Natural Disasters Occurrence, Reduction, and Restoration in Mountain Regions

Natural Disasters Occurrence, Reduction, and Restoration in Mountain Regions

Editor

Su-Chin Chen

MDPI

Basel • Beijing • Wuhan • Barcelona • Belgrade • Novi Sad • Cluj • Manchester

Editor
Su-Chin Chen
Innovation and Development
Centre of Sustainable Agriculture
National Chung Hsing University
Taichung
Taiwan

Editorial Office
MDPI
St. Alban-Anlage 66
4052 Basel, Switzerland

This is a reprint of articles from the Special Issue published online in the open access journal *Water* (ISSN 2073-4441) (available at: www.mdpi.com/journal/water/special_issues/disasters_restoration_landslides).

For citation purposes, cite each article independently as indicated on the article page online and as indicated below:

Lastname, A.A.; Lastname, B.B. Article Title. *Journal Name* **Year**, *Volume Number*, Page Range.

ISBN 978-3-7258-0858-8 (Hbk)
ISBN 978-3-7258-0857-1 (PDF)
doi.org/10.3390/books978-3-7258-0857-1

Cover image courtesy of Su-Chin Chen

Contents

About the Editor

Su-Chin Chen

Professor Chen has studied river morphology, sediment transport, debris flow, and mountainous disaster mitigation. He has served as a distinguished professor in the Department of Soil and Water Conservation at National Chung Hsing University, Taiwan. Professor Chen is the author of more than 200 scientific journal papers and 20 patents. He also received the Ministry of Science and Technology's Outstanding Research Award, the highest honor in Taiwan's academic community.

Preface

Mountain regions are critical because of their diverse geological conditions, dynamic changes, and the multiple natural hazards that often occur. Mountains are high-risk environments that can experience a variety of natural hazards, since initiated hazards often trigger secondary, cascading hazards, having a significant impact not only on the area of occurrence but often also on upstream and downstream regions. High economic losses and human casualties are caused by geophysical (rockfalls, earthquakes, and volcanic activities), hydrological (floods, avalanches, and dammed-lake outbursts), and sediment-related hazards (landslides, driftwood, debris/mud flows, and surface erosion). Under the impacts of global warming and climate change, the spatiotemporal patterns of rainfall and other weather events have become more unevenly distributed, often with a more extreme magnitude and/or intensity of events. The complexity of mountainous regions, in addition to the continued changes in climate and land use, have made it more challenging to predict mountainous hazards and their impacts on communities. Based on the countless efforts made worldwide on natural hazards in mountain regions, tight international collaboration is strongly required to answer questions related to the causes of disasters, the monitoring of hazardous phenomena, predicting disasters, and the effective reduction of hazardous consequences.

The *INTERPRAEVENT 2023* International Conference took place in Taichung, Taiwan, and for the first time, in partnership with the '*WATER*' journal, a Special Issue titled "*Occurrence, Reduction, and Restoration of Natural Disasters in Mountain Regions*", was curated, dedicated to the dissemination of scholarly research in this domain. The conference program was comprehensive, addressing a multitude of topics ranging from phenomenological investigations and event monitoring to risk analysis, policy development, resilience enhancement, and strategies for disaster response and recovery, thereby encapsulating a holistic approach to disaster preparedness.

Su-Chin Chen
Editor

water

Editorial

Natural Disasters Occurrence, Reduction, and Restoration in Mountain Regions

Su-Chin Chen

Innovation and Development Centre of Sustainable Agriculture (IDCSA), National Chung Hsing University, 145 Xingda Road, Taichung 40227, Taiwan; scchen@nchu.edu.tw

1. Introduction

The Intergovernmental Panel on Climate Change's Sixth Assessment Report [1] posits that climate change is likely to exacerbate the severity and frequency of water- and sediment-related disasters on a global scale. Under the impacts of global warming and climate change, the spatiotemporal patterns of rainfall and other weather events have become more unevenly distributed, often with a more extreme magnitude and/or intensity of events. Empirical studies conducted within the European Alps have elucidated a suite of mechanisms by which climate change is implicated in the exacerbation of slope instability, with a notable escalation in meteorological and climatic conditions that predispose terrains to the initiation of debris flows [2]. Recent decades have witnessed a correlation between heightened debris flow activity and intensified pluvial events within this region [3,4]. In Asia, one practical example is Typhoon Morakot, which struck southern Taiwan in August 2009 [5], which incurred unprecedented pluviometric extremes exceeding 3000 mm within four days, culminating in the catastrophic Shiaolin landslide with over 400 fatalities. Similarly, Japan has experienced significant losses due to slope failures, with an average of 1241 sediment disasters recorded annually from 1990 to 2019, culminating in 678 fatalities and disappearances between 2000 and 2019 [6] (Contribution 1). The comparative inefficacy of search and rescue operations for hydro-meteorological disasters relative to seismic events [7] highlights the compounded challenges of such disasters. Furthermore, mountainous regions are particularly vulnerable to a spectrum of compound disasters, including droughts, flash floods, forest fires, debris flows, and landslides, with the vicissitudes of climate change amplifying the risk profile for physical, ecological, and socio-economic systems [8,9]. Accordingly, this Special Issue endeavors to elucidate the emergent characteristics of disasters concomitant with extreme rainfall events and to survey the recent progress in state-of-the-art in situ observational technologies, early warning models, and disaster risk-reduction strategies.

2. Review of New Advances

Mountain regions are recognized as critical areas of study due to their heterogeneous geological conditions [10], the dynamism inherent in their environmental transformations (Contributions 2–4), and their propensity for frequent natural hazards. High economic losses and human casualties are caused by geophysical- (rockfalls, earthquakes, and volcanic activities), hydrological- (floods, avalanches, and dammed-lake outbursts), and sediment-related hazards (landslides, driftwood, debris/mud flows, and surface erosion). Such environments are high-risk due to the propensity of an initial hazard event to induce subsequent and potentially more destructive secondary hazards, thus affecting not only the immediate vicinity but also far-reaching effects on upstream and downstream areas (Contributions 5,6). The complexity of mountain regions and the continued changes in the climate (Contribution 7) and land use (Contributions 8,9) have made it more challenging to predict mountainous hazards and their subsequent socio-economic impacts. Hence, the

Citation: Chen, S.-C. Natural Disasters Occurrence, Reduction, and Restoration in Mountain Regions. *Water* **2024**, *16*, 313. https://doi.org/10.3390/w16020313

Received: 12 December 2023
Accepted: 12 January 2024
Published: 17 January 2024

strategic management of sediment disasters, particularly those exacerbated by climatic change, necessitates focused academic inquiry and meticulous intervention strategies.

Mitigation strategies for mountainous disasters encompass a spectrum of structural and non-structural interventions [11,12]. Regarding structural measures, protective measures involve constructing or reinforcing embankments and levees along rivers and establishing sediment-control structures, such as check dams, which can mitigate the impact of debris flows. In areas prone to rockfalls or landslides, retaining structures can stabilize slopes, thereby consequently diminishing landslide incidence. Complementing structural approaches, non-structural strategies (Contributions 10–22) engage in delineating hazard potential zones (Contributions 16–18), wherein the probabilistic risk assessment informs spatial planning. The advancement and deployment of disaster-detection and -monitoring technologies are integral to this paradigm, enabling the establishment of early warning systems essential for prompt disaster response (Contributions 19–21).

Chiang et al. (Contribution 16) applied a deep learning algorithm to ascertain the distribution of rainfall thresholds for landslides in a potentially high-risk area. Their study further encompassed evaluating the distribution of recurrence intervals through probability density functions. The primary goal was to aid decision-makers in implementing early responses to landslides and mitigating the risk of large-scale events. Complementary to this approach, Tsai et al. (Contribution 19) gathered pertinent information from satellite imagery, field investigations, major event reports, and seismic data spanning 2004 to 2016 in the mountainous regions of Taiwan. The aim was to establish a specific relationship between large-scale landslides (LSL) and triggering rainfall to enhance early warning predictions for LSL events. By integrating real-time rainfall forecasts, the dynamic variation in statistical indicators offers trend information, contributing to an increased response time for relevant evacuation operations.

Land-use management and ecological conservation are also vital aspects (Contributions 22,23), particularly in maintaining the vegetation of mountainous areas [13], including trees and grasslands, to mitigate soil erosion and reduce the risks of landslides and debris flow. Educational initiatives targeting community residents are instrumental in cultivating an understanding of the risks associated with mountainous regions and in disseminating knowledge regarding appropriate safety protocols during such geohazard events. Moreover, infrastructural resilience is bolstered through the strategic construction of shelters designed to withstand natural disasters and the meticulous planning of evacuation routes. These measures are essential in safeguarding human life by facilitating timely and orderly evacuations in the event of a disaster (Contribution 21).

Restoration research has also been the focus of research in recent years to minimize the impact of structures on nature (Contributions 24–26). Within this domain, Chiu et al. (Contribution 24) pioneered the application of Nature-based Solutions (NbS) [14] within the Huyuan Stream watershed in southern Taiwan. This initiative represents a tripartite collaboration between industrial sectors, governmental bodies, and academic institutions. The project aimed to enhance ecosystem services, providing diverse aquatic habitats, leisure sites for urban residents, and support for local agriculture. Additionally, it integrated local culture, environmental education, and professional development. Under the aegis of research leadership, this pilot study championed a holistic management paradigm that engaged multiple stakeholders in addressing the dual challenges posed by urban expansion and climatic fluctuations, concurrently advancing the quality of ecosystem services. Concurrently, there has been a nascent shift in the appraisal of mountain stream facilities, transitioning from focusing solely on safety to a broader consideration of landscape aesthetics and environmental congruence (Contributions 25,26). Peng et al. (Contribution 25) have been at the forefront of this shift, employing visual language translation as a novel methodological approach in qualitative landscape assessment. Their research has culminated in developing a model that synergizes visual harmony and aesthetic preferences, underpinned by an extensive suite of physical indicators. Such a model promises to refine the visual impact of hydrological engineering projects and enhance the evaluative processes thereafter.

3. International Conference

Global initiatives have increasingly underscored the necessity for rigorous international collaboration to elucidate the etiology of natural disasters, particularly in mountainous regions, and to develop robust frameworks for monitoring, predicting, and mitigating the impacts of such hazards. From its inception in 1967, the INTERPRAEVENT International Society, with its origins in the alpine landscapes of Austria, is esteemed for its pivotal contributions to research on slope disasters and is a vanguard interdisciplinary body that synergizes efforts from academia, governance, and industry. The INTERPRAEVENT 2023 International Conference [15] took place in Taichung, Taiwan, and, for the first time in partnership with the 'WATER' journal, has curated a Special Issue titled "Occurrence, Reduction, and Restoration of Natural Disasters in Mountain Regions," dedicated to the dissemination of scholarly research in this domain. The conference program was comprehensive, addressing a multitude of topics ranging from phenomenological investigations and event monitoring to risk analysis, policy development, resilience enhancement, and strategies for disaster response and recovery, thereby encapsulating a holistic approach to disaster preparedness.

There has been a concerted shift in disaster governance strategies in response to this recognition. The focus has transitioned from the traditional approach of imposing resilience on the environment to fostering a societal ethos that seeks to thrive in synergy with ecological systems. This approach suggests that the sustainability of human societies is intrinsically linked to their ability to adapt and integrate within the natural dynamics of the environment.

4. Conclusions

The Special Issue presents the necessity for a holistic and sustainable approach to disaster governance that is transparent. It requires the establishment of a resilient infrastructure capable of withstanding environmental calamities and, more importantly, societal transformation. This transformation would cultivate a culture that respects and aligns with ecological principles, ensuring that society and the environment can sustainably prosper together. Such an alignment represents the optimal pathway for mitigating the risks associated with mountain disasters in the face of climate change and human development.

Conflicts of Interest: The author declares no conflict of interest.

List of Contributions:

1. Hayashi, S.I.; Kunitomo, M.; Mikami, K.; Suzuki, K. Recent and Historical Background and Current Challenges for Sediment Disaster Measures against Climate Change in Japan. *Water* **2022**, *14*, 2285. https://doi.org/10.3390/w14152285.
2. Yoshino, T.; Katsura, S. Perennial Groundwater Zone Formation Processes in Thin Organic Soil Layers Overlying Thick Clayed Mineral Soil Layers in a Small Serpentine Headwater Catchment. *Water* **2022**, *14*, 3122. https://doi.org/10.3390/w14193122.
3. Wada, T.; Mishima, H.; Takemura, J.; Kobayashi, K.; Miwa, H. Transition Indices of Sediment-Transport Modes on a Debris Flow Resulting from Changing Streambed Gradients. *Water* **2022**, *14*, 1810. https://doi.org/10.3390/w14111810.
4. Chen, Y.C.; Hsu, Y.C.; Zai, E.O. Streamflow Measurement Using Mean Surface Velocity. *Water* **2022**, *14*, 2370. https://doi.org/10.3390/w14152370.
5. Noviandi, R.; Gomi, T.; Kharismalatri, H.S.; Sidle, R.C.; Ritonga, R.P.; Shiraki, K. The Mobility of Landslides in Pumice: Insights from a Flume Experiment. *Water* **2022**, *14*, 3083. https://doi.org/10.3390/w14193083.
6. Chen, J.C.; Huang, W.S. Evaluation of Rainfall-Triggered Debris Flows under the Impact of Extreme Events: A Chenyulan Watershed Case Study, Taiwan. *Water* **2021**, *13*, 2201. https://doi.org/10.3390/w13162201.

7. Tuong, V.; Van Hoang, T.; Chou, T.Y.; Fang, Y.M.; Wang, C.T.; Tran, T.D.; Tran, D.D. Extreme Droughts Change in the Mekong River Basin: A Multidisciplinary Analysis Based on Satellite Data. *Water* **2021**, *13*, 2682. https://doi.org/10.3390/w13192682.
8. Kasahara, N.; Gonda, Y.; Huvaj, N. Quantitative Land-Use and Landslide Assessment: A Case Study in Rize, Turkey. *Water* **2022**, *14*, 1811. https://doi.org/10.3390/w14111811.
9. Lusiana, N.; Shinohara, Y. The Role of Citrus Groves in Rainfall-Triggered Landslide Hazards in Uwajima, Japan. *Water* **2022**, *14*, 2113. https://doi.org/10.3390/w14132113.
10. Lo, C.M.; Lai, Y.-S.; Chu, C.H. Investigation of Rainfall-Induced Failure Processes and Characteristics of Wedge Slopes Using Physical Models. *Water* **2023**, *15*, 1108. https://doi.org/10.3390/w15061108.
11. Wang, C.L.; Ko, C.J.; Wong, H.K.; Pai, P.H.; Tai, Y.C. An Approach for Preliminary Landslide Scarp Assessment with Genetic Algorithm (GA). *Water* **2022**, *14*, 2400. https://doi.org/10.3390/w14152400.
12. Wu, C.H. Using the Spatiotemporal Hot Spot Analysis and Multi-Annual Landslide Inventories to Analyze the Evolution and Characteristic of Rainfall-Induced Landslide at the Subwatershed Scale in Taiwan. *Water* **2023**, *15*, 1355. https://doi.org/10.3390/w15071355.
13. Nguyễn, T.T.; Dong, J.J.; Tseng, C.H.; Baroň, I.; Chen, C.W.; Pai, C.C. Three-Dimensional Engineering Geological Model and Its Applications for a Landslide Site: Combination of Grid- and Vector-Based Methods. *Water* **2022**, *14*, 2941. https://doi.org/10.3390/w14192941.
14. Wan, S.; Yeh, M.L.; Ma, H.L.; Chou, T.Y. The Robust Study of Deep Learning Recursive Neural Network for Predicting of Turbidity of Water. *Water* **2022**, *14*, 761. https://doi.org/10.3390/w14050761.
15. Wu, C.Y.; Lin, S.Y. Performance Assessment of Event-Based Ensemble Landslide Susceptibility Models in Shihmen Watershed, Taiwan. *Water* **2022**, *14*, 717. https://doi.org/10.3390/w14050717.
16. Chiang, J.L.; Kuo, C.M.; Fazeldehkordi, L. Using Deep Learning to Formulate the Landslide Rainfall Threshold of the Potential Large-Scale Landslide. *Water* **2022**, *14*, 3320. https://doi.org/10.3390/w14203320.
17. Huang, Y.M.; Lu, S.W. The Effect of Temporal Characteristics on Developing a Practical Rainfall-Induced Landslide Potential Evaluation Model Using Random Forest Method. *Water* **2021**, *13*, 3348. https://doi.org/10.3390/w13233348.
18. Thanh, L.N.; Fang, Y.M.; Chou, T.Y.; Van Hoang, T.; Nguyen, Q.D.; Lee, C.Y.; Wang, C.L.; Yin, H.Y.; Lin, Y.C. Using Landslide Statistical Index Technique for Landslide Susceptibility Mapping: Case Study: Ban Khoang Commune, Lao Cai Province, Vietnam. *Water* **2022**, *14*, 2814. https://doi.org/10.3390/w14182814.
19. Tsai, T.T.; Tsai, Y.J.; Shieh, C.L.; Wang, J.H.C. Triggering Rainfall of Large-Scale Landslides in Taiwan: Statistical Analysis of Satellite Imagery for Early Warning Systems. *Water* **2022**, *14*, 3358. https://doi.org/10.3390/w14213358.
20. Catelan, F.T.; Bossi, G.; Marcato, G. Self-Sustaining Landslide Mitigation Strategy through Long-Term Monitoring. *Water* **2022**, *14*, 3824. https://doi.org/10.3390/w14233824.
21. Chiba, M.; Furuido, H.; Shibasaki, S.; Haga, K. Analysis of Socio-Economic Factors That Influence Loss of Life in Sediment-Related Disasters. *Water* **2022**, *14*, 2408. https://doi.org/10.3390/w14152408.
22. Tsai, S.C.; Lee, S.H. Disaster-Resilient Communities on Flood Plains and Their Agricultural Regeneration: A Case Study in Meinong Plain, Taiwan. *Water* **2022**, *14*, 1736. https://doi.org/10.3390/w14111736.
23. Wu, C.H. Evaluating the Landslide Stability and Vegetation Recovery: Case Studies in the Tsengwen Reservoir Watershed in Taiwan. *Water* **2021**, *13*, 3479. https://doi.org/10.3390/w13243479.
24. Chiu, Y.Y.; Wu, Y.H.; Fu, K.L.; Lai, T.C.; Chen, H.E.; Chen, S.C. Nature-Based Solutions for Disaster Reduction and Improving Ecosystem Services in the Hutoubi Watershed, Taiwan. *Water* **2023**, *15*, 2527. https://doi.org/10.3390/w15142527.
25. Chen, J.C.; Huang, C.L.; Chen, S.C.; Tfwala, S.S. Visual Harmony of Engineering Structures in a Mountain Stream. *Water* **2021**, *13*, 2201. https://doi.org/10.3390/w13233324.
26. Peng, S.H.; Huang, C.Y.; Chen, S.C. Visual Language Translation Analysis and Scenic Beauty Estimation of Mountain Stream Facilities. *Water* **2022**, *14*, 3605. https://doi.org/10.3390/w14223605.

References

1. IPCC. Sixth Assessment Report—Summary for Policymakers. Available online: https://www.ipcc.ch/report/ar6/wg1/downloads/report/IPCC_AR6_WGI_SPM.pdf (accessed on 13 July 2022).
2. Rebetez, M.; Lugon, R.; Baeriswyl, P.-A. Climatic Change and Debris Flows in High Mountain Regions: The Case Study of the Ritigraben Torrent (Swiss Alps). *Clim. Change* **1997**, *36*, 371–389. [CrossRef]
3. Dietrich, A.; Krautblatter, M. Evidence for Enhanced Debris-Flow Activity in the Northern Calcareous Alps since the 1980s (Plansee, Austria). *Geomorphology* **2017**, *287*, 144–158. [CrossRef]
4. Schlögel, R.; Kofler, C.; Gariano, S.L.; Van Campenhout, J.; Plummer, S. Author Correction: Changes in Climate Patterns and Their Association to Natural Hazard Distribution in South Tyrol (Eastern Italian Alps). *Sci. Rep.* **2020**, *10*, 10477. [CrossRef] [PubMed]
5. Wu, C.H.; Chen, S.C.; Feng, Z.Y. Formation, Failure, and Consequences of the Xiaolin Landslide Dam, Triggered by Extreme Rainfall from Typhoon Morakot, Taiwan. *Landslides* **2014**, *11*, 357–367. [CrossRef]
6. Junichi, K.; Naoki, I. Outline of Measures for Sediment Disaster by the Sabo Department of MLIT, Japan. *Landslides* **2020**, *17*, 2503–2513. [CrossRef]
7. Chiu, Y.Y.; Omura, H.; Chen, H.E.; Chen, S.C. Indicators for Post-Disaster Search and Rescue Efficiency Developed Using Progressive Death Tolls. *Sustainability* **2020**, *12*, 8262. [CrossRef]
8. Sekhri, S.; Kumar, P.; Fürst, C.; Pandey, R. Mountain Specific Multi-Hazard Risk Management Framework (MSMRMF): Assessment and Mitigation of Multi-Hazard and Climate Change Risk in the Indian Himalayan Region. *Ecol. Indic.* **2020**, *118*, 106700. [CrossRef]
9. Schneiderbauer, S.; Fontanella Pisa, P.; Delves, J.L.; Pedoth, L.; Rufat, S.; Erschbamer, M.; Thaler, T.; Carnelli, F.; Granados-Chahin, S. Risk Perception of Climate Change and Natural Hazards in Global Mountain Regions: A Critical Review. *Sci. Total Environ.* **2021**, *784*, 146957. [CrossRef] [PubMed]
10. Antonelli, A.; Kissling, W.D.; Flantua, S.G.A.; Bermúdez, M.A.; Mulch, A.; Muellner-Riehl, A.N.; Kreft, H.; Linder, H.P.; Badgley, C.; Fjeldså, J.; et al. Geological and Climatic Influences on Mountain Biodiversity. *Nat. Geosci.* **2018**, *11*, 718–725. [CrossRef]
11. Piton, G.; Recking, A. Design of sediment traps with open check dams. I: Hydraulic and deposition processes. *J. Hydraul. Eng.* **2016**, *142*, 04015045. [CrossRef]
12. Segoni, S.; Piciullo, L.; Gariano, S.L. A Review of the Recent Literature on Rainfall Thresholds for Landslide Occurrence. *Landslides* **2018**, *15*, 1483–1501. [CrossRef]
13. Ma, S.; Qiao, Y.P.; Wang, L.J.; Zhang, J.C. Terrain Gradient Variations in Ecosystem Services of Different Vegetation Types in Mountainous Regions: Vegetation Resource Conservation and Sustainable Development. *For. Ecol. Manag.* **2021**, *482*, 118856. [CrossRef]
14. Albert, C.; Brillinger, M.; Guerrero, P.; Gottwald, S.; Henze, J.; Schmidt, S.; Ott, E.; Schröter, B. Planning Nature-Based Solutions: Principles, Steps, and Insights. *Ambio* **2021**, *50*, 1446–1461. [CrossRef]
15. INTERPRAEVENT 2023 International Symposium; Natural Disasters Occurrence, Reduction, and Restoration in Mountain Regions. Available online: https://interpraevent2022.nchu.edu.tw/ (accessed on 17 April 2023).

water

MDPI

Review

Recent and Historical Background and Current Challenges for Sediment Disaster Measures against Climate Change in Japan

Shin-ichiro Hayashi [1,*], Masaru Kunitomo [2], Koso Mikami [2] and Keisuke Suzuki [3]

1 SABO & Landslide Technical Center, 2-7-5, SABO-kaikan, Hirakawa-Cho, Tokyo 102-0093, Japan
2 SABO Department, Ministry of Land, Infrastructure, Transport and Tourism, 2-1-3, Kasumigaseki, Tokyo 100-8918, Japan; kunitomo-m82ac@mlit.go.jp (M.K.); mikami-k2rg@mlit.go.jp (K.M.)
3 Chugoku Regional Development Bureau, Ministry of Land, Infrastructure, Transport and Tourism, 6-30, Kami-hachobori, Hiroshima 730-8530, Japan; suzuki-k2sh@mlit.go.jp
* Correspondence: hayashi@stc.or.jp

Abstract: Japan is a country with one of the highest incidences of sediment disasters, which will become more severe and more frequent as a result of climate change. This paper reviews the recent occurrence of sediment disasters caused by heavy rainfall affected by climate change in recent years, challenges in adaptation measures, and recent policies targeting such sediment disasters in Japan. The Ministry of Land, Infrastructure, Transport and Tourism (MLIT) has been conducting non-structural and structural measures based on legislation. Recently, climate change has resulted in more severe and frequent disasters along with damage caused by sediment movement phenomena that are not covered by the present system for warning and evacuation. Efforts to establish assessment methods concerning the risk of these phenomena are shown as examples of current challenges of climate change.

Keywords: sediment disaster prevention; SABO; adaptation measure; climate change; Japan

Citation: Hayashi, S.-i.; Kunitomo, M.; Mikami, K.; Suzuki, K. Recent and Historical Background and Current Challenges for Sediment Disaster Measures against Climate Change in Japan. *Water* **2022**, *14*, 2285. https://doi.org/10.3390/w14152285

Academic Editor: Zhi-jun Dai

Received: 29 April 2022
Accepted: 19 July 2022
Published: 22 July 2022

1. Introduction

Globally, sediment disasters are natural disasters that cause enormous human and property damage [1–3]. In the Intergovernmental Panel on Climate Change (IPCC) Sixth Assessment Report, sediment disasters are treated as one of the phenomena that could be affected by climate change and have socioeconomic impacts [4]. Climate change is expected to cause more severe and frequent sediment disasters worldwide [5–8], and there are concerns about the effects of such disasters in Japan, where 1241 sediment disasters have occurred on average from 1990 to 2019, and caused 678 dead and missing from 2000 to 2019 (approximately 21% of the number of dead and missing except those caused by the Great East Japan Earthquake) [9].

The Japanese government's Ministry of the Environment published a report in 2020 entitled, "Assessment of Impacts of Climate Change in Japan and Future Challenges" [10], mandated by the "Climate Change Adaptation Act" [11]. Sediment disaster is rated highest in all evaluation categories of "Severity", "Urgency" and "Confidence". In 2020, the Cabinet Office surveyed public awareness of "concerns about water-related problems due to the effects of climate change" and "what you would like the government to focus on water-related problems due to the effects of climate change". According to the survey results, "Frequent floods and sediment disasters due to climate instability" is the most worrisome problem (percentage of respondents: 85.6%), and "Construction and improvement of flood and sediment disaster prevention facilities" are the most desired things about which the government focuses on outcomes, for the government to focus on (percentage of respondents: 78.5%) [12].

Despite the high interest from both Japanese government agencies and the public, and the increase in sediment disasters due to climate change, few reports on global climate change related to sediment disasters in Japan have been published [6,8]. In this review, the characteristics of recent sediment disasters caused by heavy rainfall under the influence of climate change in Japan are summarized, and the challenges in advancing adaptation measures against sediment disasters are identified and reviewed.

2. Recent and Historical Background

2.1. Non-Structural and Structural Measures against Sediment Disasters in Japan

The reduction measures conducted by the Ministry of Land, Infrastructure, Transport and Tourism (MLIT) consist of non-structural measures and structural measures [9,13]. The non-structural measures such as disseminating sediment disaster risk information, issuing "Sediment Disaster Alert" [14], and land use regulation and building restrictions at the areas at risk, have been implemented to designate "Sediment Disaster Alert Areas" [15] (the number of Sediment Disaster Alert Areas: 214,000 for debris flow, 15,691 for deep-seated landslides and 442,728 for slope failure; as at the end of December 2021) [16] based on the Sediment Disaster Prevention Act [17]. The structural measures were implemented based on three laws, the SABO Act [18], Landslide Prevention Act [19] and Steep Slope Failure Prevention Act [20], to address three types of sediment disasters: debris flows, deep-seated landslides and steep slope failures (including rock fall); these three types of sediment disaster are classified using the original Japanese classification method in a precautionary approach. The SABO Act deals mainly with debris flow [21] including sediment-laden floods [22] (also referred to as sediment and flood damage [23], flood flow with active sediment transport [24]). The Landslide Prevention Act is designed to prevent disasters caused by slow-moving deep-seated landslides at slopes with landslide features in order to make the slopes available to residents for productive activities such as agriculture or forestry. The Steep Slope Failure Prevention Act stipulates structural measures against slope failures initiated at slopes with an angle of 30 degrees or steeper.

As an emergency response (non-structural measure) for large-scale sediment disaster based on the Sediment Disaster Prevention Act, "Sediment Disaster Emergency Information", is disseminated to the governors of prefectures and municipalities, as well as the public in the predicted affected area for evacuation of residents. Landslide dams [25,26] and debris flows due to volcanic eruptions [25,27] are assessed by the MLIT and deep-seated landslides are assessed by the prefectural government [25,28]. Numerical simulations are used to predict the affected area of debris flows caused by landslide dam breaks (1D riverbed variation calculation and 2D inundation calculation) and volcanic eruptions (rainfall-runoff analysis considering ash accumulation and 2D inundation calculation). [25,27,29,30]

2.2. Recent Major Sediment Disasters and Response

The JMA (Japan Meteorological Agency) reported that the observed changes to date include increases in the frequency of heavy rain and short-duration intense rainfall events. Moreover, at the end of the 21st century, the number of annual days with daily precipitation of 200 mm or more will increase by about 1.5 times; and the frequency of 1 h precipitation of 50 mm or more increases by about 1.6 times based on the 2 °C rise scenario (RCP2.6) of IPCC [31]. Figure 1 shows the relationship between the number of sediment disasters increasing from 1980 to 2019 with an increase in the number of days with precipitation of 100 mm or more [32]. In recent years, the number of sediment disasters in Japan has increased in accordance with the increase in heavy rainfall. In the torrential rains in July 2018, the MLIT reported 2581 cases of sediment disasters over a wide area, about 2.5 times the annual average, mainly in western Japan. In the following year 2019, Typhoon Hagibis, its heavy precipitation enhanced by historical warming [33], caused widespread sediment disasters in the Kanto and Tohoku regions [34], where sediment disasters had been relatively rare [35]. A total of 952 sediment disasters occurred, the highest number recorded for a single typhoon disaster since 1982. In this disaster, human casualties resulted from disasters that occurred in areas that were not designated as "Sediment Disaster Alert Areas" because the topographical conditions did not meet the criteria for designation in municipalities where heavy rainfall exceeding the criteria for the announcement of a JMA emergency warning was recorded [36]. Specifically, in Marumori, Miyagi Prefecture, causalities and property damage were caused by sediment-laden floods (Figure 2a) [37], and casualties and property damage were caused by debris flows in an area where the valley topography was unclear in Mawarikura (Figure 2b) [38], and casualties and property damage were caused by deep-seated landslides in an area where there was no clear deep-seated landslide topography in Tomioka, Gunma Prefecture (Figure 2c) [39]. In addition, these human casualties occurred in areas that did not meet the criteria for designation as Sediment Disaster Alert Areas. This suggests that, due to the intensification of heavy rainfall caused by climate change, the number of sediment disasters in areas where the number of such disasters has been small in the past is increasing, and that sediment disasters caused by sediment movement phenomena, which have been infrequent in the past, are becoming more apparent.

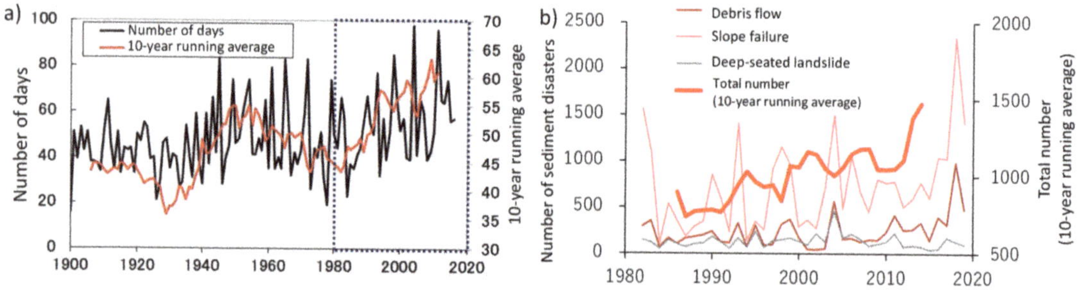

Figure 1. The relationship between the number of sediment disasters increasing from 1980 to 2019 with an increase in the number of days with precipitation of 100 mm or more: (**a**) The numbers of days with precipitation of 100 mm or more and 10-year running average in 51 observation stations in Japan (based on Uchida et. al., 2017 [32]), (**b**) the number of sediment disasters and 10-year running average, reported by the MLIT, from 1980 to 2019 in Japan.

(**a**) (**b**) (**c**)

Figure 2. Photos of sediment disasters caused by Typhoon Hagibis 2019: (**a**) sediment-laden flood, (**b**) debris flows in an area where the valley topography was unclear, (**c**) deep-seated landslides in an area where there was no clear deep-seated landslide topography (courtesy of Pasco Corporation).

In the inner belt of Southwest Japan (IBSJ), where granitic geology is widely distributed, sediment disasters that cause major damage are often caused by the simultaneous occurrence of slope failures and debris flows due to torrential rains during the rainy season, while in the Pacific side of the outer belt of Southwest Japan (OBSJ), where accretionary geology is widely distributed, sediment disasters tend to be caused by deep-seated catastrophic (rapid) landslides due to typhoons [40]. Figure 3 shows 48 major sediment disasters from April 1998 to September 2020, in which 40 disasters caused five fatalities or more in one municipality, five deep-seated catastrophic (rapid) landslides [41–44] and three sediment-laden floods [45–47], in the geological structures of Northeast Japan (NJ), Fossa Magna (FN), IBSJ and OBSJ and indicates their tendency.

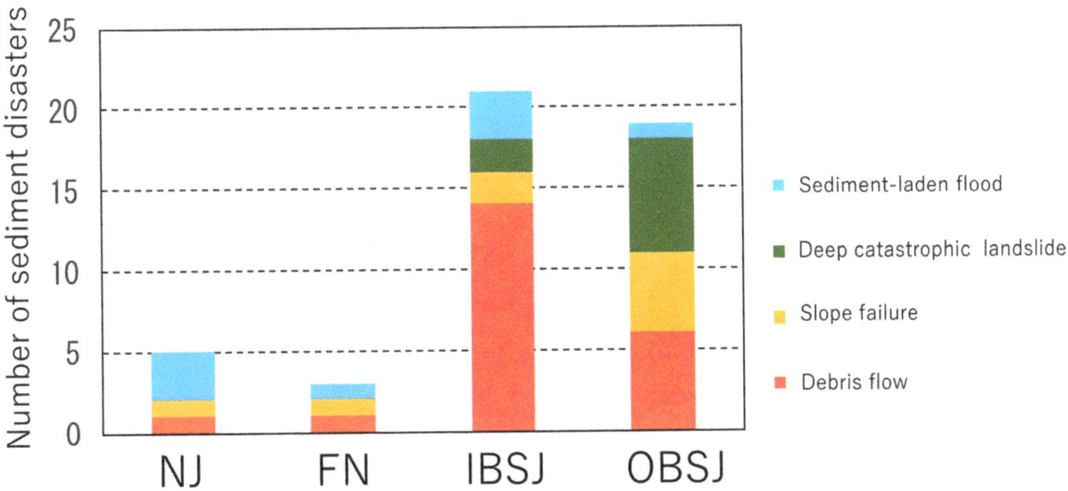

Figure 3. Type of major sediment disasters from April 1998 to September 2020 in geological structure: NJ (Northeast Japan), FN (Fossa Magna), IBSJ (the inner belt of Southwest Japan), OBSJ (the outer belt of Southwest Japan).

In the future, as climate change progresses, the influence of both typhoons and rainy season fronts will shift eastward and northward, which will further increase the amount of precipitation in western Kyushu and the Chugoku region, where many torrential rains have occurred in the past and are expected to cause torrential rains in Hokkaido, a region that has not had much precipitation in the past (refer to Figure 4 for locations of sub-region of Japan) [48,49].

In order to adapt to the increasing sediment disasters caused by climate change, it is necessary to properly assess which areas will experience more frequent sediment movement phenomena; this will become more apparent due to the changes in rainfall characteristics caused by climate change, and it is urgent to develop a new assessment method. Based on the relationship between future rainfall forecast data and the critical line method of Sediment Disaster Alert in Japan [14], studies are being conducted on where the frequency of sediment disasters will increase [50,51].

Based on the experience of recent major sediment disasters, the authors expect several outcomes as a result of climate change in Japan as follows: (1) increasing frequency of sediment disasters throughout Japan; (2) expansion of the damage area of sediment disaster; (3) occurrence of sediment disasters in regions where sediment disasters had been relatively rare; and (4) occurrence of sediment movement phenomena that would be more apparent due to climate change (such as sediment-laden flood and deep-seated landslides with no clear landslide topography on gentle slopes), which had been rare.

In particular, for sediment-laden floods and deep-seated landslides with no clear landslide topography on gentle slopes that are not covered by the present system for warning and evacuation [14,15], it is an important issue to establish a method to identify the hazardous area. Especially, as the frequent occurrence of sediment production in the mountainous area and the duration of flooding increases, the risk of damage caused by sediment-laden flood might become greater. Moreover, the frequent occurrence of simultaneous debris flows, which caused a large number of fatalities in the past due to the increase in precipitation caused by climate change, may be of concern. Recent and historical background and current challenges of sediment-laden floods and deep-seated landslides with no clear landslide topography on gentle slopes and simultaneous debris flow, are next explained in detail.

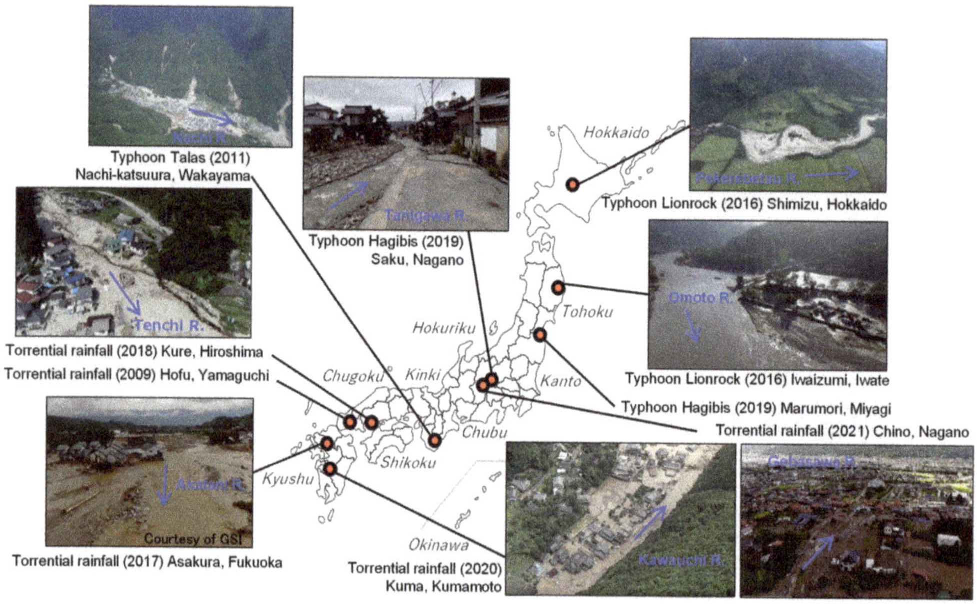

Figure 4. Distribution of recent sediment-laden floods in Japan since 2009.

2.3. *Sediment-Laden Floods and Deep-Seated Landslides with No Clear Landslide Topography on Gentle Slope and Simultaneous Debris Flows*

The affected area and damage of sediment-laden floods are more extensive than those caused by debris flows, as they have recently occurred in urban areas with relatively gentle slopes, such as alluvial fans and valley plains in Japan [23,52]. Sediment-laden floods have often caused serious damage in the past in Japan. Moreover, the frequency of sediment-laden floods has been increasing from the 2010s (Figure 4) [37,43,46,47,53–58]. Sediment-laden floods occurred when active surface erosion and small-scale shallow landslides in the bare mountains and hills caused sediment to be constantly produced and deposited on slopes and streambeds due to rainfall before Japan's rapid economic growth period (approximately from 1954 to 1973), such as the Great Hanshin Flood (1938) [59]. Sediment-laden floods and debris flows could also be enhanced by sediment production affected by the temporary loss of the mechanical soil binding effect of forest root systems due to the expansion of afforestation after World War II [60–62].

In such areas, for example, the damage caused by sediment-laden floods and debris flows was greatly reduced during the torrential rains in July 2018, even though the amount of precipitation around Mt. Rokko was higher than that of the Great Hanshin Flood [63]. The construction of hillside works, including afforestation, at the source of sediment production, and SABO dams and ground sills, in the middle and lower reach of streams, are considered to reduce erosion of hillside slopes and the discharge of sediment deposited on stream beds, to reduce sediment transport within the watershed and to prevent damage [64].

As tree felling decreased and vegetation recovered on bare mountains and hills after the energy revolution during Japan's rapid economic growth period, the frequency of surface erosion and small-scale shallow landslides at mountains and hills decreased, and the development of forest soils and weathered bedrock layers increased in all regions of Japan.

As above, the sediment production patterns have changed dramatically compared to the time when bare mountains and hills were widespread. In recent years, sediment-laden floods that cause serious damage have been caused by the simultaneous occurrence of slope failures and debris flows even in forested watersheds. In unprecedented torrential rainfalls, sediment-laden floods caused by simultaneous slope failures and debris flows might occur on forested hillslopes, even in watersheds where structural measures for preventing sediment transport have been implemented and where there is little sediment transport in the small and medium magnitude of precipitation. When a sediment-laden flood occurs, slope failures and debris flow often produce large amounts of driftwood from hillsides, which might increase the damage.

Deep-seated landslides with no clear landslide topography on gentle slopes tend to occur where subsurface flow is prone to occur on bedrock, and are related to falling pyroclastic deposits that can be a slip surface due to weathered pumice layer [39,65]. Reviews of past incidents and on-site investigations are very important to understanding the generation mechanism and flow process of deep-seated landslides with no clear landslide topography on gentle slopes, and to establish risk assessment methods.

Simultaneous debris flows caused a large number of fatalities, especially in urban and suburban areas of Japan [56,66,67]. In sediment disasters caused by the torrential rains in July 2018, approximately 70% of the streams where fatalities were caused by sediment disasters were small streams with a basin area of 0.05 km^2 or less. The percentage of streams with a basin area of 0.05 km^2 or less is equivalent to about 35% of the approximately 180,000 streams of "Debris Flow Prone Torrents" (not legal designations, identification of risk against debris flow in Japan before designating the Sediment Disaster Alert Areas). Moreover, only three streams had SABO dams before the disasters in 33 streams where fatalities occurred. In response to this disaster, the report of the "Sediment disaster prevention study committee to Ensure Effective Evacuation" [68] established by the MLIT stated that "effective and efficient measures against debris flow in small streams should be promoted" as one of the measures to be taken in the future.

3. Challenges for Sediment Disaster Measures against Climate Change

3.1. Current Challenges of Sediment-Laden Floods

In order to take appropriate adaptation measures against sediment-laden floods, it is necessary to identify watersheds that produce large amounts of sediment due to slope failures and debris flows caused by changes in rainfall characteristics associated with climate change, which are prone to being transported and deposited near residential areas in downstream. However, it is difficult to establish a method for identifying such watersheds; thus, (1) watersheds where sediment-laden floods have occurred in the past; and (2) watersheds that have the same characteristics as watersheds, which recently experienced sediment-laden floods such as topography and potential sediment production, should be considered.

In Figure 5, relationships between catchment area and volume of sediment production of rivers that have experienced sediment-laden floods in recent years (from 2009 to 2018; 11 events in which sediment budgets in watersheds were measured by LiDAR) are shown. Sediment-laden floods tend to occur in watersheds with a catchment area of 3 km^2 or more and more than 100,000 m^3 of sediment production. In Figure 6, relationships between characteristics of damaged houses and river gradient; the Akadani river in heavy rainfall in northern Kyushu, 2017; the Sozu, Tenchi and Oyaokawa rivers in the torrential rains in July 2018, and the Gofukuya river in heavy rainfall caused by Typhoon Hagibis, 2019, are shown. At a riverbed gradient of 1/150~200 or steeper, the majority of severe damage was caused to houses, which were washed away in some cases [69,70]. Based on these tendencies, the MLIT published, "Draft guideline of investigation for watershed at risk of severe damage due to sediment and flood damage (Trial version)" [71] in March 2022, for the prefectural government as a risk assessment method against sediment-laden floods.

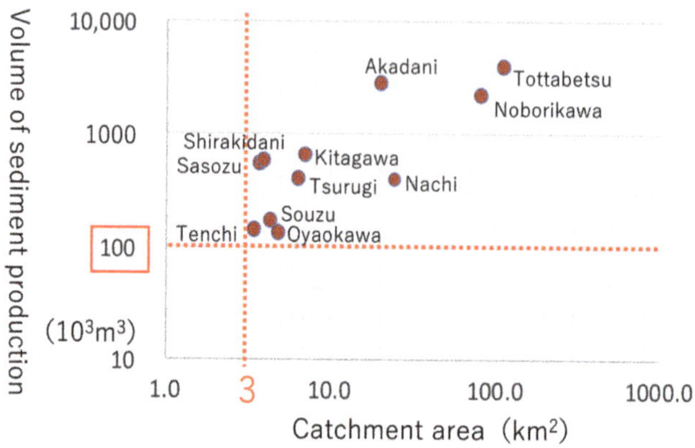

Figure 5. Relationships between catchment area and volume of sediment production of rivers that have experienced sediment-laden floods from 2009 to 2018 (cited from Sakai et al., 2021 [69]).

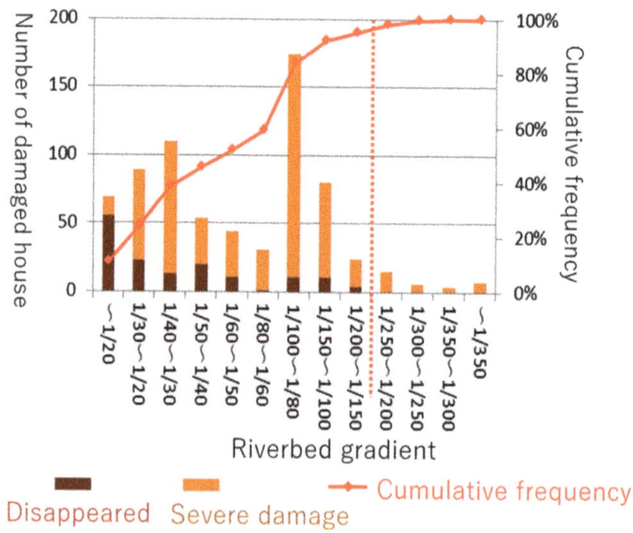

Figure 6. Relationships between characteristics of damaged houses and river gradient (cited from Sakai et al., 2021 [70]).

For the planning phase of the structural measures, Technical Criteria for River and SABO Works Basic Planning Part [25] was revised in 2019 to use planning methodologies based on riverbed variation calculations against sediment-laden floods. In 2021, for the MLIT projects in the Kizugawa River and the Kii mountains, which changed project plans based on riverbed variation calculations, continued implementation of the projects was approved by project evaluation [72,73], mainly based on cost–benefit analysis [74]. The MLIT projects against sediment-laden floods will continue to be reviewed [75], and the prefectural government will establish structural measure plans against sediment-laden floods [76] using riverbed variation calculations.

Driftwood brought by sediment-laden floods causes serious damage in downstream areas as a result of heavy rainfall in northern Kyushu, 2017 [43], in which the estimated volume of driftwoods was approximately 210,000 m^3 [77]. The MLIT and Forestry Agency started to collaborate quantitatively on structural measures against driftwood, for example sharing the data of standing timbers surveyed by LiDAR and a measure plan [78].

3.2. Current Challenges of Deep-Seated Landslides with No Clear Landslide Topography on Gentle Slope

In order to establish an assessment method to clarify the risk of deep-seated landslides with no clear landslide topography on gentle slopes, Sugimoto et al. collected literature on 26 disasters and 71 cases of deep-seated landslides in Japan with no clear landslide topography on gentle slopes due to rainfall [79]. In their study, first, as an analysis of rainfall conditions for the occurrence of landslides, the ratio (hereafter R) of the maximum daily rainfall at the time of disasters to the 100-year probable daily rainfall (using the 100-year probable rainfall at the nearest point in the JMA website's list [80] from the location of landslide) was calculated. $R \geq 1$ accounts for 80% of the cases (57 cases), indicating that many of the cases occurred in rare heavy rains. In addition, as a topographical analysis of the occurrence of landslides, whether or not the landslide topography existed before the disaster in the literature was surveyed. The results show that 86% of the cases (61 cases) had no description of landslide topography, suggesting that many of the cases occurred on gentle slopes where no clear landslide topography was observed before the disaster. As an analysis of geology and geological structure, the ratios of geology and geological structure based on descriptions in the literature were surveyed. It was found that the number of deep-seated landslides with no clear landslide topography on gentle slopes occurring on slopes where pyroclastic fall deposits (Quaternary) formed dip slope was 15/71 cases (21%) and 5/26 disasters (19%), including the landslide in Takumi, Tomioka, Gunma in this category, and the number that occurred on slopes where marine sedimentary rocks (from Quaternary to Cretaceous) formed dip slopes was 15/71 cases (21%) and 9/26 disasters (35%). They reported that these two types of geology and geological structure might have relatively high risks of deep-seated landslides with no clear landslide topography on the gentle slopes.

3.3. Current Challenges of Simultaneous Debris Flows

Downstream of streams with small catchment areas in urban and suburban areas, developed housing areas are often located directly below their valley outlet, and once a debris flow occurs, it could cause extensive damage. Moreover, because narrow roads in crowded housing areas complicate the transport of construction materials and construction machines to build an SABO dam (check dam), it is difficult to build a conventional type of SABO dam. Therefore, it is important to effectively and efficiently install measures to build facilities against debris flow in such streams [81].

The MLIT published in 2022, "Technical supplement for structural measure against debris flow in streams with small catchment area and no flowing water" [78,82] as supplemental material for the technical standard for planning and design of conventional SABO dams [83,84], to promote the introduction of a simpler structure than the conventional SABO dam. This material is to be used under the following conditions.

A stream should have two characteristics, as follows: (1) an ephemeral stream where the flow path is unclear, there is no water flow on days with no rain, and sediment transport is not expected under rainfall that does not cause a sediment disaster; and (2) a stream where the riverbed slope gradient upstream of the planned SABO dam is 10° or steeper, and the entire watershed of the upper stream of the point is a debris flow generation/runoff zone. These guidelines also apply when it is difficult to construct an SABO dam based on the conventional guidelines from the viewpoint of constructability and topographical conditions, such as (1) when a road to the construction site is very narrow; (2) when the size of the conventional SABO dam might be significantly larger than the size of the stream watershed; (3) when a large amount of soil is excavated for construction of the SABO dam; or (4) when a new road for construction cannot be prepared due to the proximity of houses, buildings or any structures.

4. Discussion

In Section 2.2, we indicate several expected outcomes of climate change in Japan. For increasing frequency and expansion of the damage area of sediment disasters, and the occurrence of sediment disasters in regions where sediment disasters had been relatively rare (outcomes: from (1) to (3)), it may be essential to predict sediment movement phenomena that will become more frequent and more apparent in each region of Japan. To this end, it is important to analyze the dominant topographical and geological factors [40] and rainfall indexes [14,40,85] that caused each past sediment movement phenomena (such as simultaneous debris flows and slope failures, sediment-laden floods, and deep-seated catastrophic (rapid) landslide), based on these results, to predict the occurrence of future sediment movement phenomena for each region using future rainfall forecast data on a nationwide scale. If the dominant topographical and geological factors and rainfall indexes could be quantified, it would improve the dissemination of sediment risk information by taking into account the effects of climate change.

For the occurrence of sediment movement phenomena that would be more apparent due to climate change, which had been rare (outcome: (4)), especially regarding sediment-laden floods, it is important to assess the potential source of sediment, and the tendencies of sediment erosion and deposition in each river system, evaluated by the riverbed variation calculation [25] and connectivity of sediment transport in rivers [86], for the risk grading of watersheds at risk of sediment-laden floods to prioritize implementation of structural measures.

Moreover, the development and implementation of techniques for estimating sediment production based on physical models, such as numerical simulations [87–90], should be expedited so that the amount of sediment production in areas with inexperienced rainfall can be appropriately estimated to implement non-structural and structural measures from the viewpoint of the estimation of the damage area caused by each sediment movement phenomenon and external forces required for the design of facilities. Since the establishment of a physical model requires further validation and technological development, the use of an empirical method to estimate the amount of sediment production according to the predicted future rainfall based on the relationship between the amount of rainfall during past sediment disasters and the amount of sediment production should be facilitated [91]. In the use of empirical methods, extrapolation to a range of rainfall that has not been experienced at all is a major problem.

Because the spatial resolution of present future rainfall forecast data (for example spatial resolution of SI-CAT DDS5TK 5 km) is often large compared to the watershed area where structural measures are conducted against sediment-laden floods, debris flow, and slope failure [92], quantitative evaluation of water and sediment discharge on a local scale affected by climate change using future rainfall forecast data is also an important issue for the planning and design of facilities.

5. Conclusions

In this review, firstly non-structural and structural measures against sediment disasters in Japan conducted by the MLIT and the prefectural government, and the characteristics of recent major sediment disasters caused by heavy rainfall under the influence of climate change in Japan, mainly describing sediment disasters caused by the torrential rains in July 2018 and Typhoon Hagibis, are summarized as a recent and historical background. Secondly, based on the characteristics, problems that need to be resolved are shown, for example, sediment-laden floods and deep-seated landslides with no clear landslide topography on gentle slopes, which are not covered by the present system for warning and evacuation. Finally, current challenges to establishing advancing adaptation measures against such sediment disasters are introduced.

To establish proper adaptation measures against sediment disasters due to climate change, the SABO Department of the MLIT has been consulting with academic experts in the "SABO technical panel for climate change" since 2019. The interim report of the panel was published in June 2020 [93]. Based on the Interim report, the MLIT plans and conducts research and technical development to update technical standards for non-structural and structural measures. The MLIT continues to exchange ideas with countries and areas affected by sediment disasters due to climate change.

Author Contributions: S.-i.H.; software, validation, formal analysis, writing—original draft preparation, visualization, project administration, M.K.; conceptualization, methodology, writing—original draft preparation, supervision, project administration, K.M.; conceptualization, writing—review and editing, supervision, project administration, K.S.; software, validation, formal analysis, writing—original draft preparation, visualization, project administration. All authors have read and agreed to the published version of the manuscript.

Funding: This research received no external funding.

Institutional Review Board Statement: Not applicable.

Informed Consent Statement: Not applicable.

Data Availability Statement: Not applicable.

Acknowledgments: The authors would like to thank the anonymous reviewers whose comments significantly improved this paper.

Conflicts of Interest: The authors declare no conflict of interest.

References

1. Petley, D. Global patterns of loss of life from landslides. *Geology* **2012**, *40*, 927–930. [CrossRef]
2. Haque, U.; Blum, P.; da Silva, P.F.; Andersen, P.; Pilz, J.; Chalov, S.R.; Malet, J.P.; Jamec Auflic, M.; Andres, N.; Poyiadji, E.; et al. Fatal landslides in Europe. *Landslides* **2016**, *13*, 1545–1554. [CrossRef]
3. Marc, O.; Stumpf, A.; Malet, J.P.; Gosset, M.; Uchida, T.; Chiang, S.H. Initial insights from a global database of rainfall-induced landslide inventories: The weak influence of slope and strong influence of total storm rainfall. *Earth Syst. Dyn.* **2018**, *6*, 903–922. [CrossRef]
4. IPCC Sixth Assessment Report—Summary for Policymakers. Available online: https://www.ipcc.ch/report/ar6/wg1/downloads/report/IPCC_AR6_WGI_SPM.pdf (accessed on 13 July 2022).
5. Stoffel, M.; Tiranti, D.; Huggel, C. Climate change impacts on mass movements—Case study from the European Alps. *Sci. Total Environ.* **2014**, *493*, 1255–1266. [CrossRef] [PubMed]
6. Gariano, S.L.; Guzzetti, F. Landslides in a changing climate. *Earth-Sci. Rev.* **2016**, *162*, 227–252. [CrossRef]
7. Gariano, S.L.; Rianna, G.; Petrucci, O.; Guzzetti, F. Assessing future changes in the occurrence of rainfall-induced landslides at a regional scale. *Sci. Total Environ.* **2017**, *596–597*, 417–426. [CrossRef]
8. Haque, U.; da Silva, P.F.; Devoli, G.; Pilz, J.; Zhao, B.; Khaloua, A.; Wilopo, W.; Andersen, P.; Lu, P.; Lee, J.; et al. The human cost of global warming: Deadly landslides and their triggers (1995–2014). *Sci. Total Environ.* **2019**, *682*, 673–684. [CrossRef]
9. Kanbara, J.; Imamori, N. Outline of measures for sediment disaster by the Sabo department of MLIT, Japan. *Landslides* **2020**, *17*, 2503–2513.
10. Ministry of the Environment—Publication of Report on Assessment of Impacts of Climate Change in Japan and Future Challenges. Available online: https://www.env.go.jp/press/108790.html (accessed on 29 April 2022).
11. e-GOV—Climate Change Adaptation Act. Available online: https://elaws.e-gov.go.jp/document?law_unique_id=430AC0000000050_20181201_000000000000000 (accessed on 29 April 2022).
12. Cabinet Office—Public Opinion Survey on Water Circulation (October 2020 Survey). Available online: https://survey.gov-online.go.jp/hutai/r02/r02-mizug.pdf (accessed on 29 April 2022).
13. Mizuyama, T. Sediment hazards and SABO works in Japan. *Int. J. Eros. Control Eng.* **2008**, *1*, 1–4. [CrossRef]
14. Osanai, N.; Shimizu, T.; Kuramoto, K.; Kojima, S.; Noro, T. Japanese early-warning for debris flows and slope failures using rainfall indices with radial basis function network. *Landslides* **2010**, *7*, 325–338. [CrossRef]
15. Uchida, T.; Nishimoto, H.; Osanai, N.; Shimizu, T. Countermeasures for sediment-related disasters in Japan using hazard maps. *Int. J. Eros. Control Eng.* **2009**, *2*, 46–53. [CrossRef]
16. Ministry of Land, Infrastructure, Transport and Tourism—Designation of Sediment Disaster Alert Areas. Available online: https://www.mlit.go.jp/mizukokudo/sabo/content/001465798.pdf (accessed on 29 April 2022).
17. e-GOV—Sediment Disaster Prevention Act. Available online: https://elaws.e-gov.go.jp/document?lawid=412AC0000000057 (accessed on 29 April 2022).
18. e-GOV—Sabo Act. Available online: https://elaws.e-gov.go.jp/document?lawid=130AC0000000029 (accessed on 29 April 2022).
19. e-GOV—Landslide Prevention Act. Available online: https://elaws.e-gov.go.jp/document?lawid=333AC0000000030 (accessed on 29 April 2022).

20. e-GOV—Steep Slope Failure Prevention Act. Available online: https://elaws.e-gov.go.jp/document?lawid=344AC0000000057_20150801_000000000000000 (accessed on 29 April 2022).
21. Mizuyama, T. Structural countermeasures for debris flow disasters. *Int. J. Eros. Control Eng.* **2008**, *1*, 38–43. [CrossRef]
22. Ministry of Land, Infrastructure, Transport and Tourism—Sediment-Laden Flood. Available online: https://www.mlit.go.jp/common/001296657.pdf (accessed on 29 April 2022).
23. Hasegawa, Y.; Nakatani, K.; Araki, Y.; Kaibori, M.; Satofuka, Y. Study on sediment disasters occurred in Tennou, Kure city, Hiroshima prefecture and considering sediment and flood damage in down stream area. *J. Jpn. Soc. Civ. Eng. Ser. B1* **2019**, *75*, 324–331. [CrossRef]
24. Harada, D.; Nagumo, N.; Nakamura, Y.; Egashira, S. Characteristics of flood flow with active sediment transport in the Sozu river flood hazards at the severe rainfall event in July 2018. *J. Disaster Res.* **2019**, *14*, 886–893. [CrossRef]
25. Ministry of Land, Infrastructure, Transport and Tourism—Technical Criteria for River and SABO Works Basic Planning Part. Available online: https://www.mlit.go.jp/river/shishin_guideline/gijutsu/gijutsukijunn/keikaku/index.html (accessed on 19 July 2022).
26. Kinki Regional Development Bureau, Ministry of Land, Infrastructure, Transport and Tourism—The Great Flood in the Kii Peninsula, The Record of Disaster Response of the Kinki Regional Development Bureau, Ministry of Land, Infrastructure, Transport and Tourism. Available online: https://www.kkr.mlit.go.jp/bousai/qgl8vl0000008ajd-att/kiihantou-kirokushi.pdf (accessed on 19 July 2022).
27. Sasaki, M. Measures of sediment disasters after 2011 eruption of Shinmoe-dake, Kirishima Volcano. *J. Jpn. Soc. Eros. Control Eng.* **2014**, *66*, 88–91.
28. Council for landslide disaster management in Niigata Prefecture. Activities of Council for landslide disaster management in Niigata Prefecture. *J. Jpn. Landslide Soc.* **2017**, *54*, 68–71. [CrossRef]
29. Uchida, T.; Yamakoshi, T.; Shimizu, T.; Yoshino, K.; Kisa, H.; Ishiduka, T. Quick analysis method for assessing debris flow hazard area induced by overtopping erosions of landslide dam and post-eruption rainstorm. *Civ. Eng. J.* **2011**, *53*, 18–23.
30. Public Works Research Institute, Japan—Development of an Open Source Program for Parallel Computation of Inundation by Stony Debris Flow, Immature Stony Debris Flow and Bed Load. Available online: https://www.pwri.go.jp/jpn/results/db/doken_kankoubutu/doken_shiryou/files/doken_shiryou_4415_00.pdf (accessed on 19 July 2022).
31. Japan Meteorological Agency—Climate Change in Japan, December 2020. Available online: https://www.data.jma.go.jp/cpdinfo/ccj/2020/pdf/cc2020_gaiyo_en.pdf (accessed on 8 July 2022).
32. Uchida, T.; Sakurai, W.; Okamoto, A. Historical Patterns of Heavy Rainfall Event and Deep-Seated Rapid Landslide Occurrence in Japan: Insight for Effects of Climate Change on Landslide Occurrence. In *Advancing Culture of Living with Landslides*; Mikoš, M., Casagli, N., Yin, Y., Sassa, K., Eds.; Springer: Cham, Switzerland, 2017; pp. 251–257.
33. Kawase, H.; Yamaguchi, M.; Imada, Y.; Hayashi, S.; Murata, A.; Nkaegawa, T.; Miyasaka, T.; Takayabu, I. Enhancement of extremely heavy precipitation induced by Typhoon Hagibis (2019) due to historical warming. *SOLA* **2021**, *17A*, 7–13. [CrossRef]
34. Ministry of Land, Infrastructure, Transport and Tourism—Problems Based on Recent Sediment Disaster Experience. Available online: https://www.mlit.go.jp/river/sabo/committee_kikohendo/200521/02shiryo.pdf (accessed on 29 April 2022).
35. Saito, H.; Korup, O.; Uchida, T.; Hayashi, S.; Oguchi, T. Rainfall conditions, typhoon frequency, and contemporary landslide erosion in Japan. *Geology* **2014**, *42*, 999–1002. [CrossRef]
36. Japan Meteorological Agency—Emergency Warning System. Available online: https://www.jma.go.jp/jma/en/Emergency_Warning/ew_index.html (accessed on 29 April 2022).
37. Irasawa, M.; Koi, T.; Tsuo, C.Y.; Kato, N.; Matsuo, S.; Arai, M.; Kaibori, M.; Yamada, T.; Kasai, M.; Wakahara, T.; et al. October 2019 sediment disaster in the Tohoku region owing to Typhoon No. 19 (Tyhpoon Hagibis). *Int. J. Eros. Control Eng.* **2020**, *13*, 48–55. [CrossRef]
38. National Institute for Land and Infrastructure Management, Japan—Report on the Investigation of Infrastructures and Building Damages Caused by Typhoon No.15 and No.19 in 2019. Available online: http://www.nilim.go.jp/lab/bcg/siryou/tnn/tnn1111pdf/ks1111.pdf (accessed on 19 July 2022).
39. Shuin, Y.; Uchida, T.; Kaibori, M.; Takeshita, W.; Tobioka, S.; Nishiwaki, A.; Yamakoshi, T. Sediment disasters in the Kanto area caused by heavy rains in October 2019, Typhoon Hagibis and locally developed front. *J. Jpn. Soc. Eros. Control Eng.* **2020**, *72*, 54–62.
40. Nishiuchi, T.; Miyase, M.; Kunita, K.; Hamaji, R.; Hayashi, S.; Matsumoto, H.; Kito, M.; Sakai, Y.; Yamada, T.; Shibata, S. Characterization of dominant predisposing and triggering factors that are highly relevant to sediment transport phenomena. In Proceedings of the Annual Meeting, Japan Society of Erosion Control Engineering, Online, 19–21 May 2021; pp. 11–12.
41. Taniguchi, Y.; Uchida, T.; Omura, H.; Ochiai, H.; Kaibori, M.; Kubota, T.; Sasahara, K.; Jitousono, T.; Shimizu, O.; Shimokawa, E.; et al. Sediment disasters caused by typhoon Nabi (T 0514) in September 2005. *J. Jpn. Soc. Eros. Control Eng.* **2005**, *58*, 46–53.
42. Sasahara, K.; Kato, H.; Sakurai, W.; Ishizuka, T.; Kaji, A. Large-scale landslides occurred at eastern part of Kochi Pref. due to Typhoon No.6, 2011. *J. Jpn. Soc. Eros. Control Eng.* **2011**, *64*, 39–45.
43. Marutani, T.; Kaibori, M.; Jitousono, T.; Mizuno, H.; Ohno, H.; Shimizu, O.; Kubota, T.; Ue, H.; Kanazawa, A.; Kawano, T.; et al. Sediment-related disasters by a heavy rainfall in the northern part of Kyushu-Island, Japan in July 2017. *J. Jpn. Soc. Eros. Control Eng.* **2017**, *70*, 31–42.

44. Satofuka, Y.; Kosugi, K.; Nakatani, K.; Masaoka, N.; Okano, K.; Kasahara, T.; Yanagisaki, G.; Yamaguchi, Y. Disasters occurred in Kansai region due to the heavy rain event of July 2018. *J. Jpn. Soc. Eros. Control Eng.* **2018**, *71*, 38–48.
45. Hashinoki, T.; Mizuyama, T.; Satoh, K.; Murakami, M. Research on sediment yield and sediment discharge which considers timing of sediment yield. *J. Jpn. Soc. Eros. Control Eng.* **2007**, *59*, 15–22.
46. Furuichi, T.; Osanai, N.; Hayashi, S.; Izumi, N.; Kyuka, T.; Shiono, Y.; Miyazaki, T.; Hayakawa, T.; Nagano, N.; Matsuoka, N. Disastrous sediment discharge due to typhoon-induced heavy rainfall over fossil periglacial catchments in western Tokachi, Hokkaido, northern Japan. *Landslides* **2018**, *15*, 1645–1655. [CrossRef]
47. Kitahara, K.; Fujimoto, K.; Fujii, T.; Osegi, H.; Tokunaga, S.; Takahashi, M.; Sato, T. Sediment-laden flood caused by Typhoon Hagibis in Saku area, Nagano Prefecture. In Proceedings of the Annual Meeting, Japan Society of Erosion Control Engineering, Online, 19–21 May 2021; pp. 25–26.
48. Osakada, Y.; Nakakita, E. Future changes of "Baiu heavy rainfall duration and accumulated precipitation" using the regional climate model verified with past real heavy rainfall events. *J. Jpn. Soc. Civ. Eng. Ser. B1* **2018**, *74*, I_19–I_24. [CrossRef]
49. Osakada, Y.; Nakamura, A.; Nakakita, E. Statistical analysis of future changes in baiu heavy rainfall duration and accumulated rainfall considering its spatio-temporal characteristics. *J. Jpn. Soc. Civ. Eng. Ser. B1* **2020**, *76*, I_7–I_12. [CrossRef]
50. Wu, Y.H.; Nakakita, E.; Kunitomo, M. Future change of rainfall-triggered landslide risk using NHRCM05 based on critical line method. *J. Jpn. Soc. Civ. Eng. Ser. B1* **2020**, *76*, I_67–I_72. [CrossRef]
51. Wu, Y.H.; Nakakita, E.; Yamaji, A. Future change of snake line pattern and its relation to sediment disasters. *J. Jpn. Soc. Civ. Eng. Ser. B1* **2021**, *77*, I_193–I_198. [CrossRef]
52. Uchida, T.; Sakai, Y. Sediment and flood damage as a characteristic form of sediment disaster and its countermeasures. In Proceedings of the Annual Meeting, Japan Society of Erosion Control Engineering, Morioka, Japan, 21–23 May 2021; pp. 7–8.
53. Furukawa, K.; Kaibori, M.; Kubota, T.; Jirousono, T.; Gonda, Y.; Sugihara, S.; Hayashi, S.; Ikeda, A.; Araki, Y.; Kashiwabara, Y. Debris disasters caused by heavy rainfall around Hofu City in Yamaguchi Prefecture on July 21, 2009. *J. Jpn. Soc. Eros. Control Eng.* **2009**, *62*, 62–67.
54. Matsumura, K.; Fujita, M.; Yamada, T.; Gonda, Y.; Numamoto, S.; Tsutsumi, D.; Nakatani, K.; Imaizumi, F.; Shimada, T.; Kaibori, M.; et al. Sediment disasters due to typhoon No.12 at 2011, occurred in Kii peninsula. *J. Jpn. Soc. Eros. Control Eng.* **2012**, *64*, 43–53.
55. Nagumo, N.; Egashira, S. Geomorphological analysis on flood and sediment hazards in the Omoto River resulting from Typhoon Lionrock in 2016. *Trans. Jpn. Geomorphol. Union* **2018**, *39*, 47–66.
56. Kaibori, M.; Hasegawa, Y.; Yamashita, Y.; Sakida, H.; Nakai, S.; Kuwada, S.; Hiramatsu, S.; Jitousono, T.; Irasawa, M.; Shimizu, O.; et al. Sediment related disaster due to heavy rainfall in Hiroshima Prefecture in July 2018. *J. Jpn. Soc. Eros. Control Eng.* **2018**, *71*, 49–60.
57. Jitousono, T.; Igura, M.; Ue, H.; Ohishi, H.; Kakimoto, T.; Kitou, K.; Koga, S.; Sakai, Y.; Sakashima, T.; Shinohara, Y.; et al. Sediment disasters in Kumamoto Prefecture by a heavy rainfall in July 2020. *J. Jpn. Soc. Eros. Control Eng.* **2020**, *73*, 41–50.
58. Sakai, Y.; Yamada, Y.; Sakata, T.; Yamakoshi, T.; Miyase, M.; Sakai, N.; Hayashi, S. Sediment disasters in the Gebasawa River, Chino, Nagano Prefecture on September 2021. *J. Jpn. Soc. Eros. Control Eng.* **2022**, *75*, 35–40.
59. Rokko SABO Office—The Great Hanshin Flood. 1938. Available online: https://www.kkr.mlit.go.jp/rokko/disaster/history/s13/s13-index.php (accessed on 29 March 2022).
60. Tada, Y.; Sanmori, T.; Daimaru, H.; Koyama, K.; Kawai, T. *Changes in Forests and Sediment Disaster over the Past 100 Years*; Japan Society of Erosion Control Engineering: Tokyo, Japan, 2010; pp. 32–33.
61. Ide, M. Looking back upon forest conservation (Chisan) measures of the last 100 years in Japan. *J. Jpn. Soc. Eros. Control Eng.* **2011**, *64*, 40–48.
62. Shuin, Y. Changes in Japan's forests and sediment disaster caused by heavy rainfall. *Sci. Fire Disaster Prev.* **2021**, *143*, 40–46.
63. Rokko SABO Office—Effects of Sabo Facilities against the Torrential Rains in July Heisei 30. Available online: https://www.mlit.go.jp/river/sabo/jirei/h30dosha/H30koukajirei_rokkou_jigyo.pdf (accessed on 29 March 2022).
64. Shinohara, Y.; Kume, T. Changes in the factors contributing to the reduction of landslide fatalities between 1945 and 2019 in Japan. *Sci. Total Environ.* **2022**, *827*, 154392. [CrossRef]
65. Sugimoto, H.; Uto, T.; Honma, H.; Takeshi, T. Features of landslides induced by the 2011 off the Pacific Coast of Tohoku Earthquake, in Shirakawa hills. *Civ. Eng. J.* **2012**, *54*, 18–21.
66. Kaibori, M.; Ishikawa, Y.; Ushiyama, M.; Kubota, T.; Hiramatsu, S.; Fujita, M.; Miyoshi, I.; Yamashita, Y. Debris flow and slope failure disasters in Hiroshima Prefecture caused by feavy rainfall in June, 1999 (Prompt Report). *J. Jpn. Soc. Eros. Control Eng.* **1999**, *52*, 34–43.
67. Kaibori, M.; Ishikawa, Y.; Satufuka, Y.; Matsumura, K.; Nakatani, K.; Hasegawa, Y.; Matsumoto, N.; Takahara, T.; Fukatsuka, K.; Yoshino, K.; et al. Sediment-related disasters induced by a heavy rainfall in Hiroshima-city on 20th August, 2014. *J. Jpn. Soc. Eros. Control Eng.* **2014**, *67*, 49–59.
68. Ministry of Land, Infrastructure, Transport and Tourism—The Report of the Sediment Disaster Prevention Study Committee to Ensure Effective Evacuation. Available online: https://www.mlit.go.jp/river/sabo/committee_jikkousei/190529/hokokusyo.pdf (accessed on 29 March 2022).
69. Sakai, Y.; Tsushima, M.; Yamakoshi, T. Sediment yield in recent events with sediment and flood damage. *Civ. Eng. J.* **2021**, *63*, 30–35.

70. Sakai, Y.; Tsushima, M.; Yamakoshi, T. Topographic analysis of damaged houses on sediment and flood damage. *Civ. Eng. J.* **2021**, *63*, 30–35.
71. Ministry of Land, Infrastructure, Transport and Tourism—Draft Guideline of Investigation for Watershed at Risk of Severe Damage due to Sediment and Flood Damage (Trial version). Available online: https://www.mlit.go.jp/river/shishin_guideline/sabo/dosyakouzuihanran_youryou_r0403.pdf (accessed on 29 March 2022).
72. Kinki Regional Development Bureau, Ministry of Land, Infrastructure, Transport and Tourism—Revaluation of Kizugawa River project. Available online: https://www.kkr.mlit.go.jp/plan/ippan/zigyohyoka/ol9a8v000003p9bu-att/no.3.pdf (accessed on 29 March 2022).
73. Kinki Regional Development Bureau, Ministry of Land, Infrastructure, Transport and Tourism—Revaluation of Kii moutains project. Available online: https://www.kkr.mlit.go.jp/plan/ippan/zigyohyoka/ol9a8v000004e27l-att/no.4.pdf (accessed on 29 March 2022).
74. Ministry of Land, Infrastructure, Transport and Tourism—Manual for Cost-Benefit Analysis of SABO Project (Draft). Available online: https://www.mlit.go.jp/river/sabo/zigyo_hyokasyuho/manualsabo.pdf (accessed on 29 March 2022).
75. Marushita, J.; Kiyotsuna, Y.; Kano, K.; Tanigawa, Y.; Ikeshima, T. A case study of planning measure against sediment and flood damage in the watershed of Tenjin river. In Proceedings of the Annual Meeting, Japan Society of Erosion Control Engineering, Miyazaki, Japan, 11–13 May 2022; pp. 481–482.
76. Kobayashi, Y.; Mori, K.; Kato, Y.; Ochi, T.; Sugiyama, S.; Mori, M.; Nakata, H. A case study of analysis against sediment and flood damage for facility layout plan. Proceedings of annual meeting, Japan Society of Erosion Control Engineering, Online, 19–21 May 2021; pp. 29–30.
77. Ministry of Land, Infrastructure, Transport and Tourism—Summary of Sediment Disasters Caused by Heavy Rainfall in Northern Kyushu, July Heisei 29. Available online: https://www.mlit.go.jp/river/sabo/h29_kyushu_gouu/gaiyou.pdf (accessed on 29 March 2022).
78. Ministry of Land, Infrastructure, Transport and Tourism—5th SABO Technical Panel for Climate Change, Material No.4 Others. Available online: https://www.mlit.go.jp/river/sabo/committee_kikohendo/220422/06siryou4.pdf (accessed on 29 March 2022).
79. Sugimoto, H.; Kamiyama, J.; Fujiwara, K.; Hayashi, S.; Matsumoto, H.; Kito, M.; Sugawara, H.; Kanai, S.; Fujimoto, R.; Okazaki, T.; et al. An investigation of inventory and conditions for occurrence of deep-seated landslide with no clear landslide topography on gentle slope. Proceedings of Annual Meeting, Japan Society of Erosion Control Engineering, Online, 19–21 May 2021; pp. 263–264.
80. Japan Meteorological Agency—Table of Probable Rainfall (51 locations). Available online: https://www.data.jma.go.jp/cpdinfo/riskmap/qt_table.html (accessed on 29 March 2022).
81. Mizuyama, T.; Wada, H.; Yoshida, K. Debris flow control structures installed in zero-order torrents—A debris flow fence. *J. Jpn. Soc. Eros. Control Eng.* **2009**, *62*, 74–76.
82. Ministry of Land, Infrastructure, Transport and Tourism—Technical Supplement for Structural Countermeasure against Debris Flow in Streams with Small Catchment Area and No Flowing Water (Trial version). Available online: https://www.mlit.go.jp/river/shishin_guideline/sabo/muryuusuikeiryuu_ryuuizikou_r0403.pdf (accessed on 29 March 2022).
83. Sabo Planning Division, National Institute for Land and Infrastructure Management, Japan. Manual of Technical Standard for Establishing Sabo Master Plan for Debris Flow and Driftwood. Available online: http://www.nilim.go.jp/lab/bcg/siryou/tnn/tnn0904.htm (accessed on 19 July 2022).
84. Sabo Planning Division, National Institute for Land and Infrastructure Management, Japan. Manual of Technical Standard for Designing Sabo Facilities against Debris Flow and Driftwood. Available online: http://www.nilim.go.jp/lab/bcg/siryou/tnn/tnn0905.htm (accessed on 19 July 2022).
85. Suzuki, M.; Kobashi, S. The critical rainfall for the disasters caused by slope failures. *J. Jpn. Soc. Eros. Control Eng.* **1981**, *34*, 16–26.
86. Shiomi, R.; Otsubo, R.; Ieta, Y.; Kiyono, M.; Nakanishi, H.; Sakai, Y.; Uchida, T. Topographic characteristics of erosion and deposition sites in watersheds caused sediment and flood damage. In Proceedings of the Annual Meeting, Japan Society of Erosion Control Engineering, Morioka, Japan, 21–23 May 2021; pp. 423–424.
87. Tomita, Y.; Mori, T.; Miya, T.; Musashi, Y.; Suzuki, T.; Mizuyama, T. Development of sediment discharge model and deformation of riverbed model for watershed-management system (WMS). *J. Jpn. Soc. Eros. Control Eng.* **2014**, *66*, 3–12.
88. Yamanoi, K.; Fujita, M. Risk estimation of multiple hazards related to sediment and water disasters occurring in heavy rainfall. *J. Jpn. Soc. Civ. Eng. Ser. B1* **2016**, *72*, I_1291–I_1296.
89. Matsushi, Y.; Toyama, M.; Matsuzaki, H.; Chigira, M. Simulation of Soil Production and Transport for Prediction of Location and Magnitude of Shallow Landslides. *Trans. Jpn. Geomorphol. Union* **2016**, *37*, 427–453.
90. Yamazaki, Y.; Egashira, S. Method to estimate landslide and their runout during heavy rainfall. *J. Jpn. Soc. Civ. Eng. Ser. B1* **2020**, *76*, I_1093–I_1098. [CrossRef]
91. Kunitoki, M.; Hayashi, S.; Matsumoto, H.; Kito, M.; Yamakoshi, T.; Sakai, Y.; Uchida, T.; Ikeda, H.; Kawai, E.; Sugihara, S. Study on the estimation method of produced sediment volume using the results of interpretation for slope failure and debris flow. In Proceedings of the Annual Meeting, Japan Society of Erosion Control Engineering, Online, 19–21 May 2021; pp. 567–568.

92. Fujita, A.; Suzuki, K.; Toda, M.; Kawabe, M.; Tanigawa, Y.; Ogawa, S.; Matsunaga, Y.; Makino, H. Basic study for setting design rainfall based on future rainfall forecast data in Uonuma region. Proceedings of Annual Meeting, Japan Society of Erosion Control Engineering, Miyazaki, Japan, 11–13 May 2022; pp. 61–62.
93. Ministry of Land, Infrastructure, Transport and Tourism—Interim Report of SABO Technical Panel for Climate Change. Available online: https://www.mlit.go.jp/river/sabo/committee_kikohendo/200521/chukan_torimatome.pdf (accessed on 29 March 2022).

water

MDPI

Article

Perennial Groundwater Zone Formation Processes in Thin Organic Soil Layers Overlying Thick Clayed Mineral Soil Layers in a Small Serpentine Headwater Catchment

Takahiko Yoshino [1,*] and Shin'ya Katsura [2,3]

1 Graduate School of Agriculture, Hokkaido University, Sapporo 060-0811, Hokkaido, Japan
2 Research Faculty of Agriculture, Hokkaido University, Sapporo 060-0811, Hokkaido, Japan
3 Center for Natural Hazards Research, Hokkaido University, Sapporo 060-8589, Hokkaido, Japan
* Correspondence: t.yoshino@eis.hokudai.ac.jp

Abstract: Groundwater zone formation in the soil layers of a headwater catchment is an important factor that controls volumetric and chemical changes in streamflow; it also induces shallow landslides. Previous studies have suggested that the groundwater zone in soil layers generally forms transiently atop low-permeability layers in response to rainfall. This study focused on an unchanneled hollow in a serpentine headwater catchment, where a semi-perennial to perennial groundwater zone was observed in thin organic soil layers (OSLs) overlying thick clay mineral soil layers (CMLs), even during dry periods. We conducted detailed observations in this catchment to clarify the formation processes of the semi-perennial to perennial groundwater zone. The results showed that water is supplied from the CMLs to the OSLs as unsaturated upward flow in areas where the OSLs are dry. This water then accumulates in the downslope hollow, which sustains the groundwater zone in the OSLs during dry periods. The frequent and long-term occurrence of upward flow can be attributed to differences in the hydraulic properties of OSLs and CMLs. This process prevents the OSLs in the hollow from drying, presumably causes volumetric and chemical changes in streamflow, and reduces the stability of OSLs.

Keywords: perennial groundwater zone; organic soil layers; clay mineral soil layers; headwater catchment; serpentine; unsaturated upward flux

Citation: Yoshino, T.; Katsura, S. Perennial Groundwater Zone Formation Processes in Thin Organic Soil Layers Overlying Thick Clayed Mineral Soil Layers in a Small Serpentine Headwater Catchment. *Water* **2022**, *14*, 3122. https://doi.org/10.3390/w14193122

Academic Editor: Su-Chin Chen

Received: 12 August 2022
Accepted: 28 September 2022
Published: 3 October 2022

1. Introduction

The groundwater zone within the soil layers of a headwater catchment has been recognized as one of the most important factors controlling volumetric [1–3] and chemical [4,5] changes in streamflow and in the induction of shallow landslides [6,7]. Clarification of the formation process of the groundwater zone within soil layers is needed to predict changes in streamflow and the occurrence of shallow landslides. Based on numerous studies involving detailed field observations in various headwater catchments, including the Maimai catchment, New Zealand [8–11]; Walker Branch watershed, U.S.A. [12]; LI1 catchment at Llyn Brianne, Wales [13]; and Hitachi Ohta experimental watershed, Japan [14], the formation process of the groundwater zone can be described as follows. Rainwater vertically infiltrates into permeable soil layers. This water reaches low-permeability layers (i.e., bedrock) and flows laterally along their surfaces, transiently forming a groundwater zone above low-permeability layers in the downslope area. This groundwater zone shrinks through the discharge of water into streams after rainfall has ceased; it generally disappears (especially in small catchments) during dry periods because of the low capacity of a small headwater catchment to store water, combined with high permeability of soil layers.

Although this formation process is widely accepted in various headwater catchments, some studies have identified a groundwater zone within the soil layers, even during dry periods. Frisbee et al. [15] observed groundwater table dynamics in two hillslope transects and calculated the water balance for them. They reported that during periods of low water table position, large-volume groundwater storage was observed in a depression located near the base of the slope that maintained streamflow. Such depressions formed atop less permeable layers are presumed to increase water storage, thereby preventing the groundwater zone from drying up. Fujimoto et al. [16] conducted hydrological observations on two types of hillslopes (valley head and side slope) in a headwater catchment; they demonstrated that the hillslope with highly convergent topography continued to sustain wet conditions through the concentration of subsurface water during dry periods. These studies highlighted the importance of hillslope topography in maintaining the groundwater zone during dry periods. Moreover, although bedrock groundwater is commonly reported to exfiltrate into channels and thus influence stream discharge and quality during baseflow periods [17–19], some studies that included detailed hydrological, hydrochemical, and hydrothermal observations have shown that bedrock groundwater can also exfiltrate into soil layers, thereby generating a perennial groundwater zone in the soil layers. Uchida et al. [20] indicated that expansion and contraction of the groundwater zone in a soil layer at a slope base during periods of normal low flow and drought were controlled by bedrock groundwater exfiltration. Katsura et al. [21] found that a groundwater zone was present within the soil layers at a slope base in a headwater catchment during dry periods; they demonstrated that the groundwater zone was maintained by bedrock groundwater exfiltration. Thus, water flow from bedrock can contribute to the formation of a groundwater zone in the soil layer during dry periods.

In this study, we observed groundwater dynamics in a small headwater catchment (0.068 ha) with gentle topography underlain by serpentine. In this catchment, a semi-perennial to perennial groundwater zone forms in thin, high-permeability organic soil layers (OSLs) in an unchanneled hollow; it remains present during dry periods. The small catchment area, lack of a large depression, low topographic convergence, and presence of thin OSLs with high permeability in this catchment suggest that the effects of catchment topography cannot explain the observed semi-perennial to perennial groundwater zone in the OSLs. A substantial contribution from bedrock groundwater exfiltration is also unlikely due to the presence of mineral soil layers that consist of thick clay mineral layers (CMLs) with low permeability above the bedrock, which inhibit bedrock groundwater flow into the OSLs. Thus, the aforementioned formation processes of a groundwater zone in soil layers, which were based on observations in catchments with no thick clay mineral layers, cannot explain the semi-perennial to perennial groundwater zone observed in this catchment. The groundwater zone observed in this catchment appeared to be controlled by another factor. Using hydrometric, hydrochemical, and thermal observations, this study explored the formation process of the semi-perennial to perennial groundwater zone observed in the OSLs of this catchment.

2. Materials and Methods

2.1. Study Site

The study was conducted in a very small forested headwater catchment (0.068 ha; 44°52′ N, 142°4′ E) on Kunneshiri Mountain in Nakagawa Experimental Forest, which is managed by Hokkaido University and located in Hokkaido, northern Japan (Figure 1). The elevation of the catchment ranges from 241 to 257 m above sea level (Figure 1c), and a spring is present at the catchment outlet. The region has a mean annual temperature of 5.9 °C (1989–2019) and precipitation of 1240 mm (1989–2019) according to data from the nearest weather station (the Japan Meteorological Agency Automated Meteorological Data Acquisition System [(AMeDAS) Nakagawa observation station, located approximately 5 km south-southwest of the catchment). The snowfall season in Nakagawa Experimental Forest is from late November through late April or early May [22], and the maximum snow depth during this season is about 1.7 ± 0.5 m [23]. The dominant tree species in Nakagawa Experimental Forest are *Abies sachalinensis*, *Picea glehnii*, *Picea jezoensis*, *Sorbus commixta*,

Quercus crispula, and *Acer pictum* [24]; the understory is densely covered with *Sasa kurilensis* and *Sasa senanensis* [25].

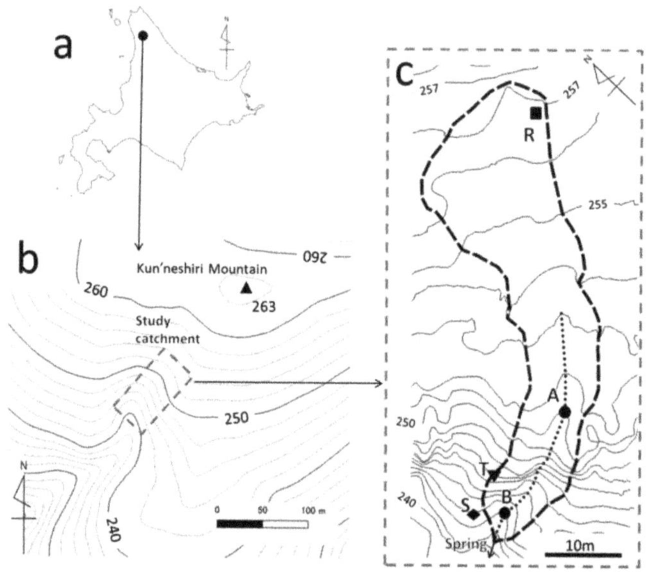

Figure 1. Maps of (**a**) Hokkaido, (**b**) the Kunneshiri Mountain area in Nakagawa Experimental Forest, and (**c**) the study catchment. In panel (**b**), black and gray contours represent intervals of 10 and 2 m, respectively. In panel (**c**), the contour interval is 1 m. Dotted and dashed lines in panel (**c**) indicate an unchanneled hollow in the study catchment and the catchment boundary, respectively.

The study catchment is underlain by serpentinite bedrock and features low elevation, gentle hillslopes, and a rounded ridge with a mean gradient of 14°. These topographic characteristics are consistent with the characteristics of other serpentinite catchments in Japan [26]. The surface soil type of the area underlain by serpentine is "Wet Iron Podzol" [27]. In this catchment, OSLs are very thin, with thickness ranging from 10 to 50 cm. Mineral soil layers derived from weathering of the serpentinite bedrock consist of clayey materials (referred to as CMLs in this study), with thickness ranging from 30 to 170 cm. Each layer is thicker in the middle of the slope and thinner in the upper and lower parts. Whereas macropores and pipes can form within CMLs and contribute to preferential flow in catchments with geological settings other than serpentinite [28,29], they are not observed in the study catchment because the CMLs consist of stiff clayey materials and the high concentration of nickel in serpentine soils inhibits root growth into the CMLs [30].

2.2. Laboratory Measurement

Figure 2 shows a flowchart of the methodology used to explore the formation process of the semi-perennial to perennial groundwater zone observed in the OSLs of this catchment. First, to measure the hydraulic properties of OSLs, CMLs, and the boundary layers between them, we collected undisturbed core samples (100 cc in volume) from a trench (Figure 3) excavated at Point S in Figure 1c. The thicknesses of the OSL and CML at this point are 10 and 33 cm, respectively. In total, three, one, and four samples were collected from the OSL, boundary layer, and CML, respectively. We confirmed that the soil and bedrock structure at this point is almost the same as the structure at other points (e.g., Points A, B, and T; mentioned later). Moreover, the area of the study catchment is very small (0.068 ha). Hence, we assumed that the collected samples were representative of each layer in the catchment.

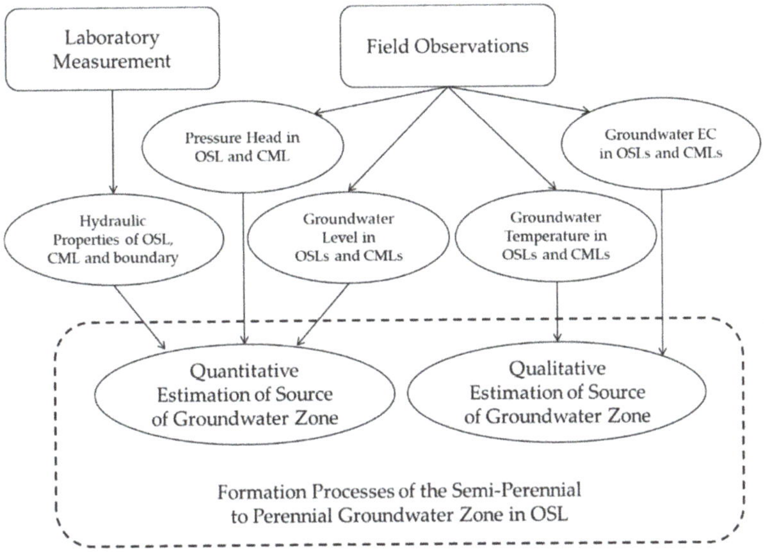

Figure 2. Flowchart of methodology of this study.

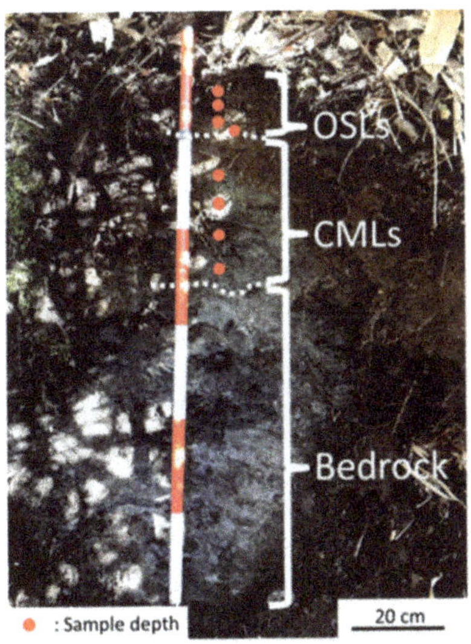

Figure 3. Photograph of the trench at Point S. White dotted lines indicate boundaries between the OSLs and CMLs and between the CMLs and bedrock. Solid red circles indicate the depths at which undisturbed core samples were collected.

We conducted water retention tests on all core samples in the laboratory. To obtain and characterize water retention curves for each sample, we fitted a lognormal model [31] to the observed relationship between volumetric water content, θ, and pressure head, ψ (cm). The lognormal model was derived by applying the lognormal distribution law to the soil pore radius distribution function, and the water retention curve is expressed as follows:

$$S_e = \frac{\theta - \theta_r}{\theta_s - \theta_r} = Q\left(\frac{ln(\psi/\psi_m)}{\sigma}\right) \tag{1}$$

where S_e represents the effective saturation; θ_s and θ_r are the saturated and residual volumetric water contents, respectively; ψ_m is the pressure head at $S_e = 0.5$; σ is a dimensionless parameter characterizing the width of the pore-size distribution; and Q is the complementary normal distribution function:

$$Q(x) = \int_x^\infty \frac{1}{\sqrt{2x}} \exp\left(-\frac{u^2}{2}\right) du \tag{2}$$

We determined the values of θ_r, ψ_m, and σ for each sample through minimization of the residual sum of squares, which was obtained from the measured and calculated values of θ at the applied ψ, with θ_s fixed at the measured value. Finally, the mean water retention curves of OSLs and CMLs were obtained using the mean parameter values of core samples collected from each layer (arithmetic means of θ_s, θ_r, and σ; geometric mean of ψ_m). The water retention curve for the OSL–CML boundary zone was obtained using parameter values determined from the OSL–CML boundary core sample.

We also measured the saturated hydraulic conductivity, K_s, for all core samples in the laboratory. The hydraulic conductivity curve was obtained using the model proposed by Kosugi [31]. The following functional relationship between hydraulic conductivity, K, and ψ was obtained by combining Equation (1) with Mualem's [32] model:

$$K(\psi) = K_s {S_e}^{1/2} \left[Q\left(\frac{In(\psi/\psi_m)}{\sigma} + \sigma \right) \right]^2 \tag{3}$$

Mean hydraulic conductivity curves for OSLs and CMLs were obtained using the geometric mean value of K_s for core samples collected from each layer with the mean parameter values of ψ_m and σ, which were determined as described above. The hydraulic conductivity curve for the OSL–CML boundary was obtained using the measured value of K_s and the parameter values of ψ_m and σ.

2.3. Field Observations

Temporal variations of ψ were measured at Point T (Figure 1c) at 60 min intervals from 15 May to 1 November 2019 using tensiometers. We selected this point because it is located at the catchment boundary in the downslope area and therefore roughly represents the average degree of wetness for the whole catchment. The OSL depth is 31 cm at this location. The tensiometer installation depths were 30 (in the OSLs), 50, and 80 cm (in the CMLs).

To separately monitor groundwater levels and temperature in the OSLs and CMLs, two pairs of observation wells (for OSLs and CMLs) were manually excavated at Points A and B along an unchanneled hollow in the study catchment (Figure 1c). All wells were constructed using 6 cm diameter polyvinyl chloride pipes perforated with numerous 5 mm diameter holes. The depths of the wells for observation of the OSLs at Points A and B (equivalent to the depths to the CML surfaces at each point) were 41 and 38.5 cm, respectively. The depths of the wells for the CMLs at Points A and B were 209 and 164 cm (equivalent to the depths of the bedrock surfaces at each point), respectively. To prevent groundwater in the OSLs from directly infiltrating into the CML wells, the upper sections of the CML wells were unperforated. The depth of the unperforated section was 60 cm at Point A and 43 cm at Point B; it was greater than the OSL depth at each point. A water-level gauge with a temperature recorder was installed at the bottom of each observation well, where it measured the groundwater level and temperature simultaneously at 60 min intervals from 15 May to 1 November 2019.

For analysis of electrical conductivity (EC; µS cm^{-1}), water samples were collected at intervals of approximately 4–5 weeks from mid-May through early November 2019. Samples of groundwater in the OSLs and CMLs were collected directly from the wells. Rainwater samples were collected using a plastic bottle equipped with a mesh-covered 21-cm-diameter funnel located at Point R in a clearing in the study catchment (Figure 1c). The measured EC values were converted to values at a standard temperature of 25 °C.

All rainfall and air temperature data were obtained from the Japan Meteorological Agency AMeDAS Nakagawa observation station.

3. Results

3.1. Hydraulic Properties

Figure 4 shows the water retention and hydraulic conductivity curves of the OSLs, CMLs, and boundary layer. The parameter values are listed in Table 1. The θ value of the OSLs is large at ψ = 0 cm (i.e., θ_s = 0.707); it shows a sharp and continuous decrease (approximately 0.204) as ψ decreases to −200 cm (Figure 4a). The decrease in θ is most prominent in the wet range of $\psi > -10$ cm (0.060). In contrast, θ in the CMLs is smaller at ψ = 0 cm (θ_s = 0.549) than in the OSLs; it decreases less sharply (0.074) from θ_s as ψ decreases to −200 cm, with no prominent decrease in the wet range. Because θ_s is equivalent to the total volume of soil pores and water drains earlier from larger pores than from smaller pores as ψ decreases, these water retention curves suggest that the OSLs contain many pores of various sizes from very large (as indicated by the prominent change in θ in the wet range) to small, whereas the CMLs have only a few large pores. The θ value of the OSL–CML boundary is similar to the θ value of OSLs at ψ = 0 cm (θ_s = 0.718), and it shows an intermediate rate of decrease (0.138) as ψ decreases to −200 cm.

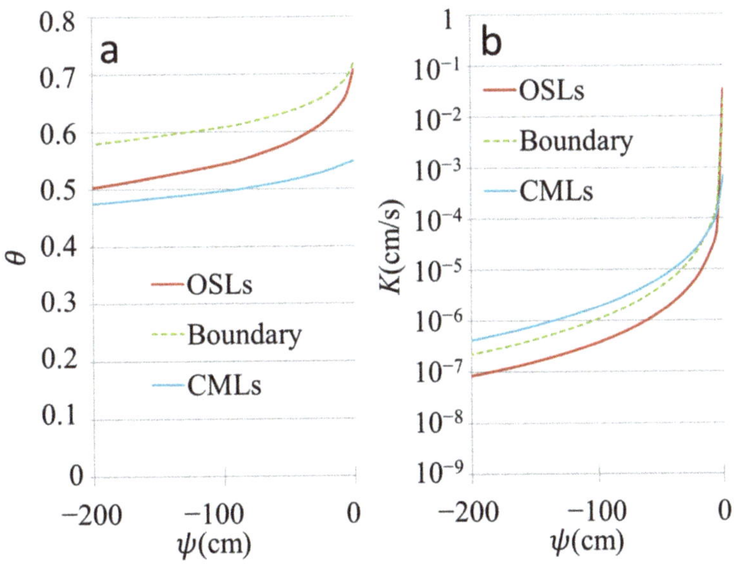

Figure 4. (**a**) Water retention curves and (**b**) hydraulic conductivity curves of the OSLs, CMLs, and boundary zone between the OSLs and CMLs.

Table 1. Parameter values for hydraulic properties.

	K_s cm s^{-1}	θ_s	θ_r	ψ_m cm	σ
OSLs	3.38×10^{-2} [a]	0.707 [b]	0.250 [b]	−303.4 [a]	3.05 [b]
CMLs	6.76×10^{-4} [a]	0.549 [b]	0.374 [b]	−303.2 [a]	2.10 [b]
Boundary	2.14×10^{-2}	0.718	0.442	−200.3	2.67

Note(s): [a] Geometric mean. [b] Arithmetic mean.

The measured K_s values of the OSLs ranged from 1.07×10^{-2} to 6.06×10^{-2} cm s^{-1}, with a geometric mean of 3.38×10^{-2} cm s^{-1}; these values were larger than the K_s of CMLs, which ranged from 2.37×10^{-4} to 6.08×10^{-3} cm s^{-1}, with a geometric mean of

6.76×10^{-4} cm s^{-1} (Table 1). Figure 4b depicts hydraulic conductivity curves for the OSLs, CMLs, and OSL–CML boundary; it shows that the K of the OSLs rapidly decreases from 3.38×10^{-2} cm s^{-1} to 2.34×10^{-5} cm s^{-1} as ψ decreases from 0 to −10 cm. In contrast, the reduction in the K values of CMLs across the same ψ range is smaller (6.76×10^{-4} cm s^{-1} to 6.58×10^{-5} cm s^{-1}). Therefore, although OSLs have a larger K_s (K at ψ = 0 cm) than CMLs, OSLs have a smaller K than CMLs in the range of $\psi < -1.5$ cm ($K < 3.00 \times 10^{-4}$ cm s^{-1}). At the OSL–CML boundary, K decreases from 2.14×10^{-2} to 7.50×10^{-5} cm s^{-1} as ψ decreases from 0 to −10 cm. In the range of $\psi < -10$ cm, OSLs, CMLs, and the OSL–CML boundary all show gentler changes in K than in the range of $\psi > -10$ cm.

3.2. Pressure Head

Figure 5 shows hourly rainfall and ψ in the OSL (30 cm depth) and CMLs (50 and 80 cm depths) at Point T. Total precipitation during the study period was 468.5 mm. Positive and negative ψ values indicate saturated and unsaturated conditions, respectively. The OSLs showed a rapid increase in ψ and often reached saturation in response to rainfall, whereas periods without rainfall resulted in ψ decreases and severe drying. Careful examination of Figure 5b demonstrates that ψ in the OSL displayed a rapid decrease from saturation to approximately −10 cm, followed by a gentler decrease as ψ decreased further. This tendency can be attributed to the water retention curve of OSLs (Figure 4a). The observed rapid decrease in ψ from saturation to $\psi = -10$ cm presumably reflects rapid drainage of water through very large pores. Based on the tendencies described here, we defined wet and dry periods in this study as periods with OSL ψ of >−10 cm and <−10 cm, respectively. The variations of ψ are shown in Figure 5b, c for wet and dry periods, respectively. The 80 cm sampling depth in the CMLs showed smaller ψ responses to rainfall and was constantly saturated in both wet and dry periods. The 50 cm sampling depth in the CMLs became saturated in response to rainfall and gradually returned to unsaturation after rainfall ceased. The ψ fluctuated less at 50 cm than at 30 cm (in the OSLs); it remained close to saturation even when ψ values at depths upward of 20 cm indicated severely dry conditions. These ψ dynamics indicate wetter conditions in the CMLs than in the OSLs, especially during dry periods.

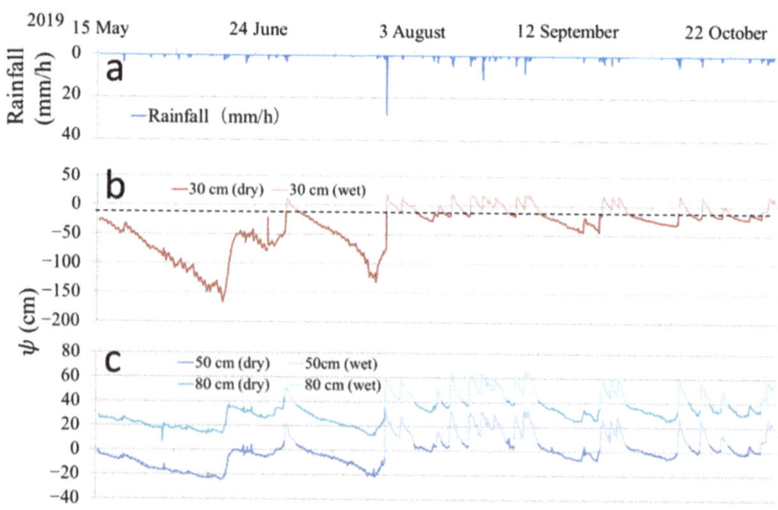

Figure 5. (**a**) Rainfall and ψ in the (**b**) OSLs (30 cm) and (**c**) CMLs (50 and 80 cm). Dashed line in panel (**b**) represents ψ = −10 cm.

3.3. Groundwater Level and Temperature

Figure 6 shows hourly rainfall and the groundwater level in the OSLs and CMLs at Points A and B. The groundwater level is presented as the depth below the ground surface, and the OSL–CML boundary is denoted with a dashed line. In the study catchment, we observed two distinct groundwater levels, in OSLs and CMLs (OSL groundwater and CML groundwater, respectively). Observation data for OSL groundwater at Point A are missing because of equipment failure from 26 May at 13:00 through 3 July at 10:00.

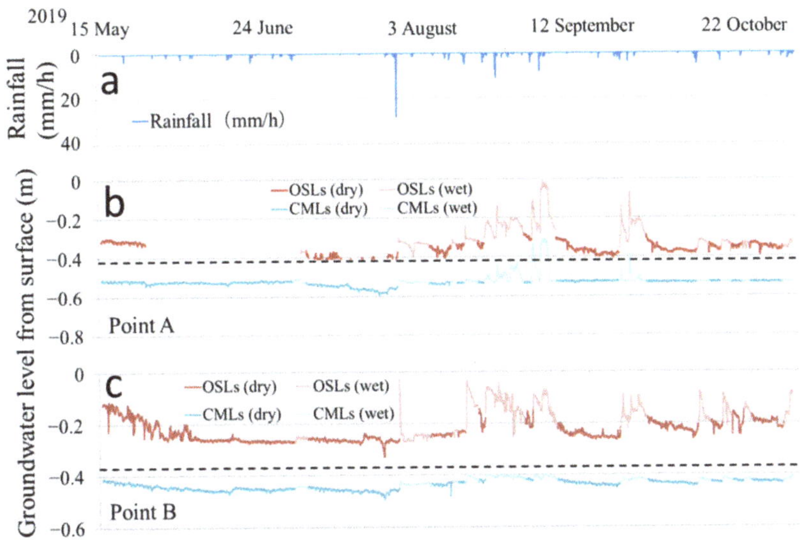

Figure 6. (**a**) Rainfall and groundwater levels at (**b**) Point A and (**c**) Point B. The groundwater level is presented as depth below the ground surface. Dashed lines in panels (**b**,**c**) indicate the boundary between the OSLs and CMLs.

At Point A, the OSL groundwater showed rapid fluctuations in response to rainfall; it was present throughout the observation period except in July, when low precipitation and high evapotranspiration during the summer drought period in Hokkaido [33] caused the groundwater level to intermittently decrease to zero (Figure 6b). Notably, OSL groundwater was sustained from 2 September through 17 September despite extremely low rainfall (5.0 mm) during this period. In contrast, the CML groundwater level was less responsive to rainfall. The CML groundwater level generally did not reach the OSL–CML boundary, although a few rainfall events increased the CML groundwater level above the OSL–CML boundary. Even during such periods, the CML groundwater level did not match the OSL groundwater level, suggesting distinct groundwater zones in the OSLs and CMLs. The minimum groundwater level in the CMLs was −0.593 m, which was observed on 22 July at 16:00 during the drought period; this level was 0.183 m below the boundary.

At Point B, the groundwater level in the OSLs was observed continuously; it showed steep increases and decreases in response to rainfall (Figure 6c). Excluding the rapid decrease and recovery of the groundwater level within the OSLs from 23 July at 1:00 to 14:00 during the summer drought period, the minimum groundwater level was −0.282 m on 17 July at 16:00 (in a dry period), which was 0.103 m above the boundary; this finding indicated that groundwater was permanently present at approximately 10 cm above the OSL–CML boundary, even during dry periods. Conversely, the CML groundwater response to rainfall was gentle and small. The CML groundwater level did not reach the OSL–CML boundary throughout the observation period. The minimum groundwater level in the CMLs was −0.492 m, which was observed on 23 July at 14:00 (in a dry period) and was

0.107 m below the boundary. Note that the groundwater level dynamics at Points A and B were similar to the pressure head dynamics at Point T (Figure 5), except that Points A and B, located in the hollow, were wetter.

Figure 7 shows rainfall, air temperature, and the temperature measured at the bottom of each observation well. Because groundwater was nearly always present in all wells, we hereafter refer to the temperature in the wells as the groundwater temperature. The groundwater temperature in the OSLs at Points A and B showed diurnal variations that can be attributed to changes in air or rainwater temperature, as well as a seasonal trend with a peak in early August. In contrast, the groundwater temperature in the CMLs at each point was minimally affected by diurnal air temperature changes or rainwater; it showed gentle seasonal variations with a peak in mid-September. The dynamics of groundwater temperature did not differ between dry and wet periods.

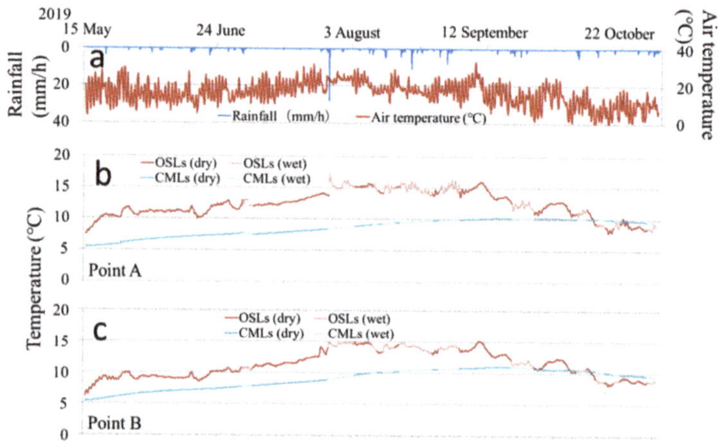

Figure 7. (**a**) Rainfall and air temperature and groundwater temperature at (**b**) Point A and (**c**) Point B.

3.4. Electrical Conductivity

Figure 8 shows the EC values of water samples. Rainwater EC (ranging from 9 to 34 μS cm^{-1}) was significantly lower than the values of other samples based on the two-tailed *t*-test ($p < 0.05$). Two water samples from the OSLs at Point A showed EC of 147 and 180 μS cm^{-1}. The EC of OSL groundwater at Point B ranged from 162 to 279 μS cm^{-1}; this overlapped the values of CML groundwater. Greater variation in groundwater EC in the CML was observed at Point B than at Point A (from 221 to 379 μS cm^{-1} and from 377 to 426 μS cm^{-1}, respectively). Overall, the OSL groundwater had significantly lower values than the CML groundwater ($p < 0.05$). Figure 8 also shows that the EC values of OSL groundwater tended to be higher during dry periods than during wet periods. No such tendency was recognized for CML groundwater.

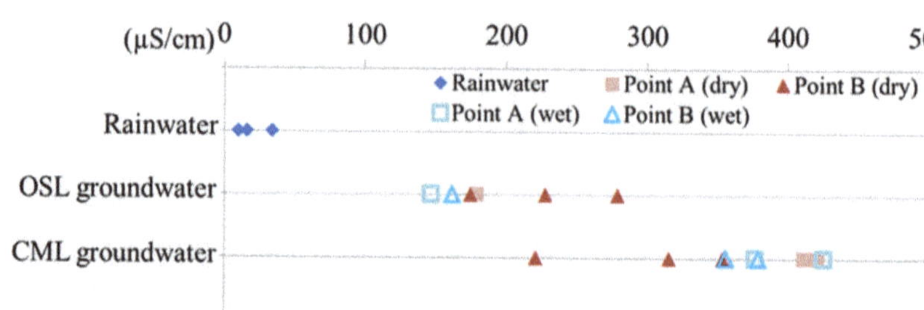

Figure 8. EC of collected water samples.

4. Discussion

4.1. Sources of Groundwater

While semi-perennial to perennial groundwater zones were present in both OSLs and CMLs within the unchanneled hollow, their fluctuation characteristics differed. The responses of CML groundwater to rainfall were small and gentle (Figure 6), presumably because of the hydraulic properties of CMLs, which have low permeability even under saturated conditions and contain small numbers of large pores (Figure 4). Additionally, ψ in the CMLs at Point T was consistently high, even during dry periods (Figure 5c), because the large number of small pores in the CMLs prevents drainage of water stored in the pores during dry periods. In contrast, the OSL groundwater level in the hollow and ψ at Point T showed rapid fluctuations in response to rainfall (Figures 5b and 6). These fluctuations during wet periods suggest rapid rainwater infiltration into and drainage from large pores, reflecting the hydraulic properties of OSLs (Figure 4). The process of groundwater generation in OSLs in the hollow during wet periods can be explained by the conventional conceptual model described in the Introduction section: rainwater vertically infiltrates into the highly permeable OSLs until it reaches the less permeable CMLs. Water then laterally flows along the CML surface, forming a groundwater zone above CMLs in the downslope area. The EC of OSL groundwater tended to be lower during wet periods than during dry periods (Figure 8), suggesting that the infiltration of rainwater with low EC (Figure 8) affected the OSL groundwater during wet periods.

Surprisingly, the groundwater zone remained present for a long period of time in the OSLs in the hollow during dry periods (Figure 6). The high permeability of the OSLs does not allow groundwater in the OSLs to be maintained consistently during dry periods. Additionally, because the study catchment is small, with gentle hillslopes and thin OSLs, no large depression or convergent topography is available to provide water storage. The EC values of OSL groundwater collected from a depth of several tens of centimeters during dry periods were high (175 to 279 $\mu S\ cm^{-1}$; Figure 8). In Japan, reported EC values in soil layers at a few centimeters directly below leaf litter were high (>300 $\mu S\ cm^{-1}$), especially in autumn when many broadleaf trees drop their leaves; these values decreased with infiltration into OSLs [34,35]. Sakuma and Sato [34] showed that the EC values of water collected from OSLs overlying pyroclastic fall deposits in Hokkaido, Japan, were in the range of approximately 30–70 $\mu S\ cm^{-1}$. Miyata et al. [36] reported that, in OSLs at the depth of 20 cm in a granitic catchment in Gifu Prefecture, Japan, the mean EC value of stored water was 17.36 $\mu S\ cm^{-1}$. In a catchment consisting of volcanic rock and sedimentary rock in Akita, Japan, the EC of water stored at approximately 20–40-cm depth in an OSL was 51 $\mu S\ cm^{-1}$ [35]. Compared with the EC values measured in these previous studies, the OSL groundwater EC values observed in the present study during dry periods were surprisingly high. This comparison suggests that the high values of OSL groundwater EC during dry periods are influenced by CML groundwater, which has high EC values (Figure 8); moreover, CML groundwater is an important source of groundwater present in the OSLs during dry periods.

One possible explanation for the contribution of CML groundwater to the groundwater zone in the OSLs involves runoff of CML groundwater into the OSLs via macropores or pipes within the CMLs. However, no such structures were observed in the study catchment. The results of temperature observations (Figure 7) suggested that such runoff of CML groundwater into the OSLs was unlikely; the groundwater temperature in the CMLs showed moderate seasonal variations and minimal impact from diurnal air or rainwater temperature, whereas water in the OSLs was sensitive to air and rainwater temperature. If macropores or pipes had contributed to water flow, the CML groundwater temperature would be affected by air or rainwater temperature, such that it would show sensitive fluctuation during wet periods; during dry periods, the OSL groundwater temperature would show moderate fluctuations driven by runoff of CML groundwater. The EC values of CML groundwater showed no clear difference between dry and wet periods (Figure 8), which is consistent with these inferences.

These findings suggest that CML groundwater contributed to maintenance of the groundwater zone in OSLs in the hollow during dry periods through a mechanism that did not involve runoff of CML groundwater into the OSLs via macropores or pipes.

4.2. Water Flow Direction and Flux Analysis across OSL-CML Boundary

4.2.1. Flow Direction Analysis

To examine the contribution of CML groundwater to the OSL groundwater zone in the hollow during dry periods, we analyzed the water flow direction between OSLs and CMLs using the ψ values recorded at Point T. We focused on the flow direction between the 30 and 50 cm tensiometers, which were located immediately above and below the OSL–CML boundary (depth: 31 cm). Figure 9 shows the temporal variations in the difference in total head, H, defined as the sum of ψ and the elevation head, between 30 and 50 cm. We used the ground surface as the reference level for calculating elevation head. A positive value of H indicates upward flow from the CMLs into the OSLs, whereas a negative value indicates downward flow from the OSLs into the CMLs. Figure 9 demonstrates that the main flow direction during wet periods was downward.

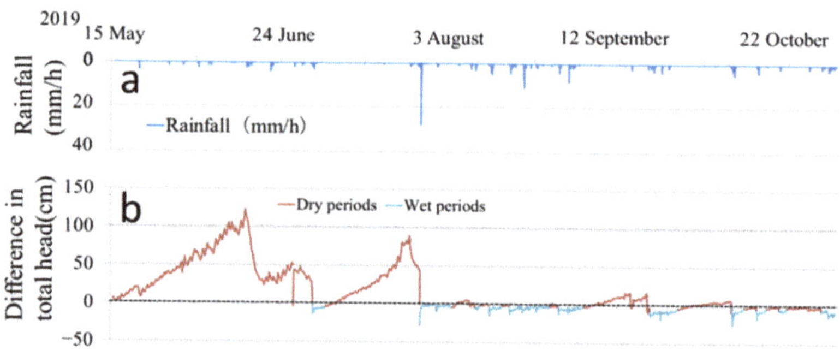

Figure 9. (**a**) Rainfall and (**b**) the difference in total head between the depths of 30 and 50 cm at Point T.

Figure 10 shows the relationship between simultaneous ψ measurements from the 30 and 50 cm tensiometers (ψ_{30} and ψ_{50}, respectively); in this plot, two groups are distinguished according to the flow direction (upward or downward). During wet periods, downward flow was observed for 1104 h, whereas upward flow was observed for only 1 h. Thus, downward flow from the OSLs to the CMLs occurred mainly during wet periods including rainfall periods, indicating that a portion of the rainwater infiltrated into the CMLs and recharged the CML groundwater. Conversely, after the cessation of rainfall, downward flow decreased and the flow direction switched from downward to upward; upward flow was observed during dry periods (Figure 9). Upward and downward flows were observed during a dry period at 2296 and 662 h, respectively (Figure 10). Thus, upward flow consistently occurred during dry periods. The OSLs, with high permeability under saturated conditions and a large number of large pores, became unsaturated and dried up during dry periods. The CMLs, with low permeability under saturated conditions and no large pores, remained nearly saturated even during dry periods. The frequent and long-term occurrence of upward flow can be attributed to these differences in hydraulic properties between OSLs and CMLs.

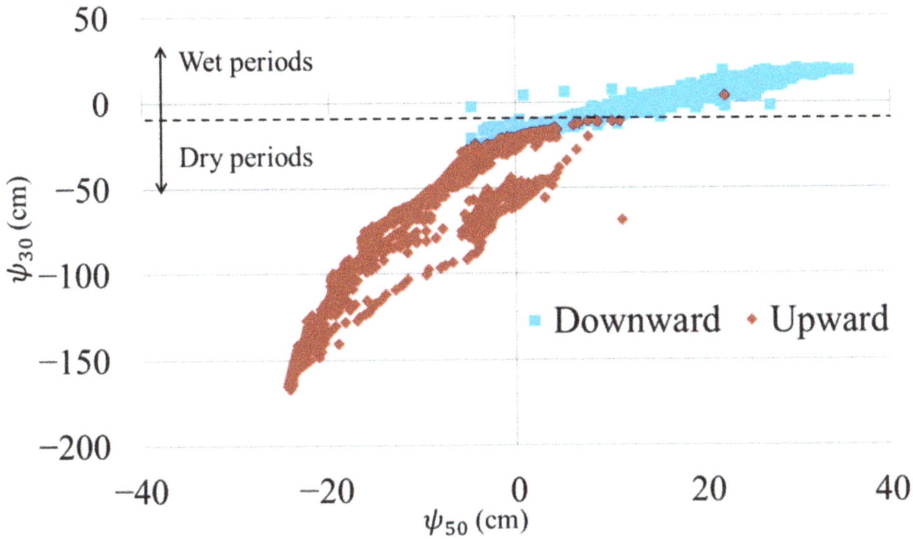

Figure 10. Relationship between simultaneous ψ_{30} and ψ_{50} observations. Dashed line represents $\psi = -10$ cm.

4.2.2. Upward Flux Analysis during Dry Periods

Next, we analyzed the vertical water flux across the OSL–CML boundary at Point T using H. According to Darcy's Law, water flux, q, at the boundary can be calculated as follows:

$$q = -K_{bo}(\psi_{31})\frac{H}{s} \tag{4}$$

where K_{bo} is the hydraulic conductivity of the OSL–CML boundary, ψ_{31} is ψ at the boundary (at 31 cm depth), and s is the distance in the direction of flow (20 cm). We computed ψ_{31} as follows:

$$\psi_{31} = \frac{\psi_{50}s_{30} + \psi_{30}s_{50}}{s} \tag{5}$$

where s_{30} and s_{50} are the vertical distances from the boundary to the 30 and 50 cm tensiometers (i.e., 1 and 19 cm), respectively. $K_{bo}(\psi_{31})$ was calculated using the hydraulic conductivity curve of the OSL–CML boundary shown in Figure 4b.

Figure 11b, c shows the temporal variations in upward and downward q, respectively. Note that downward q in Figure 11c is plotted on a logarithmic scale. The maximum upward flux during dry periods was 0.36 mm h^{-1}, observed on 1 July at 17:00, which was extremely small compared to the maximum downward flux during wet periods of 983 mm h^{-1}, which was observed on 9 October at 0:00. However, the upward flux during dry periods averaged 0.16 mm h^{-1} and totaled 360 mm, equivalent to three-quarters of the total precipitation (468.5 mm). This similarity suggests that the accumulation of water that moved upward from the CMLs into the OSLs in areas where the OSLs were dry ($\psi < -10$ cm; Figure 10) contributed to the maintenance of the groundwater zone in the OSLs along the hollow (including Points A and B) during dry periods.

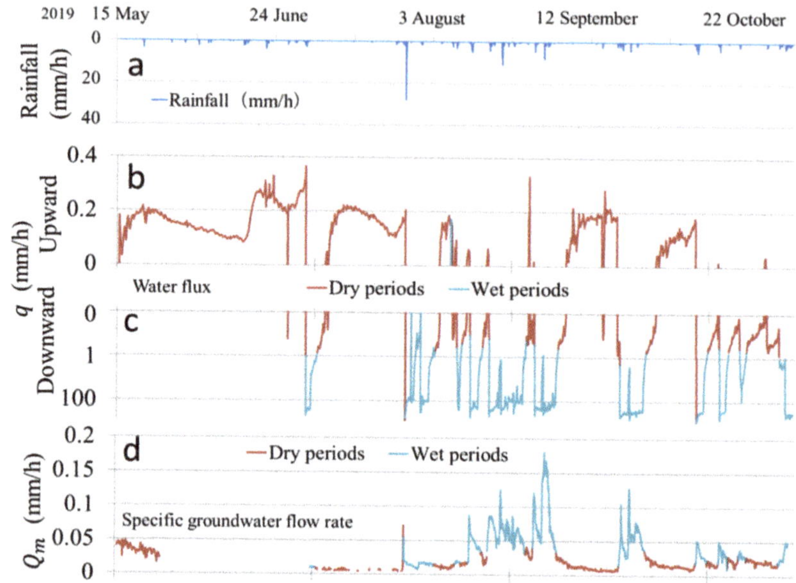

Figure 11. (**a**) Rainfall, (**b**) upward and (**c**) downward water flux at Point T, and (**d**) rate of specific groundwater flow at the midpoint between Points A and B.

To determine whether the calculated upward q can explain the observed groundwater level in the OSLs in the hollow during dry periods, we conducted a crude calculation of the specific groundwater flow rate in the OSLs at the midpoint between Points A and B. The rate of specific groundwater flow in the OSLs at the midway point, Q_m, can be approximated as follows:

$$Q_m = \frac{q_m\, A_C}{A_m} \tag{6}$$

where q_m is groundwater flux at the midway point, A_C is the cross-sectional area of the groundwater zone at the midway point calculated assuming a triangular area, and A_m is the catchment area at the midway point (0.056 ha). By applying Darcy's law to saturated groundwater flow in the OSLs between Points A and B, q_m can be approximated as follows:

$$q_m = K_{s\_OSLs} \frac{H_{AB}}{S_s} \tag{7}$$

where K_{s_OSLs} is the mean saturated hydraulic conductivity of OSLs (3.38×10^{-2} cm s^{-1}), H_{AB} is the difference in elevation of the groundwater table in the OSLs between Points A and B, and S_s is the slope length from Points A to B (19.0 m).

Figure 11d shows the temporal variations of Q_m (data are missing from 26 May at 13:00 though 3 July at 10:00 due to the lack of the groundwater level data at Point A). The maximum Q_m was 0.18 mm h^{-1}, observed on 31 August at 14:00 during a wet period. During dry periods, Q_m reached its maximum (0.072 mm h^{-1}) on 27 July at 3:00 and averaged 0.018 mm h^{-1}. The findings in Figure 11b,d indicate that upward q was larger than Q_m for nearly the entire observation period when upward flux was generated. The sums of upward q and Q_m during dry periods with upward flux from 1 July to 31 October were 172.5 and 9.76 mm, respectively. Kondo et al. [37] estimated that mean transpiration in July through October in this region (including the study catchment) was 172 mm. Comparison of these values suggests that most of the upward q is consumed by transpiration (some water that flowed only through the OSLs was presumably also consumed by transpiration), whereas the remaining q could generate the observed saturated groundwater flow in the OSL in the hollow. Thus, water supplied from the CMLs to OSLs as upward flux accumulated and contributed to the groundwater zone in the OSLs in the hollow during dry periods.

4.3. Conceptual Model of Formation Process of OSL Groundwater Zone

Based on the discussion above, Figure 12 illustrates the formation processes of the groundwater zone in the OSLs in the hollow of the study catchment. During wet periods including rainfall periods (Figure 12a), rainwater vertically infiltrates into the highly permeable OSLs until it reaches the less permeable CMLs; it then becomes perched groundwater above the OSL–CML boundary. The perched groundwater flows downslope along the OSL–CML boundary as saturated lateral flow, thereby expanding or increasing the depth of the groundwater zone in the OSLs in the hollow. In areas where the OSLs are wet ($\psi > -10$ cm), a portion of this water infiltrates into the CMLs as downward flux (Figure 10) and slowly recharges the CML groundwater.

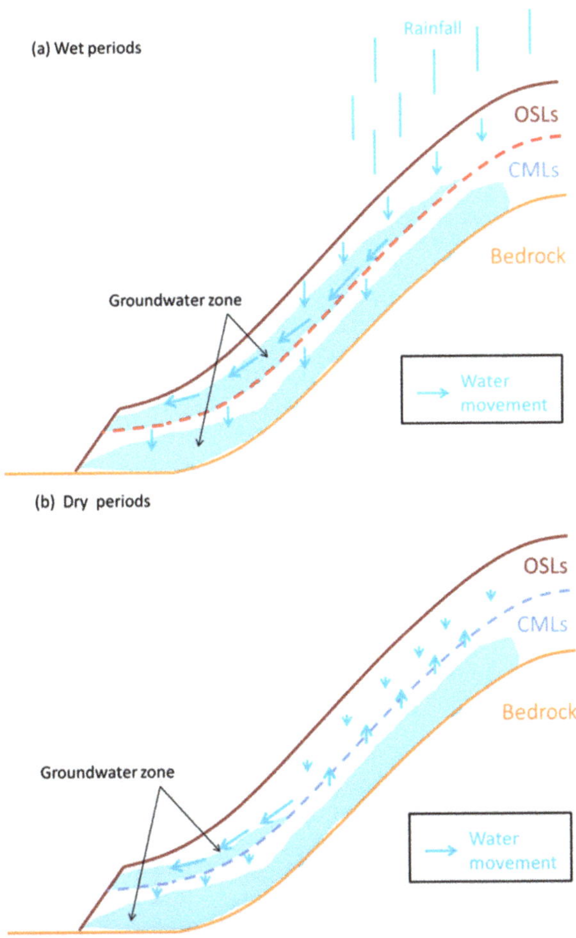

Figure 12. Formation processes of the groundwater zone within the OSLs in the study catchment during (**a**) wet periods and (**b**) dry periods.

In contrast, during dry periods (Figure 12b), the contribution of rainwater to the OSL groundwater zone decreases and upward flow from the CMLs occurs mainly in areas where the OSLs are dry ($\psi < -10$ cm) (Figure 10). This water then flows downslope along the OSL–CML boundary as unsaturated or saturated lateral flow, forming the groundwater zone in the OSLs in the hollow of the downslope area.

4.4. Importance of Water Supply from CMLs to OSLs as Upward Flux during Dry Periods

Upward flux in a headwater catchment has been reported previously [20,38,39], and two mechanisms of upward flux generation have been proposed. One mechanism is unsaturated upward flux resulting from the balance of the matrix potential of soil layers (i.e., unsaturated matrix flow), as observed in this study, which occurs as soil layers become dry and the groundwater level decreases [39–42]. The other mechanism is saturated upward flux, which occurs in association with the local exfiltration of bedrock groundwater into soil layers through fissures and cracks in the bedrock [20,38,43–45]. Although saturated upward flux is often considered as an important water supply process for soil layers, unsaturated upward flux caused by matrix flow has attracted minimal attention because of the small fluxes involved relative to saturated upward fluxes. Frisbee et al. [15] reported large-volume groundwater storage in a depression located near the base of the slope in a 0.60 ha catchment during periods of low water table position. In this case, the soil thickness of the depression was as much as approximately 4 m. From the results of this study, unsaturated upward flux may be an important process that sustains wet soil layers during dry periods in still smaller headwater catchments with thinner soil layers and without saturated upward flux from bedrock.

At Point T in the study catchment, upward flux from the CMLs to the OSLs generally occurred as unsaturated flow during dry periods. Although the maximum upward flux was small (0.36 mm h^{-1}), the total upward flux over the entire observation period was large (360 mm). This continuous upward flux contributed to the maintenance of the groundwater zone in the OSLs in the hollow of the downslope area during dry periods, which presumably caused subsequent volumetric and chemical changes in streamflow. Moreover, in a serpentinite catchment in Hokkaido, Japan, Aipassa [46] described the occurrence of surface slides (a type of slope failure) with the slip surface located at the OSL–CML boundary; such slides were reported to occur on steep slopes when rainfall is sufficiently heavy to saturate the slip surface. Although no traces of such slides were found in the study catchment, presumably because of the low gradient, continuous unsaturated upward flux can influence such slides by raising the groundwater level or expanding the groundwater zone in the OSLs of hollows during rainfall periods through the generation of a small groundwater zone before the rainfall, which reduces the stability of the OSLs. Thus, the present findings emphasize the important role of water supply from CMLs to OSLs during dry periods in the formation of a groundwater zone in OSLs in hollows, in addition to its effects on streamflow and shallow landslides.

5. Conclusions

Detailed hydrological, hydrochemical, and thermal observations in a forested serpentinite headwater catchment containing mineral soil layers composed of thick clayey materials produced by weathering of bedrock (i.e., CMLs), along with hydraulic property measurements in the laboratory, were conducted to elucidate the formation process of the semi-perennial to perennial groundwater zone observed in thin OSLs within an unchanneled hollow in the study catchment, even during dry periods. The groundwater temperature and chemistry suggested that runoff of the CML groundwater into the OSLs via macropores or pipes is unlikely. Instead, water flux analysis across the OSL–CML boundary revealed that unsaturated upward flow from the CMLs into OSLs occurs in areas where the OSLs are dry (pressure head < -10 cm). Accumulation of this upward flux can explain the groundwater zone observed in the OSLs of the hollow during dry periods. We conclude that water infiltrates from the OSLs into CMLs during wet periods (including rainfall periods) and then is supplied into the OSLs as upward flux during dry periods; this upward flux contributes to the generation of a semi-perennial to perennial groundwater zone in the OSLs in the hollow. Land users and management organizations of catchments underlain by thick CMLs should pay attention to volumetric and chemical changes in streamflow and the surface slides that could be caused by the semi-perennial to perennial groundwater zone. In future research, detailed observations of other serpentinite catchments and catchments underlain by other bedrock types with thick CMLs (e.g., mudstone and slate catchments) are needed to determine the generalizability of the present findings.

Author Contributions: Conceptualization, T.Y. and S.K.; methodology, T.Y. and S.K.; investigation, T.Y. and S.K.; data curation, T.Y.; writing—original draft preparation, T.Y.; writing—review and editing, T.Y. and S.K.; supervision, S.K.; funding acquisition, S.K. All authors have read and agreed to the published version of the manuscript.

Funding: This research was funded by Japan Society for the Promotion of Science (JSPS) KAKENHI grants (nos. JP18H03819 and JP19H02392).

Data Availability Statement: The data that support the findings of this study are available from the corresponding author upon reasonable request.

Acknowledgments: The authors thank the staff of Nakagawa Experimental Forest, Field Science Center for Northern Biosphere, Hokkaido University. They also thank their colleagues for support in the field.

Conflicts of Interest: The authors declare no conflict of interest.

References

1. Katsuyama, M.; Ohte, N. Determining the sources of stormflow from the fluorescence properties of dissolved organic carbon in a forested headwater catchment. *J. Hydrol.* **2002**, *268*, 192–202. [CrossRef]
2. Detty, J.M.; McGuire, K.J. Threshold changes in storm runoff generation at a till-mantled headwater catchment. *Water Resour. Res.* **2010**, *46*, W07525. [CrossRef]
3. Martínez-Carreras, N.; Hissler, C.; Gourdol, L.; Klaus, J.; Juilleret, J. Storage controls on the generation of double peak hydrographs in a forested headwater catchment. *J. Hydrol.* **2016**, *543*, 255–269. [CrossRef]
4. Bishop, K.; Seibert, J.; Köhler, S.; Laudon, H. Resolving the double paradox of rapidly mobilized old water with highly variable responses in runoff chemistry. *Hydrol. Process.* **2004**, *18*, 185–189. [CrossRef]
5. Gannon, J.P.; Bailey, S.W.; Mcguire, K.J.; Shanley, J.B. Flushing of distal hillslopes as an alternative source of stream dissolved organic carbon in a headwater catchment. *Water Resour. Res.* **2015**, *51*, 8114–8128. [CrossRef]
6. Van Asch, T.; Buma, T.; VanBreek, L. A view on some hydrological triggering systems in landslides. *Geomorphology* **2005**, *30*, 25–32. [CrossRef]
7. Sidle, R.C.; Bogaard, T.A. Dynamic earth system and ecological controls of rainfall-initiated landslides. *Earth-Sci. Rev.* **2016**, *159*, 275–291. [CrossRef]
8. Mosley, M.P. Streamflow generation in a forested watershed, New Zealand. *Water Resour. Res.* **1979**, *15*, 795–806. [CrossRef]
9. Mosley, M.P. Subsurface flow velocities through selected forest soils South Island, New Zealand. *J. Hydrol.* **1982**, *55*, 65–92. [CrossRef]
10. Pearce, A.J.; Stewart, M.K.; Sklash, M.G. Storm runoff generation in humid headwater catchments. 1. Where does the water come from? *Water Resour. Res.* **1986**, *22*, 1263–1272. [CrossRef]
11. McGlynn, B.L.; McDonnell, J.J.; Brammer, D.D. A review of the evolving perceptual model of hillslope flowpaths at the Maimai catchments, New Zealand. *J. Hydrol.* **2002**, *257*, 1–26. [CrossRef]
12. Wilson, G.V.; Jardine, P.M.; Luxmoore, R.J.; Zelazny, L.W.; Todd, D.E.; Lietzke, D.A. Hydrogeochemical processes controlling subsurface transport from an upper subcatchment of walker branch watershed during storm events. 2. Solute transport processes. *J. Hydrol.* **1991**, *123*, 317–336. [CrossRef]
13. Soulsby, C. Hydrological controls on acid runoff generation in an afforested headwater catchment at Llyn Brianne, Mid-Wales. *J. Hydrol.* **1992**, *138*, 431–448. [CrossRef]
14. Sidle, R.C.; Tsuboyama, T.; Noguchi, S.; Hosoda, I.; Fujieda, M.; Shimizu, T. Stormflow generation in steep forested headwaters: A linked hydrogeomorphic paradigm. *Hydrol. Process.* **2000**, *14*, 369–385. [CrossRef]
15. Frisbee, M.D.; Allan, C.J.; Thomasson, M.J.; Mackereth, R. Hillslope hydrology and wetland response of two small zero-order boreal catchments on the Precambrian Shield. *Hydrol. Process.* **2007**, *21*, 2979–2997. [CrossRef]
16. Fujimoto, M.; Ohte, N.; Tani, M. Effects of hillslope topography on hydrological responses in a weathered granite mountain, Japan: Comparison of the runoff response between the valley-head and the side slope. *Hydrol. Process.* **2008**, *22*, 2581–2594. [CrossRef]
17. Uhlenbrook, S.; Frey, M.; Leibundgut, C.; Maloszewski, P. Hydrograph separations in a mesoscale mountainous basin at event and seasonal timescales. *Water Resour. Res.* **2002**, *38*, 1096. [CrossRef]

18. Gabrielli, C.P.; McDonnell, J.J.; Jarvis, W.T. The role of bedrock groundwater in rainfall-runoff response at hillslope and catchment scales. *J. Hydrol.* **2012**, *450–451*, 117–133. [CrossRef]
19. Zhao, S.; Hu, H.; Harman, C.J.; Tian, F.; Tie, Q.; Liu, Y.; Peng, Z. Understanding of storm runoff generation in a weathered, fractured granitoid headwater catchment in northern China. *Water* **2019**, *11*, 123. [CrossRef]
20. Uchida, T.; Asano, Y.; Ohte, N.; Mizuyama, T. Analysis of flowpath dynamics in a steep unchannelled hollow in the Tanakami Mountains of Japan. *Hydrol. Process.* **2003**, *17*, 417–430. [CrossRef]
21. Katsura, S.; Kosugi, K.; Mizutani, T.; Okumura, S.; Mizuyama, T. Effects of bedrock groundwater on spatial and temporal variations in soil mantle groundwater in a steep granitic headwater catchment. *Water Resour. Res.* **2008**, *44*, W09430. [CrossRef]
22. Hiura, T.; Uejima, N.; Okuda, A.; Hojo, H.; Ishida, T.; Okuyama, S. Crowding effects on diameter growth for *Abies sachalinensis*, *Quercus crispula* and *Acer mono* trees in a northern mixed forest, Nakagawa Experimental Forest. *Res. Bull. Hokkaido Univ. For.* **1998**, *55*, 255–261. (In Japanese with English Summary)
23. Kobayashi, M.; Templer, P.H.; Katayama, A.; Seki, O.; Takagi, K. Early snowmelt by an extreme warming event affects understory more than overstory trees in Japanese temperate forest. *Ecosphere* **2022**, *13*, e4182.
24. Hiura, T.; Sato, G.; Hayashi, I. Long-term forest dynamics in response to climate change in northern mixed forests in Japan: A 38-year individual-based approach. *For. Ecol. Manag.* **2019**, *449*, 117469. [CrossRef]
25. Hiura, T.; Fujiwara, K. Density-dependence and co-existence of conifer and broad-leaved trees in a Japanese northern mixed forest. *J. Veg. Sci.* **1999**, *10*, 843–850. [CrossRef]
26. Suzuki, T. Formative processes of specific features of serpentinite mountains. *Trans. Jpn. Geomorphol. Union* **2006**, *27*, 417–460.
27. Nakata, N.; Kojima, S. Effects of serpentine substrate on vegetation and soil development with special reference to *Picea glehnii* forest in Teshio District, Hokkaido, Japan. *For. Ecol. Manag.* **1987**, *20*, 265–290. [CrossRef]
28. Tani, M. Runoff generation processes estimated from hydrological observations on a steep forested hillslope with a thin soil layer. *J. Hydrol.* **1997**, *200*, 84–109. [CrossRef]
29. Newman, B.D.; Campbell, A.R.; Wilcox, B.P. Lateral subsurface flow pathways in a semiarid ponderosa pine hillslope. *Water Resour. Res.* **1998**, *34*, 3485–3496. [CrossRef]
30. Gabbrielli, R.; Pandolfini, T.; Vergnano, O.; Palandri, M.R. Comparison of two serpentine species with different nickel tolerance strategies. *Plant Soil* **1990**, *122*, 271–277. [CrossRef]
31. Kosugi, K. Lognormal distribution model for unsaturated soil hydraulic properties. *Water Resour. Res.* **1996**, *32*, 2697–2703. [CrossRef]
32. Mualem, Y. A new model for predicting the hydraulic conductivity of unsaturated porous media. *Water Resour. Res.* **1976**, *12*, 513–522. [CrossRef]
33. Ishii, Y.; Kodama, Y.; Nakamura, R.; Ishikawa, N. Water balance of a snowy watershed in Hokkaido, Japan. In *Northern Research Basins Water Balance (Proceedings of a Workshop Held at Victoria, Canada, March 2004)*; IAHS Publication: Oxfordshire, UK, 2004; Volume 290, pp. 13–27.
34. Sakuma, T.; Satoh, F. Dynamics of inorganic elements in soils under deciduous forest (Part 1). *Res. Bull. Coll. Exp. For.* **1978**, *44*, 537–552. (In Japanese with English Summary)
35. Hayakawa, A.; Ota, H.; Asano, R.; Murano, H.; Ishikawa, Y.; Takahashi, T. Sulfur-based denitrification in streambank subsoils in a headwater catchment underlain by marine sedimentary rock in Akita, Japan. *Front. Environ. Sci.* **2021**, *9*, 664488. [CrossRef]
36. Miyata, S.; Uchida, T.; Asano, Y.; Ando, H.; Mizuyama, T. Effects of bedrock groundwater seepage on runoff generation at a granitic first order stream. *J. Jpn. Soc. Eros. Control. Eng.* **2003**, *56*, 13–19. (In Japanese with English Abstract)
37. Kondo, J.; Nakazono, M.; Watanabe, T.; Kuwagata, T. Hydrological climate in Japan (3) Evapotranspiration from forest. *J. Jpn. Soc. Hydrol. Water Resour.* **1992**, *5*, 8–18. [CrossRef]
38. McGlynn, B.L.; McDonnell, J.J.; Shanley, J.B.; Kendall, C. Riparian zone flowpath dynamics during snowmelt in a small headwater catchment. *J. Hydrol.* **1999**, *222*, 75–92. [CrossRef]
39. Harria, A.H.; Shand, P. Near-stream soil water-groundwater coupling in the headwaters of the Afon Hafren, Wales: Implications for surface water quality. *J. Hydrol.* **2006**, *331*, 567–579. [CrossRef]
40. Tsujimura, M. Behavior of subsurface water in a steep forested slope covered with a thick soil layer. *J. Jpn. Assoc. Hydrol. Sci.* **1993**, *23*, 3–18. (In Japanese with English Abstract)
41. Tsurita, T.; Yoshinaga, S.; Abe, T. Estimation of vertical water flux through forest subsoil using Buckingham-Darcy Equation. *J. Jpn. For. Soc.* **2009**, *91*, 151–158. (In Japanese with English Abstract) [CrossRef]
42. Gannon, J.P.; McGuire, K.J.; Bailey, S.W.; Bourgault, R.R.; Ross, D.S. Lateral water flux in the unsaturated zone: A mechanism for the formation of spatial soil heterogeneity in a headwater catchment. *Hydrol. Process.* **2016**, *31*, 3568–3579. [CrossRef]
43. Montgomery, D.R.; Dietrich, W.E.; Torres, R.; Anderson, S.P.; Heffner, J.T.; Loague, K. Hydrologic response of a steep, unchanneled valley to natural and applied rainfall. *Water Resour. Res.* **1997**, *33*, 91–109. [CrossRef]
44. Haga, H.; Matsumoto, Y.; Matsutani, J.; Fujita, M.; Nishida, K.; Sakamoto, Y. Flow paths, rainfall properties, and antecedent soil moisture controlling lags to peak discharge in a granitic unchanneled catchment. *Water Resour. Res.* **2005**, *41*, W12410. [CrossRef]
45. Masaoka, N.; Kosugi, K.; Yamakawa, Y.; Tsutsumi, D. Processes of bedrock groundwater seepage and their effects on soil water fluxes in a foot slope area. *J. Hydrol.* **2016**, *535*, 160–172. [CrossRef]
46. Aipassa, M.I. Landslide debris movement and its effects on slope and river channel in mountainous watershed. *Res. Bull. Coll. Exp. For.* **1991**, *48*, 375–418.

water

MDPI

Article

Transition Indices of Sediment-Transport Modes on a Debris Flow Resulting from Changing Streambed Gradients

Takashi Wada [1,*], Hirotada Mishima [1], Jin Takemura [2], Kazuki Kobayashi [3] and Hiroshi Miwa [1]

1 Department of Social Systems and Civil Engineering, Graduate School of Engineering, Tottori University, Tottori 680-8552, Japan; m22j6029b@edu.tottori-u.ac.jp (H.M.); miwa-h@tottori-u.ac.jp (H.M.)
2 Wakayama Prefectural Office, Wakayama 640-8585, Japan; takemura_j0001@pref.wakayama.lg.jp
3 Asahi Consultant Co., Ltd., Tottori 680-0911, Japan; k.kobayashi@asahic.co.jp
* Correspondence: wada-t@tottori-u.ac.jp; Tel.: +81-857-31-5284

Abstract: We conducted experiments using an experimental flume with two variable streambed gradients in the upstream and downstream parts with various debris flows, composition sizes, and supply flow rates. We investigated the transition processes of sediment transport modes along the longitudinal distances from the gradient change point using the transition mode indices, $I_{\overline{Cs}x}$, $I_{\overline{h}x}$, and $I_{\overline{U}x}$; these indices were calculated based on measurements of sediment transport concentrations, flow depths, and gravel migration velocities in the debris flow's front in the downstream part. Using these indices, we postulated that after the debris flow passed the gradient change point, the transition of the sediment transport modes progressed by changing the measured parameters to those in the steady-state condition on the gradient of the downstream parts. In addition, these indices suggested that the gravel migration velocities in the flow front interior changed most rapidly after passing the gradient change point, and that flow depths tended to change most slowly. Finally, the indices suggested that as the debris flow material became finer and the supplied flow rates became larger, the longitudinal transition sections tended to be longer because the momentum needed to transport the material was less than the total debris flow momentum.

Keywords: debris flow; sediment transport mode; transition process; changing streambed gradient; sediment transport concentration; flow depth; gravel migration velocity; flume experiment

Citation: Wada, T.; Mishima, H.; Takemura, J.; Kobayashi, K.; Miwa, H. Transition Indices of Sediment-Transport Modes on a Debris Flow Resulting from Changing Streambed Gradients. *Water* **2022**, *14*, 1810. https://doi.org/10.3390/w14111810

Academic Editor: Su-Chin Chen

Received: 30 April 2022
Accepted: 31 May 2022
Published: 4 June 2022

1. Introduction

In recent years, debris flow disasters have become more frequent in various regions of East and Southeast Asia [1–5]. To protect lives and property from these disasters, it is necessary to more precisely understand the characteristics of debris flows reaching flood plains containing residential and social structures, such as alluvial fans, and to install effective countermeasures or predict the damage caused by debris flows based on these characteristics. The important characteristics of debris flows are the flow velocity, flow depth, and sediment concentration. To investigate these parameters, the sediment transport mode of debris flow during the downflow in mountain streams must be understood.

Before a debris flow reaches the flood plains, it flows down mountain streams with continuously changing streambed gradients from steep upstream areas, such as in a valley head, to gentle downstream areas, such as the top of an alluvial fan. Therefore, the sediment transport mode changes stepwise during the downflow in mountain streams owing to the continuously changing streambed gradients. If the streambed gradient is 14–15° or higher, transported materials can be dispersed throughout the entire flow layer by the shear forces of high-density materials in the flow [6–10]. We consider this sediment transport mode the "debris flow". However, if the streambed gradient is below 14–15°, the materials cannot be dispersed throughout the flow layer and tend to settle down as the vertical downward component of gravity increases. Therefore, the flow consists of the lower layer, which is composed of high-density materials, and the upper layer, which is composed of turbulent water flow with suspended loads [11–14]. We consider this sediment transport mode the "sediment sheet flow" or "immature debris flow". The thickness of the lower layer with high-density materials decreases as the streambed gradient becomes gentler. When the streambed gradient is approximately 1–2°, the sediment transport mode changes to individual transportation; that is, bed load transport. The thickness of the lower layer with high-density materials in this transportation mode is approximately one to two times the grain diameter [14]. Considering the above, the sediment transport mode of a debris flow is a sediment sheet flow or bed load transport when the flow reaches the downstream end of a mountain stream that connects to the top of an alluvial fan, because the streambed gradient is often less than 15°. However, the flow front may be considered to maintain the "debris flow", with the materials dispersed throughout the entire flow layer, even when it reaches the stream outlet. Thus, to predict the damage caused by a debris flow on an alluvial fan more accurately, it is necessary to accurately determine the characteristics of the debris flow supplied to the fan from mountain streams based on the transition processes of sediment transport modes owing to the changing streambed gradients.

Previous studies have clarified the sediment transport mechanisms in debris flow and sediment sheet flow on a flume with a constant gradient [11–16]. However, except for the recent experimental studies on the changing of the flow characteristics for a granular sediment flow or soil mass with little moisture at the gradient change points, such as the bottom of the failure slope [17–19], only a few studies have clarified the sediment transport mechanisms in flumes with continuously changing gradients [20,21]. These studies focused only on the process of debris flow generation by a surface runoff on a streambed with steep gradients (more than 15°), while few studies have focused on the transition of sediment transport modes on the gradient changes within 2–15°, where a sediment sheet flow or bed load transport would be pronounced. The sediment flow model proposed by Egashira et al. [10,15] enables the reproduction of various modes using their constitutive laws for all streambed gradients. However, the experimental results used to verify their proposed model were also based on a flume with a constant gradient; there has not been sufficient verification of this model based on the transition processes of sediment transport modes with changing gradients. Although these conventional constitutive laws are set in the x-coordinate direction along the flow direction on the riverbed surface, the constitutive law of bed load transport in the global coordinates, which is not affected by the changing of the x-coordinate direction due to riverbed deformation, has been proposed [22,23]. These laws can estimate the sediment transport rates under non-negligible transversal and longitudinal gradients by reducing the effects of the additional laws of the erosion and deposition processes, considering the separation of the gravity vector component in the flow and transversal directions. Since these proposed constitutive laws do not deal with high-density material flow, such as debris flow, it is necessary to construct a constitutive law based on a global coordinate system for the flow and to examine whether the transition of sediment transport modes due to changing gradients can be expressed. Taking a numerical approach to sediment transport mechanisms with changing gradients, Takahama et al. [24] proposed a numerical model for a two-layer flow composed of a turbulent water flow in the upper layer and a concentrated sediment flow in the lower layer, while Suzuki and Hotta [25] proposed a numerical model using the moving particle semi-implicit (MPS) method. Their models were applied to the debris flow behaviors and depositions at changing gradient points. Since these models are based on the sediment transport mechanisms proposed by Egashira et al. [10,15], it is necessary for these models to sufficient verify the based mechanisms in the transition processes of sediment transport modes with changing gradients. Few other researchers except them have attempted numerical modeling for the transition processes of sediment transport modes.

A debris flow is composed of materials of various particle sizes, and these particles influence each other during the downflow, resulting in the occurrence of particle size segregation in the flow interior. For a stony debris flow with many boulders, the boulders become concentrated toward the flow front during the downflow in mountain streams [4,26–31].

Ashida et al. [14] conducted flume experiments on the transport mechanisms of sediment mixtures with gentle flume bed gradients of 1–8°. They clarified that although the particle migration velocities of various sizes in the flow interior are generally consistent, finer particles fall to the bottom of the flow. Furthermore, it is difficult for finer particles to migrate because the coarser particles above block them from moving; thus, the coarsening of the migrating particles becomes more pronounced. This indicates that the migrating particle size segregation (coarsening of migrating particles) also occurs in the interior of the sediment sheet flow. This was also evident in the flume experiments on debris flows composed of sand and gravel mixtures conducted at constant gradients ranging from 3 to 18° by Wada et al. [32]. Thus, to predict the damage caused by a stony debris flow on an alluvial fan more accurately, it is necessary to focus on the transitions of sediment transport modes for debris flows composed of sand and gravel mixtures resulting from changing streambed gradients.

Based on the above background, this study focused on the transition processes of sediment transport modes for the debris flow consisting of uniform-sized and mixed-sized gravel types resulting from changing streambed gradients from more than 15° to less than 15°. We conducted the flume experiments using an experimental flume consisting of two variable gradients in the upstream and downstream parts. Based on these experimental results, we calculated the transition indices of sediment transport modes and investigated the corresponding transition processes along the longitudinal distances from the gradient change point using these indices.

2. Materials and Methods

2.1. Experimental Setup and Conditions

Figure 1 shows the experimental flume and the measurement equipment used in our study. The experimental flume consisted of two variable streambed gradients in the upstream and downstream parts, with lengths of 150 cm and widths of 10 cm, with a fixed bed part for the rectification of the supplied water at the upstream end of the flume. The gradients of the upstream and downstream parts, θ_1, θ_2, were set to three types; a gentle uniform gradient ($\theta_1 = \theta_2 = 9°$), a steep uniform gradient ($\theta_1 = \theta_2 = 15°$), and a changed gradient ($\theta_1 = 15°$, $\theta_2 = 9°$). The connecting point of the upstream and downstream parts was defined as the gradient change point (at +0 cm; the positive direction was the downstream side). In addition, the length of the downstream part could be changed to 50 cm, 70 cm, or 100 cm. After the experimental materials were placed at a depth of 5 cm in the upstream and downstream parts (the total material volume including bulks equaled 15,000 cm^3), the material was eroded by the supplied water from the upstream end, resulting in the generation of debris flow. We measured the sediment transport concentration of the flow front, C_s, at the downstream end of each length flume (at +50, +80, +100, and +150 cm) using a movable sediment sampler moving in the transverse direction with respect to the flow direction. The sampler separated the debris flow front into the four boxes over the time intervals in the range of 1.0–2.0 s. Measurements were performed to determine the temporal changes in the sediment transport concentrations of the flow front in each sample. We also measured the flow front depths, h, using three ultrasonic water level sensors (E4C-DS30, OMRON Corp., Kyoto, Japan) at +10, +50, and +80 cm points, as well as the gravel migration velocities, u, in the flow front's interior using two high-speed video cameras (EXILIM PRO EX-F1, CASIO COMPUTER Co., Ltd., Tokyo, Japan) at +0 and +30 cm points.

The experimental conditions combined with the particle size compositions of the debris flow materials and inflow rates of the supplied water are listed in Table 1. The gravel sizes and debris flow depths of these conditions were set within the gravel size and gravel-size-to-flow-depth ratio ranges used in the previous experiments on debris flow with sediment mixtures performed by Wada et al. [32]. Table 1 also shows the averaged flow depth, flow velocity, Froude number, and particle Reynolds number for the "frontal part" of the debris flow for all cases. In this study, we focused on the effects of the changing gradient on the sediment transport mode at the "frontal part" of the debris flow, whereby the "frontal part" was defined as the part within 3 s from reaching the flow front at a certain measuring point. The magnitude orders of the Froude number and particle Reynolds number were consistent with those of the dimensionless parameters describing previous field and experimental debris flows, as organized by Turnbull et al. [33]. This indicated that our experiments replicated the previous field and experimental debris flows in term of the flow characteristics. The flow velocities on the gentle gradient of the downstream part were greater than those on the steep gradient of the upstream part in many cases. The reason for this may be because the debris flow velocity depends on the magnitude relation regarding the increase or decrease in the streambed gradient and sediment concentration, as shown in Equation (1) below. Under our experimental conditions, the strongest effect on the debris flow velocity is the decreasing sediment concentration rather than the decreasing streambed gradient.

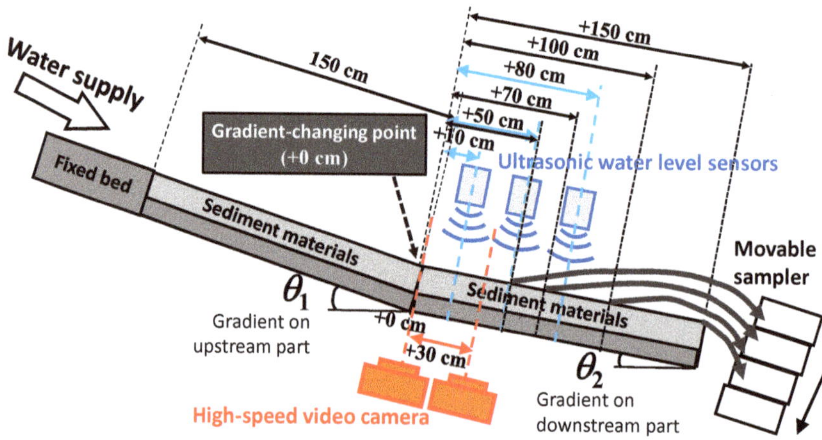

Figure 1. Experimental setup.

Table 1. Experimental cases and conditions.

Case	Debris Flow Composition		Inflow Rate [2] (cm^3/s)	Flume Gradient [2] θ_1, θ_2 (°) [3]	Measured Representative Values of the "Frontal Part" [4] of a Debris Flow with a Changing Gradient				
	d_L, d_S (mm) [1]	P_L, P_S [1]			Condition	$\overline{h_f}$ [5] (mm)	$\overline{U_f}$ [6] (mm/s)	Fr_f [7]	Re_{*f} [8]
Case 1	19.0, 7.1	20%, 80%	2000 1000	Gentle uniform gradient 9°, 9° Steep uniform gradient 15°, 15° Changed gradient 15°, 9°	Inflow rate; 2000 cm^3/s				
					Upstream part (15°)	29.767	545.45	1.028	2651.78
					Downstream part (9°)	51.300	571.43	0.811	2676.44
					Inflow rate; 1000 cm^3/s				
					Upstream part (15°)	30.000	260.87	0.489	2662.13
					Downstream part (9°)	24.533	363.64	0.746	1850.86
Case 2	7.1, non	100%, 0%			Inflow rate; 2000 cm^3/s				
					Upstream part (15°)	29.567	315.79	0.597	1979.35
					Downstream part (9°)	40.453	615.38	0.983	1780.02
					Inflow rate; 1000 cm^3/s				
					Upstream part (15°)	27.667	321.43	0.628	1914.70
					Downstream part (9°)	29.008	387.10	0.730	1507.33
Case 3	7.1, 3.0	80%, 20%			Inflow rate; 2000 cm^3/s				
					Upstream part (15°)	36.000	387.10	0.663	1931.84
					Downstream part (9°)	43.250	615.38	0.951	1627.96
					Inflow rate; 1000 cm^3/s				
					Upstream part (15°)	24.000	285.71	0.599	1577.34
					Downstream part (9°)	30.367	390.24	0.719	1364.12

Table 1. *Cont.*

Case	Debris Flow Composition		Inflow Rate [2] (cm^3/s)	Flume Gradient [2] θ_1, θ_2 (°) [3]	Measured Representative Values of the "Frontal Part" [4] of a Debris Flow with a Changing Gradient				
	d_L, d_S (mm) [1]	P_L, P_S [1]			Condition	$\overline{h_f}$ [5] (mm)	$\overline{U_f}$ [6] (mm/s)	Fr_f [7]	Re_{*f} [8]
Case 4	19.0, 3.0	20%, 80%			Inflow rate; 2000 cm^3/s				
					Upstream part (15°)	28.700	545.45	1.046	1702.92
					Downstream part (9°)	39.767	571.43	0.921	1541.15
					Inflow rate; 1000 cm^3/s				
					Upstream part (15°)	19.867	500.00	1.152	1416.83
					Downstream part (9°)	31.167	363.64	0.662	1364.36
Case 5	7.1, 3.0	20%, 80%			Inflow rate; 2000 cm^3/s				
					Upstream part (15°)	31.667	375.00	0.685	1102.11
					Downstream part (9°)	38.033	888.89	1.464	928.61
					Inflow rate; 1000 cm^3/s				
					Upstream part (15°)	19.200	181.82	0.426	858.17
					Downstream part (9°)	31.858	545.45	0.982	849.89
Case 6	3.0, non	100%, 0%			Inflow rate; 2000 cm^3/s				
					Upstream part (15°)	21.333	510.64	1.136	710.41
					Downstream part (9°)	36.747	727.27	1.219	716.84
					Inflow rate; 1000 cm^3/s				
					Upstream part (15°)	23.100	276.92	0.592	739.24
					Downstream part (9°)	28.600	631.58	1.120	632.40

Note: [1] d_1 and d_2 are the diameters of coarser and finer gravel types, respectively; P_1 and P_2 are the initial compositions of coarser and finer gravel types, respectively; [2] common conditions in all cases; [3] θ_1 and θ_2 are the gradients of the upstream and downstream parts in the experimental flume, respectively; [4] "frontal part" is the part within 3 s from reaching the flow front at a certain measuring point; [5] $\overline{h_f}$ is the averaged flow depth at the "frontal part" of a debris flow; [6] $\overline{U_f}$ is the averaged front velocity of a debris flow in the section from 50 cm upstream from the gradient change point to 50 cm downstream from the point; [7] Fr_f is the Froude number, $= \overline{U_f} / \left(g\overline{h_f}\cos\theta\right)^{0.5}$, where g is the gravitational acceleration, 9.81×10^2 mm/s^2; [8] Re_{*f} is the particle Reynolds number, $= d_m\left(g\overline{h_f}\tan\theta\right)^{0.5}/\nu$, where d_m is the mean diameter of the debris flow material and ν is the kinematic viscosity coefficient, 1.00 mm^2/s.

The particle size compositions were prepared by mixing one or two particles of the three particles with diameters of 3.0 mm, 7.1 mm, and 19.0 mm according to the ratios shown in Table 1. The average mass density of these gravel particles (σ) was 2.650 g/cm^3, the average concentration in the static sediment bed (C_*) was 0.575, and the average internal friction angle (ϕ) was 34.80°. Here, 80% of the particles in Cases 1–3 were gravel with a diameter of 7.1 mm, while 80% of the particles in Cases 4–6 were gravel with a diameter of 3.0 mm. For all conditions, the inflow rates of the supplied water were set to 1000 and 2000 cm^3/s. Note that in Cases 2, 5, and 6, the flow front depth measurements at +10 and +80 cm and the measurements of gravel migration velocities in the flow front interior at +30 cm were not taken, while in Cases 1 and 4, the measurements of sediment transport concentrations of the flow front at +50, +80, and +100 cm were also not taken. This was due to missing or oscillating measurement data caused by a malfunction of the measurement equipment.

In our experiments, no significant topographic changes due to the deposition of debris flow at the gradient change point (+0 cm) occurred during the "frontal part" passing at the point. Therefore, the effects of topographic changes on the transition of sediment transport modes for the "frontal part" were minor.

2.2. Transition Index of Sediment Transport Modes

Using the measurement results, the averaged sediment transport concentration, $\overline{C_s}$, the averaged flow depth, $\overline{h}$, and the vertical averaged gravel migration velocity, $\overline{U}$, at the "frontal part" of the debris flow were calculated. $\overline{C_s}$ is calculated using the following equation, using the volumes of water and sediment included in the samples of debris flow fronts obtained by the four boxes in the movable sampler:

$$\overline{C_s} = \sum_{i=1}^{4}(V_{sLi} + V_{sSi}) / \sum_{i=1}^{4}(V_{sLi} + V_{sSi} + V_{wi}) \tag{1}$$

where V_{sLi} and V_{sSi} are the coarser and finer sediment volumes included in the samples obtained by the i-th boxes, respectively; V_{wi} is the water volume included in the samples obtained by the i-th boxes; and i is the number of each box (i = 1–4). Note that in the cases with the debris flow consisting of uniform-sized gravel, V_{sSi} is zero. In the following, $\overline{C_s}$, $\overline{h}$, and $\overline{U}$ in the cases of uniform gradients of 15° and 9° are denoted as $\overline{C_{s15°}}$, $\overline{C_{s9°}}$, $\overline{h}_{15°}$, $\overline{h}_{9°}$, $\overline{U}_{15°}$, and $\overline{U}_{9°}$, respectively; and $\overline{C_s}$, $\overline{h}$, and $\overline{U}$ in the cases of changing gradients are denoted as $\overline{C_{sx}}$, $\overline{h}_x$, and $\overline{U}_x$ at x cm downstream from the gradient change point (+0 cm), respectively. $\overline{C_{s15°}}$, $\overline{C_{s9°}}$, and $\overline{C_{sx}}$ were calculated by averaging the sediment transport concentrations of debris flows obtained by the four samplers. Here, $\overline{h}_{15°}$ and $\overline{h}_{9°}$ were the theoretical debris flow depths on the uniform gradients of 15° and 9°, respectively, obtained using the theoretical averaged velocity equation for a stony debris flow as proposed by Takahashi [6] (Equation (2)) and the continuity equation of the flow (Equation (3)). We applied Equation (2) to the consideration of our experimental results because the scales of our experimental conditions were equivalent to Takahashi's experiments, which were conducted to confirm the equation's validity [6]. The streambed surface (i.e., the x-coordinate direction that is the basis of Equation (2)) did not change significantly at the gradient change point during the passing point of the "frontal part" in our experiments:

$$U_m = \frac{2}{5d_m}\left[\frac{g \sin\theta}{\alpha_i \sin\phi}\{C_\infty + (1 - C_\infty)\}\left(\frac{\rho_m}{\sigma}\right)\right]^{1/2}\left\{\left(\frac{C_*}{C_\infty}\right)^{1/3} - 1\right\}\overline{h}^{3/2} \tag{2}$$

$$q = U_m \overline{h} \tag{3}$$

where U_m is the vertical averaged velocity of the debris flow, d_m is the mean diameter of the material, g is the gravitational acceleration, θ is the streambed gradient, α_i is the coefficient (= 0.042), and ρ_m is the mass density of the interstitial fluid (= 1.0 g/cm^3). C_∞ is the equilibrium sediment concentration from Equation (4), which is derived from the equilibrium of the riverbed shear stress and the body forces of the debris flow on the riverbed in the dynamic equilibrium state of the flow [6]:

$$C_\infty = \frac{\rho_m \tan\theta}{(\sigma - \rho_m)(\tan\phi - \tan\theta)} \tag{4}$$

where $\overline{h}_x$ is the averaged measurement value of the flow front depth within 3 s from the time at which the flow front reaches the measuring point; $\overline{U}_{15°}$ and $\overline{U}_{9°}$ are the averaged gravel migration velocities within the "median depth" for the theoretical velocity distribution, as suggested by Takahashi et al. [34], for uniform gradients of 15° and 9°, respectively. We defined the "median depth" as the vertical range from $1/3\overline{h}$ to $2/3\overline{h}$ in the interior of the "frontal part". The theoretical velocity distribution used in this study can be adopted for the velocity distributions of both stony debris and sediment sheet flows (see [34] for details on the theoretical distribution). Note that the internal friction angles (ø) for calculating the theoretical distribution were adjusted according to the experimental results based on the uniform gradients of 15° and 9°. Here, $\overline{U}_x$ is the averaged measurement value within the "median depth" of the "frontal part".

Based on the values of the uniform gradients, $\overline{C_{s15°}}$, $\overline{C_{s9°}}$, $\overline{h}_{15°}$, $\overline{h}_{9°}$, $\overline{U}_{15°}$, and $\overline{U}_{9°}$, and the measurement results for the changing gradient, $\overline{C_{sx}}$, $\overline{h}_x$, and $\overline{U}_x$, the transition indices of sediment transport modes at x cm downstream from the gradient change point (+0 cm point), $I_{\overline{Csx}}$, $I_{\overline{h}x}$, and $I_{\overline{U}x}$, were obtained using the following equations:

$$I_{\overline{Csx}} = \frac{\overline{C_{sx}} - \overline{C_{s\theta 1}}}{\overline{C_{s\theta 2}} - \overline{C_{s\theta 1}}} \tag{5}$$

$$I_{\overline{h}x} = \frac{\overline{h}_x - \overline{h}_{\theta 1}}{\overline{h}_{\theta 2} - \overline{h}_{\theta 1}} \tag{6}$$

$$I_{\overline{U}x} = \frac{\overline{U}_x - \overline{U}_{\theta 1}}{\overline{U}_{\theta 2} - \overline{U}_{\theta 1}} \tag{7}$$

where $\overline{C_{s\theta 1}}$, $\overline{h}_{\theta 1}$, and $\overline{U}_{\theta 1}$ are the averaged sediment transport concentration, the theoretical flow depth, and the averaged gravel migration velocities within the "median depth" for the Takahashi's theoretical velocity distribution [34] at the "frontal part" of the debris flow on a uniform gradient of upstream parts, θ_1, respectively. Similarly, $\overline{C_{s\theta 2}}$, $\overline{h}_{\theta 2}$, and $\overline{U}_{\theta 2}$ are the same indicators on a uniform gradient of downstream parts, θ_2, respectively. Using these indices helps us to simply estimate the transition processes of sediment transport modes for any pattern of changed gradients from θ_1 to θ_2. In this study, we defined θ_1 = 15° and θ_2 = 9°, and the transition process of the modes on the changed gradients from 15° to 9° were discussed using these indices.

3. Results and Discussion

3.1. Transition of Sediment Transport Modes Based on Changes in Sediment Transport Concentrations of the "Frontal Part" Resulting from Changing Streambed Gradient

Figure 2 shows the sediment transport concentrations of the "frontal part" of the debris flow, C_s, for all gradient conditions in Cases 2, 5, and 6, respectively. The dotted lines in these figures show the averaged value, $\overline{C_s}$, for the sediment transport concentrations of the debris flow obtained by the four samplers. The sediment transport concentrations in the debris flow front, which were obtained by the first sampler, were significant for some cases. This reason for this was that the volumes of gravel in the debris flow front increased more than the interstitial fluid because sufficient clearance between the coarser gravel types in the flow front's interior was not enough to concentrate the coarser gravel types in the flow front. In the cases with a changing gradient, as the distance between the measuring point and the gradient change point (+0 cm) increased, the sediment transport concentrations became closer to the uniform gradient (9°) for the downstream part. Comparing the results with different inflow rates for each case, although $\overline{C_{s+50cm}}$ decreased significantly with both inflow rates, these rates were less changed at more than 50 cm downstream from the +0 cm point in the cases where the inflow rates were higher. On the other hand, in the cases where the inflow rates were lower, $\overline{C_s}$ decreased gradually as the distance from the +0 cm point increased more, and $\overline{C_{s+150cm}}$ was almost consistent with $\overline{C_{s9^\circ}}$ on the uniform gradient of the downstream part. However, in Case 6 with the finest debris flow material, $\overline{C_{s+150cm}}$ was larger than $\overline{C_{s9^\circ}}$, even in the case where the inflow rate was lower. Therefore, in the cases where the inflow rates were higher and the debris flow materials were finer, although $\overline{C_s}$ decreased significantly over a short distance from the gradient change point (+0 cm), these rates were less changed thereafter. Conversely, in the cases where the inflow rates were lower and the materials were coarser, $\overline{C_s}$ decreased more as the distance from the gradient change point (+0 cm) increased and was closer to the uniform gradient for the downstream part at +150 cm. These tendencies were confirmed by the averaged sediment transport concentrations of the "frontal part", $\overline{C_{sx}}$, for all cases, as shown in Table 2. This indicates that when the kinetic energy of a debris flow is larger, although some of the flow causes a sudden stoppage (sedimentation) by consuming the kinetic energy because the flow collides with the riverbed near the gradient change point, most of the transported sediment flows downstream because the consumed momentum is less than the total debris flow momentum. Conversely, when the kinetic energy of a debris flow is smaller, although the collision of the flow with the riverbed at the gradient change point is not significant, a firm transition in the sediment transport mode is caused because the materials are unable to maintain their dispersion in the flow's interior owing to the significant increase in the downward component of gravity on the downstream part.

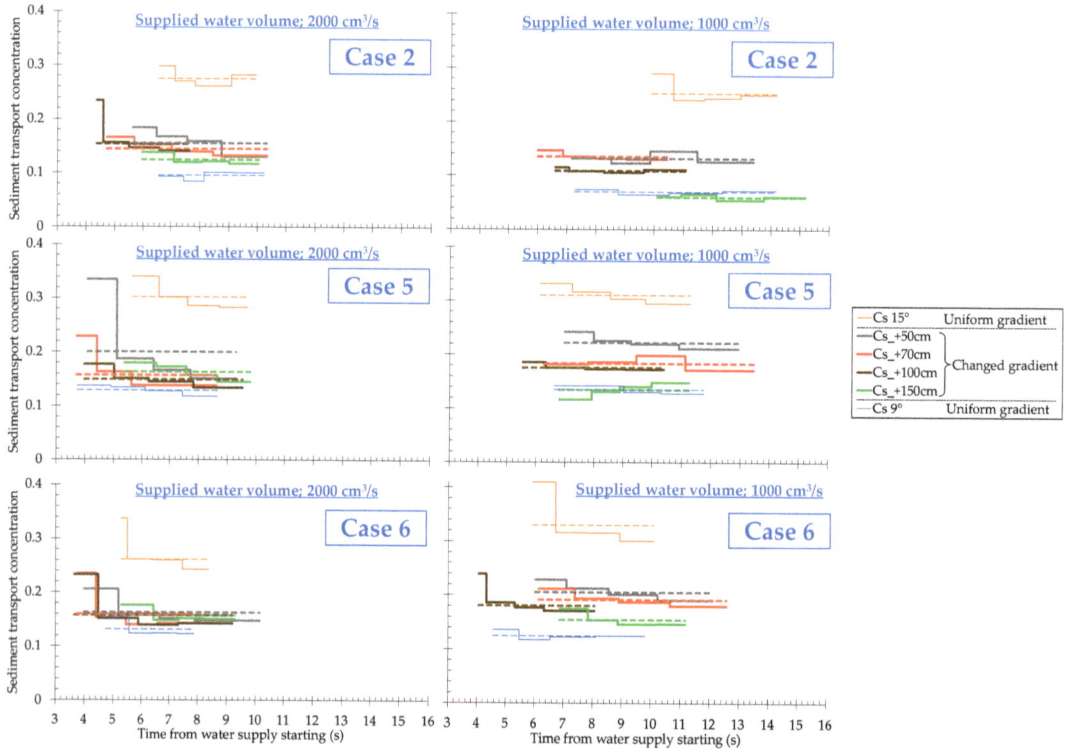

Figure 2. Time variations in sediment transport concentrations of the "frontal part" of the debris flow, C_s, for all gradient conditions in Cases 2, 5, and 6.

Table 2. Averaged sediment transport concentrations of the "frontal part" of the debris flow, $\overline{C_s}$, for all cases with two inflow rates.

Case	Inflow Rate (cm³/s)	$\overline{C_{s15^\circ}}$	$\overline{C_{s50cm}}$	$\overline{C_{s70cm}}$	$\overline{C_{s100cm}}$	$\overline{C_{s150cm}}$	$\overline{C_{s9^\circ}}$
Case 1	2000	0.260	-	-	-	0.124	0.124
Case 2		0.275	0.156	0.144	0.154	0.124	0.096
Case 3		0.250	0.164	0.156	0.148	0.127	0.110
Case 4		0.272	-	-	-	0.170	0.138
Case 5		0.302	0.200	0.157	0.149	0.163	0.129
Case 6		0.261	0.163	0.159	0.158	0.158	0.132
Case 1	1000	0.256	-	-	-	0.013	0.060
Case 2		0.253	0.132	0.135	0.109	0.060	0.070
Case 3		0.256	0.161	0.131	0.120	0.073	0.061
Case 4		0.318	-	-	-	0.130	0.060
Case 5		0.310	0.222	0.183	0.176	0.135	0.136
Case 6		0.330	0.206	0.191	0.181	0.154	0.125

Figure 3 shows the transition indices of sediment transport modes based on the sediment transport concentrations of the "frontal part" of the debris flows, $I_{\overline{Csx}}$, for all cases at the two inflow rates. Considering that $I_{\overline{Csx}} = 0$ represents the sediment transport mode at the uniform gradient for the upstream part (15°) and $I_{\overline{Csx}} = 1$ represents the mode at the uniform gradient for the downstream part (9°), it is considered that as $I_{\overline{Csx}}$ is the closer to 1, the transition to the mode for the downstream parts becomes more significant. The lower the inflow rate, the larger the change in $I_{\overline{Csx}}$ with respect to the increase in the longitudinal distance from the gradient change point (that is, the linear slope of $I_{\overline{Csx}}$).

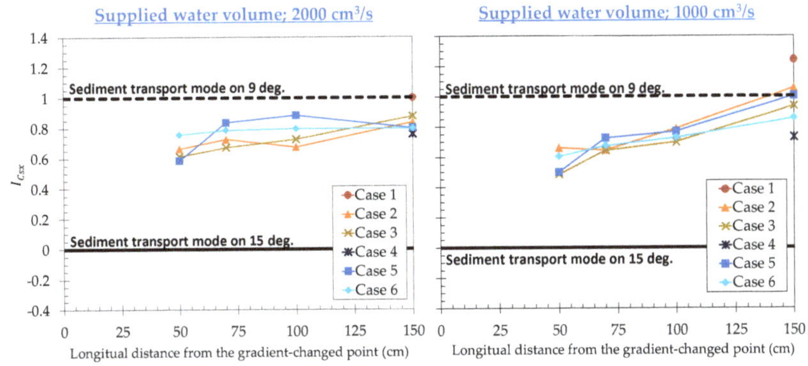

Figure 3. Transition indices of sediment transport modes based on sediment transport concentrations of the "frontal part" of the debris flow, $I_{\overline{Csx}}$, for all cases with two inflow rates.

Thus, by using the transition indices, $I_{\overline{Csx}}$, we can explicitly describe the tendencies mentioned above on sediment transport concentrations of the "frontal part" of the debris flow. However, by using $I_{\overline{Csx}}$, the similar transition tendencies for all cases (all particle size compositions of debris flow materials) also suggest that the effect of the particle size composition on the transition for sediment transport concentration is relatively small.

3.2. Transition of Sediment Transport Mode Based on Changes in Debris Flow Depths of the "Frontal Part" Resulting from Changing Streambed Gradient

Figure 4 shows the debris flow depths of the "frontal part", h, and the theoretical debris flow depths on the uniform gradients of 15° and 9°, $\overline{h}_{15°}$ and $\overline{h}_{9°}$, for all gradient conditions in Cases 1, 3, and 4, respectively. Even though the sediment transport concentrations of the "frontal part" were significant, the debris flow depths of the "frontal part" did not show remarkable increases for all cases. This may be due to the larger gravel sizes relative to the debris flow depth, whereby the supplied gravel sizes from the subsequent part by riding over the frontal part were few. The measurements of the debris flow depths for several runs in Case 3 generally showed the same trend. This suggested that our experimental results were reproducible.

At 10 cm downstream of the gradient change point, h was equal to or greater than the theoretical debris flow depth on the uniform gradient for the upstream part (15°), $\overline{h}_{15°}$, with both inflow rates. More than 50 cm downstream from the point, as the distance from the gradient change point to the measuring point increased, h was the closer to that on the uniform gradient for the downstream part (9°), $\overline{h}_{9°}$. Comparing $\overline{h}_{+80cm}$ values for different inflow rates for each case, the $\overline{h}_{+80cm}$ values in the cases with the lower inflow rates were closer to $\overline{h}_{9°}$ than that in the cases with the higher inflow rates. In addition, comparing $\overline{h}_{+50cm}$ values in the cases with lower inflow rates, $\overline{h}_{+50cm}$ in Case 1 with the coarsest debris flow material was close to $\overline{h}_{9°}$, but the $\overline{h}_{+50cm}$ values in other cases were close to $\overline{h}_{15°}$. Therefore, in the cases where the inflow rates were lower and the debris flow materials were coarser, the closeness of the debris flow depth to the theoretical depth on the uniform gradient for the downstream part was more pronounced as the distance from the gradient change point increased. When flowing at shorter distances from the gradient change point in these cases, the debris flow depths of the "frontal part" were close to $\overline{h}_{9^\circ}$. When the kinetic energy of a debris flow is smaller, a firm transition in the sediment transport mode is caused for reasons similar to those for the transitions of sediment transport concentrations by changing gradients; that is, this transition occurs because the materials are unable to maintain their dispersion in the flow's interior owing to the significant increase in the downward component of gravity on the downstream part. These tendencies were confirmed in the averaged debris flow depths of the "frontal part", $\overline{h}_x$, for all cases, as shown in Table 3.

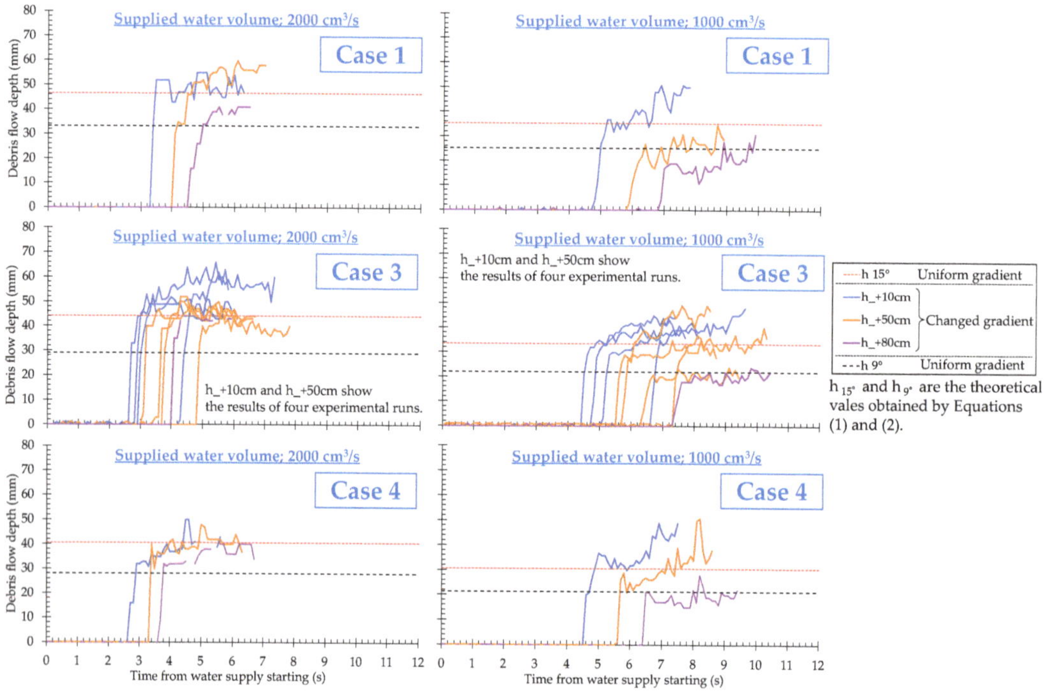

Figure 4. Time variations in debris flow depths of the "frontal part" of the debris flow, *h*, for all gradient conditions in Cases 1, 3, and 4.

Table 3. Averaged debris flow depths of the "frontal part", $\overline{h}$, and theoretical debris flow depths on the uniform gradients, $\overline{h}_{15^\circ}$ and $\overline{h}_{9^\circ}$, for all cases with two inflow rates (unit; mm).

Case	Inflow Rate (cm^3/s)	$\overline{h}_{15^\circ}$	$\overline{h}_{10cm}$	$\overline{h}_{50cm}$	$\overline{h}_{80cm}$	$\overline{h}_{9^\circ}$
Case 1	2000	46.670	48.800	51.300	35.429	33.198
Case 2		45.871	48.217	40.453	—	30.446
Case 3		44.104	50.458	43.250	41.000	29.008
Case 4		40.589	35.917	39.767	35.846	28.078
Case 5		37.428	41.100	38.033	—	23.845
Case 6		34.447	37.100	36.747	—	21.674
Case 1	1000	35.369	39.633	24.533	18.967	25.159
Case 2		34.764	40.022	29.008	—	23.074
Case 3		33.424	37.042	30.367	19.600	21.984
Case 4		30.760	35.933	31.167	18.800	21.279
Case 5		28.365	30.878	31.858	—	18.071
Case 6		26.106	31.133	28.600	—	16.426

Figure 5 shows the transition indices of sediment transport modes based on debris flow depths of the "frontal part", $I_{\overline{h}x}$, for all cases at the two inflow rates. The lower the inflow rate, the larger the change in $I_{\overline{h}x}$ with respect to the increase in the longitudinal distance from the gradient change point (that is, the linear slope of $I_{\overline{h}x}$). In addition, in the cases where the inflow rates were lower, as the debris flow materials became coarser, the linear slope of $I_{\overline{h}x}$ had a greater positive value. Thus, by using the transition indices, $I_{\overline{h}x}$, we can explicitly determine the effect of debris flow magnitudes and particle size compositions on the transition for debris flow depths of the "frontal part" by changing the gradient.

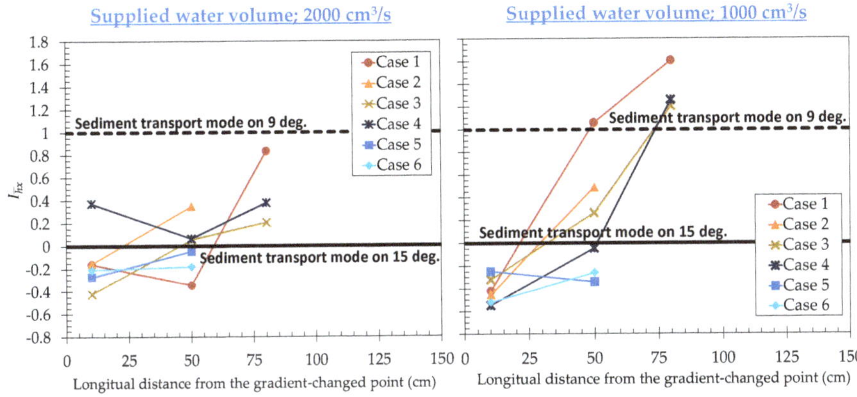

Figure 5. Transition indices of sediment transport modes based on flow depths of the "frontal part" of the debris flow, $I_{\overline{h}x}$, for all cases with two inflow rates.

3.3. Transition of Sediment Transport Mode Based on Changes in Gravel Migration Velocities in the Interior of the "Frontal Part" Resulting from Changing Streambed Gradient

Figure 6 shows the gravel migration velocity distributions in the interior of the "frontal part" of the debris flow in the cases of changing gradients and the theoretical velocity distributions on the uniform gradients of 15° and 9° in Case 3. The flow depths used to obtain these theoretical velocity distributions were the averaged measurement values in the cases of changing gradient. The main parameters used to obtain these distributions, such as the mass density of gravel types (σ), the average concentration in the static sediment bed (C_*), and the average internal friction angle (ϕ), were identified by trial and error so that the distributions were consistent with the measured distributions on the uniform gradients of 15° and 9° in Case 2 (see Figure 7). This figure also shows the averaged measured velocities within the "median depth" of the "frontal part", $\overline{U}_x$, and the averaged theoretical velocities within the "median depth", $\overline{U}_{15°}$ and $\overline{U}_{9°}$. As the debris flows flowed over longer distances from the gradient change point, $\overline{U}_x$ became closer to the theoretical velocities on the uniform gradient of the downstream part, $\overline{U}_{9°}$. In the cases where the inflow rates were lower, these tendencies were more pronounced. These trends were the similar to the aforementioned transition trends of the sediment transport concentrations and flow depths. Therefore, the transition mechanisms of the gravel migration velocities resulting from the changing gradient were estimated to be similar to the aforementioned mechanism. These tendencies were confirmed in the averaged measurement velocities within the "median depth" of the "frontal part", $\overline{U}_x$, for all cases where the inflow rates were lower, as shown in Table 4. However, in the cases where the inflow rates were higher, these tendencies were not pronounced for all cases, and in some cases there was a significant decrease in $\overline{U}_x$ near the gradient change point. The reason for this was the same as the aforementioned reason for the transition of the sediment transport concentrations. In other words, these tendencies indicated that although a part of the flow causes a sudden stop (sedimentation) by consuming the kinetic energy when the flow collides with the riverbed near the gradient change point, most of the transported sediment flows downstream because the consumed momentum is less than the total debris flow momentum.

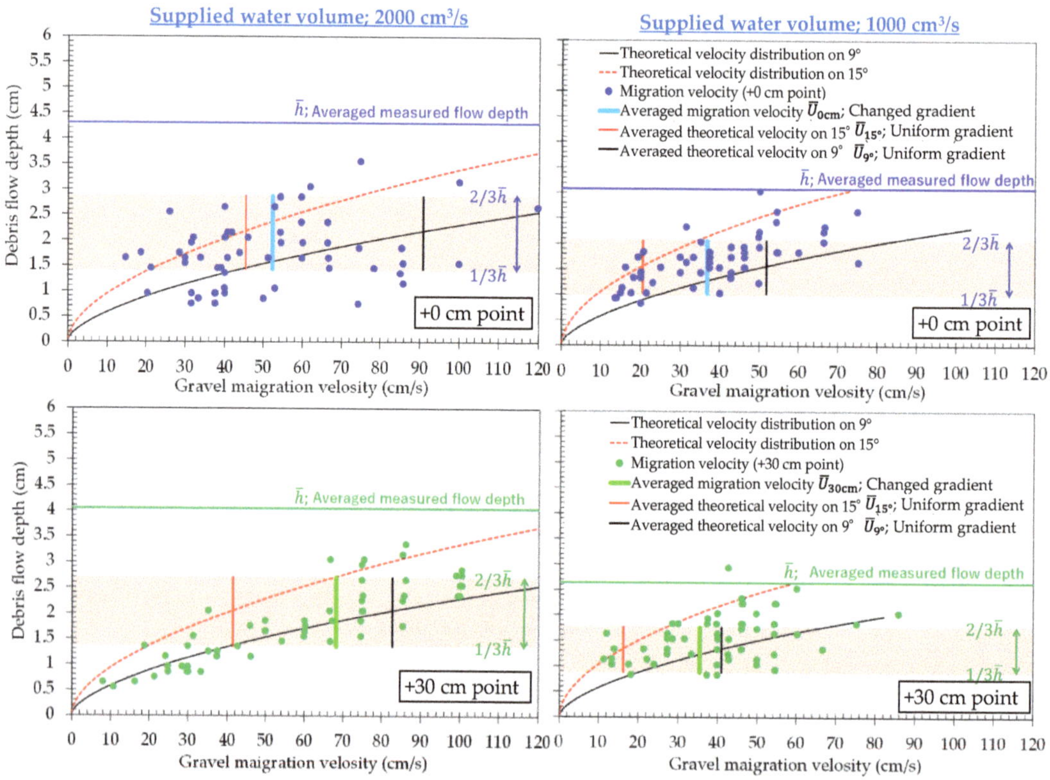

Figure 6. Measured gravel velocity distributions in the interior of the "frontal part" of the debris flow in the cases involving a changing gradient and theoretical velocity distributions on uniform gradients of 9° and 15° for Case 3.

Figure 8 shows the transition indices of sediment transport modes based on the averaged gravel migration velocities within the "median depth" of the "frontal part", $I_{\overline{U}x}$, for all cases at the two inflow rates. In the case with the lower inflow rate, the linear slope of $I_{\overline{U}x}$ was significantly positive. This demonstrates that as the debris flows flowed greater distances from the gradient change point, the front velocities were closer to the theoretical velocities on the uniform gradient of the downstream part for all cases. Conversely, in the cases with the higher inflow rate, no clear trend common to all cases is observed because the linear slope of $I_{\overline{U}x}$ is not monotonic. The possible reason for this is that a part of the flow causes a sudden stop (sedimentation) by consuming of their kinetic energy because the flow collides with the riverbed near the gradient change point. Thus, by using the transition indices, $I_{\overline{U}x}$, we can explicitly determine the effect of debris flow magnitudes on the transition for migration velocities in the interior of the "frontal part" by changing the gradient. However, it is necessary to confirm the validity of the transition indices for other pattern of changed gradients, except for the changed gradients from 15° to 9°.

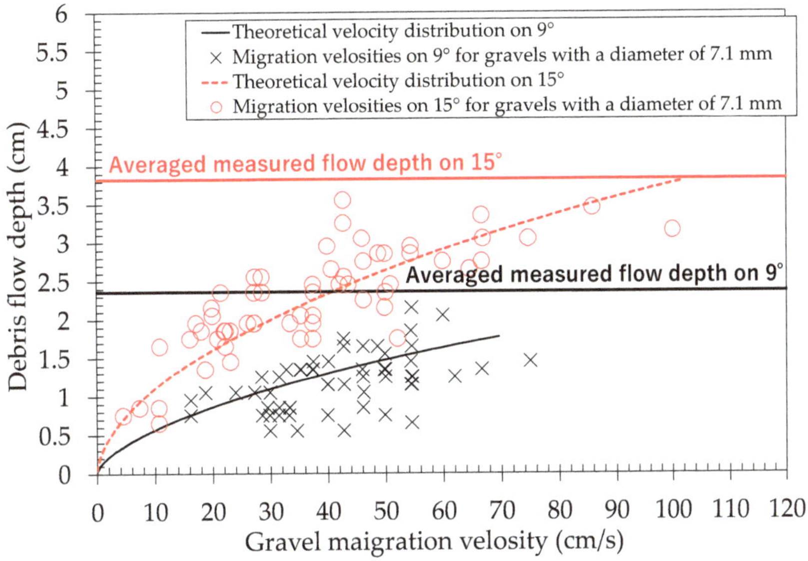

Figure 7. Comparison of theoretical velocity distribution as suggested by Takahashi et al. [34] and measured migration velocities for case 2 on uniform gradients of 15° and 9°. The main parameters used to obtain the theoretical velocity distribution were the mass density of gravel (σ) = 2650 kg/m^3, internal friction angle ($\o$) = 32°, and concentration in the static sediment bed (C_*) = 0.650.

Table 4. Averaged gravel migration velocities within the "median depth" in the interior of the "frontal part", $\overline{U}$, and theoretical migration velocities of gravel within the "median depth" on the uniform gradients, $\overline{U}_{15^\circ}$, $\overline{U}_{9^\circ}$, for all cases with two supplied water rates (unit; cm/s).

Case	Inflow Rate (cm^3/s)	Measuring Point	$\overline{U}_{15^\circ}$	$\overline{U}_{0cm}$	$\overline{U}_{30cm}$	$\overline{U}_{9^\circ}$
Case 1	2000	+0 cm	38.128	60.098	-	76.188
		+30 cm	44.759	-	60.463	89.439
Case 2		+0 cm	39.877	65.593	-	79.701
Case 3		+0 cm	45.467	52.318	-	90.860
		+30 cm	41.461	-	68.171	82.790
Case 4		+0 cm	72.187	91.512	-	131.031
		+30 cm		-	79.083	
Case 5		+0 cm	103.931	83.615	-	188.748
Case 6		+0 cm	65.018	95.179	-	118.020
Case 1	1000	+0 cm	22.102	38.424	-	15.992
		+30 cm	55.757	-	42.395	40.344
Case 2		+0 cm	22.000	42.999	-	55.472
Case 3		+0 cm	20.534	36.847	-	51.802
		+30 cm	16.269	-	35.514	41.021
Case 4		+0 cm	47.604	70.068	-	86.410
		+30 cm	44.788	-	69.550	81.298
Case 5		+0 cm	75.530	50.318	-	18.071
Case 6		+0 cm	60.641	61.651	-	16.426

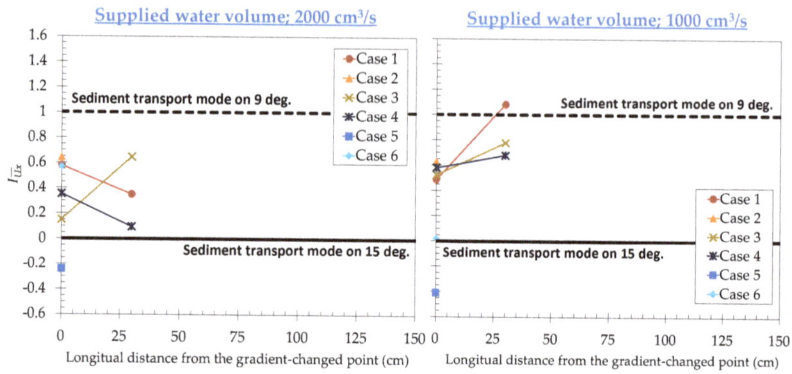

Figure 8. Transition indices of sediment transport modes based on averaged gravel migration velocities within the "median depth" in the interior of the "frontal part" of the debris flow, $I_{\overline{U}x}$, for all cases with two inflow rates.

3.4. Estimation of Transition Processes of Sediment Transport Modes Resulting from Changing Streambed Gradient Based on Transition Indices

Based on the transition indices at all measurement points in all cases and shown in Section 3.1 to Section 3.3, the transition processes of sediment transport modes for all cases were estimated, as shown in Figure 9. The transitions of sediment transport modes of the "frontal part" occurred in the section from the gradient change point (+0 cm) to +150 cm in all cases. In the cases where the inflow rates were higher or the debris flow materials were finer (i.e., the kinetic energies of the debris flow were larger), the transition was not completed at +150 cm. The materials of the debris flow with larger kinetic energy were considered to be able to maintain their dispersion in the flow interior after passing the gradient change point owing to sufficient debris flow momentum. Conversely, when the kinetic energies were smaller, the transition was completed in the shorter longitudinal transition sections. Therefore, it is important to investigate the major gravel sizes and flow magnitudes of debris flows to understand the transition processes of sediment transport modes of debris flow with various particle size compositions owing to the changes in gradient.

Comparing the longitudinal trends of the three transition indices based on the sediment transport concentrations, flow depths, and gravel migration velocities of the "frontal part", with $I_{\overline{Cs}x}$, $I_{\overline{h}x}$, and $I_{\overline{U}x}$ at the downstream part, it is suggested that $I_{\overline{Cs}x}$ significantly increases immediately after passing the gradient change point. On the other hand, $I_{\overline{h}x}$ changes most slowly. Therefore, it is inferred that the transition process of sediment transport modes resulting from changing streambed gradients is as follows. First, the changing streambed gradient causes a rapid decrease in the migration velocities for some of the gravel in the interior of the debris flow. This leads to a decrease in the sediment transport concentration. This decreases opportunities for collisions and decreases the friction of the debris flow materials in the flow interior, leading to a decrease in energy dissipation; therefore, the debris flow depth decreases. These stepwise order of the changes in hydraulic quantities is not considered in the theoretical laws of the conventional debris flow [6–16]. Therefore, the transition indices provide important findings regarding the transition of sediment transport modes due to changing gradients. However, as Zordan et al. [35] suggested, the stepwise order depends on the interaction between the changing debris flow mechanism and the streambed changes at the gradient change point. Therefore, future studies are required that focus on this interaction at the gradient change point using experimental measurements of the changing hydrodynamic quantities, such as the streambed shear stress measurements [35], as well as studies on the mechanistic approaches using the constitutive laws of the debris flow with a gradient-independent global coordinate system [22,23] or numerical approaches with the two-layer flow model or moving particle semi-implicit (MPS) method [24,25].

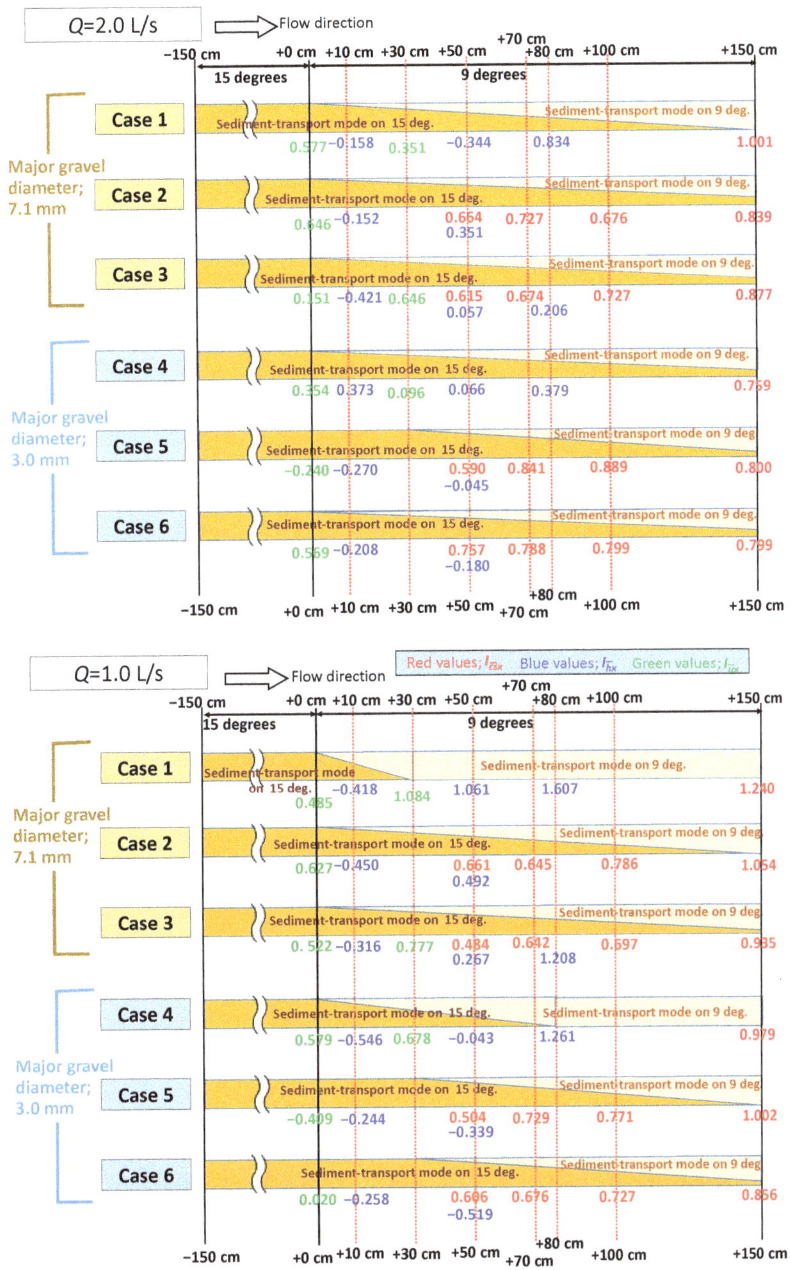

Figure 9. Estimated longitudinal transition processes of sediment transport modes of the "frontal part" of the debris flow for all cases, based on the transition indices. Red, blue, and green values in the figure are the transition indices of sediment transport modes at x cm downstream from the gradient change point for the sediment transport concentration, the averaged flow depth, and the vertical averaged gravel migration velocities at the "frontal part" of the debris flow, i.e., $I_{\overline{Csx}}$, $I_{\overline{hx}}$, and $I_{\overline{U}x}$.

4. Conclusions

In this study, using an experimental flume with two variable gradients in the upstream and downstream parts, we investigated the transition processes of sediment transport modes in debris flows composed of materials of various particle sizes owing to the changes in streambed gradients. In our discussion of the transition processes along the longitudinal distances from the gradient change point, the transition indices $I_{\overline{Cs}x}$, $I_{\overline{h}x}$, and $I_{\overline{U}x}$ were used. These were calculated based on measurements of sediment transport concentrations, flow depths, and gravel migration velocities in debris flow fronts in the downstream part. The findings of this study are as follows:

1. After a debris flow passes the gradient change point, the transition of the sediment transport modes progresses by changing the sediment transport concentrations, flow depths, and gravel migration velocities to those in the steady-state condition on the gradient of the downstream part;
2. In the cases where the debris flow magnitudes are higher and the materials are finer—that is, the kinetic energies are larger—a part of the flow will cause a sudden stop (sedimentation) because the flow will collide with the riverbed near the gradient change point. However, the transition of the sediment transport modes is less pronounced because the dispersion of the debris flow material in the flow interior is maintained after passing the gradient change point owing to the sufficient debris flow momentum. Conversely, in the cases where the inflow rates are lower and the materials are coarser, the transition is more pronounced when flowing at shorter distances from the gradient change point. Therefore, it is important to investigate the major gravel sizes and flow magnitudes of debris flows to understand the transition processes caused by the changing streambed gradient;
3. By using the three transition indices, $I_{\overline{Cs}x}$, $I_{\overline{h}x}$, and $I_{\overline{U}x}$, we can explicitly determine the effects of debris flow magnitudes and their particle size compositions on the transition processes by changing the gradient. However, it is necessary to confirm the validity of the transition indices for other patterns of changed gradients, except for the gradients that change from 15° to 9°;
4. $I_{\overline{Cs}x}$ increases significantly immediately after passing the gradient change point. In contrast, $I_{\overline{h}x}$ changes most slowly. Therefore, the transition process of sediment transport modes resulting from changing streambed gradients occurs as follows. First, the changed streambed gradient causes a rapid decrease in the migration velocities of some gravel types in the interior of the debris flow. This leads to a decrease in the sediment transport concentration. This decreases opportunities for collisions and decreases the friction of the debris flow materials in the flow interior, leading to a decrease in energy dissipation. Therefore, the debris flow depth decreases.

Our future tasks include applying the proposed concept and indices for the transition of the modes under the various conditions in terms of gradient change patterns, the flow magnitudes, and the particle size compositions of the debris flow to investigate the validity of the three transition indices, as well as to consider a general transition index that is suitable for assessing the transition under various conditions.

Author Contributions: Conceptualization, T.W. and H.M. (Hiroshi Miwa); methodology, T.W., H.M. (Hirotada Mishima), J.T., and K.K.; validation, T.W. and H.M. (Hiroshi Miwa); formal analysis, T.W. and H.M. (Hiroshi Miwa); investigation, T.W., H.M. (Hirotada Mishima), J.T., and K.K.; data curation, T.W., H.M. (Hirotada Mishima), J.T., and K.K.; writing—original draft preparation, T.W., H.M. (Hirotada Mishima), J.T., and K.K.; writing—review and editing, T.W. and H.M. (Hiroshi Miwa); visualization, T.W., H.M. (Hirotada Mishima), J.T., and K.K.; supervision, T.W. and H.M. (Hiroshi Miwa); funding acquisition, T.W. and H.M. (Hiroshi Miwa). All authors have read and agreed to the published version of the manuscript.

Funding: This research was funded by JSPS KAKENHI Grant-in-Aid for Young Scientists, Grant Number 22K14454, and Research Grants from the Electric Technology Research Foundation of Chugoku (Grant Subject: Development of a Numerical Model for Deposition Trend in a Dam

Reservoir Considering the Spatial and Temporal Distribution and Transport Processes of Various Sized Sediment Production in the Dam Reservoir Catchment).

Data Availability Statement: Not applicable.

Acknowledgments: Our great thanks go to Yasuhito Takeuchi for his aid in the conception and creation of our experimental flume.

Conflicts of Interest: The authors declare no conflict of interest.

References

1. Wang, F.; Wu, Y.H.; Yang, H.; Tanida, Y.; Kamei, A. Preliminary investigation of the 20 August 2014 debris flows triggered by a severe rainstorm in Hiroshima City, Japan. *Geoenviron. Disasters* **2015**, *2*, 17. [CrossRef]
2. Hashimoto, R.; Tsuchida, T.; Moriwaki, T.; Kano, S. Hiroshima Prefecture geo-disasters due to Western Japan Torrential rainfall in July 2018. *Soils Found* **2020**, *60*, 283–299. [CrossRef]
3. Gong, X.L.; Chen, K.T.; Chen, X.Q.; You, Y.; Chen, J.G.; Zhao, W.Y.; Lang, J. Characteristics of a Debris Flow Disaster and Its Mitigation Countermeasures in Zechawa Gully, Jiuzhaigou Valley, China. *Water* **2020**, *12*, 1256. [CrossRef]
4. Rosli, M.I.; Che Ros, F.; Razak, K.A.; Ambran, S.; Kamaruddin, S.A.; Nor Anuar, A.; Marto, A.; Tobita, T.; Ono, Y. Modelling Debris Flow Runout: A Case Study on the Mesilau Watershed, Kundasang, Sabah. *Water* **2021**, *13*, 2667. [CrossRef]
5. Wada, T.; Kodani, R.; Miwa, H. Spatial analysis on occurrence factors of multiple slope failures using topographic and rainfall indices with high spatial resolutions. *Int. J. GEOMATE* **2021**, *20*, 89–97. [CrossRef]
6. Takahashi, T. Mechanical characteristics of debris flow. *J. Hydraul. Div. Proc. Am. Soc. Civ. Eng.* **1978**, *104*, 1153–1169. [CrossRef]
7. Savage, S.B. Gravity flow of cohesionless granular materials in chutes and channels. *J. Fluid Mech.* **1979**, *92*, 53–96. [CrossRef]
8. Tsubaki, T.; Hashimoto, H.; Suetsugi, T. Grain stresses and flow properties of debris flow. *Proc. Jpn. Soc. Civ. Eng.* **1982**, *317*, 70–91. (In Japanese) [CrossRef]
9. Chen, C.L. Generalized viscoplastic modeling of debris flow. *J. Hydraul. Eng.* **1986**, *114*, 237–257. [CrossRef]
10. Egashira, S.; Miyamoto, K.; Itoh, T. Constitutive equations of debris flow and their applicability. In Proceedings of the 1st International Conference on Debris Flow Hazards Mitigation, San Francisco, CA, USA, 7–9 August 1997; pp. 340–349.
11. Takahashi, T. Study on the deposition of debris flows (3); Erosion of debris fan. *Disaster Prev. Res. Inst. Annu. Kyoto Univ.* **1982**, *25*, 327–348. (In Japanese with English abstract)
12. Hashimoto, H.; Tsubaki, T.; Hirano, M. Sediment Gravity Flow on Relatively Gentle Slopes. *Annu. J. Hydraul. Eng. Jpn. Soc. Civ. Eng.* **1986**, *30*, 235–240. (In Japanese)
13. Takahashi, T. High velocity flow in steep erodible channels. In Proceedings of the 22nd IAHR Congress, Lausanne, Switzerland, 31 August–4 September 1987; pp. 42–53.
14. Ashida, K.; Takahashi, T.; Mizuyama, T. Study on bed load equations for mountain streams. *J. Jpn. Soc. Eros. Control Eng.* **1978**, *30*, 9–17. (In Japanese with English abstract)
15. Egashira, S.; Ashida, K.; Takahama, J.; Tanonaka, S. Sediment transport formula derived from an energy dissipation model of solid-fluid mixture. *Disaster Prev. Res. Inst. Annu. Kyoto Univ.* **1990**, *33*, 293–306. (In Japanese with English abstract)
16. Mizuyama, T. Sediment transport rate in the transition region between debris flow and bed load transport. *J. Jpn. Soc. Eros. Control Eng.* **1980**, *33*, 1–6. (In Japanese with English abstract)
17. Iverson, R.M.; Logan, M.; LaHusen, R.G.; Berti, M. The perfect debris flow? Aggregated results from 28 large-scale experiments. *J. Geophys. Res.* **2010**, *115*, F03005. [CrossRef]
18. Cheng, Y.M.; Fung, I.W.H.; Li, L.; Li, N. Laboratory and Field Test and Distinct Element Analysis of Dry Granular Flows and Segregation Processes. *Nat. Hazards Earth Syst. Sci.* **2019**, *19*, 181–199. [CrossRef]
19. Chen, Z.; Yang, P.; Liu, H.; Zhang, W.; Wu, C. Characteristics analysis of granular landslide using shaking table model test. *Soil Dyn. Earthq. Eng.* **2019**, *126*, 105761. [CrossRef]
20. Takahashi, T.; Nakagawa, H.; Kuang, S. Estimation of debris flow hydrograph on varied slope bed, Erosion and Sedimentation in the Pacific Rim. In Proceedings of the Corvallis Symposium, Corvallis, OR, USA, 3–7 August 1987; Volume 165, pp. 167–177.
21. Ikeda, A.; Mizuyama, T.; Sugiura, N.; Hasegawa, Y. Study about deformation of stream bed deposit at initiation zone of debris flow. *J. Jpn. Soc. Eros. Control Eng.* **2009**, *62*, 46–51. (In Japanese with English abstract)
22. Juez, C.; Soares-Frazao, S.; Murillo, J.; García-Navarro, P. Experimental and numerical simulation of bed load transport over steep slopes. *J. Hydraul. Res.* **2017**, *55*, 455–469. [CrossRef]
23. Veronica, C.; Petrie, j.; Timbe, L.; Pacheco, E.; Astudillo, W.; Padilla, C.; Cisneros, F. Validation of an Experimental Procedure to Determine Bedload Transport Rates in Steep Channels with Coarse Sediment. *Water* **2021**, *13*, 672.
24. Takahama, J.; Fujita, Y.; Kondo, Y.; Hachiya, K. A two layer simulation model unifying debris flow and sediment sheet flow. In Proceedings of the International Congress INTERPRAEVENT 2002 in the Pacific Rim, Matsumoto, Japan, 14–18 October 2002; pp. 113–124.
25. Suzuki, T.; Hotta, N. Development of Modified Particles Method for Simulation of Debris Flow Using Constitutive Equations. *Int. J. Eros. Control Eng.* **2016**, *9*, 165–173. [CrossRef]
26. Sharp, R.P.; Nobles, L.H. Mudflow of 1941 at Wrightwood, southern California. *Bull. Geol. Soc. Am.* **1953**, *64*, 547–560. [CrossRef]

27. Curry, R.R. Observation of Alpine Mudflows in the Tenmile Range, Central Colorado. *Bull. Geol. Soc. Am.* **1966**, *77*, 771–776. [CrossRef]
28. Okuda, S.; Suwa, H.; Okunishi, K.; Nakano, M.; Yokoyama, K. Synthetic observation on debris flow part. 3, Observation at valley Kamikamihorisawa of Mt. *Yakedake. Disaster Prev. Res. Inst. Annu. Kyoto Univ.* **1977**, *20*, 237–263. (In Japanese with English abstract)
29. Ishikawa, Y. Debris flow in the Name river. *J. Jpn. Soc. Eros. Control Eng.* **1985**, *37*, 24–29. (In Japanese)
30. Teramoto, Y.; Jitousono, T.; Shimoyama, E. Flow properties of the debris flow that occurred on September 11, 1999, in the Akamatsudani River Basin at Unzen Volcano. *Res. Bull. Kagoshima Univ. Forests.* **2002**, *30*, 19–25. (In Japanese with English abstract)
31. Gray, J.M.N.T.; Kokelaar, B.P. Large particle segregation, transport and accumulation in granular freesurface flows. *J. Fluid Mech.* **2010**, *652*, 105–137. [CrossRef]
32. Wada, T.; Furuya, T.; Nakatani, K.; Mizuyama, T.; Satofuka, Y. Experimental Study on the Concentration of Coarser Particles at the Frontal Segment of a Debris Flow. *Int. J. Eros. Control Eng.* **2015**, *8*, 20–30. [CrossRef]
33. Turnbull, B.; Bowman, E.T.; McElwaine, J.N. Debris flows: Experiments and modelling. *Comptes Rendus Physique* **2015**, *16*, 86–96. [CrossRef]
34. Takahashi, T.; Satofuka, Y.; Chishiro, K. Dynamical law of debris flows in inertial regime. *Disaster Prev. Res. Inst. Annu. Kyoto Univ.* **1996**, *39*, 333–346. (In Japanese with English abstract)
35. Zordan, J.; Juez, C.; Schleiss, A.J.; Franca, M.J. Entrainment, transport and deposition of sediment by saline gravity currents. *Adv. Water Resour.* **2018**, *115*, 17–32. [CrossRef]

water

MDPI

Article

Streamflow Measurement Using Mean Surface Velocity

Yen-Chang Chen [1], Yung-Chia Hsu [2,*] and Eben Oktavianus Zai [1]

1 Department of Civil Engineering, National Taipei University of Technology, Taipei 10617, Taiwan; yenchen@ntut.edu.tw (Y.-C.C.); d91521012@g.ntu.edu.tw (E.O.Z.)
2 Department of Civil Engineering, National Central University, Taoyuan 32001, Taiwan
* Correspondence: ychsu1978@g.ncu.edu.tw; Tel.: +886-90-666-18-39

Abstract: This study developed an efficient discharge measurement method that can be applied to estimate the streamflow of natural streams and artificial channels. The conventional methods that apply current meters to measure discharge are costly, time-consuming, and labor-intensive. Owing to a shortage of observers in streamflow measurement and for the safety of hydrologists and with advances in measurement techniques, many have strongly suggested the use of non-contact methods when determining streamflow. The non-contact methods that use floats or surface velocity radar to determine the streamflow are becoming more and more popular especially during periods of high water. However, it is not easy to estimate the surface velocity coefficient of each vertical directly for determining the mean velocity in each subsection. As the relationship between the mean surface velocity and mean velocity of a stream cross-section is constant, an efficient and accurate non-contact method of streamflow measurement could be further developed. Thus, streamflow can be estimated by the constant, the mean surface velocity, and cross-sectional area of a stream. The mean velocity of a cross-section, used for parameter calibration, is usually obtained from the discharge made based on the velocity-area principle and cross-sectional area. The surface velocity was measured at the vertical that is then used to estimate mean velocity of a subsection. Once the parameter is determined, streamflow can be obtained from the surface velocity. This approach was further applied to a natural stream and an artificial channel. Measurements were made to verify the reliability and accuracy of the proposed approach. The results show that the relationship between mean channel velocity and mean surface velocity is very stable in both a natural stream and an artificial channel because the streamflow differences, given by the proposed and the conventional method, are relatively insignificant. As a result, mean surface velocity can be used to determine the streamflow quickly and provides for a reliable and accurate measurement of streamflow.

Keywords: discharge measurement; mean surface velocity; non-contact measurement; acoustic doppler flowmeter; magnetic-inductive current meter

Citation: Chen, Y.-C.; Hsu, Y.-C.; Zai, E.O. Streamflow Measurement Using Mean Surface Velocity. *Water* **2022**, *14*, 2370. https://doi.org/10.3390/w14152370

Academic Editor: Su-Chin Chen

Received: 6 June 2022
Accepted: 28 July 2022
Published: 31 July 2022

1. Introduction

Conventional methods of river discharge measurement apply the velocity-area principle [1] for field measurements while the mid-section method [2] is used to calculate the streamflow. The conventional method involves dividing a river cross-section into several subsections. In each subsection, the mean velocity and water depth are measured along the vertical to obtain the discharge of the subsection. The streamflow is the sum of the discharge measurements of all subsections. The conventional method is usually a contact method, requiring a current meter, a sounding weight, and hydrologists on site making this method costly, time-consuming, and labor-intensive. This method is not suitable for tidal streams and during the high water.

Many instruments have been invented with the goal of improving conventional methods, allowing for the rapid and accurate measurement of flow velocity and water depth. An acoustic Doppler current profiler (ADCP) which applies the Doppler effect is a relatively new instrument and has been widely considered as a method used to replace mechanical current meters for velocity measurement. Simpson [3] used a broad-band ADCP that is much faster for accurately measuring tidally affected flow than conventional methods. Boiten [4] applied ADCPs to measure discharges in open channels, according to the velocity-area principle. Costa et al. [5,6] used a boat-mounted ADCP to measure discharge for converting surface-velocity to mean velocity. Muste et al. [7] analyzed velocity profiles collected by ADCPs to propose complementary software for better support of hydraulic investigation requirements. Chauhan et al. [8] showed that the relative error of the discharge measurement is very small with the ADCP compared to the conventional method. Oberg and Mueller [9] showed that ADCP streamflow measurements are unbiased when compared to the discharges obtained by current meter, stable rating curves, salt-dilution, and acoustic velocity meter. Flener et al. [10] used the multidimensional spatial flow patterns measured by an ADCP installed on a remotely-controlled boat to monitor a spring flood. Ground penetration radar [11,12], lidar [13], pressure sensors, and sonar systems [14,15] have been developed to replace the conventional sounding weights during water depth measurement. Although these modern instruments are costlier, they can be applied to provide data when conventional instruments cannot, with the extra benefits of reducing the overall cost and time required [10,16].

It is better to not put hydrologists and equipment in contact with the water, given concerns for personal safety and efficiency. The US Geological Survey [16] has also suggested that in the future gaging stations can use remote sensing outside the flow of water to measure water stages, cross-section, and velocity, and so on. In the past and currently, floats are most often used for surface velocity measurements. The principal sources of error inherent in determining surface velocity in this way are wind and flow conditions. With the recent development of new technology, non-contact instruments [5,6] used for measuring surface flow velocity and to estimate discharge have gradually been developed, among which, surface velocity radar (SVR, [17–20]) and particle image velocimetry (PIV, [21]) are the main ones. The most important issue when applying surface velocity to estimate discharge is the choice of a surface-velocity coefficient. In a natural channel a surface-velocity coefficient of 0.85 or 0.86 is typically used to compute mean flow velocity [2]. However, the surface-velocity coefficient varies with the positions of the verticals from the river bank. Generally, the closer to the bank, the larger the surface-velocity coefficient. In addition, with higher water flow the maximum velocity will occur below the water surface, so the surface-velocity coefficient will also become larger. The velocity distributions can be used to determine the surface-velocity coefficient. The logarithmic distribution [22,23] and probabilistic velocity distribution [24] are commonly used to determine the surface-velocity coefficient. The mean velocity of the vertical can be estimated by the surface-velocity coefficient and surface velocity, and then the velocity-area principle can be used to estimate streamflow.

Although using surface velocity to estimate river discharge is quite efficient, it is still necessary to select surface-velocity coefficients for all verticals to accurately estimate the streamflow; however, it is always a difficult task to determine the best coefficients. Therefore, this study proposes a method which uses the average surface velocity and only one surface velocity coefficient to efficiently and accurately estimate streamflow.

2. Materials and Methods

2.1. Relation between Mean Velocity and Mean Surface Velocity

The mid-section method is often used to calculate streamflow. Figure 1 illustrates that in the mid-section method the mean velocity on the vertical represents the mean velocity in a subsection. The subsection area extends laterally from half the distances from the preceding vertical to half the distance to the next as shown by the hatched area in Figure 1. b_{n-1} in the Figure 1 is the distance from initial point to the n-1th vertical; d_{n-1} is the water depth at vertical n-1; $\overline{u}_{n-1}$ is the mean velocity of the n-1th vertical. Therefore, each subsection is rectangular. The subsectional discharge q_i and the subsection area a_i were calculated using (1) and (2), respectively:

$$q_i = \overline{u}_i \left(\frac{b_{i-1} + b_{i+1}}{2} \right) d_i \tag{1}$$

$$a_i = \left(\frac{b_{i-1} + b_{i+1}}{2} \right) d_i \tag{2}$$

where b_i is the distance from the initial point to vertical i; d_i is the depth of flow at vertical i; and $\overline{u}_i$ is mean velocity at vertical i. The observed discharge (Q_{obs}) and cross-sectional area (A_{obs}) can be represented as (3) and (4), respectively:

$$Q_{obs} = \sum_{i=1}^{n} q_i \tag{3}$$

$$A_{obs} = \sum_{i=1}^{n} a_i \tag{4}$$

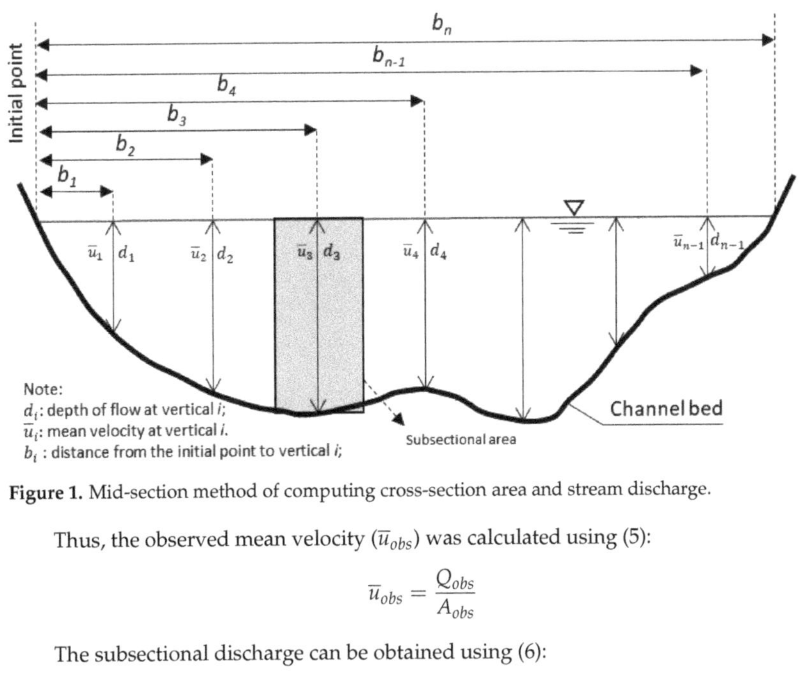

Figure 1. Mid-section method of computing cross-section area and stream discharge.

Thus, the observed mean velocity ($\overline{u}_{obs}$) was calculated using (5):

$$\overline{u}_{obs} = \frac{Q_{obs}}{A_{obs}} \tag{5}$$

The subsectional discharge can be obtained using (6):

$$q_{si} = u_{si} \left(\frac{b_{i-1} + b_{i+1}}{2} \right) d_i \tag{6}$$

where u_{si} is the surface velocity on vertical i. If the cross-section of a stream is a rectangle or close to a rectangle, and the intervals between the verticals are equal; then the area of each subsection will be equal as shown in (7):

$$q_{si} = u_{si} a \tag{7}$$

where a is the area of a subsection when the width and depth of the subsection are the same. The subsection discharge is given by (8):

$$Q_s = \sum u_{si} a = a \sum u_{si} = a n \overline{u}_s = \overline{u}_s A_{obs} \tag{8}$$

where Q_s is discharge estimated by surface velocity; and $\overline{u}_s$ is the mean surface velocity. However, (8) is not valid for estimating stream discharge. A surface-velocity coefficient must be applied to (8) to relate the data to the actual discharge amount as shown in (9):

$$Q_{obs} = \alpha Q_s \tag{9}$$

where α is surface-velocity coefficient. Thus:

$$\alpha = \frac{Q_{obs}}{Q_s} = \frac{\overline{u}A_{obs}}{\overline{u}_s A_{obs}} = \frac{\overline{u}}{\overline{u}_s} \tag{10}$$

(10) reveals that the relationship of mean cross-sectional velocity and mean surface velocity is a straight line going through the origin.

2.2. Estimation of Surface Velocity with Velocity Distribution Based on Probability

The surface velocity can be measured directly by SVR in most conditions. However, when the channel width is not enough for accommodating the instrument, one can measure the velocity profile on each vertical and use the velocity distribution equation to estimate the surface velocity. This study applied the probabilistic velocity distribution to estimate the surface velocity [24], which is shown in (11):

$$\frac{u}{u_{max}} = \frac{1}{M} ln\left[1 + \left(e^M - 1\right)\frac{\xi - \xi_0}{\xi_{max} - \xi_0}\right] \tag{11}$$

where u_{max} is the max velocity; M is a parameter; ξ is the isovel in Figure 2 [25]; u is the velocity at ξ; ξ_{max} and ξ_0 are the values of ξ at which $u = u_{max}$ and $u = 0$, respectively. In addition, a $\eta - \xi$ coordinate system can be used to describe the velocity field with a set of isovels, in which ξ and u has a one-to-one relationship, meaning that the velocities are the same on ξ, unlike the Cartesian coordinate system where the same velocity values can occur in difference locations. The ξ on the vertical line is shown in (12):

$$\xi = \frac{y}{D-h} exp\left(1 - \frac{y}{D-h}\right) \tag{12}$$

where y is the vertical distance from the channel bottom; D is the water depth; and h indicates the location where the max velocity occurs. When $h \leq 0$, the max velocity occurs on the surface; when $h \geq 0$, the max velocity occurs below water surface h. Using the velocity profile and (11), nonlinear regression can be employed to estimate the parameter in (11) and the surface velocity for each vertical.

(a)

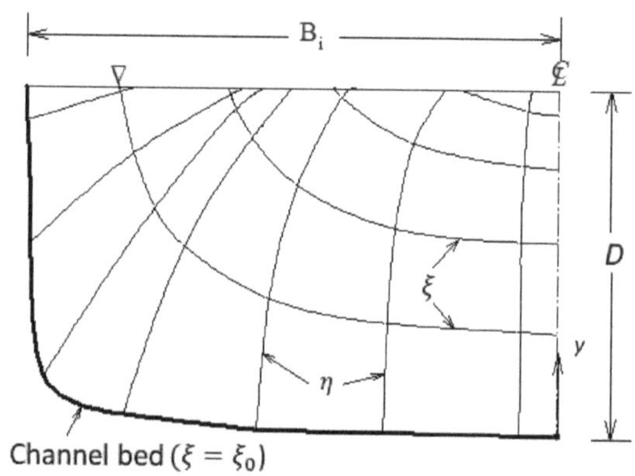

Figure 2. *Cont.*

(b)

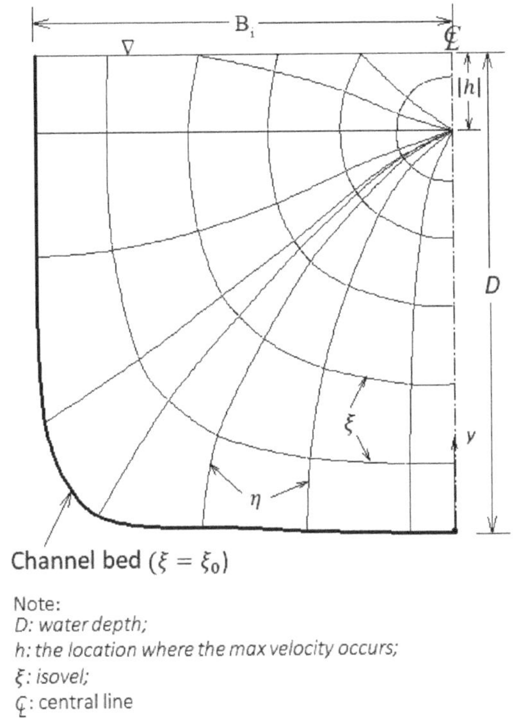

Figure 2. Velocity field in the $\eta - \xi$ coordinate system with a set of isovels. (**a**) $h \leq 0$; (**b**) $h > 0$.

2.3. Estimation of Cross Section Area and Discharge

In practice, there are many approaches for cross-sectional area estimation under different conditions. For a fixed artificial channel, the water stage is consistent enough that it can be used for estimating the cross-sectional area. For a stable channel bed without obvious erosion and sediment deposition, the relationship between the water stage and the cross-sectional area can be used to estimate the cross-sectional area of the river. For an unstable channel bed, the cross-sectional area can be estimated from the water depth of the verticals [26] using (13):

$$A_{est} = b(d - c)^e \tag{13}$$

where A_{est} is the cross-sectional area estimated by water depth; d is the water depth of a vertical; and b, c, and e are coefficients.

Once α and A_{est} are obtained, then the streamflow can be promptly evaluated from the surface velocities of the verticals.

$$Q_{est} = \alpha \overline{u}_s A_{est} \tag{14}$$

where Q_{est} is the streamflow estimated by mean surface velocity.

The proposed approach only requires a SVR to obtain the mean surface velocity for the discharge (Q_{est}) measurement. If the bed does not change too much the cross-sectional area (A_{est}) estimated by water depth may remain feasible. If not, one may also need an efficient and non-contact method, such as the GPR method [12], for fast cross-sectional measurement to update the A_{est} function. As for the conventional approach (mid-section method), it requires obtaining the area and mean velocity of each subsection in order to obtain the subsectional discharge (q_i) and the observational discharge Q_{obs}.

3. Case Study: Study Sites and Data Collection Methods

This study applied field data collected in a natural river and an artificial channel to verify the proposed approach. The study sites are located along Nankang River at Guanyin Bridge in central Taiwan and along Longen Channel in Hsinchu, northern Taiwan (Figure 3).

Figure 3. Study sites for discharge measurement.

The Nangang Creek serves as the main tributary of the Wu River, with a length of about 37.4 km and a catchment area of about 438.14 km^2. It originates from the western foothills of the Hehuan Mountain at an elevation of 3417 m. Wu River is one of the most important rivers in Taiwan, providing vast amounts of water for industrial, agricultural and domestic uses. Thus, the Third River Management Office set up a gauge station at the Guanyin Bridge for the purposes of water resources and flood management.

The Nangang Creek at the Guanyin Bridge, where the measurements were taken, is located near the geographical center of Taiwan, with water flowing from east to west. The channel of Nangang Creek near the Guanyin Bridge is divided into left and right channels. As shown in Figure 4, both the two channels are considered rectangular in cross-section, particularly during the periods of high water, while the left channel is narrower in width and shallower in depth. The collision of tectonic plates has uplifted the terrain of Taiwan over time, while floods brought by typhoons always wash the riverbeds, causing riverbeds to become unstable and change shape frequently. An electromagnetic current meter was used to estimate the water discharge at the Guanyin Bridge. The distance between two successive verticals was 3 m. The velocity observations were made based on the water depth of each vertical. When the water depth was greater than 0.6 m, the two-point method was used, while when the water depth was less than 0.6 m, the six-tenths depth method was used. This study also used a vehicle-mounted SVR unit to measure the surface velocity at each vertical (Figure 5a). A total 23 discharge measurements were made in 2018.

The water in Longen Channel comes from Longen Weir in the middle reach of the Touqian River. Its main function is to divert water from the Touqian River for irrigation and domestic use purposes. The measurement location in Longen Channel was about 1 km downstream from the water intake location. The channel at this location was formed by a box culvert with no place for mounting instruments. Therefore, the top of the box culvert was opened about 1 m wide to allow for discharge measurements. The cross-section of Lungen Channel is rectangular with a width of 2.6 m. Due to the narrow width, the SVR unit can be easily affected by the side walls. With slow water flow, the water surface is very calm, which also makes the SVR unit unable to measure the surface velocity. Therefore, an Argonaut SW acoustic Doppler Flowmeter (Sontek, San Diego, CA, USA) was used to measure the velocity profile, and then a probabilistic velocity distribution equation was used to estimate the surface velocity. The SW Flowmeter was a pulsed Doppler Current profiling system designed for measuring water velocity profiles using three acoustic beams (Figure 5b). The slanted beams, Beams 1 and 2, measure the water velocity in two dimensions, and the down-looking beam measures water depth. An SW Flowmeter is usually mounted in a channel bottom. In this study, the SW Flowmeter was installed under the sounding weight, so that the velocity profile was measured from top to bottom. The velocity observation locations in Longen Channel are shown in Figure 6. The velocity distribution was measured on eight vertical lines, the distance between the verticals was 0.3 m, and the velocity was measured at ten points on each vertical. Table 1 shows the discharge and water depth measurements of eight runs during different water stages having covered the upper and lower water supply capability of Longen Channel. Based on Figure 6, the velocity measurements of Run 4 in Table 1 are shown in Figure 7.

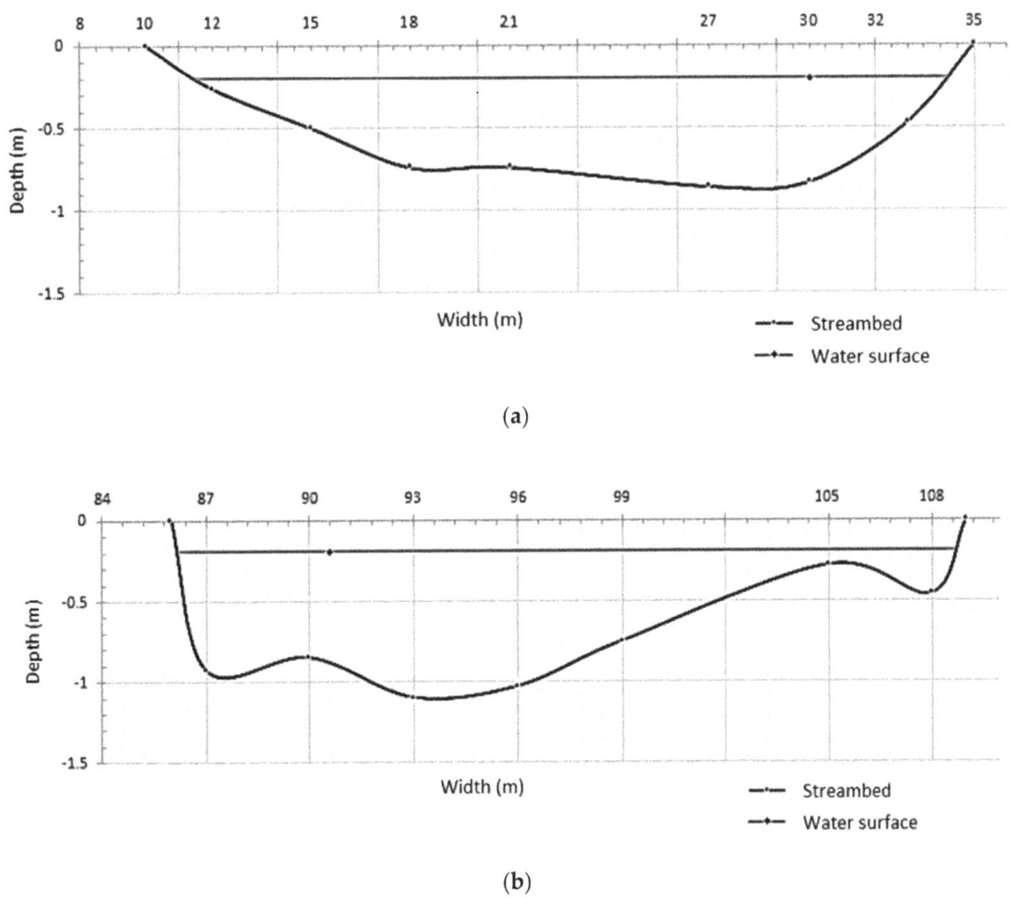

Figure 4. Channel cross sections of the Nankang River at the Guanyin Bridge; (**a**) left channel; (**b**) right channel.

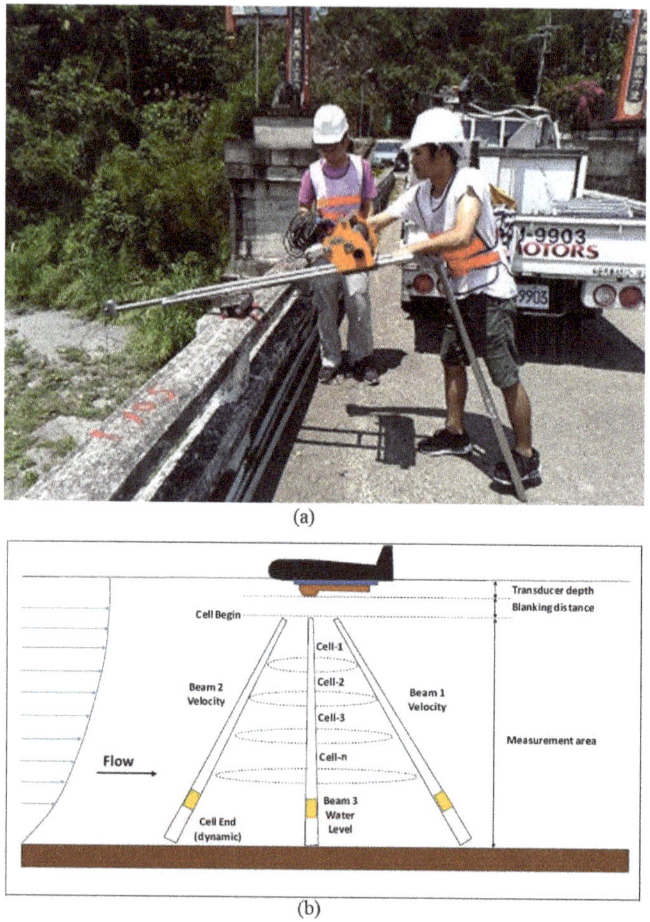

Figure 5. Instruments for measuring discharge. (**a**) A magnetic-inductive current meter and a SVR are used at the Guanyin Bridge; (**b**) A down-looking SW integrated with a sounding weight is used to measure the velocity profiles in the Longen Channel.

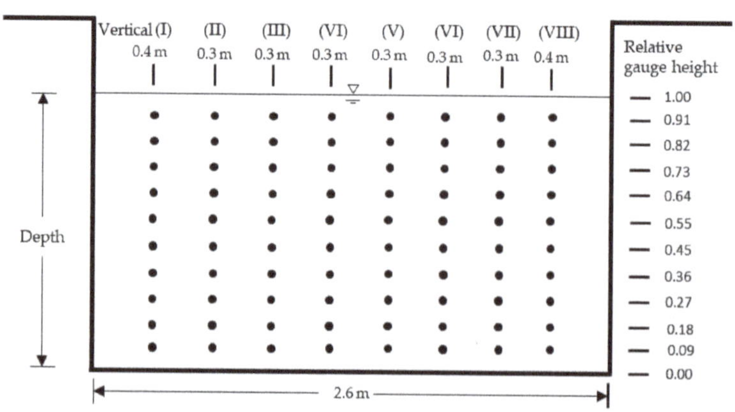

Figure 6. Discharge measurement at the Longen Channel.

Table 1. Discharge measurement of the Longen Channel.

Run	Date	Depth (m)	Q_{obs} (Observed Discharge m^3/s)
1	22 February 2019	0.87	0.43
2	6 March 2019	0.98	0.92
3	8 March 2019	1.47	3.80
4	14 March 2019	1.43	4.24
5	20 March 2019	1.15	2.04
6	25 March 2019	1.31	3.16
7	3 May 2019	1.52	3.79
8	23 May 2019	1.48	4.35

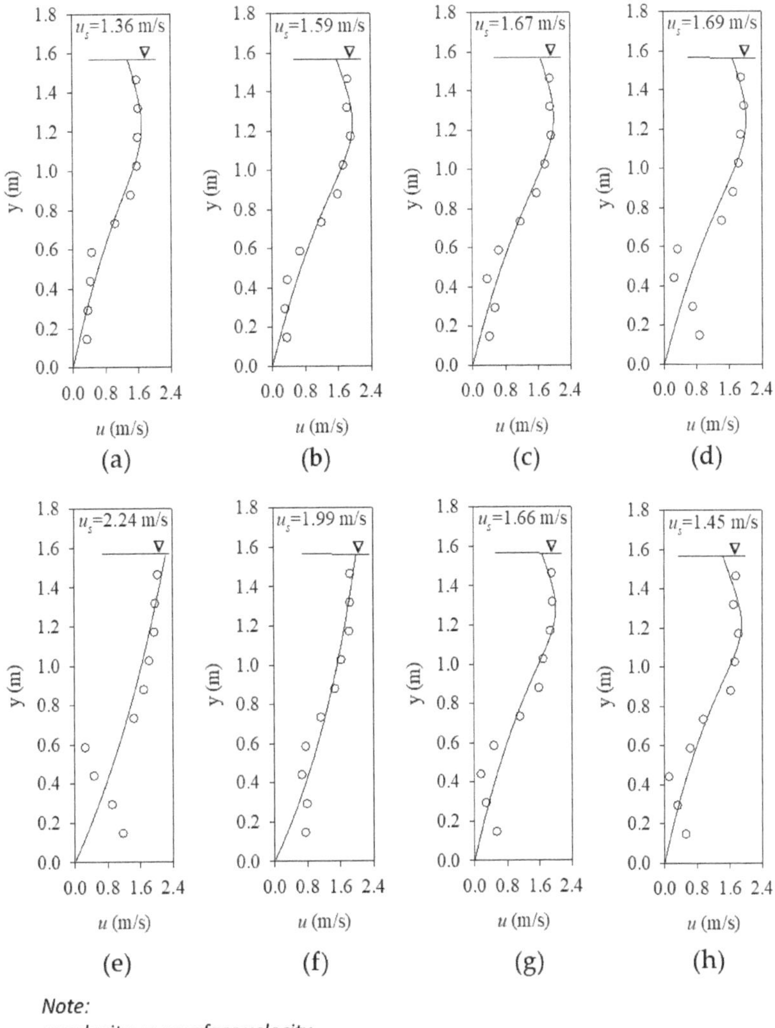

Figure 7. Subsectional velocity profile of the Run 4 in the Longen Channel. (**a**) vertical I; (**b**) vertical II; (**c**) vertical III; (**d**) vertical IV; (**e**) vertical V; (**f**) vertical VI; (**g**) vertical VII; (**h**) vertical VIII of Figure 6.

4. Results and Discussion

The discharge measurement in the natural channel at the Guanyin Bridge was taken by the conventional two-point method for the water depth and the mean vertical velocity. Meanwhile, owing to the limitations caused by the channel width, the discharge measurement at the Longen Channel was taken from the vertical velocity profile for the mean velocity. The mid-section method was then applied to both the natural and the artificial channels. The mean cross-sectional velocity can be observed by the discharge divided by the cross-sectional area. Figure 7 shows the vertical velocity profile of the Longen Channel on 8 March 2019 when the discharge was 3.80 m^3/s. The flow pattern of the Longen Channel was quite similar to a large-scale hydraulic flume in a laboratory. In Figure 7, it is obvious that the flow patterns of Longen Channel were very different from those of natural rivers. Most of the maximum velocities on the verticals occurred at a depth of about 1/4 water depth from the water surface, while the surface velocity was relatively small. It also shows that the maximum velocities of the verticals excluding verticals (e) and (f) did not occur on the water surface. Experimental studies have been shown from considerations of momentum transfer that the velocity in an open channel should decrease toward the channel bed. In a very wide channel the velocity decreases toward the bed and walls, and theoretically the maximum occurs at the water surface. The Longen Channel is a small artificial flume; therefore, depression of the maximum velocities below the water surface was observed. The flow pattern cannot be described by a logarithmic distribution. The circle in Figure 7 is the actual velocity measurement on each vertical, and the line is the velocity distributions based on (11) indicating that vertical maximum velocity does not always occur on the water surface. It also shows that the velocity profile data of the Longen Channel is difficult to describe using conventional velocity distribution theories, such as logarithm velocity distribution. However, (11) can simulate velocity profiles effectively, regardless of whether the maximal velocity occurs on or below the water surface. Therefore, the surface velocities on the verticals could also obtained precisely by using (11). In addition, using the nonlinear regression method, M, h and u_{max} can also be obtained from (11) with the vertical velocity and water depth. Thus, the mean vertical velocity ($\overline{u_i}$ *in* [1]) on each vertical can be estimated. Therefore (11) can be used to accurately estimating the mean velocity of the vertical for obtaining reliable discharge.

Figure 8 shows the relationship of the mean surface velocity and mean velocity of Nankang River at Guanyin Bridge and at Longen Channel. The surface velocity of the natural river was directly measured by SVR, while surface velocities of artificial channels were estimated by the probabilistic velocity distribution equation. All the points, including those in the left and right channels, distribute closely on the two sides of the regression. The relationships of the mean surface velocity and the mean velocity of the cross section in the left and right channels in Figure 8a do not show much difference.

When all the points were combined, all the points fall on the periphery of the regression line, and when the mean surface velocity increases, the data tends to approach the regression line. This means that as the water depth increases, the width of the cross-section increases even more making the shape of the cross-section approach the shape of a rectangle. Hence, the relationship between the mean surface velocity and the mean velocity was stable. Figure 8b shows an artificial rectangular channel with all the points falling near the regression line. This means that the relationship between $\overline{u}_s$ and $\overline{u}_{obs}$ of the artificial rectangular channel is very stable. It reveals that the mean cross-sectional velocity and mean surface velocity in both the natural and the artificial channels have correlation coefficients of 0.92 and 0.99, respectively. As the artificial channel is confined with the concrete bed and walls, the velocity is less affected by natural factors; and thus, it performs better than the natural channel. Figure 8 demonstrates that the relationship between the mean surface velocity and the mean velocity of natural rivers and artificial channels is quite stable, forming a linear relationship through the origin.

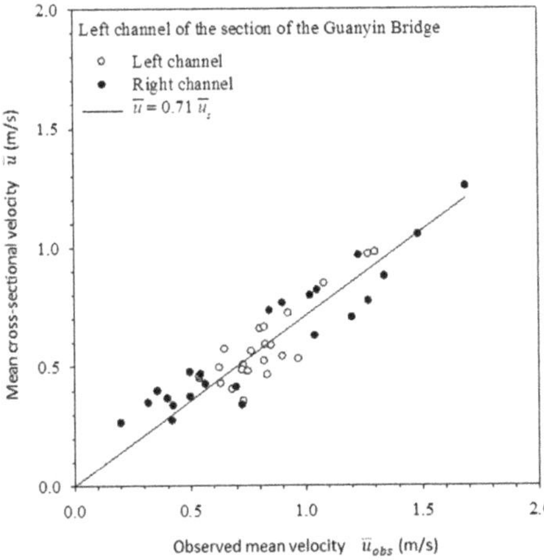

(**a**)

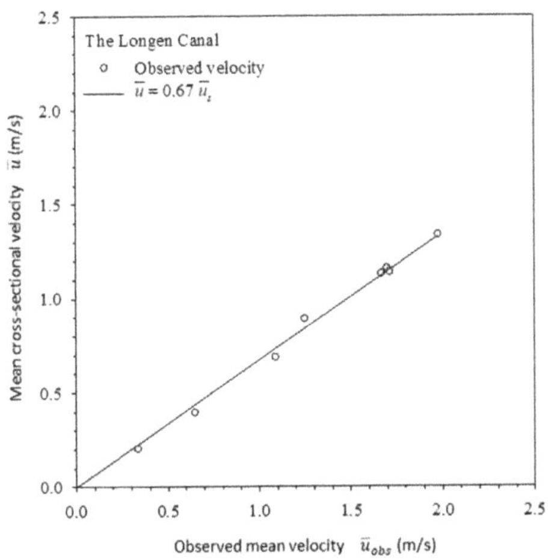

(**b**)

Figure 8. Relation between mean velocities of cross section and surface. (**a**) the Nankang River at the Guanyin Bridge; (**b**) the Longen Channel.

Nangang Creek is a mountain river at Guanyin Bridge, with steep slopes and continuous erosion and sediment deposition. When a flood occurs, the river course will initially be deepened, although the river course will be silted back in as the water recedes. However, the data from 2018 show that the water stage did not change much during the study; nevertheless, the flow and cross-sectional area of the river have changed greatly over time due to erosion and deposition along the river. Thus, it is impossible to estimate the water cross-sectional area based on the water stage. In order to quickly estimate the cross-sectional area, the water depths of the left and right channels at 30 m and 91 m from the initial point were used to establish the relationship between water depth and cross-sectional area (Figure 9). The points (dots and open circles) in the figure represent observed data in the right and left channels, respectively, while the lines represent the depth-cross-sectional area rating curves obtained by nonlinear regression. The nonlinear regression method was applied to relate the water depth and the cross-sectional area with the coefficients given from the observations. The points did not deviate too far from the regression lines. The correlation coefficients of the left and right channels were 0.94 and 0.90, respectively, suggesting that the regression equations are quite useful for making estimations of cross-sectional area by using depths.

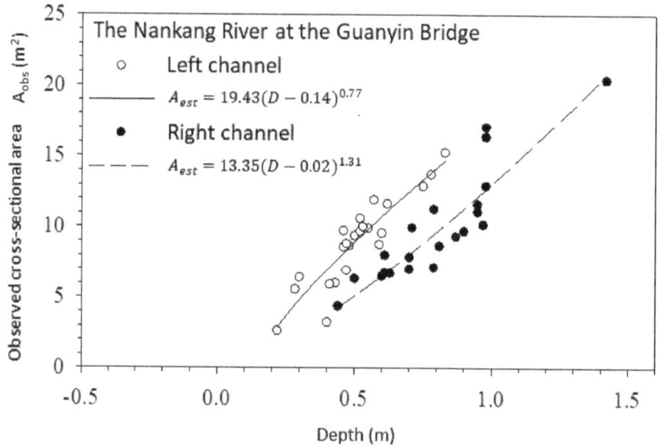

Figure 9. Relation between gage height and cross section area at the Guanyin Bridge.

As the shape of the artificial channel is rectangular, the cross-sectional area can be obtained easily from the water depth and the width of the channel. Therefore, the cross-sectional area of the Longen Canal can be quickly and accurately estimated by the water depth of each run and the width of the channel.

After the two relationships: (1) between mean cross section and mean surface velocity, and (2) between water depth and cross-sectional area were established, the estimated discharge could be quickly obtained with mean surface velocity and water depth. The estimated discharge with the mean surface velocity and water depth were:

$$Q_{est} = 13.77\overline{u}_s(D - 1.35)^{0.78} \text{ for the left channel of the Nangang River} \quad (15)$$

$$Q_{est} = 9.48\overline{u}_s(D - 0.02)^{1.31} \text{ for the right channel of the Nangang River} \quad (16)$$

$$Q_{est} = 1.74\overline{u}_s D \text{ for the Longen Canal} \quad (17)$$

Figure 10 illustrates the accuracy of the discharge measurement using the mean surface velocity. The x- and y-axes represent the discharge measured by the mean surface velocity and the conventional method, respectively. This figure also shows that the natural channel cross-section was close to a rectangle; however, the estimation of the cross-sectional area of the natural channel was not as accurate as that of the artificial channel. As a result, the measurement of the artificial channel was better than that of the natural channel, but all points fall on the 45-degree agreement line, which demonstrates that the streamflow measured by both the conventional and the proposed methods were close to each other with less than 1% error on average.

A strong correlation ($R^2 > 0.97$) between the above two methods demonstrates the accuracy and reliability of using the mean surface velocity method for discharge measurement in both the natural rivers and artificial channels. Therefore, the authors concluded the measurement of river discharge can be obtained promptly and accurately using the proposed approach, which only requires one to measure the surface velocity to obtain the mean surface velocity, and estimate the cross-sectional area based on the water depth (or water stage).

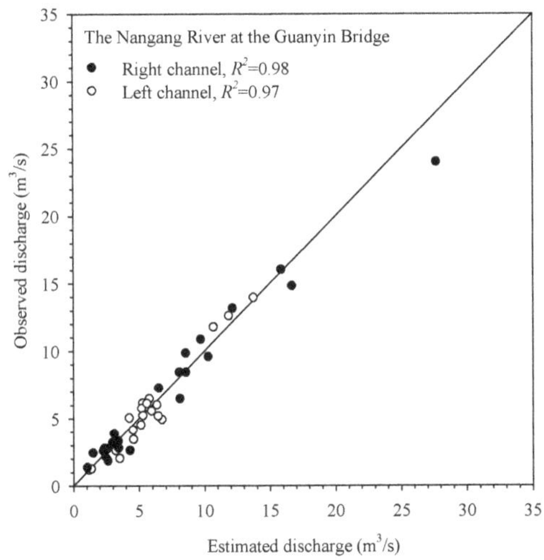

(a)

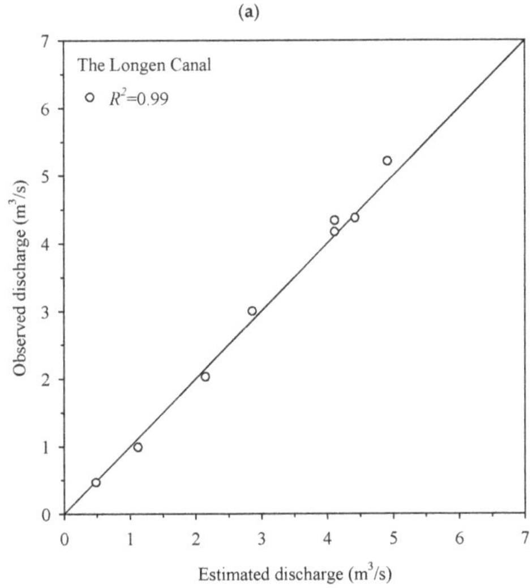

(b)

Figure 10. Accuracy of discharge measurement by surface mean velocity; (**a**) the Nankang River at the Guanyin Bridge; (**b**) the Longen Channel.

5. Conclusions

The conventional methods for discharge measurement consume more labor and time than the proposed approach. Particularly, it is more risky when taking discharge measurement during the high water. The proposed approach applies the relationship between the mean cross-sectional velocity and the mean surface velocity as a constant in each cross section if the cross section does not change too much. Once the constant of the relationship is found, the discharge can be easily estimated based on the constant, the mean surface velocity, and the cross-sectional area. The mean surface velocity can be obtained through the SVR or other means; and the cross-sectional area can be estimated by water depth (or stage) or other means. Nowadays, the surface velocity radar (SVR) has become a popular instrument, which makes the surface velocity measurement easy, reliable and accurate.

The data collection from a natural stream and an artificial channel was used to demonstrate the proposed approach. Two case studies were conducted at two locations with many session runs at different times with low to intermediate flows. Both conventional and the proposed methods were conducted for flow measurements. The results show the feasibility of measuring streamflow with an accuracy of less than 1% difference on average comparing the proposed and the conventional methods. Moreover, the strong correlation coefficient (>0.97) of the observed discharge and the estimated discharge suggests the reliability of the proposed approach for streamflow measurement in both the natural and artificial channels. In previous studies, the conventional methods required extra coefficients (velocity profile or surface velocity coefficient) of each subsection for determining the mean velocity in each subsection, involving more time and labor cost. In this study, we proposed that the mean surface velocity to estimate streamflow can provide an opportunity to improve the conventional methods in streamflow measurement and maintain the accuracy of discharge measurement.

This approach can substantially reduce the uncertainty involved in determining surface velocity coefficients that are employed in conventional methods, while maintaining the accuracy of discharge measurement. Compared with the conventional methods, the proposed approach saves more labor and time cost. As it is a non-contact method, it can also reduce the risk to human life and measuring instruments when taking measurements in the natural environment. The proposed approach can be applied in both natural and artificial channels for flow measurement. Based on the measurement sessions implemented in this study, we conclude that the proposed approach can provide reliable and accurate streamflow measurement from low to intermediate flows. The constancy of the relationship between the mean cross-sectional velocity and the mean surface velocity suggests that this approach might also be applied for high water conditions, which would need further tests in the future.

Author Contributions: All authors have read and agreed to the published version of the manuscript. Conceptualization, Y.-C.C.; Data curation, E.O.Z., Formal analysis, Y.-C.H.; Funding acquisition, Y.-C.C.; Investigation, Y.-C.H. and E.O.Z.; Methodology, Y.-C.C.; Supervision, Y.-C.C.; Validation, Y.-C.H.; Writing–original draft, Y.-C.H.

Funding: This article was based on work supported by the Ministry of Science and Technology, Taiwan (Grant no. MOST 110-2221-E-027-021-).

Institutional Review Board Statement: Not applicable.

Informed Consent Statement: Not applicable.

Data Availability Statement: The data presented in this study are available on request from the corresponding author.

Acknowledgments: The authors would like to acknowledge the Third River Management Office and the Northern Region Water Resources Office of the Water Resources Agency, Taiwan for their kind assistance in measurement sessions.

Conflicts of Interest: The authors declare no conflict of interest.

References

1. Herschy, R.W. *Streamflow Measurement*; Routledge: New York, NY, USA, 2009. [CrossRef]
2. Rantz, S.E. *Measurement and Computation of Streamflow: Volume 1. Measurement of Stage and Discharge*; USGS Water Supply Paper 2175; U.S. Government Printing Office: Washington, DC, USA, 1982. [CrossRef]
3. Simpson, M.R. *Discharge Measurements Using a Broad-Band Acoustic Doppler Current Profiler*; US Geological Survey: Sacramento, CA, USA, 2001.
4. Boiten, W. *Hydrometry*, 1st ed.; CRC Press: Lisse, The Netherlands, 2003.
5. Costa, J.E.; Spicer, K.R.; Cheng, R.T.; Haeni, F.P.; Melcher, N.B.; Thurman, E.M.; Plant, W.J.; Keller, W.C. Measuring stream discharge by non-contact methods—A proof-of-concept experiment. *Geophys. Res. Lett.* **2000**, *27*, 553–556. [CrossRef]

6. Costa, J.E.; Cheng, R.T.; Haeni, F.P.; Melcher, N.; Spicer, K.R.; Hayes, E.; Plant, W.; Hayes, K.; Teague, C.; Barrick, D. Use of radars to monitor stream discharge by noncontact methods. *Water Resour. Res.* **2006**, *42*, W07422. Available online: https://agupubs.onlinelibrary.wiley.com/doi/epdf/10.1029/2005WR004430 (accessed on 26 July 2022). [CrossRef]
7. Muste, M.; Yu, K.; Spasojevic, M. Practical aspects of ADCP data use for quantification of mean river flow characteristics; part I: Moving-vessel measurements. *Flow Meas. Instrum.* **2004**, *15*, 1–16. [CrossRef]
8. Chauhan, M.S.; Kumar, V.; Dikshit, P.K.S.; Dwivedi, S.B. Comparison of discharge data using ADCP and current meter. *Int. J. Adv. Earth Sci.* **2014**, *3*, 81–86.
9. Oberg, K.A.; Mueller, D.S. Validation of streamflow measurements made with acoustic Doppler current profilers. *J. Hydraul. Eng.* **2007**, *133*, 1421–1432. [CrossRef]
10. Flener, C.; Wang, Y.; Laamanen, L.; Kasvi, E.; Vesakoski, J.M.; Alho, P. Empirical modeling of spatial 3d flow characteristics using a remote-controlled ADCP system: Monitoring a spring flood. *Water* **2015**, *7*, 217–247. [CrossRef]
11. Chen, Y.C.; Kao, S.-P.; Wu, C.-H. Measurement of stream cross section using ground penetration radar with Hilbert–Huang transform. *Hydrol. Process.* **2014**, *28*, 2468–2477. [CrossRef]
12. Chen, Y.C.; Kao, S.P.; Hsu, Y.C.; Chang, C.W.; Hsu, H.-C. Automatic measurement and computation of stream cross-sectional area using ground penetrating radar with an adaptive filter and empirical mode decomposition. *J. Hydrol.* **2021**, *600*, 126665. [CrossRef]
13. Mitchell, S.; Thayer, J.P.; Hayman, M. Polarization lidar for shallow water depth measurement. *Appl. Opt.* **2010**, *49*, 6995–7000. [CrossRef]
14. Gupta, S.D.; Shahinur, I.M.; Haque, A.A.; Ruhul, A.; Majumder, S. Design and implementation of water depth measurement and object detection model using ultrasonic signal system. *Int. J. Eng. Res. Dev.* **2012**, *4*, 62–69.
15. Chen, Y.-C.; Yang, C.-M.; Hsu, N.-S.; Kao, T.-M. Real-time discharge measurement in tidal streams by an index velocity. *Environ. Monit. Assess.* **2012**, *184*, 6423–6436. [CrossRef]
16. U.S. Geological Survey (USGS). *Streamflow Information for the Next Century—A Plan for the National Streamflow Information Program of the U.S. Geological Survey*; USGS Open-File Report 99-456; U.S. Geological Survey: Reston, VA, USA, 1999. [CrossRef]
17. Gença, O.; Ardıçlıoğlu, M.; Ağıralioğlu, N. Calculation of mean velocity and discharge using water surface velocity in small streams. *Flow Meas. Instrum.* **2014**, *41*, 115–120. [CrossRef]
18. Chen, Y.C.; Liao, Y.J.; Chen, W.-L. Discharge estimation in lined irrigation channels by using surface velocity radar. *Paddy Water Environ.* **2018**, *16*, 857–866. [CrossRef]
19. Novak, G.; Rak, G.; Prešerena, T.; Bajcar, T. Non-intrusive measurements of shallow water discharge. *Flow Meas. Instrum.* **2017**, *56*, 14–17. [CrossRef]
20. Welber, M.; Coz, J.L.; Laronne, J.B.; Zolezzi1, G.; Zamler, D.; Dramais, G.; Hauet, A.; Salvaro, M. Field assessment of noncontact stream gauging using portable surface velocity radars (SVR). *Water Resour. Res.* **2016**, *52*, 1108–1126. [CrossRef]
21. Legleiter, C.J.; Kinzel, P.J.; Nelson, J.M. Remote measurement of river discharge using thermal particle image velocimetry (PIV) and various sources of bathymetric information. *J. Hydrol.* **2017**, *554*, 490–506. [CrossRef]
22. Hauet, A.; Morlot, T.; Daubagnan, L. Velocity profile and depth-averaged to surface velocity in natural streams: A review over a large sample of rivers. *E3S Web Conf.* **2018**, *40*, 06015. [CrossRef]
23. Corato, G.; Ammari, A.; Moramarco, T. Conventional point-velocity records and surface velocity observations for estimating high flow discharge. *Entropy* **2014**, *16*, 5546–5559. [CrossRef]
24. Chiu, C.L. Velocity distribution in open-channel flow. *J. Hydraul. Eng.* **1989**, *115*, 576–594. [CrossRef]
25. Chiu, C.L. Entropy and 2-D velocity distribution in open-channels. *J. Hydraul. Eng.* **1988**, *114*, 738–756. [CrossRef]
26. Chen, Y.C.; Chiu, C.L. A fast method of flood discharge estimation. *Hydrol Process.* **2004**, *18*, 1671–1684. [CrossRef]

water

MDPI

Article

The Mobility of Landslides in Pumice: Insights from a Flume Experiment

Rozaqqa Noviandi [1,*], Takashi Gomi [2,3], Hefryan S. Kharismalatri [3], Roy C. Sidle [3,4], Rasis P. Ritonga [1] and Katsushige Shiraki [2]

1 United Graduate School of Agricultural Science, Tokyo University of Agriculture and Technology, Fuchu, Tokyo 183-8509, Japan
2 Graduate School of Agriculture, Tokyo University of Agriculture and Technology, Fuchu, Tokyo 183-8509, Japan
3 Institute of Global Innovation Research, Tokyo University of Agriculture and Technology, Fuchu, Tokyo 183-8538, Japan
4 Mountain Societies Research Institute, University of Central Asia, Bishkek 720001, Kyrgyzstan
* Correspondence: rozaqqa.noviandi@gmail.com; Tel.: +81-42-367-5646

Abstract: Risk of landslide hazards strongly depends on how far landslide sediment travels, known as landslide mobility. Previous studies mentioned enhanced mobility of earthquake-induced landslides in volcanic deposits compared to those from other geologic/soil settings. A flume apparatus constructed at a 1:300 scale was used to examine the mobility of landslides with pumice. Four pumice samples were collected from landslides induced by the 2018 Eastern Iburi earthquake, Hokkaido, Japan. Laboratory tests confirmed the unique low specific gravity of the pumice (1.29–1.33), indicating numerous voids within pumice particles. These voids allowed pumice to absorb a substantial amount of water (95–143%), about 9–15 times higher than other coarse-grained soils. Our flume experiments using various saturation levels (0–1) confirmed the influence of this inner-particle water absorption on pumice mobility. Because a low value of specific gravity indicates a low strength of soil, grain crushing may occur on the pumice layer, causing water from the internal voids to discharge and fluidize the transported landslide mass. Our findings indicate that such earthquake-induced landslides can be as mobile as those induced by rainfall, depending on the initial water content of the pumice layers. These conditions might be associated with water accumulation from previous rainfall events and the water-holding capability on pumice layers.

Keywords: earthquake-induced landslides; landslide mobility; pumice; flume experiment

Citation: Noviandi, R.; Gomi, T.; Kharismalatri, H.S.; Sidle, R.C.; Ritonga, R.P.; Shiraki, K. The Mobility of Landslides in Pumice: Insights from a Flume Experiment. *Water* **2022**, *14*, 3083. https://doi.org/10.3390/w14193083

Academic Editor: Su-Chin Chen

Received: 30 August 2022
Accepted: 25 September 2022
Published: 30 September 2022

1. Introduction

Landslides are common natural hazards that occur in hilly and mountainous terrain due to intense rainfall and earthquakes [1]. Landslide impacts strongly depend on how far landslide sediment travels, widely known as landslide mobility [2]. With increasing mobility, landslides are more likely to affect community life systems [3]. For instance, the 2005 La Conchita landslide destroyed 13 houses within the 350 m landslide mobility path, causing more than 10 fatalities in California [4]. The 2006 Leyte landslide with a 3800 m travel distance destroyed more than 500 houses and caused 1126 fatalities on Leyte Island, Philippines [5]. The 2014 Oso earthquake-induced landslide crossed the entire 1-km-wide floodplain of the North Fork Stillaguamish River, causing more than 42 fatalities in Oso, Washington [2].

Landslide mobility is largely affected by topographic factors as these parameters alter landslide momentum. GIS analysis on 33 deep-seated landslides in Kii Peninsula, Japan, noted high variability in landslide mobility that depended on tributary junction angle and stream gradient [6]. A small-scale laboratory experiment confirmed this finding and suggested that a tributary junction angle of >60° had a high risk of landslide dam formation, whereas junction angles < 30° potentially generate debris flows [7]. Field investigation and GIS analysis reported that the 46 landslides caused by the 2008 Wenchuan earthquake traveled 347–4170 m depending on slope gradient [8]. Because landslide momentum depends strongly on landslide mass, previous studies suggested volume-dependent landslide mobility, which is widely viewed as a key mechanism for landslide sediment movement in the Earth and other planetary bodies [9–11].

In addition to topographic factors, landslide mobility is strongly controlled by soil water condition during landslide initiation [12]. Here, water facilitates the downslope fluidization of the collapsed landslide mass [13], where higher soil water content causes greater landslide mobility [14]. For instance, a numerical simulation of landslide mobility in Yining, China, suggested that landslides with saturation levels > 85.7% travelled 25% longer than those with lower water contents [15]. A flume experiment showed that a 40% increase in water content elevated landslide travel distance 2-fold due to the reduction of soil strength [16]. Similarly, a laboratory experiment showed that landslides transformed into debris flows at water contents exceeding saturated condition [7].

The mobility of rainfall and earthquake-induced landslides may differ because of the diverse triggering mechanisms, leading to a dissimilar soil-water contents during landslide initiation. In general, rainfall drives water pressure within a slope to a critical level, significantly reducing the shear resistance of slope materials [17]. In contrast, earthquakes initiate landslides by altering gravitational forces on slopes, which surpass the critical stability threshold [18]. Thus, landslides induced by earthquakes are typically less mobile than those induced by rainfall because they are less influenced by soil-water fluxes. Indeed, previous studies showed that 95% of landslides (of similar volume) induced by the 2015 Gorkha earthquake had 1.5–2 times lower mobility compared to those induced by the September 2011 Typhoon Talas [3,6,19]. Only a few of these earthquake-induced landslides in the Himalaya were as mobile as rainfall-induced landslides because of the steep topography [3]. Hence, the mobility of earthquake-induced landslides has been neglected or not strongly considered in many landslide studies.

Nevertheless, some earlier studies found exceptional mobility of earthquake-induced landslides, particularly those in volcanic deposits. For instance, landslides triggered by the 1968 Tokachi-Oki earthquake occurred in pumice and volcanic ash soils with 1.3–3 times greater mobility compared to other landslides around the globe of similar volume (e.g., [10,19,20]). This pattern also occurred in landslides during the 2001 El Salvador earthquake, transporting > 10^6 m^3 of pumice and volcanic ash [21]. A global investigation of 132 non-volcanic and 50 volcanic landslides triggered by rainfall and earthquakes suggested much higher mobility of landslides in volcanic deposits compared to those in other geologic settings [10]. Such a high mobility of these earthquake-induced landslides might be associated with water content in volcanic deposit layers [20,21].

Volcanic areas are typically covered by tephra-volcanic ejecta that contains various silicates with different sizes [22,23]. Among tephra, pumice has a vesicular texture formed due to the rapid cooling of ejected volcanic rock [22]. Pumice typically has lower density (0.2–0.9 g/cm^3) and higher porosity (50–90%) compared to non-volcanic soils (density > 0.9 g/cm^3, porosity < 60%) [22–24]. This uniqueness is associated with voids inside pumice particles [25]. These voids may also allow pumice to hold a substantial amount of water, depending on pore size and connectivity [22,23]. Because water content is one of the most influential factors for mobility [12], such unique pumice characteristics may influence landslide mobility. Thus, the presence of pumice may control the mobility of earthquake-induced landslides that failed in volcanic deposits. Nevertheless, the effect of pumice on landslide mobility has not been systematically investigated.

Numerous landslides with a density of 326 landslides/km^2 occurred during the Eastern Iburi earthquake, Hokkaido, on 6 September 2018 [26]. Approximately 6000 landslides and debris flow were reported; mobilized distances ranged from 20 to 450 m [27]. The mean distance of transported landslide sediments was 154 m, while the distances varied significantly depending on locations [28]. For instance, a 170 × 85 m^2 landslide in the Towa

River catchment travelled from 20 to 30 m [29], whereas a larger landslide (850 × 400 m^2) in the Hidaka-horonai River catchment travelled 350 m forming a landslide dam [30]. Sliding surfaces of landslides occurred along volcanic deposit layers that originated from the eruption of Mt. Tarumae (Ta-d: 9 ka) and Mt. Eniwa (En: 20 ka) [28]. Hence, materials transported by these landslides consisted primarily of volcanic deposits, i.e., pumice and volcanic ash.

Field and remote sensing studies of the influence of pumice on landslide mobility are difficult due to complex topography and other variable site factors. As such, flume experiment can be effective to consider the range of factors affecting landslide mobility within a simplified configuration. Indeed, a previous study mentioned the effectiveness of flume experiments in demonstrating landslide dam and debris flow formation for various geologic (i.e., soil types) and hydrologic (i.e., water contents) conditions (e.g., [7]). Therefore, the objectives of this study are to: (1) investigate physical characteristics affecting water content of pumice within hillslopes impacted by the Eastern Iburi earthquake and (2) examine the influence of pumice physical characteristics on landslide mobility using a flume experiment. Based on the findings of this study, we then discuss possible mechanisms controlling the mobility of landslides in pumice. This study provides insights to improve landslide risk assessment in volcanic areas.

2. Methodology

2.1. Soil Sampling

Soil samples were collected from three landslide locations in Yosino and Tomisato areas of Atsuma Town, Hokkaido, where a maximum seismic intensity of 7 was recorded (Japan Meteorological Agency scale). These sites are 5 km west of Atsuma Town (Figure 1a,b). Mean annual precipitation and temperature recorded in Atsuma Town is 997 mm and 6.7 °C, respectively (AMeDAS Atsuma based on data from 1976 to 2019). The topography of these sites is hilly with altitudes ranging from 100 to 200 m. Dominant vegetation cover is secondary deciduous forest (e.g., *Betula platyphylla*, *Quercus mongolica*) and mixed-conifer plantations, such as larch (*Larix kaempferi*) and Todo fir (*Abies sachalinensis*).

These landslide sites are surrounded by a well-developed fault system (Figure 1c). The eastern boundary fault zone is located in the Ishikari lowland that originates at Bibai and terminates at Yufutsu, located 7 km northwest and 67 km north of Atsuma town, respectively. Within this fault zone, two active faults (Yusufu and Maoi) strike from NNE/SSW to NNW/SSE [31]. Another active fault, the Karumai fault, is located southwest of the epicenter of the Iburi earthquake (Figure 1c). Two major faults (Atsuma and Biratori) are located 7 km west and next to landslide sites generating a NNW/SSE trend [31].

Thirteen geological units were identified based on a 1:200,000 geological map published by the Geological Survey of Japan (Table 1). The basement complex of these landslide sites is mainly composed of sandstone, mudstone, conglomerate, and sandstone (with tuff) from the middle to late Pliocene periods. This geology is covered by pyroclastic fall deposits, originated from the eruption of Mt. Eniwa (20 ka) and Mt. Tarumae (9 ka) [27]. Tarumae (Ta) is the most recent pyroclastic fall deposit, consisting of Ta-a (1739 A.D.), Ta-b (1667 A.D.), Ta-c (2.5 ka), and Ta-d (8.7–9.2 ka) [30]. Among all Tarumae deposits, Ta-d consists of dark red pumice with a coarse-grained texture and numerous vesicles, whereas other tephra has various size ranges. In all landslide sites, volcanic ash soil was formed between Tarumae deposits creating alternate soil layers. Total depth of surface geology (including soil) is approximately 4–5 m [32].

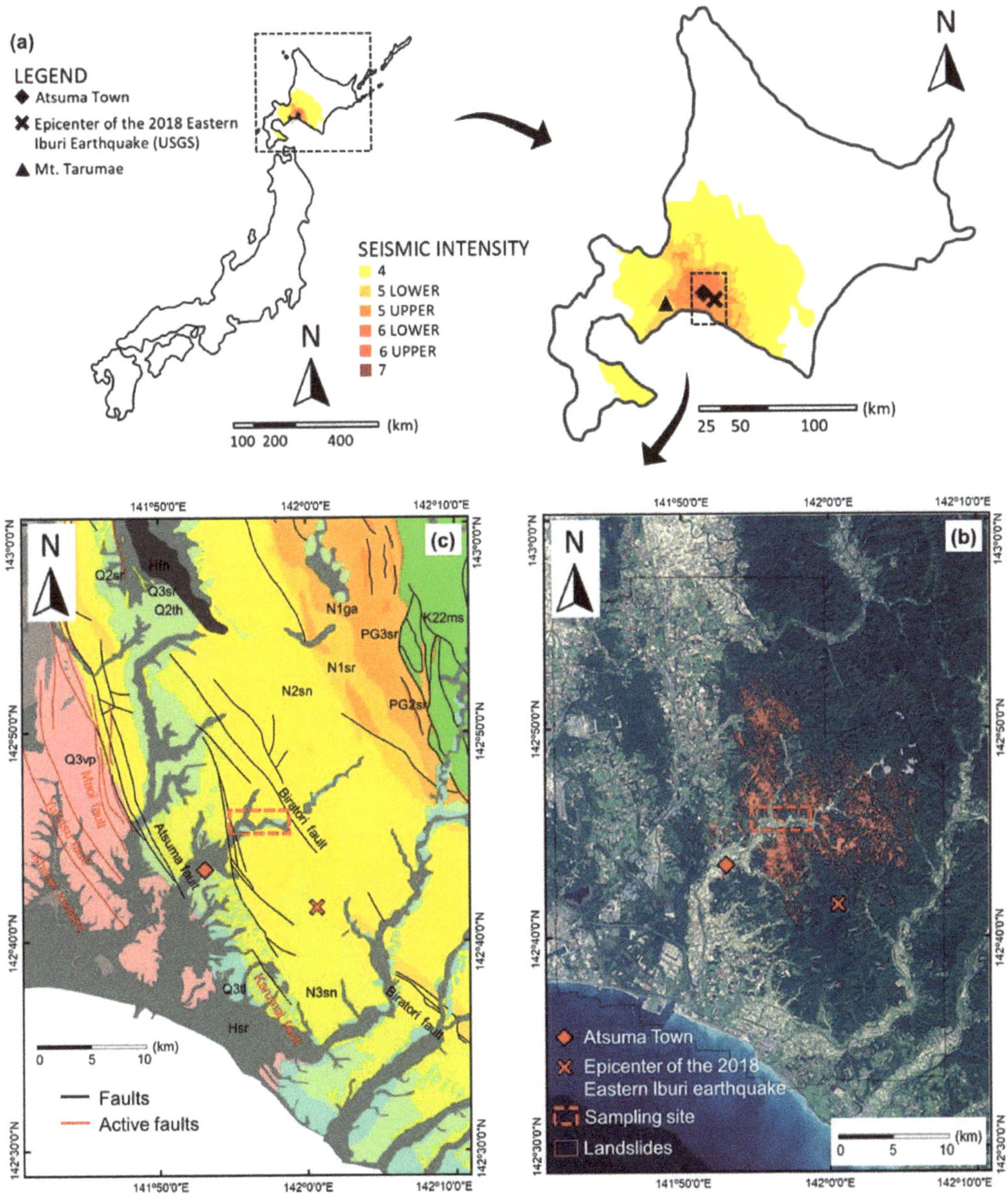

Figure 1. (**a**) Distribution of seismic intensity during the 2018 Eastern Iburi earthquake, Hokkaido, Japan (Modified from Japan Meteorological Agency [33]); (**b**) Landslide distribution provided by the Geospatial Authority of Japan; (**c**) Geological settings within the area affected by landslides derived from a 1:200,000 geological map published by the Geological Survey of Japan (adapted with permission from Ref. [31]. 2019, Shuai Zhang et al.). Geological units are described in (Table 1). Red dotted box in (**b**,**c**) indicates location where soil samples were collected.

Table 1. Geological unit classification.

Codes	Lithology	Age
N2sn	Sandstone, mudstone, conglomerate, and sandstone (with tuff)	Middle to Late Miocene
N3sn	Diatomaceous siltstone with sandstone and conglomerate	Late Miocene to Pliocene
N1sr	Mudstone, sandstone, and conglomerate (with tuff)	Early Miocene to Middle Miocene
Q2th	Mud, sand, gravel, and peat	Middle Pleistocene
Q2sr	Mud, sand, gravel, and peat	Middle Pleistocene
Q3tl	Mud, sand, gravel, peat, and volcanic materials	Late Pleistocene
Q3vp	Rhyolite pumice block, lapilli, and ash	Late Pleistocene
Q3sr	Sand and volcanic ash sand	Late Pleistocene
PG2sr	Sandstone, mudstone, and conglomerate (with coal and tuff)	Middle Eocene
PG3sr	Tuffaceous siltstone with sandstone and conglomerate	Late Eocene to Early Oligocene
Hfn	Fan deposits gravel, sand, and mud (with peat and volcanic ash)	Late Pleistocene to Holocene
Hsr	Clay, silt, sand, gravel, and peat	Late Pleistocene to Holocene
K22ms	Marine sandstone	Late Cretaceous

Field investigations showed that Ta-d (hereafter referred to as pumice) became the major slip surface of landslides [28,30]. Therefore, we collected pumice from our landslide sites. Four disturbed (400 cc volume) and two bulk (disturbed, volume unmeasured) pumice samples were collected in two periods. The first samples were collected at locations 1 and 2 on 3–4 November 2018 (Figure 2). The 30 days antecedent precipitation index (API_{30}), representing the catchment wetness based on the rainfall that occurred over 30 preceding days, was about 104 mm during this sampling dates (measured from AMeDAS Atsuma). At location 1, we collected disturbed and bulk pumice samples at a depth of 130 cm in the landslide scarp (S1) (Figure 2b). Disturbed and bulk pumice samples were also collected from the 60 cm depth in the landslide depositional zone at location 2 (S2) (Figure 2c). The second soil sampling was conducted on 14 June 2019 (API_{30} = 64 mm). During this time, two disturbed samples were collected from depths of 125 and 175 cm in the landslide scarp (location 3; S3 and S4, respectively) (Figure 2d). Since soil surfaces that were exposed after the landslides might be vulnerable to soil erosion, it is possible that the surface soil properties could change between the first and second sampling periods. Thus, the upper 20 cm of exposed surface soil was removed when collecting samples, and soil samples were also collected from excavated trenches.

Disturbed samples were collected using 400 cc stainless tubes; 11.3 cm diameter and 4 cm long. These samples were used for analyzing water content, density, particle size distribution, specific gravity, porosity, and water absorption limit of pumice. Bulk samples were collected in a 650 × 850 mm plastic bag for laboratory experiments. All collected samples were stored in closed containers to minimize the change in soil properties due to direct solar radiation. The samples were then tested one week after sampling dates, 12–16 November 2018 and 24–30 June 2019 for those collected during the first and second sampling periods, respectively. Bulk samples were oven-dried at 105 °C for 24 h and then stored in sealed plastic bags for later use in laboratory experiments.

2.2. Soil Characterization

Laboratory tests were conducted according to American Standard for Testing Materials (ASTM). Dry density and water content were examined using ASTM D7263-09 and ASTM D4643-17 by oven-drying the specimen at 105–110 °C for 24 h. Dry density was estimated based on the ratio of dry weight divided by the volume, while the gravimetric water content was based on the weight of water divided by dry sediment weight.

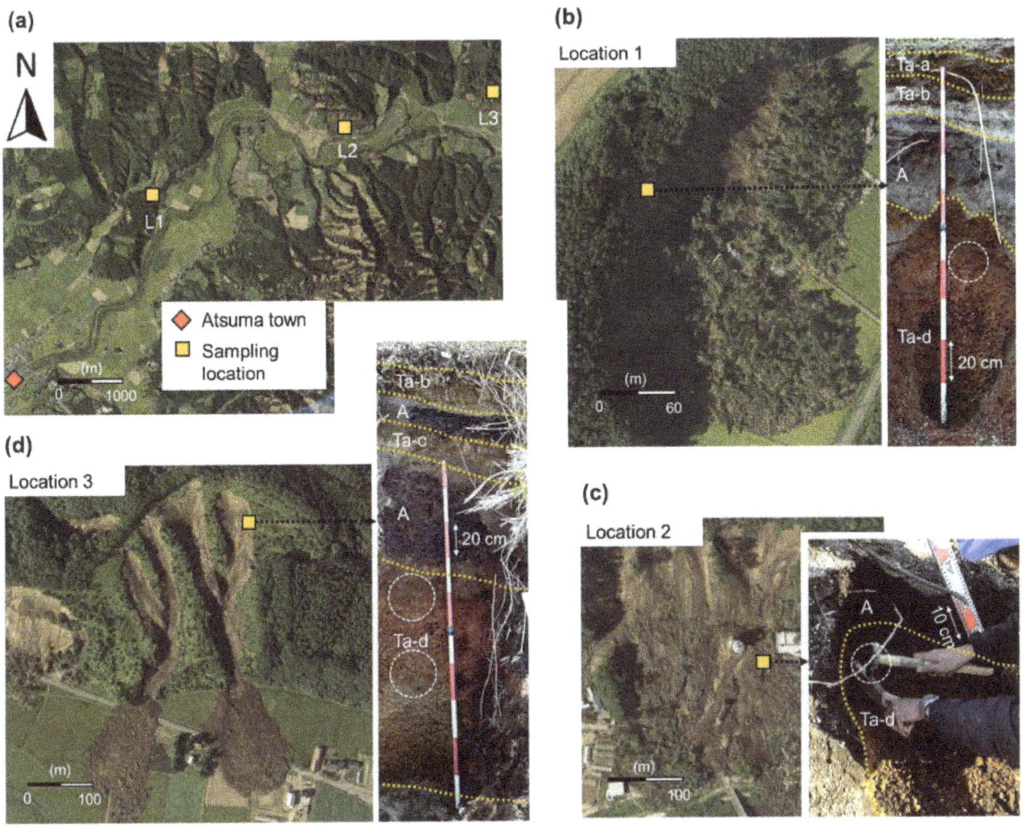

Figure 2. (**a**) General location of soil sampling. Landslide and soil conditions at (**b**) sampling location 1, (**c**) sampling location 2, and (**d**) sampling location 3. Ta is the Tarumae deposit that consists of Ta-a (1739 A.D.), Ta-b (1667 A.D.), Ta-c (2.5 ka), and Ta-d (8.7–9.2 ka). "A" represents volcanic ash soil. The white-dashed circle indicates the location where soil samples were collected.

Particle size distribution was analyzed based on ASTM D6913-04 using sieve opening sizes of 19 mm, 9.5 mm, 4.75 mm, 2 mm, 0.85 mm, 0.425 mm, 0.25 mm, 106 μm, and 75 μm. Dry soils were placed into sieves and then vibrated using a FRITSCH automatic sieve shaker for 10 min. The number of particles retained in each sieve was then weighed. For particles smaller than 75 μm, particle size distribution was analyzed by a laser diffraction particle size analyzer SALD-2300. The soil was classified based on ASTM D2487-06 and USCS Classification. The coefficient of curvature (C_c) and coefficient of uniformity (C_u) representing soil gradation were calculated based on the cumulative percentages of particle size distribution. The coefficient of curvature and coefficient of uniformity were estimated as:

$$C_c = \frac{(D_{30})^2}{D_{10} \times D_{60}} \quad (1)$$

$$C_u = \frac{D_{60}}{D_{10}} \quad (2)$$

where D_{10}, D_{30}, and D_{50} indicate particle diameters finer than 10%, 30%, and 60% of the total mass, respectively.

Specific gravity (G_s), representing particle density, was estimated based on ASTM D854-05 for particles finer than 4.75 mm and ASTM C127-07 for particles larger than

4.75 mm. This test was conducted without crushing pumice particles to maintain their original characteristics. The average specific gravity was estimated based on ASTM C127-07. Porosity (n) and saturated water content (w_{Sr}) were calculated from specific gravity. Porosity represents total voids in the soil, whereas saturated water content represents the water content when all voids are filled with water. Because pumice may contain unconnected voids [23], which could prevent water diffusion [25], we estimated the water absorption limit (W_A) based on the maximum absorbed water that generates a constant mass of pumice particles.

Since specific gravity (G_s) was measured without crushing pumice particles, the volume used in our G_s calculation includes dead-end pores (V_{vd}) and solid fraction of pumice (V_s). We then measured the specific gravity of crushed pumice particles (G_{sc}) to estimate the volume of solid fraction (V_s). By assuming that all active pores can effectively store the water, we estimated the volume of active pores (V_{va}) from the water absorption limit (W_A). Thus, the dead-end to active pore ratio (R_v), representing pore connectivity within pumice, is estimated as follows:

$$R_v = \left(\frac{1}{G_s} - \frac{1}{G_{sc}}\right)\frac{1}{W_A} \tag{3}$$

2.3. Laboratory Experiment for Examining Landslide Mobility

A flume constructed at a 1:300 scale was used to examine the mobility of landslides with pumice. The flume consisted of two segments that represent landslide initiation and deposition (Figure 3a,b). All segments were 10 cm wide, 15 cm high, and constructed with 1-cm thick acrylic material. The primary segment was set to 45° with a total length of 50 cm. This condition represented a 45° slope with a length of 150 m. Because most landslides occurred on slopes between 25° and 30° [26], we focused on a relatively steeper slope, e.g., in headwater catchment. The secondary segment had an inclination of 10° and a total length of 130 cm. This represents a valley; 30 m wide, 390 m long, and a 10° gradient. A bucket was placed at the end of the secondary channel to collect transported sediment. To quantify mobility, we divided the flume into four segments, i.e., A (20 cm), B (65 cm), C (65 cm), and D (Outlet) (Figure 3c,d).

We used pumice collected from the landslide at location 2 in the experiment to represent sediment characteristics during transport and deposition. Transported materials by the 2018 Eastern Iburi earthquake may consist not only of pumice but also other tephra and volcanic ash with different physical properties. Despite various types of soil transported during the slope failures, we assumed that soil at the landslide slip surface has the most influence on landslide mobility because it controls basal friction resistance. Because landslides occurred mostly at pumice layers [31], we used pumice to represent landslide mobility during the Eastern Iburi earthquake. The sample volume of pumice that we used was 500 cc, representing a large landslide (13,500 m^3) based on our 300× scaling. Since the Eastern Iburi earthquake generated > 6000 landslides with total sediment production of 30 million m^3 [31,34], the mean landslide volume was about 5000 m^3.

Because water content is an influential factor for landslide mobility [12], we applied eight saturation levels (S = 0.0, 0.15, 0.3, 0.45, 0.6, 0.7, 0.8, and 1.0), where $S = 0$ represents completely dry conditions and $S = 1$ represents saturated conditions. These different saturation levels represent the variability of initial water content during earthquake-induced landslides. We situated the water-sediment mixture 20 cm upstream of the channel junction behind an acrylic gate (Figure 3c). The gate was then abruptly opened to allow the sediment to flow downstream. Such an abrupt gate opening represents a condition where the slope already collapsed, and movement was mainly driven by gravity. This experimental process was also intended to lessen the variability of initial mobility, similar to a study examining the mobility of debris flows in a large-scale flume [35].

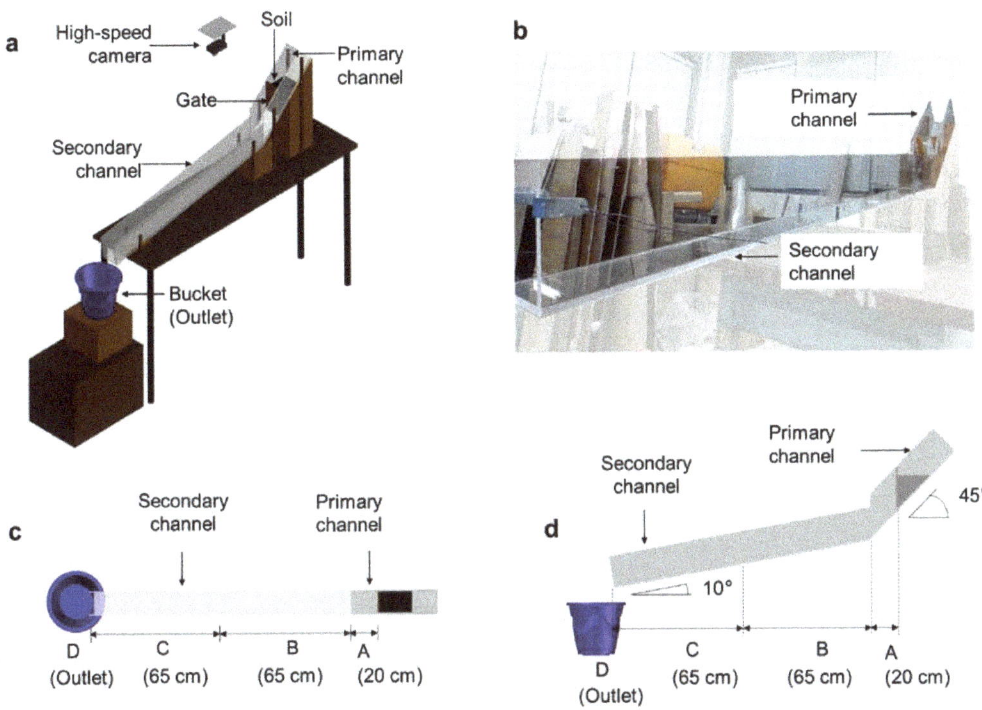

Figure 3. (**a**) Schematic illustration of the experimental setting; (**b**) Flume apparatus; (**c**) Flume segments (**top**); (**d**) Flume segments (**side**).

After the movement of sediment stopped, we estimated the L/H ratio based on the total travel distance (L) and total drop height (H) of sediment. This parameter is widely used to quantify the mobility of landslide sediment [3,9,10,20]. L/H is derived from the law of energy conservation by assuming a constant moving mass without a progressive mass loss [9]. We then collected the sediment from all segments. All collected sediment samples were oven-dried at 105 °C for 24 h and weighed. The same procedure was repeated four times for each level of saturation. A total of 32 experiments were conducted during this study, from 26 March 2019 to 13 April 2019.

3. Results

3.1. Physical Characteristics of Pumice

Dry density, water content, and specific gravity of the four pumice samples varied by sampling locations (Table 2). Dry density ranged from 0.33 to 0.45 g/cm^3; lowest in S2 and highest in S3. Gravimetric water content was highest in S2 (172%), followed by S4 (160%), S1 (147%), and S3 (139%). Specific gravity ranged from 1.29 to 1.33, with the lowest value in S1 and the highest in S3. Similar to the previously mentioned parameters, the particle size of pumice differed depending on sampling locations. The mean diameter of particle size distribution (D_{50}) ranged between 1.7 to 6.3 mm, with the lowest in S3 and the highest in S1. Since more than 50% of pumice was retained on the 75 μm sieve, pumice was classified as coarse-grained soil according to ASTM D2487-06. S1 and S2 were classified as gravel, whereas S3 and S4 were classified as sand because more than 50% of their particles passed through the 4.75 μm but were retained on the 75 μm sieves. All samples except S3 were poorly graded with a narrow size range; C_u ranged from 4.8 to 6.3, whereas C_c ranged from 0.4 to 7. S3 was classified as well-graded with $C_u > 6$ and $1 < C_c < 3$.

Table 2. Physical characteristics of pumice.

Physical Characteristics	Sample			
	S1	**S2 ***	**S3**	**S4**
Dry density (ρ, g/cm^3)	0.33	0.3	0.41	0.35
Water content (w, %)	147	172	139	160
Specific gravity (G_s)	1.29	1.3	1.33	1.19
Mean particle size (D_{50}, mm)	6.3	4.8	1.7	4.5
Coefficient of uniformity (C_c)	6.3	4.8	6.1	5.2
Coefficient of curvature (C_u)	7.0	6.1	1.2	0.4
Void ratio (e)	2.9	3.3	2.2	2.4
Porosity (n, %)	75	77	70	74
Saturated water content (w_{Sr}, %)	221	256	167	200
Water absorption limit (W_A, %)	113	143	95	118
Dead-end to active pore ratio (R_v, %)	34.0	26.5	38.0	38.1
Soil classification	GP	GP	SW	SP

Notes: * Sample used for flume experiment, GP is poorly graded gavel, SP is poorly-graded sand, and SW is well-graded sand.

Because soil particle size affects soil physical properties [36], void ratio, porosity, saturated water content, and water absorption limit of the four pumice samples differed depending on particle size distributions, yet followed a similar pattern. In general, S2 had the highest values of these parameters (e = 3.3, n = 77%, w_{Sr} = 256%, W_A = 143%), whereas the lowest ones were for S3 (e = 2.2, n = 70%, w_{Sr} = 167%, W_A = 95%). In contrast, the dead-end to active pores ratio (R_v) was the highest for S4 and the lowest for S2 (Table 2).

3.2. Characteristics of Mobility

The L/H varied depending on the saturation level (S) (Figure 4). For dry soil (S = 0), L/H ranged between 2–2.2 (mean = 2.1; SD = 0.07). As pumice became wetter, L/H progressively decreased up to S = 0.6. Then L/H began to increase above S = 0.6 and peaked at S = 1 (mean = 3.2; SD = 0). Thus, S = 0.6 marks the threshold for landslide mobility change; the lowest L/H among all conditions, ranging between 1.6 and 1.8 (mean = 1.7; SD = 0.1). Nevertheless, because the sediment entered the bucket (segment D) during the S = 1 experiment, the actual travel distance for this condition could not be measured. Hence, the L/H under S = 1 conditions was the same in all four repetitions (L/H = 3.2) and underestimates mobility. Therefore, we projected the slope of the mean L/H value at S = 0.6–0.8 to S = 0.8–0.1 to predict the mobility under saturated conditions (S = 1) (Figure 4).

3.3. Characteristics of Sediment Deposition

Sediment deposition in all segments differed depending on the saturation level (Figure 5). For all dry sediments (S = 0), deposition was highest in segment B, followed by segment A, and no sediment reached segments C and D. With increasing saturation (up to S = 0.6), sediment deposition increased in segment A and decreased in segment B. For S > 0.6, sediment deposition occurred in downstream segments C and D. Further increases in the saturation level caused a gradual increase in sediment deposition in segments C and D, whereas deposition in segments A and B gradually decreased. This increasing pattern of sediment deposition in segments C and D indicates greater sediment movement downslope.

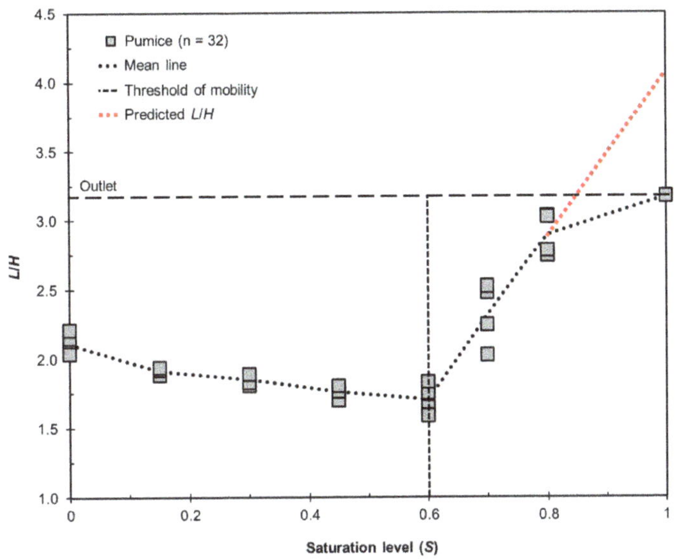

Figure 4. The L/H of pumice for various saturation levels. X-axis shows saturation levels (S), where $S = 0$ represents completely dry conditions, whereas $S = 1$ represents completely saturated conditions. Y-axis shows L/H, the ratio between total travel distance (L, cm) and total drop height (H, cm) of sediment. Both S and L/H are dimensionless parameters.

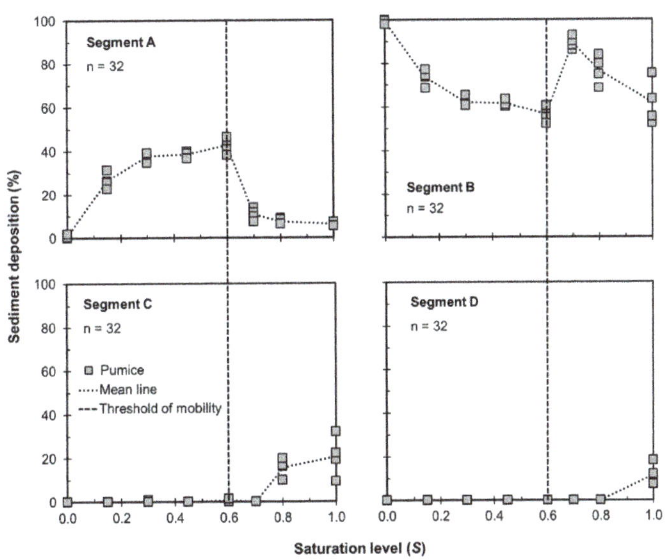

Figure 5. Sediment deposition in all flume segments. X-axis shows saturation levels, where $S = 0$ represents completely dry conditions and $S = 1$ represents completely saturated conditions. Y-axis shows the percentage of sediment deposition in all flume segments (%). Segments A, B, C, and D represent the initiation, upper deposition, lower deposition, and outlet segments, respectively.

4. Discussion

4.1. *Uniqueness of Pumice*

Pumice is characterized by a low specific gravity, representing particle density. The specific gravity of soils typically ranges between 2.6 and 2.8 [37], whereas that in the four pumice samples we collected was 1.29–1.33. Since particle density influences soil strength [38], such a low specific gravity indicates a low strength of pumice. Indeed, changes in particle size distribution after triaxial tests indicated a low shear strength of pumice [39], and grain crushing due to overburden pressure [31]. The low specific gravity of pumice might be associated with numerous voids inside these volcanic particles (e.g., [25]).

Internal voids within pumice greatly influence pumice density because soil is a three-phase material consisting of solids, voids, and water. The four pumice samples had a dry bulk densities of 0.3–0.41 g/cm^3, lower than that of typical non-volcanic soils (>0.9 g/cm^3) [22–24]. Our findings were similar to those derived from the Plinian deposit of the Minoan eruption of Santorini, Greece (0.19–0.32 g/cm^3 pumice bulk density) [22]. Similarly, pumice derived from the 79 AD Mount Vesuvius eruption had a low dry bulk density, ranging from 0.6 to 0.7 g/cm^3 [23].

The presence of internal voids in pumice also enhances porosity compared to non-volcanic soils. The porosity of the four pumice samples was 70–75%, whereas for non-volcanic soils, porosity values are typically <60% [22–24]. Our porosity values agreed with previous studies in Santa Maria, Guatemala and Medicine Lake, California (63–73% porosity) [25]. Moreover, the internal voids in pumice may result in a high matric potential (suction) [23]. Such a condition enables pumice to absorb and possibly hold a substantial amount of water. Indeed, the inner-particle water absorption limit for pumice was 95–143%, which is 9–15 times higher than that suggested by ASTM C127, with the absorption limit ranging between 8–10% for typical coarse-grained soils. Such a high-water absorption limit on pumice was also found in previous studies (e.g., [23]), ranging from 67 to 70%.

This high-water absorption limit on pumice might also be associated with its pore connectivity. Dead-end to active pore ratios (R_v) among four pumice samples ranged from 26.5% to 38.1%, indicating that voids inside pumice particles are composed of 72.4–79.1% active pores and 20.9–27.6% dead-end pores. Our finding of a predominance of active versus dead-end pores is similar to findings of pumice originating from the eruption of Montagne Pele'e volcano in France, i.e., >75% active pores inside pumice particles [40]. Previous studies at the mountaintop lake of Changbaishan, China, and Mount Mazama, USA similarly reported >80% of interconnected porosity in pumice [41,42].

4.2. *Landslide Mobility among Various Water Content*

The presence of water within soil particle contacts strongly influences soil strength [43], affecting the mobility of geophysical gravitational flows [44]. For sand-sized particles, a significant increase in cohesion occurred at water contents < 5% because of the formation of a liquid bridge at soil particle contacts [45]. Similarly, a laboratory experiment indicated that water surface tension was the highest at a water content of <20% for particles < 110 μm, attributed to inter-particle cohesion [43]. This liquid cohesion weakened when particle size exceeded 150 μm because the effect of gravitational forces was larger than water surface tension [46]. Moreover, most studies confirmed a significant reduction of shear strength at high water content (e.g., saturated conditions) because of high pore-water pressure (e.g., [10,12,43,45,47]).

Similar to previous studies (e.g., [7,15,44]), the variability of sediment mobility in our experiment might be associated with the availability of water at soil particle contacts. Indeed, completely dry conditions generated the highest L/H for conditions under $S \leq 0.6$ because lightweight materials with no liquid cohesion between particles promoted individual mobility for granular materials (e.g., [48]). The continuous decrease in L/H for $0 < S \leq 0.6$ might be associated with the liquid bridge that was generated by inter-particle cohesion [45]. Nearly saturated conditions caused much sediment transport to the outlet because of high pore-water pressure [47].

4.3. The Effect of Soil Characteristics on Landslide Mobility

Soil porosity is an important geotechnical factor for landslide mobility [2,7], because it defines water storage on soils [49]. A numerical simulation suggested that a 2% increase in soil porosity may elevate the travel distance of the 2014 Oso landslide because the porous, water-saturated media potentially liquefied during transport. A laboratory experiment confirmed this finding and found that sediment with the highest porosity generated the greatest sediment transport downstream in saturated conditions [7].

Although the porosity of pumice samples was high, the inner-particle absorption of water in these samples was also high, affecting inter-particle water availability. Compared to a previous study in the same flume configuration [7], pumice in our study had higher porosity and generated greater sediment transport to the outlet (7–18%) than their sediments under saturated conditions (used in [7]). However, this pumice is less mobile compared to weathered sedimentary rock (mean sediment transport to outlet = 61.3%) despite having 28% larger porosity. Because inner-particle water absorption (W_A) on pumice was equal to $S = 0.56$, only about 44% of saturated water was retained in inter-particle voids. Although pumice had high porosity, since their W_A is also high, the inter-particle water availability might be less than that in weathered sedimentary rock. Such conditions imply that higher porosity does not always represent higher mobility.

Water absorption within pumice particles also controls the pattern of L/H (Figure 6a). Indeed, since the water absorption limit for pumice corresponded to $S = 0.56$, water less than $S < 0.56$ may be absorbed by the internal voids with little water on particle contacts. Such a condition enhances inter-particle cohesion and thereby reduces L/H. Hence, $S = 0.6$ became the threshold of L/H, and $S > 0.6$ generated much higher sediment transport because water exceeding water absorption limit may weaken particle contacts (Figure 6b).

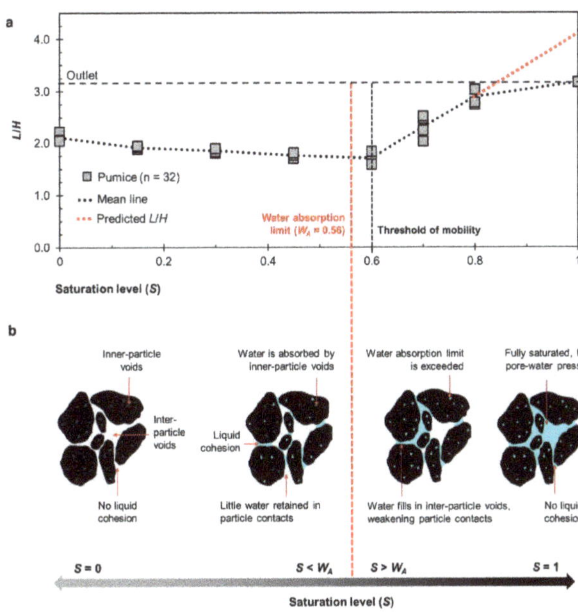

Figure 6. (**a**) Effects of inner-particle water absorption (W_A) on pumice mobility. X-axis shows saturation levels (S), where $S = 0$ represents completely dry conditions and $S = 1$ represents completely saturated conditions. Y-axis shows L/H, representing the ratio between total travel distance (L, cm) and total drop height (H, cm) of sediment. (**b**) Schematic illustration of possible soil-water interactions on pumice. Water absorption limit (W_A, %) is converted to the saturation level (S) by dividing W_A (%) by the saturated water content (w_{Sr}, %).

Thus, we concur with previous studies that porosity is influential in landslide mobility (e.g., [2,7]). However, we highlight that inner-particle water absorption is also a key factor that controls mobility.

4.4. The Mobility of Landslides in Pumice

The water content of our four pumice samples was high, with saturation level (S) ranging between 0.5 and 0.6. Based on our experiments, values of L/H using $S = 0.6$ were 1.6–1.8, with a scaled-up volume of 13,500 m^3. These values of L/H were low compared to those from field investigations in the same area that ranged from 4.2 to 5.2 for volumes of 10,000–15,000 m^3 [30]. Similarly, GIS analysis of landslides triggered during the 1968 Tokachi-Oki earthquake had higher values of L/H for landslides in pumice, ranging between 2.7–7 for similar ranges of volume [20]. This condition implies that the actual landslides were more mobile than those in our experiment, which may indicate a higher water content in the field during the actual landslides. These differences in water content might be associated with water accumulation from previous rainfall events. Indeed, API_{30} on the date when landslides occurred was 230 mm (measured from AMeDAS Atsuma), which was 2.2–3.6 times larger than that during field investigations.

These differences in L/H might also be associated with scaling effects. A previous study suggested that small-scale experiments generated much larger effects of shear resistance from liquid cohesion and smaller effects of pore-water pressure compared to the actual phenomena, which is proportional to h^3, where h represents landslide thickness [50]. As such, small scale experiments with much smaller h compared to the full-scale natural events potentially underestimate landslide mobility. Additionally, since overburden pressure depends strongly on h, the actual landslides may generate much larger overburden pressure than those in small-scale experiments, potentially crushing soil particles during transport. As soil particles are crushed, more soil mass occupies the same soil volume, elevating pore-water pressure, which consequently fluidizes the landslide mass, thus enhancing landslide mobility (L/H) [51]. Previous studies mentioned that this grain-crushing-induced fluidization not only occurs in completely saturated conditions [52], but also in partially saturated soils [53].

Although we did not observe any grain crushing in our experiments, which may be attributed to the much smaller h compared to actual landslides, our findings indicate that such unique pumice characteristics can facilitate landslide fluidization. Since a low value of specific gravity indicates a low soil strength [38], pumice can be easily crushed during transport. Indeed, the change in particle size distribution after triaxial testing indicates crushability of pumice under overburden pressure [39]. Field investigation confirmed that grain crushing occurred on landslides in pumice [21,31]. Because internal voids in pumice can absorb and possibly hold a substantial amount of water, such grain crushing may allow water to escape from internal voids thereby fluidizing the landslide mass during transport.

Therefore, our findings concur with previous studies that water content is a key factor controlling the mobility of landslides in pumice (e.g., [20]). Gravimetric water contents of pumice samples based on laboratory tests were high, ranging between 95 and 143%. However, we highlight that unique pumice characteristics such as high water absorption may facilitate water fluidization in pumice layers, enhancing landslide mobility. Variability of mobility might be associated with the variability of initial water content in pumice layers influenced by topography and water-holding capability.

5. Conclusions

We conducted a flume experiment to examine the mobility of earthquake-induced landslides in pumice soils. Pumice used in this experiment was derived from landslides induced by the 2018 Eastern Iburi Earthquake, Japan. This pumice was unique because of the low specific gravity compared to non-volcanic soils, which indicated a low soil strength [38]. Since this low value of specific gravity might be associated with numerous voids inside pumice particles, pumice can absorb and possibly hold a substantial amount of water; 9–15 times higher than for typical coarse-grained soils. Using various saturation levels, our flume experiment confirmed that this inner-particle water absorption influences pumice mobility (L/H ratio).

Since grain crushing may occur on landslides in pumice [21,31], our findings indicated that unique pumice characteristics (i.e., high water absorption) potentially enhance their mobility. Therefore, although the initiation of earthquake-induced landslides is less influenced by soil-water conditions, they can be as mobile as those induced by rainfall, depending on the initial water content of pumice layers. Since pumice can absorb a substantial amount of water, this initial water content can be influenced by topography and water-holding capability in pumice layers, as well as water accumulation during antecedent rainfall. We believe that these findings provide insights for improving landslide risk assessment in catchments underlain by volcanic deposits.

Author Contributions: Conceptualization, resources, and supervision, T.G.; Methodology, T.G., H.S.K. and R.N.; Investigation, T.G., H.S.K., R.N. and R.P.R.; Data curation and writing—original draft preparation, R.N.; Validation, T.G., R.C.S. and K.S.; Writing—review and editing, all co-authors. All authors have read and agreed to the published version of the manuscript.

Funding: This research received no external funding.

Acknowledgments: We thank the FLOuRISH Institute (Institute for Fostering Leadership to create future value through Outstanding Research, Innovation and Sustainable Higher education), Tokyo University of Agriculture and Technology, Japan, for supporting this study. We also thank Yoshiharu Ishikawa and Yohei Arata for valuable discussions and constructive feedbacks on conducting field investigations.

Conflicts of Interest: The authors declare no conflict of interest.

References

1. Sidle, R.C. Mass Movement. In *Encyclopedia of Natural Hazards, Encyclopedia of Earth Sciences Series;* Bobrowsky, P.T., Ed.; Springer: Dordrecht, The Netherlands, 2013. [CrossRef]
2. Iverson, R.; George, D.; Allstadt, K.; Reid, M.; Collins, B.; Vallance, J.; Schilling, S.; Godt, J.; Cannon, C.; Magirl, C.; et al. Landslide mobility and hazards: Implications of the 2014 Oso disaster. *Earth Planet. Sci. Lett.* **2015**, *412*, 197–208. [CrossRef]
3. Roback, K.; Clark, M.K.; West, A.J.; Zekkos, D.; Li, G.; Gallen, S.F.; Chamlagain, D.; Godt, J.W. The size, Distribution, and Mobility of Landslides Caused by the 2015 M. 7.8 Gorkha Earthquake, Nepal. *Geomorphology* **2018**, *301*, 121–138. [CrossRef]
4. Jibson, R.W. The 2005 La Conchita, California, Landslide. *Landslides* **2006**, *3*, 73–78. [CrossRef]
5. Evans, S.G.; Guthrie, R.H.; Roberts, N.J.; Bishop, N.F. The Disastrous 17 February 2006 Rockslide-debris Avalanche on Leyte Island, Philippines: A Catastrophic Landslide in Tropical Mountain Terrain. *Nat. Hazards Earth Syst. Sci.* **2007**, *7*, 89–101. [CrossRef]
6. Kharismalatri, H.S.; Ishikawa, Y.; Gomi, T.; Shiraki, K.; Wakahara, T. Collapsed material movement of deep-seated landslides caused by Typhoon Talas 2011 on the Kii Peninsula, Japan. *Int. J. Eros. Control Eng.* **2017**, *10*, 108–119. [CrossRef]
7. Kharismalatri, H.S.; Ishikawa, Y.; Gomi, T.; Sidle, R.C.; Shiraki, K. Evaluating Factors for Controlling Sediment Connectivity of Landslide Materials: A Flume Experiment. *Water* **2019**, *11*, 17. [CrossRef]
8. Guo, D.; Hamada, M.; He, C.; Wang, Y.; Zou, Y. An empirical model for landslide travel distance prediction in Wenchuan earthquake area. *Landslides* **2014**, *11*, 281–291. [CrossRef]
9. Hayashi, J.N.; Self, S. A comparison of pyroclastic flow and debris avalanche mobility. *J. Geophys. Res. Solid Earth* **1992**, *97*, 9063–9071. [CrossRef]
10. Legros, F. The Mobility of Long Run-out Landslides. *Eng. Geol.* **2002**, *63*, 301–331. [CrossRef]
11. Lucas, A.; Mangeney, A.; Ampuero, J.P. Frictional velocity-weakening in landslides on Earth and on other planetary bodies. *Nat. Commun.* **2014**, *5*, 3417. [CrossRef]
12. Legros, F. Landslide Mobility and the Role of Water. In *Landslides from Massive Rock Slope Failure, Earth and Environmental Sciences;* Evans, S.G., Mugnozza, G.S., Strom, A., Hermanns, R.L., Eds.; Springer: Dordrecht, The Netherlands, 2006; Volume 49, pp. 233–242. [CrossRef]
13. Ochiai, H.; Okada, Y.; Furuya, G.; Okura, Y.; Matsui, T.; Sammori, T.; Terajima, T.; Sassa, K. A fluidized landslide on a natural slope by artificial rainfall. *Landslides* **2004**, *1*, 211–219. [CrossRef]
14. Bessette-Kirton, E.K.; Coe, J.A.; Schulz, W.H.; Cerovski-Darriau, C.; Einbund, M.M. Mobility characteristics of debris slides and flows triggered by Hurricane Maria in Puerto Rico. *Landslides* **2020**, *17*, 2795–2809. [CrossRef]
15. Yang, L.; Wei, Y.; Wang, W.; Zhu, S. Numerical Runout Modeling Analysis of the Loess Landslide at Yining, Xinjiang, China. *Water* **2019**, *11*, 1324. [CrossRef]

16. Zhou, G.G.D.; Hu, H.S.; Song, D.; Zhao, T.; Chen, X.Q. Experimental study on the regulation function of slit dam against debris flows. *Landslides* **2019**, *16*, 75–90. [CrossRef]
17. McColl, S.T. Landslide Causes and Triggers. In *Landslide Hazards, Risks and Disasters*; Shroder, J.F., Davies, T., Eds.; Academic Press: Cambridge, MA, USA, 2015; pp. 17–42. [CrossRef]
18. Zhang, Y. *Earthquake-Induced Landslides*; Springer Nature: Singapore, 2018; pp. 1–31. [CrossRef]
19. Chigira, M.; Tsou, C.-Y.; Matsushi, Y.; Hiraishi, N.; Matsuzawa, M. Topographic precursors and geological structures of deep-seated catastrophic landslides caused by Typhoon Talas. *Geomorphology* **2013**, *201*, 479–493. [CrossRef]
20. Ling, S.; Chigira, M. Characteristics and triggers of earthquake-induced landslides of pyroclastic fall deposits: An example from Hachinohe during the 1968 M7.9 Tokachi-Oki earthquake, Japan. *Eng. Geol.* **2020**, *264*, 105301. [CrossRef]
21. Crosta, G.B.; Imposimato, S.; Roddeman, D.; Chiesa, S.; Moia, F. Small Fast-moving Flow-like Landslides in Volcanic Deposits: The 2001 Las Colinas Lanslide (El Salvador). *Eng. Geol.* **2005**, *79*, 185–214. [CrossRef]
22. Esposito, L.; Guadagno, F.M. Some special geotechnical properties of pumice deposits. *Bull. Eng. Geol. Environ.* **1998**, *57*, 41–50. [CrossRef]
23. Whitham, A.G.; Sparks, R.S.J. Pumice. *Bull. Volcanol.* **1986**, *48*, 209–223. [CrossRef]
24. Maeda, T.; Soma, K.; Warkentin, B.P. Physical and Engineering Characteristics of Volcanic Soils in Japan Compared in Other Countries. *Irrig. Eng. Rural. Plan.* **1983**, *3*, 16–31. [CrossRef]
25. Fauria, K.E.; Manga, M.; Wei, Z. Trapped bubbles keep pumice afloat and gas diffusion makes pumice sink. *Earth Planet. Sci. Lett.* **2017**, *460*, 50–59. [CrossRef]
26. Wang, F.; Fan, X.; Yunus, A.P.; Subramanian, S.S.; Alonso-Rodriguez, A.; Dai, L.; Xu, Q.; Huang, R. Coseismic landslides triggered by the 2018 Hokkaido, Japan (Mw 6.6), earthquake: Spatial distribution, controlling factors, and possible failure mechanism. *Landslides* **2019**, *16*, 1551–1566. [CrossRef]
27. Konagai, K.; Nishiyama, S.; Ohishi, K.; Kodama, D.; Nanno, Y. Large Ground Deformations Caused by the 2018 Hokkaido Eastern Iburi Earthquake. *JSCE J. Disaster Factsheets* **2018**, *FS2018-E-0003*, 1–9.
28. Ritonga, R.P.; Gomi, T.; Tarigan, S.D.; Kaswanto, R.L.; Kharismalatri, H.S.; Noviandi, R.; Arata, Y.; Ishikawa, Y. Land Cover and Characteristics of Landslides Induced by the 2018 Mw 6.7 Eastern Iburi Earthquake, Hokkaido. *Int. J. Eros. Control Eng.* **2020**, *13*, 76–83. [CrossRef]
29. Ito, Y.; Yamazaki, S.; Kurahashi, T. Geological features of landslides caused by the 2018 Hokkaido Eastern Iburi Earthquake in Japan. *Geol. Soc. London Spéc. Publ.* **2020**, *501*, 171–183. [CrossRef]
30. Osanai, N.; Yamada, T.; Hayashi, S.-I.; Kastura, S.; Furuichi, T.; Yanai, S.; Murakami, Y.; Miyazaki, T.; Tanioka, Y.; Takiguchi, S.; et al. Characteristics of landslides caused by the 2018 Hokkaido Eastern Iburi Earthquake. *Landslides* **2019**, *16*, 1517–1528. [CrossRef]
31. Zhang, S.; Li, R.; Wang, F.; Iio, A. Characteristics of landslides triggered by the 2018 Hokkaido Eastern Iburi earthquake, Northern Japan. *Landslides* **2019**, *16*, 1691–1708. [CrossRef]
32. Yamagishi, H.; Yamazaki, F. Landslides by the 2018 Hokkaido Iburi-Tobu Earthquake on September 6. *Landslides* **2018**, *15*, 2521–2524. [CrossRef]
33. Japan Meteorological Agency (JMA). Monthly Report on Earthquakes and Volcanoes in Japan September 2018. Available online: https://www.data.jma.go.jp/svd/eqev/data/gaikyo/monthly/201809/201809monthly.pdf (accessed on 20 September 2022).
34. Ministry of Land Infrastructure, Transport, and Tourism (MLIT). The Collapse Area of the 2018 Hokkaido Eastern Iburi Earthquake was the Largest Since the Meiji Era. Available online: http://www.mlit.go.jp/river/sabo/h30_iburitobu/181005_sediment_volume.pdf (accessed on 7 April 2019).
35. Iverson, R.M.; Logan, M.; Lahusen, R.G.; Berti, M. The perfect debris flow? Aggregated results from 28 large-scale experiments. *J. Geophys. Res. Earth Surf.* **2010**, *115*, F03005. [CrossRef]
36. Nicholson, P.G. *Soil Improvement and Ground Modification Methods*; Elsevier: Amsterdam, The Netherlands, 2015; pp. 231–234. [CrossRef]
37. Bowles, J.E. *Foundation Analysis and Design*, 5th ed.; The McGraw-Hill Companies, Inc.: Singapore, 1997; pp. 1–29.
38. Roy, S.; Bhalla, S.K. Role of Geotechnical Properties of Soil on Civil Engineering Structures. *Resour. Environ.* **2017**, *7*, 103–109.
39. Li, R.; Wang, F.; Zhang, S. Controlling role of Ta-d pumice on the coseismic landslides triggered by 2018 Hokkaido Eastern Iburi Earthquake. *Landslides* **2020**, *17*, 1233–1250. [CrossRef]
40. Bernard, M.-L.; Zamora, M.; Géraud, Y.; Boudon, G. Transport properties of pyroclastic rocks from Montagne Pelée volcano (Martinique, Lesser Antilles). *J. Geophys. Res.* **2007**, *112*, B05205. [CrossRef]
41. Klug, C.; Cashman, K.V.; Bacon, C.R. Structure and physical characteristics of pumice from the climactic eruption of Mount Mazama (Crater Lake), Oregon. *Bull Volcanol.* **2002**, *64*, 486–501. [CrossRef]
42. Li, X.; Yang, W.; Zou, Q.; Zuo, Y. Investigation on Microstructure, Composition, and Cytocompatibility of Natural Pumice for Potential Biomedical Application. *Tissue Eng. Part C Methods* **2010**, *16*, 427–434. [CrossRef] [PubMed]
43. Louati, H.; Oulahna, D.; de Ryck, A. Effect of the Particle Size and the Liquid Content on the Shear Behavior of Wet Granular Material. *Powder Technol.* **2017**, *315*, 398–409. [CrossRef]
44. Roche, O.; Attali, M.; Mangeney, A.; Lucas, A. On the run-out distance of geophysical gravitational flows: Insight from fluidized granular collapse experiments. *Earth Planet. Sci. Lett.* **2011**, *311*, 375–385. [CrossRef]
45. Mitarai, N.; Nori, F. Wet Granular Materials. *Adv. Phys.* **2006**, *55*, 1–45. [CrossRef]

46. Shi, H.; Mohanty, R.; Chakravarty, S.; Cabiscol, R.; Morgeneyer, M.; Zetzener, H.; Ooi, J.Y.; Kwade, A.; Luding, S.; Magnanimo, V. Effect of Particle Size and Cohesion on Powder Yielding and Flow. *KONA Powder Part. J.* **2018**, *35*, 226–250. [CrossRef]
47. Wang, G.; Sassa, K. Pore-pressure generation and movement of rainfall-induced landslides: Effects of grain size and fine-particle content. *Eng. Geol.* **2003**, *69*, 109–125. [CrossRef]
48. Staron, L.; Lajeunesse, E. Understanding how volume affects the mobility of dry debris flows. *Geophys. Res. Lett.* **2009**, *36*, L12402. [CrossRef]
49. Du, J.; Fan, Z.-J.; Xu, W.-T.; Dong, L.-Y. Research Progress of Initial Mechanism on Debris Flow and Related Discrimination Methods: A Review. *Front. Earth Sci.* **2021**, *9*, 629527. [CrossRef]
50. Iverson, R.M. Scaling and design of landslide and debris-flow experiments. *Geomorphology* **2015**, *244*, 9–20. [CrossRef]
51. Gerolymos, N.; Gazetas, G. A model for grain-crushing-induced landslides—Application to Nikawa, Kobe 1995. *Soil Dyn. Earthq. Eng.* **2007**, *27*, 803–817. [CrossRef]
52. Sassa, K.; Fukuoka, H.; Wang, F.; Wang, G. Dynamic properties of earthquake-induced large-scale rapid landslides within past landslide masses. *Landslides* **2005**, *2*, 125–134. [CrossRef]
53. Sassa, K.; Fukuoka, H.; Scarascia-Mugnozza, G.; Evans, S. Earthquake induced landslides: Distribution, motion and mechanisms. *Soils Found* **1996**, *36*, 53–64. [CrossRef]

water

MDPI

Article

Visual Harmony of Engineering Structures in a Mountain Stream

Jinn-Chyi Chen [1], Chia-Ling Huang [2], Su-Chin Chen [2,3,*] and Samkele S. Tfwala [4]

1 Department of Hydraulic Engineering, Fujian College of Water Conservancy and Electric Power, Yongan 366000, China; chenjinnchyi@gmail.com
2 Department of Soil and Water Conservation, National Chung Hsing University, Taichung 40227, Taiwan; judy820517@gmail.com
3 Innovation and Development Center of Sustainable Agriculture, National Chung Hsing University, Taichung 40227, Taiwan
4 Department of Geography, Environmental Science and Planning, University of Eswatini, Matsapha M202, Eswatini; sstfwala@uniswa.sz
* Correspondence: scchen@nchu.edu.tw

Abstract: This study uses the cognitive factor of "visual harmony" to assess the visual quality of stream engineering in a mountainous region. Images of engineering structures such as revetments and submerged dams in the mountain streams of Taiwan were collected. Three image groups with different structures invaded by vegetation were used for a questionnaire survey, which yielded 154 valid samples. We used statistical analysis to develop a model of visual harmony H with respect to the percentage of visible greenery G_R, that is, the perceived curve of vegetation change. A comparison of our data with the literature determined the upper and lower bound curves of the relationship between H and G_R. We found that the physical elements of "softscape" and "hardscape"—namely, percentage of visible water W_R, visible structure I_R, and visible natural material on the structure N_R—affected this relationship. Results show that H is equivalent to visual preference P, and both can be improved by better green visibility (increasing G_R and $G_R < 50\%$), avoiding low water visibility ($W_R < 10\%$), or increasing the amount of visible natural material ($N_R > 0.9$). High visibility of the structures ($I_R > 0.3$) may decrease H and P. We ultimately propose a visual harmony or preference model concerning a combined physical indicator that comprises G_R, W_R, I_R and N_R. Results of this study could be helpful to improve or access the aesthetics of stream engineering design.

Keywords: harmony; preference; cognitive factors; physical elements; questionnaire survey

Citation: Chen, J.-C.; Huang, C.-L.; Chen, S.-C.; Tfwala, S.S. Visual Harmony of Engineering Structures in a Mountain Stream. *Water* **2021**, *13*, 3324. https://doi.org/10.3390/w13233324

Academic Editor: Tammo Steenhuis

Received: 15 October 2021
Accepted: 19 November 2021
Published: 23 November 2021

1. Introduction

The concept of harmony is an essential focus of attention in engineering, management, art, and design. One such example is the harmony sought to balance safety (for humans), ecology, and landscape in ecological engineering within the more extensive study of soil and water conservation [1]. Other examples include human–water harmony when developing evaluation models in hydraulic engineering and water resources management [2–4], harmony in human-river encounter sites [5], harmony in ecosystem service [6], and many studies concerns about colour harmony in colour combinations [7,8], art and design [9], and urban spaces [10,11].

1.1. Visual Harmony

There are many definitions of harmony in different study fields. In colour harmony, key words associated with harmony include "completeness", "complementary/analogous", "order", "balance", "unambiguous", and "pleasing" [12,13]. "Harmony" and "preference" are often used interchangeably and are regarded equivalent in early previous analyses [13]. However, some studies showed that two terms of "preference" for and "harmony" of colour combinations may produce confusing results [8,13].

Within the study of visual landscape, harmony reflects unity and balance, or the pleasant "arrangement" of landscape attributes [14]. It is closely related to coherence [15] or unity. Harmony, coherence and unity are interchangeable terms [16]. Harmony or coherence has been associated with water features [16–18] and land-use suitability [16,19,20]. It is an important cognitive indicator [21] and may be associated with landscape physical elements. Thus, harmony was selected as an indicator in this work to investigate changes of proportions of water and vegetation, and property of engineering structures (scale and texture) in mountainous stream landscape. Visual harmony in this paper describes the perceived balance of landscape attributes to form a coherent, unified whole. Visual preference focuses on people's preference or aversion to the landscape. Visual preference and harmony were regarded as different terms in this work. They are important cognitive processes in aesthetic assessment.

1.2. Aesthetic Assessments of Rivers

Defining aesthetics has been addressed by [13]. Ideally, it refers to the study of human minds and emotions in relation to the sense of beauty. However, aesthetics, such as we usually talk about (the beauty or ugliness of an object or a scene), is subjective, and it depends on personal aesthetic experiences. Thus, the terms public perception or preference were generally used by researchers to investigate aesthetic quality or aesthetic assessment. Visual preferences reflect the degree to which people experience it as beautiful (or ugly). It is a cognitive process and a product of perception that should be considered in studies involving people's aesthetic judgments and selectivity, including perceptions of the environment [22]. Research on the aesthetic assessments or visual quality of rivers has often been addressed in many studies, such as river restoration measures [23], river with recreation infrastructures [24,25], wood in rivers [26,27], water quality [28], and engineering structures in mountainous streams [21,29]. There are a few soils and water conservation studies on the visual quality of sediment control structures in mountain streams written from the perspective of perception [21,29]. The goal of stream engineering design is safety and functionality. Yet, with more excellent environmental quality and growing public interest in landscape beauty, visual preference and visual harmony has become a part of engineering design. Chen et al. [21] proposed a model of visual preference associated with cognitive factors, such as harmony, naturalness, vividness, and closeness, in sediment control structure in mountainous streams. The property of softscape elements, i.e., the visible body of water and vegetation, associated with cognitive indicators were also presented. However, the model was developed based on cognitive factors. Studies on how visual harmony (*H*) and visual preference (*P*) are affected by the structure's texture and its proportion (namely, hardscape elements in this work), and how they are associated with softscape elements, are still lacking. Particularly, an evaluation model using softscape and hardscape elements may be more interesting to a hydraulic or landscape designer and manager than applying cognitive factors. Softscape and hardscape are direct and the designer or manager can try to adjust the physical environment elements to prompt aesthetic quality.

1.3. Objectives of This Study

Engineering structures allow for self-design by nature and considering their harmony with nature is a design norm. For examples, a stone revetment, groundsill, or dam with stone lining has voids between stones at the surface that allow vegetation to invade, increasing greenery and, thus, visual beauty [30]. However, overgrown greenery can destroy the harmony of the stream's landscape and sediment control structures [21]. Such invasion of vegetation in stone revetments and other sediment control structures is well documented in the literature [21,29]. However, a lack of study determines the impact curve nor evaluation model of visual harmony nor preference with respect to visual greenery for application. If such a model were established, the visual quality of engineering design, management, and maintenance could be improved. There also exists some evidence of an evaluation model of visual preference that discusses the influence of softscapes (water and greenery) on cognitive factors [21], but it ignores hardscapes, despite such elements being vital in design.

Visual harmony is a cognitive process resulting from a sense of visual order, unify, and coherence between and among the landform, water, vegetation, or structures visible. It allows the human eye to connect the "softscapes" and "hardscapes" easily—that is, physical elements—through engineering design. Harmony may increase scenic beauty and public acceptance [21,22,29] and it should be seriously considered in assessment visual quality of engineering structures in mountain stream [21]. This paper focuses on visual harmony of engineering structures in a mountain stream, particularly vegetation invading engineering structures. Emphasis is on how visual harmony and preference are affected by vegetation after engineering structures were completed, and how to develop a model of visual harmony with respect to the physical elements, such as the percentage of visible water, visible structure, and visible natural material on the structure. The visual harmony influenced by the proportions of water, vegetation, and engineering structures, and structure texture were studied. Visual harmony associated with preference is also presented. Understanding the visual harmony of physical elements may improve the aesthetics of stream engineering design and may help scholars develop a model for assessing visual quality.

2. Overview of Gungfunnan, Chotengkeng, and Tungsikun Creeks

Taiwan is a mountainous island. Hills cover seventy percent of its total landmass. The rivers in mountainous regions have steep slopes and flow at rapid velocities, necessitating the need for sediment control engineering to stabilize the streambed and prevent hazards related to sediment and water flow. In 2001, the Soil and Water Conservation Bureau (SWCB) in Taiwan began to promote innovation in ecological engineering [1]. Currently, the typical engineering structures in Taiwan include stone revetments, groundsills, submerged dams, and check dams.

We chose a location upstream of the Gungfunnan creek in Yilan Country and Chotengkeng creek in Taichung City in the Taiwanese mountains (see Figure 1 and Table 1). The Gungfunnan creek has a watershed area of 866 ha and a main stream length of 5.5 km. In 1997, Typhoon Sinlaku caused landslides and floods in the watershed, leading to massive sediment deposition downstream. The Chotengkeng creek has a watershed area of 1320 ha and a main stream length of 8 km. It experienced floods, landslides, and massive debris flow during Typhoon Mindulle in July 2004. Subsequently, the SWCB has constructed a series of stone groundsills, stone revetments, and submerged dams in the two creeks to control sediment flow. The stones for such ecological structures were sourced from on-site streambeds or stream sides to utilise the natural resource and increase the flow area of the creeks. Groundsills can stabilize streambeds and the toe of the slopes, but they also tend to raise the streambed, leading to vegetation invasion. Stone revetments mainly protect the toe of the streambank to avoid erosion. In these structures, the voids between the stones allow vegetation to proliferate compared with concrete revetments.

Figure 1 shows three observation points, namely, P1, P2, and P3, and their viewing scenes in the study area. The viewing scenes at P1 and P2 were taken upstream of Gungfunnan creek. Figure 1b,c show the stone revetment, groundsill works, and submerged dam. P3 (see Figure 1d) was taken from Chung-Ho bridge downstream of Chotengkeng creek, showing a declined groundsill with stone lining. Figure 1b–d were collected from the SWCB in the early stages after engineering works were completed. All images were captured from either river banks or on a bridge. They were easily viewable with the naked eye such that aerial, long-range, and panoramic shots were not necessary. These images were used to investigate the changes in vegetation on the riversides and riverbeds.

Figure 1. Study area and observation sites in the Gungfunnan creek in Yilan Country and Chotengkeng creek in Taichung city, Taiwan. Note. Figure 1b–d were collected from SWCB in the early stage after the engineering works were completed, and they were used to investigate the changes in vegetation on the riversides and riverbeds. (**a**) Maps showing the study area and observation sites P1, P2, and P3; (**b**) Viewing scene at P1(SWCB, November 2012); (**c**) Viewing scene at P2 (SWCB, September 2012); (**d**) Viewing scene at P3 (SWCB, September 2008).

Table 1. Watershed characteristics of Gungfunnan (upstream) and Chotengkeng creeks.

Location	Watershed Area (ha)	Mainstream Length (km)	Stream Width (m)	Average Riverbed Slope	Engineering Structures
Gungfunnan creek	866	5.5	6–15	Mainstream: 7.07%; Tributary: 15–17%	Stone revetments, stone groundsills, and submerged dam
Chotengkeng creek	1320	8.0	10–20	Mainstream: 2.7%; Tributary: 10–16%	Declined groundsill with stone lining

3. Methods

First, field images were collected, followed by image processing. Secondly, the physical elements of each image were calculated. Thirdly, a questionnaire survey was conducted. Participants were asked to rate selected images for their visual preferences and visual harmony. The mean value of the visual preference and harmony in each image was determined through statistical analysis. Finally, a visual harmony and preference model that includes a combined physical indicator was proposed.

3.1. Image Collection and Processing

Images were simulated using Adobe® Photoshop® CC in 2015 version. All simulated images had the same background (sky, mountain, and water); and only the vegetation on the riverbed and riversides differed. Figure 2 shows the changes of the riverside landscape in the upstream section of Gungfunnan creek. Five images C1–1, C1–2 were collected from SWCB, C1–4 from field investigation, and C1–3 and C1–5 images from image processing, referred to as images C1–2 and/or C1–4. Image C1–3 simulates the vegetation invasion in the stage between November 2012 (Image C1–2) and July 2018 (Image C1–4). After engineering was completed, six years (Image C1–4), vegetation started to naturally invade riverbanks and riverbed, compared to Image C1–2. However, excessive vegetation on banks and riverbed increased flow resistance and led to overflow. Image C1–5, simulation based on C1—2 and C1–4 images, simulates vegetation removed from riverbed after July 2018 for management requirements. This followed the background of image C1–2, and merging the vegetation on river banks of image C1–4.

Figure 2. *Cont.*

C1–4

C1–5

Figure 2. Changes in the riverside landscape from engineering works, vegetation invasion, and human maintenance in the up-stream section of Gungfunnan creek. These images were collected between 2011 and 2018. The field images include the river landscape before engineering (C1–1) and after engineering treatment (C1–2 to C1–4). Image C1–3 shows increasing vegetation after Image C1–2, and Image C1–5 shows no riverbed vegetation after Image C1–4. C1–1: Prior to engineering, with visible greenery, G_R, at 66% (SWCB, September 2011). C1–2: Early stage after the engineering complete, with visible greenery, G_R, at 24% (SWCB, November 2012). C1–3: Vegetation invasion, with visible greenery, G_R, at 31% (Simulation based on C1–2 and C1–4 images). C1–4: Vegetation invasion, with visible greenery, G_R, at 76% (July 2018). C1–5: No riverbed vegetation, with visible greenery, G_R, at 60% (Simulation based on C1–2 and C1–4 images).

Figure 3 shows four images (C2–1, C2–2, C2–3, and C2–4) with different vegetation growing over the revetment in the upstream section of Gungfunnan creek. Image C2–1 were collected from SWCB in 2012 and Image C2–4 were taken from field investigation in 2018. Images C2–2 and C2–3 were generated by image processing to simulate the changes of vegetation invasion on revetment. The simulated vegetation elements referred to the real images of C2–1 and C2–4. All images in Figure 3 showed the early stage of complete engineering (C2–1), and gradual invasion of vegetation after engineering (C2–2 to C2–4).

Figure 4 includes three images (C3–1, C3–2, and C3–3) showing changes in the riverbed vegetation at Chung-Ho bridge in the downstream section of Chotengkeng creek. Image C3–1 was collected from SWCB in 2008, and C3–2 and C3–3 images are from image processing. C3–3 is a merged image that combined the real vegetation on the riverbed from field investigation in 2018 and the background of Image C3–1. Twelve images (see Figures 2–4) were used to investigate the changes in vegetation.

C2–1

C2–2

Figure 3. *Cont.*

C2–3

C2–4

Figure 3. Changes in the revetment vegetation in the upstream section of Gungfunnan creek. Showing the early stage of complete engineering (C2–1), and the gradual invasion of vegetation after engineering (C2–2 to C2–4). Images C2–1 and C2–4 were collected from the field, respectively from SWCB in 2012 and in 2018. Images C2–2 and C2–3 were generated by image processing to simulate the changes of vegetation invasion on the revetment. C2–1: Engineering complete, with visible greenery, G_R, at 8% (SWCB, September 2012). C2–2: Vegetation invasions on revetment, with visible greenery, G_R, at 25% (Simulation). C2–3: Vegetation invasions on revetment, with visible greenery, G_R, at 41% (Simulation). C2–4: Vegetation invasions on revetment, with visible greenery, G_R, at 60% (July 2018).

C3–1

C3–2

C3–3

Figure 4. Changes in the riverbed vegetation at Chung-Ho bridge, at the downstream section of Chotengkeng creek. Image C3–1 was collected SWCB in 2008, and C3–2 and C3–3 are from images processing. C3–3 is a merged image that combined the real vegetation on riverbed from field investigation in 2018 and the background of Image C3–1. C3–1: Engineering complete, with visible greenery, G_R, at 19% (SWCB, September 2008). C3–2: Simulation based on Image C3–1, with visible greenery, G_R, at 30%. C3–3: Merged image of Image C3–1 and field image on July 2018, with visible greenery, G_R, at 45%.

3.2. Calculating the Physical Elements Using the Four Visible Indicators

All images were adjusted to the dimension of 72.25 × 54.19 cm (2048 × 1536 pixels) in Adobe® Photoshop®. Four visible indicators, namely, the percentages of visible greenery (G_R), visible water (W_R), visible structure (I_R), and visible natural material on the structure (N_R) in each image were determined as follows:

$$G_R = G/T \quad (1)$$

$$W_R = W/T \quad (2)$$

$$I_R = I/T \quad (3)$$

$$N_R = N/I \quad (4)$$

In all equations, T is the total number of pixels of an image. G, W, and I are the numbers of pixels of green vegetation, water, and engineering facilities, respectively, in the image. Engineering facilities here refer to structures of stream engineering, such as dam, groundsill, and/or revetment. A few images contain part of the road. Visible natural material, N, is the number of pixels of natural material in the engineering facilities. The natural materials in an image are all from on-site stones or boulders. For each image, pixels of green vegetation, water, engineering facilities, and natural material were judged by the naked eye, and the number of pixels were determined in Adobe® Photoshop®.

3.3. Questionnaire Survey

Visual preference is a cognitive process and a product of perception that is important for analysing people's aesthetic judgments and selectivity, including perceptions of the environment [22]. It is generally regarded as a critical index for evaluating the overall landscape quality accepted by the public. It can be affected by many cognitive factors, such as naturalness, harmony, openness or closeness, vividness, and complexity [31,32]. Visual preference positively correlates with harmony, naturalness, and vividness but negatively with openness or closeness for a landscape with sediment control structures in mountain streams. Among the four cognitive factors, visual harmony has the most substantial effect on visual preference [21]. Thus, we selected two cognitive factors, namely, visual harmony and visual preference in this study.

In the literature, questionnaire surveys are the most widely used for studying the visual perception of a landscape. An online questionnaire survey was accordingly distributed using cloud based SurveyCake to licensed engineers, academic researchers, consultant companies, and the general public through email and social media channels such as LINE, WeChat, and Facebook, who were then asked to further disseminate the questionnaire among their peers. Respondents provided their gender, age, and professional background under complete anonymity and rated their visual preferences and visual harmony for each image.

Twelve images were divided into three groups, C1, C2 and C3, as shown in Figures 2 and 4, and rated using a five-point Likert scale. That is, each image had five responses for visual preference or harmony. That is, each image had five responses for visual preference or harmony: "very high (1)" , "high (0.8)" , "normal (0.6)", "low (0.4)" and "very low (0.2)". For example, if the image scores 1, the respondent's visual preference for it is very high, and a score of 0.2 means low preference. All questionnaire samples were divided into two groups: general public and experts. An expert in this work is defined as a person with a master's degree or career experience over two years in hydraulic engineering, civil engineering, soil and water conservation, or landscape architecture. There were 154 valid respondents: 67 for experts and 87 for the general public, 8% aged under 20 years, 71% aged 20–30 years, 16% aged 30–40 years, and 5% aged 30–40 years. The scores on visual preference and visual harmony for each image were statistically analysed using Microsoft Excel and Golden Software Grapher ver 12.

3.4. Results Presentation

There are 12 images with different degrees of visible greenery in the upstream section of Gungfunnan creek (groups C1 and C2 in Figures 2 and 3 showing the changes to the riverside and revetment), and in Chotengkeng creek (group C3 in Figure 4). All images were used for the questionnaire survey. The mean value of the visual preference and harmony was determined through statistical analysis of the scores of each image. Finally, we present our results as follows:

First, the visual quality of stream engineering, given the invasion of vegetation was investigated to determine the perceived curve of vegetation changes (VCC) based on the relationship between visual harmony (or preference) and visible greenery. Secondly, the observed data from this study were compared with previous data [21,30,33], as summarized in Table 2, to determine if VCC is a reasonable measure in our study's context. Third, we discussed the relationship between visual harmony (or preference) and visible greenery, and their relationship with the softscape and hardscape (e.g., water and the textures of each structure). Finally, we proposed a visual harmony model that includes a combined physical indicator.

Table 2. Studies on visual quality of stream engineering in mountainous regions.

Researchers	Perceived Factors	Source of Images	Groups in Questionnaire	Rates in Questionnaire
Ho et al. [30]	Visual preference and naturalness	Field images show changes to stream landscape from vegetation invasion on stone revetments in the Nan-Shi-Ken stream, Northern Taiwan; includes four images, A–17 to A–20 as shown in Figure A1, at a fixed site at different times after the stream remediation	No specific classification	The values of visual preference and naturalness were divided into eight intervals from zero to one
Jiang [33]	Visual preference, naturalness, harmony, closeness, vividness, continuity, and mystery	Changes to stream landscape from vegetation invasion on riversides in Northern Taiwan; includes four images, A–21 to A–24 as shown in Figure A1, at a fixed site using image simulation	No specific classification	Five-point Likert scale from 5 (highest) to 1 (lowest); the scores are normalized (divided by 5) to 1, 0.8, 0.6, 0.4, and 0.2 in this study for comparison
Chen et al. [21]	Visual preference, naturalness, harmony, closeness, and vividness	Sediment control structures including revetments, grade control structures, submerged dams, and check dams constructed by the Soil and Water Conservation Bureau, Taiwan, in mountain streams; includes sixteen images, A–1 to A–16 as shown in Figure A1, collected from different sites	Expert and general public	

4. Results and Discussions

4.1. Harmony

Figure 5 shows the relationship between the harmony and percentage of visible greenery in the data for groups C1, C2, and C3. A comparison between the two samples from the general public and experts is presented. There are many data points overlapping among the groups, and the samples from the general public and experts are consistent for the relationship between visual harmony and visible greenery. Overall, visual harmony increases with the increase in visible greenery at the approximate value of <40% but slightly decreases for G_R > 60%. That is, it exhibits an opposite trend at a critical value of G_R. The approximate critical value of visible greenery may be between 40% and 60%, consistent with previous studies [21,33], as shown in Figure 6. Optimal harmony can be achieved when visible greenery is between 0.4 and 0.6.

As indicated by red and black circles, data from previous studies are more scattered than the current data, as indicated by red and black solid dots. This may be because most images collected from previous studies were taken from different locations and their characteristics are diverse, including different landforms and sediment control structures. Sixteen images, all from different locations and those with different scenes and engineering structures, were collected [21]. We studied only three locations and focused on changes to vegetation in the same image background.

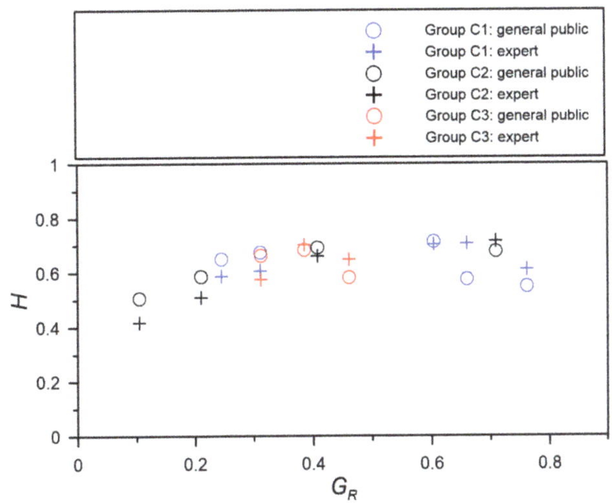

Figure 5. Relationship between the harmony H and percentage of visible greenery G_R in the current data for groups C1, C2, and C3, and comparison between the two samples for the general public and experts.

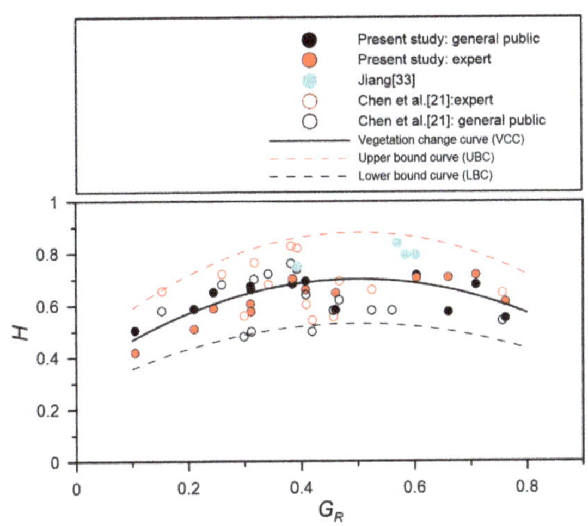

Figure 6. Relationship between the harmony H and percentage of visible greenery G_R: comparison between current and previous findings.

4.1.1. Visual Harmony and Visible Greenery

The fit line between visual harmony and visible greenery based on current data (Groups C1, C2, and C3) is shown in Figure 6. The fit equation (the solid line in the figure) with the correlation coefficient of r = 0.78 can be expressed as follows:

$$H = -1.45(G_R - 0.5)^2 + 0.7 \quad (5)$$

In Equation (5), harmony is maximum, at $H = 0.7$ when $G_R = 0.5$; that is, there is a moderate increase in the amount of vegetation. For example, when $G_R < 0.5$, the perception of harmony increased. This equation was developed based on changes in vegetation from stream engineering under the following conditions: $0.2 < G_R < 0.8$, $10\% < W_R < 30\%$, $I_R < 60\%$, and $0.1 < N_R < 1.0$. These conditions almost cover the ranges of previous studies. Thus, data from previous studies, but mostly not from vegetation variations, except [33], are scattered compared with current data. The upper and lower bounds of all data show trend similar to that in Equation (5), that is, approximately 1.25 times and 0.75 times that of Equation (5), as shown below:

$$H = -1.81(G_R - 0.5)^2 + 0.88 \text{ for upper bound curve} \quad (6)$$

$$H = -1.09(G_R - 0.5)^2 + 0.53 \text{ for lower bound curve} \quad (7)$$

Equation (5) was obtained from the fit curve of vegetation changes, or VCC. The VCC (Equation (5)) is located at the centre between the upper bound (Equation (6)) and lower bound (Equation (7))) curves. The three curves were used to discuss the changes of the visual harmony and visible greenery relationship between the two bounds, the effect on the relationship of harmony with visible greenery, and visual perception, respectively.

4.1.2. Factors Affecting the Harmony and Visible Greenery Relationship

Harmony refers to the unity and balance of the entire landscape in the visual effect. It can be enhanced or diminished by introducing or combining similar or dissimilar landscape elements and spatial arrangements [21]. Harmony, in this study, can be affected by landscape elements such as vegetation and water amount, size, form and colour of the structure, texture or material of the structure's surface, and the spatial arrangements thereof. The effect of water amount, visual proportion of structure, and materials on the structure's surface on the relationship between harmony and visible greenery are discussed as follows.

(1) Water amount

We find that a moderate amount of vegetation can improve harmony, but excessive vegetation could be detrimental to harmony. The reason may be that excess vegetation harms the overall unity of the stream landscape at $G_R > 50\%$, thereby reducing harmony. Besides vegetation, water is an important softscape element in stream landscape. Water features in the literature are related to coherence or harmony of the scene [17] and it is helpful to add the sense of orderliness in a landscape [18].

Figure 7a shows that harmony varies with the percentage of visible greenery and water amount. Three fitting curves based on various values of W_R: $20\% < W_R < 30\%$ with r = 0.85, $10\% < W_R < 20\%$ with r = 0.65, and $W_R < 10\%$ with r = 0.58, show direct proportionality between harmony and visible water, similar to G_R. For example, $H = 0.65$ with $W_R = 20$–30% is higher than $H = 0.55$ with $W_R = 10$–20% and $H = 0.45$ with $W_R < 10\%$ when $G_R = 30\%$. The softscape elements of visible greenery and visible water may also affect each other. An image with larger visible greenery may limit the space of water and reduce W_R, while an image with smaller visible greenery may promote more W_R. We find that most images with higher visible water are at $G_R < 50\%$ in Figure 7a. Generally, $W_R < 10\%$ is mainly distributed near the lower bound curve or $G_R > 50\%$, which is unfavourable to harmony. Most of the data collected show that visible water is less than 30%. Two data points with $W_R = 42\%$—samples from the general public and experts in Image A-2 in

Figure A1 of Appendix A—show that the value of visual harmony is smaller than visual harmony with W_R = 30% (Image A-13). Excessive water amount may thus destroy the balance or unity of an image and reduce harmony [21]. However, our collected data are limited to W_R < 30%. We recommend more studies to investigate the critical value of W_R.

(2) Proportion of structure

The proportion of the structure in the image affects the harmony. Figure 7b shows the relationship between visual harmony and visible greenery at various percentages of visible structure (I_R). The values of visual harmony with I_R > 0.3 are concentered between VCC and the lower bound curve, occupying 73% of the total. Visual harmony values with I_R < 0.3 are mostly distributed between the upper bound curve and VCC at G_R < 50%, occupying 78% of the total. That is, visual harmony with I_R < 30% is higher than that with I_R > 30%. Thus, harmony will be low if the proportion of the structure in the image is extremely large. In visual, relatively large artificial structures in stream or river may reduce the naturalness of the landscape, affect the unity and coordination of the overall environment, and reduce the harmony.

(3) Materials on the surface

In the past, the surface of structure was often smoothed with cement. Advances in ecology and aesthetics now demands the surface of such structures to be roughened, encouraging the use of local materials, such as boulders and stones. A structure with local stone surfaces can also save cost and resources and improve the ecological function of the structure. However, the stone is conducive to plant invasion because of its porous nature and the voids between stones. Thus, the surface of the structure gradually begins to green from vegetation invasion.

The materials of structure's surface were classified into near-natural, unnatural, and other types using the index N_R, the percentage of natural material on the structure that is visible. On-site stones or boulders are commonly used natural materials in mountain stream engineering. A structure with N_R > 0.9 is classified as "natural type," N_R < 0.1 is an "unnatural type," and 0.1 < N_R < 0.9 is "other." Images attributed to the unnatural type used concrete with an artificially smoothed or rough texture on the surface. Most data showing N_R > 0.9 are concentrated between the upper bound curve and VCC, occupied 77% of the total, and all data near the lower bound curve show N_R < 0.1. That is, most images showing natural materials on the structure's surface elicit higher harmony, while those with unnatural ones elicit lower harmony. Harmony can be regarded as visually unifying and balancing for all the hardscape elements (e.g., form, colour, texture or materials) and softscape elements (e.g., water and vegetation). Thus, engineering structures that use natural materials, such as stone revetments, are more easily integrated with the surrounding environment than materials not found in nature [21]; and it will be helpful to increase visual harmony.

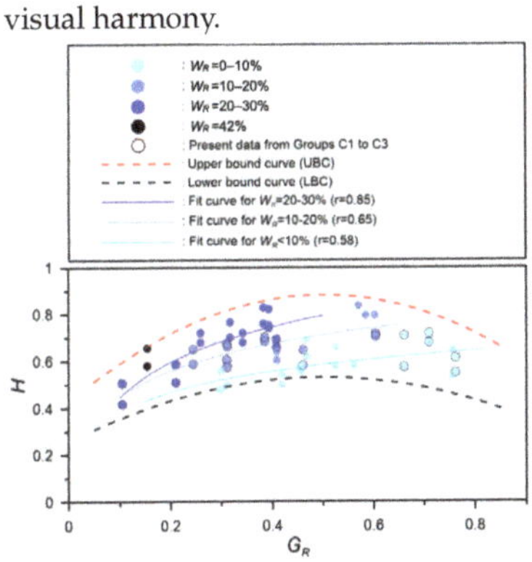

(a)

Figure 7. *Cont.*

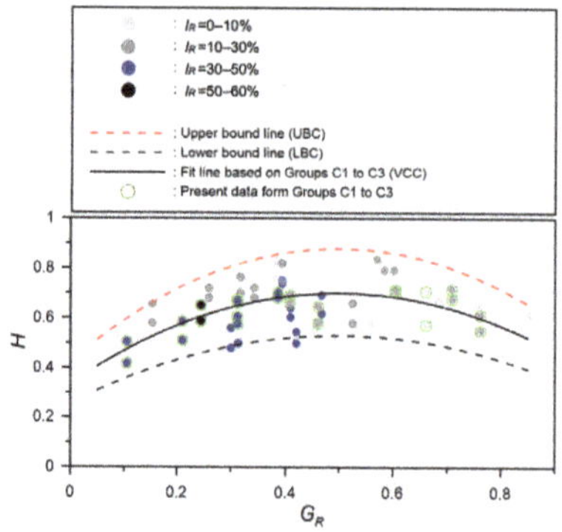

(**b**)

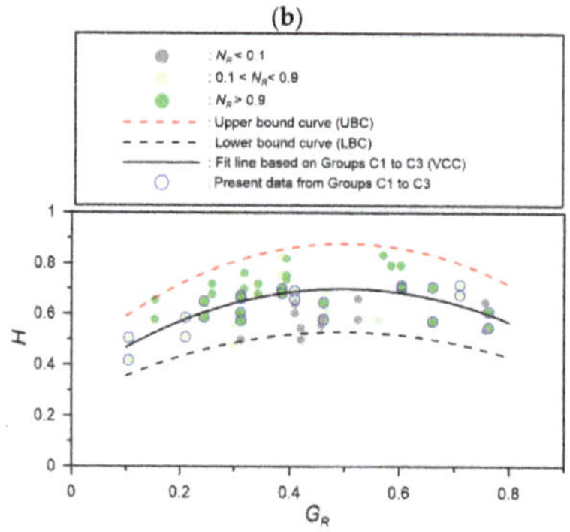

(**c**)

Figure 7. Relationship between harmony H and percentage of visible greenery G_R at various values of water amount, proportion of structure, and material used. (**a**) Various percentage of visible water (W_R); (**b**) various percentage of visible structure (I_R); (**c**) various percentage of natural material on structure in visibility (N_R).

4.2. Visual Preference

4.2.1. Relationship between Visual Preference and Harmony

The visual preference associated with visual harmony for samples of the general public and experts are shown in Figure 8. Two examples from the general public and expert are not different between visual preference and H.

Figure 8 shows that visual harmony is highly correlated to visual preference either for present or previous studies. The variations in visual harmony were approximately equivalent to the variations in P; the relationship between visual preference and visual harmony can be expressed as:

$$P = H \tag{8}$$

This result implies that visual harmony is an important indicator to evaluate visual preference and this aligns with the previous findings that P and H are often used interchangeably [13]. Chen et al. [21] also suggested that harmony has a high weighting factor on visual preferences, and it should be seriously considered for assessment of visual quality of sediment control structures in mountain streams. When visual harmony is replaced with visual preference in Equations (5)–(7), that is, $P = H$, the middle (VCC), upper bound, and lower bound curves between visual preference and visible greenery can be determined (see Figure 9, which also shows data from current and previous findings).

4.2.2. Factors Affecting the Visual Preference and Visible Greenery Relationship

Because visual preference is highly correlated to harmony, we expect similar results for visual preference in Figure 10. However, data for visible water, structure, and natural material for the relationship between visual preference and visible greenery are slightly scattered compared with the visual harmony and visible greenery relationship. Visual preference increases with increasing visible greenery (G_R at 50%) but decreases when $G_R > 50\%$. The relation between preference and visible greenery is also affected by the visibility of water, structures, and natural material. Most data points showed that higher visual preferences between the upper bound curve and VCC, usually have the following characteristics: $0.1 < W_R < 0.3$ (72% in Figure 10a); $I_R < 0.3$ (70% in Figure 10b); and $N_R > 0.9$ (occupied 70% in Figure 10c). Most data were close to the lower bound curve, implying lower visual preference, have excessive amount of greenery ($G_R > 0.5$), much less water ($W_R < 0.1$), excess of structures ($I_R > 0.3$ and $G_R < 0.5$), or use of unnatural materials ($N_R < 0.1$). Visual preference can also be increased at $W_R < 0.3$ and $G_R < 0.5$.

Vegetation and water are important softscape elements in stream landscape. The common hardscape elements include the structure's surface materials, scale, and form. Harmony and visual preference can be altered by adjusting the softscape and hardscape elements, as follows: (1) setting $G_R < 50\%$ by managing and maintaining the vegetation on streambeds and revetments; (2) constructing, for example, a weir or groundsill, to raise the water level for $W_R < 10\%$; (3) using natural materials for the structures' surface, and (4) ensuring the structure is not too large ($I_R < 30\%$) as seen from a viewpoint. These suggestions will be helpful in designing new stream structures in mountainous regions and in maintaining existing ones.

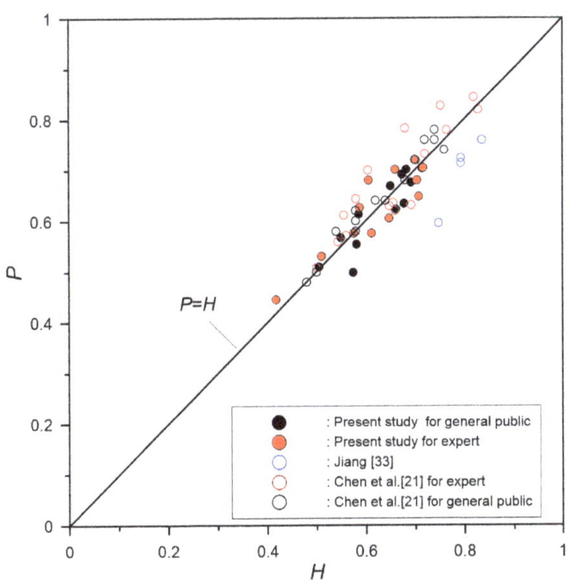

Figure 8. Visual preference P and harmony H: comparison between current and previous findings.

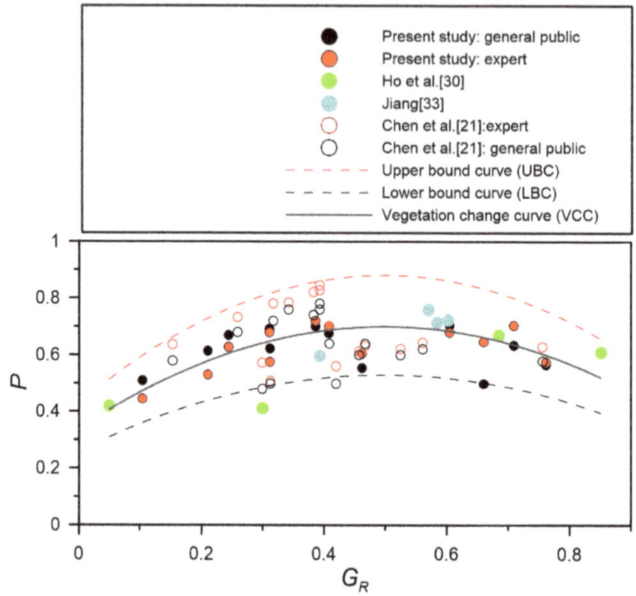

Figure 9. Visual preference P and percentage of visible greenery G_R: comparison between current and previous findings.

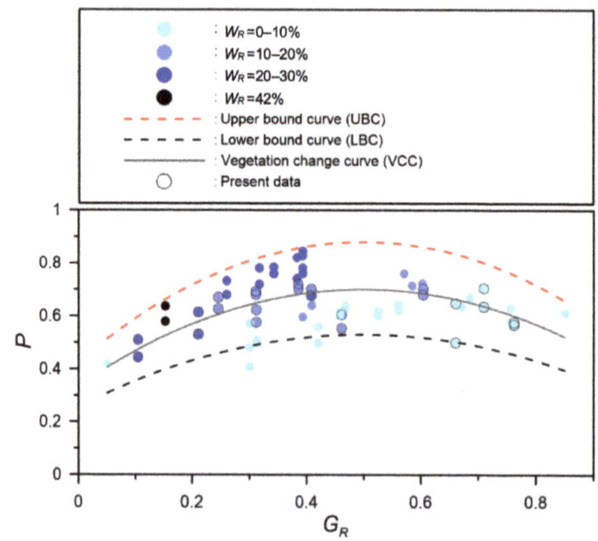

(a)

Figure 10. *Cont.*

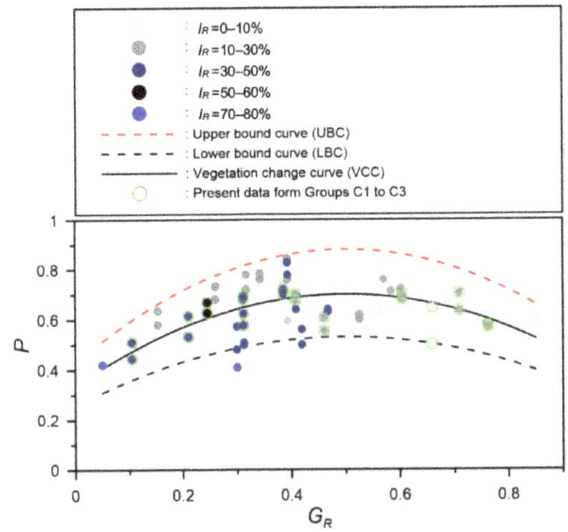

(b)

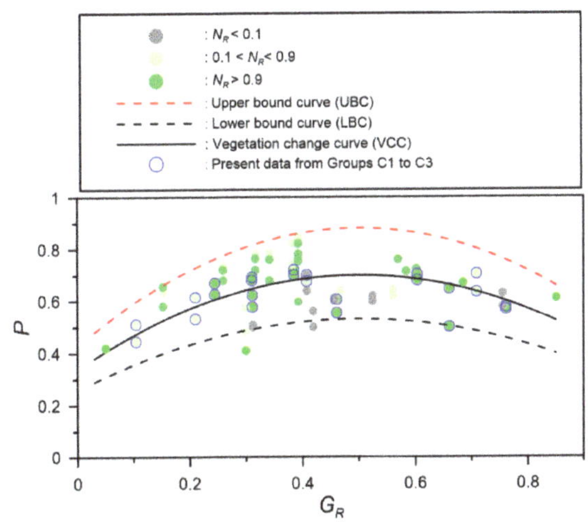

(c)

Figure 10. Visual preference P and percentage of visible greenery G_R at various values of water amount, proportion of structure, and material used. (**a**) Various percentages of visible water (W_R); (**b**) various percentages of visible structure (I_R); (**c**) various percentages of natural material on structure in visibility (N_R).

4.3. Model Using the Combined Physical Indicator M

The pixels of an image in this study are composed of visible greenery amount (G_R), water (W_R), infrastructure (I_R) and other visuals (O_R). O_R is the percentage of pixels in an image excluding G_R, W_R and I_R, that is, $O_R = 1 - G_R - W_R - I_R$; it is mainly occupied by the sky and/or sediment (soil or rock). $N_R I_R$ is the percentage of infrastructure or the structure's surface covered by natural material. A combined physical indicator M comprising all visible indicators was used to analyze the relationship between visual harmony (or preference) and M. We assume that all physical elements, G_R, W_R, $N_R I_R$, and O_R, are linearly correlated to M, and can be expressed as follows:

$$M = \beta_1 G_R + \beta_2 W_R + \beta_3 N_R I_R + \beta_4 O_R - \varepsilon_m \quad (9)$$

In Equation (9), β_1, β_2, β_3, and β_4 are the weighting coefficients for G_R, W_R, $N_R I_R$ and O_R, respectively. $\beta_1 + \beta_2 + \beta_3 + \beta_4 = 1$. ε_m is the modified term owing to excessive green amount ($G_R > 0.5$). Assume that ε_m is linearly correlated to the term of $(G_R - 0.5)$ and expressed by $\varepsilon_m = \alpha(G_R - 0.5)$ in which $\alpha = 0$ when $G_R \leq 0.5$. The empirical coefficients for β_1, β_2, β_3, β_4 and α were determined by trial and errors method to check the correlation between visual harmony or visual preference and M. With $\beta_1 = 0.45$, $\beta_2 = 0.25$, $\beta_3 = 0.2$, $\beta_4 = 0.1$, and $\alpha = 0.5$ in Equation (9), we obtain:

$$M = 0.45G_R + 0.25W_R + 0.2N_R I_R + 0.1O_R - \varepsilon_m \quad (10)$$

where $\varepsilon_m = 0.5(G_R - 0.5)$ when $G_R > 0.5$ and $\varepsilon_m = 0$ when $G_R \leq 0.5$.

Figures 11 and 12 showed that visual harmony and visual preference are positively correlated to M. The fit equations between visual harmony or visual preference and M with correlation coefficient r = 0.70 and r = 0.65, respectively, shown in the solid line in Figures 11 and 12, can be expressed as:

$$H = 1.815M + 0.178 \quad (11)$$

$$P = 1.716M + 0.198 \quad (12)$$

Two dash lines, apart from the above equations with $\pm s$ and standard deviation s = 0.095 and 0.098 for visual harmony and preference, respectively, are also shown in Figures 11 and 12. This result implies that M can be used as an indicator to evaluate visual harmony or visual preference when G_R, W_R, N_R, and I_R are provided.

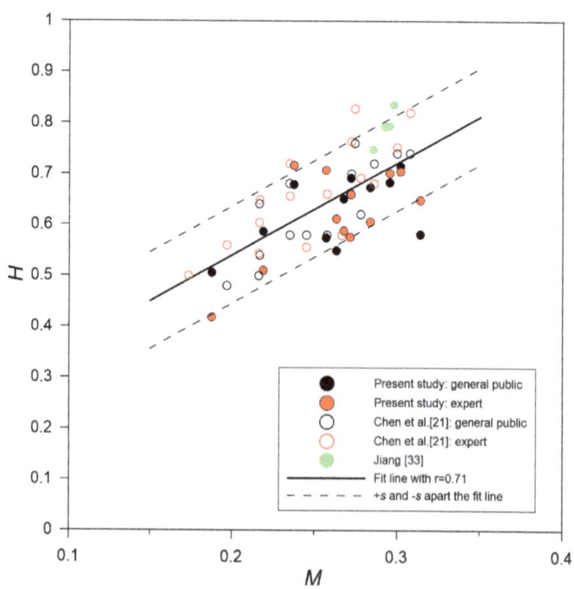

Figure 11. Relationship between harmony H and the combined physical indicator M. Note: the figure shows a fit line (the solid line, see Equation (11)) with correlation coefficient r = 0.71 and two dash lines apart, from the fit line with $\pm s$ and standard deviation s = 0.095.

A model of visual preference associated with cognitive factors (harmony, naturalness, closeness, and vividness)—a psychological based model—was proposed by [21] to evaluate the visual preference for sediment control structures in mountain streams. It was predicted using all perceived or cognitive factors. It had high correlation coefficient with $r = 0.81$. Generally, this model has higher predicated results than physical-based model because visual preference is a perceived parameter. However, the physical-based model (Equations (11) and (12)) is convenient when evaluating visual harmony (or preference) because all hardscape and softscape elements, such as G_R, W_R, N_R, and I_R, can be taken more easily from a field by an image; and harmony or preference can be achieved by ensuring that all landscape elements are considered holistically [21]. Our findings could be helpful for those who design a landscape in mountainous streams with engineering structures, which can be preferred by the public or harmonious with the surrounding environment.

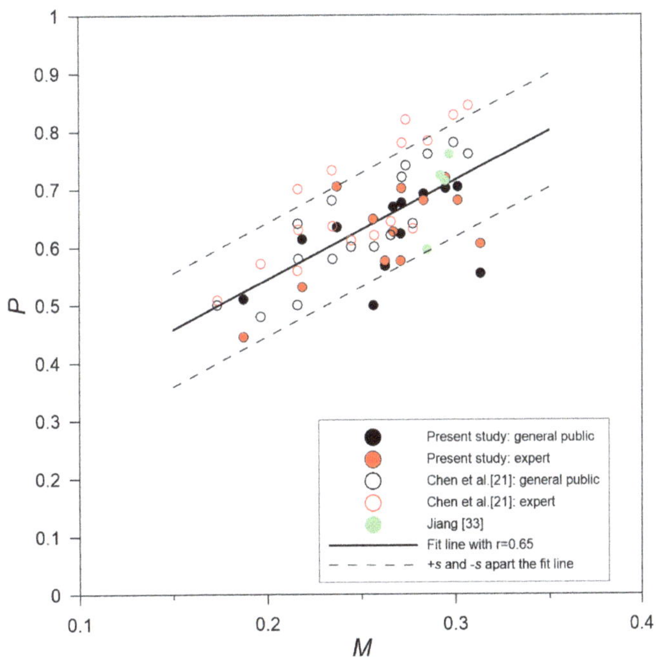

Figure 12. Relationship between visual preference P and combined physical indicator M. Note: the figure shows the fit line (the solid line, see Equation (12)) with the correlation coefficient r = 0.65 and two dashed lines, apart from the fit line with $\pm s$ and standard deviation $s = 0.098$.

5. Conclusions

This study evaluated the visual harmony of engineering structures in mountain streams. The results are as follows:

- Visual harmony H is highly correlated to visual preference P for a landscape of stream engineering in mountainous regions. The perceived curve of vegetation change (VCC) that responds to variations of visual harmony and preference owing to vegetation changes were determined. We found that 1.25 and 0.75 times of VCC can be used to describe the upper and lower bound curves for the stream landscape.
- Vegetation invasion can improve visual quality but may decrease with excess plant growth (visible greenery $G_R > 50\%$) without management and maintenance. Both H and P can be improved by cleaning excess vegetation in riverbeds or riverbanks. The

model for the relationship between H (or P) and G_R proposed herein can be used as reference for the maintenance of streams structures in mountainous regions.
- H and P can be altered by adjusting the softscape and hardscape. Both G_R and visible water W_R need to be increased to improve H and P, if $G_R < 50\%$. Generally, $W_R < 10\%$ is unfavorable to visual harmony and may reduce preference. Hardscape elements also have a positive effect if the visibility of engineering structures (I_R) is below 30% and the proportion of natural materials on structure (N_R) is over 0.9.
- The model of H (or P) and the combined physical indicator M was developed. This model is composed of visible indicators, G_R, W_R, N_R, and I_R, that can be taken by an image. It is convenient for the designer or manager to determine a stream landscape with engineering structures, which can be preferred by the public or harmonious with the surrounding environment. Results of this study could improve the visual beauty for stream engineering and environment design.
- Images analysed in this study were limited to a landscape with engineering structures in Taiwan's mountain streams. The model of visual harmony (or preference) was developed under the following conditions: stream width of 5 m to 20 m; stream gradient of 5–40%; an image showing a view with both sides of river visible at an accessible height; and the physical visual indicators in the ranges of $5\% < G_R < 85\%$, $W_R < 42\%$, $I_R < 80\%$, and $N_R < 1.0$. However, other factors such as color (seasonal changes in vegetation) and arrangement of landscape elements, such as water form and landform, were not included and should be explored in future studies.

Author Contributions: Conceptualization, J.-C.C. and S.-C.C.; methodology, J.-C.C.; formal analysis, J.-C.C. and C.-L.H.; resources, S.-C.C. and S.S.T.; writing and editing, J.-C.C., S.-C.C. and S.S.T.; funding acquisition, S.-C.C. All authors have read and agreed to the published version of the manuscript.

Funding: This research was funded by the Ministry of Science and Technology, Taiwan, under grant MOST 109-2327-B-005-004, Soil and Water Conservation Bureau, Taiwan, Grant Number 108 SWCB-10.1-C-01-06-001(6), and Sanming City Science and Technology Bureau, China, Grant Number 2020-S-81.

Data Availability Statement: The datasets generated during and/or analysed during the current study are available from the corresponding author on reasonable request.

Acknowledgments: We appreciate the image processing assistance of graduate students at Huafan University and National Chung Hsing University.

Conflicts of Interest: The authors declare no conflict of interest. The funders had no role in the design of the study; in the collection, analyses, or interpretation of data; in the writing of the manuscript, or in the decision to publish the results.

Appendix A

A–1 A–2 A–3 A–4

Figure A1. *Cont.*

Figure A1. Images collected from previous studies on mountain streams in Taiwan. Note. Images A–1 to A–16 show various sediment control structures at different locations from [21]; Images A–17 to A–20 [30] and Images A–21 to A–24 [33] are from revetments with vegetation invasion at a fixed location.

References

1. Wu, H.-L.; Feng, Z.-y. Ecological engineering methods for soil and water conservation in Taiwan. *Ecol. Eng.* **2006**, *28*, 333–344. [CrossRef]
2. Zuo, Q.T.; Zhang, Y.; Lin, P. Index system and quantification method for human-water harmony. *J. Hydraul. Eng.* **2008**, *39*, 440–447.
3. Ding, Y.; Tang, D.; Dai, H.; Wei, Y. Human-Water Harmony Index: A New Approach to Assess the Human Water Relationship. *Water Resour. Manag.* **2014**, *28*, 1061–1077. [CrossRef]

4. Zuo, Q.; Li, W.; Zhao, H.; Ma, J.; Han, C.; Luo, Z. A Harmony-Based Approach for Assessing and Regulating Human-Water Relationships: A Case Study of Henan Province in China. *Water* **2021**, *13*, 32. [CrossRef]
5. Zingraff-Hamed, A.; Bonnefond, M.; Bonthoux, S.; Legay, N.; Greulich, S.; Robert, A.; Rotgé, V.; Serrano, J.; Cao, Y.; Bala, R.; et al. Human–River Encounter Sites: Looking for Harmony between Humans and Nature in Cities. *Sustainability* **2021**, *13*, 2864. [CrossRef]
6. Ćosić-Flajsig, G.; Vučković, I.; Karleuša, B. An Innovative Holistic Approach to an E-flow Assessment Model. *Civ. Eng. J.* **2020**, *6*, 2188–2202. [CrossRef]
7. Ou, L.-C.; Luo, M.R. A colour harmony model for two-colour combinations. *Color Res. Appl.* **2006**, *31*, 191–204. [CrossRef]
8. Schloss, K.B.; Palmer, S.E. Aesthetic response to color combinations: Preference, harmony, and similarity. *Atten. Percept. Psychophys.* **2011**, *73*, 551–571. [CrossRef]
9. Ebigbagha, S. Achieving Harmonious Colour Relationship in Art/Design: Towards a Mathematical Model. *Afr. Res. Rev.* **2015**, *9*, 157. [CrossRef]
10. Rahnama, M.R.; Khavari, F. Analytical Study of Color Harmony in Urban Spaces of Mashhad, Northeast Iran. *Int. J. Sci. Eng. Res.* **2014**, *5*, 756–762.
11. Zhu, X.; Tan, S. Degree of Harmony of Urban Land Use and Economic Development: Case Study of Wuhan Metropolitan Area. *J. Sustain. Dev.* **2014**, *7*, 139. [CrossRef]
12. Judd, D.B.; Wyszecki, G. *Color in Business, Science, and Industry*, 3rd ed.; John Wiley & Sons, Ltd: New York, NY, USA, 1975; Volume 42, p. 576.
13. Palmer, S.E.; Schloss, K.B.; Sammartino, J. Visual Aesthetics and Human Preference. *Annu. Rev. Psychol.* **2013**, *64*, 77–107. [CrossRef]
14. United States Forest Service. *Landscape Aesthetics: A Handbook for Scenery Management*; U.S. Department of Agriculture, Forest Service: Washington, DC, USA, 1995.
15. Bell, S. *Landscape: Pattern, Perception and Process*; Routledge: London, UK, 2012.
16. Tveit, M.; Ode, Å.; Fry, G. Key concepts in a framework for analysing visual landscape character. *Landsc. Res.* **2006**, *31*, 229–255. [CrossRef]
17. Litton, R.B. *Water and Landscape: An Aesthetic Overview of the Role of Water in the Landscape*; Water Information Center: New York, NY, USA, 1974.
18. Kaplan, R. *Down by the Riverside: Informational Factors in Waterscape Preference*; USDA Forest Service General Technical Report: Washington, DC, USA, 1977; pp. 285–289.
19. Mansvelt, J.; Kuiper, J.J. Criteria for the humanity realm: Psychology and physiognomy and cultural heritage. In *Checklist for Sustainable Landscape Management*; Elsevier: London, UK, 1999.
20. Kuiper, J. Landscape quality based upon diversity, coherence and continuity: Landscape planning at different planning-levels in the River area of The Netherlands. *Landsc. Urban Plan* **1998**, *43*, 91–104. [CrossRef]
21. Chen, J.-C.; Cheng, C.-Y.; Huang, C.-L.; Chen, S.-C. Assessment of the Visual Quality of Sediment Control Structures in Mountain Streams. *Water* **2020**, *12*, 3116. [CrossRef]
22. Memari, S.; Pazhouhanfar, M. Role of Kaplan' Preference Matrix in the Assessment of Building façade, Case of Gorgan, Iran. *Arman. Archit. Urban Dev.* **2018**, *10*, 13–25.
23. Junker Koehler, B.; Buchecker, M. Aesthetic Preferences Versus Ecological Objectives in River Restorations. *Landsc. Urban Plan.* **2008**, *85*, 141–154. [CrossRef]
24. Stewart, W.; Larkin, K.; Orland, B.; Anderson, D. Boater preferences for beach characteristics downstream from Glen Canyon Dam, Arizona. *J. Environ. Manag.* **2003**, *69*, 201–211. [CrossRef]
25. Arnberger, A.; Eder, R.; Preiner, S.; Hein, T.; Nopp-Mayr, U. Landscape Preferences of Visitors to the Danube Floodplains National Park, Vienna. *Water* **2021**, *13*, 2178. [CrossRef]
26. Piégay, H.; Gregory, K.; Bondarev, V.; Chin, A.; Dahlstrom, N.; Elosegi, A.; Gregory, S.; Joshi, V.; Mutz, M.; Rinaldi, M.; et al. Public Perception as a Barrier to Introducing Wood in Rivers for Restoration Purposes. *Environ. Manag.* **2005**, *36*, 665–674. [CrossRef]
27. Chin, A.; Daniels, M.D.; Urban, M.A.; Piégay, H.; Gregory, K.J.; Bigler, W.; Butt, A.Z.; Grable, J.L.; Gregory, S.V.; Lafrenz, M.; et al. Perceptions of wood in rivers and challenges for stream restoration in the United States. *Env. Manag.* **2008**, *41*, 893–903. [CrossRef] [PubMed]
28. Wilson, M.I.; Robertson, L.D.; Daly, M.; Walton, S.A. Effects of visual cues on assessment of water quality. *J. Environ. Psychol.* **1995**, *15*, 53–63. [CrossRef]
29. Peng, S.-H.; Han, K.-T. Assessment of Aesthetic Quality on Soil and Water Conservation Engineering Using the Scenic Beauty Estimation Method. *Water* **2018**, *10*, 407. [CrossRef]
30. Ho, L.-C.; Chen, J.-C.; Chang, C.-Y. Changes in the visual preference after stream remediation using an image power spectrum: Stone revetment construction in the Nan-Shi-Ken stream, Taiwan. *Ecol. Eng.* **2014**, *71*, 426–431. [CrossRef]
31. Kaplan, R.; Kaplan, S. *The Experience of Nature: A Psychological Perspective*; Cambridge University Press: New York, NY, USA, 1989; p. 340.
32. Ode, Å.; Fry, G.; Tveit, M.S.; Messager, P.; Miller, D. Indicators of perceived naturalness as drivers of landscape preference. *J. Environ. Manag.* **2009**, *90*, 375–383. [CrossRef]
33. Jiang, J.-G. *Visual Preference Assessment on Streams*; Huafan University: New Taipei City, Taiwan, 2016.

water

MDPI

Article

Extreme Droughts Change in the Mekong River Basin: A Multidisciplinary Analysis Based on Satellite Data

Vo Tuong [1], Thanh-Van Hoang [2,*], Tien-Yin Chou [2], Yao-Min Fang [2], Chun-Tse Wang [2], Thanh-Danh Tran [1] and Dung Duc Tran [3]

[1] Faculty of Civil Engineering, Ho Chi Minh City Open University, Ho Chi Minh City 700000, Vietnam; voquangtuong@gmail.com (V.T.); danh.tt@ou.edu.vn (T.-D.T.)

[2] Geographic Information System Research Center, Feng Chia University, 100 Wenhwa Rd., Situn, Taichung 40724, Taiwan; jimmy@gis.tw (T.-Y.C.); frankfang@gis.tw (Y.-M.F.); james@gis.tw (C.-T.W.)

[3] Center of Water Management and Climate Change, Institute for Environment and Resources, Vietnam National University—Ho Chi Minh City (VNU-HCM), Ho Chi Minh City 700000, Vietnam; dungtranducvn@yahoo.com

* Correspondence: van@gis.tw

Abstract: This study evaluates droughts in the Mekong River Basin (MKB) based on a multidisciplinary method, mainly using the Standardized Precipitation Index (SPI) and Mann–Kendall (MK) test. Precipitation data corresponding to the seasonality of the regional climate were retrieved from Integrated Multi-satellitE Retrievals for Global Precipitation Measurement from 2001 to 2020, at a monthly temporal scale and 0.1 degree spatial resolution. Drought events and their average interval, duration, and severity were determined based on Run theory. Our results revealed the most extreme drought period was in January 2014, at the time the lowest precipitation occurred. Spatial extreme drought results indicated that Zone 2 in the upstream MKB has the highest frequency of drought, with 44 events observed during 19 years, and experiences the most severe droughts, whereas Zone 24 in the downstream MKB has the most prolonged drought duration of seven months. The periods and locations of extreme drought were identified using the SPI, corresponding to historic droughts of the MKB. Furthermore, the MK test shows an increasing trend of droughts in the lower MKB and the cluster analysis identified six clusters of times series. Overall, our study provides essential findings for international and national water resource stakeholders in identifying trends of extreme drought in the MKB.

Keywords: extreme drought; Mekong; SPI; Mann–Kendall; time series clustering; integrated multi-satellite retrievals (IMERG)

Citation: Tuong, V.; Hoang, T.-V.; Chou, T.-Y.; Fang, Y.-M.; Wang, C.-T.; Tran, T.-D.; Tran, D.D. Extreme Droughts Change in the Mekong River Basin: A Multidisciplinary Analysis Based on Satellite Data. *Water* **2021**, *13*, 2682. https://doi.org/10.3390/w13192682

Academic Editor: Su-Chin Chen

Received: 3 September 2021
Accepted: 24 September 2021
Published: 28 September 2021

1. Introduction

Drought is among those natural disasters that have caused severe damage to humanity, society, economy, and environment [1–6]. In the three recent decades, droughts have occurred in many countries across the world, such as the United States [7], Brazil [8], China [9], and Vietnam [10]. Drought has affected the socio-economy by reducing crop productivity, cultivated area, and crop yield. As a result, drought has increased agricultural production costs and decreased the income of agricultural labor. Furthermore, the prices of food could be highly influenced by droughts. Severe drought can have complex effects, ranging beyond the direct impacts on crops and livestock to an array of indirect impacts associated with sanitation, nutrition, loss of livelihood, displaced populations, and international disputes [5].

The Mekong River flows through China, Myanmar, Thailand, Laos, Cambodia, and Vietnam and its catchment basin covers an area of 797,000 km^2 with more than 60 million people. The river begins in the Lasagongma Spring in the plateaus of Tibet and flows about 4350 km southeast to the East Sea of Vietnam. The Mekong River basin is divided into two parts according to topography. The upstream is 1000 m above mean sea level, mainly in the south of China, and is the so-called upper part of the Mekong. The lower part, or the downstream, is less than 300 m compared to the mean sea level, extending from the south of Chinese-Lao border to Myanmar, Cambodia, and Vietnam. The average annual discharge of the Mekong River is about 15,000 km^3/year. The Mekong is one of the most biologically diverse areas globally, with hundreds of types of fish and river dolphins, crocodiles, and otters. Fishing is an essential activity on the river. However, the Mekong River is also used extensively for navigation. According to statistics from the World Food Organization, more than 60% of the population in the region lives mainly on agriculture and fisheries and depends on hydroelectricity [11]. The climate of the Mekong River Basin (MKB) varies by topography, from a temperate seasonal climate in the upstream, to a tropical monsoon climate in the downstream area. The rainy season usually lasts from June to November, while the dry season usually lasts from December to May. The average annual rainfall of the MKB ranges from 600 to 1700 mm [12].

Along with economic development, increasing electricity demand has led to the construction of a series of hydroelectric projects in the upper Mekong. Consequently, there are noticeable changes in climate, soil, and the hydraulic regime of sediment in the area. Pokhrel et al. [12] reviewed studies on the effects of climate change, land-use change, hydropower dam construction, and hydrological regime on the Mekong River. The study emphasizes the need to consider human factors, climate change, hydrological regime, and farming factors and at the same time find solutions for sustainable development for the region. While seeking real solutions to solve the problems caused by upstream hydropower reservoirs, climate change, and agricultural changes, assessing rainfall distribution for the area is an urgent issue. A practical assessment and forecasting of rainfall distribution will help water resource managers to develop appropriate coping strategies and solutions for damage mitigation.

Due to the extensive study catchment, metering station data is often challenging to complete. The use of satellite imagery is suitable for rain assessment for the area. Nigam, Ruiz-Barradas, and Chafik [13] used satellite imagery developed by NASA and Japan, the Tropical Rainfall Measurement Mission (TRMM), to assess hydro-meteorological changes for the entire region. Wang et al. [14] used rainfall data taken from the Tropic Rainfall Measurement Mission (TRMM 3B42V7) satellite and monitoring station data from the Mekong River Commission to build a distributed hydrological model that aims to assess hydrological processes for the area. However, recent studies have shown rain data from the Integrated Multi-satellitE Retrievals (IMERG) satellite models are more accurate [14–18] than those taken from TRMM, CHIRPS, and PERSIANN-CCS satellites.

This study aims to address the abovementioned knowledge gap. We used IMERG satellite images taken during the past two decades (2001–2020) to evaluate the rainfall distribution for the entire MKB. We hypothesized that extreme droughts are increasing in many regions of the MKB. We tested our hypothesis using satellite data over the years 2001 to 2020. These data were analyzed based on the run theory and a multidisciplinary approach using the Standardized Precipitation Index (SPI), the Zonal Statistic Model, and the Mann–Kendal test. According to Wanders and Wada [19] and Dai, Zhao, and Chen [20], drought is impacted by climate change, which affects drought characteristics, including drought severity, drought duration, and drought area. The area affected by drought is large. Therefore, the observation is difficult to obtain with gauge measurement methods. Especially in developing countries, it is insufficient to investigate climatic environmental observations. This leads to the risk of uncertainty in predicting natural disasters such as drought. Therefore, our findings are valuable for supporting international and national water resource stakeholders in determining appropriate solutions for regional development in the MKB.

2. Data Availability and Methods

This study follows an integrated framework including the three main steps presented in Figure 1. In the first step, we created an automatic program to retrieve precipitation data from IMERG (retrieved from https://gpm.nasa.gov/, accessed date: 5 May 2021). The program requires determining the spatial coverages or areas of interest and the range of temporal interest or duration of data. It also requires a temporal interval step for data retrieval. In the second step, we computed the SPI for meteorological drought estimation. The last step is to extract the characteristics of drought corresponding to each eco-watershed of the Mekong River, using zonal statistics with drought properties determined by the run theory. After that, we estimated drought characteristics and drought trends to get the comprehensive drought properties by applying the Mann–Kendall test. In these steps, the extreme droughts based on time series clustering and the monthly drought trends detected by the Mann–Kendall test are the most important results.

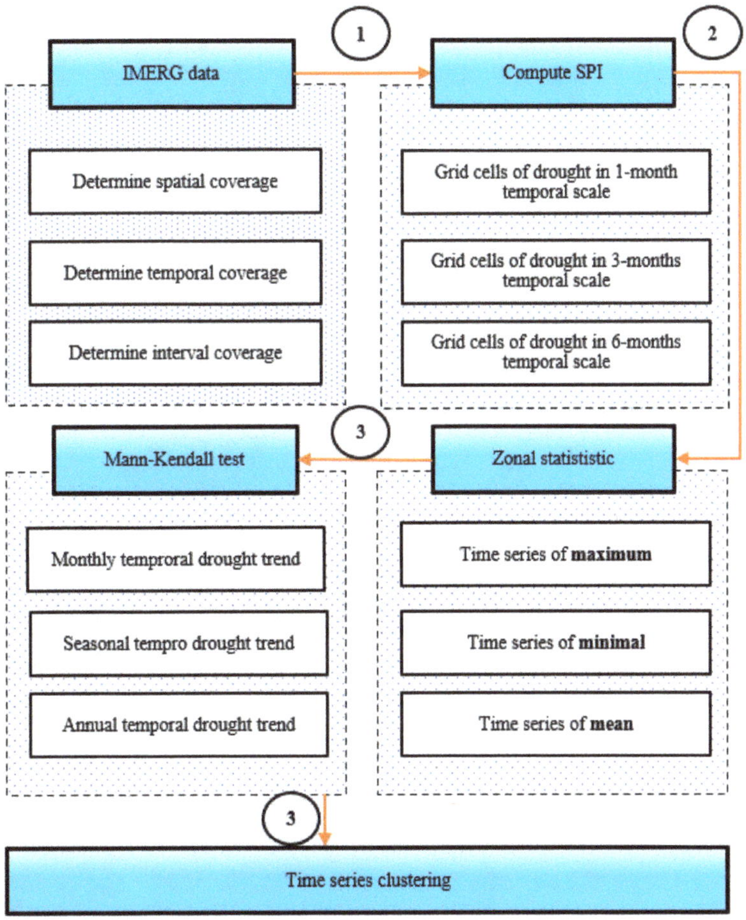

Figure 1. Methodology framework with the three main steps.

2.1. Study Area and Satellite Precipitation Data from IMERG

The study area covers the Mekong River Basin (MKB) with a latitude of 8.3–33.9 North and a longitude of 93.8–108.9 East (Figure 2). From its source on the Tibetan Plateau in China to the Mekong Delta, the main river goes throughout Myanmar, Thailand, Laos, Cambodia, and Vietnam, and is considered one of the world's great rivers. Natural disasters have caused a large number of negative influences on many people's lives in the region; of those, floods and droughts are the two most frequent phenomena that have increasing happened in recent decades. Therefore, many countries and organizations have been concerned with this river basin and finding sustainable solutions for socio-economic development.

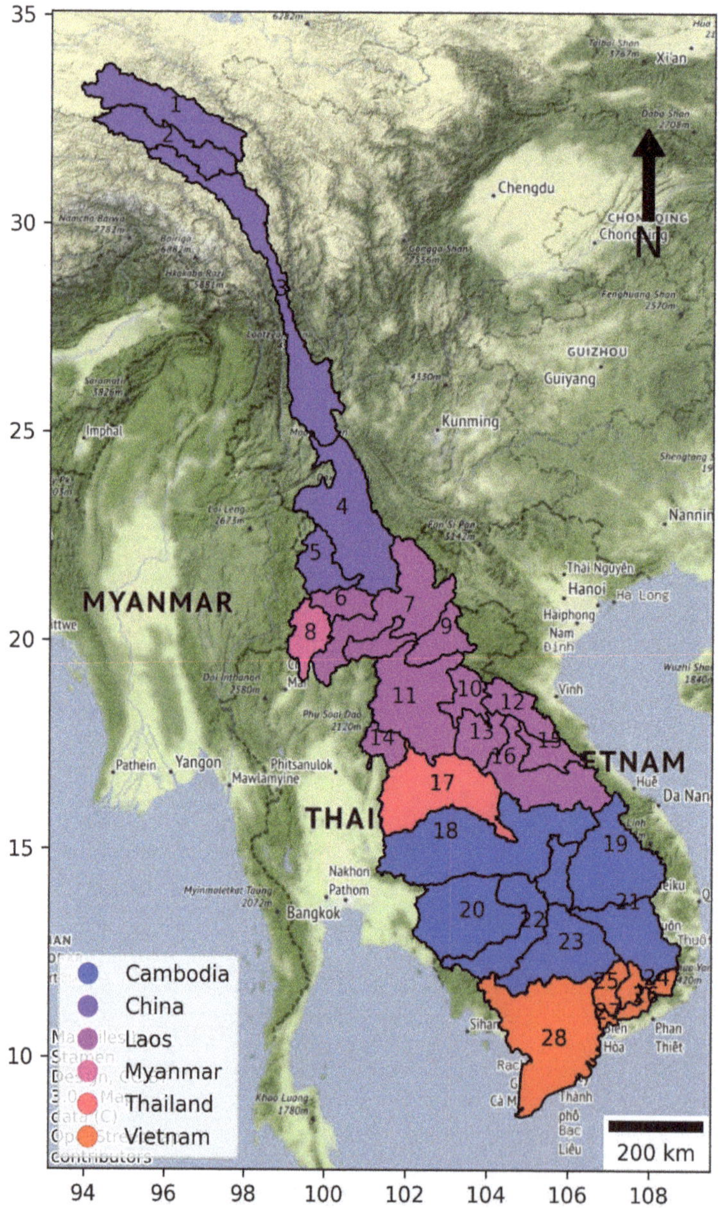

Figure 2. Study area of the Mekong River Basin.

The Mekong River Commission (MRC) annual report indicated that severe drought in the Lower Mekong River (LMB) has caused significant economic losses [21]. The report details the severe droughts in the downstream areas that cause heavy damage to crops, impacted the environment, and affected people's lives. In addition, the report also cites that the drought in 2016 caused severe economic losses in Thailand, estimated at $1.7 billion US. The MKB has been at high risk of droughts and the increasing trend continues to exacerbate the risk, as evidenced by the increasing intensity and duration of the droughts that occurred in the past two decades. This is further confirmed by the findings of several climate change studies by the MRC and other organizations, showing the LMB is likely to see more severe droughts in the next 30 to 90 years due to less precipitation, high air temperatures, and high evapotranspiration combined with increasing demand for water as a result of the growing population in the basin.

In our study, monthly precipitation was retrieved from Integrated Multi-satellitE Retrievals (IMERG). The IMERG products include early multi-satellite, late multi-satellite, and final satellite-gauge products with spatial and temporal resolutions of 1/10° and 30 min. This system used the unified U.S.-developed algorithm that provides the day-1 multi-satellite precipitation product for the U.S. [22].

The study area was divided into 28 zones as ecology catchments and colored by the six MKB countries, following Gassert et al. [23]. We numbered zones following the latitude, identified from 1 to 28. Detailed zonal characteristics are described in Figure 2 and Table 1.

Table 1. Mekong zones classified by ecological characteristics by Gassert et al. [23].

Sub-Name	Main-Name	Country	Area (km^2)	Zonal ID
Za Qu	Mekong	China	45,083	1
Ngom Qu	Mekong	China	29,981	2
Qingshuilang Shah	Mekong	China	68,623	3
Weiyuan Jiang	Mekong	China	64,201	4
Nam Loi	Mekong	China	17,551	5
Nam Pho/Nam Ngaou	Mekong	Laos	12,481	6
Nam Mae Ing	Mekong	China	52,569	7
Nam Mae Kok	Mekong	Myanmar	16,654	8
Nam Beng/Nam Ngeun	Mekong	Laos	26,321	9
Nam Nhiep/Nam Sane	Mekong	Laos	12,759	10
Nam Beng/Nam Ngeun	Mekong	Laos	52,977	11
Nam Cadinh	Mekong	Laos	16,378	12
Songkhram	Mekong	Laos	16,513	13
Huai Luang/Nam Phoung/Nam	Mekong	Laos	9601	14
Nam Kam/Nam Hinboun/Huai	Mekong	Laos	11,708	15
Nam Kam/Nam Hinboun/Huai	Mekong	Laos	39,733	16
Nam Chi	Mekong	Thailand	53,461	17
Se Bang Nouan	Mekong	Cambodia	88,451	18
Se Kong	Mekong	Cambodia	50,831	19
Upper Tonle Sap	Mekong	Cambodia	49,903	20
Huai Tomo/Tonle Repon	Mekong	Cambodia	45,589	21
St. Sen	Mekong	Cambodia	29,546	22
Siem Bok	Mekong	Cambodia	41,001	23
Lagna Da Rgna	Viet Nam, Coast	Vietnam	9920	24
Dong Nai	Viet Nam, Coast	Cambodia	8746	25
Lagna Da Rgna	Viet Nam, Coast	Vietnam	42,359	26
Song Be Delta	Viet Nam, Coast	Vietnam	1873	27
Saigon	Viet Nam, Coast	Cambodia	74,097	28

2.2. Standardize Precipitation Index (SPI)

We calculated the Standardized Precipitation Index (SPI) using the methods derived from a concept by McKee et al. [24]. In this concept, drought is initiated by a decrease in precipitation that causes a water shortage compared to water demand. The index is developed assuming that precipitation directly impacts hydrological variables (e.g., river discharge, groundwater, and soil moisture). The SPI has commonly been used to estimate the occurrence of meteorological drought due to its simplicity for calculation and applicability for various durations. The classification of drought levels based on the SPI referenced from Tsakiris and Vangelis [25] is presented in Table 2.

Table 2. Classification of SPI values.

Drought Category	Probability (%)	Values
Extreme wet	2.30	2.00 ≤ SPI
Very wet	4.40	1.99 ~ 1.50
Moderately wet	9.20	1.49 ~ 1.00
Near normal	68.20	0.99 ~ −0.99
Moderate drought	9.20	−1.00 ~ −1.49
Severe drought	4.40	−1.50 ~ −1.99
Extreme drought	2.30	−2.00 ≥ SPI

Source: Tsakiris and Vangelis (2004).

After computing several temporal scales, 1-month, 3-month, and 6-month, we proposed using 3 months (SPI3) as a typical drought index for later analyses.

2.3. *Zonal Statistic Model*

Zonal statistic methods were used to summarize and aggregate the raster values intersecting a vector geometry. For instance, zonal statistics provide the mean precipitation or maximum elevation of an administrative unit. Additionally, functions are provided to query a raster and get an interpolated value rather than the simple nearest pixel. The values within the zone were assumed to be normalized. The percentile was computed based on mean and standard deviation. For the extreme value of zone, the percentile method Q1 from Hyndman and Fan [26] was used for a query. The extreme drought values are determined when the non-linear integrated drought index (NDI) equals or is below the 10th percentile or corresponds to $z_{score} = -1.282$. The mean (μ), standard deviation (σ), z_{score} and percentile values of zone (x_p) were computed following Equations (1)–(4) below:

$$\mu = \frac{1}{N}\sum_{i}^{N} x_i \tag{1}$$

$$\sigma = \sqrt{\frac{1}{N}\sum_{i}^{N} (x_i - \mu)^2} \tag{2}$$

$$z_{score} = \frac{x - \mu}{\sigma} \tag{3}$$

$$x_p = z_{score} \times \sigma + \mu \tag{4}$$

In the next steps, the Mann–Kendall test, the empirical decomposition model, Moran's I statistic, and the Gi* statistic are used to estimate tempo-spatial trends of extreme drought.

2.4. *Drought Properties*

Drought events and drought characteristics are determined using the run theory [27], presented in Figure 3. Drought events begin when the SPI value is lower than the threshold (−1 means moderate drought, and −2 means extreme drought) and continue until it is higher than the threshold. Drought severity (DS) is defined by the total value of the SPI [28]. Drought duration (DD) is defined by the number of months in the event [29]. Drought interval is the time between the beginning of a drought event and the beginning of the next drought event [30].

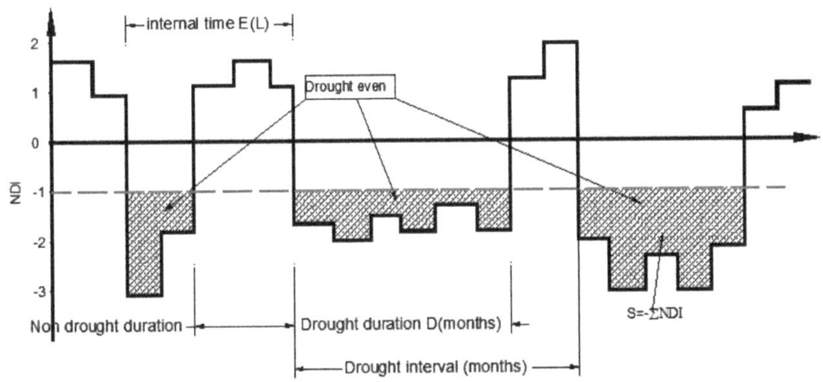

Figure 3. Drought characteristic based on the run theory.

2.5. Mann–Kendall Test

The Mann–Kendall test was used to detect the trends of historical droughts in the MKB. This method considers data distribution and can cope with outliers [31]. The processing of Mann–Kendall tests follows Equations (5) to (9) below:

$$Z_R = \frac{R - \frac{2N_1N_2}{N_1+N_2} + 1}{\sqrt{\frac{2N_1N_2(2N_1N_2-N)}{N^2(N-1)}}} \tag{5}$$

where Z_R is run homogeneous test result, R is run number, N_1 is the number of values lower than medium, and N_2 is the number of values higher than the median.

If Z_R value corresponds to 5% significance level or below, then the data is non-homogenous. Only homogeneous data are used to determine trend conditions [32].

The Mann–Kendall test is computed following Equations (6)–(9) below.

$$Z_{MK} = \left\{ \begin{array}{c} \frac{S-1}{V(S)} \text{ for } S > 0 \\ 0 \quad \text{for } S = 0 \\ \frac{S+1}{V(S)} \text{ for } S < 0 \end{array} \right\} \tag{6}$$

$$V(S) = \frac{n(n-1)(2n+5)}{18} \tag{7}$$

$$S = \sum_{k=1}^{n-1} \sum_{j=k+1}^{n} \text{sgn}(x_j - x_k)) \tag{8}$$

$$\text{sgn}(x_j - x_k) = \left\{ \begin{array}{c} +1 \text{ if } (x_j - x_k) > 0 \\ 0 \text{ if } (x_j - x_k) = 0 \\ -1 \text{ if } (x_j - x_k) < 0 \end{array} \right\} \tag{9}$$

The Z_{MK} having a positive value means an increasing trend, a negative value means a decreasing trend, and a zero value means no trend. V(S) is the variance and S is the Kendall sum statistic. The difference between each consecutive value is computed as positive (+1), negative (−1), and neural (0). x_j and x_k are value of time series at time j among n observational values. The trend is considered significant if Z_{MK} is higher than the significant levels. For instance, a confidence level of 90% has $\alpha = 10\%$ ($Z_{MK} \geq Z_{\alpha/2} = \pm 1.645$).

2.6. Time Series Clustering

We used K-Means and Hierarchical Clustering algorithms to group the dataset. K-Means clustering is a method that aims to partition n observation into k clusters. The cluster is built to follow the nearest mean (cluster centers or cluster centroid). K-Means clustering minimizes within-cluster variances (squared Euclidean distances). The term "K-Means" was first used in study [33]. The theory of the K-Means clustering method lets (Ω, A) be a probability space. Suppose we have a sequence of independent copies Z_1 Z_2, .. Z_n of random vector Z. The aim is to partition Ω into a finite number k of clusters $\Omega_1, \ldots \Omega_k$.

3. Results

3.1. Retrieved Precipitation IMERG Data

Based on the precipitation data retrieved from IMERG, we determined spatial distribution for precipitation in the MKB from 2001 to 2020 (Figure 4). Overall, the IMERG data performed well for the precipitation characteristics of the MKB, which were found to be similar compared with those from study [34]. The average monthly precipitation fluctuates in a range of lower than 10 mm to over 350 mm. Of those, the precipitation event in November 2014 had the lowest precipitation over the period, with the average monthly precipitation at 2.97 mm (Figure 4). The results also show that rainfall increases from January to August and decrease gradually from August to December. The most rainfall is in summer (JJA) and autumn (SON). Meanwhile, winter (DJF) and spring (MAM) rainfall are lower.

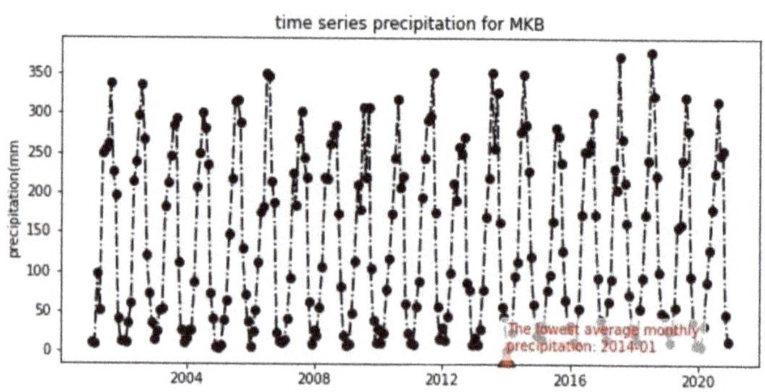

(a)

Figure 4. *Cont.*

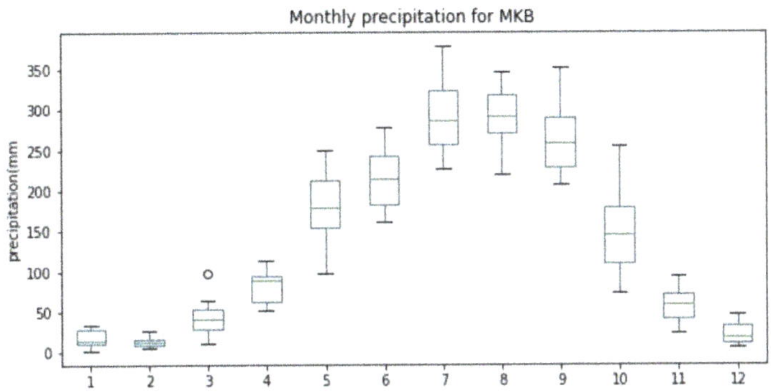

(b)

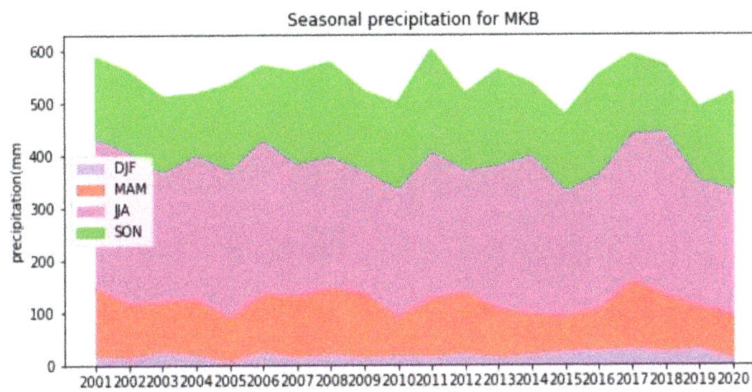

(c)

Figure 4. The lowest average monthly precipitation (**a**), monthly (**b**), and seasonal (**c**) precipitation of the MKB with DJF (December, January, February), MAM (March, April, May), JJA (June, July, August), and SON (September, October, November).

Rainfall in the MKB is not distributed differently between the upstream and downstream (Figure 5). However, in general, the climate in the MKB can be divided into two seasons: the rainy season and the dry season. The rainy season is from June to November, and the dry season is from November to May. This result is similar to the study by Lee and Dang [35]. Based on the analysis results, the IMERG satellite data were applied to the meteorological spatial analysis and spatial analysis for the MKB area.

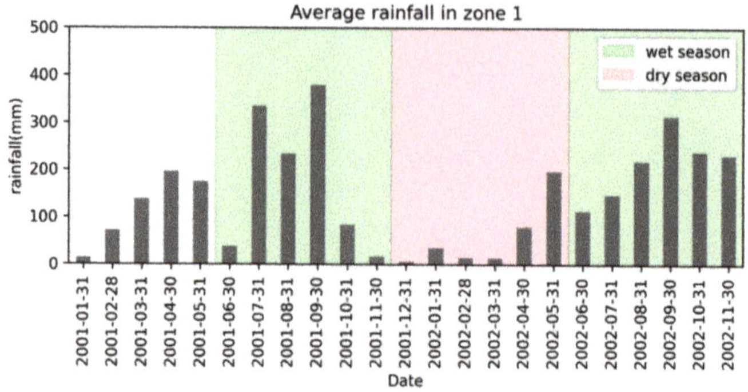

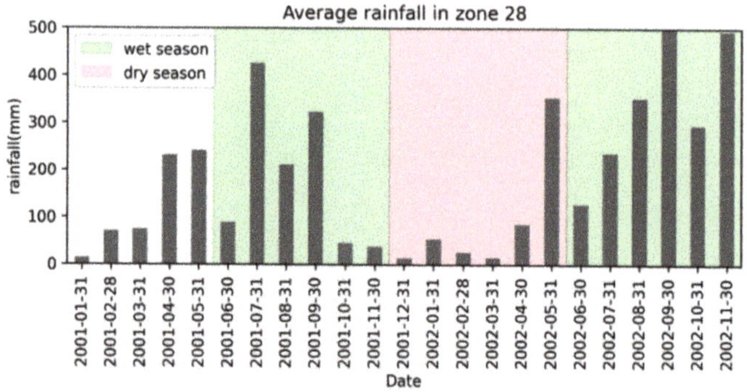

Figure 5. Average rainfall distribution in Zone 1 and Zone 28.

3.2. Simulation Results of Historical Droughts of the MKB Using SPI

To evaluate the spatial distribution of drought, we calculated the SPI index for the entire region before analyzing the max, min, mean, and standard derivation (std) values at each zone. Figure 6 shows the results of zonal statistical analysis in 2001–2002 with an example for Zone 1. Various raster pixel values of SPI in the same zone were compared using a zonal statistical model. Except for the std value, the max, min, and mean values tend to fluctuate quite similarly in Zone 1. Therefore, we chose the min value, because it is the one that causes the most extensive drought and poses the riskiest situation.

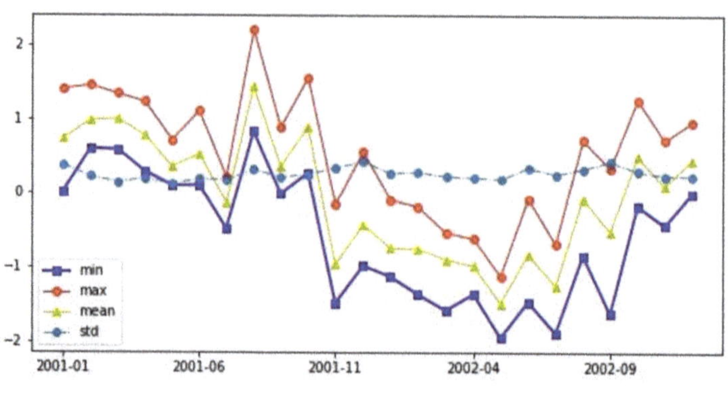

Figure 6. Zonal statistical calculations for Zone 1.

We computed the SPI to explore extreme drought severity events in the MKB. The analysis of historic drought determined that drought phenomena are complicated in the study area. The analysis results determined the spatial distribution and the dry time for the entire basin. The results show the differences in each zone's drought characteristics, including the number of droughts that occurred, the duration, the severity, and the time interval between two consecutive droughts. According to the run theory, Zone 2 has the most frequent drought events (44 extreme drought events). Zone 26 has a minor drought, with eight extreme drought events (Figure 7). The circles in Figure 7's graphs are the outliers that mainly occur when the drought duration in the region is more complex and unpredictable than the normal drought level. However, Zone 24 has the highest average drought duration of 7 months, while severe extreme droughts occurred under two months. The interval drought events, or the time between two drought events, indicate that the average interval of extreme drought is around ten months in the MKB.

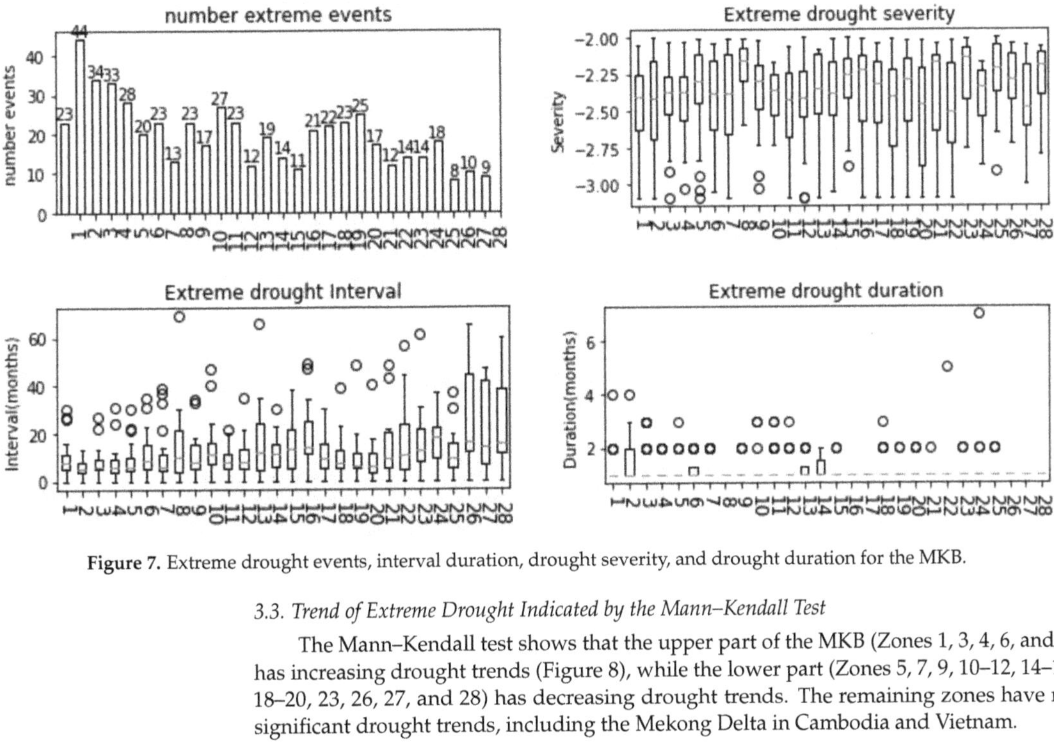

Figure 7. Extreme drought events, interval duration, drought severity, and drought duration for the MKB.

3.3. *Trend of Extreme Drought Indicated by the Mann–Kendall Test*

The Mann–Kendall test shows that the upper part of the MKB (Zones 1, 3, 4, 6, and 8) has increasing drought trends (Figure 8), while the lower part (Zones 5, 7, 9, 10–12, 14–16, 18–20, 23, 26, 27, and 28) has decreasing drought trends. The remaining zones have no significant drought trends, including the Mekong Delta in Cambodia and Vietnam.

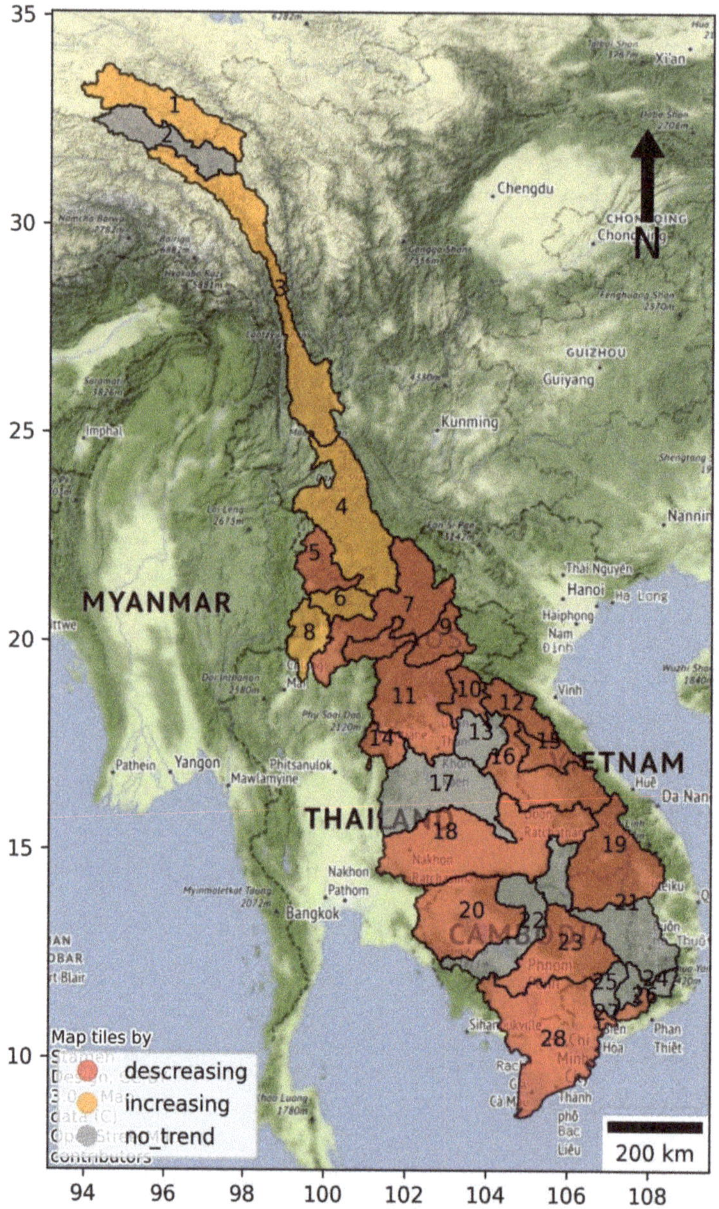

Figure 8. Trend of droughts in the MKB.

The results of the drought trend test show that most of the SPI indicators in the lower Mekong region tend to increase, or meteorological droughts have decreased. However, for a small area in the upper Mekong basin, SPI values decreased, or, in other words, meteorological drought increased (Figure 9). For example, Zone 1 in the upper Mekong tends towards increased drought (SPI decrease) with a rate of 0.002% in the past 20 years. Meanwhile, drought in the downstream area tends towards decrease (SPI increase) to a similar degree. Details of MBK's long-term trend analysis are presented in Table 3 and Figure 10 for slope changes in the zones.

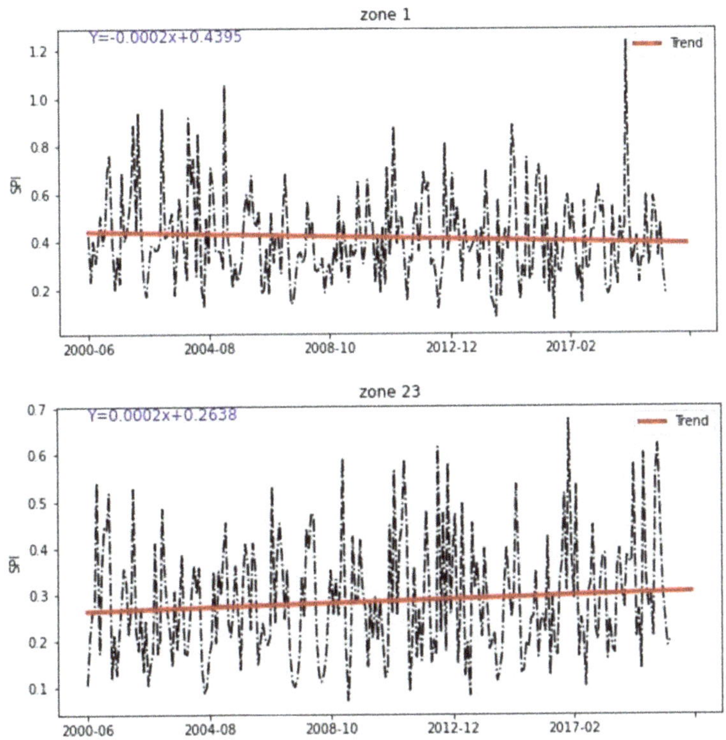

Figure 9. Trend of drought for Zone 1 and Zone 23 in the MKB.

Table 3. Drought trend and equation slopes of 28 zones within the MKB.

Zone	Trend	Equation	Zone	Trend	Equation
1	increasing	Y = −0.0002x + 0.4395	15	decreasing	Y = 0.0002x + 0.2421
2	no_trend	Y = −0.0x + 0.6357	16	decreasing	Y = 0.0002x + 0.2486
3	increasing	Y = −0.0003x + 0.5832	17	no_trend	Y = 0.0x + 0.468
4	increasing	Y = −0.0002x + 0.5522	18	decreasing	Y = 0.0003x + 0.4116
5	decreasing	Y = 0.0002x + 0.3545	19	decreasing	Y = 0.0004x + 0.4464
6	increasing	Y = −0.0001x + 0.3857	20	decreasing	Y = 0.0004x + 0.4423
7	decreasing	Y = 0.0002x + 0.4203	21	no_trend	Y = −0.0x + 0.2823
8	increasing	Y = −0.0001x + 0.207	22	no_trend	Y = 0.0x + 0.2624
9	decreasing	Y = 0.0002x + 0.4526	23	decreasing	Y = 0.0002x + 0.2638
10	decreasing	Y = 0.0002x + 0.3552	24	no_trend	Y = 0.0x + 0.2664
11	decreasing	Y = 0.0004x + 0.46	25	no_trend	Y = 0.0x + 0.2834
12	decreasing	Y = 0.0002x + 0.381	26	decreasing	Y = 0.0001x + 0.123
13	no_trend	Y = 0.0x + 0.2479	27	decreasing	Y = 0.0001x + 0.276
14	decreasing	Y = 0.0001x + 0.3633	28	decreasing	Y = 0.0003x + 0.1946

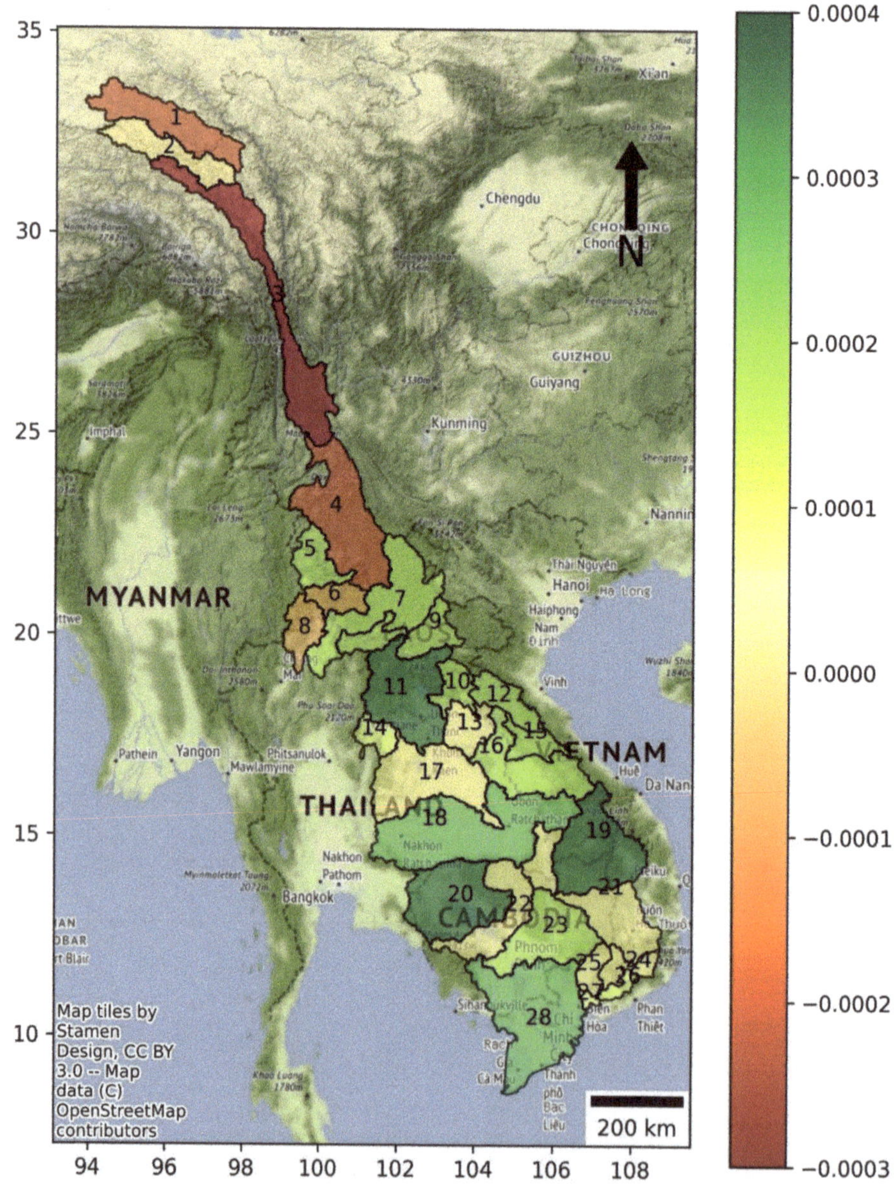

Figure 10. Slope of drought trend in the MKB.

3.4. Clustering Time Series

K-Means Clustering shows six groups, each of which has similar zone's characteristics (Figure 11). The largest, cluster 2, has six zones (1, 6, 7, 21, 22, 24), and the smallest, cluster 4, has three zones (2, 3, 4). Most of the zones are located in the middle area of the MKB. The remaining clusters have four to five zones, of which cluster 3, with all zones, is situated in the lowest part of the basin. The results show that the meteorological limit is relatively uniform in the upper Mekong region, with slight fluctuation. All basins stretching from latitude 32 degrees north to 22 degrees north (Zone 2, Zone 3, Zone 4) share the same characteristics. In contrast, in the lower part of the Mekong, the meteorological drought is highly variable. Especially in Zones 10 to 16, the characteristics of meteorological drought appear in this area. In addition, neighboring regions may have different meteorological drought characteristics (Zone 1 and Zone 2). In that climate, the locations are very far apart but still have the same meteorological drought characteristics (Zone 1 and Zone 24).

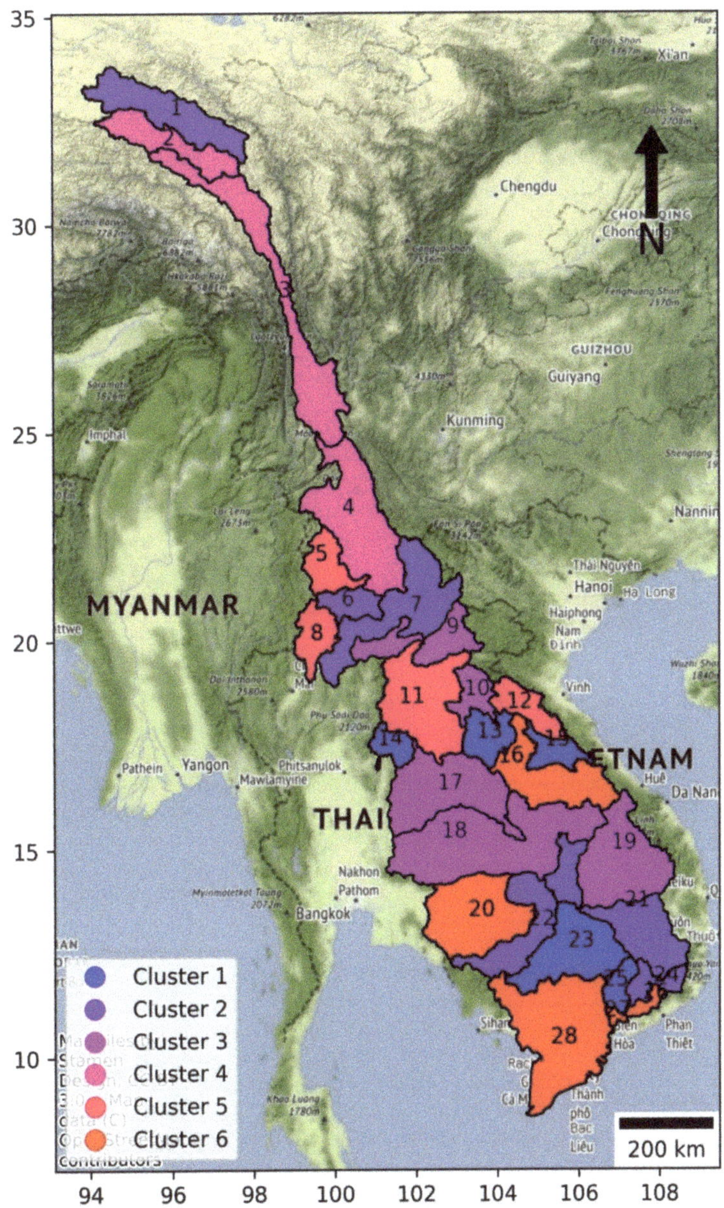

Figure 11. Clustering time series.

4. Discussion

4.1. Drought and Its Overall Impacts in the Mekong River Basin

This study has analyzed data retrieved from Integrated Multi-satellitE Retrievals for Global Precipitation Measurement from 2001 to 2020. The analysis findings of the same seasonal pattern and the lowest precipitation occurred in November during the extreme drought event across the MKB countries. The most rainfall drops in summer (JJA) and autumn (SON), while winter (DJF) and spring (MAM) rainfall are lower. In addition, average monthly precipitation in the range of lower than 10 mm to over 350 mm in the period, and rainfall increases from January to August and gradual decreases from August to December, have implied that the data are reliable for applying SPI calculations and using the Mann-Kendall trend detection test [36–38].

The SPI calculations based on the run theory have revealed 44 extreme drought events as the most frequent occurrence in Zone 2 during the 19 analysis years. It means this zone (which belongs to China) should consider droughts in the coming years. This zone is located close to Zones 1, 3, and 4 (also belonging to China), which were detected as having increasing drought trends under the Man–Kendall test. Besides, the lowest extreme drought event (8 extreme events) was found in Zone 25 (which belongs to Cambodia) over the 19 years (Figure 7). Next to this zone, however, is Zone 24, which has the highest average drought duration of seven months compared to the remaining zones. Decreasing drought trends were found in Zones 5, 7, 9, 10–12, 14–16, 18–20, 23, 26, 27, and 28 in the lower part of the MKB (Myanmar, Thailand, Laos, Cambodia, Vietnam), which may indicate the increasing precipitation from 2001 to 2020, while the upper part (Zones 1, 3, 4, 6, and 8, located mainly in China) has increasing drought trends.

The Mann–Kendall test shows that the lower part of the MKB has decreasing drought trends, while the upper part has to deal with increasing drought trends. These results imply hydrological conditions of precipitation decreases downstream which are the same as those found by Dang et al. [39]. The remaining zones with no significant drought trends, such as those in the Mekong Delta (Cambodia and Vietnam) may be due to the relatively balanced water conditions in the area or insufficient data availability for the data we used for our analysis [35]. The results were also supported by the drought trend test, with increased SPI indicators found in the lower Mekong region, providing the decreasing trend of meteorological drought. However, our clustering results based on the K-Means Clustering method indicate that drought characteristics vary across the MKB, indicated by zone distributions; of those, only cluster 3 has all zones located in the lowest section at the same time. We realized that the meteorological limit is relatively uniform in the upper Mekong region, with slight fluctuation. Precipitation in the Mekong basin is subject to the precipitation oscillation explained by Räsänen and Kummu, (2013) [40]. However, the episodes of drought do not reveal any proven periodicity. This may explain the apparent low spatial coherence of the number of clusters, which may reflect local climatic conditions.

4.2. Policy Implication

The significant variability of precipitation across the zones of the countries in the MKB has been recognized in our study. Based on the findings, we recommend that countries with historical drought zones and those with increasing drought trends consider these phenomena in their water governance systems. These countries should enhance the capabilities developed for drought forecasting, warning, and preparedness systems. Although the member countries that belong to the MKB have long acknowledged the challenges associated with drought and natural disasters, the changing climate and other environmental pressures could exacerbate the phenomenon. Our study recommends developing a policy that focuses more on identifying solutions that minimize drought impact and exposure and maximize precipitation water storage for water use in the dry seasons.

4.3. Limitation and Future Outlook

Although this study has put great efforts into detecting drought trends in the MKB using various satellite image analysis methods, two main limitations are partially associated with the chosen method of calculating drought indexes for a large-scale area such as the MKB. First, the not-too-long period of observing satellite data (2001–2020) may partly influence the analysis results. This study uses statistical methods of analyzing satellite data, so the longer the time series data, the better the calculation to determine the extreme values. Second, the model could be improved if more observational data from the ground stations were available to adjust or assimilate into the models.

Based on the scope of this study, we only evaluated increasing and decreasing drought trends. Predicting the degree of increase and decrease of drought, such as intensity and time extreme, could be further studied. The Mann–Kendall test is a non-parametric method of trend assessment that does not consider serial correlation or seasonality effects. However, it is one of the simplest and most popular methods for analyzing and evaluating time-series-based trends. This study can be further extended using other trend analysis methods such as the Modified Mann–Kendall test [41], the Yue and Wang Modified M–K Test [36], or the Seasonal Sen's Slope Estimator [41] to calculate serial autocorrelation and seasonality.

5. Conclusions

In this study, a multidisciplinary analysis based on satellite data is used to detect droughts and their trends in the Mekong River Basin (MKB) regarding how their historical dynamics could be explained for seeking potential adaptation and mitigation measures in the coming years. To our knowledge, this is one of the first studies integrating IMERG satellite images analyzed by various methods. In this study, we have therefore shown the trajectory of drought dynamics, the highest and lowest frequent droughts, and the locations and durations of droughts across the MKB. We argue that the method used by this study has broad transferability, offering the potential for drought assessments in regions worldwide facing similarly intense changes, particularly in the context of climate change.

The findings of this study reveal that even rainfall data are sufficient for drought detections, and that these types of data, particularly in terms of satellite images, present a dynamic drought process across the MKB. First, the zones with high rainfall in the upper basin show decreasing drought trends, which are not the same for those in downstream areas. Second, it should be noted that the lowest and highest severe droughts changed across the river basin with different durations, implying that future drought trends are complex. These would be indicators of a movement towards unsustainability, which should be taken into account by the countries in the MKB.

Author Contributions: V.T. developed the methodology and concept, performed data analysis and structured and wrote the first draft; T.-V.H. and D.D.T. reviewed literature, wrote the draft version, edited the manuscript and hydrographs, and managed correspondence; T.-D.T. and C.-T.W. analyzed data and provided maps, T.-Y.C. and Y.-M.F. reviewed the state-of-the-art and supervised the team. All authors have read and agreed to the published version of the manuscript.

Funding: This work was funded by Ho Chi Minh City Open University under grant number E2020.05.01.

Institutional Review Board Statement: Not applicable.

Informed Consent Statement: Not applicable.

Data Availability Statement: Not applicable.

Acknowledgments: This work was funded by Ho Chi Minh City Open University under grant number E2020.05.01. We also thank the Geographic Information Research Center—Feng Chia University Taiwan for their valuable comments and support.

Conflicts of Interest: The authors confirm that there is no conflict of interest.

References

1. Nguyen, N.A. Historic drought and salinity intrusion in the Mekong Delta in 2016: Lessons learned and response solutions. *Vietnam J. Sci. Technol. Eng.* **2017**, *59*, 93–96. [CrossRef]
2. Development, M. of A. and R. *The Drought and Salwater Intrusion in the Highlands, South Middle Region, and Vietnamese Mekong Delta in 2016 and Proposed Actions 2016*. Available online: https://reliefweb.int/report/viet-nam/vietnam-drought-and-saltwater-intrusion-emergency-plan-action-epoa-dref-operation-n (accessed on 3 March 2021).
3. Thao, N.T.T.; Khoi, D.N.; Xuan, T.T.; Tychon, B. Assessment of Livelihood Vulnerability to Drought: A Case Study in Dak Nong Province, Vietnam. *Int. J. Disaster Risk Sci.* **2019**, *10*, 604–615. [CrossRef]
4. Ahmadalipour, A.; Moradkhani, H. Multi-dimensional assessment of drought vulnerability in Africa: 1960–2100. *Sci. Total Environ.* **2018**, *644*, 520–535. [CrossRef]
5. Barlow, M.; Zaitchik, B.; Paz, S.; Black, E.; Evans, J.; Hoell, A. A Review of Drought in the Middle East and Southwest Asia. *J. Clim.* **2016**, *29*, 8547–8574. [CrossRef]
6. Tre, B.; Vinh, T.; Giang, K. *Vietnam. The Drought and Salinity Intrusion in the Mekong River Delta of Vietnam*; CGIAR Research Centers in Southeast Asia Centers: Gelderland, The Netherlands, 2016.
7. Alexander, L. V Global observed long-term changes in temperature and precipitation extremes: A review of progress and limitations in IPCC assessments and beyond. *Weather Clim. Extrem.* **2016**, *11*, 4–16. [CrossRef]
8. Ribeiro, I.O.; Andreoli, R.V.; Kayano, M.T.; Sousa, T.R.; Medeiros, A.S.; Godoi, R.H.M.; Godoi, A.F.L.; Junior, S.D.; Martin, S.T.; Souza, R.A.F. Biomass burning and carbon monoxide patterns in Brazil during the extreme drought years of 2005, 2010, and 2015. *Environ. Pollut.* **2018**, *243*, 1008–1014. [CrossRef] [PubMed]
9. Pescaroli, G.; Alexander, D. Understanding Compound, Interconnected, Interacting, and Cascading Risks: A Holistic Framework. *Risk Anal.* **2018**, *38*, 2245–2257. [CrossRef] [PubMed]
10. Le, P.V.V.; Phan-Van, T.; Mai, K.V.; Tran, D.Q. Space-time variability of drought over Vietnam. *Int. J. Climatol.* **2019**, *39*, 5437–5451. [CrossRef]
11. Frenken, K. Irrigation in Central Asia in Figures: AQUASTAT Survey-2012. *FAO Water Rep.* **2013**, *39*, 1020–1203.
12. Pokhrel, Y.; Burbano, M.; Roush, J.; Kang, H.; Sridhar, V.; Hyndman, D.W. A review of the integrated effects of changing climate, land use, and dams on Mekong river hydrology. *Water* **2018**, *10*, 266. [CrossRef]
13. Nigam, S.; Ruiz-Barradas, A.; Chafik, L. Gulf Stream excursions and sectional detachments generate the decadal pulses in the Atlantic multidecadal oscillation. *J. Clim.* **2018**, *31*, 2853–2870. [CrossRef]
14. Wang, W.; Lu, H.; Yang, D.; Sothea, K.; Jiao, Y.; Gao, B.; Peng, X.; Pang, Z. Modelling hydrologic processes in the Mekong River Basin using a distributed model driven by satellite precipitation and rain gauge observations. *PLoS ONE* **2016**, *11*, e0152229. [CrossRef]
15. Fang, J.; Yang, W.; Luan, Y.; Du, J.; Lin, A.; Zhao, L. Evaluation of the TRMM 3B42 and GPM IMERG products for extreme precipitation analysis over China. *Atmos. Res.* **2019**, *223*, 24–38. [CrossRef]
16. Zhang, Z.; Tian, J.; Huang, Y.; Chen, X.; Chen, S.; Duan, Z. Hydrologic evaluation of TRMM and GPM IMERG satellite-based precipitation in a humid basin of China. *Remote Sens.* **2019**, *11*, 431. [CrossRef]
17. Ma, Q.; Li, Y.; Feng, H.; Yu, Q.; Zou, Y.; Liu, F.; Pulatov, B. Performance evaluation and correction of precipitation data using the 20-year IMERG and TMPA precipitation products in diverse subregions of China. *Atmos. Res.* **2020**, *249*, 105304. [CrossRef]
18. Jiang, S.; Wei, L.; Ren, L.; Xu, C.-Y.; Zhong, F.; Wang, M.; Zhang, L.; Yuan, F.; Liu, Y. Utility of integrated IMERG precipitation and GLEAM potential evapotranspiration products for drought monitoring over mainland China. *Atmos. Res.* **2020**, *247*, 105141. [CrossRef]
19. Wanders, N.; Wada, Y. Human and climate impacts on the 21st century hydrological drought. *J. Hydrol.* **2015**, *526*, 208–220. [CrossRef]
20. Dai, A.; Zhao, T.; Chen, J. Climate change and drought: A precipitation and evaporation perspective. *Curr. Clim. Chang. Rep.* **2018**, *4*, 301–312. [CrossRef]
21. Mekong River Commission, M. Annual Report: 2019. Available online: http://interactive.mrcmekong.org/mrc-annual-report-2019/homepage/ (accessed on 23 September 2021).
22. Gan, F.; Gao, Y.; Xiao, L. Comprehensive validation of the latest IMERG V06 precipitation estimates over a basin coupled with coastal locations, tropical climate and hill-karst combined landform. *Atmos. Res.* **2021**, *249*, 105293. [CrossRef]
23. Gassert, F.; Rai, P.; Reig, P.; Luck, M. Mekong river basin study. *World Resour. Inst. Work. Pap.* **2012**. Available online: https://www.wri.org/data/mekong-river-basin-study (accessed on 23 September 2021).
24. McKee, T.B.; Doesken, N.J.; Kleist, J. The relationship of drought frequency and duration to time scales. In Proceedings of the 8th Conference on Applied Climatology, Anaheim, CA, USA, 17–22 January 1993; Volume 17, pp. 179–183.
25. Tsakiris, G.; Vangelis, H. Towards a Drought Watch System based on Spatial SPI. *Water Resour. Manag.* **2004**, *18*, 1–12. [CrossRef]
26. Hyndman, R.J.; Fan, Y. Sample quantiles in statistical packages. *Am. Stat.* **1996**, *50*, 361–365.
27. Guerrero-Salazar, P.L.A.; Yevjevich, V.M. Analysis of Drought Characteristics by the Theory of Runs. Ph.D. Thesis, Colorado State University, Libraries, Fort Collins, CO, USA, 1975.
28. Wu, R.; Zhang, J.; Bao, Y.; Guo, E. Run Theory and Copula-Based Drought Risk Analysis for Songnen Grassland in Northeastern China. *Sustainability* **2019**, *11*, 6032. [CrossRef]

29. Montaseri, M.; Amirataee, B.; Nawaz, R. A Monte Carlo Simulation-Based Approach to Evaluate the Performance of three Meteorological Drought Indices in Northwest of Iran. *Water Resour. Manag.* **2017**, *31*, 1323–1342. [CrossRef]
30. Ayantobo, O.O.; Li, Y.; Song, S. Multivariate Drought Frequency Analysis using Four-Variate Symmetric and Asymmetric Archimedean Copula Functions. *Water Resour. Manag.* **2019**, *33*, 103–127. [CrossRef]
31. Ali, R.; Kuriqi, A.; Abubaker, S.; Kisi, O. Long-Term Trends and Seasonality Detection of the Observed Flow in Yangtze River Using Mann-Kendall and Sen's Innovative Trend Method. *Water* **2019**, *11*, 1855. [CrossRef]
32. Güçlü, Y.S. Improved visualization for trend analysis by comparing with classical Mann-Kendall test and ITA. *J. Hydrol.* **2020**, *584*, 124674. [CrossRef]
33. MacQueen, J. Some methods for classification and analysis of multivariate observations. In Proceedings of the Fifth Berkeley Symposium on Mathematical Statistics and Probability, Oakland, CA, USA, 1 January 1967; Volume 1, pp. 281–297.
34. Thilakarathne, M.; Sridhar, V. Characterization of future drought conditions in the Lower Mekong River Basin. *Weather Clim. Extrem.* **2017**, *17*, 47–58. [CrossRef]
35. Lee, S.K.; Dang, T.A. Spatio-temporal variations in meteorology drought over the Mekong River Delta of Vietnam in the recent decades. *Paddy Water Environ.* **2019**, *17*, 35–44. [CrossRef]
36. Yue, S.; Wang, C. The Mann-Kendall Test Modified by Effective Sample Size to Detect Trend in Serially Correlated Hydrological Series. *Water Resour. Manag.* **2004**, *18*, 201–218. [CrossRef]
37. Hundertmark, W. Building drought management capacity in the Mekong River basin. *Irrig. Drain.* **2008**, *57*, 279–287. [CrossRef]
38. Dang, V.H.; Tran, D.D.; Cham, D.D.; Hang, P.T.; Nguyen, H.T.; Truong, H.V.; Tran, P.H.; Duong, M.B.; Nguyen, N.T.; Le, K.V.; et al. Assessment of Rainfall Distributions and Characteristics in Coastal Provinces of the Vietnamese Mekong Delta under Climate Change and ENSO Processes. *Water* **2020**, *12*, 1555. [CrossRef]
39. Räsänen, T.; Kummu, M. Spatiotemporal influences of ENSO on precipitation and flood pulse in the Mekong River Basin. *J. Hydrol.* **2013**, *476*, 154–168. [CrossRef]
40. Hamed, K.H.; Ramachandra Rao, A. A modified Mann-Kendall trend test for autocorrelated data. *J. Hydrol.* **1998**, *204*, 182–196. [CrossRef]
41. Hipel, K.W.; McLeod, A.I. (Eds.) Chapter 23 Nonparametric Tests for Trend Detection. In *Time Series Modelling of Water Resources and Environmental Systems*; Elsevier: Amsterdam, The Netherlands, 1994; Volume 45, pp. 853–938. ISBN 0167-5648.

water

MDPI

Article

Quantitative Land-Use and Landslide Assessment: A Case Study in Rize, Türkiye

Natsuki Kasahara [1], Yutaka Gonda [2,*] and Nejan Huvaj [3]

1 Graduate School of Science and Technology, Niigata University, Niigata 950-2181, Japan; f21e084g@mail.cc.niigata-u.ac.jp
2 Faculty of Agriculture, Niigata University, Niigata 950-2181, Japan
3 Civil Engineering Department, Middle East Technical University, Ankara 06800, Türkiye; nejan@metu.edu.tr
* Correspondence: gonda@agr.niigata-u.ac.jp

Abstract: Currently, many studies have reported that many landslides occur in tea or rubber plantation areas. In these areas, it is important to make a landslide susceptibility map and to take necessary measures to mitigate landslide damage. However, since historical landslide distribution data and land use data are not available, quantitative landslide assessment measurements have not been made in many countries. Therefore, in this study, landslide distribution maps and land use maps are created with worldwide available satellite imagery and Google Earth imagery, and the relationship between landslides and land use is analyzed in Rize, Türkiye. The results show that landslides are 1.75 to 5 times more likely to occur in tea gardens than in forests. It was also found that land use has the highest contribution to landslides among the landslide conditioning factors. The landslide assessment, using a simple landslide detection method and land use classification method with worldwide available data, enabled us to quantitatively reveal the characteristics of landslides. The results of this study reveal that quantitative landslide assessments can be applied in any location, where relatively high resolution satellite imagery and Google Earth imagery, or its alternatives, are available.

Keywords: land use; quantitative landslide assessment; Rize; satellite images; tea garden

Citation: Kasahara, N.; Gonda, Y.; Huvaj, N. Quantitative Land-Use and Landslide Assessment: A Case Study in Rize, Türkiye. *Water* **2022**, *14*, 1811. https://doi.org/10.3390/w14111811

Academic Editor: František Petrovič

Received: 28 April 2022
Accepted: 2 June 2022
Published: 4 June 2022

1. Introduction

Rainfall-induced landslides occur worldwide, and they cause economic losses and human casualties every year [1]. Intense or prolonged rainfall induces landslides. Lithology, slope gradient, slope aspect, elevation, vegetation cover, and proximity to drainage line are considered to be influential physical parameters in the occurrence of rainfall-induced landslides [2,3]. Regarding vegetation cover, it is well known that improper land use change affects the landslide frequency [1,4–6]. Some studies in Türkiye [7,8], India [9], and Sri Lanka [10,11] have reported that landslides frequently occur in areas where the land use has converted into tea garden or rubber plantation areas. The contributing factors for increasing landslide susceptibility are thought to be: (1) shallower roots of tea or rubber plantations than the original trees [7,9], (2) improper drainage systems around the plantation garden [12,13], and (3) excessive use of fertilizers [9,14].

Effective land use planning and management based on landslide risk assessment is necessary to mitigate damages caused by landslides [15]. Landslide susceptibility can be assessed qualitatively or quantitatively [16]. Qualitative landslide assessment is a subjective assessment based on experience and the knowledge of the experts employed [16–19]. On the other hand, quantitative landslide risk assessment is an objective and reproducible assessment based on comprehensive historical data [16–18,20]. Quantitative landslide risk assessment is useful because it provides a basis for the prioritization of measurement and mitigation actions [16]. However, quantitative landslide risk assessment is still inadequate in many countries [16,17,21–23], because financial resources and personnel for allocating baseline information to landslide assessment (e.g., landslide historical data, land use data, rainfall data, data regarding the distribution of soil properties) are limited [24,25].

Remote sensing information is the best alternative source for collecting these limited data [26]. Current satellite images make it possible to obtain accurate time-series information on the earth's surface over wide areas. If satellite images before and after landslide occurrence are available, it is a powerful tool for detecting landslide areas for landslide susceptibility assessment in many countries [26,27]. Detecting landslides using satellite images has the advantage of reducing the time and cost compared to conventional methods of using field surveys and aerial photos [27–29]. Recently, high resolution and highly revisited frequency satellite images have become more easily available, and they are sometimes available for free. For example, the Education and Research Program of Planet Labs provides access to the PlanetScope Imagery and the RapidEye Archive to anyone who belongs to the university [30]. In addition, Google Earth images is a very useful alternative for any user.

To take effective measures to mitigate landslide damages in the area, where quantitative landslide risk assessment is insufficient, it is very important to present an easy method (without a high level of techniques and knowledge) to generate the data necessary for landslide risk assessment, even if there is some compromise in the spatial resolution accuracy of the data obtained.

The objective of this study is to generate the data that is necessary for landslide risk assessment (such as landslide distribution map and the land use map) using globally available satellite images and Google Earth images, to quantitatively assess the relationship between landslide, land use, and topography, and to demonstrate the usefulness of this series of methods.

2. Study Area and Data

The Black Sea region of Türkiye is a highly landslide-prone area, due to its steep topography and heavy annual precipitation [3]. Many rainfall-induced landslides cause damage in the Rize region, which is located in the Eastern Black Sea region, and there have been 1 to 32 casualties every year from 1973 to 2010 [8]. A significant upward trend of fatal landslides was observed in Rize from 1952 to 2019 [31]. Rize is a very famous area for tea production, and tea is an economically valuable plant and essential for farmers [32]. The alder forests were converted to tea gardens, especially in the last 50 years, and the tea plantation area has increased by 32 times between 1940 and 2010 [4,7,33]. Many studies have reported that landslide incidents increase as tea garden areas increase [4,7,8,13]. Although Karsli et al. [7] studied the effect of land use changes on landslides, studies quantitatively evaluating the relationship between land use and landslide in Rize are rather limited.

The study area covers 683.1 km^2 in Guneysu district, Derepazari district, Merkez district, and a part of Cayeli district in Rize city, which is between a latitude of 40°82′66″ and 41°08′38″ N, and a longitude of 40°38′86″ and 40°79′67″ E (Figure 1). The altitude of the study area ranges from 0 to 2453 m above the mean sea level. The terrain is steep and mountainous, except for the coastal area (Figure 2). The flat coastal areas are populated with houses, while villages and tea gardens spread out in the mountainous areas despite the steep slopes. The geological map of the study area is shown in Figure 3.

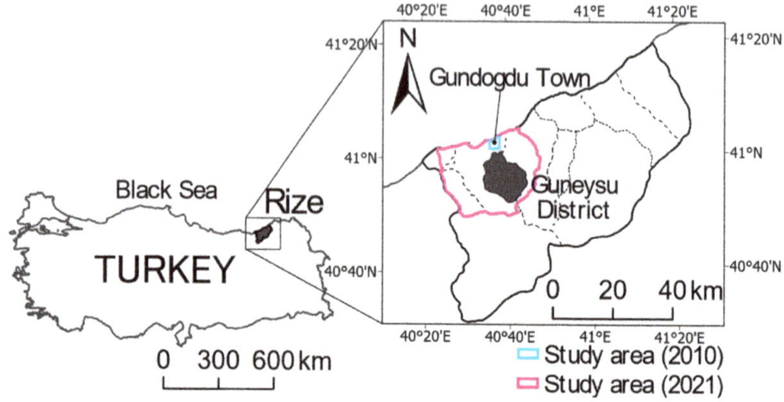

Figure 1. Location map of study area.

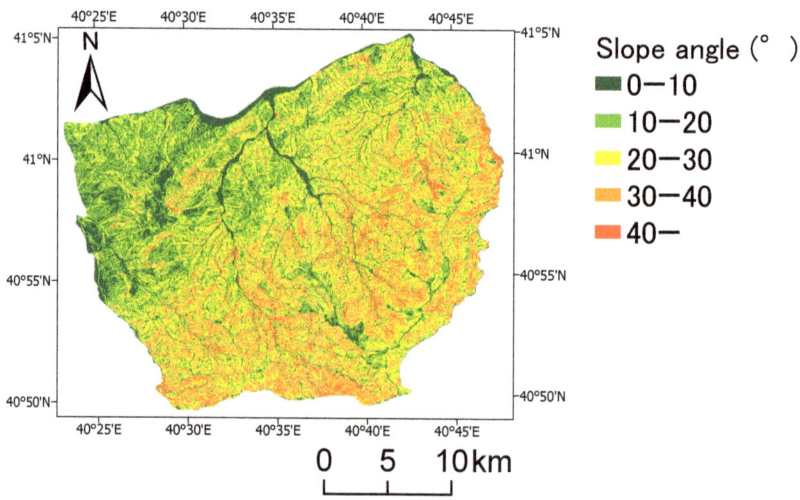

Figure 2. Distribution of the slope angle in the study area.

Heavy rainfall on 14 July 2021 triggered many landslides, mainly in Guneysu district (Figure 4), which resulted in six casualties [34]. The average July monthly rainfall was 152 mm according to the Turkish State Meteorological Service (TSMS) of the Ministry of Agriculture and Forestry [35]. However, 220 mm of rainfall was received only in 7 h [36] and it caused these landslides.

In addition, heavy rainfall on 26 August 2010 also triggered many landslides in Gundogdu town (Figure 4), which resulted in 14 casualties [37]. The average monthly rainfall in August is 197 mm [35]. However, 219 mm of rainfall was received within the day and on the day before the disaster [8], causing these landslides. For these landslides, only the concentrated area concerning the landslide, covering 10.6 km^2 and located between a latitude of 41°02′40″ and 41°05′64″ N, and a longitude of 40°59′39″ and 40°63′48″ E, is used for analysis (Figure 1).

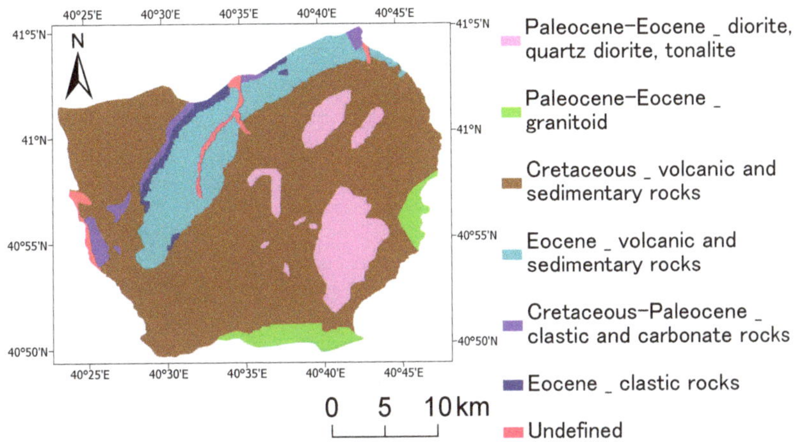

Figure 3. Geological map of the study area (adapted from Turkish General Directorate of Mineral Research and Exploration [38]).

(**a**) (**b**)

Figure 4. Heavy rainfall-induced landslides in Rize city in Türkiye. (**a**) Landslide in 2010 (photo from Turkish Disaster and Emergency Management Directorate [39]). (**b**) Landslide in 2021 (photo from Turkish Disaster and Emergency Management Directorate [34]).

3. Methodology

RapidEye imagery, PlanetScope imagery [30], and Google Earth imagery were used in order to detect the landslide area and classify the land use. The details of the satellite images used in this study are shown in Table 1. In addition, the Shuttle Radar Topography Mission (SRTM) DEM with a 30 m resolution was used to obtain topographic parameters such as slope gradient. The image analysis tool in ArcGIS Pro 2.6 was used for creating landslide distribution maps and land use, and for analyzing the relationship between landslide, land use, and other landslide-contributing factors. The rainfall data were acquired at 17 rainfall stations in Rize, and these were provided by TSMS [40]. The spatial variation of the cumulative daily rainfall the day before and the day of the landslide in 2021 (06:00, 13 July 2021–06:00, 15 July 2021 (UTC)) were interpolated using ordinary kriging.

Table 1. Satellite data used for landslide detection and land use classification.

Landslide Event	Mission	Resolution after Processing	Sensor Type	Spectral Bands (nm)	Acquisition Date	
					Pre-Event	Post-Event
26 August 2010	RapidEye	5 m	MS	B (440–510) G (520–590) R (630–685) Red Edge (690–730) NIR (760–860)	22 Jun 2010	16 May 2011
14 July 2021	PlanetScope	4.1 m	MS	B (455–515) G (500–590) R (590–670) NIR (780–860)	3 July 2021	19 July 2021 27 August 2021

3.1. Landslide Detection

Landslides were detected by visual interpolation or semi-automatic extraction using Google Earth images and satellite images. Although visual interpolation is the most common method for landslide mapping, it requires experience and time, since the landslides are mapped manually by experts [28]. Therefore, semi-automatic landslide detection is important for quick and easy-to-implement analysis. A change detection method, which is a semi-automatic method, was used. This method is the most common landslide detection method since it is simple and easy to apply [41].

In this study, landslides were detected using the change detection method from satellite images once, and then each landslide area was checked visually using Google Earth images or other satellite images. This method reduces the required skill needed compared with the conventional visual interpolation method. It also improves the accuracy of the change detection method. Therefore, it improved the efficiency of landslide detection. Only landslides of more than 100 m^2 in area were targeted for detection, since it is relatively difficult to identify small landslide areas.

The supervised classification method of the maximum likelihood classification algorithm was used to detect landslides in 2010. Training samples were gathered from Google Earth imagery. The landslide distribution map was created after visual checking using Google Earth images.

The GSI (Grain Size Index) is the index that has a positive correlation with fine sand content, and the GSI analysis was originally used for detecting desertification [42]. However, in this study, GSI was used to detect landslides in 2021. GSI is calculated as follows:

$$GSI = (R - B)/(R + B + G) \quad (1)$$

where, R, B, and G are the reflectance of the red, blue, and green bands of the satellite images. The GSI value was close to 0 in vegetated and water areas, and high in bare soil surfaces [42]. In this study, the landslide area was defined as GSI < 0.2 in the satellite image before the landslide, and GSI $\geq$ 0.2 in the satellite image after the landslide. The landslide distribution map was created after visual checking with satellite images that were taken after the landslide.

3.2. Land Use Classification

In this study, three land use classes were established: tea garden, forest, and road/house/stream. Land use before the 2010 landslide and the 2021 landslide were classified using a Random Forest (RF) machine learning technique using satellite images. RF is an ensemble learning algorithm that generates multiple decision trees based on random subsets of training data [43], and it is one of the most accurate machine learning algorithms for land use classification [44,45]. The accuracy of land use classification in a landslide area immediately before a landslide occurs has a significant impact on the reliability of the quantitative landslide risk assessment. Therefore, after land use classification by the RF method, land use in a landslide area was checked visually using Google Earth images.

The landslide spreading area in the 2010 rainfall event was relatively small (approximately 10 km^2) and covered the coastal area, while the landslide spreading area in the 2021 rainfall event was large (approximately 370 km^2) and covered coastal and steep mountainous areas. It is difficult to classify land use in large and heterogeneous landscape areas [46]. Furthermore, the shadows effect on rugged terrain can lead to classification errors [47,48]. Thus, the satellite images were divided into inland and coastal areas when classifying land use in 2021. In addition, each divided image was divided again into sunny parts and shaded parts, using the slope aspect and the azimuth of the sun at the time that satellite image was taken. The land use of each image was classified individually.

3.3. Analysis of the Relationship between Landslide, Land Use, and Other Landslide Contributing Factors

The relationship between landslide and land use was analyzed statistically using a created landslide distribution map and a land use map. The landslide area ratio, which is the ratio of landslide area to the total area, was used for quantitative assessment. (e.g., Landslide area ratio in tea gardens = Landslide area in tea gardens (m^2)/Total area of tea gardens (m^2) $\times$ 100). In general, the ground surface of the ground that experienced landslides was disturbed. For the 2021 landslides, previous landslides could be detected because Google Earth images from before the landslides are available. Therefore, follow-up landslides were excluded from the analysis.

To clarify the difference in landslide characteristics between the tea gardens and the forests, the relationship between landslides and the other factors (such as rainfall amount, geology, elevation, slope aspect, slope angle, plane curvature, profile curvature, flow accumulation, distance to the road, distance to stream, distance to first- or second-order drainage lines) were analyzed in tea gardens and forests, individually.

To reveal the dependency of land use on the occurrence of landslides, the dependence of each landslide conditioning factor on the occurrence of landslides was analyzed using Hayashi's quantification theory type II. Approximately 680,000 points were extracted from 10 km^2 of landslide concentrated area, and Hayashi's quantification theory type II was applied. Hayashi's quantification theory type II is a method of multivariate discrimination analysis [49]. Except for rainfall amount, four major landslide conditioning factors (items) that were revealed from preliminary analysis were adopted as landslide contributing factors. Each item is divided into some categories, and the contribution of each item is expressed as category scores and item range. A positive value of category score indicates that the corresponding category will promote occurrences of landslides. On the other hands, negative values indicate that the corresponding category will restrain landslides. The order of contribution to landslide occurrence was judged from the item range. The larger the item range, the more contribution to occurrences of landslides.

4. Result

4.1. Landslides

The landslides detected in 2010 and 2021 are shown in Figure 5. Landslides were concentrated in a small area during the 2010 landslides, while landslides were spread over a large area in the 2021 landslides.

In the 2010 landslides, 415 landslides were detected using the maximum likelihood method, and 348 landslides, after visual checking. In the 2021 landslides, 13,409 landslides were detected using GSI analysis, and 910 landslides, after visual checking (Table 2). It took one person 15 days to visually check the landslide areas. A total of 240 landslides were follow-up landslides, and 670 landslides were new ones. The number of actual landslides was less than 7% of the detected landslides when using GSI analysis in the 2021 landslides, because satellite images were taken immediately (5 days) after landslide occurrences, and sediment flowed areas were also recognized as landslide areas.

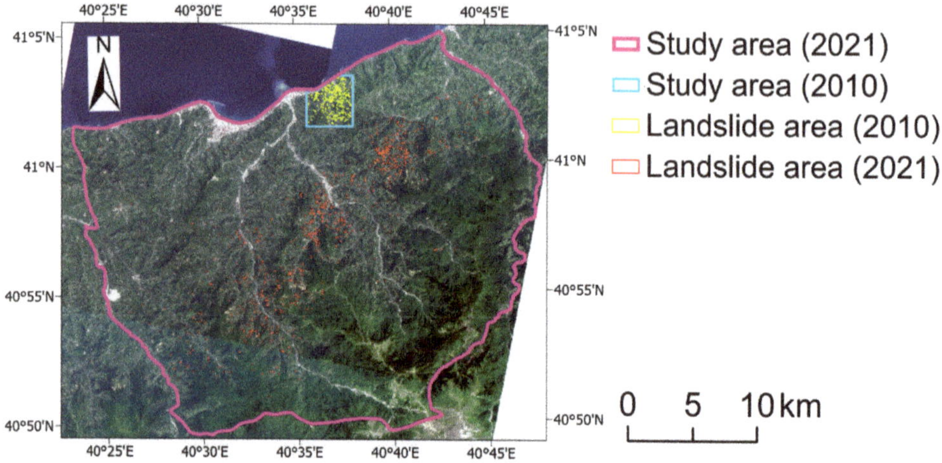

Figure 5. Landslide distribution map (yellow and red lines indicate each landslide area; satellite image is acquired from PlanetScope [30]).

Table 2. Landslides that occurred in 2010 and 2021.

		26 August 2010	14 July 2021
Number of landslides	Before visual check	415	13,409
	After visual check	348	910
Landslide area (m^2)		291,556	648,229
Minimum landslide area (m^2)		104.9	104.1
Maximum landslide area (m^2)		15,720.7	8378.2
Study area (km^2)		10.65	683.06

4.2. Land Use

The classified land use in 2010 and 2021 is shown in Figure 6. The land use classification accuracy assessment was performed in the multiple test area. Google Earth images were used for creating the ground truth data. To verify the land use classification accuracy, the homologous points in the land use map and in Google Earth images were compared directly. Confusion matrices were created to calculate the overall accuracy, user's accuracy, producer's accuracy, and kappa coefficient (Tables 3 and 4). These statistical accuracy assessments provide a measure of confidence in the satellite image classification. Each column of the confusion matrix represents a ground truth data, and the values in the column correspond to the land use classification of the ground truth points. The results indicated overall accuracies of 78.6% and 75.1%, and a kappa coefficient of 0.61 and 0.53 for the classified images of 2010 and 2021, respectively. Although the accuracies were lower than the recommended level [50], it was considered sufficiently accuracy to determine the percentages of each land use type in a large area.

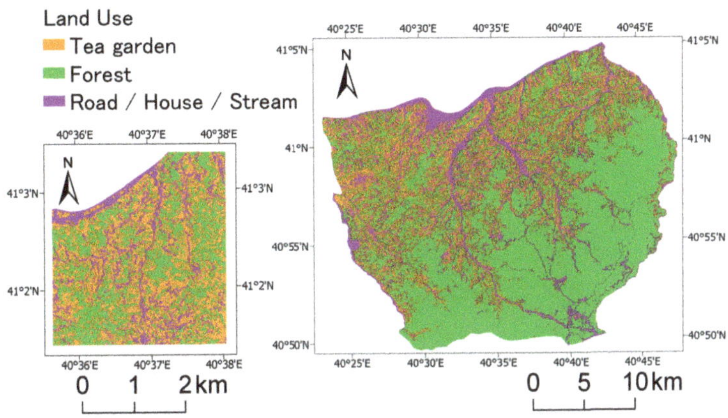

Figure 6. Land use map (**a**) in 2010 and (**b**) 2021.

Table 3. Confusion matrix between ground truth and classified land use in 2010.

Classified	Ground Truth Data			SUM	User's Accuracy
	Tea Garden	**Forest**	**Road/House/Stream**		
Tea garden	3663	589	205	4457	82.2%
Forest	371	1246	8	1625	76.7%
Road/House/Stream	287	84	747	1118	66.8%
SUM	4321	1919	960	7200	
Producer's accuracy	84.8%	64.9%	77.8%		
Overall accuracy	78.6%				
Kappa coefficient	0.61				

Table 4. Confusion matrix between ground truth and classified land use in 2021.

Classified	Ground Truth Data			SUM	User's Accuracy
	Tea Garden	**Forest**	**Road/House/Stream**		
Tea garden	16,051	5110	1403	22,564	71.1%
Forest	7780	35,244	921	43,945	80.2%
Road/House/Stream	1809	806	2419	5034	48.1%
SUM	25,640	41,160	4743	71,543	
Producer's accuracy	62.6%	85.6%	51.0%		
Overall accuracy	75.1%				
Kappa coefficient	0.53				

4.3. Relationship between Landslide, Land Use, and Other Landslide Contributing Factors

The landslide area ratio in the tea gardens was 1.75 times higher than in forests in the 2010 landslides, and five times higher than in forests, in the 2021 landslides (Table 5). Landslides were more likely to occur in the tea gardens than in the forests, regardless of rainfall events.

Table 5. Landslides in each land use area in 2010 and in 2021.

		Total Area (ha)	Landslide Area (ha)	Landslide Area Ratio (%)
2010	Tea garden	366.08	13.57	3.71
	Forest	223.20	4.73	2.12
	Road/House/Stream	94.11	0.41	0.43
2021	Tea garden	17,748.30	29.11	0.16
	Forest	42,879.02	14.23	0.03
	Road/House/Stream	9647.04	2.58	0.03

The relationship between the slope angle and landslides was analyzed in the tea gardens and forest (Table 6). In the tea garden, the landslide area ratio increased as the slope angle increased. The landslide area ratio reached a peak value at a slope angle of 20–30° in the 2010 landslide, and 30–40° in the 2021 landslide. This result was the same as Yalcin [51], which indicated that almost all of the landslides occurred at angles ranging over 10–40° in the eastern part of Rize. However, in the forest, there was no relation between the slope angle and the landslide area ratio. Therefore, in the slope range of 30–40°, the landslide area ratio in the tea gardens was 3.5 times more than in the forest in the 2010 landslide, and 9.1 times more than in the forest in the 2021 landslide.

Table 6. Landslides and slope angles in each land use area in 2010 and in 2021.

		Area (ha)		Landslide Area (ha)		Landslide Area Ratio (%)	
	Slope Angle (°)	Tea Garden	Forest	Tea Garden	Forest	Tea Garden	Forest
2010	0–10	52.6	12.2	1.36	0.40	2.584	3.303
	10–20	222.5	93.1	7.18	2.08	3.226	2.236
	20–30	236.8	173.5	10.39	4.11	4.386	2.367
	30–40	57.4	67.9	2.22	0.75	3.875	1.109
	40–	1.30	1.30	0.02	0.02	1.566	1.793
2021	0–10	3197.3	2767.3	1.28	0.93	0.040	0.034
	10–20	10,505.3	13,170.5	9.59	4.23	0.091	0.032
	20–30	9925.1	27060.3	23.61	10.09	0.238	0.037
	30–40	3698.8	21283	10.15	6.40	0.274	0.030
	40–	336.1	2550.3	0.75	0.54	0.222	0.021

Landslide distribution and two-days rainfall distribution in the 2021 landslides are shown in Figure 7. Rainfall data were lacking in the western part of the study area, so it was not possible to interpolate the accumulated rainfall. Figure 7 shows that the landslide concentrated area and the rainfall concentrated area were almost the same. With less than 120 mm 48-hr rainfall, the landslide area ratio in both the tea gardens and the forest was very low (<0.02%; Table 7). It is only in the tea gardens that the landslide area ratio increased significantly when 48-hr rainfall exceeded 120 mm, and the landslide area ratio increased with increasing rainfall. On the other hand, in the forest, the landslide area ratio increased significantly when the 48-hr rainfall exceeded 160 mm. There was a smaller number of rainfall-induced landslides in the tea gardens compared with the forest.

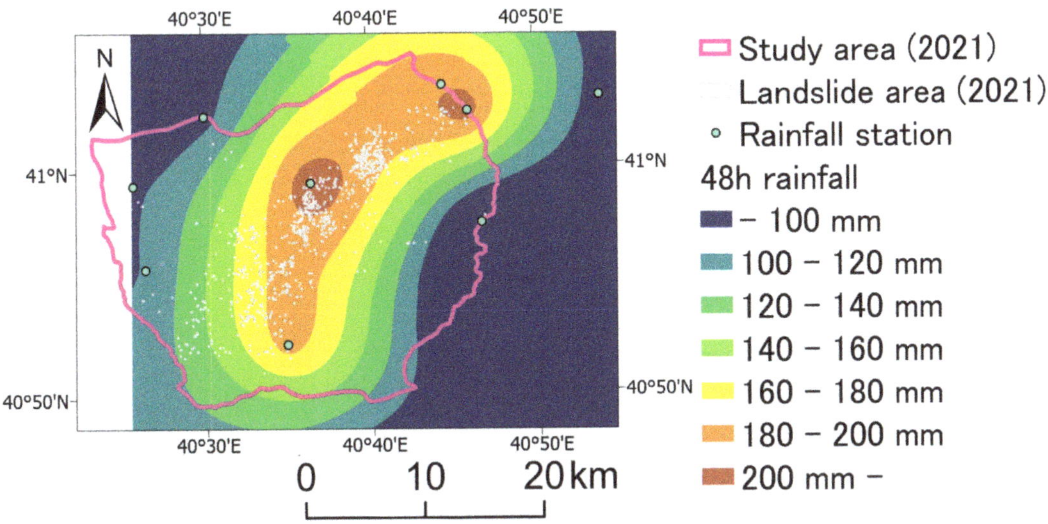

Figure 7. Landslide distribution and two-days rainfall distribution in 2021 (data from TSMS).

Table 7. Landslide area ratio and cumulative rainfall in 2021 (data from TSMS).

	Landslide Area Ratio (%)	
48-hr Rainfall (mm)	**Tea Garden**	**Forest**
–100	0.003	0.000
100–120	0.006	0.009
120–140	0.050	0.011
140–160	0.062	0.009
160–180	0.188	0.045
180–200	0.346	0.096
200–220	0.561	0.162

The dependence of each landslide conditioning factor on the occurrence of landslides was analyzed in the area shown in Figure 8. Our preliminary analysis revealed that land use, slope angle, distance to the first- or second-order drainage line, and flow accumulation (i.e., how much area drains water to a given point) are major factors for landslide occurrence. Therefore, these four factors were chosen as items (landslide contributing factor) for analysis using quantification theory type II. The result of quantification theory type II is shown in Table 8. Each landslide conditioning factor can be quantified on the basis of the category score and the item range.

The discriminative ratio was 58.3%. Judging from the range of the category scores, it is clear that land use was the most critical factor for landslide occurrence.

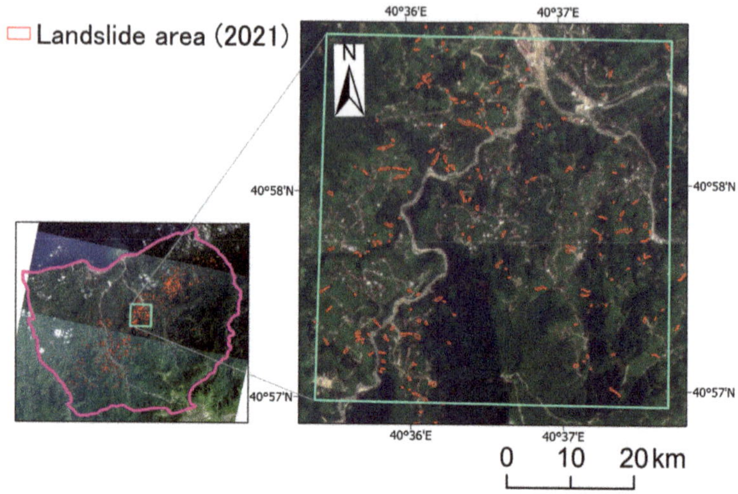

Figure 8. Quantification theory type II analysis of the area (satellite image is acquired from PlanetScope [30]).

Table 8. Result of quantification theory type II analysis.

Item	Category	Category Score	Range
Land use	Tea garden	1.066	1.602
	Forest	−0.536	
	Road/House/Stream	−0.511	
Slope angle	0°–10°	−0.847	1.276
	10°–20°	−0.454	
	20°–30°	0.429	
	30°–	−0.010	
Flow accumulation	0	−0.162	0.924
	0–5	−0.076	
	5–10	0.036	
	10–20	−0.062	
	20–100	0.227	
	100–	0.762	
Distance to first or second drainage line	-20 m	0.308	0.423
	20–40 m	−0.115	
	40–60 m	−0.110	
	60–80 m	−0.070	
	80–100 m	−0.051	
	100 m–	0.008	

5. Discussion

5.1. Landslide Detection and Land Use Classification

Regarding landslide detection, landslide detection using the maximum likelihood method gave much fewer mis-detections than GSI analysis. However, it is necessary to provide training data of the landslide area to detect landslides when using the maximum likelihood method. Immediately after a landslide occurs, it is not possible to obtain Google Earth images that clearly show the landslide area that are necessary for creating training data. Therefore, to detect landslides soon after a disaster, it is a more effective method to detect landslides using GSI analysis and then to select actual landslide areas using visual checking with another satellite image.

In this study, landslide assessment was conducted without a field survey. The accuracy of landslide detection using the change detection method and land use classification using the RF method were not high. However, visual verification after these measurements enabled us to efficiently generate the landslide distribution map and the land use map.

In countries where quantitative landslide assessment is still inadequate, it is important to objectively reveal the characteristics of landslides, even while using a less accurate method. In addition, to develop studies on landslides, it is essential to accumulate landslide historical data, land use data, and rainfall distribution data, and to make them available to the public [52].

5.2. Landslides and Tea Gardens

It is clear that in tea gardens landslides are more likely to occur in Rize. Some differences might relate to landslide occurrences between tea gardens and forests:

- Tea roots are approximately 50 cm in depth [53]. On the other hand, field research on rainfall-triggered landslides in another district of Rize (Kaptanpasa) with a similar geology as that of this study revealed that the mean depth of landslides is 1.05 m, and 98% of the landslides were less than 3 m in depth [54]. The depth of the slip surface of typical landslides around the study area tends to be deeper than the depth of the tea roots. Even if the tea roots penetrate the slip surface of a landslide, they would have little effect on retaining the landslide occurrence.
- Yüksek et al. [33] indicates that the average saturated hydraulic conductivity from the surface to 50 cm depth is 8.5 mm/h in the tea garden, and 24.7 mm/h in the forest [33]. Üyetürk [55] indicates that the saturated hydraulic conductivity in the tea gardens are in range of 0.54 to 3.96 mm/h. Therefore, it is easy to saturate the surface soil layer in the tea gardens, and greater surface flow can occur in the tea gardens than in the the forest. Yalcin [56] mentions that surface runoff is one of the primary factors leading to a landslide. In addition, some studies have reported that in order to reduce landslide incidences in tea gardens, it is important to make a proper drainage system that can collect rainfall and prevent its infiltration into the soil [8,13,14,56].
- In terms of the soil properties of the tea gardens and the forest, Yüksek et al. [33] indicates that saturated hydraulic conductivity, porosity, soil organic matter, plant available water, and total N are significantly different. The introduction of cultivation techniques using fertilizers has affected soil properties.

However, it is still not comprehensively clear how these differences affect landslide occurrence in the tea gardens. Clarification on the mechanism of landslides in tea gardens is required for future studies.

5.3. Landslides in Tea Gardens in Rize and Other Regions

Japan is also a tea-producing country, and one of the most landslide-prone countries in the world. Heavy rainfalls occur frequently and induce landslides. Heavy rainfall over 11 August 2021–15 August 2021 induced more than 80 landslides in tea gardens in Ureshino city, Saga prefecture (33°06′2″ N and 130°03′31″ E). The accumulated rainfall was 1178 mm, which was four times higher than the average August monthly precipitation [57]. In addition, heavy rainfall on 14 July 2012 induced some landslides in the tea gardens, and more than 40 ha of tea gardens were damaged in Yame city, Fukuoka prefecture (33°21′19″ N latitude and 130°55′79″) [58]. The daily rainfall was 415 mm, which was the highest record in Yame city [59]. However, landslides in forests have been reported much more frequently than in tea gardens in Japan [60,61]. Some differences might relate to the landslide occurrence between the tea gardens in Rize and in Japan:

- A total of 72% of tea gardens were on ground that was inclined more than 15° in Rize. On the other hand, only 32% of tea gardens were on slopes steeper than 15° in Japan [62]. In short, many more tea gardens are on a steep slope in Rize. In addition, the steeper the slope angle, the higher the landslide area ratio, both for Rize and for Japan. However, the landslide area ratio achieves a peak value of around 25–30° in tea gardens in Rize, while it is generally 30–35° in Japan [63]. This means that more landslides tend to occur on smaller slope angle in Rize, as compared to Japan. The synergistic effects of these two differences between Rize and Japan might be the reason for why a collapse is more likely to occur in tea gardens in Rize than in Japanese tea gardens.

- In Japan, it is common practice to pile up stone walls in tea gardens with steep slopes to prevent soil erosion and to stabilize slopes. However, in Türkiye, it seems that such landslide prevention measures are not thoroughly implemented in tea gardens.

However, it is still not clear how the tea gardens in Rize differ from those in Japan, and what the crucial difference are for landslide occurrence. It would be useful, in considering effective landslide mitigation measures in Türkiye, to clarify the differences between tea gardens in Rize and in Japan.

6. Conclusions

This study aimed to quantitatively assess the relationships between land use and landslides, using globally available data. The landslide distribution map and land use map in Rize were prepared using globally available satellite images and Google Earth images. Additionally, it was found that landslides were 1.75 to 5 times more likely to occur in the tea gardens than in the forest. In addition, less rainfall triggers landslides in the tea gardens than in the forest. In addition, the landslide area ratio dramatically increases when the 48-hr rainfall exceeds 120 mm in the tea gardens and 160 mm in the forest. Additionally, in steep sloped areas (where the slope angle is 30–40°), landslides were 3.5 to 9.1 times more likely to occur in the tea gardens than in the forest.

Even if there are no landslide historical records, it is possible to create a landslide distribution map and to quantitatively assess landslide susceptibility using rather high resolution satellite images and Google Earth images, or alternative images. Therefore, there is a possibility for conducting landslide assessments quantitatively in any location, where those images mentioned above are available.

Author Contributions: Conceptualization, Y.G.; investigation, N.K. writing—original draft preparation, N.K.; writing—review and editing, Y.G. and N.H.; supervision, Y.G. funding acquisition, Y.G. All authors have read and agreed to the published version of the manuscript.

Funding: This research was funded by the Asia Air Survey Co., Ltd., Japan.

Acknowledgments: This study was originally initiated as part of the short-term scientific visits via MEXT Re-Inventing Japan Project (Türkiye) GLocal Age 2020, "Global Disaster Risk and Recovery Program". We appreciate the international collaboration atmosphere provided with this project.

Conflicts of Interest: The authors declare no conflict of interest.

References

1. Froude, M.; Petley, D. Global Fatal Landslide Occurrence from 2004 to 2016. *Nat. Hazards Earth Syst. Sci.* **2018**, *18*, 2161–2181. [CrossRef]
2. Dai, F.C.; Lee, C.F.; Li, J.; Xu, Z.W. Assessment of Landslide Susceptibility on the Natural Terrain of Lantau Island, Hong Kong. *Environ. Geol.* **2001**, *40*, 381–391. [CrossRef]
3. Ercanoglu, M.; Gokceoglu, C. Assessment of Landslide Susceptibility for a Landslide-Prone Area (North of Yenice, NW Türkiye) by Fuzzy Approach. *Environ. Geol.* **2002**, *41*, 720–730. [CrossRef]
4. Reis, S. Analyzing Land Use/Land Cover Changes Using Remote Sensing and GIS in Rize, North-East Türkiye. *Sensors* **2008**, *8*, 6188–6202. [CrossRef]
5. Chen, C.Y.; Huang, W.L. Land Use Change and Landslide Characteristics Analysis for Community-Based Disaster Mitigation. *Environ. Monit. Assess.* **2013**, *185*, 4125–4139. [CrossRef]
6. Carrión-Mero, P.; Montalván-Burbano, N.; Morante-Carballo, F.; Quesada-Román, A.; Apolo-Masache, B. Worldwide Research Trends in Landslide Science. *Int. J. Environ. Res. Public Health* **2021**, *18*, 9445. [CrossRef]
7. Karsli, F.; Atasoy, M.; Yalcin, A.; Reis, S.; Demir, O.; Gokceoglu, C. Effects of Land-Use Changes on Landslides in a Landslide-Prone Area (Ardesen, Rize, NE Türkiye). *Environ. Monit. Assess.* **2009**, *156*, 241–255. [CrossRef]
8. Filiz, M.; Avcı, H.; Usta, P. Heyelanların Yerleşim Alanlarına Etkilerinin İncelenmesi (Rize-Gündoğdu Örneği). *E-J. New World Sci. Acad. Eng. Sci.* **2011**, *6*, 1200–1211. (In Turkish)

9. Yunus, A.P.; Fan, X.; Subramanian, S.S.; Jie, D.; Xu, Q. Unraveling the Drivers of Intensified Landslide Regimes in Western Ghats, India. *Sci. Total Environ.* **2021**, *770*, 145357. [CrossRef]
10. Senanayake, S.; Pradhan, B.; Huete, A.; Brennan, J. Assessing Soil Erosion Hazards Using Land-Use Change and Landslide Frequency Ratio Method: A Case Study of Sabaragamuwa Province, Sri Lanka. *Remote Sens.* **2020**, *12*, 1483. [CrossRef]
11. Perera, E.N.C.; Jayawardana, D.T.; Ranagalage, M.; Jayasinghe, P. Spatial Multi Criteria Evaluation (SMCE) Model for Landslide Hazard Zonation in Tropical Hilly Environment: A Case Study from Kegalle. *Geoinfor. Geostat. Overv.* **2018**, S3. [CrossRef]
12. Vasantha Kumar, S.; Bhagavanulu, D.V.S. Effect of Deforestation on Landslides in Nilgiris District—A Case Study. *J. Indian Soc. Remote Sens.* **2008**, *36*, 105–108. [CrossRef]
13. Hacisalihoglu, S.; Gumus, S.; Kezik, U. Land Use Conversion Effects Triggered by Tea Plantation on Landslide Occurence and Soil Loss. *Fresenius Environ. Bull.* **2018**, *27*, 2933–2942.
14. Yalcin, A. Environmental Impacts of Landslides: A Case Study from East Black Sea Region, Türkiye. *Environ. Eng. Sci.* **2007**, *24*, 821–833. [CrossRef]
15. Dai, F.C.; Lee, C.F.; Ngai, Y.Y. Landslide Risk Assessment and Management: An Overview. *Eng. Geol.* **2002**, *64*, 65–87. [CrossRef]
16. Corominas, J.; van Westen, C.; Frattini, P.; Cascini, L.; Malet, J.P.; Fotopoulou, S.; Catani, F.; Van Den Eeckhaut, M.; Mavrouli, O.; Agliardi, F.; et al. Recommendations for the Quantitative Analysis of Landslide Risk. *Bull. Eng. Geol. Environ.* **2014**, *73*, 209–263. [CrossRef]
17. Lee, E.M. Landslide Risk Assessment: The Challenge of Estimating the Probability of Landsliding. *Q. J. Eng. Geol. Hydrogeol.* **2009**, *42*, 445–458. [CrossRef]
18. Aleotti, P.; Chowdhury, R. Landslide Hazard Assessment: Summary Review and New Perspectives. *Bull. Eng. Geol. Environ.* **1999**, *58*, 21–44. [CrossRef]
19. Quesada Román, A. Landslides and Floods Zonation Using Geomorphological Analyses in a Dynamic Basin of Costa Rica. *Rev. Cart.* **2021**, *102*, 125–138. [CrossRef]
20. Quesada-Román, A. Landslide Risk Index Map at the Municipal Scale for Costa Rica. *Int. J. Disaster Risk Reduct.* **2021**, *56*, 102144. [CrossRef]
21. van Westen, C.J.; van Asch, T.W.J.; Soeters, R. Landslide Hazard and Risk Zonation—Why Is It Still so Difficult? *Bull. Eng. Geol. Environ.* **2006**, *65*, 167–184. [CrossRef]
22. Castellanos Abella, E.A.; Van Westen, C.J. Qualitative Landslide Susceptibility Assessment by Multicriteria Analysis: A Case Study from San Antonio Del Sur, Guantánamo, Cuba. *Geomorphology* **2008**, *94*, 453–466. [CrossRef]
23. Gaprindashvili, G.; Guo, J.; Daorueang, P.; Xin, T.; Rahimy, P. A New Statistic Approach towards Landslide Hazard Risk Assessment. *Int. J. Geosci.* **2014**, *5*, 38–49. [CrossRef]
24. Harp, E.L.; Reid, M.E.; McKenna, J.P.; Michael, J.A. Mapping of Hazard from Rainfall-Triggered Landslides in Developing Countries: Examples from Honduras and Micronesia. *Eng. Geol.* **2009**, *104*, 295–311. [CrossRef]
25. Đurić, D.; Mladenović, A.; Pešić-Georgiadis, M.; Marjanović, M.; Abolmasov, B. Using Multiresolution and Multitemporal Satellite Data for Post-Disaster Landslide Inventory in the Republic of Serbia. *Landslides* **2017**, *14*, 1467–1482. [CrossRef]
26. Hong, Y.; Adler, R.; Huffman, G. Use of Satellite Remote Sensing Data in the Mapping of Global Landslide Susceptibility. *Nat. Hazards* **2007**, *43*, 245–256. [CrossRef]
27. Tofani, V.; Segoni, S.; Agostini, A.; Catani, F.; Casagli, N. Use of Remote Sensing for Landslide Studies in Europe. *Nat. Hazards Earth Syst. Sci.* **2013**, *13*, 299–309. [CrossRef]
28. Guzzetti, F.; Mondini, A.C.; Cardinali, M.; Fiorucci, F.; Santangelo, M.; Chang, K.T. Landslide Inventory Maps: New Tools for an Old Problem. *Earth-Science Rev.* **2012**, *112*, 42–66. [CrossRef]
29. Casagli, N.; Fanti, R.; Nocentini, M.; Righini, G. *Assessing the Capabilities of VHR Satellite Data for Debris Flow Mapping in the Machu Picchu Area (C101-1)*; Sassa, K., Fukuoka, H., Wang, F., Wang, G., Eds.; Landslides; Springer: Berlin/Heidelberg, Germany, 2005. [CrossRef]
30. Planet Team. Planet Application Program Interface: In Space for Life on Earth. Available online: https://api.planet.com (accessed on 15 March 2022).
31. Görüm, T.; Fidan, S. Spatiotemporal Variations of Fatal Landslides in Türkiye. *Landslides* **2021**, *18*, 1691–1705. [CrossRef]
32. Ercisli, S. The Tea Industry and Improvements in Türkiye. In *Global Tea Breeding*; Advanced Topics in Science and Technology in China; Springer: Berlin/Heidelberg, Germany, 2012. [CrossRef]
33. Yüksek, T.; Göl, C.; Yüksek, F.; Yüksel, E.E. Long-Term Effects of Conversion From Natural Forest to Tea Cultivation on Soil Properties: Humid Northern Black Sea Region. *Afr. J. Agric. Res.* **2018**, *4*, 665–674. [CrossRef]
34. AFAD. Rize'de Meydana Gelen Aşırı Yağışlar Hakkında. Turkish Disaster and Emergency Management Directorate. Available online: https://www.afad.gov.tr/rizede-meydana-gelen-asiri-yagislar-hakkinda# (accessed on 17 July 2021).
35. TSMS. İllere Ait Mevsim Normalleri (1991–2020). Turkish State Meteorological Service of Ministry of Agriculture and Forestry. Available online: https://www.mgm.gov.tr/veridegerlendirme/il-ve-ilceler-istatistik.aspx?m=RIZE (accessed on 18 January 2022).
36. AFAD. Ministers Soylu, Kurum and Karaismailoğlu Visit Areas Hit by Flood and Landslide in Riz. Turkish Disaster and Emergency Management Directorate. Available online: https://en.afad.gov.tr/ministers-soylu-kurum-and-karaismailoglu-visit-areas-hit-by-flood-and-landslide-in-rize (accessed on 4 August 2021).

37. Yılmaz, S. *Rize Bölgesel Heyelan ve Sel Afeti Müdahale ve Iyileştirme Çalışmalar*; Unpublished presentation: Rize, Türkiye, 2010. (In Turkish)
38. Turkish General Directorate of Mineral Research and Exploration. Geological Map of Türkiye TRABZON. Available online: https://www.mta.gov.tr/en/maps/geological-500000 (accessed on 8 December 2021).
39. AFAD. Heyelan Albümü. Turkish Disaster and Emergency Management Directorate. Available online: https://heysemp.afad.gov.tr/tr/26149/Heyelan-Albumu (accessed on 22 April 2022).
40. TSMS. Gerçekleşen Toplam Yağışlar. Turkish State Meteorological Service of Ministry of Agriculture and Forestry. Available online: https://www.mgm.gov.tr/sondurum/toplam-yagis.aspx (accessed on 17 July 2021).
41. Li, Z.; Shi, W.; Myint, S.W.; Lu, P.; Wang, Q. Semi-Automated Landslide Inventory Mapping from Bitemporal Aerial Photographs Using Change Detection and Level Set Method. *Remote Sens. Environ.* **2016**, *175*, 215–230. [CrossRef]
42. Xiao, J.; Shen, Y.; Tateishi, R.; Bayaer, W. Development of Topsoil Grain Size Index for Monitoring Desertification in Arid Land Using Remote Sensing. *Int. J. Remote Sens.* **2006**, *27*, 2411–2422. [CrossRef]
43. Breiman, L. Random Forests. *Mach. Learn.* **2001**, *45*, 5–32. [CrossRef]
44. Akar, Ö.; Güngör, O. Classification of Multispectral Images Using Random Forest Algorithm. *J. Geod. Geoinf.* **2012**, *1*, 105–112. [CrossRef]
45. Talukdar, S.; Singha, P.; Mahato, S.; Pal, S.; Liou, Y.A.; Rahman, A. Land-Use Land-Cover Classification by Machine Learning Classifiers for Satellite Observations—A Review. *Remote Sens.* **2020**, *12*, 1135. [CrossRef]
46. Abdullah, A.Y.M.; Masrur, A.; Gani Adnan, M.S.; Al Baky, M.A.; Hassan, Q.K.; Dewan, A. Spatio-Temporal Patterns of Land Use/Land Cover Change in the Heterogeneous Coastal Region of Bangladesh between 1990 and 2017. *Remote Sens.* **2019**, *11*, 790. [CrossRef]
47. Giles, P.T. Remote Sensing and Cast Shadows in Mountainous Terrain. *Photogramm. Eng. Remote Sens.* **2001**, *67*, 833–839.
48. Sood, V.; Singh, S. Analytical Analysis of Shadow Removing Algorithms over Land Use and Land Cover Classification. *Himal. Geol.* **2018**, *39*, 223–232.
49. Jiang, Y.; Wang, C.; Zhao, X. Damage Assessment of Tunnels Caused by the 2004 Mid Niigata Prefecture Earthquake Using Hayashi's Quantification Theory Type II. *Nat. Hazards* **2010**, *53*, 425–441. [CrossRef]
50. Foody, G.M. Status of Land Cover Classification Accuracy Assessment. *Remote Sens. Environ.* **2002**, *80*, 185–201. [CrossRef]
51. Yalcin, A. GIS-Based Landslide Susceptibility Mapping Using Analytical Hierarchy Process and Bivariate Statistics in Ardesen (Türkiye): Comparisons of Results and Confirmations. *Catena* **2008**, *72*, 1–12. [CrossRef]
52. Reis, S.; Bayrak, T.; Yalçın, A.; Atasoy, M.; Nişancı, R.; Ekercin, S. Rize Bölgesinde Yağış Heyelan İlişkisi. *Jeodezi JeoInf. Arazi Yönetimi Derg.* **2008**, *2*, 5–9. (In Turkish)
53. Uyeturk, C.E.; Huvaj, N.; Bayraktaroglu, H.; Huseyinpasaoglu, M. Geotechnical Characteristics of Residual Soils in Rainfall-Triggered Landslides in Rize, Türkiye. *Eng. Geol.* **2020**, *264*, 105318. [CrossRef]
54. Muhammet, D. Physically-Based Modeling of Rainfall-Triggered Landslides in Kaptanpasa, Rize. Master's Thesis, Middle East Technical University, Çankaya/Ankara, Türkiye, 2022.
55. Üyetürk, C.E. Geotechnical Characterization of Soils Prone to Rainfall-Induced Landslides in Rize (Northern Türkiye). Master's Thesis, Middle East Technical University, Çankaya/Ankara, Türkiye, 2019.
56. Yalcin, A. The Effects of Clay on Landslides: A Case Study. *Appl. Clay Sci.* **2007**, *38*, 77–85. [CrossRef]
57. Saga Local Meteorological Office, JMA. Heavy Rainfall in Saga Prefecture 11/8/2021–19/8/2021. Available online: https://www.jma-net.go.jp/fukuoka/chosa/saigai/20210906_saga.pdf (accessed on 19 May 2022).
58. Asahiro, K.; Kanekiyo, H.; Tani, M. Research into the distribution of damage to farmland in Yame City, Fukuoka Prefecture, from the heavy rains in the northern part of Kyushu in 2012, and difficulties for the recovery. *J. Jpn. Inst. Landsc. Archit.* **2014**, *77*, 649–654. [CrossRef]
59. Hukuoka Region Headquarters, JMA. North Kyushu Heavy Rainfall Event of July 2012. Available online: https://www.jma-net.go.jp/fukuoka/chosa/kisho_saigai/20120711-14.pdf. (accessed on 19 May 2022).
60. Miyabuchi, Y.; Torii, M. Landslide Disaster Triggered by July 2020 Heavy Rainfall in the Southern Part of Kumamoto Prefecture, Southwestern Japan. *J. Geogr.* **2021**, *130*, 107–116. [CrossRef]
61. Kaibori, M.; Hasegawa, Y.; Yamashita, Y.; Sakida, H.; Nakai, S.; Kuwada, S.; Hiramatsu, S.; Jitousono, T.; Irasawa, M.; Shimizu, O.; et al. Sediment Related Disaster Due to Heavy Rainfall in Hiroshima Prefecture in July, 2018. *J. Jpn. Soc. Eros. Control Eng.* **2018**, *71*, 49–60. [CrossRef]
62. Sugawara, J. Landslides in Tea Plantation Fields in Shizuoka, Japan. *Int. J. GEOMATE* **2013**, *4*, 495–500. [CrossRef]
63. Yajima, Y.; Sasaki, Y.; Kurahashi, T. Analysis of Disaster Features Using Slope Disaster Database of National Roads. *J. Jpn. Soc. Eng. Geol.* **2008**, *48*, 304–311. [CrossRef]

water

MDPI

Article

The Role of Citrus Groves in Rainfall-Triggered Landslide Hazards in Uwajima, Japan

Novia Lusiana [1,2] and Yoshinori Shinohara [3,*]

[1] Interdisciplinary Graduate School of Agriculture and Engineering, University of Miyazaki, Miyazaki 889-2192, Japan; novialusiana@ub.ac.id
[2] Environmental Engineering Study Program, Faculty of Agricultural Technology, Universitas Brawijaya, Malang 65145, Indonesia
[3] Faculty of Agriculture, University of Miyazaki, Miyazaki 889-2192, Japan
* Correspondence: shinohara@cc.miyazaki-u.ac.jp; Tel.: +81-985-58-7939

Abstract: Landslides often cause deaths and severe economic losses. In general, forests play an important role in reducing landslide probability because of the stabilizing effect of the tree roots. Although fruit groves consist of trees, which are similar to forests, practical land management, such as the frequent trampling of fields by laborers and compression of the terrain, may cause such land to become prone to landslides compared with forests. Fruit groves are widely distributed in hilly regions, but few studies have examined their role in landslide initiation. This study aims at filling this gap evaluating the predisposing and triggering conditions for rainfall-triggering landslides in part of Uwajima City, Japan. A large number of landslides occurred due to a heavy rainfall event in July 2018, where citrus groves occupied about 50% of the study area. In this study, we combined geodata with a regression model to assess the landslide hazard of fruit groves in hilly regions. We developed maps for five conditioning factors: slope gradient, slope aspect, normalized difference vegetation index (NDVI), land use, and geology. Based on these five maps and a landslide inventory map, we found that the landslide area density in citrus groves was larger than in forests for the categories of slope gradient, slope aspect, NDVI, and geology. Ten logistic regression models along with different rainfall indices (i.e., 1-h, 3-h, 12-h, 24-h maximum rainfall and total rainfall) and different land use (forests or citrus groves) in addition to the other four conditioning factors were produced. The result revealed that "citrus grove" was a significant factor with a positive coefficient for all models, whereas "forest" was a negative coefficient. These results suggest that citrus groves have a higher probability of landslide initiation than forests in this study area. Similar studies targeting different sites with various types of fruit groves and several rainfall events are crucial to generalize the analysis of landslide hazard in fruit groves.

Keywords: landslide hazard; land use evaluation; multiple logistic regression; rainfall variability

Citation: Lusiana, N.; Shinohara, Y. The Role of Citrus Groves in Rainfall-Triggered Landslide Hazards in Uwajima, Japan. *Water* **2022**, *14*, 2113. https://doi.org/10.3390/w14132113

Academic Editor: Stefano Luigi Gariano

Received: 27 April 2022
Accepted: 29 June 2022
Published: 1 July 2022

1. Introduction

Landslides are frightening and destructive natural disasters. The Global Fatal Landslide Database recorded 55,997 fatalities from 4862 landslide events from 2004 to 2016 [1], and the number of landslides has increased over time [2,3]. For example, the Centre for Research on the Epidemiology of Disaster and United Nations Office for Disaster Risk Reduction [4] reported a 48% increase in landslide events from 1999 to 2019. Although several approaches for landslide mitigation exist, understanding the hazards in areas prone to landslides is a fundamental approach.

Geology, morphology, vegetation, and human activities are factors affecting landslide hazards [5,6], and land cover is a key determining factor. Forests and trees play important roles in reducing landslide risk mainly due to shear strength enhancement by roots on soil particles of slip surfaces [7,8]. A comparison of landslide distribution among land use types has been conducted by numerous studies: a lower landslide area density (LAD) has been found in forests than in bare land [9], grasslands, and shrublands [10,11].

Fruit groves, such as citrus groves, apple orchards, and pear orchards, are composed of trees categorically similar to forests. Therefore, fruit groves are easily founded in hilly regions [12,13]. However, land treatment of fruit groves differs greatly from forests. In fruit groves, laborers frequently trample fields, and compressed terrain is formed due to irrigation efficiency, soil wash prevention, and cultivation. Land treatment of fruit groves, in turn, may impact landslide initiation and distribution. In Iran, for example, a higher landslide density has been reported for fruit groves than in forests and agricultural lands [11,14]. In Thailand, landslides frequently occurred in forest areas converted to fruit groves [15], especially in mountainous areas where the largest area destroyed by landslides occurred in a region with fruit groves [16]. Landslides in fruit groves cause huge economic losses; for instance, in Southern Thailand, the total economic damages are estimated at approximately $300 million a year [17]. Landslide sensitivity determines whether a danger becomes a disaster [18]. Thus, reducing activities that impair sensitivity is an effective approach to prevent damage. Until now, few studies have evaluated the role of fruit groves in landslide initiation. However, such evaluation is essential for policymakers to develop a land use plan in regions where fruit production thrives.

Landslide hazard has previously been evaluated using landslide inventory maps (LIMs) as well as maps for the effects of triggering factors [6]. Rainfall is the most common trigger for landslides [19,20], although earthquakes and volcanic activities also frequently initiate landslide events. Landslide modeling and prediction of rainfall-triggered landslides requires rainfall distribution data [21]. These data can be obtained from the global rainfall database, with a resolution of >1 km [22,23] or developed by spatial interpolation of ground-based observations [24,25]. Therefore, although rainfall is highly heterogeneous, rainfall distribution data with high resolution are not generally available. Recently, rainfall distribution data with 250-m resolution, observed in X-band multiparameter (MP) radar networks, have become available in Japan. Thus, this rainfall data would be efficient for evaluating the landslide triggering conditions in fruit groves in the study area.

This study aimed to evaluate the impact of fruit groves on landslide initiation in part of Uwajima City, Japan. This region has widespread tangerine citrus groves in hilly areas that were affected by many landslides in July 2018 due to heavy rainfall. The objectives of the study were to (1) examine landslide distributions and conditioning factors, including several rainfall indices; (2) compare the landslide density among land use types for each category of slope gradient, slope aspect, NDVI, and geology; and (3) develop the logistic regression models to evaluate the significance of citrus groves in the occurrence of landslides.

2. Materials and Methods

2.1. Site Descriptions

This study was conducted in the northern part of Uwajima City, Ehime prefecture, Japan (Figure 1). The total area is 30.28 km^2, occupying 6.5% of Uwajima City. The annual rainfall from 2001 to 2020 was recorded at the Uwajima rainfall station of the Japan Meteorological Agency, which is 4.4 km south of the study area; the minimum and maximum annual rainfall were between 1049.5 mm in 2002 and 2305.0 mm in 2004 (mean, 1767 mm). The mean annual temperature was 17.1 °C. There were seasonal variations in rainfall; rainfall from June to September was heavier than that from November to February. The elevation of the study area is 5–620 m.a.s.l (mean, 61 m.a.s.l). Areas with elevations $\leq$200 m.a.s.l occupy over 70% of the study area. The study area is dominated by gentle terrains; 79% of the study area has a slope gradient of $\leq$24°. For land use, citrus groves occupied about 50% of the study area. The area mainly comprised sandy turbidite, muddy turbidite, and turbidite soil. More detailed topography, geology, and land use of the study area are shown in the results section.

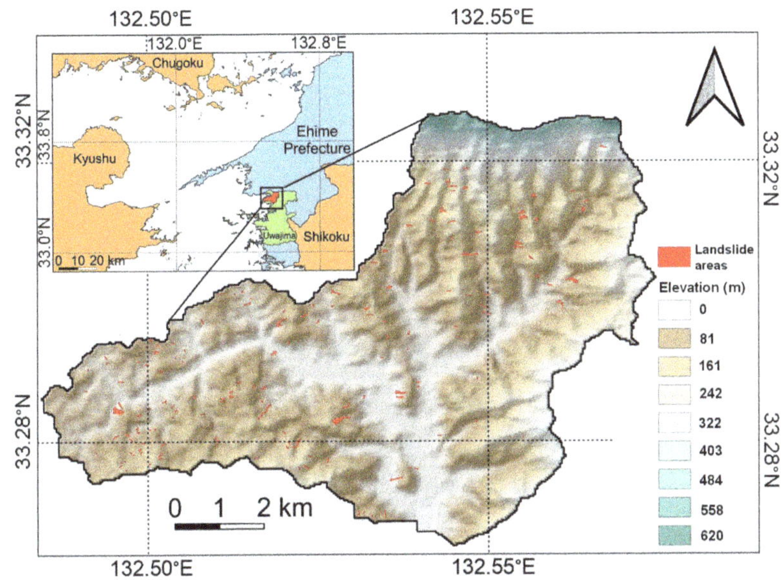

Figure 1. Location of the study area (in Uwajima City, southwest part of Ehime prefecture) and landslide distribution in July 2018. The height of elevation means height above sea level.

2.2. The Rainfall Event in July 2018

Heavy rainfall events occurred from 1 to 8 July 2018. At the Uwajima station, the total rainfall was 578.0 mm; the maximum hourly rainfall observed from 06:00 to 07:00 on 7 July was 49.0 mm (Figure 2). The rainfall event induced numerous landslides in several cities in Ehime prefecture, particularly in Uwajima City. Uwajima City suffered the greatest landslide disaster on 7 July 2018; in total, 51 debris flows and 102 steep slope failures affected houses and public facilities [26].

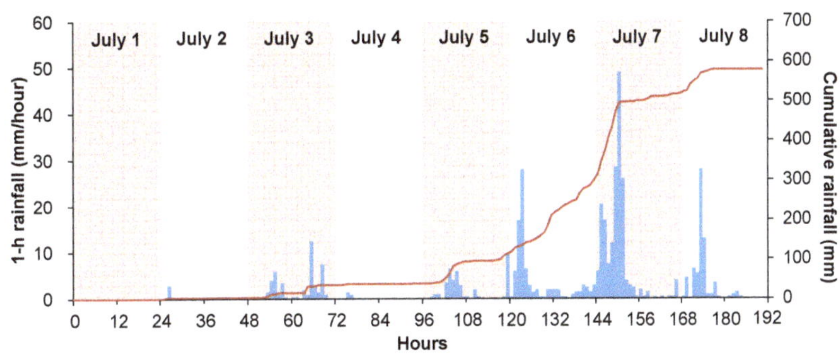

Figure 2. Hourly rainfall (blue bars) and cumulative rainfall (a red line) on 1–8 July 2018, at the Uwajima rainfall station.

2.3. Data Preparation

We used a LIM, distributions of slope gradient, slope aspect, NDVI, land use, and geology as conditioning factors, and rainfall for analysis. Data of the conditioning factors were generally selected based on the abundance, accessibility, region characteristics, and research purposes [27,28]. Because we aimed to evaluate the effects of citrus groves on landslide initiation, we considered land use as a primary conditioning factor. We also included morphological factors such as slope gradient and aspect, as well as NDVI and geology, as conditioning factors. The number of conditioning factors may not be sufficient for a landslide susceptibility study. However, they were appropriate for the hazard assessment in the area according to the study purposes. Table 1 summarizes the datasets used. We used a LIM provided by the Geospatial Information Authority of Japan (GSI) [29], which was produced using aerial photographs taken both before and shortly after the rainfall event. The LIM included 162 landslides and occupied 0.53% of the study area (Figure 1). The slope gradient and aspect were calculated using a digital elevation model (DEM). ASTER Global DEM (GDEM) was used for the calculations [30], and NDVI was calculated from the reflectance of the near-infrared wavelength (NIR) and red wavelength (RED):

$$\mathrm{NDVI} = \frac{\mathrm{NIR} - \mathrm{RED}}{\mathrm{NIR} + \mathrm{RED}} \tag{1}$$

Herein, we used the NIR and RED of Landsat 8, collection 2, level-2, measured on 20 April 2018 [31], which was a clear day just before the landslide events. Landsat series collection 2, level-2, was generated from collection 2, level-1 inputs. The level-1 consisted of quantized and calibrated scaled digital number values representing multispectral image data, whereas level-2 data consisted of pixel values estimating the spectral reflectance measured at the Earth's surface without atmospheric scattering or absorption effects [32]. Land use and geological maps were provided by the Ministry of Land, Infrastructure, Transport, and Tourism in vector format (polygon coverage) [33]; these were converted into raster data with a resolution of 100 m. The land use map included six categories (i.e., "forest," "other agricultural land (land where vegetables and fruits are cultivated)," "field (agriculture)," "building site," "road," and "water"). Within the study area, the category "other agricultural land" mainly included citrus groves. Although the original classification was used when examining the landslide density for each category, these were further classified as "forest," "citrus grove," and "the others", which represented the field, building site, road, and water categories. The purpose of the new classification was to highlight the difference in landslide initiation between "forest" and "citrus grove." The geological map in this area included five geological units: "sandy turbidite," "muddy turbidite," "turbidite," "limestone block," and "chert block."

Table 1. Summary of datasets.

Datasets	Data Type	Scale/ Resolution	Source
Landslide inventory maps (LIM)	Polygon coverage	1:25,000	[29]
Digital Elevation Model (DEM)	Raster	30×30 m	[30]
Normalized Difference Vegetation Index (NDVI)	Raster	30×30 m	[31]
Land use	Polygon coverage	1:200,000	[33]
Geology	Polygon coverage	1:200,000	[33]
Rainfall	Raster	258×258 m	[34]

For rainfall analysis, rainfall distributions based on the X-band MP radar network were used [34]. We downloaded the data that were recorded every 10 min on 1–8 July, and we calculated the 1-h, 3-h, 12-h, and 24-h maximum rainfall and total rainfall during the study period.

2.4. Analysis

First, maps for the one triggering and five conditioning factors were developed. For rainfall, maps for the 1-h, 3-h, 12-h, and 24-h maximum rainfall and the total rainfall during the study period were developed. Second, we calculated the landslide area proportional to the entire study area, and the LAD for each category was calculated. The LAD was calculated as the ratio of the area with landslides to the total area of each category [35,36]. We established eight categories for slope aspect, and six categories with the same data range were set for the slope gradient and NDVI. For the five rainfall indices, four categories with the same data range were set. Third, the LAD was compared in the new land use categories for each category of slope gradient, slope aspect, NDVI, and geology. Finally, multivariate analyses were performed to comprehensively examine the effects of each factor on landslide initiation. For this analysis, we applied logistic regression although random forest has also been frequently used recently. However, according to a previous review [37], logistic regression has been the most commonly used for landslide susceptibility modeling. The logistic regression can be expressed as follows [38]:

$$z = b_0 + b_1x_1 + b_2x_2 + \cdots + b_nx_n \tag{2}$$

where z is a dependent variable, b_0 is the constant, b_i is ith regression coefficient, x_i is the ith independent variables, i is the number of independent variables, and n (= 6 in this study) is the number of independent variables.

Landslides and non-landslide points are required to predict the probability of landslide occurrences in the logistic regression model and evaluating the model accuracy by true- or false- positive and negative rates. Landslide points at pixels with the highest elevation for each landslide were set [39,40]. Non-landslide points were randomly selected in areas without landslides using the "random points inside polygons" function in QGIS. We set the minimum distance to 10 m, and the observed mean distance was 262.3 m as a result. Following the previous literature [38,41], the number of non-landslide and landslide points was the same. Both landslide and non-landslide points were divided into training sets for model development and testing sets for model validation. According to the suggestions in previous studies [38,42–44], the ratios of the training and testing sets were 80% and 20%, respectively.

The slope gradient, slope aspect, NDVI, land use, geology, and rainfall were considered as independent variables. Dummy variables were used for categorical factors of land use and geology [45] (see Appendix A Tables A1 and A2). We developed ten models with different rainfall indices, namely 1-h, 3-h, 12-h, and 24-h maximum rainfall and total rainfall and different land use categories, namely "forest" and "citrus grove" (Appendix A Table A3). To investigate the linearity among independent variables [46], the variance inflation factor (VIF) was used, which is widely used for multicollinearity analysis. VIF indicates the effect of the determination coefficients on the variance of the estimated regression coefficients [45]. A high VIF value indicates that the model is not feasible due to multicollinearity among independent variables [45,46]. We set an acceptable VIF of 2.5, because several studies stated that VIF $\geq$ 2.5 indicates considerable collinearity [47,48], and VIF > 5 or 10 indicates a serious collinearity problem. The two land use types were separately calculated because the VIF values were >2.5 without separation.

The independent variables for each model were reduced stepwise based on the akaike information criterion (AIC). A stepwise regression automatically assigned the best subset of independent variables throughout several stages [49]. Each conditioning factor was evaluated one-by-one at every step using t statistics to determine the variable coefficients. To ensure acceptable model accuracy, the final models were evaluated using the area under the curve (AUC), which was obtained by constructing a receiver operating characteristic (ROC) curve comparing true- and false-positive rates, an F1 score by Equation (3) [50], and the overall accuracy by Equation (4) [51]. The all-accuracy score takes a value between 0% and 100%, and a higher score indicates higher model accuracy [46,51,52]. All geoprocessing was performed with QGIS (ver. 3.10), and logistic regression analysis was performed using R (ver. 4.1.0) programming language.

$$F1\ score = \frac{TP}{TP + \frac{1}{2}(FP + FN)} \tag{3}$$

$$Overall\ Accuracy = \frac{TP + TN}{TP + FP + TN + FN} \tag{4}$$

where TP is the true positive, TN is the true negative, FP is the false positive, and FN is the false negative.

In the evaluations, we set the threshold of 0.5 between the occurrence and nonoccurrence of landslides.

3. Results

3.1. Landslide Area Density on Each Conditioning Factor

Figure 3 shows the maps of the five conditioning factors. The proportional area and LAD for each category in all conditioning factors are shown in Figure 4. The center of the LAD distribution was a slope gradient of 16–24°. The LAD was smaller when the slope gradient was smaller or larger than 16–24° (Figure 4a). There were no considerable differences in the proportional area among the eight slope aspects. The LADs in the southern- and northern-facing slopes tended to be larger than those in the western- and eastern-facing slopes (Figure 4b). The NDVI for the study area was 0.05 and 0.83; over 70% of the area had an NDVI of >0.68 (Figure 4c). For land use, "citrus grove" was the dominant (52.74%), followed by "forest" (35.80%) and "building site" (7.31%) (Figure 4d). The highest LAD was found on "citrus grove" (0.41%), followed by "forest" (0.11%). The LADs of the other four land use types were 0.01% (Figure 4d). Limestones and chert blocks occupied <2% (Figure 4e). The LAD of sandy turbidites was more than twice as large as that of muddy turbidites and turbidite.

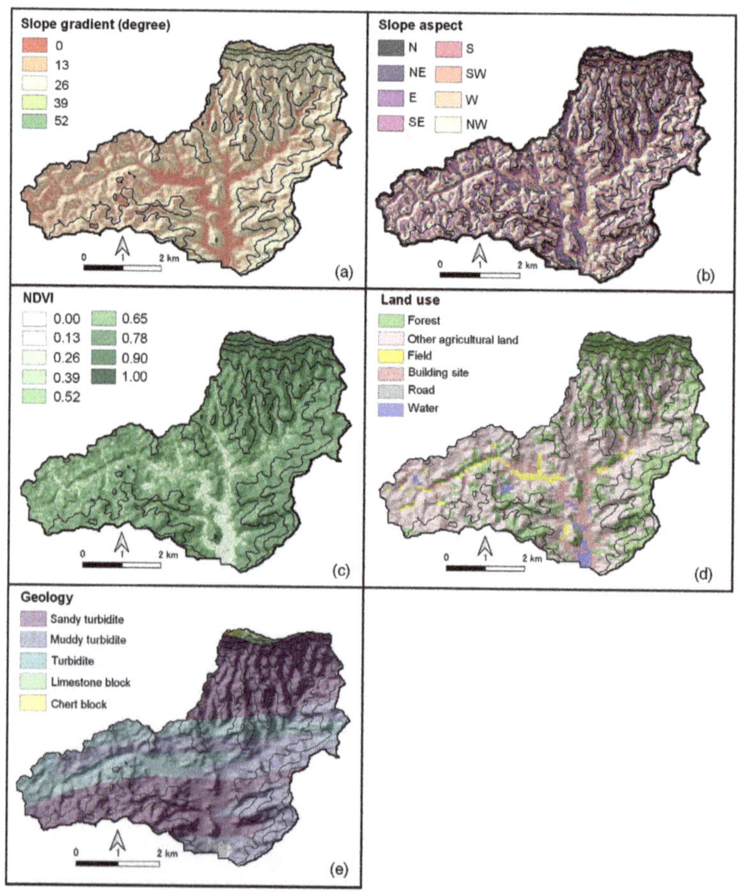

Figure 3. The distributions of the landslide conditioning factors: (**a**) slope gradient, (**b**) slope aspect, (**c**) normalized difference vegetation index, (**d**) land use, and (**e**) geology. The solid lines in the maps indicate contour lines with an interval of 100 m.

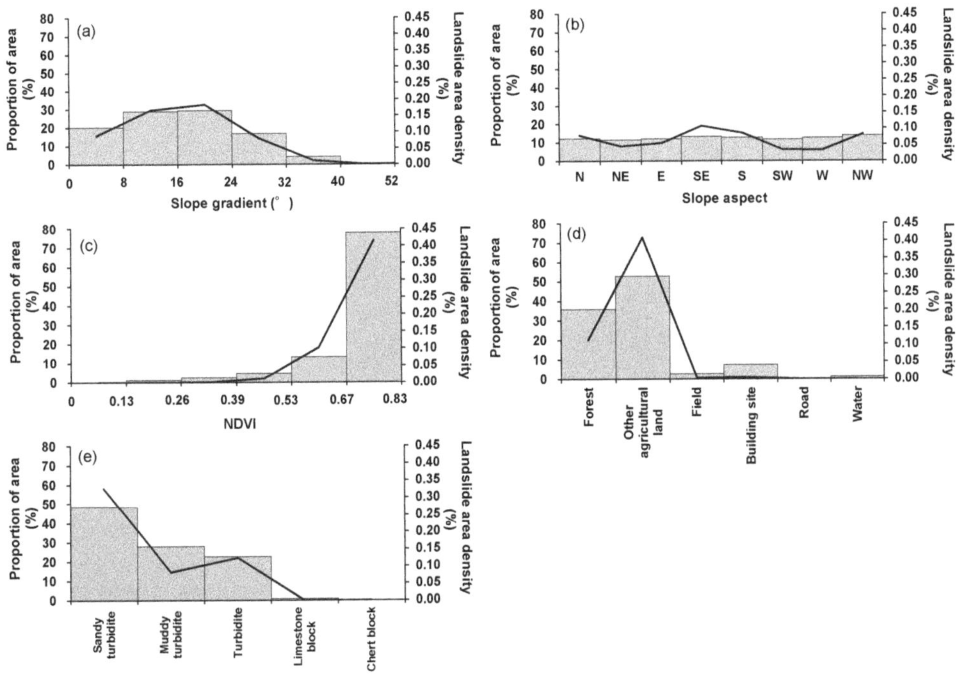

Figure 4. Proportion of area (bars) and landslide area density (lines) for each category of the five conditioning factors: (**a**) slope gradient; (**b**) slope aspect; (**c**) normalized difference vegetation index; (**d**) land use; and (**e**) geology.

3.2. Landslide Area Density Based on Rainfall Indices

Areas with a high 1-h maximum rainfall (35–63 mm) were distributed in the western part of the study area (Figure 5a). In contrast, areas with a high 3-h maximum rainfall were distributed in the northern part of the study area (Figure 5b). The distributions of the 12-h (Figure 5c) and 24-h (Figure 5d) maximum rainfall were similar to that of the total rainfall (Figure 5e); these were higher in the northeast to southwest areas and lower in the southeastern and southwestern areas. The LAD increased with an increase in the 1-h maximum rainfall (Figure 5f). The pattern of the 3-h maximum rainfall was different from that of the 1-h maximum rainfall. The middle range had the highest LAD (Figure 5g). The LADs for the 12-h (Figure 5h) and 24-h (Figure 5i) maximum rainfall and the total rainfall (Figure 5j) were comparable due to the similarity in rainfall distributions. The LAD steeply increased from the first to the second categories, and it gradually increased from the second to the fourth categories.

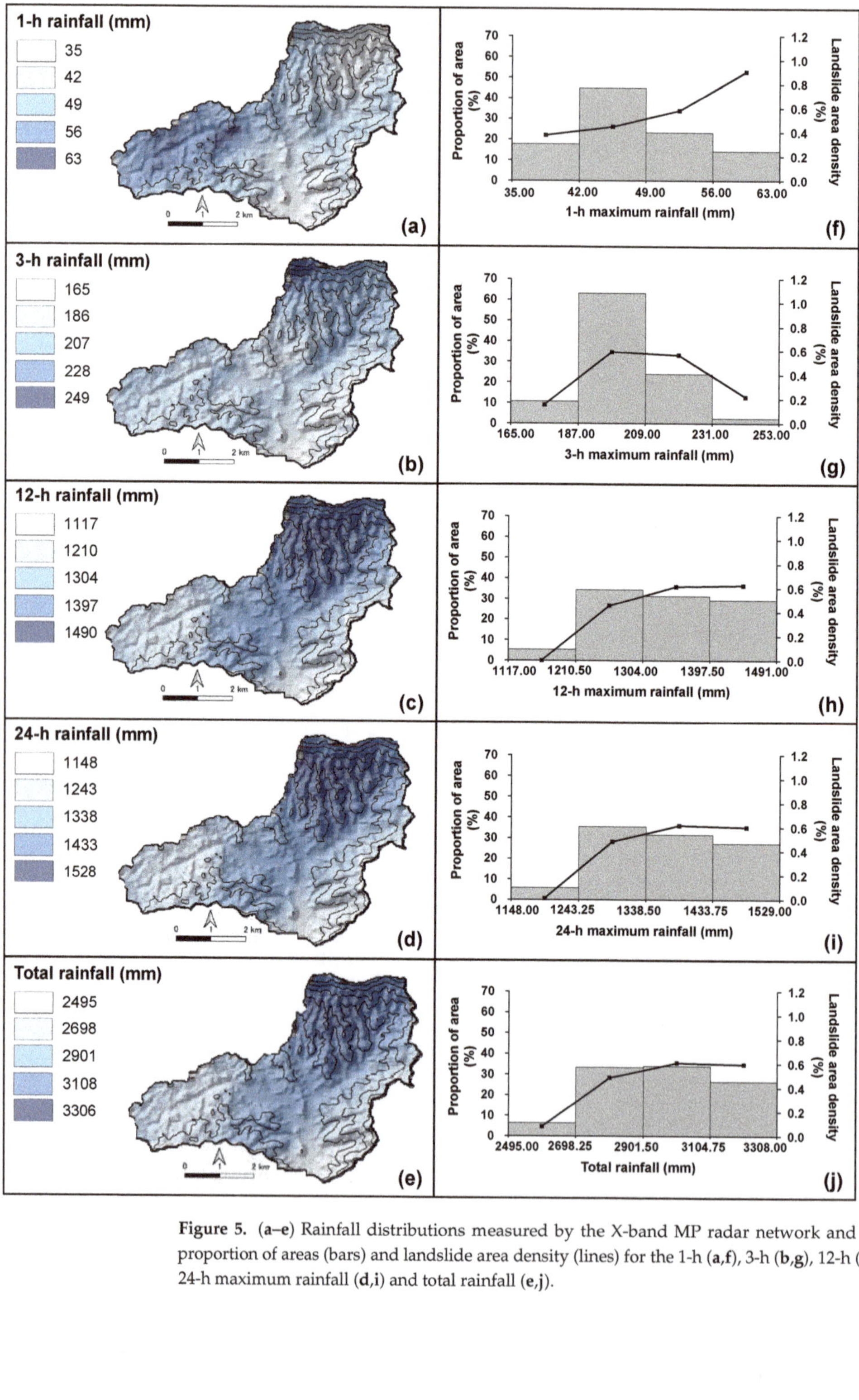

Figure 5. (**a–e**) Rainfall distributions measured by the X-band MP radar network and (**f–j**) the proportion of areas (bars) and landslide area density (lines) for the 1-h (**a**,**f**), 3-h (**b**,**g**), 12-h (**c**,**h**), and 24-h maximum rainfall (**d**,**i**) and total rainfall (**e**,**j**).

3.3. Land Use Effect to Landslide Density Distribution

Figure 6 compared the LADs of "forest," "citrus grove," and "the others" for each category of slope gradient, slope aspect, NDVI, and geology. For the slope gradient (Figure 6a), the LAD for citrus groves was larger than that for forests on gentle slopes (<24°); in contrast, it was an inconsiderable difference on steep slopes (>32°). For the slope aspect (Figure 6b), NDVI (Figure 6c), and geology (Figure 6d), the LAD for "citrus grove" was larger than that for "forest" in all categories.

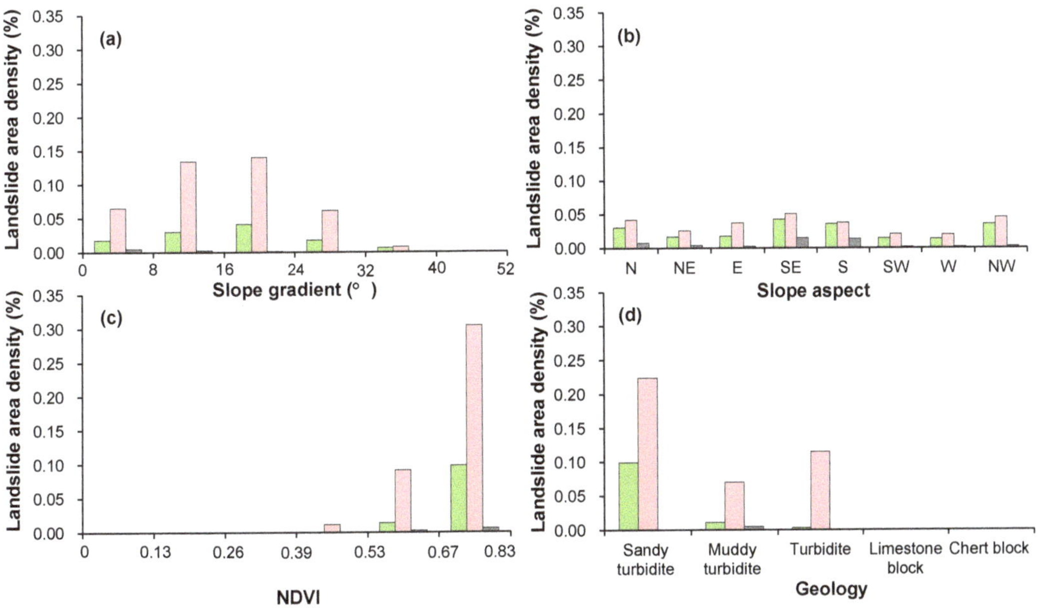

Figure 6. Landslide area density of "forest" (green), "citrus grove" (pink), and "the others (gray)" for each category of conditioning factors: (**a**) slope gradient; (**b**) slope aspect; (**c**) normalized difference vegetation index, and (**d**) geology.

3.4. Significant Effects of Landslide Conditioning Factors

All the independent variables for the ten models had a VIF of <2 (Appendix A Tables A4 and A5), indicating that all the variables could be simulated together. Table 2 summarized the selected factors; after the stepwise processes of logistic regression analysis, NDVI, "forest," and "citrus grove" were selected as significant factors. Because rainfall was not selected for models 2–5 and 7–10, the same result was obtained for the eight models. For models 1–10, NDVI and slope gradient were selected with negative coefficients, and "sandy turbidites" (Dummy 3) was selected with positive coefficients. In models 1–5, "forest" (Dummy 1) was selected with negative coefficient, and "citrus grove" (Dummy 2) was selected with positive coefficients in models 6–10. Additionally, rainfall was selected with a positive coefficient for models 1 and 6. The factors with a positive coefficient indicated an increasing value (or the presence) of factors that increase the landslide hazard. The AIC value for model 1 and 6 (321.1 and 311.7, respectively) were lower than those for models 2–5 and 7–10 (323.9 and 314.2, respectively). Figure 7 shows the ROC of all models; the AUC for models 1 and 6 were larger than that for models 2–5 and 7–10 as well as the F1 scores and the overall accuracy for models 1–5. However, the same accuracy values for models 6–10 were found in the F1 scores and the overall accuracy (Table 3).

Table 2. The coefficients of selected independent variables by logistic regression analysis for models 1–10.

Variable	Coefficient		Variable	Coefficient	
	Model 1	Models 2–5		Model 6	Models 7–10
Slope gradient	–	–0.01	Slope gradient	–	–0.005
Slope aspect	–	–	Slope aspect	–	–
NDVI	–3.27 ***	–3.48 ***	NDVI	–3.10 ***	–3.29 ***
Dummy 1	–0.14 *	–0.15 *	Dummy 2	0.22 ***	0.23 ***
Dummy 3	0.13	0.12	Dummy 3	0.13	0.13
Dummy 4	–	–	Dummy 4	–	–
Rainfall	0.01	–	Rainfall	0.01	–

–: eliminated variables by stepwise regression process; *** $p < 0.001$, ** $p < 0.01$, * $p < 0.05$.

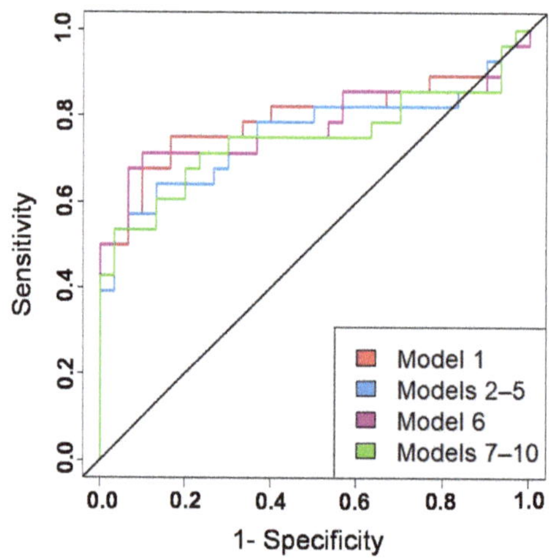

Figure 7. Receiver operating characteristic (ROC) curves with the value of the area under the curve (AUC) for all models developed by logistic regression analysis.

Table 3. The values of the AUC, F1 scores, and overall accuracy for models 1–10 developed by logistic regression analysis.

Method	Model 1	Models 2–5	Model 6	Models 7–10
AUC	79.40	75.95	77.86	74.64
F1 score	77.78	67.92	72.73	72.73
Overall accuracy	79.31	70.69	74.14	74.14

4. Discussion and Conclusions

The land use category "citrus grove" occupied >50% of the study area (Figure 4). The LAD for "citrus grove" was the highest among the six land use categories (Figure 4). When LAD was compared with "forest", "citrus grove," and "the others" for each category of conditioning factors, the LAD for "citrus grove" was larger than that for "forest" and "the others" in most cases (Figure 6). Specifically, the LAD of citrus groves on gentle terrain was higher than that of forests on steeper terrains (32–40° of gradient) (Figure 6). This result may be attributed to the study location, which is predominantly covered by gentle terrain (78.46%), and "citrus grove" was more distributed on gentle terrain than steep terrain with 42.82%, and 9.92% of the area, respectively. Furthermore, "citrus grove" was selected as a significant factor with positive coefficients in the logistic regression analysis for all models.

Thus, citrus groves contributes more to increasing the landslide hazard than forests in the study area.

Some studies [11,14] also reported a higher landslide density of fruit groves to landslides compared with forests, although there are few extensive studies examining the differences in landslide hazard between forest lands and fruit groves. However, there are several possible reasons for fruit groves contributing more to the landslide hazard. First, fruit groves are often developed by cutting the upper side of the slope and leveling the lower side of the slope to build pits for transplanting the seedlings. Generally, manmade slopes are more susceptible to landslides than natural slopes [53]. Manmade slopes and leveled natural slopes tend to change the physical properties of the soil. The main cause of hazard in manmade slopes is excavation [54,55]; geomechanical studies of soft rock slopes state that large deformation of the slope may occur on layered excavations without retaining the structure, due to the unloading of stress, which can lead to slope instability [55,56]. Second, irrigation in fruit grove areas can maintain a high soil water content. Third, fruit groves are composed of a single species. Areas with low biodiversity may have a weak root network system in terms of soil-binding capacity [57]. In addition, soil compaction by trampling may also alter soil physical properties to influence landslide initiation.

This study used radar-based rainfall data with high spatial and temporal resolutions. The result showed 1-h maximum rainfall significantly affected the landslide occurrence (Table 2). The AIC and AUC of model 1 and 6, including 1-h maximum rainfall, were lower and larger than models 2–5 and models 7–10, respectively, where rainfall indices were excluded. In a previous study [58] that compared rainfall data from XRAIN radar with those based on rain gauges, the reported differences were 2% for normalized bias, 22% for normalized error, 0.96 for correlation coefficient, and 3.5 mm for root mean square error. High-quality radar-based rainfall measurement can improve the evaluations of landslide hazard.

The results of this study suggest that fruit groves should not be developed on slopes near houses or public facilities, where people can be harmed and assets destroyed. This may apply to other areas with similar topographic and geological conditions. The results of this study (i.e., higher contribution in landslide initiation and distribution in fruit groves than forests) are identical to previous studies [11,14]. However, previous studies have only examined landslide initiation among land use types without considering variations in rainfall distributions; this study examined landslide hazard in several steps while considering several rainfall distributions, which expands upon previous work. This is an improvement over the previous studies. However, this study had some limitations. First, this study only included one rainfall event, several events would have been desirable for generalization. Second, historical land use changes and landslides were not considered, which may affect the current landslide hazard [59,60]. Third, the mechanism that leads to higher landslide initiation in fruit groves compared with forests could not be clarified. Examining and comparing the root system as well as hydrological aspects of fruit groves with forests will further increase the understanding of these mechanisms. Finally, other data inputs (i.e., natural vs. manmade slopes, soil characteristics, and root types) for the logistic regression model will help to better understand the landslide hazard of fruit groves. Our final goal is to be able to generalize landslide hazard in fruit groves, and the study presented in this paper is a step in this direction.

Author Contributions: Conceptualization, Y.S.; methodology, N.L.; formal analysis, N.L.; data curation, N.L.; writing—original draft preparation, N.L.; writing—review and editing, Y.S.; visualization, N.L.; funding acquisition, Y.S. and N.L. All authors have read and agreed to the published version of the manuscript.

Funding: This work was supported by the Japan Society for the Promotion of Science (JSPS) KAKENHI Grant Number JP21K04590, and the APC was funded by the Japan Science and Technology Agency (JST) through "Support for Pioneering Research Initiated by the Next Generation".

Institutional Review Board Statement: Not applicable.

Informed Consent Statement: Not applicable.

Data Availability Statement: Not applicable.

Conflicts of Interest: The authors declare no conflict of interest.

Appendix A

Table A1. Rule of dummy variables on land use.

	Forest	Citrus Grove	Others
Dummy 1	1	0	0
Dummy 2	0	1	0

Table A2. Rule of dummy variables on geology.

	Sandy Turbidite	Muddy Turbidite	Others
Dummy 3	1	0	0
Dummy 4	0	1	0

Table A3. Independent variables of models 1–10.

Model	Factor										
	Slope Gradient	Slope Aspect	NDVI	Forest	Citrus Grove	Geology	1h-Max Rainfall	3h-Max Rainfall	12h-Max Rainfall	24h-Max Rainfall	Total Rainfall
1	√	√	√	√		√	√				
2	√	√	√	√		√		√			
3	√	√	√	√		√			√		
4	√	√	√	√		√				√	
5	√	√	√	√		√					√
6	√	√	√		√	√	√				
7	√	√	√		√	√		√			
8	√	√	√		√	√			√		
9	√	√	√		√	√				√	
10	√	√	√		√	√					√

√: including model.

Table A4. Variance inflation factors (VIF) for the five models (i.e., models 1–5).

Landslide Conditioning Factors	Model 1	Model 2	Model 3	Model 4	Model 5
Slope gradient	1.27	1.08	1.11	1.11	1.10
Slope aspect	1.01	1.01	1.01	1.01	1.01
NDVI	1.23	1.14	1.18	1.18	1.18
Dummy 1	1.10	1.13	1.15	1.15	1.15
Dummy 3	1.83	1.83	1.83	1.83	1.83
Dummy 4	1.77	1.78	1.77	1.77	1.77
Rainfall	1.37	1.05	1.18	1.18	1.16

Table A5. Variance inflation factors (VIF) for the five models (i.e., models 6–10).

Landslide Conditioning Factors	Model 6	Model 7	Model 8	Model 9	Model 10
Slope gradient	1.27	1.09	1.11	1.11	1.10
Slope aspect	1.01	1.01	1.01	1.01	1.01
NDVI	1.23	1.15	1.18	1.18	1.18
Dummy 2	1.11	1.16	1.20	1.20	1.77
Dummy 3	1.81	1.80	1.80	1.80	1.80
Dummy 4	1.77	1.78	1.77	1.77	1.77
Rainfall	1.38	1.08	1.22	1.22	1.20

References

1. Froude, M.J.; Petley, D.N. Global fatal landslide occurrence from 2004 to 2016. *Nat. Hazards Earth Syst. Sci.* **2018**, *18*, 2161–2181. [CrossRef]
2. Petley, D.N.; Hearn, G.J.; Hart, A.; Rosser, N.J.; Dunning, S.A.; Oven, K.; Mitchell, W.A. Trends in landslide occurrence in Nepal. *Nat. Hazards* **2007**, *43*, 23–44. [CrossRef]
3. Carrión-Mero, P.; Montalván-Burbano, N.; Morante-Carballo, F.; Quesada-Román, A.; Apolo-Masache, B. Worldwide research trends in landslide science. *Int. J. Environ. Res. Public Health* **2021**, *18*, 9445. [CrossRef] [PubMed]
4. UNDRR/CRED. Human Cost of Disasters. In *An Overview of the Last 20 Years 2000–2019*; UN Office for Disaster Risk Reduction: Geneva, Switzerland, 2020; pp. 6–7.
5. Dahal, B.K.; Dahal, R.K. Landslide hazard map: Tool for optimization of low-cost mitigation. *Geoenvironmental Disasters* **2017**, *4*, 8. [CrossRef]
6. Van Westen, C.J.; Castellanos, E.; Kuriakose, S.L. Spatial data for landslide susceptibility, hazard, and vulnerability assessment: An Overview. *Eng. Geol.* **2008**, *102*, 112–131. [CrossRef]
7. Wu, W.; Sidle, R.C. A Distributed slope stability model for steep forested basins. *Water Resour. Res.* **1995**, *31*, 2097–2110. [CrossRef]
8. Schwarz, M.; Lehmann, P.; Or, D. Quantifying lateral root reinforcement in steep slopes—From a bundle of roots to tree stands. *Earth Surf. Process. Landforms* **2010**, *35*, 354–367. [CrossRef]
9. Schmaltz, E.M.; Steger, S.; Glade, T. The influence of forest cover on landslide occurrence explored with spatio-temporal information. *Geomorphology* **2017**, *290*, 250–264. [CrossRef]
10. Shu, H.; Hürlimann, M.; Molowny-Horas, R.; González, M.; Pinyol, J.; Abancó, C.; Ma, J. Relation between land cover and landslide susceptibility in Val d'Aran, Pyrenees (Spain): Historical aspects, present situation and forward prediction. *Sci. Total Environ.* **2019**, *693*, 133557. [CrossRef]
11. Joybari, J.; Kavian, A.A.; Mosaffaie, J. Effect of land use on landslide movement in the Tavan District, Qazvin. *Watershed Manag. Res. J.* **2017**, *30*, 29–39.
12. Jia, Q.; Wang, Y.P. Relationships between leaf area index and evapotranspiration and crop coefficient of hilly apple orchard in the Loess Plateau. *Water* **2021**, *13*, 1957. [CrossRef]
13. Shi, D.; Tan, H.; Rao, W.; Liu, Z.; Elenga, H.I. Variations in water content of soil in apricot orchards in the western hilly regions of the Chinese Loess Plateau. *Vadose Zone J.* **2020**, *19*, e20034. [CrossRef]
14. Mousavi, S.Z.; Kavian, A.; Soleimani, K.; Mousavi, S.R.; Shirzadi, A. GIS-based spatial prediction of landslide susceptibility using logistic regression model. *Geomatics Nat. Hazards Risk* **2011**, *2*, 33–50. [CrossRef]
15. Sangrawee, P.; Rangsipanich, A.; Suvachananonda, T. Analyzing the effects of land use changes for landslide susceptibility assessment: A case study of LabLae District, Uttradit Province, Thailand. In Proceedings of the 38th Asian Conference on Remote Sensing in Space Applications Touching Human Lives, ACRS, New Delhi, India, 3–7 October 2017.
16. Boonyanuphap, J. Cost-benefit analysis of vetiver system-based rehabilitation measures for landslide-damaged mountainous agricultural lands in the Lower Northern Thailand. *Nat. Hazards* **2013**, *69*, 599–629. [CrossRef]
17. Miyagi, T.; Prasad, G.B.; Tanavud, C.; Potichan, A.; Hamasaki, E. Landslide risk evaluation and mapping—Manual of aerial photo interpretation for landslide topography and risk management. *Rep. Natl. Res. Inst. Earth Sci. Disaster Prev.* **2004**, *66*, 75–137.
18. World Meteorological Organization. Integrated Flood Management Tools Series-Management of Sediment-Related Risks; Issue 12; 2011. Available online: https://library.wmo.int/doc_num.php?explnum_id=4274) (accessed on 10 April 2022).
19. Iverson, R.M. Landslide triggering by rain infiltration. *Water Resour. Res.* **2000**, *36*, 1897–1910. [CrossRef]
20. Saito, H.; Uchiyama, S.; Hayakawa, Y.S.; Obanawa, H. Landslides triggered by an earthquake and heavy rainfalls at Aso Volcano, Japan, Detected by UAS and SfM-MVS photogrammetry. *Prog. Earth Planet. Sci.* **2018**, *5*, 1–10. [CrossRef]
21. Chang, K.T.; Chiang, S.H. An integrated model for predicting rainfall-induced landslides. *Geomorphology* **2009**, *105*, 366–373. [CrossRef]
22. Chang, K.; Chiang, S.-H.; Lei, F. Analysing the relationship between typhoon-triggered landslides and critical rainfall conditions. *Earth Surf. Process. Landforms* **2008**, *33*, 1261–1271. [CrossRef]
23. Fick, S.E.; Hijmans, R.J. WorldClim 2: New 1-km spatial resolution climate surfaces for global land areas. *Int. J. Climatol.* **2017**, *37*, 4302–4315. [CrossRef]
24. Sun, Q.; Miao, C.; Duan, Q.; Ashouri, H.; Sorooshian, S.; Hsu, K.L. A review of global precipitation data sets: Data sources, estimation, and intercomparisons. *Rev. Geophys.* **2018**, *56*, 79–107. [CrossRef]
25. Karger, D.N.; Wilson, A.M.; Mahony, C.; Zimmermann, N.E.; Jetz, W. Global daily 1 km land surface precipitation based on cloud cover-informed downscaling. *Sci. Data* **2021**, *8*, 307. [CrossRef] [PubMed]
26. Mori, S.; Ono, K. Landslide disasters in Ehime prefecture resulting from the july 2018 heavy rain event in Japan. *Soils Found.* **2019**, *59*, 2396–2409. [CrossRef]
27. Lee, C.T.; Huang, C.C.; Lee, J.F.; Pan, K.L.; Lin, M.L.; Dong, J.J. Statistical approach to storm event-induced landslides susceptibility. *Nat. Hazards Earth Syst. Sci.* **2008**, *8*, 941–960. [CrossRef]
28. Dou, J.; Yamagishi, H.; Pourghasemi, H.R.; Yunus, A.P.; Song, X.; Xu, Y.; Zhu, Z. An Integrated artificial neural network model for the landslide susceptibility assessment of Osado Island, Japan. *Nat. Hazards* **2015**, *78*, 1749–1776. [CrossRef]
29. Geospatial Information Authority of Japan (GSI). Map/Aerial Photo Viewing Display. 2022. Available online: https://mapps.gsi.go.jp/maplibSearch.do;jsessionid=DAD5878D8589927635F982E6AAFE3A9F#1 (accessed on 10 April 2022).

30. Tachikawa, I.T.; Hato, M.; Kaku, M.; Iwasaki, A. The Characteristics of ASTER GDEM Version 2. IGRASS 2011. Available online: https://www.jspacesystems.or.jp/ersdac/GDEM/J/ (accessed on 10 April 2022).
31. United State Geological Survey (USGS). Earth Explorer. 2022. Available online: https://earthexplorer.usgs.gov/ (accessed on 10 April 2022).
32. Pinto, C.T.; Jing, X.; Leigh, L. Evaluation analysis of landsat level-1 and level-2 data products using in situ measurements. *Remote Sens.* **2020**, *12*, 2597. [CrossRef]
33. Ministry of Land, Infrastructure, Transport and Tourism (MLIT) of Japan. National Land Numerical Information Download Service. 2022. Available online: https://nlftp.mlit.go.jp/ksj/ (accessed on 10 April 2022).
34. Ministry of Land, Infrastructure, Transport and Tourism (MLIT) of Japan. X/C Band MP Radar (XRAIN) Real-Time Rainfall Information by Data Integration and Analysis System Program (DIAS). 2022. Available online: http://apps.diasjp.net/xband/ (accessed on 10 April 2022).
35. Dai, F.C.; Xu, C.; Yao, X.; Xu, L.; Tu, X.B.; Gong, Q.M. Spatial distribution of landslides triggered by the 2008 ms 8.0 Wenchuan earthquake, China. *J. Asian Earth Sci.* **2011**, *40*, 883–895. [CrossRef]
36. Keefer, D.K. Statiscal analysis of an earthquake-induced landslide distribution—The 1989 Loma Prieta, California event. *Eng. Geol.* **2000**, *58*, 231–249. [CrossRef]
37. Reichenbach, P.; Rossi, M.; Malamud, B.D.; Mihir, M.; Guzzetti, F. A review of statistically-based landslide susceptibility models. *Earth-Sci. Rev.* **2018**, *180*, 60–91. [CrossRef]
38. Bai, S.B.; Wang, J.; Lü, G.N.; Zhou, P.G.; Hou, S.S.; Xu, S.N. GIS-based logistic regression for landslide susceptibility mapping of the Zhongxian segment in the Three Gorges Area, China. *Geomorphology* **2010**, *115*, 23–31. [CrossRef]
39. Danjo, T.; Ishizawa, T.; Kimura, T. Spatial analysis of the landslide characteristics caused by heavy rainfall in the Northern Kyushu District in July 2017 using topography, geology, and rainfall levels. *J. Disaster Res.* **2018**, *13*, 832–845. [CrossRef]
40. Di Napoli, M.; Di Martire, D.; Bausilio, G.; Calcaterra, D.; Confuorto, P.; Firpo, M.; Pepe, G.; Cevasco, A. Rainfall-induced shallow landslide detachment, transit and runout susceptibility mapping by integrating machine learning techniques and gis-based approaches. *Water* **2021**, *13*, 488. [CrossRef]
41. Tien Bui, D.; Pradhan, B.; Lofman, O.; Revhaug, I. Landslide susceptibility assessment in Vietnam using support vector machines, decision tree, and nave bayes models. *Math. Probl. Eng.* **2012**, *2012*, 974638. [CrossRef]
42. Nefeslioglu, H.A.; Gokceoglu, C.; Sonmez, H. An assessment on the use of logistic regression and artificial neural networks with different sampling strategies for the preparation of landslide susceptibility maps. *Eng. Geol.* **2008**, *97*, 171–191. [CrossRef]
43. Pradhan, B.; Lee, S. Delineation of landslide hazard areas on Penang Island, Malaysia, by using frequency ratio, logistic regression, and artificial neural network models. *Environ. Earth Sci.* **2010**, *60*, 1037–1054. [CrossRef]
44. Gautam, P.; Kubota, T.; Aditian, A. Evaluating underlying causative factors for earthquake-induced landslides and landslide susceptibility mapping in upper Indrawati watershed, Nepal. *Geoenviron. Disasters* **2021**, *8*, 30. [CrossRef]
45. Gujarati, D.N.; Porter, D.C. *Multicollinearity: What Happens If the Regressor Are Correlated?* Basic Econometrics, 5th ed.; McGraw-Hill Irwin: New York, NY, USA, 2013; pp. 328–330.
46. Corsini, A.; Mulas, M. Use of ROC curves for early warning of landslide displacement rates in response to precipitation (Piagneto Landslide, Northern Apennines, Italy). *Landslides* **2017**, *14*, 1241–1252. [CrossRef]
47. Midi, H.; Sarkar, S.K.; Rana, S. Collinearity diagnostics of binary logistic regression model. *J. Interdiscip. Math.* **2010**, *13*, 253–267. [CrossRef]
48. Johnston, R.; Jones, K.; Manley, D. Confounding and collinearity in regression analysis: A cautionary tale and an alternative procedure, illustrated by studies of british voting behaviour. *Qual. Quant.* **2018**, *52*, 1957–1976. [CrossRef]
49. Hengl, T.; Heuvelink, G.B.M.; Stein, A. A generic framework for spatial prediction of soil variables based on regression-kriging. *Geoderma* **2004**, *120*, 75–93. [CrossRef]
50. Hand, D.J.; Christen, P.; Kirielle, N. F*: An interpretable transformation of the F-measure. *Mach. Learn.* **2021**, *110*, 451–456. [CrossRef] [PubMed]
51. Chen, W.; Xie, X.; Wang, J.; Pradhan, B.; Hong, H.; Bui, D.T.; Duan, Z.; Ma, J. A comparative study of logistic model tree, random forest, and classification and regression tree models for spatial prediction of landslide susceptibility. *Catena* **2017**, *151*, 147–160. [CrossRef]
52. Xing, Y.; Yue, Y.; Guo, Z.; Chen, Y.; Hu, J.; Travé, A. Large-Scale landslide susceptibility mapping using an integrated machine learning model: A case study in the Lvliang Mountains of China. *Front. Earth Sci.* **2021**, *9*, 622. [CrossRef]
53. Rupke, J.; Huisman, M.; Kruse, H.M.G. Stability of man-made slopes. *Eng. Geol.* **2007**, *91*, 16–24. [CrossRef]
54. Zhang, F.; Liu, G.; Chen, W.; Liang, S.; Chen, R.; Han, W. Human-induced landslide on a high cut slope: A case of repeated failures due to multi-excavation. *J. Rock Mech. Geotech. Eng.* **2012**, *4*, 367–374. [CrossRef]
55. Li, Q.; Wang, Y.M.; Zhang, K.B. Case study of cutting slope with progressive failure due to continuous excavation. In Proceedings of the 8th Japan-China Geotechnical Symposium Challenges to Breakthrough Geotechnics, Kyoto, Japan, 28–29 September 2020; pp. 436–440.
56. Ayalew, L.; Moeller, D.; Reik, G. Geotechnical aspects and stability of road cuts in the Blue Nile Basin, Ethiopia. *Geotech. Geol. Eng.* **2009**, *27*, 713–728. [CrossRef]

57. Ziegler, A.D.; Bruun, T.B.; Guardiola-Claramonte, M.; Giambelluca, T.W.; Lawrence, D.; Thanh Lam, N. Environmental consequences of the Demise in Swidden cultivation in Montane Mainland Southeast Asia: Hydrology and geomorphology. *Hum. Ecol.* **2009**, *37*, 361–373. [CrossRef]
58. Shakti, P.C.; Misumi, R.; Nakatani, T.; Iwanami, K.; Maki, M.; Maesaka, T.; Hirano, K. Accuracy of quantitative precipitation estimation using operational weather radars: A case study of heavy rainfall on 9–10 September 2015 in the east Kanto region, Japan. *J. Disaster Res.* **2016**, *11*, 1003–1016. [CrossRef]
59. Bernardie, S.; Vandromme, R.; Thiery, Y.; Houet, T.; Grémont, M.; Masson, F.; Grandjean, G.; Bouroullec, I. Modelling landslide hazards under global changes: The case of a Pyrenean Valley. *Nat. Hazards Earth Syst. Sci.* **2021**, *21*, 147–169. [CrossRef]
60. Tyagi, A.; Tiwari, R.K.; James, N. *GIS-Based Landslide Hazard Zonation and Risk Studies using MCDM BT—Local Site Effects and Ground Failures*; Springer: Singapore, 2021; pp. 251–266.

water

MDPI

Article

Investigation of Rainfall-Induced Failure Processes and Characteristics of Wedge Slopes Using Physical Models

Chia-Ming Lo [1], Yu-Sen Lai [1,*] and Chen-Han Chu [2]

[1] Department of Civil Engineering, National Yang Ming Chiao Tung University, Hsinchu 30010, Taiwan
[2] Department of Civil and Disaster Prevention Engineering, National United University, Miaoli 360302, Taiwan
* Correspondence: nycu409612010.c@nycu.edu.tw

Abstract: In this study, we conducted small-scale physical modeling tests to consider the impact of the infiltration of rainfall in order to investigate the processes involved in wedge slope deformation and failure. The experiments were conducted under controlled conditions of the intersection angle and half-wedge angle. The observations obtained during each stage of deformation and failure were used to explain how gravity deformation varies on wedge slopes and infer how rainfall influences slope failure. The results indicated that half-wedge angle is a crucial factor in the deformation failure of slopes. The failure mechanisms of small-intersection angle slopes (sliding model) differ considerably from those of large-intersection angle slopes (free falling or toppling model). The infiltration of surface water can have a significant influence on rock layer deformation and the speed of failure. Details of the failure characteristics of wedge slope models were discussed in this paper.

Keywords: physical modeling; rainfall; wedge slope; the intersection angle; half-wedge angle

1. Introduction

Extreme weather events in recent years have greatly increased the amount of rainfall in the mountains of Taiwan. Substantial rainfall affects rock slope safety by infiltrating rock slopes via weak planes, thereby weakening the shear strength of the slope body, raising the water table, and increasing the weight and downslide stress of the soil. This, in turn, causes parts of the rock slopes to deform and collapse. The fragile geological environment and heavy rainfall conditions in Taiwan make for frequent landslide disasters, mostly around mountain roads, villages, and valley banks. Large-scale rockslide disasters in the past include the complex landslide in Tuchang Village, Hsinchu in 2004, the collapse event at 3.1 km on National Freeway No. 3 in 2010, and the complex landslide in Zhongzhi Village, Wulai in 2016, all of which greatly threatened protected targets downstream. Rock slope failure is generally determined by the discontinuities between strata, such as bedding, faults, foliation, and joints [1]. Prior to rock slope failure and creep deformation, tension cracks can generally be observed developing at the top of the slope, affecting it or causing the rocks on the slope toe to overturn or bulge. The damage gradually extends upwards toward important public facilities or villages, and if the scope of the creep damage is not determined and protective measures are not taken in a timely manner, major disasters may occur. In addition to enhancing emergency responses to landslides, disaster warnings for evacuation are also crucial to minimize losses. Model tests, numerical simulations, and real-time deformation monitoring are thus employed to grasp rock slope deformation, the timing of failure, and the hazard ranges of rock failure and deposits. These are all important issues in rock slope disaster prevention [2–8].

Wedge failures are a common type of failure in rockslides, and mostly occur where the strata have an oblique attitude [9]. Most wedge failures take place in rock masses via the following methods [10]: (a) sliding along the line of intersection of the two planes, (b) sliding along only a single plane accompanied by rock mass fracturing, (c) rotation and

Citation: Lo, C.-M.; Lai, Y.-S.; Chu, C.-H. Investigation of Rainfall-Induced Failure Processes and Characteristics of Wedge Slopes Using Physical Models. *Water* **2023**, *15*, 1108. https://doi.org/10.3390/w15061108

Academic Editors: Renato Morbidelli and Su-Chin Chen

Received: 8 January 2023
Revised: 25 February 2023
Accepted: 9 March 2023
Published: 14 March 2023

sliding along only a single plane accompanied by rock mass fracturing, and (d) if the rock mass has a high joint density, gradual disintegration of the rock mass along two planes. A number of studies have examined the issue of rock wedge stability [1,2,4,5,9,11–16] and also pointed out that wedge failure is related to its intersection angle and half-wedge angle, but they did not consider the three-dimensional rock slope state. Hoek and Bray [9] proposed the rigid wedge method (RWM), which assumes that the wedge is a rigid body and ignores shear forces that intersect perpendicular to the discontinuities. It is assumed to be zero in the slope stability analysis, and RWM does not consider the deformation of the rock mass itself. Kovari and Fritz [17] considered the influence of static and dynamic loads on wedge models and used model tests and numerical methods to assess the applicability of the limit equilibrium method to wedge slopes. Riquelmea et al. [6] used Slope Mass Rating (SMR) to characterize the wedge failure mechanism found through 3D point cloud analysis. Grenon and Hadjigeorgiou [3] proposed a methodology for linking a fracture system model to limit equilibrium and analyze the wedge failure. Smith and Arnhardt [18] proposed the circle method, which can assess the sliding and toppling mechanisms of individual planes and pairs of planes forming wedges. Model verifications were conducted using real-world data from Turkey and Japan, and the influence of dynamic loads on slope stability in earthquake-prone areas was analyzed and discussed using parameter back-calculation. Kusmar et al. [14] designed four types of test models in their study: static tests, dynamic tests, dry tests, and wet tests. Although these tests conformed to the limit equilibrium method, they could not accurately reflect the behavior of wedge slopes for the following reasons:

(a) The two-dimensional limit equilibrium method cannot describe the behavior of real three-dimensional wedge slope failure;
(b) Rock wedges are composite materials with both joints and weak planes, so an analysis with a single rigid body model cannot completely describe its physical behavior;
(c) The material parameters and variability of rock masses are complex and cannot be described using a single material; thus, the limit equilibrium method cannot express the influence of weak planes in the slope on failure characteristics.

To fill these gaps, this study examined the distributions of discontinuities and changes in intersection planes and angles in various types of oblique slopes, developed moderately simplified physical models, and used these models to investigate the deformation of wedge slopes until failure under conditions of rainfall infiltration. The factors of wedge failure caused by rainfall include rise in groundwater level (increase in water pressure), decrease in effective stress and frictional resistance, failure caused by seepage of weak surface of rock mass, etc. The softening of weak interlayers is also an important factor in rainfall-induced landslides. Xu et al. [8] used nanoindentation experiments and simulations to study the softening of weak interlayers during landslides. The results of the study found that the friction angle of shale was almost unchanged after immersion in water, while the elastic modulus and cohesion decreased significantly. The shear strength of the weak interlayer decreased significantly, and the plastic zone developed along the weak interlayer, eventually forming a landslide. However, because of the complexity all the factors, the establishment of more complex physical models or numerical models has its limitations, and it is difficult to deeply understand the characteristics influencing each factor. Therefore, in this study, we only moderately simplified the wedge failure model of the rock slope, and first discussed the shear strength weakening characteristics of the discontinuity of the rock mass, assuming that the shear strength weakening of the discontinuity occurs in the entire rock mass during the rainfall period. In order to initially grasp the failure behavior under different wedge slope conditions, other factors will be gradually explored in the future.

2. Methodology

2.1. Introduction to Study Area

The study area is located between 81.7 km and 81.9 km and at 83.9 km of Provincial Highway No. 2 in Nanyali of Ruifang District, New Taipei City, Taiwan. In terms of geological conditions, the section from 81.7 km to 81.9 km belongs to the Dapu section of the Kueichoulin Formation, which mainly consists of muddy sandstone and white sandstone and is on the west wing of the Pitou Syncline, which strikes northeast–southwest. The 83.9 km location belongs to the Erjiu section of the Kueichoulin Formation, which mainly consists of muddy sandstone and shale (Figure 1). The abundance of vertical joints has created steep rock faces and, on the northeast coast, many of the rock faces are concave due to differential erosion caused by the ocean wind in the muddy sandstone. The shale is often eroded, leaving the sandstone in the upper portion suspended without support. Consequently, the Ruifang section of Provincial Highway No. 2 is prone to rockfalls and wedge sliding after heavy rain or earthquakes, which severely threatens the safety of road users (Figure 2). According to field surveys, there are three primary sets of joint planes in Nanyali. Stereographic projections show that two of the weak planes intersect, and the exposure of intersection lines on the slope surface can easily lead to wedge failure. We found six intersection lines in the study area, and four intersection points were exposed. The plunge at the points where Joint 1 intersects with Joints 2 and 3 is between 60° and 64°, and the plunge of the intersection point of Joints 2 and 3 is approximately 38° (Figure 3). On-site observations revealed that the average slope below Provincial Highway No. 2 was around 30°, whereas most of the upper slopes were steep and near 90°. Due to the three joint sets, many collapses have left V-shaped rock formations with talus cone deposits below them. According to Figure 1, the geological composition of this area is a combination of muddy sandstone and white sandstone. Because of insufficient support from below, wedge-shaped masses formed by Joints 2 and 3 slide along the intersection line. The intersection angle seen onsite is roughly 40°, and triangular rock masses are left behind (Figure 2). Field surveys show that approximately 30% of the wedge slopes are controlled by Joints 1 and 2; roughly 30% are controlled by Joints 1 and 3, and about 40% are controlled by Joints 2 and 3. The dip angles of the wedge slopes mainly controlled by Joint 1 are steeper, at about 60°. The intersection angles of the wedge slopes controlled by Joints 2 and 3 are close to 40°. Thus, this study simplified the site conditions and designed 7 physical models to carry out related experiments. The intersection angles include 40°, 60°, and 70°, and the half-wedge angles include 25°, 45°, 60°, 75°, eccentricity, etc., and their conditions are similar to the field wedge rock mass.

2.2. Design of the Physical Model

We simplified on-site conditions to design simulation tests. A stainless-steel platform was used to simulate the intersection angle and wedge angle of the wedge masses. Both Zhang et al. [7] and Bowa et al. [19] pointed out that the main cause of wedge instability is the water pressure caused by gravity. When the rock slope is affected by rainfall infiltration, the uplift forces acting along the joint planes increase, and can lead to rapid deterioration of the stability of rocky slopes and wedge sliding. Thus, a sprinkler was installed above the platform to simulate rainfall. The rainfall caused surface water that gradually infiltrated the rock mass, affecting the stability of the model and ultimately leading to slope failure. The physical model of the rock mass comprised grinding stone and resin to simulate sandstone and shale, respectively, that weaken upon encountering water. Although these models do not represent complex full-scale cases, they are sufficient for the analysis of wedge slope failure.

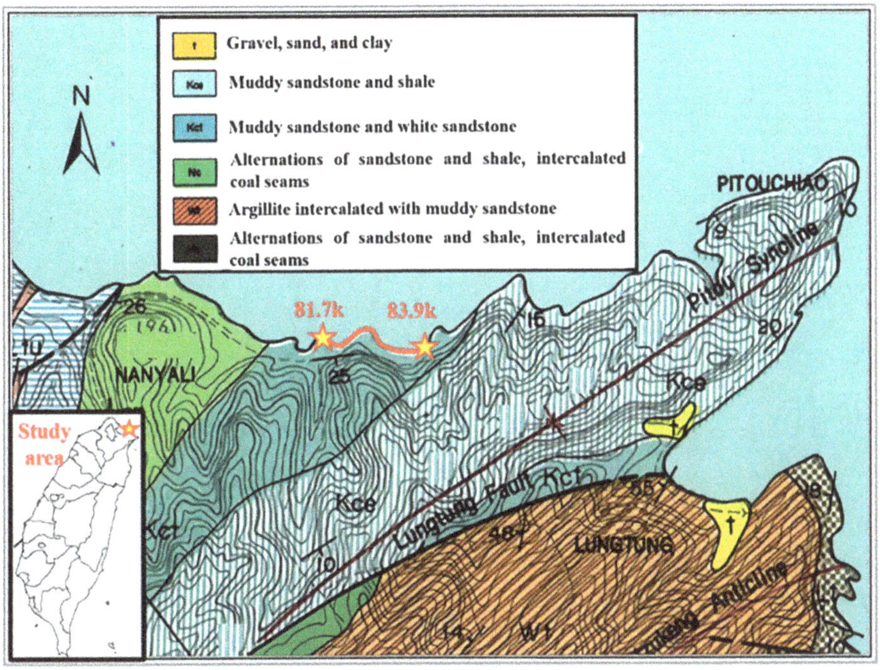

Figure 1. The 1/50,000-scale geological map of the study area.

Figure 2. Wedge failure characteristics on the slope at 81.8 km of Provincial Highway No. 2. Red, green and blue frames means joint plane 1, 2 and 3 respectively.

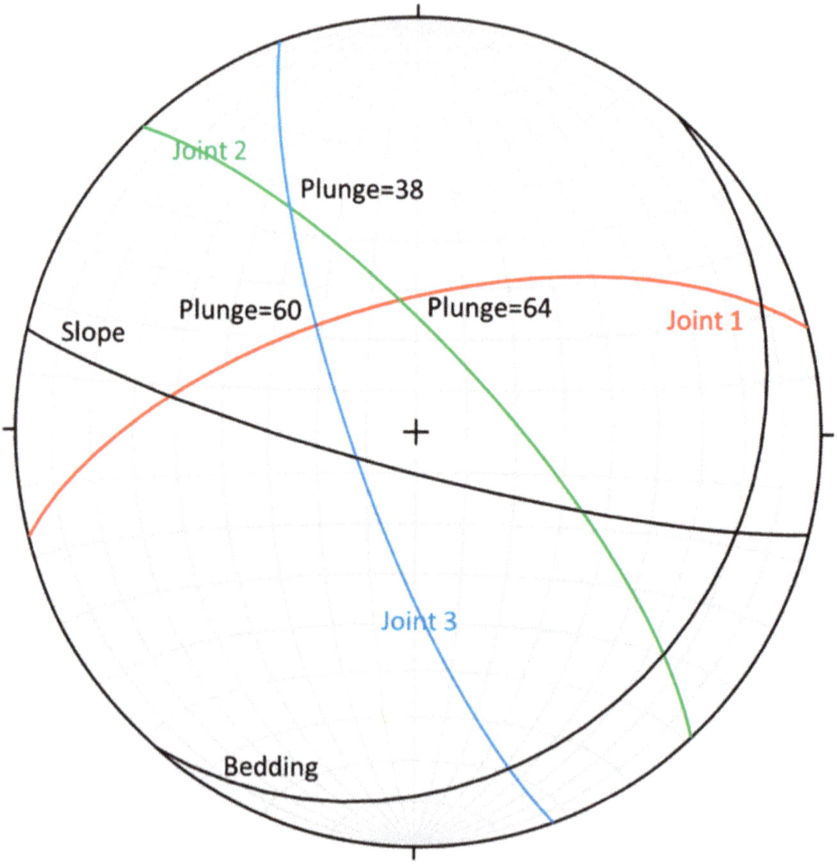

Figure 3. Attitudes of primary weak planes and stereographic projection in the study area.

In the design of the physical models, we first investigated and generalized the morphology of wedge slope failures in Nanyali. Based on the obtained morphological characteristics, we determined the adjustable variables needed for the platform, including the half-wedge angle (ω) between two discontinuities and the wedge intersection angle (i_a). The platform has two major parts that can be adjusted (Figure 4). The first part is the half-wedge angle. The angles between the two discontinuity steel plates on the sides and the normal line can be adjusted from 0° to 90°. The steel plates on the sides can move freely, which means that the angles between the two steel plates and the normal line do not have to be identical; simulations can have designs in which $\omega_1 \neq \omega_2$ (unique conditions in which the two half-wedge angles are not the same). The second part is the wedge intersection angle. Two apparatuses on the sides can be used to adjust the intersection angle, the range of which is between 10° and 90°, to simulate gentle to steep slopes.

To ensure that the influence of the rainfall simulated by the sprinkler remained consistent for all the tests, we did not adjust the height, position, or support of the sprinkler and fixed the rainfall intensity at 150 mm/h.

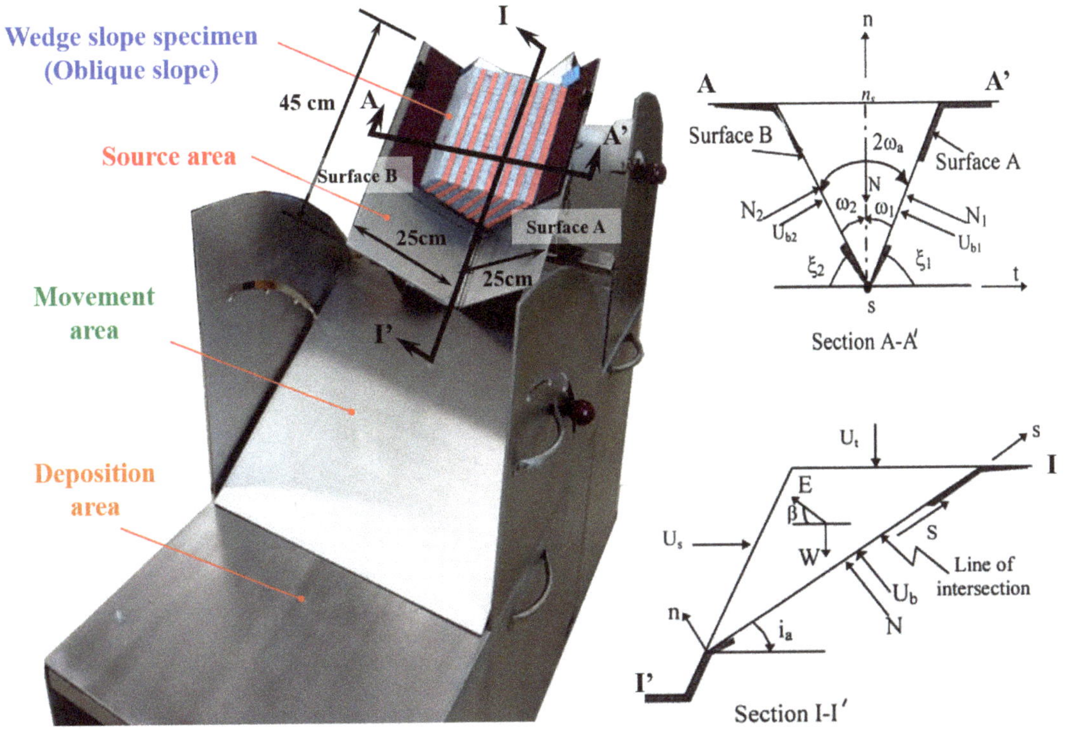

Figure 4. Setup of the physical model.

2.3. Simulation Rock Materials

The rock slopes in the study area near Nanyali mainly consist of thick white-gray layers or fine-to-medium-grained blocks of sandstone occasionally interbedded with thin layers of shale. Most of the rocks deposited onsite are shaped like triangular prisms. Thus, after comparing on-site rock wedges with our physical model, we found that grinding stones in the shape of triangular prisms were similar to the on-site sandstone. Thus, we chose triangular grinding stones to represent the sandstone material. Resin has characteristics similar to those of the interbedded shale or muddy material. When it encounters water, it weakens the strength of the simulated rock mass, accelerating the failure of the model. This helped us simulate the behavior of rocks infiltrated by rain. In the test models, we arranged layers of triangular grinding stones and then stacked them to form a wedge. Polyvinyl acetate (resin) was applied between the grinding stones and layers as an adhesive to create an oblique slope structure (Figure 5). We kept the amount of resin in the rock mass to 8 mm^3 per layer and 2 mm^3 between layers. This ensured the water weakened the planes between layers in the oblique slope to reflect the failure behavior of discontinuities. The physical model thus reflects the failure characteristics of discontinuities (weaker strength). Additionally, we assumed that the intact rock was the undamaged material during the test, and the rock mass failure mainly occurred at the discontinuous position between the rock blocks. The two supports in the upper portion of the platform were both stainless-steel plates, which have smooth surfaces and little friction. We therefore glued the model to the test platform using sandpaper and clamped the sandpaper to the surfaces of the stainless-steel plates to increase the friction between the two. The bottom bases of the models were kept within 15 cm × 25 cm. Before each test, the triangular grinding stones were kept fixed and dry; the resin in the model was left to dry for 5 to 7 days. The purpose was to keep the sizes and strength properties of the models consistent. The details about the rock slope material properties are listed in Table 1.

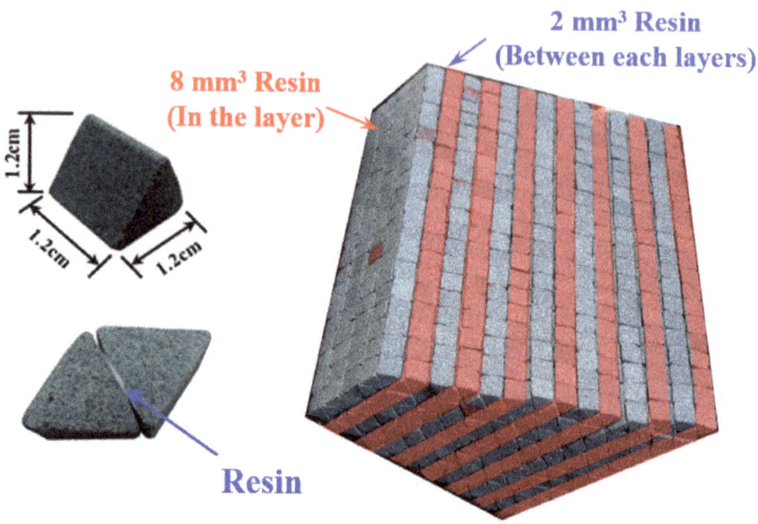

Figure 5. Rock wedge (oblique slope) model comprising grinding stones and resin.

Table 1. Material properties for the physical models.

Properties	Value
Specific gravity (intact artificial rock)	3.65
Density (intact artificial rock)	2600 kg/m^3
Friction angle (intact artificial rock)	32°
Permeability (intact artificial rock)	1.5×10^{-9}–2.3×10^{-9} cm/s
Permeability (artificial rock discontinuity)	1.2×10^{-3}–1.7×10^{-3} cm/s
Compressive strength (intact artificial rock)	7.46 MPa
Compressive strength (the resin in a dry condition)	0.15 MPa
Modulus of elasticity (intact artificial rock)	3 GPa

2.4. *Physical Model Test Items*

Based on field surveys and inductive analysis, we used physical models to change the characteristics of ω and i_a and designed different test items to examine the process and characteristics of wedge failure. We focused on three variables (angle i_a, angle ω, and the eccentricity of angle ω) to construct 7 physical models (see Table 2). We also altered the direction of the oblique slope for preliminary observation of the influence of the oblique direction on rock wedge failure.

Table 2. Experiment parameters for the physical models.

Experiment Numbering	i_a	ω_1	ω_2	Oblique Direction
A	40°	25°	25°	left
B	40°	45°	45°	right
C	40°	60°	60°	right
D	40°	75°	75°	left
E	70°	45°	45°	right
F	70°	25°	25°	right
G	60°	10°	45°	right

3. Failure Processes of Physical Wedge Models

Rainfall intensity was controlled at 15 mm/h for constant surface water infiltration conditions. We varied i_a, ω, the eccentricity of ω, and the oblique direction while observing the processes and behavioral characteristics of wedge failure in the oblique slope under rainfall and slope toe daylight conditions. The primary conditions of Model A were $i_a = 40°$ and ω1 = ω2 = 25°, and the duration of the test was 12 h 27 min 55 s. The test procedure is depicted in Figure 6. In the initial stage, the initial resting state before the rainfall began, the oblique slope strata were inclined towards the left plate, and none of the rocks showed any signs of displacement or deformation. After the test started, the model began to soften from the water, and then, because of its own weight, further softened and deformed. The rock at locations No. 6 and No. 7 (close to the left plate) showed signs of dislocation and deformation. The model gradually slid downwards along the weak plane along the left plate. Some of the rock where the slope toe daylighted (locations No. 9 and No. 10) fell. This stage is mainly characterized by local slope failure (00:00:00–02:36:44) followed by the development of wedge failure (02:36:45–08:02:20). Because of surface water infiltration and the weakening of the strata at locations No. 4–10, approximately 17% of the lower portion of the entire rock mass slid downward, and tension cracks appeared in the upper portion of locations No. 5–9. The tension cracks allowed surface water to infiltrate the model even more easily, thereby accelerating the collapse of the lower portion of the rock. Once some of the rock in the lower portion of the model collapsed, the daylighting of the slope toe caused the upper portion of the rock mass to slide down along the intersection angle. The tension cracks near the top of the slope gradually progressed towards the source, enlarging the wedge failure range. The final stage is the accelerated development of rock mass sliding (08:02:21–12:27:55). Following the collapse of the lower portion of the rock mass, the surface water continued to infiltrate deep into the rock mass, causing tension cracks to form at the source and deeper within the rock. Thus, the lower portion of the rock mass continued to deform and collapse. With progressive wedge failure, tension cracks began appearing in multiple places along the left and right plates and developing along the sides. Ultimately, around 30% of the center of the rock wedge and fell along the intersection line, and only a small-scale collapse took place instead of complete wedge failure.

The duration of the Model B ($i_a = 40°$ and ω1 = ω2 = 45°) test was 12 h 26 min 23 s (Figure 7). At first (initial slight deformation stage at 00:00:00–01:29:00), surface water infiltrated the model via weak planes, softening the model and causing slight deformation. At locations No. 2 and No. 3, tension cracks could be seen at the slope toe. In this stage, the model only presented slight deformation toward the left plate with no significant displacement. At 01:29:01–04:32:30, the entire model gradually moved downhill. Tension cracks appeared at the slope toe at location No. 2, and after the rainfall continued for an hour, the slope toe at locations No. 1–4 slowly slid, creating local tension cracks. Apparent bulges could be seen where the slope toe daylighted, and deformation was more significant at the slope toe on the left plate of the model than that on the right plate (continued development of slight deformation). During this stage (slope stabilization at 04:32:31–12:26:23), the rocks at locations No. 6–12 stabilized and showed no signs of deformation. The tension cracks at the slope toe of locations No. 2 and No. 3 continued to develop, and small-scale collapses took place. However, there were no significant deformations or tension cracks in the overall rock mass. Before the end of the test, the rock wedge had not slid significantly and remained stable after 05:56:13.

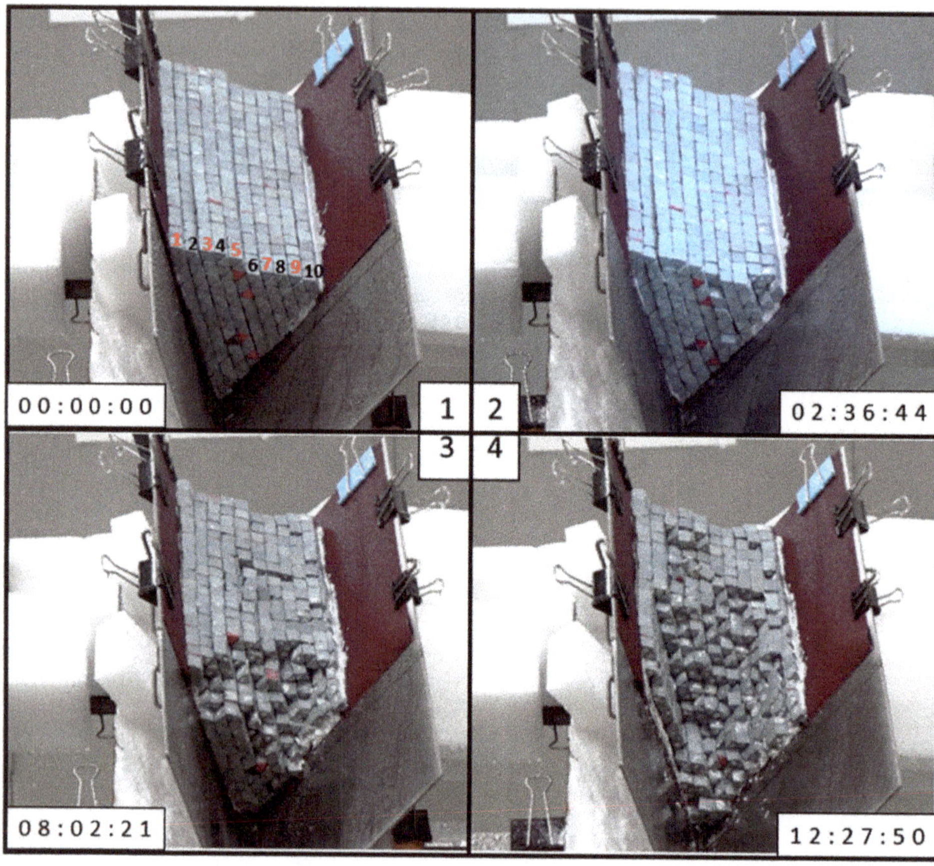

Figure 6. Physical model A (i_a = 40° and $\omega1 = \omega2$ = 25°).

Model C (i_a = 40° and $\omega1 = \omega2$ = 60°) was tested for about 10 h 53 min 51 s (Figure 8). During the initial deformation stage (00:00:00–06:15:57), the model began to soften from the infiltrating surface water, and then, because of its own weight, began to deform, and some tension cracks appeared on the surface. Because of the progression of tension cracks, the daylighting of the slope toe at locations No. 8–10 caused some of the rock to fall. Because of the continued influence of surface water infiltration and gravity, the rock wedge slowly slid along the intersection line. Local tension crack development continued on the slope, accelerating the infiltration of surface water and weakening the strength of the rock. This stage is characterized by the sliding of the entire rock mass (06:15:58–08:26:51). As the duration of surface water infiltration increased, the resin at the bottom on the left and right plates gradually lost its shear strength. Ultimately, the entire rock wedge began to slide along the intersection line because of its own weight, and only 5% of the rock remained on the slope (08:26:52–10:53:51). The test conditions of model D (i_a = 40° and $\omega1 = \omega2$ = 75°) and model C were similar (Figure 9), both $\omega1$ and $\omega2$ exceeded 50 degrees, and only the failure time of the test model was different. In the initial sliding stage (00:00:00–00:56:55), the model began to soften from the infiltrating surface water, and then, because of its own weight, the entire rock wedge began sliding toward the slope toe. As the surface water continued to infiltrate the rock wedge, the surface of the rock showed no apparent tension cracks. However, the strata in the oblique slope were inclined toward the left plate, which caused most of the surface water that had infiltrated the rock mass to flow through the strata on the right plate toward the left plate. Consequently, the rock on the right plate of the mode gradually weakened, resulting in twisting, deformation, and sliding (00:56:56–01:56:51). Then, the bottom of the rock wedge gradually lost its shear strength. Ultimately, the entire rock wedge began to slide along the intersection line and the two intersection planes because of its own weight, and only 10% of the rock remained on the slope (01:56:51–02:37:39).

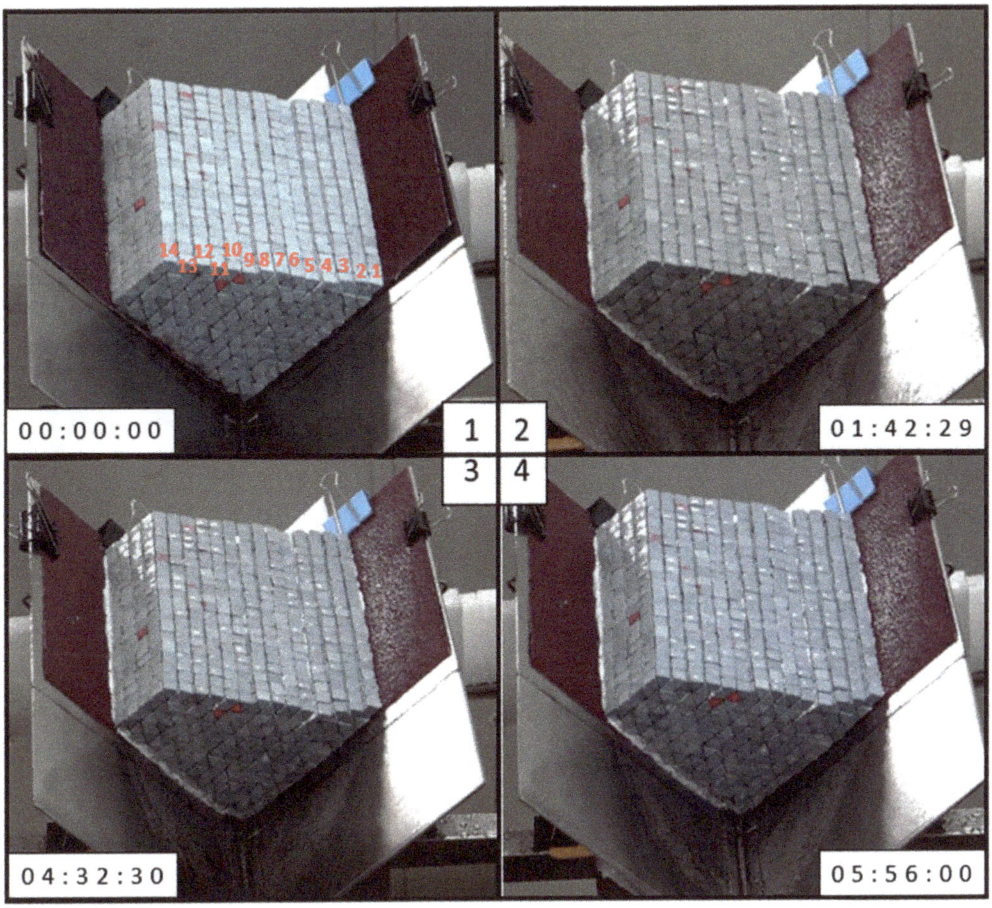

Figure 7. Physical Model B (i_a = 40° and $\omega 1 = \omega 2 = 45°$).

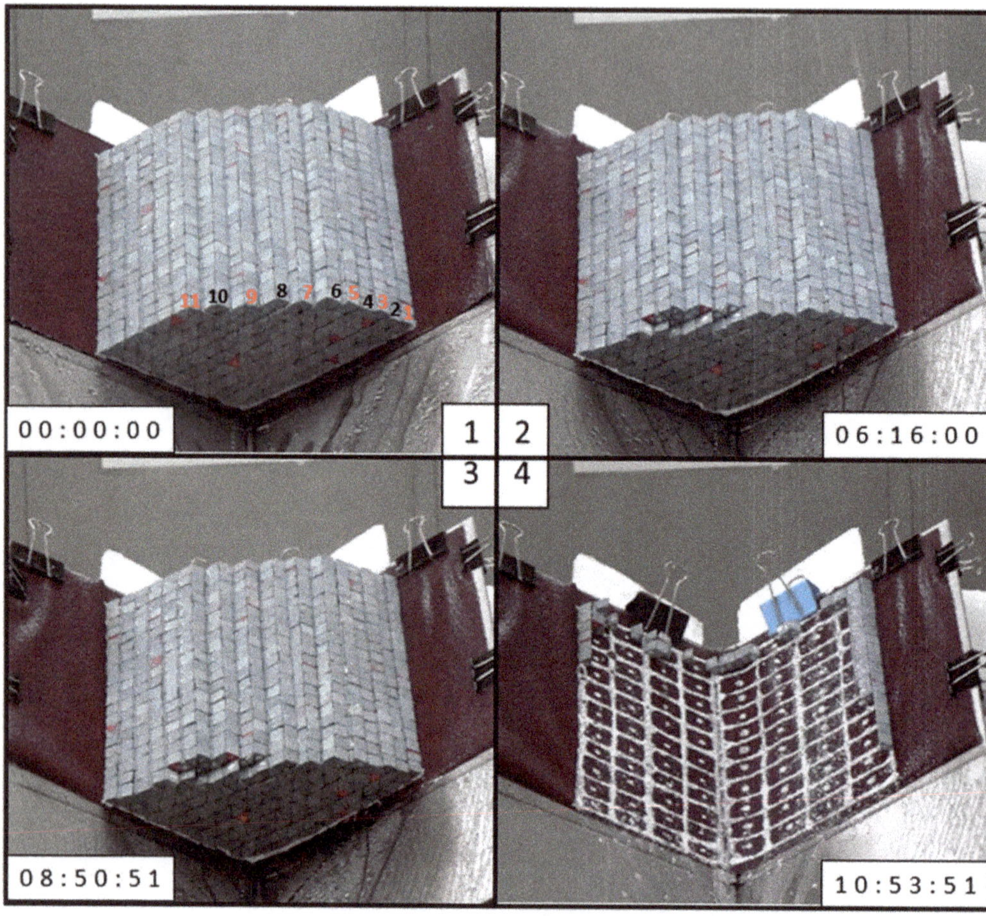

Figure 8. Physical test process of Model C (i_a = 40° and ω1 = ω2 = 60°).

The primary conditions of Model E were i_a = 70° and ω1 = ω2 = 25°, and the duration of the test was 3 h 24 min 03 s (Figure 10). At the first stage (the initial deformation stage; 00:00:00–01:44:39), the model softened because of rainfall infiltration, and the greater intersection angle meant that gravity exerted a greater impact on the model. Significant deformation was observed after 20 min of rainfall. The weak planes in the oblique slope were inclined towards the right plate, and, as a result, soon after the surface water had infiltrated the model, the surface layer at the slope toe on the right plate (locations No. 7–10) showed local collapses and toppling failure. During the second stage (the local rock mass failure stage; 01:44:40–03:14:12), as some of the rock at the slope toe on the right plate collapsed, the rock on the right plate also became unstable, gradually resulting in wedge failure. Approximately 30% of the rock wedge slid and collapsed along the slope, greatly reducing the support below the rock mass on the right plate. Losing its support on the right, the entire rock wedge relied on the rock on the left plate and the intersection line for support. Then, as the duration of surface water infiltration increased, gravity and the weakening of the rock near the left led to gradual leaning and deformation. Subsequently, the rock in the back collapsed and toppled on a large scale along the intersection line, and only 5% of the rock remained on the slope. The duration of the Model F (i_a = 70° and ω1 = ω2 = 45°) test was 7 h 49 min 14 s (Figure 11). At the first stage (the initial slope deformation stage; 0:00:00–1:30:44), the model softened because of surface water infiltration, and the greater intersection angle meant that gravity exerted a greater impact on the model. The model thus deformed and shifted toward the slope toe, and some of the rock at location No. 14 collapsed. The weak planes in the oblique slope were inclined towards the right plate, and the surface water could easily infiltrate and reach the weak places from above, resulting in weakening and failure on the surface on the right plate. At the local rock mass failure stage (1:30:45–5:37:36), rocks continued to fall at location No. 14. Because of gravity and surface water infiltration, the center of the model deformed and tilted forward. Local collapses were caused at locations No. 1–3, causing the deforming and forward-tilting slope toe at locations No. 3–9 to lose some of its support at the wings. After an hour of rainfall, the rock at locations No. 3–9 continued to deform and tilt forward. After four hours of rainfall, the forward-tilting rock became an independent mass on the verge of collapse. With the continuing infiltration of rainfall and the steep intersection angle, this mass eventually toppled. After the rock at locations No. 3–9 lost its support on the sides, the mass toppled. The remaining rock on the right plate also gradually toppled towards the intersection line, ultimately leaving only about 20% of the rock on the slope (5:37:37–07:49:14).

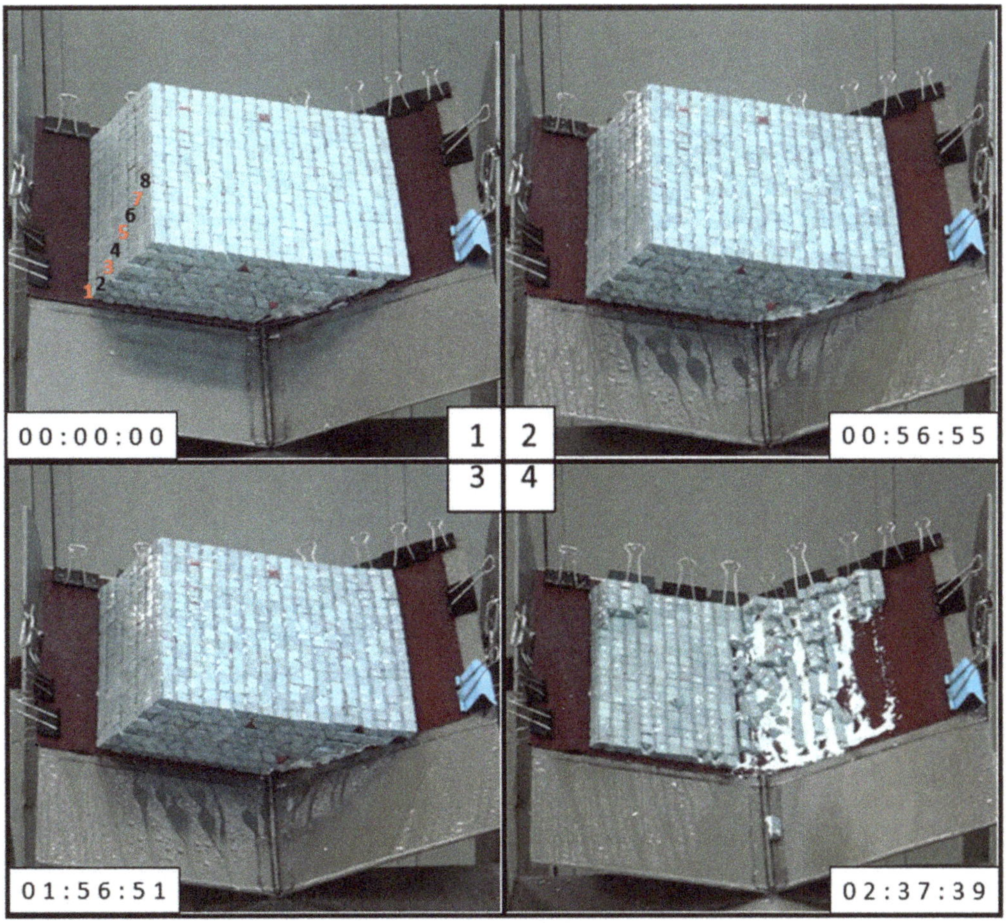

Figure 9. Physical test process of Model D (i_a = 40° and ω1 = ω2 = 75°).

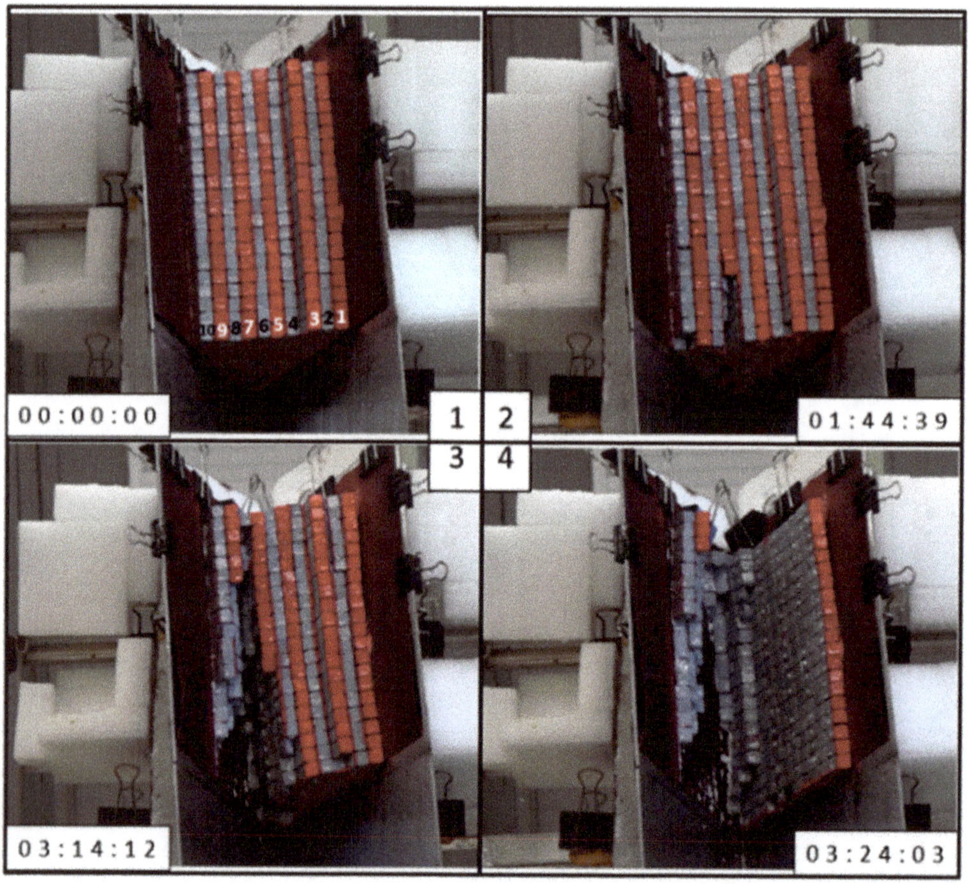

Figure 10. Physical Model E (i_a = 70° and $\omega 1 = \omega 2$ = 25°).

The primary conditions of Model G were i_a = 60°, $\omega 1$ = 40°, and $\omega 2$ = 10°, with two eccentric half-wedge angles, and the duration of the test was 3 h 24 min 03 s (Figure 12). At the initial deformation stage under conditions of surface water infiltration (00:00:00–02:19:25), the model's center of gravity was on the left plate (the side with a gentler slope), where ω was greater. In addition, the greater intersection angle meant that gravity exerted a greater impact on the model from the onset. As the surface water continued to infiltrate the model, it began to weaken and gradually deform and slide toward the toe on the left plate, leaving most of the rock on the left plate under compression, while most of the rock on the steep right plate was under tension. Before the midpoint of the test, rock continued to fall from the bottom of the model, resulting in local vertical torsion and displacement at locations No. 2 and No. 3. Weakened by water, the entire tock mass gradually slid and deformed. Next, the overall model reached the stage of local rock mass failure (02:19:26–02:33:32). Vertical torsion and displacement continued at locations No. 2 and No. 3 until tension cracks developed at location No. 8 (the cracks parallel to the intersection angle). The rock at locations No. 3–7 then lost its support on the sides, causing the rock mass to slide downwards along the intersection line, and only approximately 35% of the rock remained on the slope. The last stage is characterized by the rock wedge sliding and acceleration of the failure (02:33:33–04:00:51). After the collapse of the center rock mass, the rock at locations No. 9 and No. 10 remained on the slope. However, due to the loss of the rock on the intersection line, the rock remaining on the right plate gradually toppled toward the intersection line. The remainder of the rock was also softened by the infiltrating surface water and collapsed. Afterwards, approximately 15% of the rock remained on the right plate. The failure behaviors displayed by the entire rock wedge included sliding (gentle slope on left plate), collapsing, and toppling (steep slope on right plate).

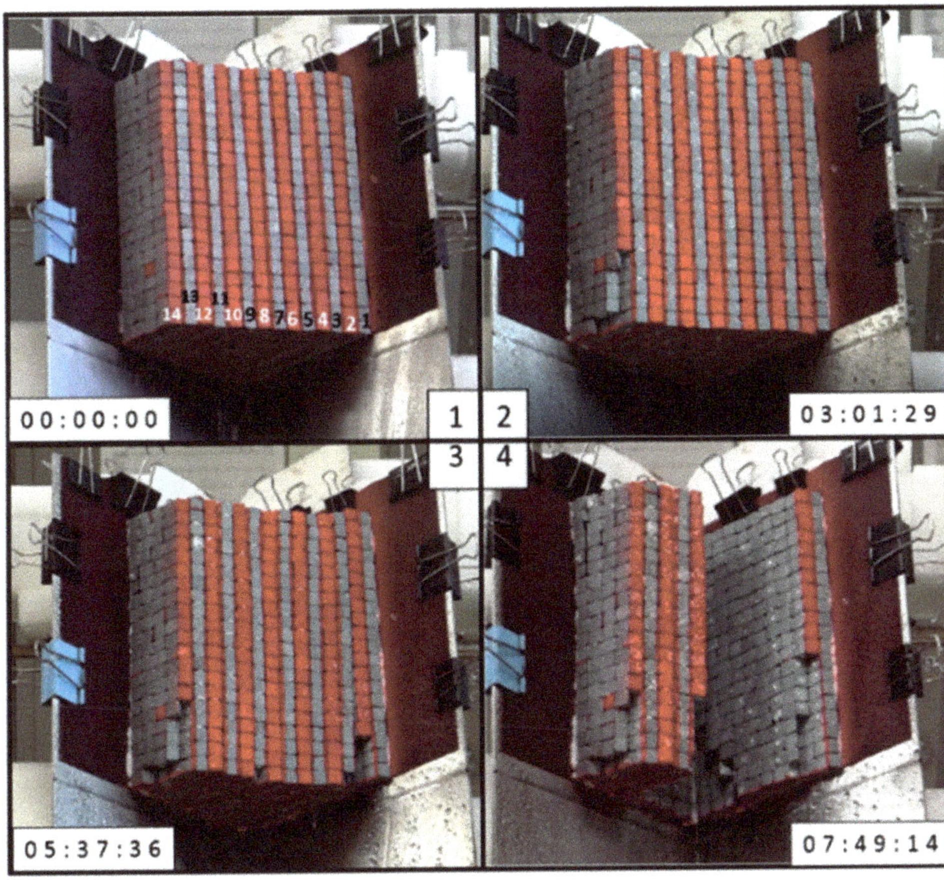

Figure 11. Physical Model F (i_a = 70° and $\omega 1 = \omega 2 = 45°$).

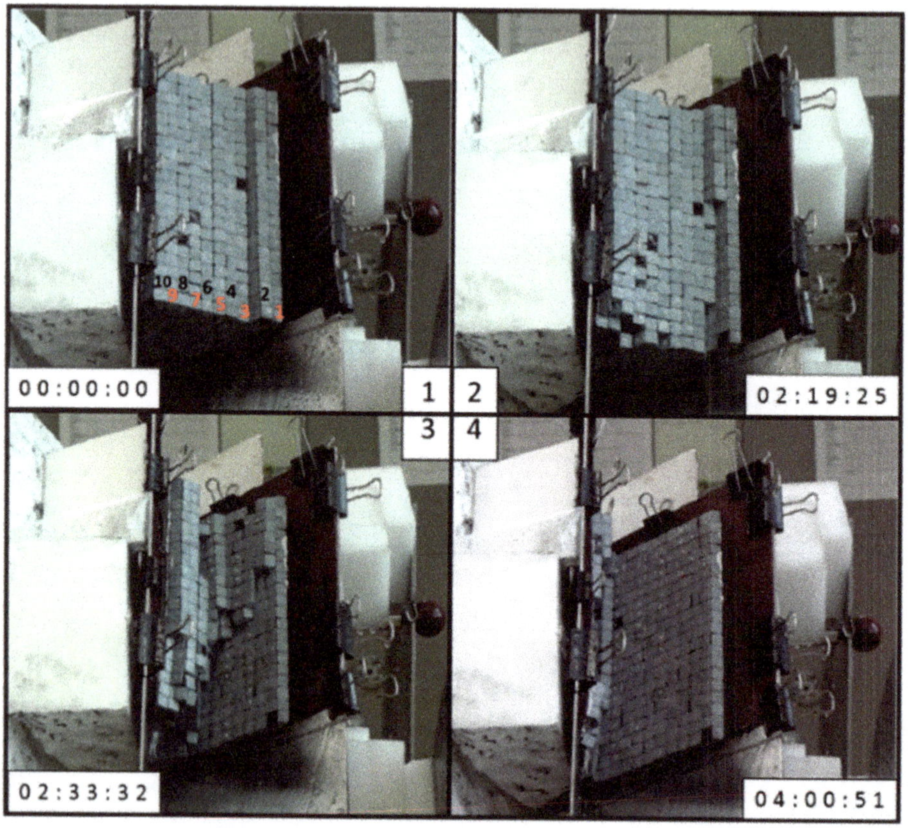

Figure 12. Physical Model G (i_a = 60°, ω1 = 40° and ω2 = 10°).

4. Discussion

4.1. Comparison of Physical Models

4.1.1. Influence of Small i_a and Changing ω

Comparison of the test results with i_a fixed at 40° and different ω (Figure 13) revealed that the failures in the models with smaller ω mainly began at the slope toe and gradually progressed towards the source until wedge failure took place. The final failures were shallower in depth. In the model with ω = 60°, collapses initially only took place where the slope toe daylighted, and the surface of the rock showed deformation and tension crack development. In the end, the entire rock wedge collapsed. In the model with ω = 75°, the rock displayed no significant deformation or crack development. Total rock wedge failure took place within a short period of time, and the characteristics were similar to those of planar failure.

A comprehensive comparison of the three tests showed that with a small i_a, the failure behavior resulting from a smaller ω produced relatively higher stability and shallower failure depth because of the two discontinuities on the sides, and this is categorized as progressive wedge failure. In contrast, a larger ω resulted in greater failure depth and a larger sliding rock wedge. Additionally, most of the sliding followed the intersection line and planes, and the wedge failure behavior was similar to planar sliding. Clearly, under small i_a, the influence of ω on wedge failure characteristics was significant, and failure behavior varied significantly with ω. From a mechanical point of view, the change in the half-wedge angle mainly affected the mechanical properties of wedge rock mass confining pressure. The lower half-wedge angle led to larger confining pressure in the wedge rock mass, which made the resin between the discontinuities in the rock mass more easily damaged by soaking and softening under rainfall conditions. Additionally, the failure behavior of the rock mass was similar to the progressive failure: the damage was first concentrated at the slope toe of the wedge rock mass, and then gradually failed towards the top of the slope. Therefore, under the condition of high confining pressure (the smaller half-wedge angle condition), the failure behavior of this kind of wedge rock mass is characterized by discontinuity, which mainly controls its failure characteristics. However, under the condition of larger half-wedge angles (smaller confining pressure), since the boundary constraints of the wedges on both sides are relatively small, the interface (or discontinuity at the bottom) of the wedge rock mass will dominate its overall failure characteristics. Therefore, its failure behavior is very similar to translation sliding.

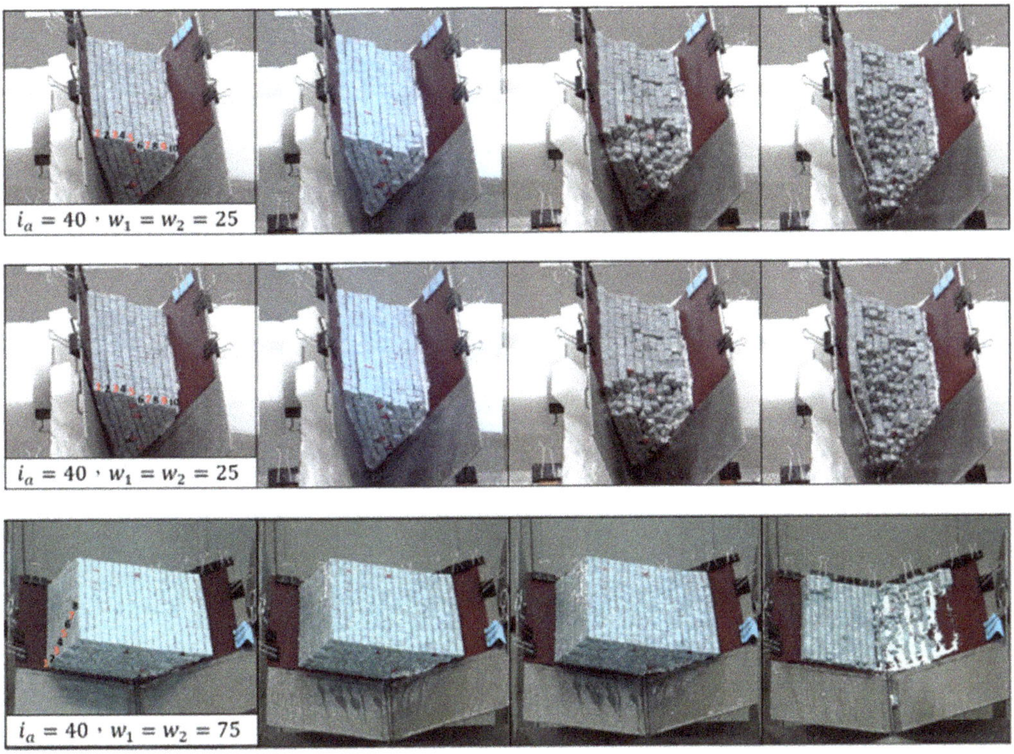

Figure 13. Comparison of tests with i_a = 40° and different ω.

4.1.2. Influence of Large i_a and Changing ω

Comparison of the test results with i_a fixed at 70° and ω equal to 50° and 90° (Figure 14) revealed that once some of the rock at the toe of the model with a smaller ω collapsed, the steep terrain caused the rock wedge to gradually deform and tilt forwards. After the partial collapse on the right plate and the loss of support, the rock on the left plate was also weakened by infiltrating surface water, resulting in the loss of support on both sides and inducing toppling failure. The model with the greater ω also gradually lost support on both sides, resulting in some of the rock in the middle deforming and tilting forward. In the end, the continuing rain led to toppling failure, and after the rock in the middle toppled over, the rock on the right plate along the intersection line also toppled over. Comparison of the two tests showed that because of the steeper inclination angle, the failure depths were significant in both tests, and the failures were similar in type. Clearly, under large i_a, the influence of ω on wedge failure characteristics is smaller, and failure behavior is mainly determined by the extent of discontinuity erosion and the toppling and deformation characteristics of the rock. From the perspective of its mechanical mechanism, when the intersection angle reached 70 degrees, the change of its half-wedge angle (that is, the change of the confining pressure of the wedge rock mass) did not cause a significant impact on the failure mechanism. Both of them exhibited toppling failure characteristics under this condition, and both occurred in the intersection line range. Therefore, when the wedge rock mass reaches the topographical condition of the toppling, the discontinuous strength of the rock mass within the intersection line range is the main factor controlling the toppling failure.

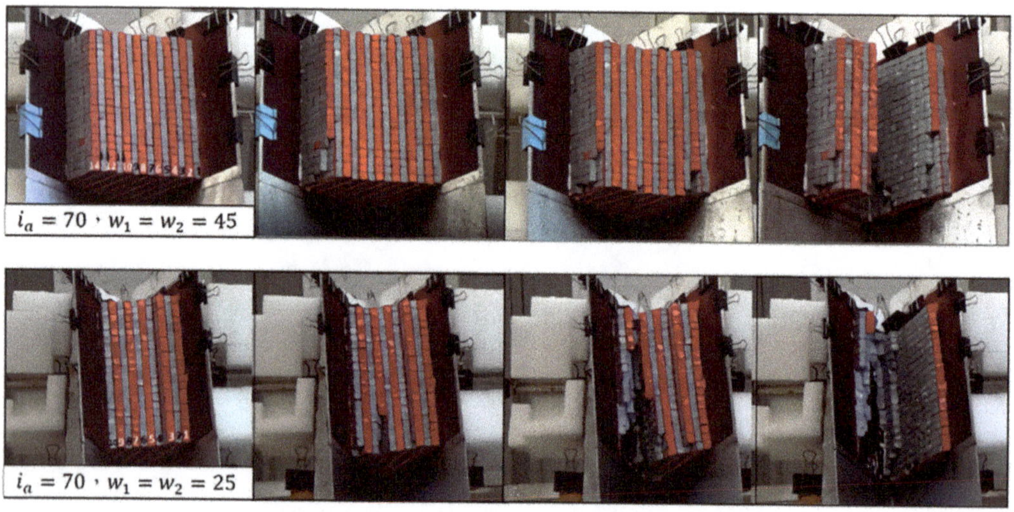

Figure 14. Comparison of tests with $i_a = 70°$ and different ω.

4.1.3. Influence of Large ω, Small ω and Changing i_a

Comparison of the test results with ω fixed at 45° and two i_a (Figure 15) revealed that after the model with a greater i_a was weakened by water, tension cracks developed swiftly, causing the rock in the middle of the model to deform and ultimately topple. Only a small portion of the rock fell in the model with a smaller i_a, and no significant deformation or wedge failure took place during the test. Comparison of the two tests showed that toppling or falling was more likely with a greater i_a due to gravity, whereas a smaller i_a produced greater stability. In addition, most of the failure behavior involved the rock wedge sliding along the intersection line or plane. Comparison of the test results with ω fixed at 25° and two i_a (Figure 16) revealed that the model with a smaller i_a took longer to reach wedge failure. The tension cracks gradually developed towards the source, and the failure in the end was shallower. The result was similar to progressive wedge failure. Gravity exerted a greater impact on the model with a greater i_a, and the common failure types were toppling and falling. A comparison of the two tests revealed a shallower failure depth in the model with a smaller i_a, longer time to wedge failure, and greater overall stability than in the model with a greater i_a. As far as its mechanical mechanism is concerned, when the wedge rock mass is in a small confining pressure state, the change of its intersection angle causes a significant difference in the failure mode of the rock slope, that is, sliding along the intersection angle and half-wedge face (at intersection angles less than 60 degrees, the failure is dominated by the discontinuity at the bottom), or toppling (at intersection angles greater than 60 degrees, the failure is dominated by the discontinuity strength near the intersection angle) failure behavior. Additionally, when the wedge rock mass is in a state of high confining pressure, it also has the same mechanical properties. That is, progressive failure occurs along the slope toe to the top of the rock slope (at intersection angles less than 60 degrees, the failure is mainly controlled by the discontinuity of the rock mass), or overturning (at intersection angles greater than 60 degrees, the failure is mainly controlled by the discontinuity strength near the intersection angle) failure behavior.

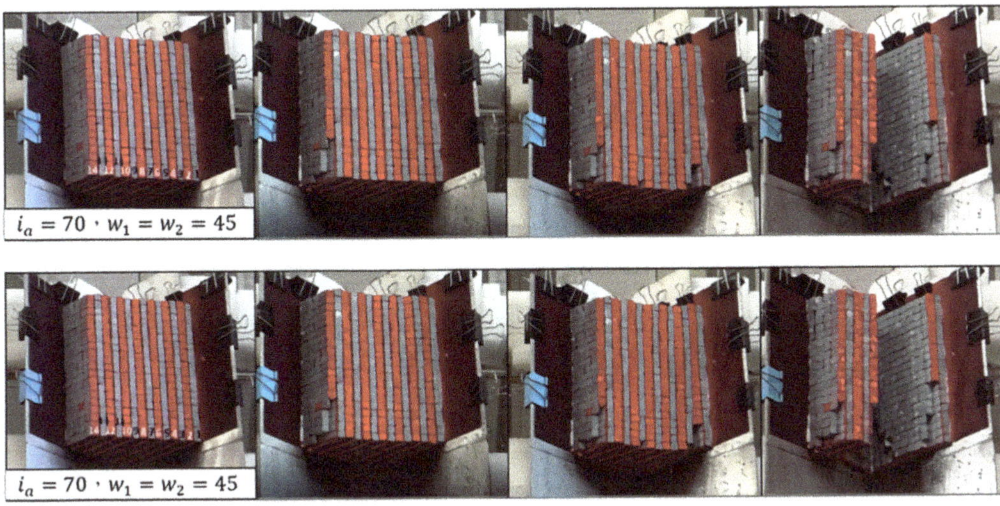

Figure 15. Comparison of tests with ω = 45° and different i_a.

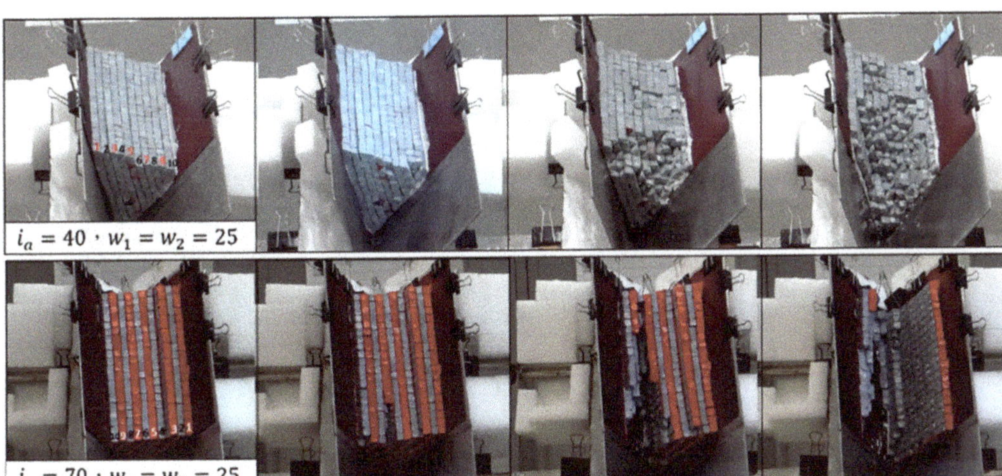

Figure 16. Comparison of tests with ω = 25° and different i_a.

4.1.4. Influence of ω Eccentricity

Comparison of the test results with different ω (eccentric) and identical ω (Figure 17) revealed that the initial failure with identical ω is determined by the distribution of weak planes in the oblique slope and that the type of the failure is determined by i_a. In contrast, the initial failure in the eccentric model with different ω is determined by the distribution of weak planes in the oblique slope as well as the slope of the terrain on the left and right plates. The side with the gentler slope was mostly compressed, most of the rock mass was characterized by toppling/sliding failure along this side, and it was the primary sliding zone of the wedge failure. The side with the steeper slope was mostly under tension, and the stability of most of the rock mass on this side was controlled by the steep terrain. In the end, toppling or failing failure occurred on this side. In terms of its mechanical behavior, when the half-wedge angles were different, the wedge rock mass more easily failed along the smaller half-wedge angle position. The main reason for this is that when the wedge rock mass is eccentric, the main stress is concentrated on the relatively gentle half-wedge surface, resulting in sliding or toppling failure first occurring on this side. After the smaller half-wedge angle is damaged to a certain extent, the other side decompresses, and failure behaviors such as falling or overturning occur.

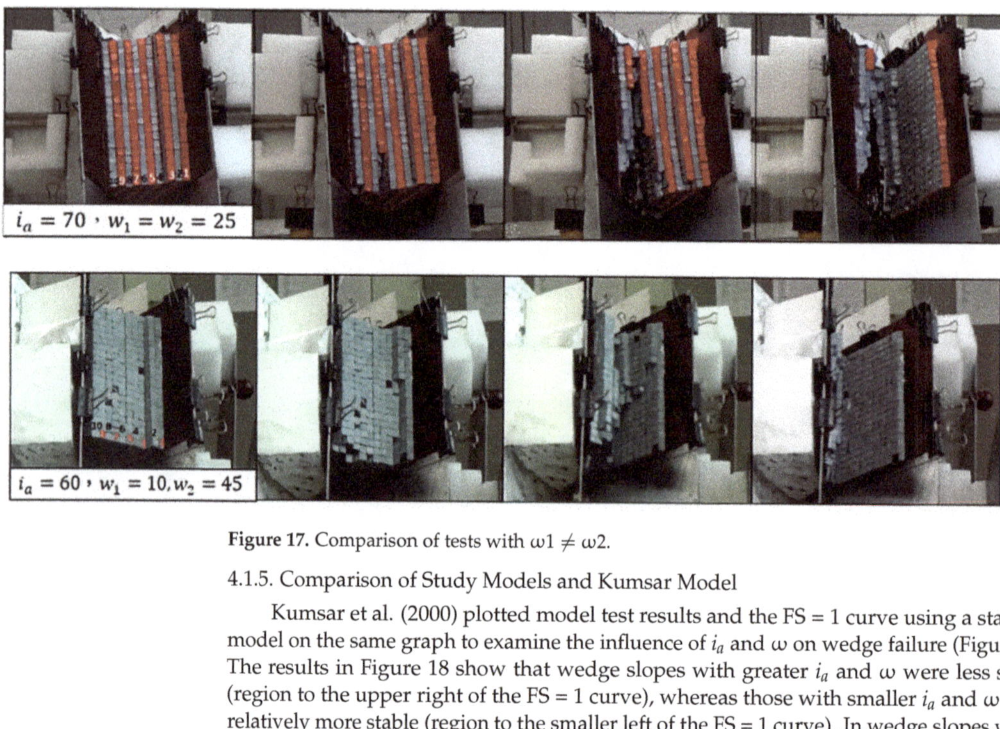

Figure 17. Comparison of tests with $\omega 1 \neq \omega 2$.

4.1.5. Comparison of Study Models and Kumsar Model

Kumsar et al. (2000) plotted model test results and the FS = 1 curve using a stability model on the same graph to examine the influence of i_a and ω on wedge failure (Figure 18). The results in Figure 18 show that wedge slopes with greater i_a and ω were less stable (region to the upper right of the FS = 1 curve), whereas those with smaller i_a and ω were relatively more stable (region to the smaller left of the FS = 1 curve). In wedge slopes with a greater i_a, only a smaller ω can create more stable conditions, whereas if ω is greater, then i_a must be smaller for the sake of stability. These results demonstrate the importance of i_a and ω in wedge slopes.

The marks of the physical models in Figure 18 show that Models E and F, in which $i_a = 70°$ and $\omega = 25°$ and $45°$, respectively, were both located in the unstable region. This means that a greater i_a tended to result in wedge failure (Figure 18). Model E had a smaller ω than Model F; it was thus closer to the FS = 1 curve and had greater stability. Comparison of the wedge failure time of the two models also confirmed that Model F was likely to succumb to total wedge failure sooner than Model E. In contrast, Models A and B were situated in the stable region (wedge failure took place in Model A); however, the stability formula and test only considered rigid wedge failure models, and the formula did not consider the impact of continuous rainfall on the wedge mass. Our models took both the rock material and rainfall infiltration into account, and after the continuous infiltration of surface water weakened the strength of the rock material, progressive wedge failure took place. This demonstrates that these two factors also influence wedge slope stability. The conditions for Models C and D were similar. Although i_a was only $40°$, the overall wedge failure was similar to planar failure when ω was greater than $50°$, and total failure tended to occur swiftly following continuous surface water infiltration. This further demonstrates that the wedge failure of rock material and the continuous infiltration of surface water are suitable predictors for conventional stability analysis. We next marked the parameters of the eccentric model (Model G, where ω = 10° and 45°) in Figure 19 and found that with ω = 45° (G2), Model G fell in the unstable region, whereas with ω = 10° (G1), it fell in the stable region. However, the test results revealed wedge failure occurring within a short period of time (indicating instability), which means that when ω on the two slides of the slope differ, using the greater ω for stability analysis is the best approach (Figure 19).

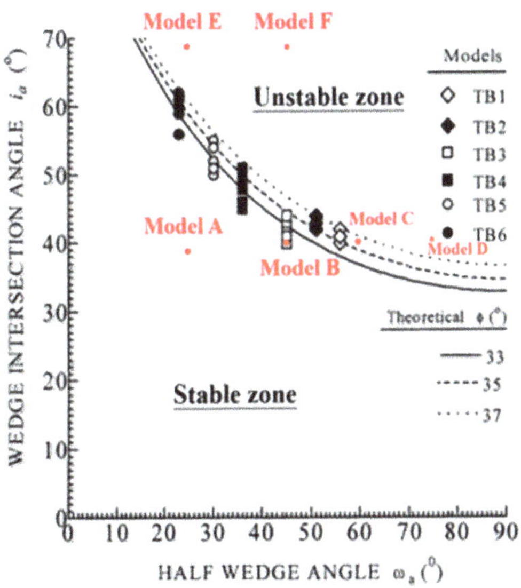

NO.	i_a	ω_a
A	40°	25°
B	40°	45°
C	40°	60°
D	40°	75°
E	70°	25°
F	70°	45°

Figure 18. Comparison of the study models and Kumsar's model.

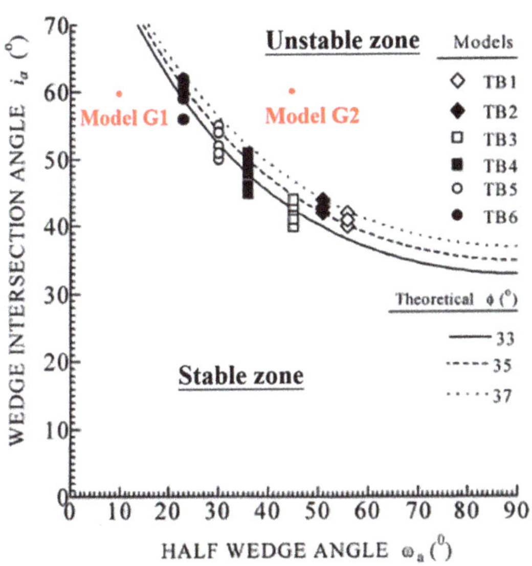

NO.	i_a	ω_a
G1	60°	10°
G2	60°	45°

Figure 19. Comparison of the eccentric model and Kumsar's model.

5. Conclusions

Our field surveys revealed that the wedge slopes along Provincial Highway No. 2 in Nanyali are mainly controlled by three sets of joints. Approximately 40% of the wedge slopes are controlled by Joints 2 and 3, and their intersection angles mostly range from 40° to 60°. We thus designed our physical models accordingly. The results of the physical tests revealed that most of the initial failure took place where the slope toe daylighted and that rockfalls or toppling were the most common. After some of the rock at the slope toe collapsed, tension cracks gradually developed toward the source, and surface water continued to infiltrate the rock mass via tension cracks, which weakened the shear strength within weak planes, reduced the overall stability of the rock wedge, and ultimately led to wedge failure.

The physical tests also revealed that i_a had the greatest impact on overall rock wedge stability: i_a less than 50° tended to result in sliding-type failure, and i_a greater than 90° tended to result in toppling- or falling-type wedge failure. The variable with the second greatest impact was ω; when ω was less than 40°, forces on the rock mass from both sides produced progressive wedge failure characteristics. The failure began at the slope toe which the oblique slope was inclined towards and then gradually developed. Following local wedge failure, tension cracks formed in the rock above, and this cycle continued. The failure behavior resulting from ω greater than 45° was similar to that of planar rock sliding; there were fewer tension cracks, and the sliding mass was relatively whole. Most of the failure took place in deeper locations, thereby demonstrating that the characteristics of wedge failure are mostly controlled by i_a and ω. Comparisons of the physical models and stability analysis models revealed that greater ω and i_a increase the probability of wedge failure; however, existing stability formulas overlook the influence of rock material and continuous rainfall infiltration on wedge slope stability. This should be a focal point of future improvements to stability formulas. In wedge slopes where ω on the two slides of the slope differ, using the greater ω for stability analysis is the best approach.

Our results indicated that wedge slopes with ω less than 40° in Nanyali should be inspected for progressive wedge failure, and any tension cracks on the slope surface or wedge failure at the slope toe should be noted. Wedge slopes with ω greater than 45° should be inspected for slope toe daylighting to prevent large-scale wedge sliding. In eccentric wedge slopes ($\omega 1 \neq \omega 2$), a greater i_a means a greater chance of large-scale sliding on gentler slopes, which is important information for engineers.

Author Contributions: Conceptualization, C.-M.L. and Y.-S.L.; Methodology, C.-H.C.; Validation, C.-M.L. and C.-H.C.; Formal analysis, C.-M.L.; Investigation, Y.-S.L.; Data curation, C.-H.C.; Supervision, C.-M.L. and Y.-S.L. All authors have read and agreed to the published version of the manuscript.

Funding: This research received no external funding.

Data Availability Statement: Not applicable.

Conflicts of Interest: The authors declare no conflict of interest.

References

1. Wyllie, D.C.; Mah, C.W. *Rock Slope Engineering*, 4th ed.; Spon Press: London, UK, 2004.
2. Wang, Y.J.; Yin, J.H.; Chen, Z.; Lee, C.F. Analysis of wedge stability using different methods. *Rock Mech. Rock Eng.* **2004**, *37*, 127–150. [CrossRef]
3. Grenon, M.; Hadjigeorgiou, J. A design methodology for rock slopes susceptible to wedge failure using fracture system modelling. *Eng. Geol.* **2008**, *96*, 78–93. [CrossRef]
4. Alejano, L.R.; Ferrero, A.M.; Ramirez-Oyanguren, P.; Alvarez Fernandez, M.I. Comparison of limit-equilibrium, numerical and physical models of wall slope stability. *Int. J. Rock Mech. Min. Sci.* **2011**, *48*, 16–26. [CrossRef]
5. Havaej, M.; Stead, D. Investigating the role of kinematics and damage in the failure of rock slopes. *Comput. Geotech.* **2016**, *78*, 181–193. [CrossRef]
6. Riquelmea, J.A.; Tomas, R.; Jewell; Abellan, A. Characterization of Rock Slopes through Slope Mass Rating using 3D point clouds. *Int. J. Rock Mech. Min. Sci.* **2016**, *84*, 165–176. [CrossRef]

7. Zhang, S.; Xu, Q.; Peng, D.; Zhu, Z.; Li, W.; Wong, H.; Shen, P. Stability analysis of rock wedge slide subjected to groundwater dynamic evolution. *Eng. Geol.* **2020**, *270*, 105528. [CrossRef]
8. Xu, J.; Tang, X.; Wang, Z.; Feng, Y.; Bian, K. Investigating the softening of weak interlayers during landslides using nanoindentation experiments and simulations. *Eng. Geol.* **2020**, *277*, 1–12. [CrossRef]
9. Hoek, E.; Bray, J.W. *Rock Slope Engineering*, 3rd ed.; Institute of Mining and Metallurgy: London, UK, 1981; p. 358.
10. Piteau, D.R.; Martin, D.C. Mechanics of rock slope failure. In *Stability in Surface Mining Volume 3, Proceedings of the Third International Conference on Stability in Surface Mining, Vancouver, BC, Canada, 1–3 June 1982*; Society of Mining Engineers of the American Institute of Mining, Metallurgical, and Petroleum Engineers: New York, NY, USA, 1982; pp. 113–169.
11. Adhikary, D.; Dyskin, A.V.; Jewell, R.J.; Stewart, D.P. A study of the mechanism of flexural toppling failure of rock slopes. *Rock Mech. Rock Eng.* **1997**, *30*, 75–93. [CrossRef]
12. Bobet, A. Analytical Solutions for Toppling Failure. *Int. J. Rock Mech. Min. Sci.* **1999**, *36*, 971–981. [CrossRef]
13. Park, H.; West, T.R. Development of a probabilistic approach for rock wedge failure. *Eng. Geol.* **2001**, *59*, 233–251. [CrossRef]
14. Kumsar, H.; Aydan, O.; Ulusay, R. Dynamic and static stability assessment of rock slopes against wedge failures. *Rock Mech. Rock Eng.* **2000**, *33*, 31–51. [CrossRef]
15. Yoon, W.S.; Jeong, U.J.; Kim, J.H. Kinematic analysis for sliding failure of multi-faced rock slopes. *Eng. Geol.* **2002**, *67*, 51–61. [CrossRef]
16. Ahangar-Asr, A.; Faramarzi, A.; AJavadi, A. A new approach for prediction of the stability of soil and rock slopes. *Eng. Comput.* **2010**, *27*, 878–893. [CrossRef]
17. Kovari, K.; Fritz, P. Stability analysis of rock slopes for plane and wedge failure with aid of a programmable pocket calculator. In Proceedings of the 16th US Symposium on Rock Mechanics (USRMS), Minneapolis, MN, USA, 22–24 September 1975; pp. 25–33.
18. Smith, J.V.; Arnhardt, C. A New Assessment Method for Structural-Control Failure Mechanisms in Rock Slopes—Case Examples. *AIMS Geosci.* **2016**, *2*, 214–230. [CrossRef]
19. Bowa, V.; Samiselo, W.; Manda, E.; Lei, Y.; Zhou, W.; Shane, A.; Chinyanta, S. Wedge Failure Analysis of the Slope Subjected to Uplift Forces by Analytical Method at Chingola Open Pits F & D. *J. Min. Environ.* **2021**, *12*, 941–952.

water

MDPI

Article

An Approach for Preliminary Landslide Scarp Assessment with Genetic Algorithm (GA)

Chih-Ling Wang, Chi-Jyun Ko †, Hock-Kiet Wong, Pei-Hsin Pai and Yih-Chin Tai *

Department of Hydraulic and Ocean Engineering, National Cheng Kung University, Tainan 70101, Taiwan; monkey93135@gmail.com (C.-L.W.); ws612531@gmail.com (C.-J.K.); ikhowong93@hotmail.my (H.-K.W.); paipei14916@smail.nchu.edu.tw (P.-H.P.)
* Correspondence: yctai@ncku.edu.tw
† Current address: Institute of Applied Mechanics, National Taiwan University, Taipei 10617, Taiwan.

Abstract: For the investigation of landslide mass movement scenarios through numerical simulation, a well-defined released mass and a precise initial source area are required as prerequisites. In the present study, we present a genetic algorithm-based approach for preliminarily assessing the landslide scarp when the local field data are limited, using an ellipse-referenced idealized curved surface (ER-ICS)—a smooth surface constructed with respect to an ellipse. According to a specified depth at the center, there are two distinct curvatures along the major and minor axes, respectively. To search for the most appropriate ICS, the reference ellipse is translated, rotated, and/or side-tilted to achieve the optimal orientation for meeting the best fitness to the assigned condition (delineated area or failure depths). The GA approach may significantly enhance the efficiency, by reducing the number of candidate ICSs and notably relaxing the searching ranges. The proposed GA-ER-ICS method is examined and shown to be feasible, by mimicking the source area of a historical landslide event and through application to a landslide-prone site. In addition to evaluating the fitness of the ICS-covered area with respect to the source scarp, the impacts of various ICSs on the flow paths are investigated as well.

Keywords: genetic algorithm (GA); landslide-prone area; landslide scarp assessment; ellipse-referenced idealized curved surface (ER-ICS); flow paths; scenario investigation

Citation: Wang, C.-L.; Ko, C.-J.; Wong, H.-K.; Pai, P.-H.; Tai, Y.-C. An Approach for Preliminary Landslide Scarp Assessment with Genetic Algorithm (GA). *Water* **2022**, *14*, 2400. https://doi.org/10.3390/w14152400

Academic Editor: Olga Petrucci

Received: 31 May 2022
Accepted: 25 July 2022
Published: 2 August 2022

1. Introduction

From the viewpoint of hazard assessment or disaster mitigation, the core concerns are the plausible threats to residents and potential damage to infrastructure. Over the years, researchers have proposed various empirical laws (e.g., [1–3]) and physics-based numerical simulation tools (e.g., [4–7]) in order to delineate landslide-susceptible hazard zones. With either of these approaches, the volume of the released mass and the location of the possible failure surface are prerequisites for the evaluation process. However, predicting the released landslide volume and estimating the failure surface are highly challenging, due to high uncertainty caused by the complexity of spatial geological and hydrological variations. At the same time, weathering effects and the material composition at sites are also generally inhomogeneous (e.g., [8,9]).

With the rapid development of UAVs and modern remote sensing techniques (e.g., LiDAR, UAV-LiDAR, SAR, InSAR, and UAV-SAR), high-resolution digital elevation models (DEMs) have become popular, in which detailed topographic features can be well-recognized. Furthermore, with expensive and time-consuming geological field surveys, the scar boundaries of landslide-prone areas can be estimated and delineated, with respect to some specific features (e.g., crowns, bulges, trenches, or fissures). Based on the DEM, a geometric interpretation method—namely, the Sloping Local Base Level (SLBL) method—has been suggested by Jaboyedoff et al. [10,11] for the approximation of a 3D failure surface, which possesses a constant second derivative in the down-slope direction (i.e., a parabolic curve in section view). Besides the SLBL approach, Reid et al. [12,13] have proposed a spherical surface for analyzing the slope stability of a 3D stratovolcano edifice. Considering the spatial geological structure and groundwater patterns, the concept of spherical failure surfaces has been adopted and extended in the open-source software Scoops3D (see, e.g., [14,15]). Tun et al. [16] have further applied Scoops3D for calculating the probability of multiple failures, where a genetic algorithm (GA) with the first-order reliability method (FORM) was utilized.

For a well-defined landslide scarp area, Kuo et al. [17] have proposed a smooth minimal surface (SMS) method to approximate the failure surface by a smooth surface, where the constructed fracture surface is determined according to the minimal surface area, with the prerequisites of a given landslide volume and a convex polygon-outlined region. Instead of fitting the scarp boundary, Tai et al. [18] suggested the concept of using an idealized curved surface (ICS) to mimic the fracture surface for numerical simulation, where the ICS is defined by two distinct curvatures in the down-slope and cross-slope directions, respectively (cf. Figure 1b). As the ICS approach does not request an exact fitness to the assigned area (source area or delineated area), a search process is needed to find the most appropriate ICS. Motivated by the ellipticity of landslide shapes [19], and for ease of proceeding with the search process, Ko et al. [20] have utilized an ellipse with a specified depth to represent the corresponding ICS (ellipse-reference ICS; ER-ICS), where the depth is used to determine the landslide volume. In the search process, the optimal reference ellipse is selected, with respect to the best fitness to the assigned area, through translation, rotation, and/or side-tilting (cf. Figure 2). The ER-ICS search process is an exhaustive method considering all the candidate ICSs, where the associated depth is determined by the assigned volume. Due to the non-trivial topography, the reference ellipse-covered area varies for different orientations, such that the determination of depth should be repeated for each candidate ICS. Therefore, the search process is rather time-consuming, limiting the search range. Taking the ICS-D in Ko et al. [20] as an example, the search range covers 7 × 7 grids for the top and bottom vertices, yielding 10,633 candidate ellipses with corresponding ICSs to be evaluated. Hence, an efficient method, which either reduces the number of candidate ellipses or enhances the computational performance, is highly desirable. Accordingly, a genetic algorithm (GA) [21–24] can provide an optimal solution for reducing the number of candidate ellipses.

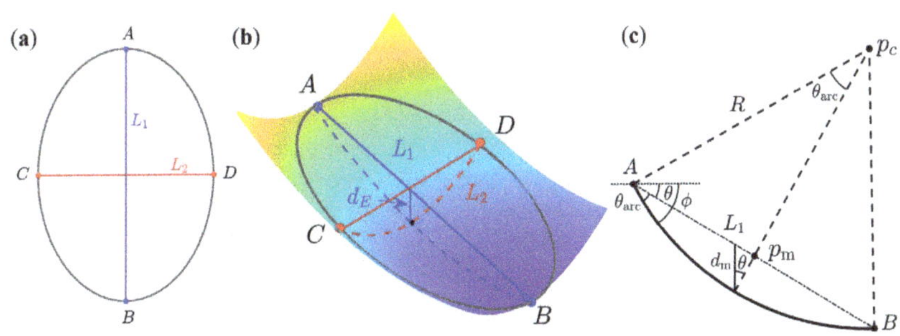

Figure 1. Ellipse-referenced idealized curved surface (ER-ICS): (**a**) The reference ellipse with major axis L_1 and minor axis L_2; (**b**) The constructed ICS with respect to a depth d_E below the middle point; (**c**) Section view of the ICS along the major axis, where $d_{\mathrm{m}} \cos\theta = d_E$ with inclination angle θ.

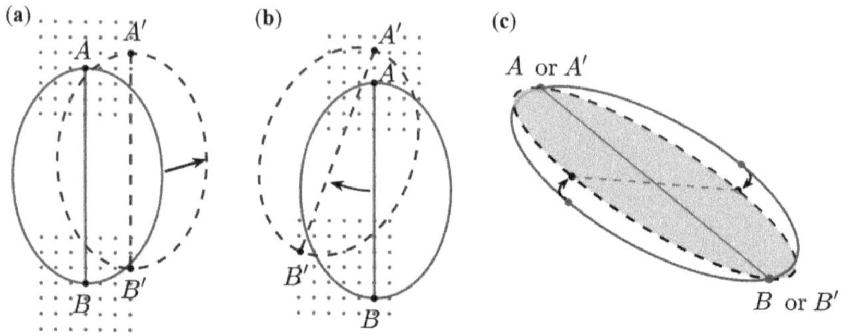

Figure 2. To search for the optimal ellipse, the reference ellipse is (**a**) translated, (**b**) rotated, and/or (**c**) side-tilted.

From the viewpoint of engineering applications, a genetic algorithm (GA) is an optimization tool utilizing an iterative process. It delivers solutions through the use of evolution operators such as mutation, crossover, and selection [24]. In the present study, the "canonical genetic algorithm" (cf. [21,23,24]) is adopted for the search process; that is, the initial population is generated with the first reference ellipse mapping within the assigned search ranges of parameters randomly. The parameters determining the orientation of the reference ellipse comprise the genes. Each constructed ICS is evaluated based on the fitness function, and then assigned a fitness value. The selection process follows the method of *roulette wheel selection* [24], where the possibility of being selected is proportional to fitness (also known as *fitness proportionate selection*). After the selection process, recombination (crossover) of the genes (parameters determining the orientation of the reference ellipse) is conducted for breeding the populations of the next generation, where mutation is allowed with a slight possibility (generally $\leq$1%).

In addition to the GA approach, a manipulation treatment is introduced to isolate some redundant portion(s) of the ICS-covered area, as the complex topography may induce the inclusion of unexpected regions in the neighborhood of the target area. The redundant part can be automatically isolated without additional manual operation, through utilization of *Erosion* and *dilation* operations in morphological image processing (e.g., [25]). This manipulation treatment may retain an ICS whose main portion has good fitness, while removing the redundant part(s).

The feasibility and applicability of the GA-ER-ICS approach are investigated through the validation of a historical landslide event and application to a plausible failure surface, based on the measured failure depths in a landslide-prone area. In terms of the fitness of the target area, the convergence of the employed GA approach is examined, considering the number of generations. The application to landslide-prone areas demonstrates another advantage of the GA-ER-ICS approach, in that the failure surface can be mimicked in a flexible manner for various scenarios. All of the selected ICSs with the corresponding released volumes of landslide mass are integrated into a GPU-accelerated simulation tool (MoSES_2PDF [26]), in order to investigate the impacts of various ICSs on their consequent flow paths.

The remainder of this manuscript is structured as follows. In Section 2, the construction of ellipse-referenced ICS is reviewed, where the manipulation approach for isolating the redundant portion of the ICS-covered area is detailed. In Section 3, the employment of the GA procedure is elaborated. The procedure of the GA-ER-ICS searching process is given in Section 4. Numerical investigations and the application to a landslide-prone area are discussed in Section 5. The key features of the proposed GS-ER-ICS approach and its potential for engineering applications are summarized and highlighted in the concluding remarks.

2. Ellipse-Referenced Idealized Curved Surface (ER-ICS)

2.1. Construction of the ER-ICS

The ICS concept was first suggested to replicate the failure surface of a sliding-type landslide in Tai et al. [18]. As the ICS is characterized by two constant curvatures in the down- and cross-slope directions, respectively, Ko et al. [20] used an ellipse with a specified depth to construct the ICS; that is, the Ellipse-Referenced Idealized Curved Surface (ER-ICS). As shown in Figure 1a, $\overline{AB}$ (with length L_1) denotes the major axis of the ellipse along the down-slope direction, while $\overline{CD}$ (with length L_2) represents the minor axis. Once L_1 and L_2 are fixed, the ICS can then be constructed with respect to a specified depth, d_E, where the corresponding curvature radii (R_1 and R_2) are determined accordingly (cf. panels b and c). In general, point A sits at the upper part of the scarp area, point B is at the lower part, and the length L_2 of $\overline{CD}$ determines the width of the site. Once the ICS is constructed, the landslide volume can be calculated based on the DEM; that is, the reference ellipse defines the orientation and location of the ICS, while the depth determines the two curvatures and the associated landslide volume.

For each reference ellipse, the corresponding depth is determined in accordance with an assigned prerequisite, such as the released volume of mass or a specific failure depth at some location. Hence, it takes time to find the covariant depth for each reference ellipse. There may be thousands of reference ellipses for one site, where each ellipse yields an ICS and serves as a candidate. To construct the most appropriate ICS, Ko et al. [20] have suggested trying various locations/orientations of the ellipse (i.e., translating, rotating, or side-tilting the reference ellipse; cf. Figure 2), where four methods (methods A–D) were considered and evaluated. In the present study, method D in Ko et al. [20] is employed, in which, in addition to translation, rotation, and/or side-tilting, the RE is allowed to stretch or shrink slightly, while keeping the area of the RE invariant. Even though the area of the ellipse is fixed, thousands of candidate ICSs still have to be constructed. If not additionally specified, the most appropriate ICS is the one with the minimal deviation index, calculated as:

$$\Lambda_S = \frac{|A_{\rm ICS} - A_{\rm ICS\cap sa}| + |A_{\rm sa} - A_{\rm ICS\cap sa}|}{A_{\rm sa}}, \tag{1}$$

which indicates the deviation of the ICS-covered area from the source (target) area (cf. [20]). In (1), $A_{\rm ICS}$ denotes the ICS-covered area, $A_{\rm sa}$ is the source (target) area, and $A_{\rm ICS\cap sa}$ represents their intersection.

2.2. Manipulation of the ER-ICS-Covered Area

Due to the complex topography, the constructed ICS might intersect the neighboring hill and include some additional unexpected area. An example can be seen in Figure 3a, in which the blue area indicates the source (target) area and the red dot line depicts the outline of the ICS-covered area. A sizable unexpected area, indexed by "II", can be seen on the right-hand side. Figure 3b shows the section view of the dashed line in panel a, where the portion in tawny color indicates the ICS-determined landslide body of interest, while the intersected part of the unexpected portion (II) is marked in yellow. This unexpected portion can be seen as redundant and is usually isolated manually (e.g., as in [20]), which is highly time-consuming and could differ between individuals. Here, a manipulation process (morphological process), on the basis of the OpenCV software [27], is introduced in order to segment and isolate the redundant section(s) automatically.

The manipulation process consists of two operations: Erosion and Dilation. The *Erosion* process shrinks the area through a local minimum over the area of a given kernel $\mathcal{K}$ (e.g., an $n \times n$ matrix or a circle). It replaces the image pixels (grid) in the anchor point (center of $\mathcal{K}$) with the local minimum. The *Dilation* process extends the area shrunk in the *Erosion* process by determining the local maximum for the anchor point over the kernel-covered area. In the present study, the kernel $\mathcal{K}$ is set using a 3×3 matrix, which is scanned over the whole DEM in the erosion and dilation operations (cf. Figure 3). Letting the grids within the ICS-covered area be indexed by 1 and 0 set for those in the rest of the area (i.e., as a mask), one can divide the connected areas into several blocks by carrying out the Erosion operation several times, in order to isolate the redundant one(s). After that, the resultant mask for the major scarp area is retained and recovered through several runs of the Dilation operation. Figure 3c depicts the eroded area(s) after five operations, where the redundant portion (II) is separated from the primary area (I). The recovered major scarp area is illustrated in Figure 3d, where six Dilation operations have been conducted. It should be noted that the resultant scarp area is valid only when the ICS sits below the topographic surface locally. The additional Dilation operation (versus the five Erosion operations) is conducted to alleviate the cliffs possibly sitting at the boundary of the resultant mask, if they exist.

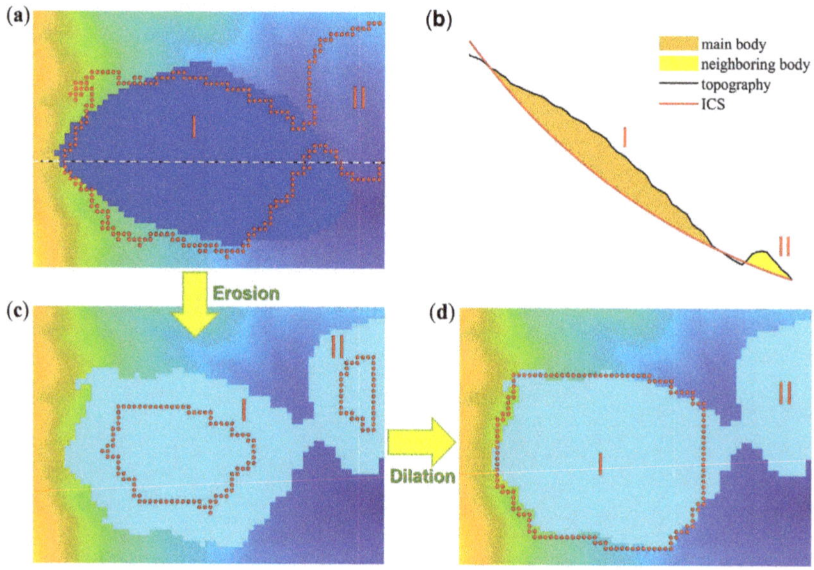

Figure 3. Manipulation of the ICS-covered area. The blue area represents the source (target) area, and the preliminary ICS-covered area is shown in aqua-blue. (**a**) The initial ICS-covered area is marked by the red dotted line and divided into two portions (I and II); (**b**) Section view along the white dashed line in panel (**a**); (**c**) Outlines of the two portions after Erosion operations; (**d**) The resultant mask of the target portion after Dilation operations.

3. Genetic Algorithm Approach

For high efficiency in the search process, a genetic algorithm (GA) is employed (see, for example, [22–24,28]). Together with the GA approach, determination of the failure depth (i.e., d_E, as indicated in Figure 1b) is implemented based on the CUDA structure (cf. [29]) for high-performance GPU computation. A GA is a search process, which is generally used to generate an optimal solution under a given context. The approach used here is composed of two parts: the first consists of decoding the genes which determine the candidate solutions (i.e., the orientation of the reference ellipse; RE), while the second part involves the fitness function, which is used to evaluate the corresponding performance. In this study, the genes used for constructing the RE are $(\delta x, \delta y, \delta\theta, \delta L_1)$, where δx and δy denote translations, $\delta\theta$ denotes rotation, and δL_1 represents stretching/shrinking of the reference ellipse. Each reference ellipse is associated with a depth, in order to construct the ICS and meet the assigned landslide volume. If not additionally specified, the deviation index, Λ_S—defined in (1)—represents the fitness (i.e., serves as a fitting function) between the ICS-covered area and the target area.

In the GA search process, thirty ICSs (population size) are constructed in each generation. During each successive generation, the better-fitted ICSs (i.e., those with smaller deviation index) are selected to breed the next generation through genetic crossover and mutation. The DNA size is 12, and the rates for crossover and mutation are 0.8 and 0.01, respectively. Figure 4 illustrates the framework of the GA-ER-ICS search process. Each ICS is constructed in the GPU section, where the depth is determined based on the assigned volume and the given genes for the orientation of the RE. The GA procedure in the CPU section evaluates the fitness and provides the genes through crossover and mutation operations. We experimentally determined that a plateau of fitness was reached after circa 10 generations. As this result might depend on the size of the search range of genes, we set 10 and 15 generations as the termination criteria for GA-ER-ICS construction in the following numerical investigation and site application, respectively.

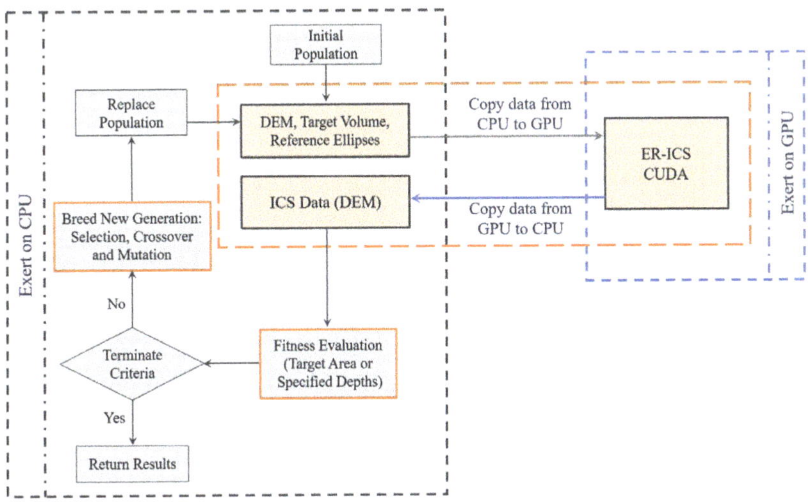

Figure 4. The framework of the GA-ER-ICS search process, in which the ICSs are constructed in the GPU section and the GA is operated in the CPU section.

4. Procedure of the GA-ER-ICS Search Process

The search process consists of four main stages: (1) Preparation; (2) Input and parameter setting; (3) ER-ICS construction; and (4) GA process, termination, and results output.

Stage 1: Preparation and determination of the initial reference ellipse

Based on the source area (post event) or the delineated area of the landslide-susceptible zone, four starting reference points are assigned to determine the initial reference ellipse (RE) on the DEM (cf. Figure 5a). In general, these four reference points indicate the length and width of the target area. The highest and lowest points (in elevation) compose the major axis of the initial RE, while the length of the minor axis is determined by regression for a minimal root mean square (RMS), with respect to the other two reference points.

Stage 2: Input and parameter setting

In this stage, the ranges of the genes $(\delta x, \delta y, \delta \theta, \delta L_1)$ are given, and the size of the population in each generation, as well as the termination criterion (the number of generations), must be assigned. With the initial RE determined in Stage 1, the initial population is randomly generated within the range of the genes.

Stage 3: ER-ICS construction

For each RE in the initial population (or the replaced population of the evolution), the individual depth is determined according to the assigned released volume, which can be given by assignment or through the use of a volume–area empirical relation, as suggested in, for example, [17,18,30]. As each RE is associated with one most appropriate ICS, this process (including the manipulation process introduced in Section 2.2) is highly time-consuming. It is, therefore, conducted in the GPU section (cf. Figure 4). Once all the ICSs are constructed, the GPU section returns the corresponding sets of genes and DEMs to the CPU section.

Stage 4: GA process, termination, and results output

The fitness of all determined ICS (DEMs) is evaluated based on the deviation index Λ_S given in (1), in the case where the source/target area is available. On the other hand, when the evaluation is based on the specified failure depth(s), the fitness is computed based on the root mean square (RMS) between the ICS and the specified depth(s). In both cases, a smaller value of the fitness index (i.e., Λ_S or RMS) indicates better fitness. As illustrated in Figure 4, the next generation is bred as the new population, until the termination criterion (i.e., the maximum number of generations) is fulfilled. The best-fitted ICS in the last generation is considered the most appropriate ICS in the GA search process.

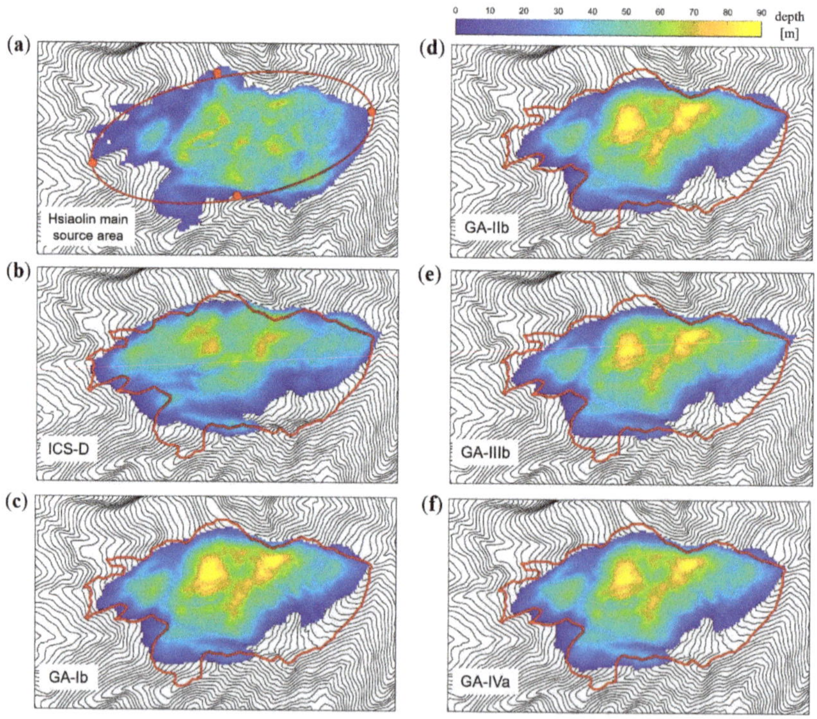

Figure 5. The shape and depth distribution of the initial landslide mass: (**a**) The measured data in the main source area, where the four red points stand for the starting reference points and the red line represents the initial reference ellipse; (**b**) The ICS constructed by the method-D in [20]; (**c**) The best-fitted ICS in condition GA-I (i.e., GA-Ib); (**d**) The best-fitted ICS in condition GA-II (i.e., GA-IIb); (**e**) The best-fitted ICS in condition GA-III (i.e., GA-IIIb); (**f**) The best-fitted ICS in condition GA-IV (i.e., GA-IVa).

5. Numerical Investigation and Application in a Landslide-Prone Area

The performance of the proposed GA-based ER-ICS (GA-ER-ICS) method, in terms of constructing a plausible failure surface, was investigated against a historical large-scale landslide event, and a trial application was arranged in a landslide-prone area. The historical event was the Hsiaolin landslide, triggered during typhoon Moratok in 2009 in southern Taiwan. Due to its huge scale, severe damage (more than 450 victims), and unique characteristics, the 2009 Hsiaolin landslide has become one of the most representative deep-seated landslides worldwide, for which ample well-documented data are available. For further details, the readers are referred to [31–37]. The DEM and initial conditions used in [32,38] were employed in the numerical investigation. Note that only the main source area listed in Kuo et al. [32] was taken into account when mimicking the failure surface, as it provides more than 94% of the total landslide mass.

For application of the GA-ER-ICS method to a landslide-prone area, we assigned a potential large-scale landslide area named Kuanghua-T002, located in Taoyuan county, northern Taiwan. As creeping movements have been detected in this area for years, three inclinometers (indexed by K18-1BW, K18-2BW, and K19-1BW) had been installed. Both K18-1BW and K18-2BW were broken on 31 May 2018, while K19-1BW was broken on 3 December 2019, due to local mass movements (cf. Chen [39]). The plausible outline of the potential deep-seated landslide has been suggested and delineated in Chen [39].

The ranges of genes used in the GA procedure to search for the most appropriate ICSs are listed in Table 1. Due to the complex topography in the neighborhood of the target area, redundant areas were found and, so, the manipulation process was employed for site Kuanghua-T002. For reference landslide volumes, we took the measured amount (21,180,535 m^3) in the Hsiaolin case and 348,581 m^3 (estimated by the volume–area empirical relation suggested in [30]) for the Kuanghua-T002 case. The consequent flow paths of the released landslide masses were computed using a GPU-accelerated simulation tool (MoSES_2PDF in [26]). The values of the material parameters used for computation were identical to those presented in [20], as collected in Table A1. The following index:

$$\Lambda_{\text{path}} = \frac{|A_{\alpha}^{\text{path}} - A_{\alpha\cap\beta}^{\text{path}}| + |A_{\beta}^{\text{path}} - A_{\alpha\cap\beta}^{\text{path}}|}{A_{\alpha\cap\beta}^{\text{path}}}, \tag{2}$$

was introduced for quantitative investigation of the discrepancy of flow paths between two scenarios. In (2), $A_{\alpha\cap\beta}^{\text{path}}$ denotes the intersection (overlapped area) of the flow paths between Scenarios α and β. In the scenario campaigns, the index Λ_S quantifies the fitness of the constructed ICS, while Λ_{path} evaluates the difference between two ICSs (scenarios).

Table 1. Range of genes used in the GA for plausible ICS searches in the 2009 Hsiaolin event and site Kuanghua-T002 cases.

	$(\delta x, \quad \delta y, \quad \delta\theta, \quad \delta L_1)$	**Manipulation**	**Volume (m^3)**
Hsiaolin event	$(\pm 8\Delta x, \pm 8\Delta y, \pm 15^\circ, \pm 8\Delta x)$	no	21,180,535
Kuanghua-T002	$(\pm 5\Delta x, \pm 5\Delta y, \pm 10^\circ, \pm 5\Delta x)$	yes	348,581

5.1. *Numerical Investigation (The 2009 Hsiaolin Landslide)*

With respect to the main source area of the 2009 Hsiaolin landslide, the most appropriate ICS was selected through the GA-ER-ICS search process. Here, the feasibility and applicability are investigated in terms of three aspects: (a) Convergence with respect to the number of generations; (b) the effectiveness of the side-tilting operation; and (c) the impacts of the different coverages between ICSs on the consequent flow paths. The resolution of the used digital elevation map (DEM) was 10 m ($\Delta x = \Delta y = 10$ m), and the computational domain was 3700×2210 m^2, where the projection of the main scarp on the horizontal plane covered an area of 624,900 m^2.

Four conditions for the GA-ER-ICS search process were arranged, and three runs for each situation were carried out (cf. Table 2). The impact of the number of generations (i.e., 10 or 15 generations) on the GA process was also investigated. The search processes with side-tilting are indexed by GA-I and GA-II, in which the tilting angle was determined by the inclination of the minor axis of the reference ellipse to the horizon (i.e., the inclination angle of $\overline{CD}$, shown in Figure 1b). GA-I and GA-III used ten generations, while GA-II and GA-IV used 15 generations. The constructed ICSs were compared with the main source area. In addition, the most appropriate ICS (ICS-D) in Ko et al. [20], as determined by method D, was included in the campaigns for comparison. We first focused on the fitness of the landslide source area and the sensitivity of the ICSs constructed by the proposed GA-ER-ICS approach under various conditions. After that, the consequent flow paths were computed, with respect to the best-fitted ICSs. The flow paths were determined using the moving mass-covered areas of 61 sets of results from 0.0 s to 181.83 s with an interval of 3.0305 s, where only areas with flow thickness greater than 10 cm were considered.

Table 2. Fitness of the ICSs under various conditions for the main source area in the 2009 Hsiaolin event.

Condition	Run	Generation	Side-Tilting	Λ_S
ICS-D		-	-	28.72%
GA-I	a	10	yes	30.05%
	b *	10	yes	25.01%
	c	10	yes	25.83%
GA-II	a	15	yes	24.08%
	b *	15	yes	24.05%
	c	15	yes	24.24%
GA-III	a	10	no	25.76%
	b *	10	no	24.23%
	c	10	no	29.12%
GA-IV	a *	15	no	24.15%
	b	15	no	25.52%
	c	15	no	24.82%

* The best-fitted among the three runs.

5.1.1. Fitness to the Main Source Area

With respect to the starting reference points (the four red points marked in Figure 5a), three runs were performed for each condition (GA-I to GA-IV). The performance of the selected ICS was evaluated with respect to the fit to the measured source area, as indexed by the value of Λ_S defined in (1); see Table 2. A smaller value of Λ_S indicates a better fit. As the GA process does not guarantee identical results, the Λ_S value varied with each run. It was found that ten generations (GA-I and GA-III) did not provide satisfactory convergence of fitness; however, a plateau of fitness is reached when the evolution terminated at 15 generations. Despite the small difference, inclusion of the side-tilting operation slightly improved the fitness (see GA-I vs. GA-III and GA-II vs. GA-IV). Furthermore, it was found that 10 of the resultant ICSs, among the 12 runs of the GA search process, presented better fitness than ICS-D. The better performance of the GA approaches is possibly due to the more considerable translation distance (i.e., ± 8 grids in the GA approach versus ± 3 grids in the method of exhaustion). Together with its high efficiency, the proposed GA approach is apparently superior to the exhaustive method.

The best-fitted ICSs in the four conditions (GA-Ib, GA-IIb, GA-IIIb, and GA-IVa) are exhibited in Figure 5. The measured depth distribution of the released landslide mass from the primary source area is shown in panel a, and its outline is depicted in all the other panels for comparison. Panel b displays the ICS-D result. All the ICSs, including ICS-D, yielded higher thickness than the measured data (cf. Figure 5). This phenomenon is suspected to have been induced by two causes. The first is the smaller ICS-covered area, such that a higher thickness of the initial landslide body is required to retain the constant reference volume. The second reason could be that the ICS is a smooth surface, differing from the natural failure surface, for which the local geological conditions may play a crucial role. Another interesting finding is that the GA-selected ICSs were rather similar, for which the value of Λ_S ranged from 24.05% (GA-IIb) to 25.01% (GA-Ib). It was also found that the side-tilting operation did not significantly improve the performance with either 10 or 15 generations, if there were three runs for selection in each condition. Hence, the operation of side-tilting was not employed for the application in a landslide-prone area (Section 5.2).

5.1.2. Impacts on Flow Paths

Ko et al. [20] have investigated the consequent flow paths of the ICS by comparing the results computed with ICS-D against the results calculated using the measured landslide scarp. The area discrepancy between the measured failure surface (FS) and ICS-D was 26.82% at the initial stage of initiation, which reduced to circa 9.5% for the computed results at the rest state. Here, we focus on the discrepancy of the flow paths computed with the selected ICSs. The differences between the initial areas and the discrepancy of the flow paths are quantitatively denoted by the indices Λ^{init}_{path} and Λ_{path}, respectively (see Table 3). In agreement with the illustration in Figure 5, where all the GA process-selected ICSs can be seen to be rather similar, the values of Λ^{init}_{path} were in the range from 4.13% to 6.90% (cf. Table 3). With 15 generations, no significant discrepancy between GA-IIb (with side-tilting) and GA-IVa (without side-tilting) could be identified, with respect to the initial ICS-covered area and the consequent flow paths, where $\Lambda^{init}_{path} = 4.13\%$ and $\Lambda_{path} = 3.87\%$.

Table 3. Comparison of the flow paths under various conditions for the 2009 Hsiaolin event.

Impacts	**ICS**	Λ^{init}_{path}	Λ_{path}
side-tilting	GA-Ib vs. GA-IIIb	6.90%	10.52%
	GA-IIb vs. GA-IVa	4.13%	3.87%
generation	GA-Ib vs. GA-IIb	5.38%	6.06%
	GA-IIIb vs. GA-IVa	5.70%	7.77%
Exhaustion vs. GA	ICS-D vs. GA-Ib	15.08%	14.24%
	ICS-D vs. GA-IIb	14.77%	12.63%
	ICS-D vs. GA-IIIb	12.84%	13.97%
	ICS-D vs. GA-IVa	14.53%	13.13%

Among the four selected best-fitting ICSs (GA-Ib, GA-IIb, GA-IIIb, and GA-IVa), the results computed with GA-Ib had the most significant discrepancy from the results, compared with ICS-D, where $\Lambda^{init}_{path} = 15.08\%$ and $\Lambda_{path} = 14.24\%$. Figure 6 depicts the associated distinctions at the initial stage (panel a) and the flow paths (panel b). The four red markers in panel a are the initial reference points in the search process, and the red line outlines Hsiaolin village. It was found that most of the differences took place around the source area, due to the shape discrepancy in the initial stage. For a thorough overview, the flow paths of ICS-D, GA-Ib, GA-IIb, GA-IIIb, and GA-IVa were all collected, and are displayed in a comparable way in Figure 7, in which the blue zone represents the shared paths. The individual routes yielded by ICS-D, GA-Ib, GA-IIb, GA-IIIb, and GA-IVa are shown in red, magenta, cyan, green, and yellow, respectively. Due to the minor distinction between the source areas, only one of the two outlines is given in each comparison scenario. We refer the readers to Figure A1 in Appendix B for the details of the outlines between the source areas, in accordance with the sequence of Figure 7. As has already been elaborated and illustrated in Figure A1e, the flow paths of GA-IIb and GA-IVa were very close, with $\Lambda_{path} = 3.87\%$. Both of them were constructed using 15 generations in the GA process, revealing that, with a sufficient number of generations for evolution, the effectiveness of the side-tilting operation becomes insignificant. The data listed in Table 3 indicate that the GA-ER-ICS method may deliver the selected ICSs with good convergence, in terms of the flow paths. This convergence might benefit from the repetition of operations and a sufficient number of generations for evolution in the GA search process.

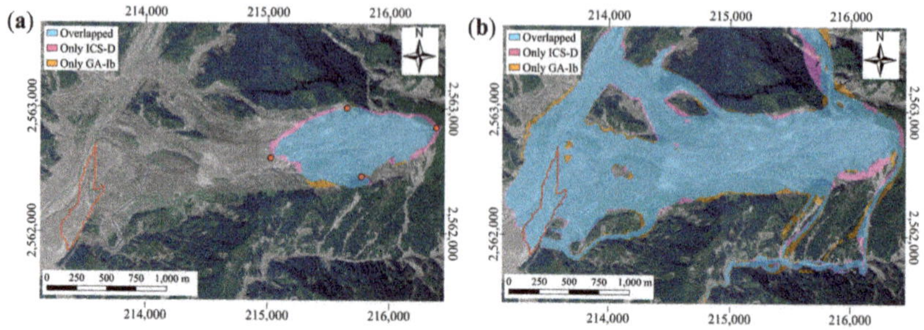

Figure 6. Orthophoto and landslide mass-covered areas in the computation, where cyan indicates the overlapped area, the pink area is covered only by the results computed with the ICS-D, and the area in dark-yellow is occupied only in the condition of GA-Ib. The red line denotes Hsiaolin village, and the four red markers in panel a represent the initial reference points in the GA-ER-ICS search process. (**a**) Initial stage; (**b**) Flow paths (Orthophoto: Courtesy of Serial Survey Office, Forestry Bureau, Taiwan).

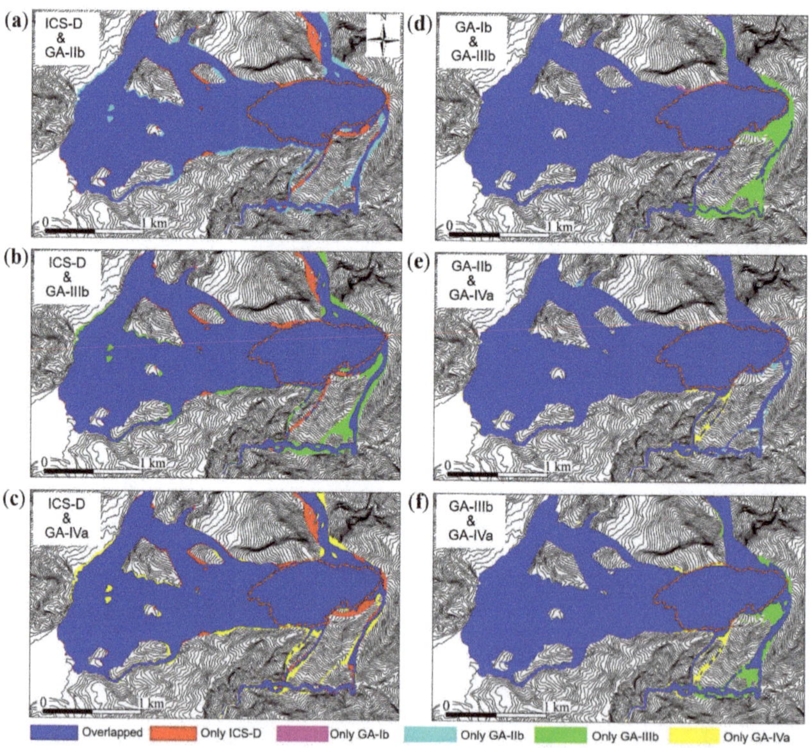

Figure 7. Flow paths in various conditions, where the blue represents the overlapping area. The red, magenta, cyan, green, and yellow areas are covered only by results computed with ICS-D, GA-HL-Ib, GA-HL-IIb, GA-HL-IIIb, and GA-HL-IVa, respectively. The brown dash-dotted lines outline the source areas: (**a**) ICS-D; (**b**) GA-IIIb; (**c**) GA-IVa; (**d**) GA-Ib; (**e**) GA-IIb; (**f**) GA-IVa.

5.2. Application to a Landslide-Prone Area (Kuanghua-T002)

Based on the records of the three installed inclinometers (K18-1BW, K18-2BW, and K19-1BW), three scenarios (A, B, and C) were designed, as listed in Table 4. In Scenario A, the fitness evaluation is based on the failure depths recorded by gaging wells K18-1BW and K18-2BW. Only K19-1BW is considered in Scenario B, while the failure depths of all three gaging wells are taken into account in Scenario C. In this application, the DEM has a resolution of 5 m and the horizontal projection of the delineated source area is 32,325 m^2. This is much smaller than the main source area in the Hsiaolin case, so we set $\delta x = \delta y = \delta L_1 = \pm 5\Delta x$ in the GA-ER-ICS search process.

Table 4. Scenarios for the Kuanghua-T002 landslide-prone area.

Scenario	Referenced Gaging Well	ICS-Determined Failure Depth (m)			Λ_S	$V_{Target}/V_{Guzzetti}$
		K18-1BW	K18-2BW	K19-1BW		
A	K18-1BW & K18-2BW	5.68	24.98	(38.71)	23.36%	1.5
B	K19-1BW	(none)	(14.08)	27.35	29.47%	1.0
C	K18-1BW, K18-2BW & K19-1BW	4.63	19.27	30.71	36.58%	1.0
Inclinometer records	-	5	25	27	-	-

Note: Values in brackets are not for the fitness evaluation in the GA search processes.

Although local movement and failures have been detected by the inclinometers, they did not take place at the same time, and the landslide body has not yet been ultimately released. Hence, the exact volume of the landslide mass is not available for determining the depth, as well as constructing the ICS. The reference volume for constructing the ICS can basically be approximated by empirical laws based on the delineated source area, such as Guzzetti's empirical volume–area relation [30] or other similar laws, as detailed in [17,18]. As the ICS-determined failure surface is required to be close to the records of the inclinometers, with as sound a fitness to the delineated area (e.g., $\Lambda_S \leq 40\%$) as possible in the GA-search process, one may not always obtain satisfactory results. In such circumstances, we can relax the constraint of the reference (target) volume; for example, the reference volume V_{Target} in Scenario A was suggested as $1.5 \times V_{Guzzetti}$ for a more satisfactory result. The best-fitting ICSs in Scenarios A, B, and C are listed in Table 4, where the values in brackets were not taken into account when evaluating the fitness in the GA search processes.

The best-fitting ICSs in Scenarios A, B, and C are illustrated in the left panels of Figure 8. The yellow line depicts the outline of the delineated area, while the ICS-covered regions are marked in an emerald-green color. In Figure 8a, the four red markers represent the reference points for determining the first RE in the GA-ER-ICS search process. As reported in Table 4, the discrepancy between the ICS-determined source area and the delineated one was indexed by Λ_S, whose value ranged from 23.36% to 36.58%. It is interesting to find that the best-fitting ICS in Scenario A was found with a reference volume $V^A_{Target}/V_{Guzzetti} = 1.5$. At the same time, it delivered the best fitness ($\Lambda_S = 23.36\%$) among these three scenarios.

The flow paths under the three scenarios were computed in accordance with the ICS-determined source area. The computational domain covered 1275×740 m^2 and the DEM had a resolution of $\Delta x = \Delta y = 5$ m. The simulation period was 101.0153 s, and the flow paths were determined using the moving mass-covered areas of 51 sets of results from 0.0 s to 101.0153 s, with an interval of 2.0203 s. Similar to the illustration of the flow paths in the Hsiaolin event, only the areas covered by flow thickness of more than 10 cm were taken into account when determining the flow paths. Figure 8d–f present the flow paths under the three scenarios. The turquoise color indicates the overlapping area, while the green, orange, and pink areas are covered only by the results computed in Scenarios A, B, and C, respectively. As reported in Table 5, although the discrepancy between the ICS-covered areas Λ^{init}_{path} was more significant (ranging from 22.16% to 40.12%), the discrepancy for the whole flow paths Λ_{path} was reduced to less than 10%. We suspect that the channelized topography diminished the discrepancy at the early stage.

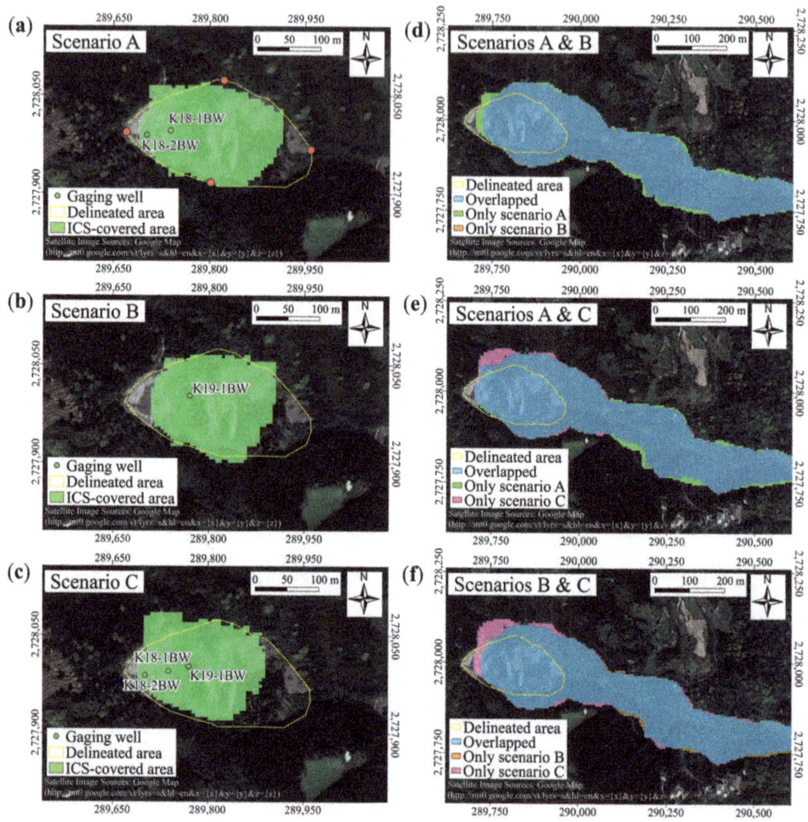

Figure 8. The delineated area (outlined by a yellow line), the ICS-covered area, and the consequent flow paths, where the four red points in panel (**a**) are the initial vertices of the GA-ER-ICS search process. The emerald-green indicates the ICS-covered area, the turquoise represents the overlapping area, and the green, orange, and pink regions are covered only by the results computed under Scenarios A, B, and C, respectively: (**a**) ICS-covered area in Scenario A; (**b**) ICS-covered area in Scenario B; (**c**) ICS-covered area in Scenario C; (**d**) Computed flow paths in Scenarios A and B; (**e**) Computed flow paths in Scenarios A and C; and (**f**) Computed flow paths in Scenarios B and C. (Satellite image: http://mt0.google.com/vt/lyrs=s&hl=en&x=x&y=y&z={z} (accessed on 4 May 2022)).

Table 5. Comparison of flow paths between various scenarios.

Kuanghua-T002	$\Lambda_{\text{path}}^{\text{init}}$	Λ_{path}
Scenarios A vs. B	22.16%	6.80%
Scenarios A vs. C	30.42%	8.93%
Scenarios B vs. C	40.12%	9.72%

Among the scenarios, the released mass in Scenario A was 50% more than those in the other two scenarios. Still, the corresponding values of Λ_{path} were not as notable as the volume difference. In this regard, the volume of released mass seems not to be the critical factor for the flow paths. On the other hand, the released volumes in Scenarios B and C were identical, but the values of the associated $\Lambda_{\text{path}}^{\text{init}}$ and Λ_{path} were dominant. Nevertheless, the discrepancy in the campaign between Scenarios B and C reduced from 40.12% at the initial stage to 9.72% for the whole flow-flushed region, indicating that the channelized topography in the downstream area plays a significant role in the consequent flow paths (see also [18]); that is, a considerable discrepancy at the initial stage may be attenuated during the movement in a channelized topography. These findings reveal that the flow paths are not sensitive to various ICSs under a channelized topography. In the three scenarios, the simulated flow paths did not touch the houses sitting in the lower right corner of Figure 8d–f. Although a more detailed study concerning various parameter sets is needed, these results indicate that the buildings do not lie in the center of the flow paths and, so, are at low risk. Such flow path investigations, using constructed ICSs under various scenarios, might provide an excellent representative hazard assessment model for landslide-prone areas.

6. Discussion and Concluding Remarks

In the present work, we outlined an efficient methodology integrating a GA approach with the reference ellipse-ICS method, in order to mimic the plausible failure surface, as well as estimate the scarp, for a landslide-prone area. The proposed method does not aim to extract a precise estimation but, instead, to provide a preliminary assessment; especially when no detailed field data are available. The reference ellipse-ICS method [20] utilizes the reference ellipse to construct the ICS. However, the goodness-of-fit to the assigned targets (e.g., shape of the area, failure depths) highly depends on the location and orientation of the reference ellipse. As a matter of course, there are thousands of orientations (candidate ICSs) to be evaluated for a target site, in order to determine the most appropriate ICS. The GA process reduces the number of candidate ICSs by preserving the critical characteristics of the ICS through the concept of evolution, such that one can practically employ a broader range of parameters (treated as genes) in the search process.

For each assigned condition/scenario, the search operation was repeated three times, and the best-fitted one was selected for simulation of the associated flow path. The fitting ability of the constructed ICSs to the main source area of the 2009 Hisiaolin landslide was numerically studied, and the effectiveness of the side-tilting operation and the convergence against the number of generations were examined. It is interesting to note that all of the GA-selected ICSs exhibited similar shapes, with deviation index $\Lambda_S \in [0.2405, 0.3005]$. The results also revealed that the side-tilting operation did not have a remarkable impact on the fitness performance. In addition, it was found that a satisfactory ICS could be found after ten generations among the three runs, although a more stable plateau of fitness was obtained with 15 generations. In the investigation of flow paths, most of the discrepancy took place in the early stage, and the presence of a channelized topography may attenuate the impact of the difference between ICSs on the flow path.

Application of the proposed method to a landslide-prone area (Kuanghua-T002) exhibited a representative example, concerning the utilization of the GA-ER-ICS method for a preliminary hazard assessment, through the delineation of a potential zone without requiring detailed geological structure or hydrological conditions as prerequisites. Based on the records of installed inclinometers, three scenarios (Scenarios A, B, and C) were considered. The GA-ER-ICSs were constructed and selected in accordance with the recorded failure depths in the scenarios. It was found that simultaneously meeting the fitness of the target area, failure depth(s), and the assigned volume of released mass is highly challenging. Hence, some compromises (e.g., the fitness to the delineated area or the landslide volume) may be needed. In the three scenarios, all of the ICS-covered areas deviated from the delineated region, and were distinct from each other. Despite the clear distinction among the ICS-covered zones, simulation of the landslide routes revealed that the consequent flow paths were not sensitive under a channelized topography.

It is worth noting that the ICS is a preliminary approximation to the landslide scarp and should be used for the purpose of scenario investigation by numerical simulation when only limited field data are available. It is intended to make up for a deficiency, instead of replacing conventional slope stability analyses. We should admit that there is inevitably a notable discrepancy between the smooth ICS and the non-trivial natural failure surface. Nevertheless, our investigations support the fact that this discrepancy does not significantly influence the consequent flow paths (see also [18,20]). The integration of the GA approach with the ER-ICS method has significantly enhanced the efficiency of the searching process. In comparison with the method of exhaustion (ER-ICS), the proposed GA-ER-ICS may reduce the computational time from more than 20 hrs (for ICS-D) to circa 13.5 mins (e.g., for GA-IIb with 15 generations) with a PC (i7-9700 CPU@3.00 GHz×8, 64 GB memory, Linux OS Ubuntu 18.04), when a GPU (NVIDIA GeForce TRX 2080Ti) is utilized. It should be noted that, in the GA-ER-ICS process, a significantly more extensive translation range (17 × 17 grids) is taken into account in the search process, compared to that in the exhaustive method (7 × 7 grids for the ICS-D in [20]). The proposed GA-ER-ICS method and the GPU-accelerated simulation tool [26] facilitate a highly efficient hazard assessment system, which is currently under development. We intend to report updates on the system in due time.

Author Contributions: Conceptualization, Y.-C.T. and P.-H.P.; methodology, Y.-C.T.; software, C.-L.W. and C.-J.K.; validation, C.-L.W., H.-K.W. and Y.-C.T.; formal analysis, C.-L.W. and H.-K.W.; investigation, H.-K.W. and Y.-C.T.; resources, Y.-C.T.; data curation, H.-K.W.; writing—original draft preparation, Y.-C.T.; writing—review and editing, Y.-C.T.; visualization, H.-K.W. and P.-H.P.; supervision, Y.-C.T.; project administration, Y.-C.T.; funding acquisition, Y.-C.T. All authors have read and agreed to the published version of the manuscript.

Funding: This research was partly founded by the Ministry of Science and Technology, Taiwan (MOST 110-2221-E-006-045-MY2) and the Soil and Water Conservation Bureau, Council of Agriculture, Taiwan (SWCB-111-035).

Institutional Review Board Statement: Not applicable.

Informed Consent Statement: Not applicable.

Data Availability Statement: The data that support the findings of this study are available from the corresponding author upon reasonable request.

Acknowledgments: The authors are grateful to Chun-Wei Lu for his support and discussions related to the GA method.

Conflicts of Interest: The authors declare no conflict of interest.

Abbreviations

The following abbreviations are used in this manuscript:

CUDA	Compute Unified Device Architecture
DEM	Digital Elevation Model
ER-ICS	Ellipse-Referenced ICS
GA	Genetic Algorithm
GA-ER-ICS	Genetic Algorithm-selected Ellipse-Referenced ICS
GPU	Graphics Processing Unit
ICS	Idealized Curved Surface
RE	Reference Ellipse
SLBL	Sloping Local Base Level
UAV	Unmanned Aircraft System

Appendix A

In the GPU-accelerated simulation tool (MoSES_2PDF [26]), there are five material parameters $(\alpha_\rho, \delta_b, C_d, N_R, \vartheta_b)$ and one initial concentration, ϕ_0^s, for the flow body to be set (cf. Tai et al. [38]). Here, $\alpha_\rho = \rho^f/\rho^s$ is the density of the interstitial fluid to the solid constituent, δ_b denotes the angle of basal friction of the solid constituent, C_d represents the drag coefficient between the interstitial fluid and the solid constituent, N_R is proportional to the inverse viscosity (similar to the Reynolds number), and ϑ_b denotes the fluid friction coefficient at the basal surface. It should be noted that the values of the material parameters depend on the composition of the moving mass. The determination of their values requires further study, and is beyond the current scope of this paper. To enable a comparison with the results provided in previous studies [18,20,38], identical values were adopted in the investigation of the consequent flow paths, which are listed in Table A1.

Table A1. Material parameters and initial volume fraction used for computing flow paths.

Parameter	α_ρ	δ_b	c_D	N_R	ϑ_b	ϕ_0^s
Used value	1.42/2.6	16°	6.0	268	5.0	0.5

Appendix B

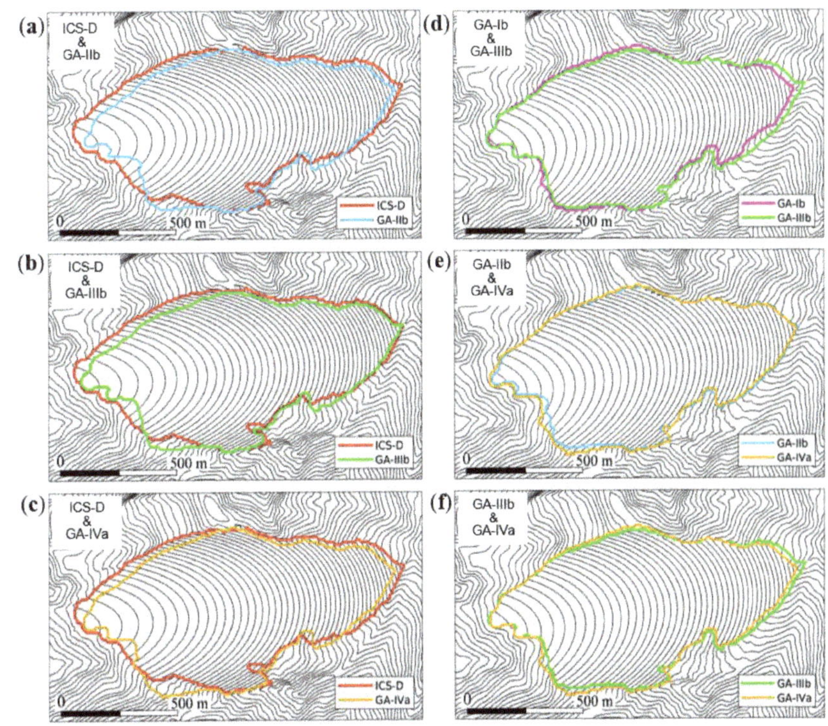

Figure A1. Outlines of the ICS-determined failure (source) areas used for computing the flow paths in the corresponding panels in Figure 7.

References

1. Hsu, K.J. Catastrophic debris streams (sturzstroms) generated by rockfalls. *Geol. Soc. Am. Bull.* **1975**, *86*, 129–140. [CrossRef]
2. Davies, T.R. Spreading of rock avalanche debris by mechanical fluidization. *Rock Mech.* **1982**, *15*, 9–24. [CrossRef]
3. Legros, F. The mobility of long-runout landslides. *Eng. Geol.* **2002**, *63*, 301–331. [CrossRef]
4. Savage, S.B.; Hutter, K. The motion of a finite mass of granular material down a rough incline. *J. Fluid Mech.* **1989**, *199*, 177–215. [CrossRef]
5. Gray, J.; Wieland, M.; Hutter, K. Gravity-driven free surface flow of granular avalanches over complex basal topography. *Proc. R. Soc. Lond. Ser. A Math. Phys. Eng. Sci.* **1999**, *455*, 1841–1874. [CrossRef]
6. Pudasaini, S.P.; Hutter, K. Rapid shear flows of dry granular masses down curved and twisted channels. *J. Fluid Mech.* **2003**, *495*, 193. [CrossRef]
7. Tai, Y.C.; Kuo, C.Y.; Hui, W.H. An alternative depth-integrated formulation for granular avalanches over temporally varying topography with small curvature. *Geophys. Astrophys. Fluid Dyn.* **2012**, *106*, 596–629. [CrossRef]
8. Jaboyedoff, M.; Chigira, M.; Arai, N.; Derron, M.H.; Rudaz, B.; Tsou, C.Y. Testing a failure surface prediction and deposit reconstruction method for a landslide cluster that occurred during Typhoon Talas (Japan). *Earth Surf. Dyn.* **2019**, *7*, 439–458. [CrossRef]

9. Yeh, C.H.; Dong, J.J.; Khonevisan, S.; Juang, C.H.; Huang, W.C.; Lu, Y.C. The role of the geological uncertainty in a geotechnical design–A retrospective view of Freeway No. 3 Landslide in Northern Taiwan. *Eng. Geol.* **2021**, *291*, 106233. [CrossRef]
10. Jaboyedoff, M.; Baillifard, F.; Couture, R.; Locat, J.; Locat, P.; Rouiller, J. Toward preliminary hazard assessment using DEM topographic analysis and simple mechanic modeling. In *Proceedings of the Landslides Evaluation and Stabilization*; Lacerda, W., Ehrlich, M., Fontoura, A., Sayo, A., Eds.; Balkema: Rotterdam, The Netherlands, 2004; pp. 191–197.
11. Jaboyedoff, M.; Couture, R.; Locat, P. Structural analysis of Turtle Mountain (Alberta) using digital elevation model: Toward a progressive failure. *Geomorphology* **2009**, *103*, 5–16. [CrossRef]
12. Reid, M.E.; Christian, S.B.; Brien, D.L. Gravitational stability of three-dimensional stratovolcano edifices. *J. Geophys. Res. Solid Earth* **2000**, *105*, 6043–6056. [CrossRef]
13. Reid, M.E.; Sisson, T.W.; Brien, D.L. Volcano collapse promoted by hydrothermal alteration and edifice shape, Mount Rainier, Washington. *Geology* **2001**, *29*, 779–782. [CrossRef]
14. Reid, M.E.; Christian, S.B.; Brien, D.L.; Henderson, S. Scoops3D–Software to Analyze Three-Dimensional Slope Stability Throughout a Digital Landscape. In *US Geological Survey Techniques and Methods*; Volcano Science Center: Menlo Park, CA, USA, 2015.
15. Alvioli, M.; Lee, G.; An, H.U. Three-dimensional, time-dependent modeling of rainfall-induced landslides over a digital landscape: A case study. *Landslides* **2018**, *15*, 1071–1084.
16. Tun, Y.W.; Llano-Serna, M.A.; Pedroso, D.M.; Scheuermann, A. Multimodal reliability analysis of 3D slopes with a genetic algorithm. *Acta Geotech.* **2019**, *14*, 207–223. [CrossRef]
17. Kuo, C.Y.; Tsai, P.W.; Tai, Y.C.; Chan, Y.H.; Chen, R.F.; Lin, C.W. Application Assessments of Using Scarp Boundary-Fitted, Volume Constrained, Smooth Minimal Surfaces as Failure Interfaces of Deep-Seated Landslides. *Front. Earth Sci.* **2020**, *8*, 211. [CrossRef]
18. Tai, Y.C.; Ko, C.J.; Li, K.D.; Wu, Y.C.; Kuo, C.Y.; Chen, R.F.; Lin, C.W. An idealized landslide failure surface and its impacts on the traveling paths. *Front. Earth Sci.* **2020**, *8*, 313. [CrossRef]
19. Taylor, F.E.; Malamud, B.D.; Witt, A.; Guzzetti, F. Landslide shape, ellipticity and length-to-width ratios. *Earth Surf. Process. Landf.* **2018**, *43*, 3164–3189. [CrossRef]
20. Ko, C.J.; Wang, C.L.; Wong, H.K.; Lai, W.C.; Kuo, C.Y.; Tai, Y.C. Landslide Scarp Assessments by Means of an Ellipse-Referenced Idealized Curved Surface. *Front. Earth Sci.* **2021**, *9*, 862. [CrossRef]
21. Holland, J.H. *Adaptation in Natural and Artificial Systems*; University of Michigan Press: Ann Arbor, MI, USA, 1975.
22. De Jong, K.A. *An Analysis of the Behavior of a Class of Genetic Adaptive Systems*; University of Michigan: Ann Arbor, MI, USA, 1975.
23. Holland, J.H. *Adaptation in Natural and Artificial Systems: An Introductory Analysis with Applications to Biology, Control, and Artificial Intelligence*; MIT Press: Cambridge, MA, USA, 1992.
24. Whitley, D. A genetic algorithm tutorial. *Stat. Comput.* **1994**, *4*, 65–85. [CrossRef]
25. Van Den Boomgaard, R.; Van Balen, R. Methods for fast morphological image transforms using bitmapped binary images. *CVGIP Graph. Model. Image Process.* **1992**, *54*, 252–258. [CrossRef]
26. Ko, C.J.; Chen, P.C.; Wong, H.K.; Tai, Y.C. MoSES_2PDF: A GIS-Compatible GPU-accelerated High-Performance Simulation Tool for Grain-Fluid Shallow Flows. *arXiv* **2021**, arXiv:2104.06784.
27. OpenCV Modules. Available online: https://docs.opencv.org/3.2.0/index.html (accessed on 18 February 2021).
28. Grefenstette, J.J. Optimization of control parameters for genetic algorithms. *IEEE Trans. Syst. Man Cybern.* **1986**, *16*, 122–128. [CrossRef]
29. CUDA C++ Programming Guide. Available online: https://docs.nvidia.com/cuda/archive/11.4.0/cuda-c-programming-guide/index.html (accessed on 17 August 2021).
30. Guzzetti, F.; Ardizzone, F.; Cardinali, M.; Rossi, M.; Valigi, D. Landslide volumes and landslide mobilization rates in Umbria, central Italy. *Earth Planet. Sci. Lett.* **2009**, *279*, 222–229. [CrossRef]
31. Dong, J.J.; Li, Y.S.; Kuo, C.Y.; Sung, R.T.; Li, M.H.; Lee, C.T.; Chen, C.C.; Lee, W.R. The formation and breach of a short-lived landslide dam at Hsiaolin village, Taiwan—Part I: Post-event reconstruction of dam geometry. *Eng. Geol.* **2011**, *123*, 40–59. [CrossRef]
32. Kuo, C.; Tai, Y.C.; Chen, C.; Chang, K.; Siau, A.; Dong, J.; Han, R.; Shimamoto, T.; Lee, C. The landslide stage of the Hsiaolin catastrophe: Simulation and validation. *J. Geophys. Res. Earth Surf.* **2011**, *116*. [CrossRef]
33. Lo, C.M.; Lin, M.L.; Tang, C.L.; Hu, J.C. A kinematic model of the Hsiaolin landslide calibrated to the morphology of the landslide deposit. *Eng. Geol.* **2011**, *123*, 22–39. [CrossRef]
34. Tsou, C.Y.; Feng, Z.Y.; Chigira, M. Catastrophic landslide induced by typhoon Morakot, Shiaolin, Taiwan. *Geomorphology* **2011**, *127*, 166–178. [CrossRef]
35. Chang, P.Y.; Chen, C.C.; Chang, S.K.; Wang, T.B.; Wang, C.Y.; Hsu, S.K. An investigation into the debris flow induced by Typhoon Morakot in the Siaolin Area, Southern Taiwan, using the electrical resistivity imaging method. *Geophys. J. Int.* **2012**, *188*, 1012–1024. [CrossRef]
36. Kuo, Y.S.; Tsai, Y.J.; Chen, Y.S.; Shieh, C.L.; Miyamoto, K.; Itoh, T. Movement of deep-seated rainfall-induced landslide at Hsiaolin Village during Typhoon Morakot. *Landslides* **2013**, *10*, 191–202. [CrossRef]
37. Wu, J.H.; Lin, W.K.; Hu, H.T. Post-failure simulations of a large slope failure using 3DEC: The Hsien-du-shan slope. *Eng. Geol.* **2018**, *242*, 92–107. [CrossRef]

38. Tai, Y.C.; Heß, J.; Wang, Y. Modeling Two-Phase Debris Flows With Grain-Fluid Separation Over Rugged Topography: Application to the 2009 Hsiaolin Event, Taiwan. *J. Geophys. Res. Earth Surf.* **2019**, *124*, 305–333. [CrossRef]
39. Chen, C.W. *Investigation and Mitigation Strategies Planning for the Large Scale Landslide on Shiuwunau Area and Kuanghua Area (2019)*; Technical Report; Rep.-No.: SWCB-109-200; Soil and Water Conservative Bureau, Council of Agriculture: Nantou, Taiwan, 2020.

water

MDPI

Article

Using the Spatiotemporal Hot Spot Analysis and Multi-Annual Landslide Inventories to Analyze the Evolution and Characteristic of Rainfall-Induced Landslide at the Subwatershed Scale in Taiwan

Chunhung Wu

Department of Water Resources Engineering and Conservation, Feng Chia University, Taichung 40724, Taiwan; chhuwu@fcu.edu.tw; Tel.: +886-4-24517250-3231

Abstract: This study used rainfall and annual landslide data for the 2003–2017 period in Taiwan to determine the long-term evolution of landslides and conducted a spatiotemporal analysis of landslides at the subwatershed scale. The historically severe landslide induced by Typhoon Morakot in 2009 was mainly distributed in the central mountainous region and southern Taiwan. The Mann–Kendall trend test revealed that in 2003–2017, 13.2% of subwatersheds in Taiwan exhibited an upward trend of landslide evolution. Local outlier analysis results revealed that the landslide high–high cluster was concentrated in the central mountainous region and southern Taiwan. Moreover, the spatiotemporal analysis indicated 24.2% of subwatersheds in Taiwan in 2003–2017 as spatiotemporal landslide hot spots. The main patterns of spatiotemporal landslide hot spots in 2003–2017 were consecutive, intensifying, persistent, oscillating, and sporadic hot spots. The recovery rate in the first two years after the extreme rainfall-induced landslide event in Taiwan was 22.2%, and that in the third to eighth years was 31.6%. The recovery rate after extreme rainfall-induced landslides in Taiwan was higher than that after major earthquake-induced landslides in the world, and the new landslides were easily induced in the area of rivers and large landslide cases after Typhoon Morakot in 2009.

Keywords: landslide evolution; landslide recovery rate; local outlier analysis; Mann–Kendall trend test; spatiotemporal analysis; sub-watershed scale; Taiwan

Citation: Wu, C. Using the Spatiotemporal Hot Spot Analysis and Multi-Annual Landslide Inventories to Analyze the Evolution and Characteristic of Rainfall-Induced Landslide at the Subwatershed Scale in Taiwan. *Water* **2023**, *15*, 1355. https://doi.org/10.3390/w15071355

Academic Editor: Su-Chin Chen

Received: 23 February 2023
Revised: 25 March 2023
Accepted: 28 March 2023
Published: 1 April 2023

1. Introduction

The rainfall variability under the present and future climate is examined as a function of global warming [1]. The trend of extreme rainfall events has been rising during the last three decades [2], and flood events are sensitive to global climate changes [3]. Rainfall-induced landslides are a common geological disaster in Taiwan that jeopardizes the safety of human life and property. The most severe rainfall-induced landslide disasters have been recorded in Taiwan in the past three decades, especially in 2001–2010 [4–7]. During 2001–2010, deaths caused by landslide disasters in Taiwan exceeded 1000, including 214 in 2001 caused by Typhoon Toraji, 45 in 2004 caused by Typhoon Mindulle, 33 in 2004 caused by Typhoon Aere, 21 in 2008 caused by Typhoon Sinlaku, and 699 in 2009 caused by Typhoon Morakot. Debris flow and sediment deposition in rivers are secondary geological disasters in watershed units caused by extreme rainfall-induced landslides [8,9]. The dense distribution of landslides and high sediment deposition in rivers have been the two main driving forces for the long-term geomorphologic evolution of landslides in mountainous watersheds in Taiwan in the past decade. Strategies for predicting landslides based on the characteristics of long-term landslide evolution must be developed to mitigate the effects of landslide disasters and secondary geological disasters.

Landslides caused by large-scale earthquakes or extreme rainfall events are usually clustered in mountainous areas near the earthquake epicenter [10–12] or with abundant rainfall during extreme rainfall events [6,13]. The spatial-temporal analysis method has been widely used to explore the long-term evolution and distribution of specific phenomena, including water use efficiency [14], the long-term GDP development trend [15], groundwater storage [16], observation of crop growth [17], and urban sprawl [18]. To effectively analyze the temporal distribution and evolution of landslides, a spatiotemporal landslide database should be developed by combining multitemporal landslide inventories [8]. Multitemporal high-resolution and NDVI satellite imagery has been used to build multitemporal landslide inventories for identifying and analyzing multiple landslides [10,19,20]. Satellite images captured by FORMOSAT-2 with a spatial resolution of 2 m during 2003–2017 were used to develop annual landslide inventories in Taiwan. Several serious landslide disaster events occurred during 2003–2017 in Taiwan [21,22]. The spatiotemporal landslide database can be used to analyze the long-term evolution, distribution, and characteristics of landslides in Taiwan [8,9,23].

Landslide evolution analysis has been used to observe the long-term deformation of specific large landslide cases [24,25] or the spatial and temporal distribution of landslides on a watershed scale [8,10–12]. Long-term landslide evolution analysis needs multi-temporal remote sensing data, including SPOT images, interferometric synthetic aperture radar (InSAR), and light detection and ranging (LIDAR), to produce landslide inventories, NDVI images, or a digital elevation model for deformation observation [8–12,19–21,26–31]. The long-term evolution of landslides on a watershed scale has been increasingly studied in the past decade, especially in areas that experience major earthquake events (2005 Kashmir earthquake in Pakistan, 2008 Wenchuan earthquake in China, and 2015 Gorkha earthquake in Nepal) or extreme rainfall events (2009 Typhoon Morakot in southern Taiwan). Many approaches have been used to analyze the long-term evolution of landslides, including the evaluation of the spatial and temporal distribution of landslides using annual data of landslide areas and the number of landslides [10,11,19,26,27], evaluation of the spatial distribution of new, enlarged, and recovered landslides [19,26,27], assessment of the relationship between the spatial distribution of annual landslides and geomorphological factors [10,11,19,26,28], and assessment of the spatial and temporal distribution of landslide activity [10,11,19,26]. The temporal stability of landslides at the subwatershed scale has been examined by evaluating the relationship between annual rainfall [17], monthly rainfall [10], or daily rainfall [19] and the annual number of landslides or area of landslides. Images taken at multiple time points, including those obtained from Google Earth, aerial photographs, and field investigation images, have also been used to study temporal differences in landslide stability and recovery [19,20,26]. The landslide area frequency distribution have been used to examine the composition of recurring and new landslide areas in landslide inventories [20], and the β coefficient of the landslide area frequency distribution was used as a basis for comparison with other earthquake-induced landslide events [28].

The analysis of landslide evolution usually requires more than 10 years of climate data and landslide inventories [5,27]. In this study, we used the hot spot analysis tool to evaluate the characteristics of long-term landslide evolution and distribution in Taiwan [32]. Typhoon Morakot induced the most severe rainfall-induced landslide events in the last two decades in Taiwan, especially in southern Taiwan. Taiwan is one of the most landslide-prone regions in the world [33], and the number of landslides and areas affected in Taiwan reached a historical peak during 1999–2017. The analysis of the long-term evolution of landslides in Taiwan at the subwatershed scale should help determine the evolution characteristics of rainfall-induced landslides.

The purpose of this study was to explore and explain the long-term evolution characteristics of rainfall-induced landslides at the subwatershed scale in Taiwan using annual landslide inventories from 2003 to 2017 and spatiotemporal hot spot technology. Knowing the long-term evolution characteristic of landslides can contribute to planning the correct engineering in the correct place to prevent secondary geological hazards, especially in a watershed with dense landslide distribution. The long-term rainfall records from 16 rainfall stations for the period 2003–2017 were collected, and the relationship between rainfall and landslide evolution was analyzed. Two spatiotemporal landslide cube models were developed in this study. The first was based on the annual landslide inventories from 2003–2017 and was used for exploring the long-term evolution of landslides in Taiwan. The other model was based on the annual landslide inventories from 2008–2017 and used to explore the long-term evolution of landslides caused by Typhoon Morakot in 2009. Moreover, the complete set of hot spots was statistically analyzed. We used spatial cluster and outlier analysis to explore the spatial and temporal distribution of landslides in Taiwan and the emerging hot spot analysis tool to explore the spatial and temporal hot spot patterns of landslides in Taiwan. We also determined the spatial and temporal distribution of landslide recovery after extreme rainfall-induced landslide events and compared it with that after major earthquake-induced landslide events.

2. Study Area

The research area, Taiwan, is located in East Asia. The elevation in Taiwan ranges from 0 m to 3952 m, and the average elevation is around 738 m (Figure 1a). The slope in Taiwan ranges from 0° to 89.8°, and the average slope is around 21.8°. The area with a slope >30° occupies 38.3% of the total area in Taiwan. Based on the delineation of sub-watersheds proposed by the Soil and Water Conservation Bureau in 2021 [34], there were 839 sub-watersheds in Taiwan (Figure 1b). The average slope of the 839 sub-watersheds is 22.5°, and 42.1% of the 839 sub-watersheds have a slope >30°.

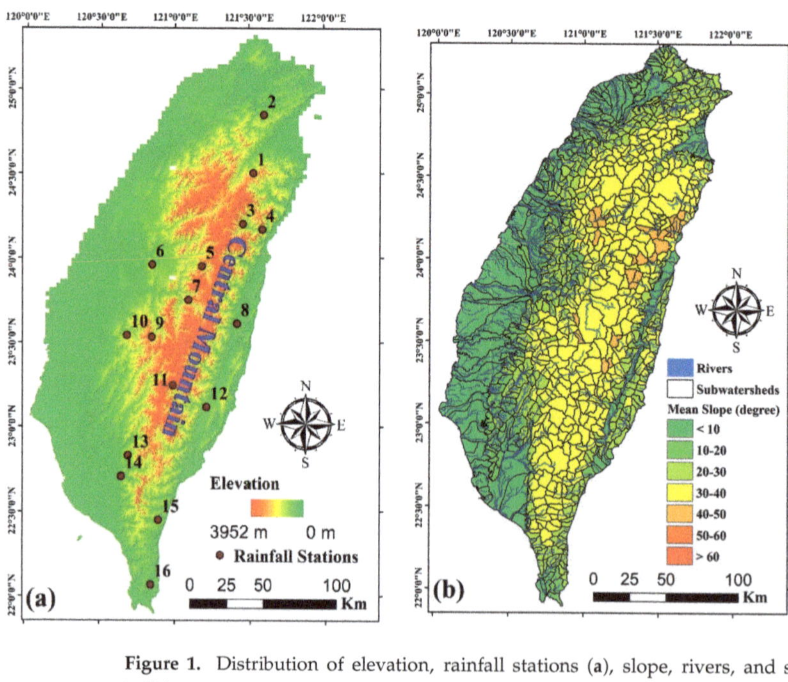

Figure 1. Distribution of elevation, rainfall stations (**a**), slope, rivers, and sub-watersheds (**b**) in Taiwan.

The long-term average annual rainfall in Taiwan was estimated at 2510 mm, but that in 2013–2020 was estimated at 2454 mm based on the open data information on the Water Resources Agency website [35]. The characteristic of annual rainfall in the past 2 decades in Taiwan was obvious oscillating. The average annual rainfall in 2014 and 2016 in Taiwan was 1921 mm and 3278 mm, and the difference between the annual rainfall in 2014 and 2016 was 1357 mm, i.e., 54.0% of the long-term average annual rainfall. The geological settings in Taiwan are shown in Figure 2. The terrace deposits (21.7%), alluvium (8.8%), and

Sanhsia group and its equivalents (7.7%) were the three main geological settings in Taiwan. The main stratigraphical formations in the northern central mountain region include the Oligocene-era Aoti formation, Oligocene-era Kankou formation, Eocene-era Szeleng sandstone, and Oligocene-era Tatungshan formation, and those in the southern central mountain region include Eocene-era Hsitsun Formation, Hsinkao Formation, Miocene-era Lushan Formation, late Miocene-era Sanhsia group and its equivalents, and late Paleozoic-era Tananao schist. Most landslides were concentrated in central and southern Taiwan, especially in the five landslide-prone stratigraphical formations, including the Changchihkeng Formation, Chaochou Formation, Pilushan Formation, Lushan Formation, and Tayuling Formation. The occupied percentage of the five landslide-prone stratigraphical formations in Taiwan is <30%, but the landslide area within the five landslide-prone stratigraphical formations after 2009 Typhoon Morakot occupied >70% of the all landslide areas in Taiwan. The main composition in the five landslide-prone stratigraphical formations are sandstone, shale, and slate. Detailed information of geological settings in Taiwan is also shown in Supplementary Materials.

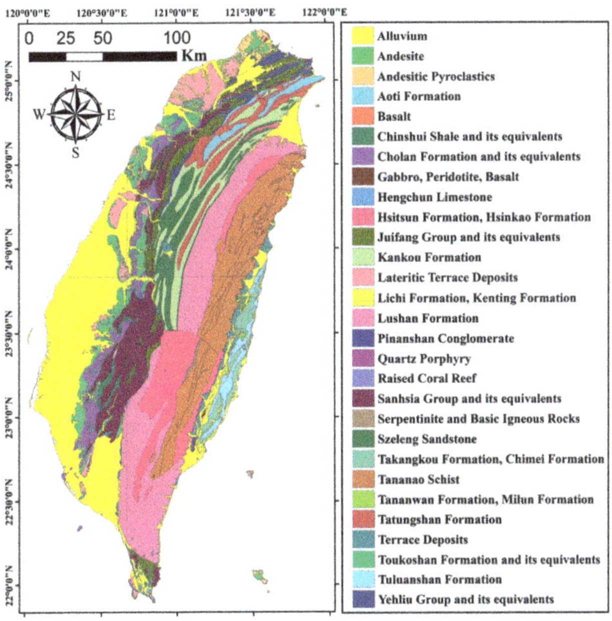

Figure 2. Geological settings in Taiwan (Scale 1:250,000).

3. Materials and Method

The flow chart of this research is shown in Figure 3.

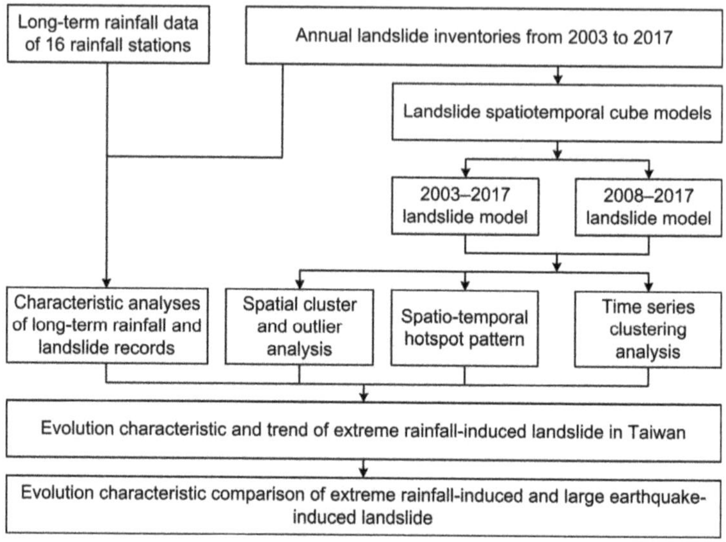

Figure 3. The flow chart in the study.

3.1. Materials

The rainfall data for the period 2003–2017 used in this study were collected from 16 rainfall stations (Figure 1a). Eight rainfall stations, namely stations 1, 3, 5, 7, 9, 11, 13, and 15, were located in the central mountainous region, which is the main spatiotemporal landslide hot spot in Taiwan, and the other eight rainfall stations, namely stations 2, 4, 6, 8, 10, 12, 14, and 16, were located outside the central mountainous region and not within the main spatiotemporal landslide hot spot. The data obtained from the 16 rainfall stations are detailed in the Supplementary Materials.

A digital elevation model with a resolution of 5 m was used in the study to conduct geomorphologic analyses. The delineation of subwatersheds in Taiwan (Figure 1b) was based on the proposed delineation from the Soil and Water Conservation Bureau [34], which is commonly used in research in Taiwan. The total area of 839 subwatersheds in Taiwan ranges from 0.39 to 575.8 km^2, with an average area of 42.9 km^2. Multiannual landslide inventories in 2003–2017 were generated by the Forestry Bureau in Taiwan by selecting images captured by the FORMOSAT-2 satellite with a spatial resolution of 2–5 m in January every year to identify landslides; this entails using a supervised classification method to develop landslide inventories for the previous year [18]. The production process of the annual distribution of landslides in 2003–2017 in Taiwan is shown in the Supplementary Materials. The large-scale landslide case considered in this study had a landslide area of >100,000 m^2.

3.2. Rainfall Seasonality Index (RSI)

Climatic factors are key to the recovery and long-term evolution of landslides, and the centralization of rainfall during the rainy seasons is a crucial factor that influences the recovery of rainfall-induced landslides in Taiwan [8,9,23]. The rainfall seasonality index (RSI) was used to measure the centralization of monthly rainfall in a year [36] and to assess the relationship between rainfall centralization and the recovery of earthquake-induced landslides in post-seismic periods [27]. The RSI is given [37] as follows:

$$RSI = \frac{1}{\bar{R}}\sum_{n=1}^{12}\left|\bar{X}_n - \frac{\bar{R}}{12}\right| \quad (1)$$

where $\bar{R}$ is the average annual rainfall in the research period and $\bar{X}_n$ is the average monthly rainfall of the nth month in the research period. The centralization of monthly rainfall based on the RSI is described in Table 1 [38].

Table 1. The explanation of the centralization of monthly rainfall based on the rainfall seasonality index [36].

RSI	Rainfall Regime
$\leq$0.19	Very equable
0.20–0.39	Equable but with a definite wetter season
0.40–0.59	Rather seasonal with a short drier season
0.60–0.79	Seasonal
0.80–0.99	Markedly seasonal with a long drier season
1.00–1.19	Most rain in 3 months or less
$\geq$1.20	Extreme, almost rain in 1–2 months

Note: The table was reprinted with permission from Ref. [37]. 1981, Walsh, R.P.D. and Lawler, D.M.

3.3. Landslide Spatiotemporal Cube Models

The annual landslide inventories in 2003–2017 and 2008–2017 in Taiwan were used to construct the two spatiotemporal landslide cube models in this study, namely the 2003–2017 spatiotemporal landslide model and the 2008–2017 spatiotemporal landslide model. The 2003–2017 spatiotemporal landslide model was used to analyze the long-term evolution of landslides in Taiwan, and the 2008–2017 spatiotemporal landslide model was used to analyze the evolution of landslides caused by Typhoon Morakot. The basic unit in the spatiotemporal cube models was data from the 839 subwatersheds based in Taiwan. The x and y coordinates in the spatiotemporal cube model denote the x and y coordinates of the centroid in the subwatershed scale image and represent the spatial dimensions in a specific year. The term t denotes the time dimension. The landslide ratio in the subwatershed for a specific year was calculated using the spatiotemporal cube model and was defined as the ratio of the landslide area to the total subwatershed area. The process of building the spatiotemporal landslide cube models is shown in the Supplementary Materials. The spatiotemporal landslide cube models were developed using ArcGIS Pro software Ver 1.2, and the data were converted to the NetCDF data format. The spatiotemporal hot spot analyses were conducted using the NetCDF data set.

The Mann–Kendall test was used to measure the temporal trend of landslide ratio in a specific basic unit in the two spatiotemporal cube models. The temporal trend meant that the landslide ratio in a specific period had been increasing or decreasing over time. The purpose of the Mann–Kendall test was to compare the landslide ratio in the specific subwatershed between the current year and the previous year. A positive value, negative value, or value of 0 in the Mann–Kendall test indicated an increasing trend, a decreasing trend, or no obvious change, respectively, over time for the annual landslide distribution.

3.4. Spatial Cluster and Outlier Analysis

The spatial landslide pattern is key to identifying landslide-prone regions and determining landslide evolution. The Anselin Local Moran's I index [38] was used to explore the spatial clusters and outliers of landslides in space and time dimensions in the study. Based on the calculations obtained using the Anselin Local Moran's I index, the spatial landslide pattern can be classified into six types, including a statistically significant cluster of high values (only high–high cluster), a statistically significant cluster of low values (only low–low cluster), an outlier in which a high landslide ratio value was surrounded by low landslide ratio values (only high–low outlier), outlier in which a low landslide ratio value was surrounded by high landslide ratio values (only low–high outlier), multiple types, and never significant.

3.5. Spatiotemporal Hot Spot Patterns

Spatiotemporal hot spot patterns are useful in explaining landslide evolution and can be obtained by analyzing the spatiotemporal landslide cube models using the Space Time Pattern Mining toolbox in Arc GIS Pro software Ver 1.2. This tool can be used to analyze the distribution and patterns of multiannual landslide data in terms of both space and time. A basic unit in the spatiotemporal landslide cube model represents the landslide ratio in a specific subwatershed in a specific year, and all basic units were input into the emerging hot spot analysis tool. The statistically significant values of each basic unit were used to identify the types and spatiotemporal distribution of hot spots.

The hot spot analysis tool estimated the Getis-Ord Gi statistic for each basic unit in the spatiotemporal models [39]. The pattern of the spatiotemporal landslide hot spots was identified using the landslide ratio of the basic unit and the neighboring basic units in both space and time. If the landslide ratios exhibited high (low) significant clustering, the basic unit was identified as a hot (cold) spot. The statistical significance of each basic unit in the spatiotemporal landslide models was determined based on the z-scores and p values for the Getis-Ord Gi statistic [39]. If a basic unit had a high landslide ratio and was surrounded by other basic units with high landslide ratios, the basic unit was identified as a significant hot spot. Based on the z-score and p value of each basic unit in the spatiotemporal landslide model, eight patterns of hot and cold spots, or no pattern, were detected. The eight hot or cold spot patterns were new, consecutive, intensifying, persistent, diminishing, sporadic, oscillating, and historical. The original definition of 8 patterns had been described on the ESRI website [40], and the adjusted definition of 8 patterns which was suitable to characterize the pattern of landslide evolution is shown in Supplementary Materials. The new, consecutive, intensifying, persistent, diminishing, sporadic, oscillating, and historical hot spots were abbreviated as NHS, CHS, I, PHS, DHS, SHS, OHS, and HHS, respectively, in this study.

The 2003–2017 spatiotemporal landslide model contained 12,585 basic units, and the 2008–2017 spatiotemporal landslide model contained 8390 basic units. The spatial neighborhood distance and time step were set as 13.6–14.2 km and 5 years, respectively.

4. Results

4.1. Long-Term Spatial and Temporal Distribution of Rainfall and Landslide Data in 2003–2017

The annual rainfall distribution data for 2003–2017 were obtained from the 16 rainfall stations (Figure 1a) in Taiwan and are presented in Figure 4. The complete data obtained from the 16 rainfall stations are presented in the Supplementary Materials. The average annual rainfall in 2003–2017 (2993.3 mm) was slightly higher (by 56.9 mm) than that in 2008–2017 (2930.5 mm). The average annual rainfall at the eight rainfall stations in the central mountainous region in 2003–2017 (3158.7 mm) was slightly higher (by 330.8 mm) than that at the eight rainfall stations outside the central mountainous region (2827.9 mm). The average ratios of the accumulated rainfall in the rainy season (from May to October) to the annual rainfall in 2003–2017 at the eight rainfall stations inside and outside the central mountainous region were 79.0% and 80.7%, respectively. The data indicated that the average accumulated rainfall in the rainy season from 2003–2017 at the eight rainfall stations in the central mountainous region was higher by 213.3 mm than that at another eight rainfall stations outside the central mountainous region. The three largest differences in average annual rainfalls recorded in 2003–2017 between the rainfall stations in and outside the central mountainous region were 1803 mm (No. 11 and No. 12 rainfall stations located in southeastern Taiwan), 1172 mm (No. 13 and No. 14 rainfall stations located in southwestern Taiwan), and 464.3 mm (No. 1 and No. 2 rainfall stations located in northern Taiwan). The distribution of annual rainfall recorded at the 16 rainfall stations in 2003–2017 in Taiwan was oscillating, especially in southern Taiwan. The annual rainfall recorded at rainfall station No. 13 in 2005 and 2011 was 7012 and 1832 mm, respectively, and the difference in annual rainfall between 2005 and 2011 was 5180 mm, 1.21 times the average annual rainfall. Furthermore, the temporal distribution of rainfall in the central mountainous region in southern Taiwan in 2003–2017 was obviously different from that outside the central mountainous region.

The annual number of landslides and their temporal distribution in 2003–2017 in Taiwan are shown in Figure 5. The annual landslide area in 2003–2017 in Taiwan ranged from 216.1 km^2 in 2008 to 623.1 km^2 in 2009, and the average annual landslide area was estimated to be 424.1 km^2. The average landslide ratio (i.e., the ratio of landslide area to total area) in Taiwan in 2003–2017 was 1.15%. Based on the analysis of landslides in Taiwan, areas with a landslide ratio of >1.0% can be considered areas with serious disaster occurrence [3]. The annual number of landslides in Taiwan ranged from 19,041 in 2008 to 51,898 in 2013, and the average annual number of landslides was estimated to be 32,070. The average landslide density in Taiwan in 2003–2017 was estimated to be 0.87/km^2. The annual number of large-scale landslides in 2003–2017 in Taiwan ranged from 279 in 2003 to 1036 in 2009, and the average number of large-scale landslides was estimated to be 638.5. Thus, for 2003–2017, Taiwan is regarded as a region with a serious disaster occurrence, especially for the period of 2009–2017.

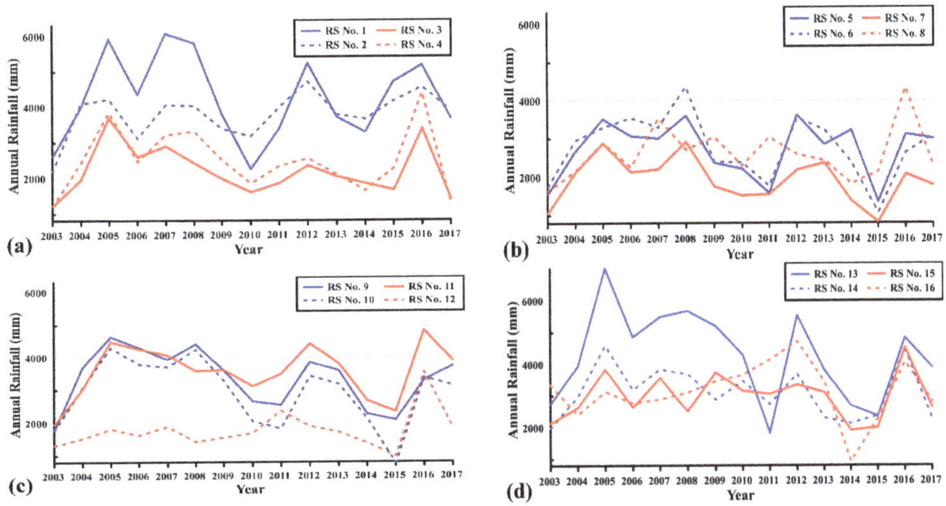

Figure 4. Temporal distribution of annual rainfall in the rainfall stations no. 1–4 (**a**), no. 5–8 (**b**), no. 9–12 (**c**), and no. 13–16 (**d**) in 2003–2017 in Taiwan.

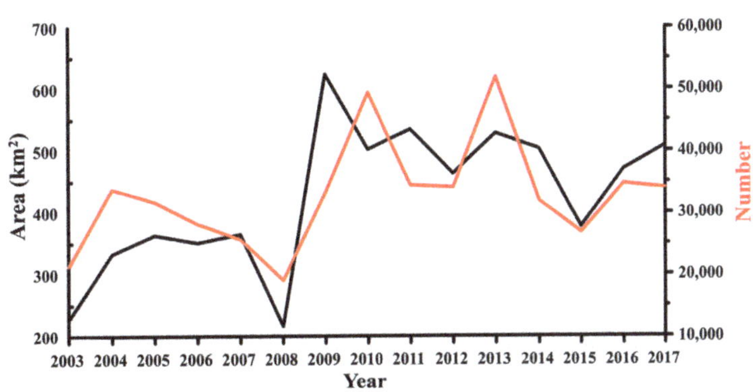

Figure 5. Temporal distribution of landslide area (**black line**) and number (**red line**) in 2003–2017 in Taiwan.

Taiwan has a nonuniform temporal distribution of annual and monthly rainfall, especially in southern Taiwan. The RSI was used to quantify the centralization of monthly rainfall at the 16 rainfall stations. Table 2 presents the average RSI values at the 16 rainfall stations in 2003–2017 in Taiwan. The rainfall stations located in the central mountainous region that recorded the top three highest RSI values were stations No. 13 (RSI = 0.78), No. 15 (RSI = 0.74), and No. 9 (RSI = 0.73). These three rainfall stations recorded relatively dry seasons (Table 1). High RSI values enable faster remobilization of sediment yield after severe landslide events [37]. The monthly rainfall regime in 2003–2017 in western, southwestern, and southern Taiwan exhibited obvious centralization, which implies that recovery after landslides in these three regions is more difficult and slower than in other regions in Taiwan.

Table 2. The rainfall seasonality index value in each region in Taiwan from 2003–2017.

No. of RS [1]	Region	RSI	No. of RS [1]	Region	RSI
1	Northern	0.62	9	Western	0.73
2	Taiwan	0.36	10	Taiwan	0.84
3	Eastern	0.47	11	Southeastern	0.76
4	Taiwan	0.65	12	Taiwan	0.70
5	Western	0.56	13	Southwestern	0.78
6	Taiwan	0.78	14	Taiwan	0.94
7	Eastern	0.64	15	Southern	0.74
8	Taiwan	0.45	16	Taiwan	0.64

Note: 1 means the number of rainfall stations in Figure 1a.

4.2. *Landslide Spatialtemporal Cube Model and Trend*

The 2003–2017 landslide model contained 12,585 basic units, and the 2008–2017 landslide model contained 8390 basic units. The results of the Mann–Kendall trend test for the two models are presented in Figure 6. In the 2003–2017 landslide model, 13.2%, 11.6%, and 75.2% of the subwatersheds accounted for upward, downward, and nonsignificant trends, respectively, whereas in the 2008–2017 landslide model, the corresponding percentages were 9.3%, 11.9%, and 78.8%, respectively. The subwatersheds exhibiting the upward trend in the 2003–2017 and 2008–2017 landslide models were concentrated in the central mountainous region of Taiwan, especially in southern Taiwan. This region was also severely affected by Typhoon Morakot. The subwatersheds exhibiting the downward trend in the 2003–2017 and 2008–2017 landslide models were spread across Taiwan and located outside the central mountainous region. The data indicated that the long-term evolution of landslides over 2003–2017 exhibited no obvious change after Typhoon Morakot in 2009, and that the landslide-prone subwatersheds were concentrated in the central mountainous region of Taiwan (i.e., upstream of the river watershed).

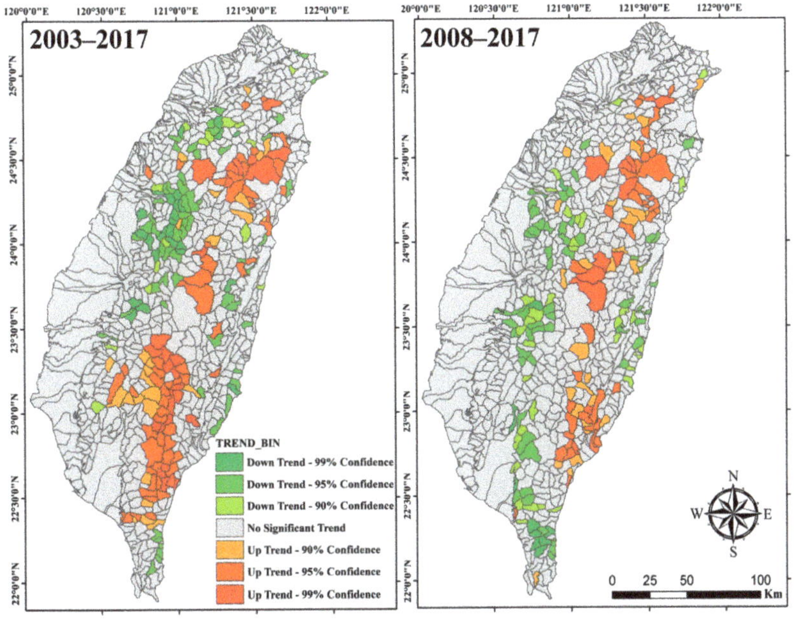

Figure 6. The Mann–Kendall trend test for the landslide cube model in 2003–2017 (**left**) and 2008–2017 (**right**).

4.3. Results of Spatial Clusters and Outliers

The distributions of the spatial landslide clusters and outliers in the 2003–2017 and 2008–2017 landslide models were analyzed using the local outlier analysis tool with a neighborhood time step of 5 years and 499 permutations in ArcGIS Pro software Ver 1.2. The results are shown in Figure 7. The high–high landslide cluster was still concentrated in the central mountainous region, whereas the low–low cluster was concentrated in the plains in west and north Taiwan. The statistical results of the spatial landslide cluster and outlier analyses are shown in Table 3.

A total of 447 and 502 subwatersheds were identified as significant spatial clusters (only high–high and only low–low) in the 2003–2017 and 2008–2017 landslide models, respectively. The only high–high cluster was distributed in the central mountainous region of south Taiwan in 2003–2017 and 2008–2017. This finding indicates that the central mountainous region in southern Taiwan was an originally landslide-prone region. The number of subwatersheds identified as having multiple types of patterns was 214 in 2003–2017 and decreased to 137 in 2008–2017; by contrast, the number of subwatersheds designated as never significant increased from 101 to 142, and the number designated as only the low–low cluster type increased from 241 to 305. The subwatersheds designated as multiple types in 2003–2017 and later identified as never significant, and only the low–low cluster type in 2008–2017 were mainly distributed in northern Taiwan. The number of subwatersheds designated as only the high–high cluster type in the 2008–2017 landslide model in north Taiwan was smaller than that in the 2003–2017 landslide model. This result indicated that the landslide recovery rate in northern Taiwan was gradually improving because the sediment deposition disaster caused by Typhoon Morakot was mainly distributed in southern Taiwan (see the annual landslide distribution in 2009–2017 in Supplementary Materials). The number of subwatersheds marked as only high–low outliers declined from 2003–2017 to 2008–2017, and these subwatersheds were mainly distributed in the piedmont regions across Taiwan. Only nine and five subwatersheds were designated as only low–high outliers in the 2003–2017 and 2008–2017 landslide models, respectively, and these subwatersheds were mainly distributed in southern Taiwan.

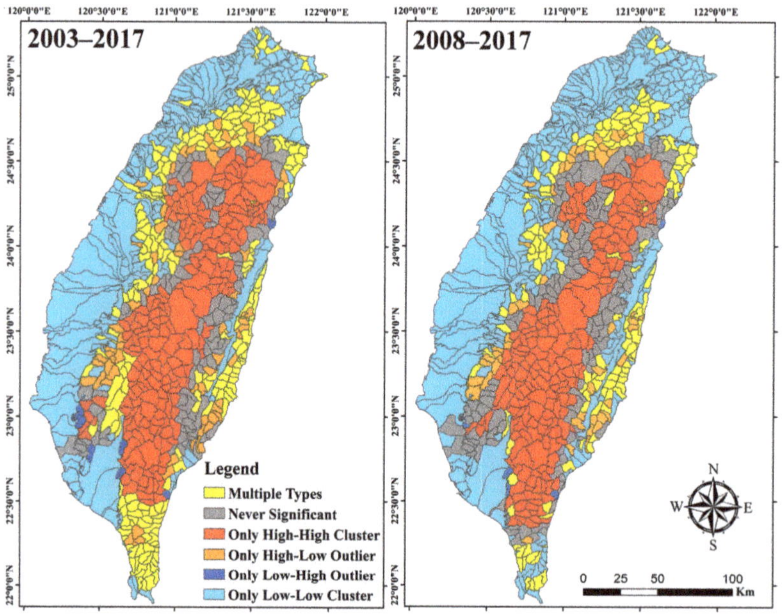

Figure 7. Distribution of landslide spatial cluster and outlier using the Anselin Local Moran's I index.

Table 3. Statistical data of landslide spatial clusters and outlier analyses in Taiwan.

Type	Numbers of Sub-Watersheds	
	2003–2017	2008–2017
Mul. *	214	137
NS *	101	142
HH *	206	197
HL *	68	53
LH *	9	5
LL *	241	305

Note: * The Mul., NS, HH, HL, LH, and LL mean multiple, never significant, only high-high cluster, only high-low outlier, only low-high outlier, only low-low cluster types.

4.4. Landslide Spatiotemporal Hot Spot Analyses

The distribution and pattern of the spatiotemporal landslide hot spots in the 2003–2017 and 2008–2017 landslide models obtained using the emerging hot spot analysis tool in ArcGIS Pro software are shown in Figure 8. The statistical data of the spatiotemporal landslide hot and cold spots in Taiwan are presented in Table 4. The percentages of subwatersheds marked as hot spots in the 2003–2017 and 2008–2017 landslide models were 24.2% and 22.6%, and those of subwatersheds marked as cold spots were 50.1% and 52.8%. The landslide hot spots in the 2003–2017 and 2008–2017 landslide models were concentrated in the central mountainous region, but the pattern of spatiotemporal landslide hot spots in the 2003–2017 landslide model was obviously more concentrated than that in the 2008–2017 landslide model. The spatiotemporal hot spot pattern of subwatersheds accounting for >1.0% in the 2003–2017 landslide model included the consecutive hot spot (13.2%), intensifying hot spot (6.0%), persistent hot spot (1.8%), oscillating hot spot (1.5%), and sporadic hot spot (1.0%), but the pattern in the 2008–2017 landslide model included only the consecutive hot spot (21.9%).

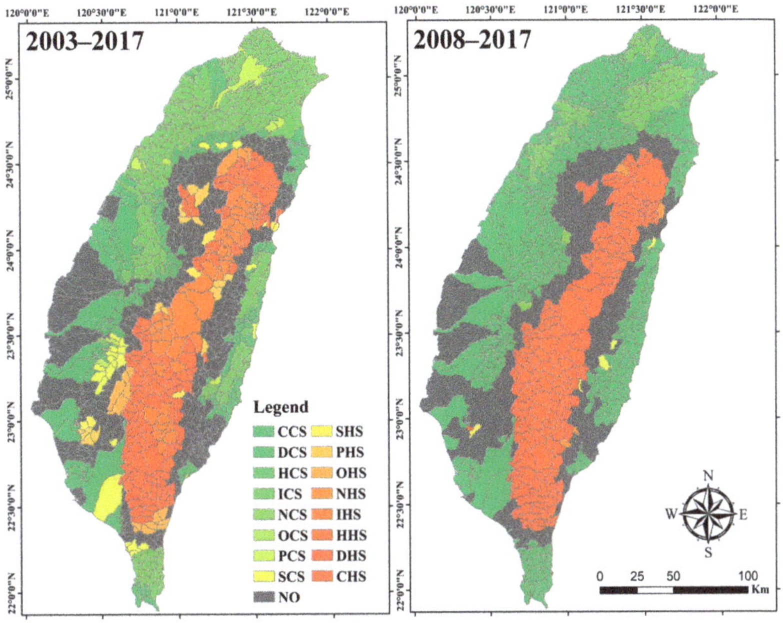

Figure 8. Distribution and pattern of landslide spatiotemporal hot spots in the 2003–2017 and 2008–2017 landslide models.

Table 4. Occupied percentage of landslide spatiotemporal hot spots and cold spots in 2003–2017 and 2008–2017 in Taiwan.

Pattern	Occupied Percentage		Pattern	Occupied Percentage	
	2003–2017	2008–2017		2003–2017	2008–2017
Cold spot	50.1	52.8	Hot spot	24.2	22.6
CCS	15.5	44.0	CHS	13.2	21.9
DCS	0.0	0.0	DHS	0.4	0.0
HCS	0.0	0.0	HHS	0.0	0.0
ICS	30.5	8.2	I HS	6.0	0.0
NCS	0.0	0.2	NHS	0.4	0.5
OCS	0.0	0.0	OHS	1.5	0.0
PCS	1.2	0.0	PHS	1.8	0.0
SCS	2.9	0.4	SHS	1.0	0.2
No pattern	25.7	24.6			

The trend and pattern of the distribution of landslide cold spots in the 2003–2017 and 2008–2017 models were similar to those of the distribution of landslide hot spots. The spatiotemporal cold spot pattern of subwatersheds accounting for >1.0% in the 2003–2017 landslide model included the intensifying cold spot (30.5%), consecutive cold spot (15.5%), sporadic cold spot (2.9%), and persistent cold spot (1.2%), but the pattern in the 2008–2017 landslide model only included the consecutive cold spot (44.0%) and intensifying cold spot (8.2%). The percentage of subwatersheds marked as having no pattern was 25.7% and 24.6%, respectively, in the 2003–2017 and 2008–2017 models. This result indicated that the spatiotemporal landslide hot spots were still primarily distributed in the central mountainous region of Taiwan.

5. Discussion

A comparison of the spatial and temporal distribution of landslide stability after severe landslide events induced by large-scale earthquakes and extreme rainfall can help clarify the characteristics of landslide recovery. Some studies [10,12,19,26] have elucidated the spatial and temporal distribution of landslide stability using the landslide recovery rate after large-scale earthquake-induced landslide events. The landslide recovery rate was defined as the ratio of the difference between the landslide area in a specific year and the landslide area in an earthquake-induced or rainfall-induced landslide year to the landslide area in an earthquake-induced or rainfall-induced landslide year. A comparison of the annual landslide recovery rate between earthquake-induced and extreme rainfall-induced landslide areas can reveal the characteristic differences between earthquake-induced and rainfall-induced landslide recovery.

The annual landslide recovery rate in northern Sichuan Province, China, in the first two years after the 2008 Wenchuan earthquake, was estimated to be 13.45%, and that in the third to eighth years was estimated to be 10.56% [41]. The landslide recovery rates in northern Pakistan in the fifth, ninth, eleventh, and thirteenth years after the 2005 Kashmir earthquake were 9.5%, 43.8%, 50.2%, and 61.3%, respectively [10]. Typhoon Morakot in 2009 was the most severe rainfall-induced landslide event during 2003–2017 in Taiwan. The number of landslides and the area affected in 2008 in Taiwan were estimated to be 19,041 and 167.1 km^2, respectively, and those in 2009 were estimated to be 33,274 and 502.2 km^2, respectively. The mean annual landslide ratio of 839 subwatersheds in 2009 in Taiwan was estimated to be 1.33%, and that in the first to second years and third to eighth years after Typhoon Morakot ranged from 1.01% to 1.06% and 0.71% to 1.01%, respectively. The landslide recovery rate in the first to second years after Typhoon Morakot in the 839 subwatersheds was estimated to be 22.2%, and that in the third to eighth years was estimated to be 31.6%.

The landslide recovery rate in the landslide-affected subwatersheds in the study was estimated to assess the actual landslide recovery after Typhoon Morakot in Taiwan. Subwatersheds were defined as "landslide affected" if their landslide ratio in 2009 was larger by at least 0.3% than that in 2008. The 278 landslide-affected subwatersheds were distributed in southern Taiwan (Figure 9). The mean landslide ratio in the 278 landslide-affected subwatersheds in 2009 was larger by 2.86% than that in 2008. The annual landslide recovery rates of the landslide-affected subwatersheds from 2010 to 2017 are shown in Figure 9. The landslide recovery rate of 278 landslide-affected subwatersheds in the first to second years after Typhoon Morakot was 31.6%, and that in the sixth to eighth years was 47.7%. The recovery rate of landslides induced by extreme rainfall events in Taiwan was larger than that of landslides induced by large-scale earthquake events in the world [10,42]. The landslide recovery rate of the subwatersheds with a landslide ratio >3.0% in 2009 was larger than that of all landslide-affected subwatersheds. By contrast, the landslide recovery rate of the subwatersheds with a landslide ratio <3.0% in 2009 was lower than that of all landslide-affected subwatersheds. Furthermore, the landslide recovery rate of the subwatersheds with a landslide ratio >3.0% in 2009 in the first two years after Typhoon Morakot was 35.4%, and that in the sixth to eighth years was 53.7%. The trend of the landslide recovery rate increased after Typhoon Morakot, but the landslide recovery rates in 2013, 2016, and 2017 showed a declining trend.

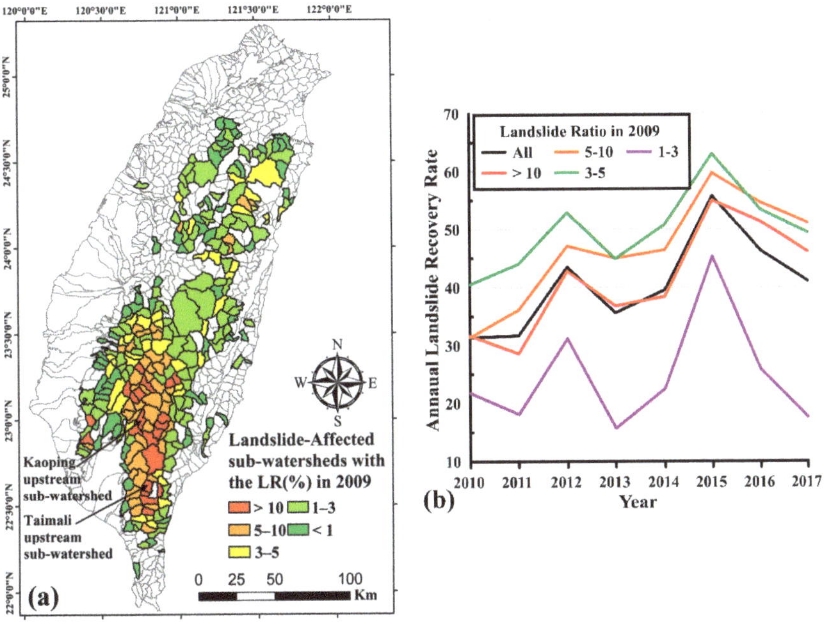

Figure 9. Distribution (**a**) and the annual landslide recovery rate (**b**) of 278 landslide-affected subwatersheds after the 2009 Typhoon Morakot in Taiwan.

Two subwatersheds, namely the Taimali upstream subwatershed and the Kaoping upstream subwatershed (Figure 9a), were selected as examples to assess the spatial distribution of landslide recovery after Typhoon Morakot in 2009. The number of landslides and landslide ratios after Typhoon Morakot in the Taimali upstream subwatershed were 353 and 24.5%, respectively, and those in the Kaoping upstream watershed were 165 and 15.9%, respectively. The Taimali upstream subwatershed and Kaoping upstream subwatershed were listed among the top five subwatersheds with the most severe landslide disasters after Typhoon Morakot in Taiwan. Landslide frequency [9] was defined as the total number of landslides that occurred from 2010 to 2017 in the study. The distribution of landslide frequency in the two subwatersheds from 2010 to 2017 is presented in Figure 10. The mean frequency of landslides in the Taimali and Kaoping upstream subwatersheds induced by Typhoon Morakot was 4.32 and 1.58 years, respectively, and outside the landslide area was 2.24 and 2.41 years, respectively. This result indicates that the landslide recovery time inside the area of the landslide induced by Typhoon Morakot was longer than that outside the landslide area (i.e., the new or enlarged landslide area from 2010 to 2017). Most areas with a high landslide frequency, especially for large-scale landslides, were located in the source area of the landslide induced by Typhoon Morakot. Landslides induced by Typhoon Morakot in the Taimali and Kaoping upstream watersheds accounted for 96.9% and 73.5% of the high landslide frequency grids (i.e., landslide frequency > 5), respectively, inside the landslide area. Moreover, landslides induced by Typhoon Morakot in the Taimali and Kaoping upstream watersheds accounted for 69.2% and 58.9%, respectively, of the low landslide frequency grid (i.e., landslide frequency < 3) inside the landslide area. The grid with a landslide frequency value outside the area of the landslide induced by Typhoon Morakot covered the bank area of sinuous reaches, gully areas, and the boundary of the original landslide area. This result indicated that the area along the sediment transport path was the main region of occurrence of secondary geohazards after extreme rainfall-induced landslide events.

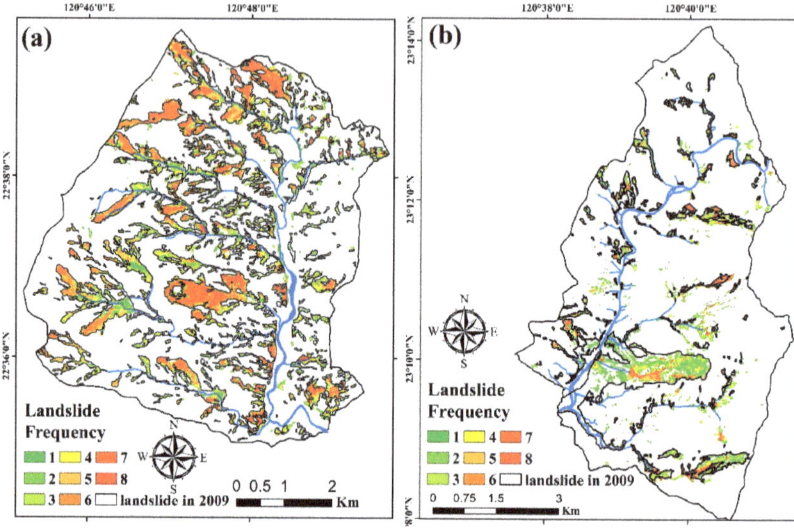

Figure 10. Distribution of landslide frequency from 2010 to 2017 in the Taimali upstream sub-watershed (**a**) and Kaoping upstream sub-watershed (**b**).

The recovery of most of the subwatersheds with a landslide ratio >5% in 2009 in southern Taiwan was difficult in the landslide hillslope, and this was related to four factors, namely landslide-prone geological settings, abundant rainfall during Typhoon Morakot, temporal nonuniform monthly rainfall, and sediment yielded from numerous landslides being deposited on the landslide hillslope or transported to the rivers. Most of the sub-watersheds with a landslide ratio of >5% in 2009 were located in the Lushan Formation, Histsun Formation, Hsinkao Formation, and Sanshia Group and its equivalents (Figure 2). The aforementioned three sediment formations contain shale, sandstone, and slate and are prone to landslides. During Typhoon Morakot in southern Taiwan, the rainfall total was estimated to be >2000 mm in 3 days [3]. The abundant rainfall at these landslide-prone geological formations may have resulted in a severe landslide disaster in 2009 in southern Taiwan, and huge amounts of sediment from this landslide area were deposited on the landslide hillslope or transported to creeks or rivers. This observation can be verified by the finding that the core areas of large-scale landslide cases or the vicinity of creeks or rivers had high landslide frequencies in 2010–2017 (Figure 10). The RSI values at the rainfall stations for the subwatersheds with a landslide ratio >5% in 2009 (Figure 9a) were >0.70 in 2010–2017, which indicated that most of the rainfall occurred in the rainy season. The seasonal rainfall regime also affects the stability of sediment in southern Taiwan. The sediment in these areas was unstable, and rainfall events with low accumulated rainfall could induce landslides on the hillslope or transport sediment to the rivers. Therefore, the core areas and boundaries of large-scale landslides and the vicinity of creeks and rivers had high landslide frequencies. This was also why the consecutive and intensifying pattern of spatiotemporal landslide hot spots was observed in southern Taiwan. The spatiotemporal analysis of landslide evolution helped characterize the patterns of landslide hot spots and revealed the reason for the difficult and slow recovery after landslides in these regions.

The composition change of soil grain before and after the 2009 Typhoon Morakot in the upstream watershed was also obvious. Most sediment yield from landslides during 2009 Typhoon Morakot was deposited in the down hillslope or transported into the rivers in the upstream watersheds. However, the fine sediment in the sediment deposition in the down hillslope or in the riverbed was gradually transported downstream after several heavy or flooding events after the 2009 Typhoon Morakot [43]. The composition of sediment deposition in the down hillslope or in the river became coarse. The coarse sediment deposition in the river also became an armor layer.

The mechanisms and triggering factors of secondary geological hazard, and the characteristics of long-term evolution can be explored by using the spatiotemporal hot spot analysis method based on the results in this study. The findings in this study can be referred to or used in other countries that had faced a serious landslide disaster in the past few years, such as the serious landslide disaster caused by heavy rainfall events in July 2020 in southern Japan, including Kumamoto, Nagano, and Kagoshima Prefectures.

The main reasons for the dense landslide caused by the 2009 Typhoon Morakot in Taiwan were extreme rainfall and subsequent flooding. Besides designing engineering to reduce or prevent secondary geological hazards, it is also important to reduce the flooding impact caused by extreme rainfall events in Taiwan [43,44]. Planning flood detention ponds and increasing the riverbed roughness coefficient have been proven to lower the impact of flooding on the safety of human life and property [45,46]. It is also important to examine the drainage ability of each reach or river in the upstream watershed in Taiwan under possible climate change conditions [47].

6. Conclusions

The study period (1990 to 2020) recorded the highest landslide density in the history of Taiwan. The Chi-Chi earthquake in 1999 and several extreme rainfall events resulted in severe landslide disaster events in Taiwan. Therefore, investigating the long-term evolution of landslides and secondary geological disasters is valuable. Based on the analysis using long-term rainfall records and annual landslide inventories for 2003–2017, recovery after landslides was the most difficult in the central mountainous region and southern Taiwan. These regions were also the densest landslide areas during Typhoon Morakot in 2009. The monthly rainfall regimes in the central mountainous region and southern Taiwan are obviously seasonal, making the sediment in these areas unstable. The Mann–Kendall trend test indicated that half of the subwatersheds that exhibited an upward trend of landslide evolution in 2003–2017 were concentrated in the central mountainous region and southern Taiwan. Furthermore, the subwatersheds with the high–high landslide cluster type were also located in the central mountainous region and southern Taiwan based on the spatial cluster and outlier analyses. The main patterns of spatiotemporal landslide hot spots in the central mountainous region and southern Taiwan were consecutive and intensifying. A comparison of landslide recovery rates between regions with large-scale earthquake-induced landslides and extreme rainfall-induced landslides showed that the recovery rate of regions affected by extreme rainfall-induced landslides in Taiwan was higher than that of regions affected by large-scale earthquake-induced landslides in the world. Rainfall-induced landslides were widely distributed in unstable areas of the central mountainous region and southern Taiwan during Typhoon Morakot in 2009 and gradually moved to the core and boundaries of large-scale landslide hillslopes and the vicinity of creeks and rivers.

Supplementary Materials: The following supporting information can be downloaded at: https://www.mdpi.com/article/10.3390/w15071355/s1. References [48,49] are cited in the Supplementary Materials.

Funding: This research was funded by the Ministry of Science and Technology of Taiwan (R.O.C.), grant number MOST 111-2625-M-035-001.

Data Availability Statement: Data sharing not applicable. Most data used in the study can be downloaded from https://data.gov.tw/.

Conflicts of Interest: The authors declare no conflict of interest.

References

1. Fowler, H.J.; Ali, H. Analysis of extreme rainfall events under the climatic change. In *Rainfall Modeling, Measurement and Applications*; Morbidelli, R., Ed.; Elsevier: Amsterdam, The Netherlands, 2022; pp. 307–326.
2. Panagoulia, D.; Bárdossy, A.; Lourmas, G. Diagnostic statistics of daily rainfall variability in an evolving climate. *Adv. Geosci.* **2006**, *7*, 349–354. [CrossRef]
3. Panagoulia, D.; Dimou, G. Sensitivity of flood events to global climate change. *J. Hydrol.* **1997**, *191*, 208–222. [CrossRef]
4. Cheng, J.D.; Huang, Y.C.; Wu, H.L.; Yeh, J.L.; Chang, C.H. Hydrometeorological and landuse attributes of debris flows and debris floods during typhoon Toraji, July 29–30, 2001 in central Taiwan. *J. Hydrol.* **2005**, *306*, 161–173. [CrossRef]
5. Milliman, J.D.; Lee, T.Y.; Huang, J.C.; Kao, S.J. Impact of catastrophic events on small mountainous rivers: Temporal and spatial variations in suspended- and dissolved-solid fluxes along the Choshui River, central western Taiwan, during typhoon Mindulle, July 2–6, 2004. *Geochim. Cosmochim. Acta* **2017**, *205*, 272–294. [CrossRef]
6. Wu, C.H.; Chen, S.C.; Chou, H.T. Geomorphologic characteristics of catastrophic landslides during typhoon Morakot in the Kaoping Watershed, Taiwan. *Eng. Geol.* **2011**, *123*, 13–21. [CrossRef]
7. Wu, C.H.; Chen, S.C.; Feng, Z.Y. Formation, failure, and consequences of the Xiaolin landslide dam, triggered by extreme rainfall from Typhoon Morakot, Taiwan. *Landslides* **2014**, *11*, 357–367. [CrossRef]
8. Wu, C.; Lin, C. Spatiotemporal hotspots and decadal evolution of extreme rainfall-induced landslides: Case studies in southern Taiwan. *Water* **2021**, *13*, 2090. [CrossRef]
9. Wu, C.H. Evaluating the landslide stability and vegetation recovery: Case studies in the Tsengwen reservoir watershed in Taiwan. *Water* **2021**, *13*, 3479. [CrossRef]
10. Shafique, M. Spatial and temporal evolution of co-seismic landslides after the 2005 Kashmir earthquake. *Geomorphology* **2020**, *362*, 107228. [CrossRef]
11. Yang, W.; Qi, W.; Wang, M.; Zhang, J.; Zhang, Y. Spatial and temporal analyses of post-seismic landslide changes near the epicentre of the Wenchuan earthquake. *Geomorphology* **2017**, *276*, 8–15. [CrossRef]
12. Rosser, N.; Kincey, M.; Oven, K.; Densmore, A.; Robinson, T.; Pujara, D.S.; Shrestha, R.; Smutny, J.; Gurung, K.; Lama, S.; et al. Changing significance of landslide Hazard and risk after the 2015 Mw 7.8 Gorkha, Nepal Earthquake. *Prog. Disaster Sci.* **2021**, *10*, 100159. [CrossRef]
13. Chen, C.W.; Chen, H.; Oguchi, T. Distributions of landslides, vegetation, and related sediment yields during typhoon events in northwestern Taiwan. *Geomorphology* **2016**, *273*, 1–13. [CrossRef]
14. Qin, S.; Ding, J.; Ge, X.; Wang, J.; Wang, R.; Zou, J.; Tan, J.; Han, L. Spatio-Temporal Changes in Water Use Efficiency and Its Driving Factors in Central Asia (2001–2021). *Remote Sens.* **2023**, *15*, 767. [CrossRef]
15. Zhao, Z.; Tang, X.; Wang, C.; Cheng, G.; Ma, C.; Wang, H.; Sun, B. Analysis of the Spatial and Temporal Evolution of the GDP in Henan Province Based on Nighttime Light Data. *Remote Sens.* **2023**, *15*, 716. [CrossRef]
16. Wang, S.; Cui, G.; Li, X.; Liu, Y.; Li, X.; Tong, S.; Zhang, M. GRACE Satellite-Based Analysis of Spatiotemporal Evolution and Driving Factors of Groundwater Storage in the Black Soil Region of Northeast China. *Remote Sens.* **2023**, *15*, 704. [CrossRef]
17. Aeberli, A.; Phinn, S.; Johansen, K.; Robson, A.; Lamb, D.W. Characterisation of Banana Plant Growth Using High-Spatiotemporal-Resolution Multispectral UAV Imagery. *Remote Sens.* **2023**, *15*, 679. [CrossRef]
18. Mansour, S.; Ghoneim, E.; El-Kersh, A.; Said, S.; Abdelnaby, S. Spatiotemporal Monitoring of Urban Sprawl in a Coastal City Using GIS-Based Markov Chain and Artificial Neural Network (ANN). *Remote Sens.* **2023**, *15*, 601. [CrossRef]
19. Chen, M.; Tang, C.; Xiong, J.; Shi, Q.Y.; Li, N.; Gong, L.F.; Wang, X.D.; Tie, Y. The long-term evolution of landslide activity near the epicentral area of the 2008 Wenchuan earthquake in China. *Geomorphology* **2020**, *367*, 107317. [CrossRef]
20. Tang, C.; Van Westen, C.J.; Tanyas, H.; Jetten, V.G. Analysing post-earthquake landslide activity using multi-temporal landslide inventories near the epicentral area of the 2008 Wenchuan earthquake. *Nat. Hazards Earth Syst. Sci.* **2016**, *16*, 2641–2655. [CrossRef]
21. Liu, C.C.; Liu, J.G.; Lin, C.W.; Wu, A.M.; Liu, S.-H.; Shieh, C.-L. Image processing of formosat-2 data for monitoring south Asia tsunami. *Int. J. Remote Sens.* **2007**, *28*, 3093–3111. [CrossRef]
22. Liu, C.C.; Ko, M.H.; Wen, H.L.; Fu, K.L.; Chang, S.-T. Instability Index Derived from a Landslide Inventory for Watershed Stability Assessment and Mapping. *ISPRS Int. J. Geo-Inf.* **2019**, *8*, 145. [CrossRef]
23. Wu, C. Certainty Factor Analyses and Spatiotemporal Characteristics of Landslide Evolution: Case Studies in the Chishan River Watershed in Taiwan. *ISPRS Int. J. Geo-Inf.* **2022**, *11*, 382. [CrossRef]
24. Ma, S.; Qiu, H.; Zhu, Y.; Yang, D.; Tang, B.; Wang, D.; Wang, L.; Cao, M. Topographic Changes, Surface Deformation and Movement Process before, during and after a Rotational Landslide. *Remote Sens.* **2023**, *15*, 662. [CrossRef]
25. Kuang, J.; Ng, A.H.-M.; Ge, L.; Metternicht, G.I.; Clark, S.R. Joint Use of Optical and Radar Remote Sensing Data for Characterizing the 2020 Aniangzhai Landslide Post-Failure Displacement. *Remote Sens.* **2023**, *15*, 369. [CrossRef]
26. Li, C.; Wang, M.; Liu, K. A decadal evolution of landslides and debris flows after the Wenchuan earthquake. *Geomorphology* **2018**, *323*, 1–12. [CrossRef]
27. Kincey, M.E.; Rosser, N.J.; Robinson, T.R.; Densmore, A.L.; Shrestha, R.; Pujara, D.S.; Oven, K.J.; Williams, J.G.; Swirad, Z.M. Evolution of Coseismic and Post-seismic Landsliding After the 2015 Mw 7.8 Gorkha Earthquake, Nepal. *J. Geophys. Res. Earth Surf.* **2021**, *126*, e2020JF005803. [CrossRef]
28. Roback, K.; Clark, M.K.; West, A.J.; Zekkos, D.; Li, G.; Gallen, S.F.; Chamlagain, D.; Godt, J.W. The size, distribution, and mobility of landslides caused by the 2015 Mw7.8 Gorkha earthquake, Nepal. *Geomorphology* **2018**, *301*, 121–138. [CrossRef]

29. Zhang, X.; Chen, L.; Zhou, C. Deformation Monitoring and Trend Analysis of Reservoir Bank Landslides by Combining Time-Series InSAR and Hurst Index. *Remote Sens.* **2023**, *15*, 619. [CrossRef]
30. Medhat, N.I.; Yamamoto, M.-Y.; Ichihashi, Y. Inclinometer and Improved SBAS Methods with a Random Forest for Monitoring Landslides and Anchor Degradation in Otoyo Town, Japan. *Remote Sens.* **2023**, *15*, 441. [CrossRef]
31. Conforti, M.; Mercuri, M.; Borrelli, L. Morphological Changes Detection of a Large Earthflow Using Archived Images, LiDAR-Derived DTM, and UAV-Based Remote Sensing. *Remote Sens.* **2021**, *13*, 120. [CrossRef]
32. Lin, S.C.; Ke, M.C.; Lo, C.M. Evolution of landslide hotspots in Taiwan. *Landslides* **2017**, *14*, 1491–1501. [CrossRef]
33. Science. A Massive Experiment in Taiwan Aims to Reveal Landslides' Surprising Effect on the Climate. Available online: https://www.science.org/content/article/massive-experiment-taiwan-aims-reveal-landslides-surprising-effect-climate (accessed on 25 January 2023).
34. Government Open Data. Available online: https://data.gov.tw/dataset/140111 (accessed on 24 September 2022).
35. Hydrological Data in Taiwan. Available online: https://www.wra.gov.tw/cp.aspx?n=3168 (accessed on 24 September 2022).
36. Walsh, R.P.D.; Lawler, D.M. Rainfall seasonality: Description, spatial patterns and change through time. *Weather* **1981**, *36*, 201–208. [CrossRef]
37. Tanyas, H.; Kirschbaum, D.; Görüm, T.; van Westen, C.J.; Tang, C.; Lombardo, L. A closer look at factors governing landslide recovery time in post-seismic periods. *Geomorphology* **2021**, *391*, 107912. [CrossRef]
38. Anselin, L. Local indicators of spatial association-LISA. *Geogr. Anal.* **1995**, *27*, 93–115. [CrossRef]
39. Getis, A.; Ord, J.K. The analysis of spatial association by use of distance statistics. *Geogr. Anal.* **1992**, *24*, 189–206. [CrossRef]
40. ESRI Website. Available online: https://pro.arcgis.com/en/pro-app/2.8/tool-reference/space-time-pattern-mining/emerginghotspots.htm (accessed on 24 September 2022).
41. Liu, J.; Mason, P.J.; Bryant, E.C. Regional assessment of geohazard recovery eight years after the Mw7.9 Wenchuan earthquake: A remote-sensing investigation of the Beichuan region. *Int. J. Remote Sens.* **2018**, *39*, 1671–1695. [CrossRef]
42. Tsai, S.R. Spatial Variation of Soil Properties in Kaoping River Basin Prior and Post Typhoon Morakot. Master's Thesis, National Pingtung University of Science and Technology, Taiwan, 2011; p. 52.
43. Lousada, S.; Cabezas, J.; Castanho, R.A.; Gómez, J.M.N. Hydraulic Planning in Insular Urban Territories: The Case of Madeira Is-land—Ribeira Brava. *Water* **2021**, *13*, 2951. [CrossRef]
44. Lousada, S.; Gonçalves, L.; Atmaca, A. Hydraulic Planning in Insular Urban Territories: The Case of Madeira Island—São Vicente. *Water* **2022**, *14*, 112. [CrossRef]
45. Lousada, S.; Castanho, R.A. GIS-Based Assessment of Morphological and Hydrological Parameters of Ribeira Dos Socorridos and Ribeira Do Vigario Basins, Madeira Island, Portugal. *Curr. World Environ.* **2021**, *16*, 408–426. [CrossRef]
46. Lousada, S.A.N.; Moura, A.D.S.; Gonçalves, L.B. Numerical modelling of the flow rate in artificial water channels: Application to Ribeira Brava's stream. *Rev. Bras. Planej. Desenvolv.* **2020**, *9*, 39–59. [CrossRef]
47. Lousada, S.; Cabezas, J.; Castanho, R.A.; Gómez, J.M.N. Land-Use Changes in Insular Urban Territories: A Retrospective Analysis from 1990 to 2018. The Case of Madeira Island—Ribeira Brava. *Sustainability* **2022**, *14*, 16839. [CrossRef]
48. Liu, C.-C.; Chen, P.-L. Automatic extraction of ground control regions and orthorectification of formosat-2 imagery. *Opt. Express* **2009**, *17*, 7970–7984. [CrossRef] [PubMed]
49. Liu, C.-C.; Shieh, C.-L.; Wu, C.-A.; Shieh, M.-L. Change detection of gravel mining on riverbeds from the multi-temporal and high-spatial-resolution formosat-2 imagery. *River Res. Appl.* **2009**, *25*, 1136–1152. [CrossRef]

water

MDPI

Article

Three-Dimensional Engineering Geological Model and Its Applications for a Landslide Site: Combination of Grid- and Vector-Based Methods

Thanh-Tùng Nguyễn [1], Jia-Jyun Dong [1], Chia-Han Tseng [2,*], Ivo Baroň [3], Chao-Wei Chen [4] and Chao-Chin Pai [5]

[1] Graduate Institute of Applied Geology, National Central University, No. 300, Zhongda Rd., Zhongli District, Taoyuan City 320317, Taiwan
[2] Department of Geology, Chinese Culture University, No. 55, Hwa-Kang Rd., Yang-Ming-Shan, Taipei City 11114, Taiwan
[3] Department of Engineering Geology, Institute of Rock Structure and Mechanics, The Czech Academy of Sciences, v.v.i., V Holesovickach 94/41, 182 09 Prague, Czech Republic
[4] LAND Engineering Consultants, Inc., Taipei City 115020, Taiwan
[5] Taipei Branch, Soil and Water Conservation Bureau, Council of Agriculture, Executive Yuan, No. 10, Jingzhong Rd., Xindian District, New Taipei City 23153, Taiwan
* Correspondence: zjh17@ulive.pccu.edu.tw

Citation: Nguyễn, T.-T.; Dong, J.-J.; Tseng, C.-H.; Baroň, I.; Chen, C.-W.; Pai, C.-C. Three-Dimensional Engineering Geological Model and Its Applications for a Landslide Site: Combination of Grid- and Vector-Based Methods. *Water* **2022**, *14*, 2941. https://doi.org/10.3390/w14192941

Academic Editors: Maria Mimikou and M. Levent Kavvas

Received: 25 June 2022
Accepted: 16 September 2022
Published: 20 September 2022

Abstract: A three-dimensional engineering geological model (EGM), which provides an approximation of the geological conditions, is a key element in any engineering project. The slope at Huafan University, Mt. Dalun, in the Western Foothills of northern Taiwan, is a dip slope that has been assumed to be unstable. The bedrock is mainly composed of intercalated sandstone and shale, where the thickness of the sandstone varies from thin to massive, interbedded with shale from the Miocene age. By interpolating the thickness of the colluvium derived from borehole data and analyzing the contours of the interpolation surface result, we find that the landslide material accumulates at the slope foot, towards the southwest in the direction of movement. Due to tectonic control—in particular, considering the two local faults that pass through the study area—the strata's orientation significantly changes over the studied slope. As a basis for the 3D EGM, polynomial surface fitting is applied for detailed analysis of the sub-surface geological structure, as well as to compute the regressive orientation of the bedding plane derived from the borehole data. Based on the calculated regression plane passing through the elevations of the geological interface (key bed), the results indicate that the regression plane's direction is consistent with the outcrop measurements. Moreover, several cross-sectional profiles are considered to visualize and clarify the 3D EGM. Finally, surface and subsurface monitoring data are compared with the result, in order to refine the 3D EGM. The proposed geological model is expected to contribute to the comprehensive understanding of gravitational slope deformation, and may serve as a guideline to minimize potential disasters.

Keywords: landslides; 3D engineering geological model; grid- and vector-based; surface and subsurface displacement monitoring; failure mechanisms; geotechnical engineering design

1. Introduction

An engineering geological model (EGM) is, according to the IAEG C25, an approximation of the geological conditions created for the purpose of solving an engineering problem [1], and it is an element of utmost importance in an engineering project [2].

Traditionally, geological maps and geological profiles are used as geological models, representing the sub-surface 3D structure in a two-dimensional manner [3], in order to communicate complex spatial geological information [4]. However, it is quite difficult for a broader audience without extensive geological knowledge to conceive of and comprehend sub-surface geological information through a two-dimensional (2D) view. Although skilled geologists know how to translate 2D information into three-dimensional (3D) information qualitatively or conceptually, inaccurate and sometimes incorrect translation may still occur [5]. Moreover, limited geological profiles are not necessarily coincident with the critical cross-sections that are important to a given engineering project. The shortcomings of this traditional presentation mode include: (1) Being unfavorable for communication between geologists and engineers, due to the knowledge gap; (2) when hard data (e.g., outcrops or boreholes) are lacking, it is difficult to obtain the geological profiles required for engineering design; (3) difficulties in constructing the 3D geometry for 3D numerical simulations; and (4) the monitoring results may be hard to interpret. Therefore, a 3D EGM may essentially contribute to better sub-surface understanding in engineering projects such as natural hazard assessment, underground work planning, hydrogeological analysis for aquifer recharge assessment, and so on [6]. Such 3D EGMs play a significant role in visualizing and communication, synthesizing all available data, and providing more accurate representations of specific geological conditions or simulating stratified geology [7–9], as well as allowing engineering geologists to test the validity of geological assumptions or hypotheses.

Three-dimensional EGMs have widely been used over the past 40 years, mainly in the hydrocarbon and mining industries [10]. They have also been used extensively for groundwater flow modelling (see, e.g., [11–13], especially for Quaternary sedimentary rocks. Although quite a few studies have applied 3D EGMs for civil engineering purposes in urban areas (e.g., [6]), 2D EGMs are still used for most landslide studies. The more efficient 3D EGMs, however, have recently been increasingly used in landslide research; for example, to reconstruct the spatial structure of a landslide [14,15], or to integrate the 3D geology of sub-surface geometry and properties to visualize landslide susceptibility and evaluate the risk of natural landslides [16–18]. Moreover, Gu et al. [18] have calculated the stability of each unit after establishing a 3D EGM to generate regional slope stability. Li et al. [19] have created a 3D EGM of a landslide, in order to assess the relationship between landslide deformation and groundwater level change beneath the sliding surface. The application of a 3D EGM in geologically vulnerable areas that might lead to disaster landslides is necessary, as the potential presence of peculiar geological properties influenced by the geometry of the bedrock and overlying soil can create unstable slopes, leading to a reduction in the resistance forces that retain them. The geological features and structures are the most important factors in the movement development of a rock mass [20]. Landslide deformation patterns and the processes that cause them can be determined through analysis of the geometry of the bedrock, the interior layers, and the slip surfaces [21,22], which can be detected from a 3D EGM. However, studies considering the monitoring and mitigation of landslides based on a critical 3D EGM remain rare, and determination of the surface or sub-surface displacement of slope mass is still a difficulty. Therefore, there is not yet any meaningful cohesiveness in the relevant literature.

Three methods can be used to model a 3D EGM of sedimentary rocks [3]: (1) Using borehole observations to create a triangle defining a surface. In geographic information system (GIS) terms, this is a triangulated irregular network (TIN); (2) by applying any of several surface-generation and contouring procedures to the borehole observations; or (3) by developing a series of parallel or intersecting interpretive cross-sections between boreholes, in order to better control the surface geometries. These surfaces can be modelled using grid-based methods [13] or vector-based methods [23]. Generally, the geological data derived from borehole or stratigraphic sections are processed to form discrete points. To make full use of these points, a grid-based method with geological contour can be carried out in the map [24]. Grid-based methods use a mesh that is either regularly spaced (triangular, square, or rectangular), or has angular grids that are consistently spaced. Grid layering is used to simulate planes (e.g., paleo-depositional surfaces or bedding planes, which should possess continuity). In particular, a GIS grid-based data model may be used to describe slope-related items such as strata, sliding surfaces, and groundwater-surface in 3D slope models in slope stability research [25]. Additionally, the bounding surface in vector-based methods computes the elevations of the contour surfaces utilizing complicated geo-modelling computations. This means that each contour line must be properly digitized, in the form of (X, Y) pairs along each contour line of specified elevation [26] (Moore et al., 1993). The vectorial foundation of the model structure is very adaptable and reasonably simple to modify computationally. Hence, a complete modelling system that integrates grid- and vector-based methodologies to stimulate the natural geological structure is required. These combined capabilities provide the essential framework and a degree of trust for a 3D EGM.

To build a reliable 3D model, all accessible geological data should be reconciled [27]. Borehole data can clearly reveal the stratification of the underlying rock, allowing for direct observation of samples [3], and are currently utilized to create 3D EGMs [7,13]; however, the construction of a 3D EGM is still a challenging task for engineering geologists.

In this study, an example of an ancient, recently re-activated landslide is presented, in order to argue the necessity of a 3D EGM for its comprehensive understanding. The 3D model building process calls for complex feedback between the interpretation of the data (observational model approach) and the model (conceptual model approach). In the interpretation, the 3D framework forces us to make interpretive decisions that would be left on the side in map-based or cross-section interpretations [5]. The slope on which Huafan University, in the Mt. Dalun region of Taiwan, is located, has been identified as a possible unstable dip slope. Numerous site investigation projects have been carried out, in order to understand the sub-surface geology and possible landslide mechanisms. There are almost 50 drilling boreholes from diverse sources. Moreover, extensive monitoring systems including surface and sub-surface displacement measurements, groundwater table measurements, and rainfall measurements, indicating a creep phenomenon in the study area [28–30]. However, the monitoring data are mainly compared with geological profiles or plan view of the landslide boundary in a two-dimensional manner. Although 3D numerical modelling has been conducted [29], the sliding surface determined from inclinometer was used to define the model boundary without incorporating a 3D EGM. The design of drainage wells, which has been carried out by a local consultant company, was also based on two-dimensional information. This indicates that previous studies have not provided a comprehensive geological model for the entire area, in the form of a 3D EGM. Therefore, the 3D EGM presented in this paper provides a comprehensive geological model for the sloping area, which can help to determine the potential landslide extent. This is crucial for anticipating the failure mechanism and assessing the slope's stability; it may also serve as a guideline for future civil engineering and geotechnical methods aimed at preventing landslides.

2. Study Area

Mt. Dalun is located in northern Taiwan's Western Foothills belt. The Huafan University campus is built on the upper part of the mountain, and it is elevated between 240 and 580 m above sea level. This location is the headstream of the Pengshan River (Figure 1).

According to the 1:50,000 geological map published by the Central Geological Survey of Taiwan [31], the main exposed strata near the study area are those of the Mushan Formation (Ms) in the early Miocene age. One of the prominent lithological features is "white sandstone" (orthoquartzite). These sandstones are mainly composed of quartz, indicating a continental shelf sedimentary environment. The sandstones in the Mushan Formation are fine- to coarse-grained, mostly in thick or massive layers with thicknesses of 8–20 m, and usually have an obvious cross-bedding sedimentary structure. Grayish black shale or thin beds of sandstone are well-developed, often alternating with thick sandstones to form interbeds [28]. The bedding plane with the strike of the rock is roughly in the east–west direction, and is inclined between 10° and 20° to the south, being consequent to the topographical slope [29,31].

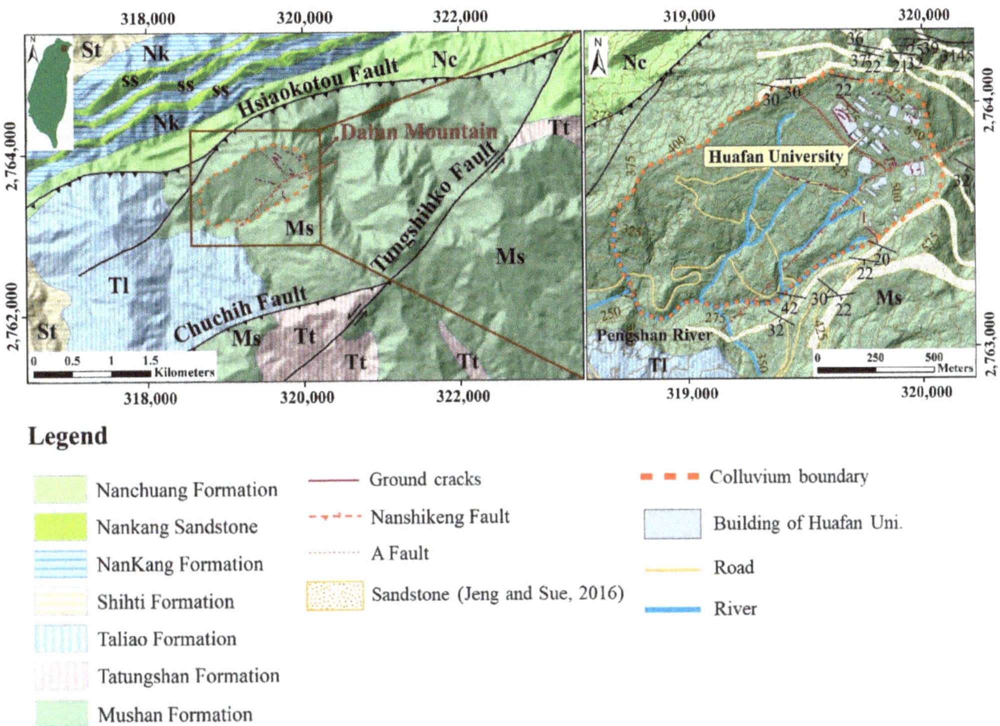

Figure 1. Location and geological map of the Mt. Dalun area. The attitudes of bedding plane are derived from the study of Jeng and Sue [28]. The orange dotted line indicates the area where the main colluvium is distributed; potential sliding blocks are estimated to occur within this area.

There are three major faults in the study area: The Hsiaokotou Fault, the Chuchih Fault, and the Tungshihko Fault. The Hsiaokotou and Chuchih Faults are thrust faults with a strike along ENE–WSW, trending and dipping to the southeast. The fault traces are intercepted by the left-lateral, strike-slip Tungshihko Fault with NE trend.

The geological structure and local lithology have determined the topographical features in the study area. The topography is defined by mountain ranges in the form of a bowl that trends northeast–southwest and has a surface slope of 20–25°. The catchment area in Mt. Dalun is represented by the mountain ridgelines that resemble the shape of a bowl on the terrain. The surface runoff within the mountain ridgelines flows down to streams, consisting of nature erosion gullies (marked as blue lines in Figure 1) on the slope surface.

By means of thorough site investigations, Jeng and Sue [28] have documented two minor faults which influence the local geological structure: (1) The Nanshihkeng Fault is a thrust fault with trace trending northeast and with a fault plane dipping 60–70° to the southeast; and (2) the A Fault is an oblique-slip fault truncated by the Nanshihkeng Fault. The strike is roughly NW–SE and dipping to the south with a dip angle of 65–70°. The A Fault is likely a normal fault with a left-lateral slip component, with its footwall relatively throwing up. Two sets of joints cut the bedrock, whose strikes are average NW–SE and NNE–SSW, thus being roughly parallel to the two local aforementioned faults [29].

Based on the results of the site investigation [28], the bedrock of the study area is overlaid by colluvium (slope wash), thick (more than 40 m) ancient landslide deposits (to the southeast of the study area), and artificial fill materials (ranging from 10 to 20 m). Due to the thick colluvium and fill materials, rock outcrops in the study area can only be found in eroded creeks or on the artificially excavated surfaces. In areas with thick landfill (e.g., the northeastern part of the study area), such as the sports ground and basketball court of Huafan University, surface displacement is significant due to inadequate compaction of the filled land or sliding along the bottom of the landfill [28,30]. These creeping areas revealed multiple listric sliding surfaces [30].

Due to the unstable slope and existing landslide potential in Mt. Dalun region, a monitoring system has been in operation since 2000, in order to determine the slope failure mechanism and to enhance the effective risk management. This includes inclinometer casings, standpipes for groundwater pressures, rainfall gauges, and a nail network for surficial deformation of the slope. To measure the sub-surface displacement and possible depth of sliding surfaces, every inclinometer casing in the borehole is recorded every month. Every six months, a network of monitoring points on the slope surface enables the measurement and monitoring of surficial displacement. It is possible to link changes in slope displacement to changes in the groundwater table and the magnitude of earthquakes by monitoring these variables after heavy rainfall or earthquakes. In addition, a countermeasure system has also been implemented in order to diminish the slope moment and to prevent the failure of the slope, including a drainage system and ground anchors. The installed drainage systems, comprising catch pits and trenches for collecting groundwater, have been carried out by a local consultant company, in order not to allow the water table to increase and submerge the sliding surface. In the future, it may be necessary to expand the number of drainages to ensure the safety of the slope.

3. Methodology to Develop a 3D EGM

The data used to develop the 3D EGM included: (1) field works; (2); interpretations of borehole cores; and (3) airborne LiDAR data. The field data included existing ground-surface geological map provided by numerous geologists, publicly available data collected by companies over dozens of years [28,30], and field investigation data obtained during field trips Mt. Dalun. The approach is described in detail below, following the flow chart depicted in Figure 2.

3.1. Topography Analysis

Based on a high-resolution (1 m) LiDAR DEM, a topographic map of the study area was generated with contour line interpretation. Moreover, the large-scale topographic features, such as geological structures, scarp, hummocky topography, and gully paths could be detected. We also analyzed the terrain in an attempt to anticipate the distribution of sandstone outcrops.

3.2. Field Investigation

The field investigation was a critical step to identify the interface between the colluvium and bedrocks, the boundaries of rock units, and bedding planes' orientation. We divided the field investigation into two stages. The first stage was a general overall view of the slope, including a preliminary assessment of terrain features and landforms and collecting outcrop information along the road in the eastern part (Figure 1). We examined classifications of lithological unit, discontinuity features, rock masses, and deposits on outcrops. In the second stage, we investigated the fields along the Pengshan River, near the toe of the slope, in order to evaluate the rock outcrop lithology for certain characteristics of rock unit classification, and investigated the dipping direction of the exposures of sedimentary strata on both sides of the riverbed. This work aims to confirm whether the type of exposure belongs to the in situ rocks or moving blocks that were deposited after ancient landslides in the past. The information gained from the field investigations and the geological map provide basic data for the establishment of a geological model in the study area. In addition, the field investigation aimed to re-examine and compare our geological records with those obtained in earlier investigations.

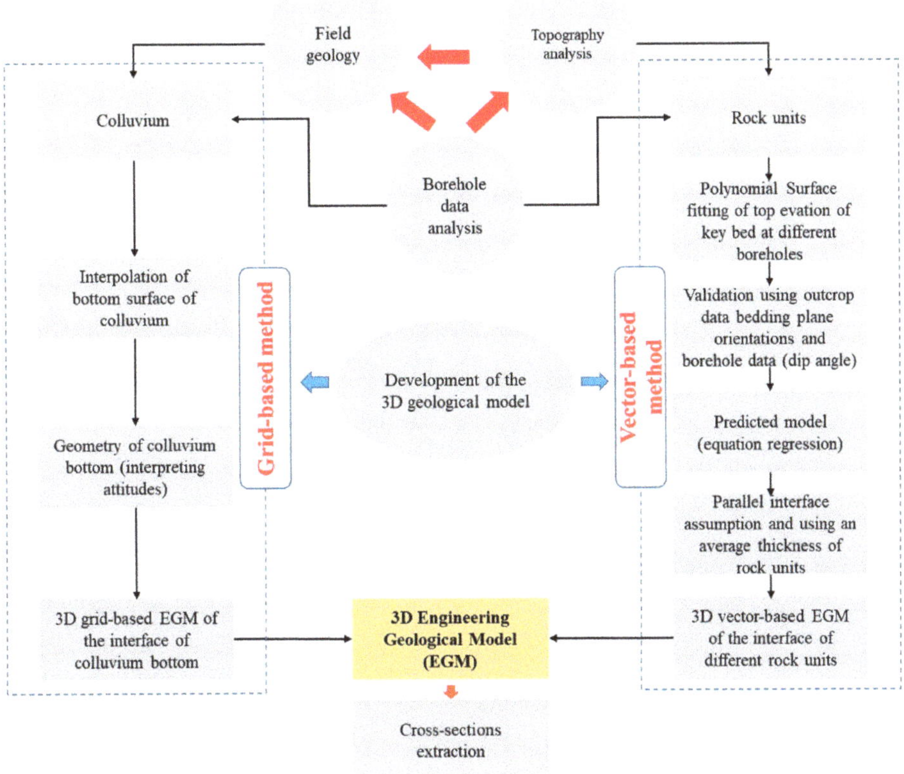

Figure 2. Workflow for development of a 3D engineering geological model (EGM).

3.3. Borehole Data Analysis

The 3D EGM developed in this study was mainly based on the stratigraphic column interpreted from rock cores. Drilling data from 50 boreholes of various depths were utilized for the analysis. The main process of borehole data analysis was as follows:

(1) Borehole data quality check. To begin, we thoroughly checked the reported locations and elevations of the boreholes, in order to ensure the correctness of the basic information. If the X and Y input values were inaccurate, the elevation of each borehole will be changed significantly, such that the elevation of the rock unit boundary will also be misjudged. The TWD97 TM2 (i.e., Taiwan Datum 1997, 2-degree Transverse Mercator) coordinate system was employed in this work, in order to ensure data coherence. As a result, this step is critical for further analysis. We used the LiDAR DEM to obtain the elevations (i.e., Z values) of the boreholes at given horizontal coordinates (i.e., X for E–W direction and Y for N–S direction), and compared them with the documented elevations of boreholes. In addition, the quality of the rock core must not have been significantly impacted by weathering (i.e., preserved in fresh bedrock) or shearing or fracturing by tectonic processes. We preferred boreholes with an adequate depth, which permit comparisons between a sequence of units in different boreholes. Only the boreholes that passed this checking step were used for subsequent 3D EGM development. In this way, we ensured that only reliable borehole data were considered prior to the following steps.

(2) Rock unit classification. The fundamental interpretation consisted of lithology identification and rock unit boundary determination. Sandstone (SS) and thin alternating layers of sandstone and shale (SS and SH, respectively) constitute the main lithology in the study area. The sedimentary features of rock units, such as whitish sandstone, particularly form thick or massive layers. The grayish-black shale and thin interbeds of sandstone and shale are also well-developed, often alternating with thick sandstones [28]. The sedimentary sequences were considered when we analyzed the rock cores; in particular, we considered a sequence of continuous rock units throughout the borehole, as opposed to focusing on a single rock unit (i.e., homogeneous in a sequence of rock units). Lithostratigraphic bodies were distinguished by distinct lithological minor structures, such as lamina and flaser structures, as a criterion for determining the sedimentary sequence. The proportion of sandstones and shales was used to classify the alternated layers of sandstone and shale into different rock units. The descriptive reference proportions are:

(I) Interlayer 1:1 $\leq$ Mj:Mn $\leq$2:1,
(II) Interbedded 2:1 $\leq$ Mj:Mn $\leq$ 4:1,
(III) Occasionally 4:1 $\leq$ Mj:Mn $\leq$ 8:1,
(IV) Main lithology 8:1 $\leq$ Mj:Mn,

where Mj denotes major lithology and Mn denotes minor lithology.

We summarized all the descriptive properties of the rock, and then split the observation data into two separate parts: colluvium on the top, and sedimentary rock beneath each drilling borehole.

(3) Comparison between rock units from different boreholes. In the advanced step, we conducted a comparison of the lithology from the 26 complete boreholes among the 50 boreholes in the study area, in order to prepare the necessary material for the building of the geological model. To make the comparisons accessible and reliable, we started with the boreholes that were close to each other. We grouped the boreholes together and compared the lithology of the strata. We assumed that the features of the strata lithology of these boreholes should be the same; however, the boreholes were distributed sporadically and also outside the campus on the downward slope. Both petrographic properties and the distribution of boreholes had changed slightly. When the rock cores were compared to each other, this aspect also had been carefully considered. There are comparison criteria that may be depended upon to ensure reliability. In this context, sedimentary characteristics, including the texture and internal structures created by bioturbation and bio-erosion, are essential for classifying rock units to build correlatively connected boreholes. Furthermore, every objective layer's thickness must have minimal change; that is, the standard deviation of the thickness should be small. Furthermore, a structural dip was recognized in the borehole description and the rock core photos, handled in laminated or thinly bedded to bedded shale and sandstone sequences. Such a structural dip is best-identified from cross-bedded sections thought to have been horizontally stratified originally, which we call cross-bedding, accordingly. As a result, a side-by-side comparison of lithological borehole data was conducted.

(4) Selection of key marker bed. The thick sandstone was selected as the key bed of the rock unit, as it was readily distinguishable. To obtain a reliable result, the thick sandstone must be identifiable over a large number of boreholes (i.e., be unique and widespread), ensuring that it exists at most locations in the study area through these boreholes. The sedimentary sequences are significant, as they help us to identify which sandstone is the proper one among the boreholes. This is based strictly on the distinguished lithology in the sedimentary sequences, as mentioned above (e.g., the laminated structure in the alternating sandstone and shale units). In addition, the key bed must fulfill the requirements for the comparison of rock units from different boreholes, as described in step (3) above.

The elevation of the colluvium bottom (top of the bedrocks), in combination with the borehole elevation and recorded depth, can be obtained in each borehole. Based on the rock unit classification results, the elevations of the top and bottom of each rock unit can also be identified. Meanwhile, the average thickness of each rock unit can be determined considering the thicknesses obtained from different boreholes.

3.4. Development of the 3D EGM

3.4.1. Interpolation of Colluvium Bottom Boundary (Depth of Bedrock Top Surface): Grid-Based Method

A dataset including the X, Y, and Z values of the interface between the colluvium and underlying bedrocks from 50 boreholes were utilized for interpolation. In addition, we assumed that the elevation of the bottom boundary of the colluvium was equal to the elevation along the Pengshan River and the water divides (ridges) and the bedrock outcrops along these lineaments. The elevation data points along these lineaments were used to interpolate the elevation of the bottom boundary of the colluvium (i.e., the bedrock top surface). These lineaments are marked with the orange dotted line in Figure 1. The 3D software Surfer 16.0 [32] was used to interpolate the boundary surface elevation at each grid with 12 × 12 m resolution using the Kriging method. Equation (1) describes the process of calculating the estimated value:

$$Z_A = \sum_{i=1}^{n} W_i Z_i, \quad (1)$$

where Z_A is the estimated value, Z_i refers to the value at position i, W_i expresses the weight at position i, and n identifies the number of neighboring data values utilized in the estimation (derived from the Golden software).

We gridded the datasets using the default properties of the Kriging interpolation. From unevenly spaced datasets, this approach creates structural contour maps which represent the colluvium bottom. The data were acquired from a total of 50 available boreholes for the whole study area. To make the result of interpolation reliable, we attempted to exclude the extrapolation when there was no data outside the borehole network in the study area. Consequently, visualization of the bedrock depth distribution in the study area was achieved (see Section 4.2).

3.4.2. Determining the Polynomial Surfaces Representing the Rock Boundaries: Vector-Based Method

Polynomial surface fitting is a mathematical approach that involves identifying and measuring trends in the characteristics of the dispersed observed data, in which the sum of the squared deviations from the trend surface is minimized. With this analysis, our purpose was to develop a fitting surface for the boundaries of each rock unit, representing a geological boundary to create the spatial geometry of the interfaces between the rock units. Thus, we decided to use a prediction model which creates an interpolated geological contact surface extending in a continuous layer from the data (i.e., the geological observations obtained from the boreholes). This is a type of regression analysis expressing each initial observation, using the product of a deterministic polynomial function of the geographic location with a general Function (2):

$$Z_A = f\,(X_A, Y_A) = b_0 + b_1 X + b_2 Y + b_3 X^n + b_4 XY + b_3 Y^n + \ldots, \quad (2)$$

where Z_A is the value of a property at location A, X_A and Y_A are the coordinate values at location A, f is the regression function, b_i is a polynomial constant, and n is the order of the polynomial [33–36]. The dataset collection step is heavily reliant on stage 2 of the borehole data analysis, including X and Y representing the longitude and latitude of boreholes, respectively, and Z is the elevation value of the top of the thick sandstone, the so-called key bed (i.e., the occurrence depth of thick sandstone).

We began our spatial distribution analysis procedure with the aforementioned 26 data points corresponding to the 16 boreholes (Table 1) which were irregularly distributed over the research region, spanning from the middle to the lower part of the slope, as well as the 10 data points distributed around the campus (Table 2). The top of the key bed in these boreholes was extracted with respect to elevation values (i.e., *Z*). Then, these values were used to carry out regression-based surface fitting, in order to get the best fit. We began by employing a polynomial function of order one, and gradually increased the order until the result fit the data well. The prediction methods for spatial computing data were then established. According to the goodness-of-fit statistics, the best polynomial surface regression model was selected, with respect to the data, in order to describe the surface trend, thus simulating the spatial distribution of the rock units. This function then could be used as a critical model to evaluate the vertical spatial distribution of the geological sub-surface at any location with geographic coordinates inside the study area. To display the surface of the interfaces (i.e., geological contacts) or the top of the thick sandstone (key bed) obtained from the model, the contour map was processed as a resultant map. This structural map guaranteed smooth boundaries and fit the data well. In order to obtain a reliable resultant interface of rock units, the contour map was validated with respect to the bedding plane orientation from outcrops. In addition, validation of the borehole data (structural dip angle) from rock core logging was significant. We assumed that all rock units were parallel to one another (i.e., the principle of sedimentary rock), and then used the average thickness of each rock unit to generate the other interfaces, based on the fitting interface of the key bed. This means that every interface of the rock units was introduced into a polynomial function. This led to creating of a so-called 3D vector-based grid of the interfaces of different rock units. Finally, cross-sections were used to visualize the spatial distribution of the rock units.

Table 1. Error of the first- and second-order polynomial regression models in comparison with the real data from the boreholes in Zone 1. Z represents the elevation of the sandstone top (key bed).

Borehole	X	Y	Z	Error	
				2nd	1st
W-4	319,803.7	2,763,737.5	413	4.10	7.00
19-7	319,722.7	2,763,660.8	390.6	5.46	2.20
19-3	319,743.2	2,763,686.7	399	3.95	2.53
DH-5	319,800.2	2,763,697.0	411	−3.14	2.79
DH-7	319,740.0	2,763,699.7	403.8	1.95	1.23
DH-4	319,792.5	2,763,743.6	422	−3.75	−0.98
DH-1	319,758.2	2,763,763.9	423	−1.00	1.57
SIS-36	319,761	2,763,769.8	427	−3.40	−0.53
18-2	319,443.5	2,763,803.5	418	−2.21	−2.63
19-1	319,456.7	2,763,780.4	414	−4.36	−4.38
18-1	319,572.3	2,763,871.3	440	4.90	3.16
17-7	319,614.1	2,763,867.3	444.4	1.03	0.32
17-3	319,273.8	2,763,519.9	314.3	4.06	8.91
17-1	319,643.2	2,763,641.1	394	−7.46	−12.01
19-2	319,177.7	2,763,324.5	262	−1.79	−0.88
21-1S	319,431.5	2,763,541.8	342.5	1.58	−2.70

Table 2. Difference between actual and predicted data using first-order regression model in Zone 2.

Borehole	X	Y	Z	Error
W-3	319,631.7	2,763,871.4	443.8	3.68
17-8	319,696.5	2,763,972	472.8	−2.5
19-8	319,691.3	2,763,949.3	466.2	−1.22
BH-1	319,823.9	2,763,774.3	428.6	−5.38
19-6	319,804.5	2,763,789.9	424.8	2.21

Table 2. *Cont.*

Borehole	X	Y	Z	Error
W-2	319,938.1	2,763,849.8	440.3	−0.67
18-4	319,941.3	2,763,915.9	452.8	2.11
SIS-14B	319,919.3	2,763,880	446.9	−0.01
SIS-26A	319,929.6	2,763,913	453.7	0.7
SIS-11B	319,848	2,763,752	416.7	1.1

4. Results of 3D EGM Development

4.1. Boundary of Paleo Landslide Deposits

Widely sporadic outcrops are exposed along both sides of the Pengshan River (Figure 3). The Mushan Formation is characterized by a preponderance of whitish fine- to medium-grained sandstone, which is easy to identify and track along the outcrops. The orientations measured at the outcrops almost differ from the background orientations (i.e., the Mushan Formation). In some places, the orientations are almost toward the south with orientation groups of steep dip angles, ranging from 49° to 55°. Particularly, in the confluence of the two creeks, there is mudstone exposure which differs from the previous rock description. In contrast, the orientation measured not far to the southeast of the intersection is almost equal to the orientations of the background, with a dip angle of about 20°. Thus, they look like the Mushan Formation, indicating that the strata in this area were not affected by tectonic activities.

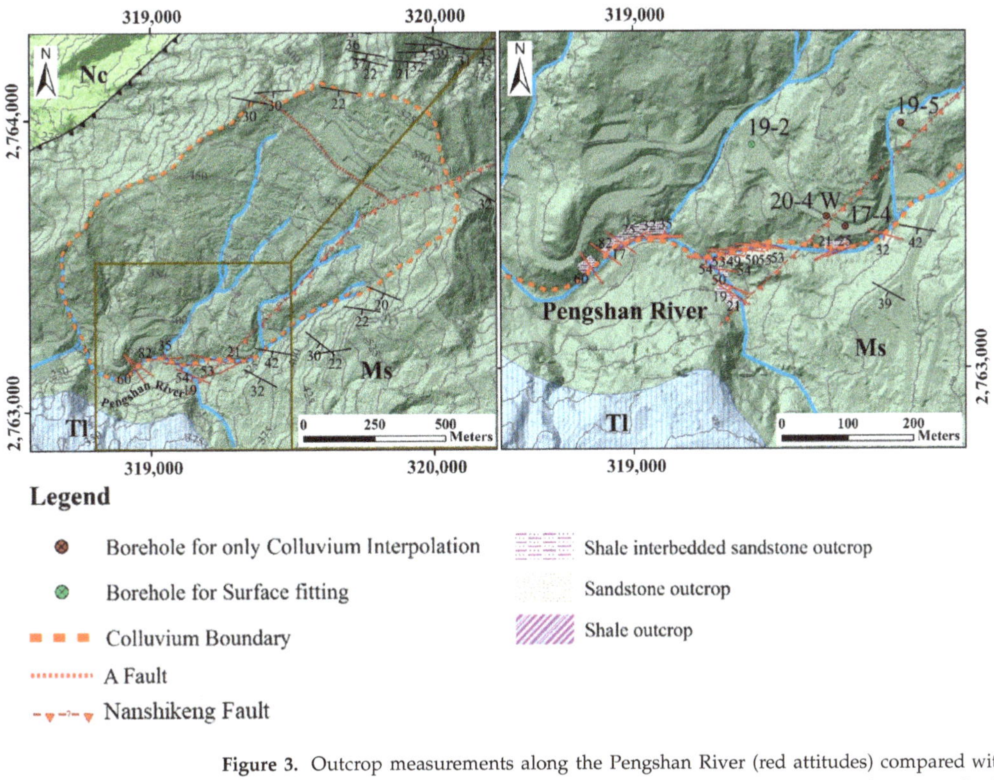

Figure 3. Outcrop measurements along the Pengshan River (red attitudes) compared with the background orientations (black attitudes). The authors measured the attitudes shown in red, whereas the attitudes shown in black were gathered from previous research.

Therefore, the outcrop measurements can be categorized into two groups, with a change in orientation. The first, representing a majority of the outcrops, is positioned on the northern bank of the river, with orientation differing from that of the Mushan Formation. As boreholes 20-4W and 17-4 near creeks indicate 36 and 40 m of colluvium, this rock may originate from ancient landslides. As a result, a boundary for the paleo landslide deposits is proposed in this study, shown as the orange dotted line in Figure 3. On the other side (i.e., on the southern bank of the river), the dip angle of the bedding is steep in the same direction, indicating that the geology in the area must be influenced by tectonic forces. There is a possibility that the Nanshihkeng Fault passes through this area, but there is not much evidence to support this claim. In addition, there are still normal orientations (i.e., background orientation) that should belong to the Mushan Formation.

4.2. Grid-Based Interface between Colluvium and Bedrock

The large-scale terrain interpretation derived from the LiDAR DEM indicated that the eastern, northern, and western sides of the Mt. Dalun area have relatively high altitudes, and the topography is bowl-shaped with collapsed cliffs. Several hummocky areas dominate the topography of the study region, characterized by many small hills and closed depressions. The contour lines, in particular, indicate that the crown is dense while the toe of the avalanche protrudes with a V or U shape. The surface of the study area was previously formed by an ancient landslide debris and by erosion caused by typhoon events. The slope surface is in a current state of equilibrium of material accumulation (Figure 4a). Figure 4b additionally depicts a Kriging map showing the elevation of the interface between the colluvium and the bedrock, which represents the depth of the colluvium. This result incorporates topographical features, outcrops, and borehole data as well. The areas constrained by the rock outcrops, such as steep terrain and the paleo-landslide boundary (cf. Section 4.1), are denoted on the map by red solid crosses (i.e., the thickness of loose materials, including the colluvium and artificial filled land, is assumed to be 0 m). This map may even reveal that the geometry of the bedding plane underneath the surface is covered by loose materials, which dips toward the southwest with an inclination of approximately 15°. Furthermore, the relationship between the colluvium and the bedrock is viewed, through several profiles, in the following sections.

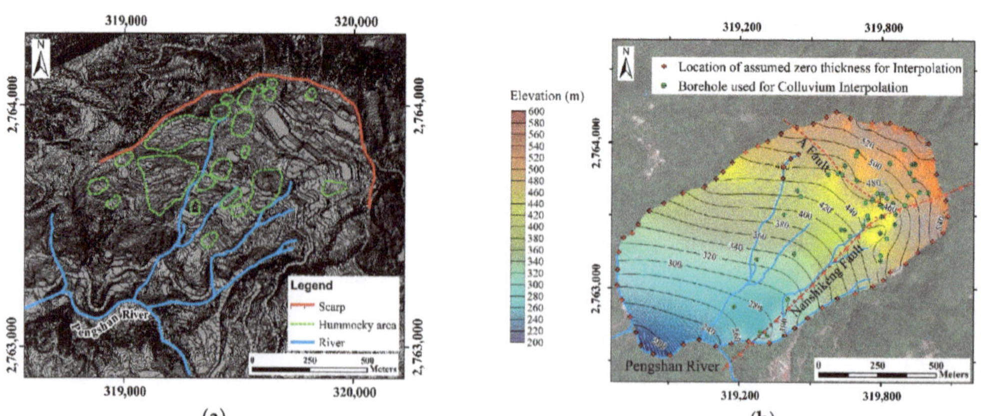

Figure 4. Topography-constrained Kriging map of the lower colluvium boundary. Data were synthesized from the boreholes and topographical analysis: (**a**) Hummocky terrain in the study area; and (**b**) Kriging map of elevation of the interface between the colluvium and the bedrock.

4.3. Geologically Uniform Zones

According to previous studies, two small-scale faults may exist in the study area: The Nanshihkeng Fault and A Fault. The potential Nanshihkeng Fault striking in NE–SW can be observed from the drilling data, in which the fault plane is inclined about 60° to the southeast and cuts through the bedrock in the slope [28]. The hanging wall dominates the eastern portion, with steep terrain to the southeast representing a reversal topography caused by the uplift phenomenon of the two thick sandstone layers in the surrounding geology. In additional, Jeng and Sue [28] have analyzed 2D resistivity data to hypothesize that a probable NW–SE striking oblique-slip fault (i.e., A Fault) intersecting the Nanshihkeng Fault might be characterized by high angle dipping. However, this fault is unclear, without any difference in topographical features.

Based on rock core analysis of the boreholes around the A Fault on the map, extensively fragmented rock and a non-cohesive crush zone as a shear zone were revealed. The shear zone is made up of gouges and rock pieces, as well as clay-rich material. This information provides evidence supporting the existence of the A Fault structure in the study area. The orientation of the A Fault can be detected from the three-point method, which corresponds to 55° NW, 75° N, suggesting that the borehole data applied in this study are consistent with the existence of the A Fault from previous studies. Notably, in following additional rock core data analysis of the borehole 20-4W around the foot part of the slope to the southwest, a piece of especially significant evidence for the presence of this fault was revealed. Figure 5a shows the shear zone like a crush and scratches, as well as a change in the bedding dip angle along with the depth of the borehole, which is a steep bedding dip angle at a depth of over 47 m. The mean inclination value ranges between 60–70°, and seems to change gradually from this depth to become a gentle bedding dip angle toward bottom of this borehole. Even though borehole 17-4 is almost filled by colluvium, the high bedding dip angle track is also indicated obviously. This implies that both boreholes 20-4W and 17-4 pass through the fault plane of the Nanshihkeng Fault, which is evolving in a drag fold pattern. Hence, the existence of these two faults is reasonable, and their distribution might dominate the bedrock orientation in a local area.

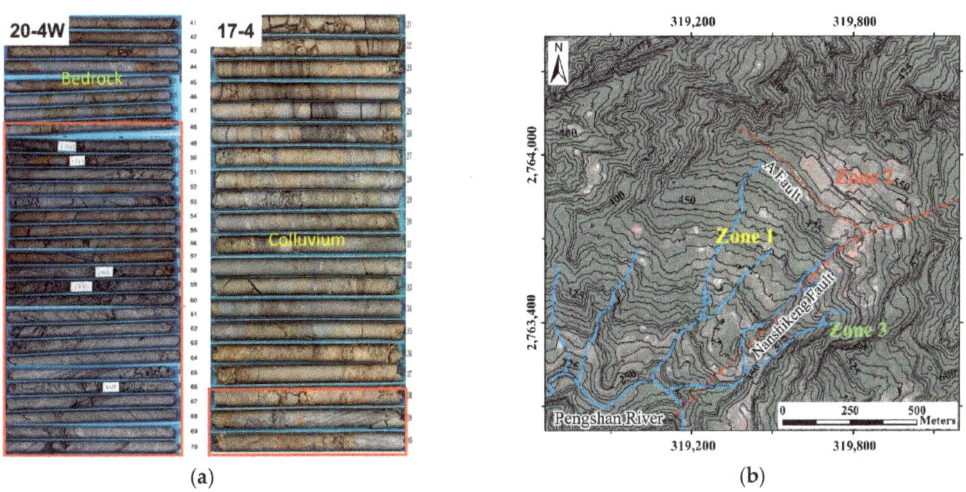

Figure 5. Geological uniform zones of the study area: (**a**) Fracture and shear zones within the red rectangles of the boreholes 20-4W and 17-4, showing a steep bedding dip angle composed of rock fragments as well as clay-rich material. Borehole 20-4W reveals that the damage zone and shear gouge may be found at depths of 56.5–59.6 m, 60.6–62.6 m, and 64.7–66.2 m, with partially crushed gouge having a ~3 cm thick bending shear plane; and (**b**) stratigraphy of the Mushan Formation, separated into the three zones by the Nanshihkeng Fault and A Fault.

However, the lithology of the sedimentary rock in the study area did not change, as revealed from the rock core analysis and field investigation. One of the lithological features was white sandstone and alternation of sandstone and shale, characteristic of the Mushan Formation, which can be easily found from the rock core logging and the outcrops. The bedding plane revealed a gentle (about 20°) dip angle for most of the rock units (borehole data). This finding was also consistent with that provided by the CGS geological survey [31]. The evolution of the Nanshihkeng Fault and the A Fault will result in a change in the spatial distribution of the bedrock; even so, the geological characteristics were the same over the whole region. Based on these conditions, the research area was divided into three zones—Zone 1, Zone 2, and Zone 3—with the two faults serving as boundaries for geological uniform zones of the study area (Figure 5b).

4.4. Development of Geological Model in Zones 1 and 2

4.4.1. Classification of Rock Units

In the Mt. Dalun area, a detailed rock core description was revealed through lithological sediment analysis of the 26 boreholes. We attempted to separate these boreholes into two groups, based on the above-mentioned fault division (cf. Section 4.3): one for the boreholes located to the north of the A Fault and another for the boreholes located to the south of the A Fault. In other words, the boreholes indicated in the western portion of the research area (Zone 1) and the boreholes around the Huafan University campus (Zone 2).

Nine rock units could be distinguished with various types of rock, correspondingly shown in Figure 6. Of these, the key bed in the geological feature was the thick sandstone (L4), with thickness ranging from 16 to 22 m. This rock unit was critical for interpolation and surface fitting in constructing the 3D EGM, as mentioned in the methodology section. In Zone 1, according to the rock core interpretation, eight rock units were revealed, excluding the colluvium layer. The group of boreholes 17-1, 17-3, 17-7, 18-1, 18-2, 19-1, 19-2, and 21-1S covered L1 to L6 in the rock unit, while the other boreholes (19-3, 19-7, W-4, DH-5, DH-7, DH-4, DH-1, and SIS-36) penetrated through L4 and above. In Zone 2, nine rock units could depict the northern stratigraphy of the research region. The borehole BH-1 had a depth of 100 m and penetrates through nine rock units. As an outcome, the stratigraphy in both Zones 1 and 2, as well as the study area overall, could be connected continuously based on the borehole BH-1. Their standard deviation was within an acceptable range, with not much difference between the boreholes in Zone 1 and Zone 2. The characteristics of the colluvium and each rock unit are described in the following.

1. Colluvium (Co). The cover material layer consists of loose and unconsolidated sediments characterized by weathered sandstone pieces and blocks with brownish color and gray matrices, fragments, debris, and soil. The fragments differ in size, shape, and crack density. In some boreholes, the fill material (dark yellowish, fine-grained) is fairly thick, which is due to the constructions, especially in the group of boreholes within the university campus. The thickness of the cover material varied, with the thickest section reaching approximately 40 m (borehole 17-4) towards the toe of the slope located to the southwest of the mountain. Based on the geological processing in the Mt. Dalun area, as well as topographical analysis, the colluvium may have originated from two sources: Loose material belonging to the ancient deep-seated landslides and so-called slope wash (i.e., debris slide).

Rock unit	Rock core sample	Description
L9 (SH/SS)		Shales occasionally interbedded with sandstones
L8 (SS-2)		Grayish thick sandstone
L7 (SS-SH)		Sandstones interlayer with shales (sandstone with laminated sandstone and shale)
L6 (SS/SH)		Sandstones occasionally interbedded with shales
L5 (SH/SS)		Shales occasionally interbedded with sandstones
L4 (SS-1)		Whitish thick sandstone
L3 (SS-SH)		Sandstones interlayer with shales (Interbedded of sandstone and shale with a dense thickly laminated bed)
L2 (SS/SH)	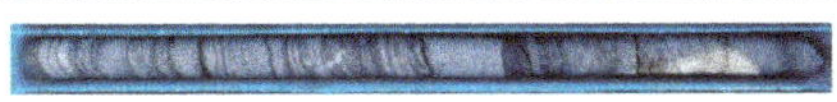	Sandstones occasionally interbedded with shales (sandstone intercalated occasionally medium shale bed)
L1 (SH/SS)		Shales occasionally interbedded with sandstones (thinly laminated bed sandstone and mainly dominat shale)

Figure 6. Typical rock cores for the nine rock units and description of the characteristics of each rock unit. The thick sandstone (L4) was used as a key bed for the polynomial regression of surface fitting.

2. Sandstone (SS). Two rock units were identified as sandstone: L4 (SS-1) and L8 (SS-2). The lithology in L4 was mostly composed of thick, medium-grained whitish sandstone intercalated with thin dark black shale layers scattered in a few places in the massive sandstone. A shear zone containing mud was found in some rock cores. The rock core contains a number of joints. As previously stated, L4 is a key bed identified in the sedimentary sequence, and the overlaying and underlying units are L5 and L3, respectively, characterized by laminated alternations of sandstone and shale. As for L8, grayish sandstone occasionally intercalated with laminated shale is the main lithology. L8 had a higher average thickness than L4, about 25 m and 16 m, respectively. Therefore, the L4 was comparatively identifiable from L8.
3. Alternations of sandstone and shale (SS, SH). This type of rock unit included L1 (SH/SS), L2 (SS/SH), L3 (SS-SH), L5 (SH/SS), L6 (SS/SH), L7 (SS-SH), and L9 (SH/SS). The ratio of sandstone and shale proportion in the recovered rock core significantly varied among these rock units. Each rock unit is described, respectively, in the following:
    a. Rock unit SH/SS (Shales occasionally interbedded with sandstones). This rock is characterized by black to gray shale interbedded with fine sandstone, very thin-bedded sandstone and shale, and mainly dominant shale percentage. It contains sandstone and shale, with shale constituting up to 70–90% of the rock core.
    b. Rock unit SS-SH (Sandstones interlayered with shales). The lithology is composed of interbedded sandstone and black to gray shale, with the thickness ranging from thinly laminated to thickly laminated, and extremely thinly bed-

ded. The sandstone content throughout the portion is approximately 50%, with beds containing parallel lamination and, occasionally, ripple cross-stratification.

c. Rock unit SS/SH (Sandstones occasionally interbedded with shales). The lithology composed of gray to whitish sandstone is the dominant lithology, with black shale interspersed and occasionally locally interbedded sandstone and shale. The sandstone beds range from medium bedded to very thickly bedded, accounting for about 70–80% of the composition.

According to the rock core interpretation, the dip angle of the bedding plane of the sedimentary rock in the above regions mostly ranged from 10° to more than 20°, implying that the bedding plane and slope topography are easily compatible. The results were also supported by the topographical features, field survey, and observations.

4.4.2. Upper Boundary of Key Bed

The application of the polynomial surface fitting method for modelling was applied to determine the geometry of the interface between L4 and L5. In further analyses, this sandstone (L4) top surface was used to estimate the boundaries of other rock units, where the boundaries were assumed to be parallel to each other. For Zone 1, the spatial distribution of the strata was created using both first- and second-order polynomial regression. The first-order polynomial was formulated as shown in Equation (3):

$$Z = -810,375.5140 + 0.065242\ X + 0.285820\ Y. \quad (3)$$

This equation denotes that assumed bedding is planar; that is, based on a linear equation. Meanwhile, the top elevation of the key bed using the second-order polynomial regression equation, which had high accuracy (R^2 = 0.99) is shown in Equation (4):

$$Z = a + bX + cY + dX^2 + eY^2 + fXY, \quad (4)$$

where $a = 1.18 \times 10^9$; $b = 4.83 \times 10^2$; $c = -9.07 \times 10^2$; $d = -7.07 \times 10^{-5}$; $e = 1.73 \times 10^{-4}$; and $f = -1.58 \times 10^{-4}$.

The error between the actual data obtained directly from the boreholes and the values from the regression processing is critical to validate the effective model fitting. In order to study the difference between the actual and interpolated data, Table 1 shows the errors obtained when using first- and second-order models. It can be seen that the maximum errors with the second-order model were lower than those with the first-order model, implying that the uncertainties in spatial formation were lower. Then, a cross-section A-A′ was plotted to visualize the errors in Table 1 for a borehole group, when using the different models in both orders (Figure 7). The sandstone layer is highlighted in yellow color in the boreholes. The red line, which indicates the geological boundary, was obtained by using regression values of the top elevation of the sandstone in the cross-section. The curved line represents the physical shape of the second-order surface polynomial regression, which was more likely to satisfy and better fit the surface than the first-order model, which is a straight line (or linear resultant). For example, borehole 17-1 displays the most obvious case of a comparison between the first- and second-order models with a decreased mismatch between the borehole and regressed data with the latter. The red line (i.e., second-order model) is more consistent with the top of the sandstone in this borehole, implying that the error was improved significantly.

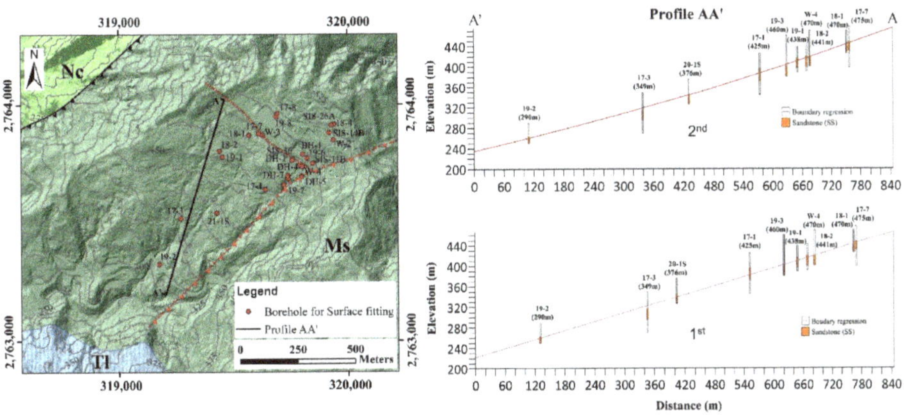

Figure 7. Cross-section A-A′ showing the resulting surface interpolation polynomials of the first (red straight line) and second (red curved line) order, for the top of the thick sandstone (L4) based on the borehole data. Each polynomial function has a different arrangement of boreholes on the profile. The deviation from the top sandstone with the second-order model is smaller than that with the first-order model.

For Zone 2, we only used first-order regression to generate the bedding distribution model, as there was not enough data in this region to perform second-order polynomial regression surface fitting, given that the majority of the boreholes did not reach the interface between L4 (the key bed) and L5. As a result, only the first order was used as the function of the polynomial regression for the geological model in Zone 2. The linear surface fitting regression equation is as follows:

$$Z = -636796.2239 - 0.009296X + 0.231637\,Y. \quad (5)$$

The root-mean-square (RMS) error values, which can demonstrate the difference between the actual and predicted data in regression, were also computed. This number was high ($R^2 = 0.98$), indicating that the estimated coefficients in the regression equation were credible. Moreover, after determining the surface equation of the fitted bedding, the regression results demonstrated that the first-order model were fairly consistent with the real data, with the error being acceptable and not changing significantly (Table 2). Except for two boreholes—BH-1 and W-3—the majority of the observed data had low error.

Based on the results of the surface fitting regression, a structural contour map of the thick sandstone (key bed) was created, in order to visualize the pattern (trend) of the sedimentary rock in the two zones. To ensure the continuity of the geological structure map of the thick sandstone (key bed, L4), Figure 8 illustrates a structural contour map comprised of two groups: Zone 1 and Zone 2. It can be seen that the shape and direction of the contours had a high association with the inferred structures. To the north of the research area, the largest portion of the location is Zone 2, with the bedding plane striking almost in the E–W direction and dipping to the south with a gentle inclination of approximately 13°. This implies that the geological structure underneath the Huafan University campus is a dip slope in the literally strictest sense. Then, progressively moving towards the downslope direction, the contour seems likely to become bent, turning towards the southwestern direction. The tangent line of these contours gradually changes, implying that the strike of the bedding plane adjusts as well, with a steeper inclination towards the southwest. Based on this contour map, a proposed geological structure can be detected with a left-lateral strike-slip fault, and the strike-slip movement on the map does not have vertical displacement of stratum. However, the displacement becomes progressively unclear along the southeastern direction of the fault, and appears to be somewhat of a slip at the intersection with the Nanshihkeng Fault. The evidence for this assumption needs to be investigated through supplementary data in the future. Furthermore, the structural contour map was consistent with the attitude of the geology (the purple attitudes) and, in particular, it fit well in the dip direction of the bedding plane, which has also been recorded by Jeng and Sue [28] and measured in the field survey (Figure 8). These orientation data (i.e., strike and dip) may be expressed as vectors that are orthogonal to the strike in their locality. Therefore, the polynomial surface fitting method procedure was used to represent the geometry of the geological contact surface between strata in the Mt. Dalun area, which was considered reasonable.

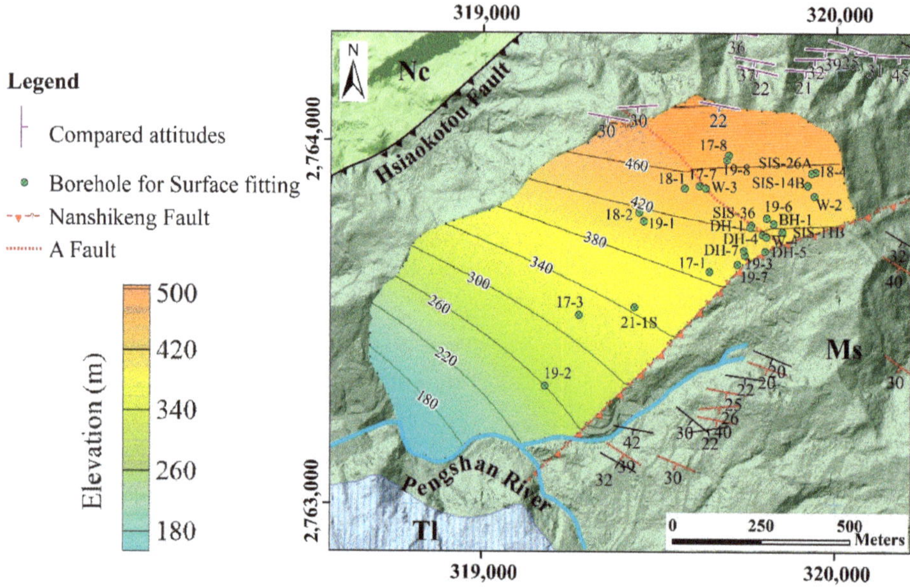

Figure 8. Structural contour of the top surface of the thick sandstone (L4, the key bed). By comparison with the outcrop measurements, a contour map with the contour interval of 2 m was created. The green circles indicate the locations of the boreholes used for the regression. The purple symbols of bedding attitudes can be utilized as a basis for comparison.

4.4.3. The Average Thickness of Each Rock Unit in Zones 1 and 2

The thickness of the rock units varied with the distribution of the boreholes. The stratigraphical sedimentary rock was constituted of a succession of rock units, which were constructed according to their sequence. These rock units are characterized by a series of boundary horizons, which represent the geological contacts between the various layers, shown as a stratigraphic column in Figure 9.

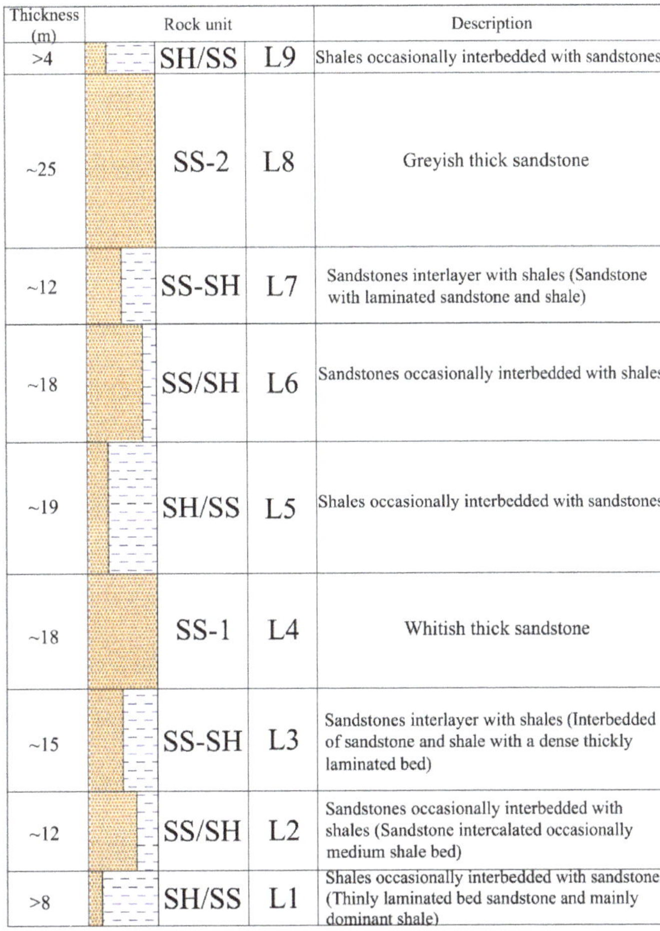

Figure 9. Stratigraphic column of the sedimentary rock in Mt. Dalun area, based on rock core analysis. The thickness of each rock unit was derived as the average thickness from the data of 26 boreholes.

4.5. *Development of Geological Model in Zone 3*

Zone 3 represents the eastern section of the study area, which lacked boreholes to analyze the underground stratigraphical conditions. According to the study of Jeng and Sue [28], two layers of thick sandstone and thin interbedded shale and sandstone shale can be found in this area. The thicknesses of these two thick sandstones are 30 m and 10 m, respectively (Figure 1). However, new outcrop measurements and geological descriptions have been carried out, such that the boundary of these two sandstone layers should be modified. Several attitudes in the outcrops were confirmed along the road to the southeast, where the topography is steep. Most of the exposures were discovered only by excavation on a construction road, which provided easy access to the outcrops. Exposures mostly included interbedded shale and sandstone. These were moderately to heavily weathered, covered by vegetation, and dip 20–40° to the southwest. The Nanshihkeng Fault may have changed the bedding plane's orientation, resulting in a steep oblique slope in this area. The result of this work is consistent with that of previous studies. A total of 15 measurements were recorded during this survey, and the statistical results indicated the orientation of the bedding plane in the stereonet (Figure 10a). The average attitude of the representative bedding plane was selected to draw a structural contour map (68° NW, 31° SW). We observed the real contact exposure between the sandstone and interbedded sandstone; shale, in particular, may be seen extending from 30 to 40 m, having a light gray tint (i.e., red circle), in Figure 11. In addition, a high-resolution slope inclination map derived from the LiDAR DEM was constructed, in order to determine and deduce the reasonable boundaries of the rock units. Then, the proposed geological map for Zone 3 was redrawn. Figure 10b exhibits the distribution of the two sandstone layers, as well as the attitudes of bedding planes. We named the thicker sandstone sandstone A, while the other was sandstone B. The distribution of sandstone A on the geological map was corroborated by the contact observed in the field. Even though the outcrop of sandstone B was not exposed in the field trip, the thick sandstone should be exposed to become the ridge. This seems to be compatible with the distribution of the thinner sandstone layer observed in the study of Jeng and Sue [28]. As a result, the outcrops drawn by Jeng and Sue [28] around the ridge are realistic, working in conjunction with the average orientation to illustrate the distribution of these two thick sandstones, particularly sandstone B. These two sandstone layers on the southeastern oblique slope are distributed along NE–SW striking steep slopes or ridges, and may turn gradually to the westward direction and terminate at the Nanshihkeng Fault. For terrain compatibility, it can be inferred that sandstone A lies above sandstone B, meaning that the top of sandstone B will be the bottom of sandstone A in the spatial distribution. This can be seen in the geological profiles provided in the following section.

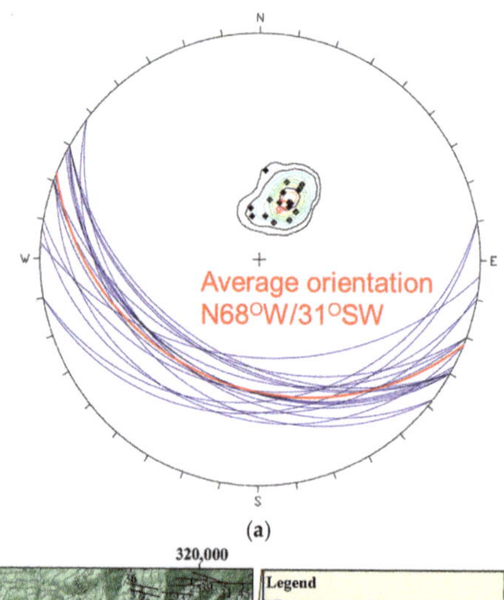

(a)

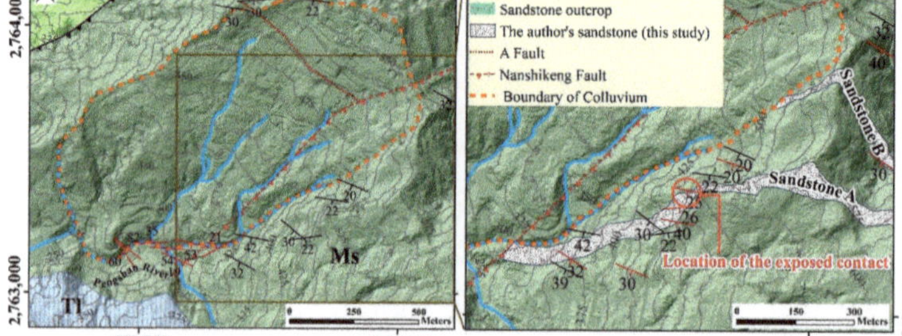

(b)

Figure 10. Geological model combining the field exposures and topographical analysis in Zone 3.

(**a**) Stereonet visualization presents the outcrop measurements; and (**b**) outcrops measured during the field survey and redistribution of two sandstone layers in the southeastern part.

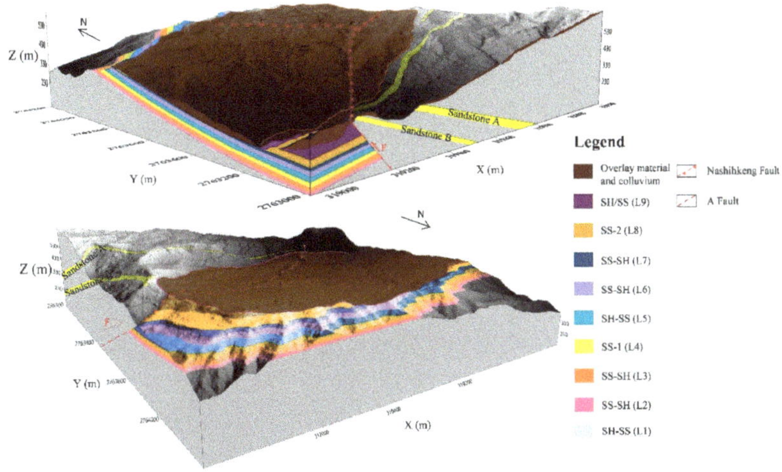

Figure 11. Three-dimensional engineering geological model of the study area visualizing the strata dip slope distribution in Zones 1 and 2, and the high oblique sandstones A and B in Zone 3. The surface was derived from the high-resolution LiDAR DEM (1 m resolution).

4.6. Three-Dimensional Engineering Geological Model and Geological Profiles

The three-dimensional engineering geological model, derived from the integration of the grid-based method (i.e., colluvium bottom elevation) and vector-based method (i.e., the geological interface of rock units) mentioned above, as well as field outcrop geological data in the research region, is shown in Figure 11. Cross-sections were also generated to interpret the 3D engineering geological model (Figure 12). The colluvium and bedrock were characterized by the rock units identifiable through the geological contact (interface) of strata. Two geological profiles (A-A′ and B-B′) from the northeastern to the southwestern part of the slope, along with their geological explanations, are detailed in Figure 12. It can be observed that the profile B-B′ passes through the eastern part of the Huafan University campus, revealing that the covered colluvium in the upper part around the campus is thin. However, their thickness becomes progressively larger toward the southwest. In particular, at the bottom of this slope where the riverbank is located to the north, there is a massive and thick colluvium accumulation area. For the sedimentary rock, the strata are parallel to each other, which is compatible with the topography. There is a thick sandstone layer (SS-2) of approximately 25 m thickness beneath the interbedded sandstone and shale (SH/SS) cover, to the north of the A Fault. However, this sandstone layer disappears in a portion in the profile A-A′, meaning that it had been eroded due to erosion of the exposed surface. Through two profiles extracted from the geological model, the interface between the colluvium and the underlying rock was found to be almost parallel to the geological boundary of the strata. The thick colluvium layer with the maximum depth lies roughly on the top of the sandstone (SS-1), which is acceptable as the sandstone at that altitude in those profiles is quite persistent and difficult to break. While other layers are mostly composed of narrow interbeds of shale and sand, this material has partly transformed into the colluvium due to long-term weathering and erosion. The two profiles also show that these outcrops of the bedrock on the ground surface here form the steps of a staircase in the topography. Therefore, the sources of colluvium almost entirely originated from this bedrock of alternative shale and sandstone (SS, SH). A cross-section in the northwestern to the southeastern direction (C-C′) was additionally drawn (Figure 13). It can be observed that the bedding was nearly consistent with the topographical surface and interface between the colluvium and itself. To visualize the spatial distribution of the two thick sandstones in the southeastern part (Zone 3), a profile D-D′ in an almost N–S direction passing across the A and Nanshihkeng Faults was also created (Figure 13). Even though the rock units of alternating sandstone and shale cannot be distinguished in the profile, the steep terrain can be explained by the uplift of the Nanshihkeng Fault as a reverse fault, based on this profile. In the southern part of the profile, the thickness of the colluvium is extremely thin, and the bedrock is virtually exposed around this part.

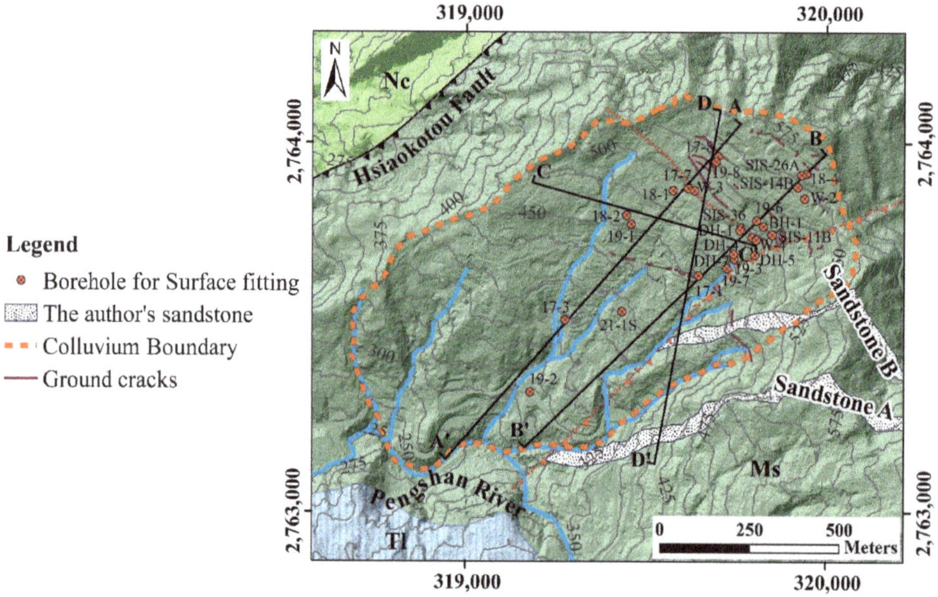

Figure 12. Distribution diagram of the four geological cross-section profiles and the borehole data. These borehole sources were gathered from available existing borehole logs in the study area. The deepest borehole was 100 m (BH-1), while the shallowest borehole was 25 m deep (SIS-26A). The red boreholes are projected along the contour line at the top of the key bed in the profiles (i.e., sandstone).

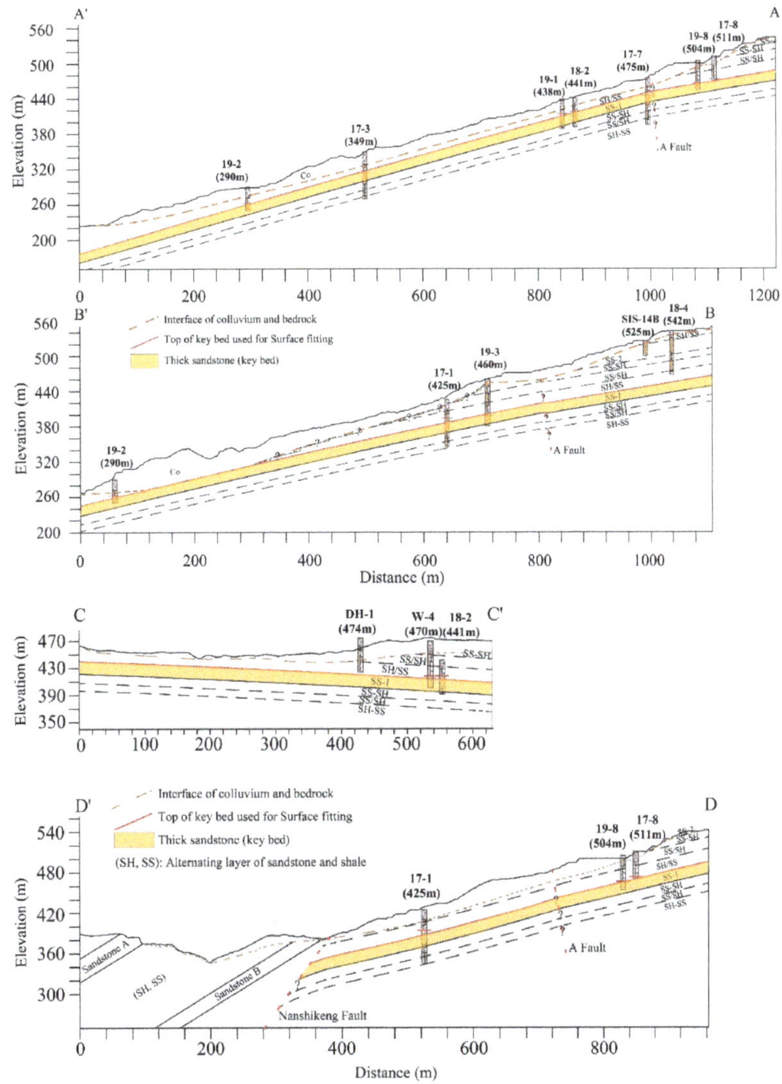

Figure 13. Geological cross-section profiles within the Mt. Dalun area. The colluvium boundary is plausibly compatible with the terrain (i.e., parallel to the topographical surface), and the deepest part of it does not cross over the thick sandstone layer (L4). The thickness of the colluvium is the largest in the lower half of the slope, and becomes increasingly thinner with topographical elevation. Two faults in the region contribute to the alteration of the spatial distribution of geological strata.

5. Applications of the 3D EGM

5.1. Surface and Sub-Surface Displacement Monitoring

5.1.1. Surficial Displacement Interpretation

According to the study of Tseng et al. [30], hundreds of displacement monitoring points were established and recorded in the study area from April 2010 to July 2017. This is a part of the long-term monitoring of surface and underground movements based on time-series data. On the map, Figure 14a displays the displacement tracks of each observation point within the Huafan University campus, from which the quantity and direction of horizontal displacement at each monitoring point throughout each period can be seen. This surficial displacement can be classified into three groups, as shown on the map. One is a larger displacement, with the main direction being down to the slope in the south-southwestern direction, denoted by red color (group 2); the second is towards the northwestern part, revealed by the black, as well as a smaller displacement movement toward almost the south (group 1). The displacement monitoring points outside the intersection part of the A Fault and the Nanshihkeng Fault are marked by dark blue color, in which the western components of displacement are more visible (group 3).

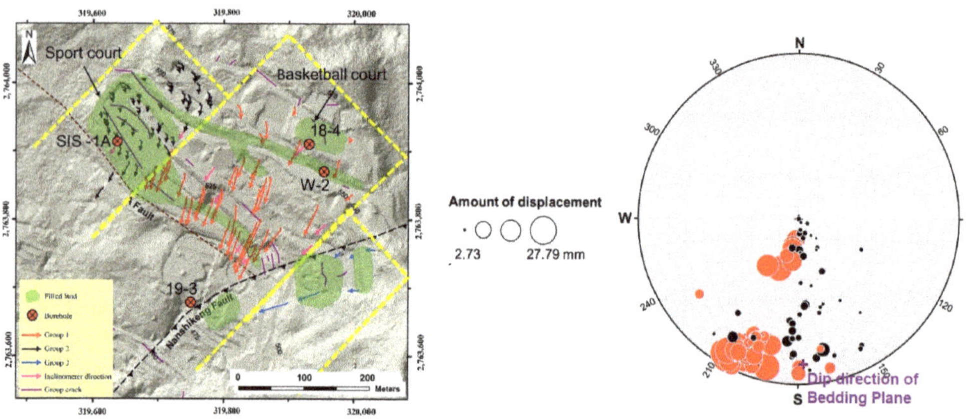

(a) (b)

Figure 14. Groups of surficial displacement at the Huafan University campus: (**a**) horizontal displacement tracks, revealed by surficial monitoring within the measurement period and the group's distribution. The displacement vectors represent the movements from the beginning to the end during the measurement period. (**b**) The geometry of displacements, i.e., the displacement vectors presented in a stereonet. The diameter scale shows magnitudes of displacement, and the solid purple cross represents the dip direction of the bedding plane in Zone 2.

To obtain detailed movement behavior of the multiple sliding blocks within the campus, the displacement vectors are indicated using a stereonet from the beginning to the ending of the monitoring period. The displacement trend was a two-dimensional resultant vector in the E and N horizontal components. To show three-dimensional surficial displacement, horizontal and vertical displacements were combined to compute the sliding surface's spatial dip, also displayed using a stereonet (Figure 14b). The red and black solid circles represent surface monitoring data groups 1 and 2, respectively. They are indicated with different diameters, according to displacement amount and direction at each monitoring site. If the vertical displacement is larger than the horizontal, the displacement vector will approach the stereonet's center, implying that the campus may have multiple sliding surfaces with different characteristics. In the results, the group 1 displacement vectors can be seen as being dispersed in the stereonet. Smaller solid circles represent less displacement. This movement is not happening in a pre-determined direction, nor is it similar to sliding on the same weak layer inside the bedrock. For further insight, the surficial displacement in the northwest (group 1) may be separated into two smaller groupings. As each small group's stereonet displacement vector is scattered from inside to outside, each group should be represented by a separate sliding block, comprising Blocks 1 and 2, respectively (Figure 15). In addition, these surface displacement vectors were also compared with the shape of the interpolated colluvium bottom (i.e., the colluvium–bedrock interface), as shown in Figure 15. The primary slope angle of the colluvium was approximately equivalent to the slope surface angle. The majority of horizontal displacements ware almost downslope or deviated from the tangent line of the colluvium contour line at a small angle. Moreover, the displacement direction of the inclinometer SIS-1A coincided well with the dip direction of the interface of colluvium and bedrock (Figure 15). This suggests that the direction of the sliding blocks should be similar to the tangential direction of the colluvium plane. Particularly, in the southwest part of the map in Figure 15 or the south portion of the A Fault, the dip direction of two displacement vectors is similar to the direction of displacement indicated by inclinometer SIS-1A. These two displacement vectors almost correlate with the dip direction of the colluvium and bedrock interface. In the stereonet of Block 2, these two displacement vectors are denoted as hollow triangles pointing southwest, even though their dip angle deviated from the dip angle of the interface (i.e., colluvium and bedrock) by about 15–20° and 20–30° in trend, respectively (Figure 15). The reason for this difference is that the shape of the contact between the colluvium and bedrock may be influenced by the outcome of interpolating this interface. In the stereonet, the green displacement vector in each sliding block denotes the head of the block when the vertical displacement is larger than horizontal displacement, producing a steep angle, while the orange vector supposedly represents a gradual change in the dip angle from the main body to the toe of the sliding block. Overall, the displacement vectors in both blocks almost follow the downward trend in the shape of the interface between the colluvium and bedrock. However, it can be seen that Block 2 has a number of displacement vectors turning toward the southeast. It can be speculated that these displacements were influenced by other blocks nearby, as an extrusion of the slope slide would have affected the opposite direction component when Block 1 slid downslope. This displacement would progressively deviate from the downward direction. One more probable reason is the shifting shape of the sliding blocks, which resemble a curved sliding boundary at the edge of the filled land distribution with the original topographical surface. This led Block 2 to slide on the simplest sliding surface, as long as it could slide under the driving force on the slope. Variation of the surface displacement vector with respect to the common sliding direction may reflect stress and other slope deformations. Block 1, with a high altitude without ground cracks nearby, could only exist at the contact between the colluvium and the underneath bedrock. This possible sliding surface could be one of multiple bottom colluvium surfaces connecting Block 1 to Block 2 in profile 3-3′, due to the apparent dip angles of surface displacements on the profile switching from gentle (18°) in Block 1 to steep (75°) at the head of Block 2. Block 2 was also identified as having a sliding block head around the filled land's distributed border, and the possibility that the sliding surface occurs at the filled land–colluvium interface is remarkable. It could be parallel to the topographical surface and plane; the red dashed lines in profiles 1-1′ and 2-2′ indicate the sliding surface geometry for the inferred interface between filled land and colluvium (Figure 16). The upward termination of these sliding surfaces on the ground surface also presents with ground cracks. However, a sliding surface may only appear at the boundary between the colluvium and bedrock in the southwest, where the two displacement vectors as mentioned above are not greatly impacted by the thick filled land in Block 2. Similar to Block 1, the geometry of the sliding surfaces changes slightly along Block 2 in the profiles, almost reflecting the apparent dip angles of surface displacements on profiles (i.e., the surface sliding parallel to apparent surface displacement angles). For the sliding surface at the colluvium–bedrock interface, the interface seems to be the consequence of a steeply vertical, stepped joint system. The depth of the sliding surface should be proportional to the depth of joints formed in the bedrock and the advent of weathering (Figure 16). These vertical joints might create blocks of mass slope failure on the surface of the bedrock, as the bedrock platform typically makes it impossible to develop "daylight" on the slope without a significant amount of the maximum principal stress; particularly in persistent, thick sandstone.

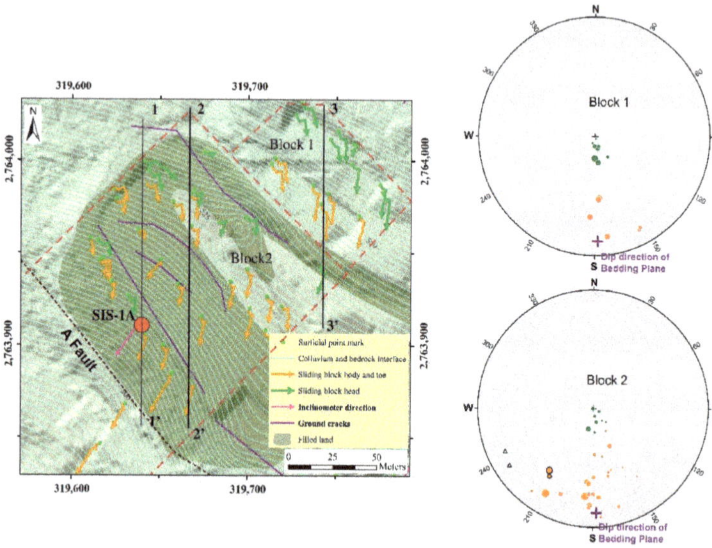

Figure 15. Geometry of each sliding block in the northwestern part (group 1) shown using stereonets. Each stereonet illustrates a block in motion, with the green dots representing the head and the orange ones representing the main body to the toe of the sliding mass. The hollow triangles on the stereonets represent the dip direction of the orange solid circles, with respect to the two surface displacements bordered by black circles in the southwest portion of the map view. The contour line interval for the colluvium–bedrock interface is 1 m.

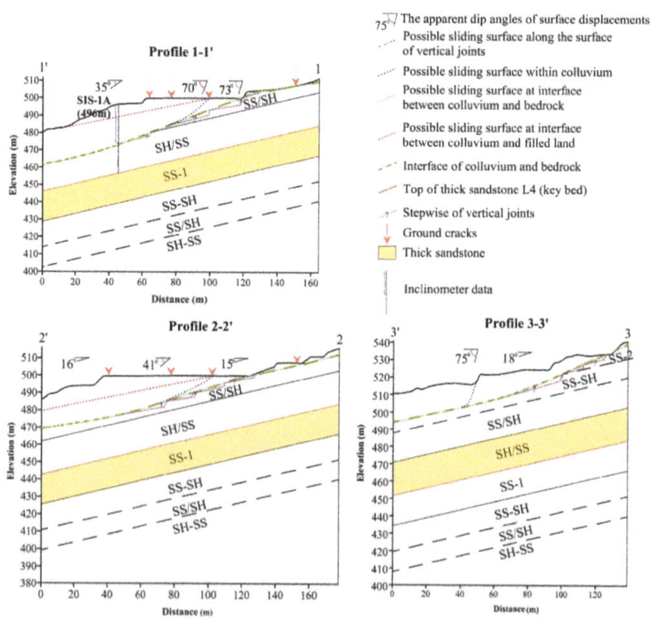

Figure 16. Cross-sectional profiles with potential sliding surfaces in the western part of the campus. Geological profiles are approximately parallel to the vectors representing the horizontal surface displacement (see Figure 15).

As the stereonet in Figure 14b shows, the displacement vectors for group 2 had significantly larger diameter. The vertical displacement for this group was almost smaller than the horizontal displacement, as they were distributed densely at the edge of the stereonet (i.e., far away from the center). In addition, it can be seen that group 2 displayed a relatively similar amount of displacement. However, there were still some displacement vectors with dominant vertical displacement, meaning that the sliding surface was nearly vertical elsewhere in group 2. One possible explanation is that the sliding surface near the head of the sliding block had greater vertical displacement, being relatively consistent with these displacement vectors towards the center of the stereonet. Considering this phenomenon, it can be assumed that the entire surficial monitoring data of group 2 were recorded for one sliding block. However, from the stereonet (Figure 14b), the dip direction of bedding (dark purple solid cross) proposed by the 3D EGM within the campus was not consistent with the surficial displacement vectors.

5.1.2. Sub-Surface Displacement Interpretation

Based on the 3D EGM indicated by the profiles provided in Section 4.6 and inclinometer measurements, the correlations between the sliding surface and geological features were revealed. The profile A-A′ suggests that the sliding surface was located approximately 15–22 m deep, with multiple surfaces (green dashed line in Figure 17). The inclinometer data of SIS-1A, 17-7, 17-2, and 17-3 indicated that the sliding surface was the interface between the colluvium and the underlying bedrock, with the inclinometer SIS-1A also having been correlated to the bottom of colluvium in the study of Tseng et al. [30]. The profile B-B′ passes through the main Huafan University campus in the eastern part, near the intersection of the Nanshihkeng and A Faults. In the upper part, the inclinometer set in the borehole 18-4 indicated a sliding surface (purple dashed line) deep in the sandstone layer (SS-2). In many instances, a break in persistent sandstone is rare; consequently, this deep sliding surface is a so-called local slope failure (e.g., a thin, soft interlayer of shale and mudstone inside sandstone, which easily slips). With reference to the underground monitoring evidence revealed by the inclinometers in boreholes 19-3 and SIS-27, it was possible to infer the presence of a sliding surface inside the bedrock at a specific depth of 38 m (purple dashed line). The sliding surface developed continuously in the rock mass at the SS/SH unit. In cooperation of the inclinometer monitoring of the boreholes 19-5 and 17-4 near the toe of the slope, the sliding surfaces at these locations were found to possibly occur deeper, at depths of 25–45 m, likely in a curved shape (green dashed line). The depth of the sliding surface was not found to lie on a plane, meaning that the sliding surface is curved.

5.1.3. Slope Instability Mechanisms Based on 3D EGM

The slope failure mechanism depends on the geological structure and spatial distribution of the objective geology of the slope. Multiple mechanisms of slope instability may occur at different locations, as revealed by the monitoring of surface and sub-surface displacements and the 3D EGM.

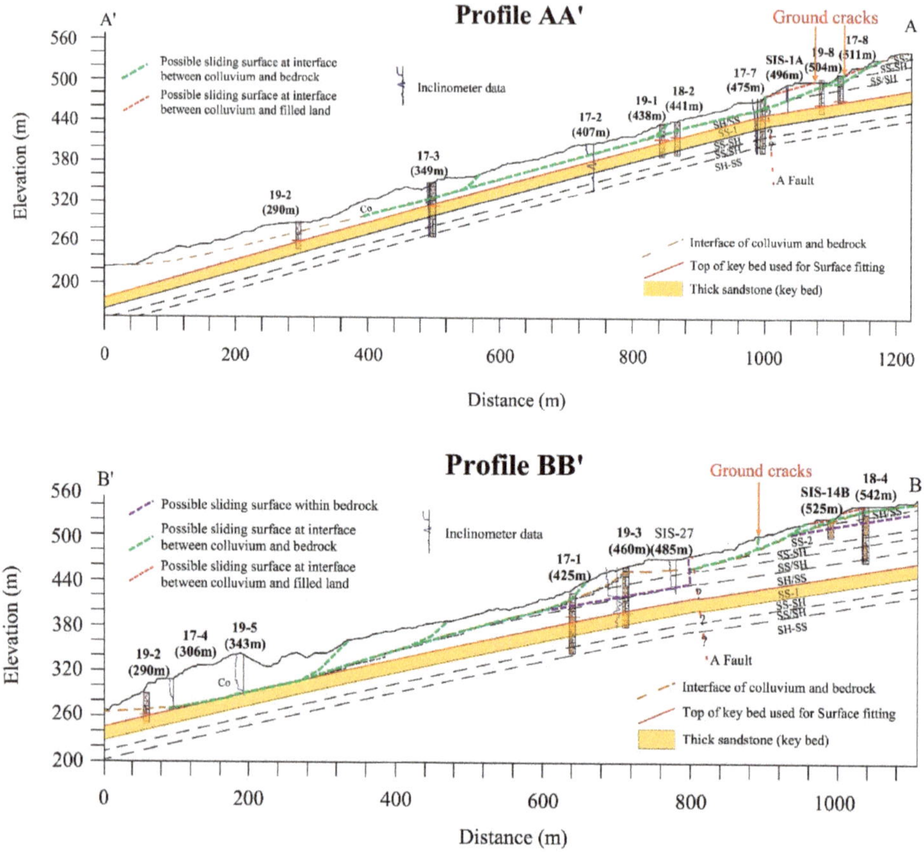

Figure 17. Potential sliding surfaces and projected inclinometer casing positions inside the slope in the profiles A-A′ and B-B′. The dashed green and dashed purple curves are inferred to be at the bottom of the colluvium and within the bedrock, respectively, based on rock cores and inclinometer measurements. The dashed red curve represents the sliding surface at the interface between the colluvium and the fill, revealed by considering the behavior of surficial displacements and topographical features.

According to the study of Jeng and Sue [28], the sports court and the western part of the campus overlie thick artificial fill, from 2 to 20 m, used for the construction of Huafan University (Figure 14a). The surficial displacement in this area (i.e., the western part of the campus) should not only have happened at the interface between the colluvium and the bedrock, but also at the interface between the fill and the colluvium (i.e., on the original topographical surface before the construction). Similarly, considering the green curving dashed lines indicating the lower portion of the slope (i.e., the location where a large volume of colluvium has accumulated), the sliding surface was determined to be within the colluvium or at the interface between the colluvium and the bedrock. These sliding blocks suggest that, at the interface between the colluvium and the bedrock, the fill and the colluvium may exhibit creeping behavior. This can be explained by the fact that, due to the unconsolidated and permeable nature of the colluvium, when the infiltrated water flows along the interface, the resistance force at the interface decreases and sliding may be triggered. The colluvium has progressively slid downstream, due to a combination of gravity, erosion, and rainfall, resulting in the emergence of several sliding surfaces. Another proposed explanation is that the heterogeneous material between the colluvium and the fill, where the heterogeneities resulting from their uneven elasticity gradually create tiny cracks within the interface. Water infiltrates and then separates the layers through a sliding surface. The cracks on the ground surface might be the consequence of the movement of fill on the original surface (i.e., the interface between the fill and the colluvium).

In addition, the inclinometer results based on the 3D EGM suggested that the sliding surface occurred inside the bedrock. The eastern slope might act under this particular mechanism for slope failure. This circumstance suggests, again, that the sliding block advances downslope, and behaves like a translational dip-slope landslide on a planar sliding surface developing parallel to the bedding plane. That means the sliding surface can develop continuously at an interbedded layer with weak strength in the rock mass, such as sandstones interbedded with shales. The vertical joints inside the slope may cause the sliding surface to break up into a series of block-like slides. Then, the entire slope toe daylights, leading to the interlayer sliding within the bedding plane, creating tension cracks at the head. Once the deformation of the bedding plane surpasses a certain threshold, surface ruptures spread along the deformation face and combine to produce a sliding surface.

In summary, based on the scenario of development shown through analysis of the 3D EGM and sub-surface/surface displacement monitoring, sliding surfaces can be detected, which cause the whole slope to become increasingly unstable. Slope failure can occur on multiple sliding surfaces through a variety of depths and mechanisms. In this case, three sliding surfaces were inferred, including both at the interface between the colluvium and the bedrock, and at the interface between the fill and the colluvium. Finally, a remarkable sliding surface mechanism was also located within the bedrock.

5.2. Application of the 3D EGM to Drainage Well Design

To improve slope stability, the installation of sub-surface drainage wells is the most common approach. As a result, we attempted to optimize the number and spacing of horizontal drains. An effective design for sub-surface drainage utilizes horizontal drains capable of capturing groundwater flow before it infiltrates an unstable region susceptible to landslides [37]. Moreover, a reasonable drainage well site can only be achieved by first understanding the potential sliding surface, as well as the ground water flow and flow pattern, which can be accomplished by using a 3D EGM. The 3D EGM proposed in this study, therefore, can also be used to optimize the drainage efficiency of the considered slope in the future. Figure 18 depicts the position of the vertical drainage well, T1, that has already been constructed in the middle of the Huafan University campus by a local consultant company, in order to illustrate the spatial distribution of geological characteristics that interact with it through the three-dimensional profiles considered in this work.

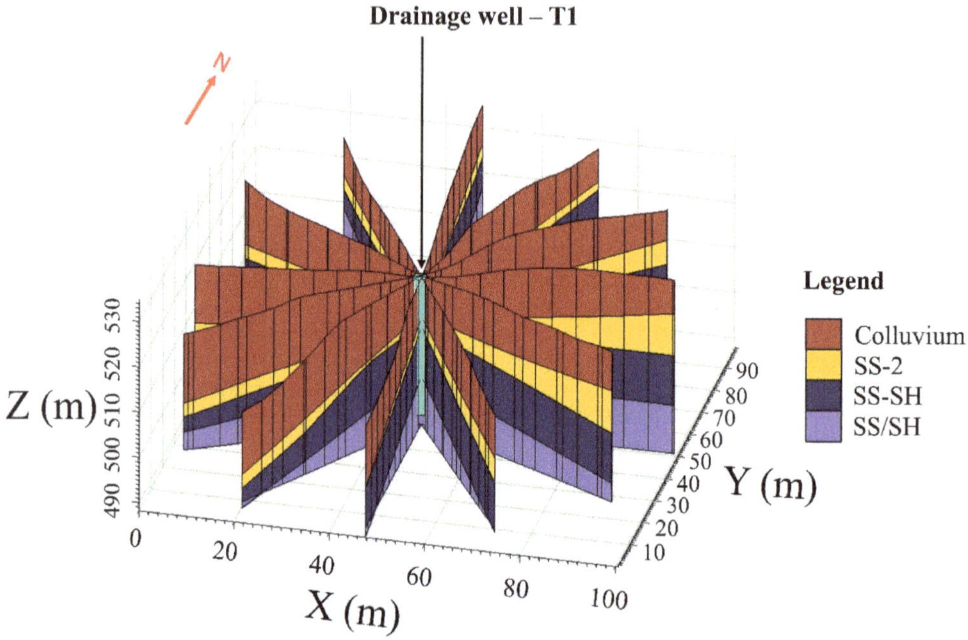

Figure 18. Three-dimensional profiles based on the 3D EGM and the drainage well T1 within the Huafan University campus. There are six profiles at 30-degree intervals. T1 has a depth of 30 m is positioned in the center of each profile.

6. Conclusions

In this study, we attempted to construct a 3D EGM of the dip slope located in the Mt. Dalun area, incorporating field geology and rock core data analysis. The spatial distribution of the colluvium and bedrock stratum interface was determined using interpolation and surface regression approaches, based on polynomial functions. The following conclusions were reached in this study:

(1) The study area can be divided into three uniform geological zones belonging to the Mushan Formation. Two faults—the Nanshihkeng Fault and the A Fault—formed the boundaries of these three zones.
(2) The interpretation of the borehole cores provided a clear overall stratigraphy of the sedimentary rocks in the area. In particular, there were nine rock units that could be distinguished, including thick sandstone and a variable proportion of interbedded shale and sandstone for Zones 1 and 2, in the following order of stratigraphic history: L1 (SH/SS), L2 (SS/SH), L3 (SS-SH), L4 (SS-1), L5 (SH/SS), L6 (SS/SH), L7(SS-SH), L8 (SS-2), and L9 (SH/SS). The thick sandstone (L4) identified as the key bed is dominated by thick medium-grained whitish sandstone, intercalated with thin dark black shale layers dispersed in a few spots in the massive sandstone. The top boundary of this rock unit was observed from the borehole samples, following which this key bed was used for surface regression. In Zone 3, the spatial distribution of the two thick sandstones was proposed in the 3D EGM model as the consequence of outcrop field geology measurements and topographical analysis, by applying a high-resolution DEM.
(3) Loose material covered the slope with varying thicknesses. It was most densely distributed in the southwestern slope, at approximately 50 m thick, and became thinner in the upper elevations along the mountain ridge. According to the topographical features, the colluvium may consist of material from various sources, such as an-

cient landslides or bedrock deterioration. The geometry of the bedding plane was calculated, based on its distribution, as a plane generally dipping 15° to the southwest.

(4) Along with the 3D EGM, the structural contour map obtained by bedding surface fitting using a second-order model revealed that the geometry of the geological interface (i.e., the top surface of the key bed) gradually changes its direction and dip inclination (Zone 1). Even though the bedding in Zones 2 and 3 proposed in this study was found to be planar, its orientation changed significantly in space. Therefore, the underground information in both zones needs to be supplemented in future research.

(5) The possibility of the future dip slope failure might be assessed through the use of a combination of surficial/underground monitoring. The geological model that incorporated the time-series of the displacement of the slope surface, revealed three sliding mechanism types. The first one comprises the sliding surface at the interface between the fill and the colluvium. The second one involves creeping on the contact surface between the colluvium and the underlying bedrock. The third mechanism is sliding occurring within the interlayer as a bedding plane. The development of deep-seated landslides through gully erosion was also proposed. However, the slope failure occurs only as local sliding and it is spatially very limited.

The three-dimensional engineering-geological model was approved as a powerful tool for further landslide hazard and risk mitigation.

Author Contributions: Conceptualization, J.-J.D.; Resources, C.-W.C. and C.-C.P.; Writing—original draft, T.-T.N.; Writing—review & editing, C.-H.T. and I.B. All authors have read and agreed to the published version of the manuscript.

Funding: This research was funded by the Ministry of Science and Technology of Taiwan and grant numbers are MOST 111-2923-M-008-006-MY3, MOST 110-2123-M-008-001, MOST 111-2123-M-008-001 and MOST 110-2116-M-034-005; and by Czech Science Foundation and grant number is GAČR 22-24206J.

Acknowledgments: We would like to thank colleagues at the Taipei Branch of Soil and Water Conservation Bureau and the Huafan University for providing their data for analyses and valuable comment and assistance in the field work. The research was supported by the international bi-lateral project "Earthquake-triggered landslides in recently active and stabilized accretionary wedges" of the Taiwanese Ministry of Science and Technology (MOST 111-2923-M-008-006-MY3) and the Czech Science Foundation (GAČR 22-24206J). We would like also to thank the support by the MOST projects: MOST 110-2123-M-008-001; MOST 111-2123-M-008-001; MOST 110-2116-M-034-005.

Conflicts of Interest: The authors declare no conflict of interest. The funders had no role in the design of the study; in the collection, analyses, or interpretation of data; in the writing of the manuscript; or in the decision to publish the results.

References

1. Parry, S.; Baynes, F.J.; Culshaw, M.G.; Eggers, M.; Keaton, J.F.; Lentfer, K.; Paul, D. Engineering geological models: An introduction: IAEG commission 25. *Bull. Eng. Geol. Environ.* **2014**, *73*, 689–706. [CrossRef]
2. Fookes, P.G. Geology for engineers: The geological model, prediction and performance. *Q. J. Eng. Geol. Hydrogeol.* **1997**, *30*, 293–424. [CrossRef]
3. Turner, A.K. Challenges and trends for geological modelling and visualisation. *Bull. Eng. Geol. Environ.* **2006**, *65*, 109–127. [CrossRef]
4. Culshaw, M.G. From concept towards reality: Developing the attributed 3D geological model of the shallow subsurface. *Q. J. Eng. Geol. Hydrogeol.* **2005**, *38*, 231–284. [CrossRef]
5. Caumon, G.; Collon-Drouaillet, P.L.C.D.; De Veslud, C.L.C.; Viseur, S.; Sausse, J. Surface-based 3D modeling of geological structures. *Math. Geosci.* **2009**, *41*, 927–945. [CrossRef]
6. De Rienzo, F.; Oreste, P.; Pelizza, S. Subsurface geological-geotechnical modelling to sustain underground civil planning. *Eng. Geol.* **2008**, *96*, 187–204. [CrossRef]
7. Lemon, A.M.; Jones, N.L. Building solid models from boreholes and user-defined cross-sections. *Comput. Geosci.* **2003**, *29*, 547–555. [CrossRef]
8. Wu, Q.; Xu, H.; Zou, X. An effective method for 3D geological modeling with multi-source data integration. *Comput. Geosci.* **2005**, *31*, 35–43. [CrossRef]

9. Zhu, L.; Zhang, C.; Li, M.; Pan, X.; Sun, J. Building 3D solid models of sedimentary stratigraphic systems from borehole data: An automatic method and case studies. *Eng. Geol.* **2012**, *127*, 1–13. [CrossRef]
10. Kessler, H.; Mathers, S.; Sobisch, H.G. The capture and dissemination of integrated 3D geospatial knowledge at the British Geological Survey using GSI3D software and methodology. *Comput. Geosci.* **2009**, *35*, 1311–1321. [CrossRef]
11. D'Agnese, F.A.; Faunt, C.C.; Turner, A.K.; Hill, M.C. *Hydrogeologic Evaluation and Numerical Simulation of the Death Valley Regional Ground-Water Flow System, Nevada and California*; US Department of the Interior; US Geological Survey: Denver, CO, USA, 1997; Volume 96, p. 4300.
12. Logan, C.; Sharpe, D.R.; Russell, H.A. *Regional Three-Dimensional Stratigraphic Modelling of the Oak Ridges Moraine Area, Southern Ontario*; Natural Resources Canada; Geological Survey of Canada: Ottawa, ON, Canada, 2001.
13. Tremblay, T.; Nastev, M.; Lamothe, M. Grid-based hydrostratigraphic 3D modelling of the Quaternary sequence in the Châteauguay River watershed, Quebec. *Can. Water Resour. J.* **2010**, *35*, 377–398. [CrossRef]
14. Bistacchi, A.; Massironi, M.; Superchi, L.; Zorzi, L.; Francesca, R.; Giorgi, M.; Genevois, R. A 3D geological model of the 1963 Vajont landslide. *Ital. J. Eng. Geol. Environ.* **2013**, *6*, 531–539.
15. Wang, J.; Schweizer, D.; Liu, Q.; Su, A.; Hu, X.; Blum, P. Three-dimensional landslide evolution model at the Yangtze River. *Eng. Geol.* **2021**, *292*, 106275. [CrossRef]
16. Thierry, P.; Prunier-Leparmentier, A.M.; Lembezat, C.; Vanoudheusden, E.; Vernoux, J.F. 3D geological modelling at urban scale and mapping of ground movement susceptibility from gypsum dissolution: The Paris example (France). *Eng. Geol.* **2009**, *105*, 51–64. [CrossRef]
17. Thapa, P.B. An approach towards integrated modelling of 3D geology and landslide susceptibility in the Lesser Himalaya of central Nepal. *J. Nepal Geol. Soc.* **2014**, *47*, 65–76. [CrossRef]
18. Gu, T.; Wang, J.; Fu, X.; Liu, Y. GIS and limit equilibrium in the assessment of regional slope stability and mapping of landslide susceptibility. *Bull. Eng. Geol. Environ.* **2015**, *74*, 1105–1115. [CrossRef]
19. Li, Z.; Zhang, F.; Gu, W.; Dong, M. The Niushou landslide in Nanjing City, Jiangsu Province of China: A slow-moving landslide triggered by rainfall. *Landslides* **2020**, *17*, 2603–2617. [CrossRef]
20. Varnes, D.J. Slope Movements And Types And Processes. Landslides Analysis and Control. *Transp. Res. Board Spec. Rep.* **1978**, *176*, 11–33.
21. Schulz, W.H.; McKenna, J.P.; Kibler, J.D.; Biavati, G. Relations between hydrology and velocity of a continuously moving landslide—Evidence of pore-pressure feedback regulating landslide motion? *Landslides* **2009**, *6*, 181–190. [CrossRef]
22. Crozier, M.J. Landslide geomorphology: An argument for recognition, with examples from New Zealand. *Geomorphology* **2010**, *120*, 3–15. [CrossRef]
23. Ross, M.; Parent, M.; Lefebvre, R. 3D geologic framework models for regional hydrogeology and land-use management: A case study from a Quaternary basin of southwestern Quebec, Canada. *Hydrogeol. J.* **2005**, *13*, 690–707. [CrossRef]
24. Tang, W.; Tang, H. Grid-based three-dimensional geological models of the update method. In Proceedings of the 2011 International Conference on Multimedia Technology, IEEE, Hangzhou, China, 26–28 July 2011; pp. 1086–1089.
25. Xie, M.; Esaki, T.; Qiu, C.; Wang, C. Geographical information system-based computational implementation and application of spatial three-dimensional slope stability analysis. *Comput. Geotech.* **2006**, *33*, 260–274. [CrossRef]
26. Moore, I.D.; Turner, A.K.; Wilson, J.P.; Jenson, S.K.; Band, L.E. GIS and land-surface-subsurface process modeling. *Environ. Modeling GIS* **1993**, *20*, 196–230.
27. Lindsay, M.D.; Aillères, L.; Jessell, M.W.; de Kemp, E.A.; Betts, P.G. Locating and quantifying geological uncertainty in three-dimensional models: Analysis of the Gippsland Basin, southeastern Australia. *Tectonophysics* **2012**, *546*, 10–27. [CrossRef]
28. Jeng, C.J.; Sue, D.Z. Characteristics of ground motion and threshold values for colluvium slope displacement induced by heavy rainfall: A case study in northern Taiwan. *Nat. Hazards Earth Syst. Sci.* **2016**, *16*, 1309–1321. [CrossRef]
29. Tseng, C.H.; Chan, Y.C.; Jeng, C.J.; Hsieh, Y.C. Slip monitoring of a dip-slope and runout simulation by the discrete element method: A case study at the Huafan University campus in northern Taiwan. *Nat. Hazards* **2017**, *89*, 1205–1225. [CrossRef]
30. Tseng, C.H.; Chan, Y.C.; Jeng, C.J.; Rau, R.J.; Hsieh, Y.C. Deformation of landslide revealed by long-term surficial monitoring: A case study of slow movement of a dip slope in northern Taiwan. *Eng. Geol.* **2021**, *284*, 106020. [CrossRef]
31. Central Geological Survey of Taiwan (CGS), 2000. Explanatory Text of the Geological Map of Taiwan, scale 1:50,000, Sheet 32. Hsintien. (In Chinese). Available online: https://twgeoref.moeacgs.gov.tw/GipOpenWeb/imgAction?f=/2000/20000110/EAB.PDF (accessed on 24 June 2022).
32. Golden software, llc, Surfer 16. Golden software 2016, llc, Po box 281, Golden, CO 80402-0281 USA. Available online: https://www.goldensoftware.com/blog/surfer-16-is-here (accessed on 24 June 2022).
33. Krumbein, W.C. Confidence intervals on low-order polynomial trend surfaces. *J. Geophys. Res.* **1963**, *68*, 5869–5878. [CrossRef]
34. Chorley, R.J.; Haggett, P. Trend-surface mapping in geographical research. *Trans. Inst. Br. Geogr.* **1965**, *37*, 47–67. [CrossRef]
35. Mei, S. Geologist-controlled trends versus computer-controlled trends: Introducing a high-resolution approach to subsurface structural mapping using well-log data, trend surface analysis, and geospatial analysis. *Can. J. Earth Sci.* **2009**, *46*, 309–329. [CrossRef]

36. Bamisaiye, O.A. Subsurface mapping: Selection of best interpolation method for borehole data analysis. *Spat. Inf. Res.* **2018**, *26*, 261–269. [CrossRef]
37. Cook, D.I.; Santi, P.M.; Higgins, J.D. 2007 AEG Student Professional Paper: Graduate Division: Horizontal Landslide Drain Design: State of the Art and Suggested Improvements. *Environ. Eng. Geosci.* **2008**, *14*, 241–250. [CrossRef]

water

MDPI

Article

The Robust Study of Deep Learning Recursive Neural Network for Predicting of Turbidity of Water

Shiuan Wan [1], Mei-Ling Yeh [2,*], Hong-Lin Ma [3] and Tein-Yin Chou [4]

[1] Department of Information Technology, Ling Tung University, Taichung 40851, Taiwan; shiuan123@teamail.ltu.edu.tw
[2] GIS Research Center, Director, Feng Chia University, Taichung 40724, Taiwan
[3] GIS Research Center, Feng Chia University, Taichung 40724, Taiwan; elmer@gis.tw
[4] GIS Research Center, Dean, Feng Chia University, Taichung 40724, Taiwan; jimmy@gis.tw
* Correspondence: milly@gis.tw

Abstract: Water treatment is an important process, as it improves water quality and makes it better for any end use, whether it be drinking, industrial use, irrigation, water recreation, or any other kind of use. Turbidity is one of the fundamental measurements of the clarity of water in water treatment. Specifically, this component is an optical feature of the amount of light on scatter particles when light is shined on a water sample. It is crucial in water reservoirs to provide clean water, which is difficult to manage and predict. Hence, this study focuses on the use of robust deep learning models to analyze time-series data in order to predict the water quality of turbidity in a reservoir area. Deep learning models may become an alternative solution in predicting water quality because of their accuracy. This study is divided into two parts: (a) the first part uses the optical bands of blue (B), green (G), red (R), and infrared (IR) to build a regression function to monitor turbidity in water, and (b) the second part uses a hybrid model to analyze time-series turbidity data with the recursive neural network (RNN2) model. The selected models' accuracies are compared based on the accuracy using the input data, forecasting level, and training time. The analysis shows that these models have their strengths and limitations under different analyzed conditions. Generally, RNN2 shows the performance regarding the root-mean-square error (*RMSE*) evaluation metric. The most significant finding is that the RNN2 model is suitable for the accurate prediction of water quality. The *RMSE* is used to facilitate a comparison of the accuracy of the sampling data. In the training model, the training data have an *RMSE* of 20.89, and the testing data have an *RMSE* of 30.11. The predicted R-squared values in the RNN2 model are 0.993 (training data) and 0.941 (testing data).

Keywords: water quality; deep learning; recursive neural network

Citation: Wan, S.; Yeh, M.-L.; Ma, H.-L.; Chou, T.-Y. The Robust Study of Deep Learning Recursive Neural Network for Predicting of Turbidity of Water. *Water* **2022**, *14*, 761. https://doi.org/10.3390/w14050761

Academic Editor: Su-Chin Chen

Received: 25 January 2022
Accepted: 25 February 2022
Published: 28 February 2022

1. Introduction

Mountain areas can store large amounts of water in various ways for human beings. In Taiwan, watershed areas can be affected by different man-made developments; abuse; or natural factors, such as typhoons, heavy rains, and other soil and sand disasters. For instance, landslides are often directly or indirectly influenced by the turbidity of the water source and can cause reservoir siltation, seriously threatening water quality. Moreover, the variations in rainfall and soil/rock characteristics in reservoirs are also dominant factors of water quality [1–3]. A smart image management platform is required for water resources, and it would also help to strengthen water conservation and land use management. Past studies that have monitored the environment can be divided based on the two types of systems used: a monitoring system or a decision support system. Most of these systems are built based on the consideration of the factors of the water body and the land use surrounding the reservoirs [4]. Although water quality comprises a very broad area, such as hydro-chemical and physicochemical properties, this study only focuses on the turbidity value of water, but other influencing factors of water quality that can also be predicted by our proposed method.

The Water Resources Department (WRD) in Taiwan effectively and constantly monitors the variation in water quality. Most turbidity problems are caused by landslides after heavy rain or illegal land use in the watershed area. However, developing intelligent image recognition technology is well accepted for water conservation and land management in the watershed area. Accordingly, the goal of this study is to construct a prediction-based image system to gain a better understanding of reservoir areas and water quality protection areas. This type of system safeguards the ecological conservation of water collection areas to ensure the safety and cleanliness of national drinking water, which hopefully ensures the sustainable use of water resources [5]. Regarding the effective management and control of image information, it can easily determine the location of actual violations and understand the results of image interpretation. Hence, the Water Resources Department (WRD) built the "Water Resource Conservation Smart Image Management Platform" in 2018. It is more effective in improving the convenience of the overall operation of image management, strengthening the functionality of the platform. The spatial management platform that we are developing will use the advantages of geographic information systems in relational analysis [6,7]. This imaging system should contain two major functions: (a) the ability to transfer image data to compute turbidity and (b) the ability to build a predictive water system to monitor turbidity in a reservoir. Specifically, the imaging system has an interface with related demand areas, it is connected to electronic maps and uses aerial images as base maps, and it constructs spatial displays to provide different images and real-time information references from the Water Resources Department.

However, the variation in water quality is hard to predict, and it may also be influenced by climate factors [8]. Turbidity in water is a major factor, and it is purified by water treatment plants in downstream in the water system. Thus, if an imaging system providing real-time data can transfer information to govern water quality, it will be of great help in understanding the current situation of the water [4]. In addition, if a proper predictive water system can be constructed, it will offer a better understanding of the near feature of turbidity in water. To achieve the prediction of water variation, a time-series neural network is adopted in this study.

As previously mentioned, the observations and predictions of water components (such as turbidity) seem to be a crucial problem for water treatment plants and should be a precaution when handling the current stage and next stage of the water body. Accordingly, in the current stage, one can consider using regression analysis to solve the problem. However, prediction involves a large amount of know-how in data analysis. Deep learning methods show promise for time-series predictions, for example, the automatic learning of temporal dependence and the automatic handling of temporal structures, such as trends and seasonality. With clear explanations, standard Python libraries (Keras and TensorFlow 2) have many useful analysis tools that can be used to carry out solutions for many fields, such as the prediction of PM 2.5 [9]. Furthermore, long short-term memory (LSTM) is another artificial recurrent neural network (RNN) architecture used in the field of deep learning. Currently, LSTM networks are applied to classify, process, and make predictions based on time-series data, since there can be lags of unknown duration between important events in an analysis of time-series data [10–12]. Therefore, using these tools to govern step-wise data and to train a neural network could be a possible way to resolve turbidity in water. The water quality of reservoirs, such as Qingtan Weir, is affected by heavy rains. This study plans to develop a boosting neural network to calculate the huge changes in turbidity under sudden conditions. The measured turbidity value is established in each sub-routine to correct the prediction of the drastic change in water turbidity of RNN2.

The two main types of analytical neural networks in the time series are (a) recurrent neural network (RNN1) and (b) recursive neural network (RNN2) [10–12]. The connections between the neurons of the time recurrent neural network form a directed graph, while the structural recurrent neural network uses a similar neural network structure to recursively construct a more complex deep network. Time recursive neural networks can describe dynamic time behavior, and they are unlike feedforward neural networks, which accept a more specific structure input. RNN2 cyclically transmits the state in its network. That is, it can accept a wider range of time-series structures. For example, handwriting recognition was the first research result to successfully use RNN2. In general, the accuracy of the classifier can usually be improved by averaging the decisions of the classifier set. In general, when the various classifiers are different and accurate, a greater improvement can be expected. It was decided to obtain the results by adopting a basic learning algorithm and using it in multiple time domains on different training sets [13].

To achieve the goal of this study, we used the image data of the turbidity of Qingtan Weir, Taiwan. The sensor under the water attained the real value of the turbidity. Moreover, a UAV took an image to acquire the band value of the water body at the sensor. The regression model was built, and the time-series data were used to predict the turbidity change in Qingtan Weir after heavy rains or after soil condition changes through the RNN2 model. The paper is structured as follows: (1) Introduction, (2) Materials and Methods, (3) Methods, and (4) Discussion of the Results. The last part is divided into two steps: the first step generates the regression model, and the second step uses the neural network with the RNN2 approach.

2. Materials and Methods

2.1. Study Area

Reservoir areas are important for soil water conservation areas. Therefore, most reservoir areas are involved in many types of land use and other induced water quality issues. There are currently 95 reservoirs in Taiwan. The first stage of this study considered a total water area of about 700,000 m^3 of a reservoir, which is an appropriate size for analysis of water quality in a specific area. The effective capacity is 150,000 m^3, and the effective size is 716.8 km^2. Hence, the study selected Qingtan Weir (Figure 1) as an example to demonstrate the application on monitoring and prediction. The longitude and latitude of the Qingtan Weir are located at 24.948N and 121.545E, respectively. More specifically, this study took the Qingtan Weir water area sampling as a means to analyze monthly the past series of image data by a UAV, to monitor the water quality from images, and to produce the acquisition of UAV images. Then, the study built a new monitoring and prediction model based on an intelligent technology system, which is a computer system platform using a regression approach [8]. However, since one series of data was collected, the RNN model was applied to predict water quality in time series.

Due to the complicated components of the water body, this study decided to use the different changes in the color of the surface water of remote images to calculate the turbidity value of the water body through the reflectance of different wavebands [13]. Figure 1 displays the observation points, namely, A, B, and C, that were selected. The underwater sensor was mounted at those points, and time-series data were collected stably and automatically. In the first stage, we compared these three points and selected point B, which displayed the best outcomes of water quality. Due to points A and C being near to a sandy shore, some unpredictable values may have occurred as a result of rainfall. This study used multiple regression analysis methods to predict the turbidity value of water quality. The first trial used regression analysis. Then, it established the dependent variable (Y; turbidity) as a function of the independent variable (X; B, G, R, and IR) model and then estimated the parameters of the function based on the data obtained from the sample. The purpose of this was to present the image bands of the data to form a correlation of the turbidity of the water body.

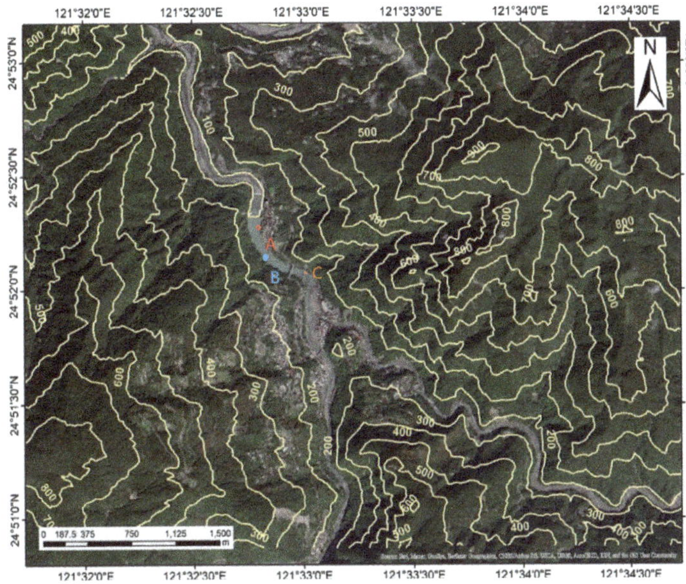

Figure 1. The contour elevation line of Qingtan Weir.

2.2. Study Plan

As previously mentioned, we received the image data from the UAV at point B in Figure 1. The water quality of turbidity was used as a series of outcomes to build the regression function by considering the image bands of B, G, R, and IR as input variables. The rainfall database was also used for calibration to amend the regression function. Figure 2 presents the research steps for analysis. The prediction of turbidity is quite difficult, especially considering the rainy hours of emergency notice for water treatment plants. The novelty this paper is that we used two neural network systems:

(a) A regular neural network to handle no-rain hours.
(b) The boosting model (RNN2) to handle rainy hours.

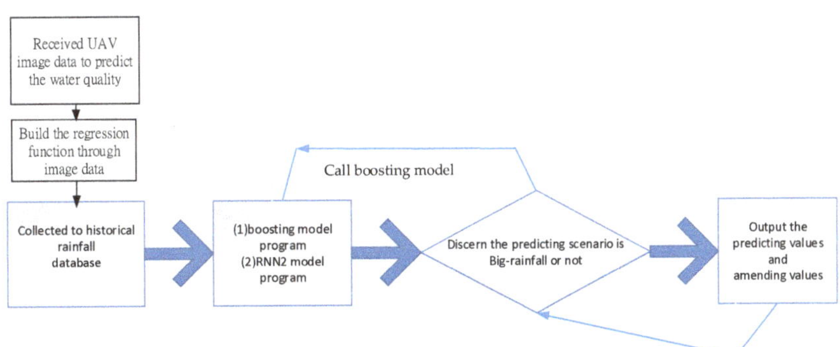

Figure 2. Steps for research analysis.

2.3. Image Data

The plan of the study was to use the acquisition image data at 10 a.m. of the selected study area in cloudless weather for a standard image format. Since the optical scanning operations are performed simultaneously, the image data will be operated to meet the requirements of a scanning height of 15 m above the ground and FOV of 40° to obtain the ground resolution as 0.4 m images. At least 72 time intervals were recorded (locations are refer points A, B, and C in Figure 1) for the time-series data, which were used as an element to identify image features and make a prediction in future steps.

3. Methods

3.1. Neural Network (NN) Model

Neural networks are constructed based on the structure of the human brain, and they have contributed to current machine learning technology. They provide a simplified mathematical model to solve various nonlinear problems. In the past, researchers reviewed the ANN model's predictions of solar energy [14] and groundwater levels [15]. Past studies showed that the ANN model has a more accurate prediction than other conventional models, such as the Angstrom, conventional, linear, nonlinear, and fuzzy logic models [16].

The traditional neural network model comprise input, hidden, and output layers with auxiliary components, such as neurons, weight, bias, and activation functions [16]. Figure 3 shows basic neural network architecture with a multilayer perceptron. The input layers may receive input values from the signal or data, and the hidden layer analyzes the input values. The output layer collects the data from the hidden layer and decides the output. In the learning process, the neural network modifies its structure to obtain the same reference or set point as the supervisor. The training process is repeated until the difference between the neural network output and the supervisor lies within an acceptable range.

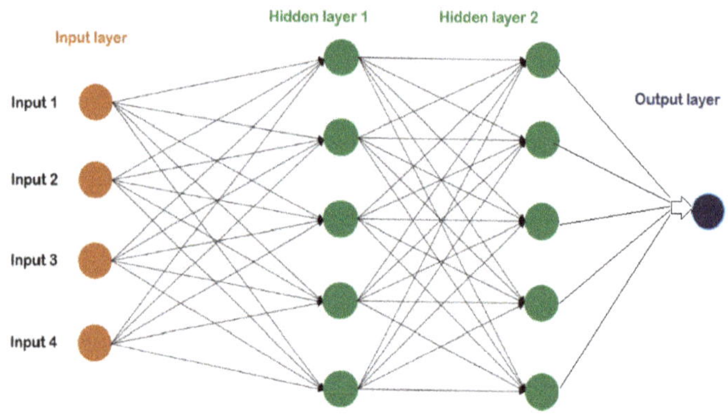

Figure 3. The traditional neural network model.

3.2. Recurrent Neural Network; RNN1

A recurrent neural network (RNN) is a class of artificial neural networks where connections between nodes form a directed graph along a temporal sequence. Previous RNN studies have drawn great attraction for water quality predictions [17–23]. This allows it to exhibit temporal dynamic behavior. Derived from feedforward neural networks, RNNs can use their internal state (memory) to process variable-length sequences of inputs. This makes them applicable to tasks such as unsegmented, connected handwriting recognition or speech recognition.

RNN models (see Figure 4) in particular are designed to analyze sequential data. They have been successfully used in fields such as speech recognition, machine translation, and image captioning. RNN processes sequence data by elements, and they preserve a state to represent the information at time steps. The traditional neural network assumes that all units of the input vectors are independent. Accordingly, the traditional neural network is ineffective for prediction when using sequential data. The architecture of RNNs comprise three main components (input, hidden neuron, and activation function). The previous hidden state (h_t) can be formulated as

$$h_t = \tanh(\mathrm{U}\cdot \mathrm{x}_t + \mathrm{W}\cdot h_{t-1})$$

where x_t is the input at time t, h_t is the hidden neuron at time t, U is the weight of the hidden layer, and W is the transition weights of the hidden layer. The input and previous hidden states are combined to produce information as the current and previous inputs go through a tanh function. Then, the output is the new hidden state, performing as the neural network memory because it holds information from the previous network. Training regular RNNs results in a series of vanishing and exploding gradient problems. In the case of the exploding gradient, the problem is resolved after backpropagation, which is closed at a certain point. However, the results may not be optimal because all the weights are not updated. In the case of the vanishing gradient, it can be adjusted by initializing the weights to reduce the possibility of a vanishing gradient [24].

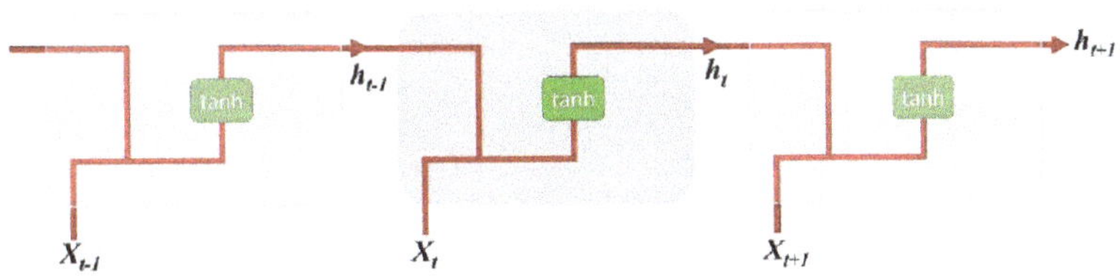

Figure 4. RNN model.

3.3. *Recursive Neural Network; RNN2*

A recursive neural network is a type of deep learning neural network. It is usually created by applying the same set of weights recursively over structured data for inputs. Then, it produces a structured prediction over variable-size input structures, or a scalar prediction on it, by traversing a given structure in topological order. Recursive neural networks have been successfully resolved of the prediction in learning sequence data [20–22]. A recursive neural network is a type of deep neural network produced by a set of weights recursively over a time series of structured inputs to produce consecutive predictions over variable-size input structures, or a scalar prediction on it [23–25]. It traverses a given structure in topological order. Recursive neural networks, sometimes abbreviated as RNN2 (see Figure 5), have been successful, for instance, in learning sequence and tree structures in natural language processing. RNN2 was first introduced to learn distributed representations of structure, such as logical terms.

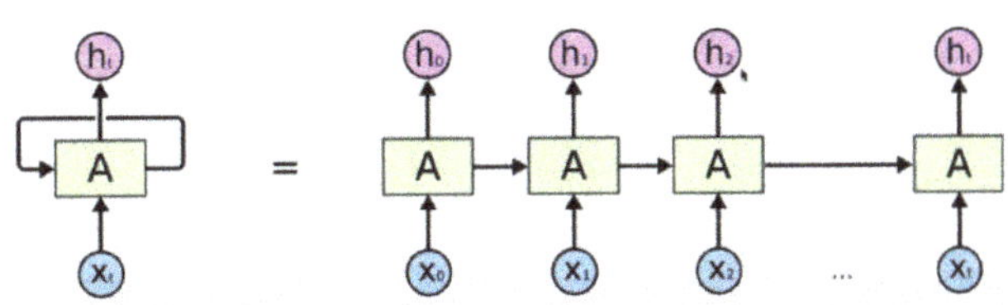

Figure 5. RNN2 model.

Although recursive neural network (RNN2) models demonstrate advantages in system robustness and time-series data computation, they have been widely applied to redundant data optimization. Existing RNN2 models suffer from local minimum and do not have planning completeness. Therefore, many techniques are used to solve the redundant data in time-series data acquisition [25,26]. Recursive neural networks usually operate on any hierarchical structure, transforming child representations into parent representations. In general, stochastic gradient descent (SGD) is used to train this type of network. The gradient is calculated using backpropagation on varying data through a time series. In detail, every node in a given layer is connected to every other node in the next successive layer through a directed (one-way) connection. Each node (neuron) has a time-varying real-valued activation. Each connection (synapse) has a real-valued weight that can be modified. Nodes are either input nodes (receive data from outside the network), output nodes (produce results), or hidden nodes (modify data in the process from input to output). For supervised learning in discrete-time data, a sequence of real-valued input vectors arrives at the input node, one vector at a time. In any given time step, each non-input unit calculates its current activation (result) as a non-linear function of the weighted sum of the activations of all units connected to it [27,28]. The target activation given by the supervisor is applied to certain output units at a specific time step [29,30].

4. Discussion of the Results

Step 1: Build the regression function using image data.

In the first stage, the goal was to monitor turbidity in water. This study used multiple regression analysis to predict the turbidity value of water quality. B, G, R, and IR are four variables taken from image data, and the images were taken by a UAV. The turbidity values were extracted from sensor B. One series of the regression function was obtained. One of them can be formulated as

$$T = -1.238 \times R + 3.616 \times G - 2.991 \times B - 0.048 \times IR + 391.448, R^2 = 0.931 \quad (1)$$

where T is the turbidity.

Part of the regression error is plotted in Figure 6. The average error is about 0.07%. The Y-axis is the real value of turbidity, and the X-axis is the value predicted by regression.

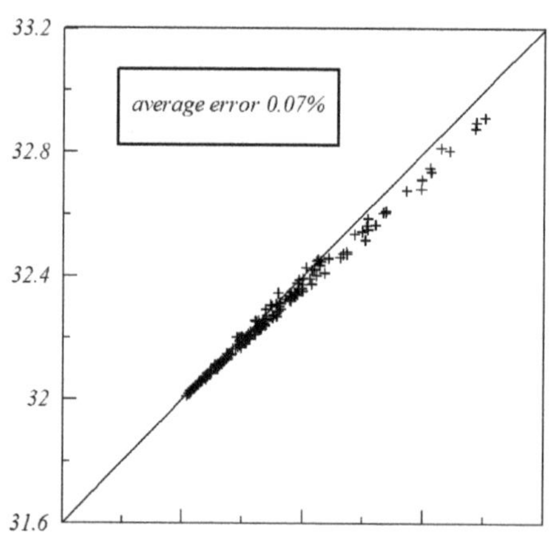

Figure 6. Prediction error of regression model (R^2 = 0.931).

Step 2: Use the time-series data of turbidity to build the RNN2 model.

The architecture of our RNN2 model was built based on a neural network framework, and it is shown in Figure 7. There is one input layer of neurons and one single output of neurons. Four hidden layers are constructed for computation. Two parts are constructed as (a) RNN2 parts and (b) NN parts. A total of 72 inputs of the neuron are set up to receive the signal of the water quality, considering the turbidity. Layer 2 is the hidden layer, and the received neurons of this layer take turns to transmit the data to the NN part of the system.

There are 72 input neurons of 12 h, with each hour taking 6 sampling rates. We selected four hidden layers, and the first two are in the RNN part and the last two are connected to the NN part. To accelerate and consider the convergence speed of the network, we decided to use the same number of neurons as in the input layer. The rectified linear unit (ReLu) was used as the activation function, which is used in the calculation of the core of the NN and RNN systems. Since the ReLu converts all the negative and zero input values to zero, using ReLu as an activation function for RNN2 is a legitimate concern. Srivastava et al. [31] suggested dropout to feedforward neural networks and RBMs and noted the probability of dropout being around 0.5 for hidden units and 0.2 for inputs, which worked well for a variety of tasks. There is a dropout ratio of 20%, which was set up in the first three layers. The last hidden layer was set up for 50% to accelerate the convergence of iterations.

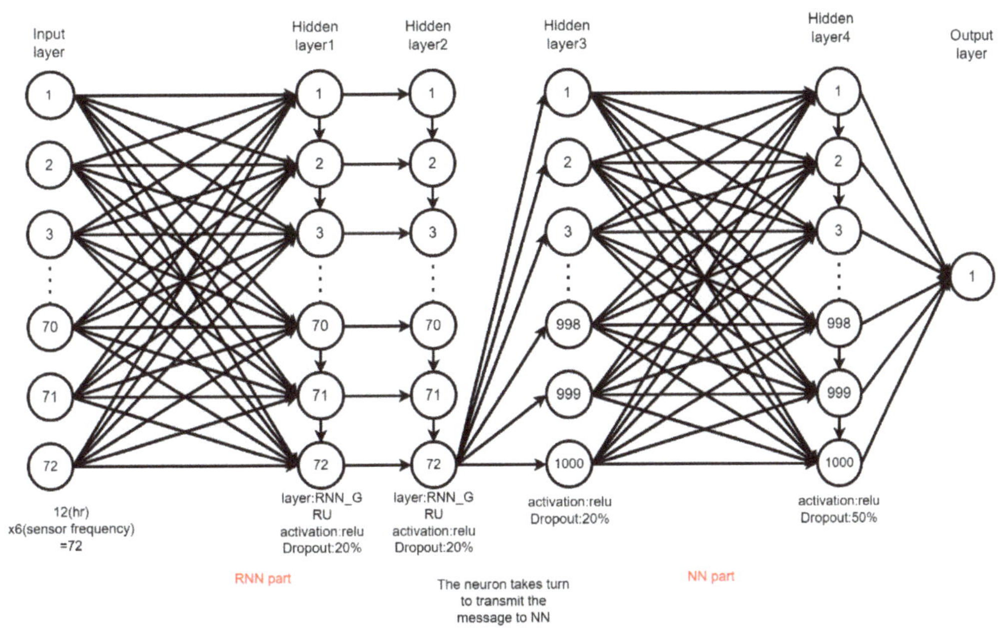

Figure 7. Flowchart of RNN2 design.

Step 3: The indicators of accuracy.

(1) *RMSE*: Root-mean-squared error is a frequently used measure of the differences between values predicted by an estimator and the values observed. It can be formulated as

$$RMSE = \sqrt{\frac{\Sigma(y - \hat{y})^2}{T}}$$

where T represents the different predictions, $\hat{y}$ is the value predicted by an estimator, and y is the real value.

(2) *MAE*: Mean absolute error (*MAE*) is a measurement of errors between paired observations with the absolute value.

(3) R^2 is a statistical measure that represents the proportion of the variance for a dependent variable.

(4) *CV*: Coefficient of variation is a statistical measure of the relative dispersion of data points in a data series around the mean. It can be formulated as

$$CV = \frac{\sigma}{\mu}$$

where σ is the standard deviation, and μ is the mean.

4.1. Training Model

This study collected the training data in October 2019. The RNN2 model was built for two different cases of (a) accumulated rainfall corresponding to 0 and (b) accumulated rainfall greater than 0. The training dataset comprises 0.7 of the entire dataset, and the remainder is used as the testing dataset for the self-testing of the model. Figure 8 presents the prediction error distribution diagram.

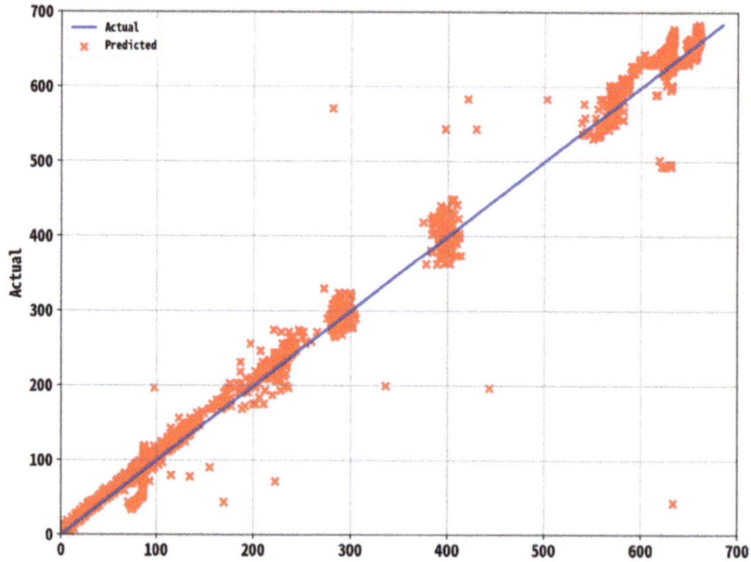

Figure 8. Prediction error of RNN2 (R^2 = 0.947).

4.2. Prediction Model for Convergence

RNN2 converges quickly, which is illustrated in Figure 9a for the case of no-rain days. It varies dramatically at the initial stage. The accuracy becomes stable at 16 epochs. The *RMSE* of the training dataset descends from 300,000. The test data move downward after the fifth epoch. Figure 9b shows the case of no-rain days of *RMSE* vs. epochs. The network converges at 14 epochs. The variations in Figure 9a,b are very similar.

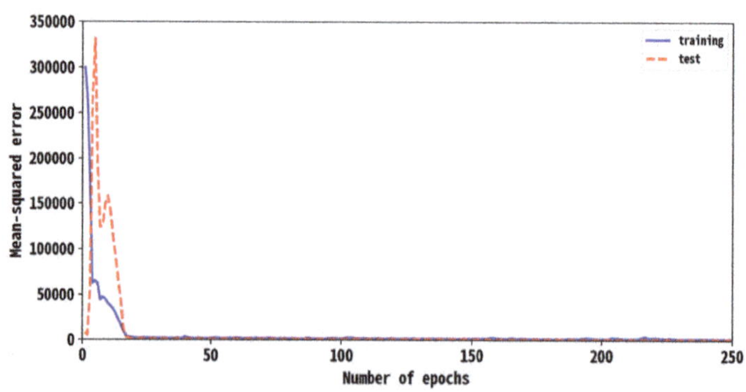

(**a**) Epochs vs. RMSE for no-rain days.

Figure 9. *Cont.*

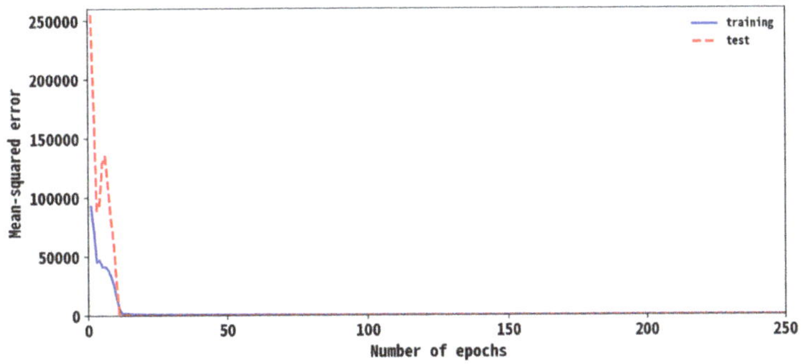

(**b**) Epochs vs. RMSE for rainy days.

Figure 9. Epochs of (**a**) no-rain days and (**b**) rainy days.

4.3. Prediction for Time–History

Time–history analysis is used to analyze the performance of RNN2 models. Accordingly, the turbidity of water can apply RNN2 for the step-wise analysis of temporal data. In our study, a portion of data is extracted to display the physical meaning of the analysis. The X-axis is the time and date, and the Y-axis is the density of turbidity. We used October 2018 to build the RNN2 model. Figure 10a shows the time–history of the accuracy of real data (blue) and verification (red) for October 2018. Figure 10b shows the time–history of accuracy of real data (blue) and prediction (red) for November 2018. The *RMSE* of October is 20.89, which is smaller than that of November (36.11); this makes sense as the prediction outcomes are greater. However, the differences between October and November are sufficiently small to explain the prediction of outcomes, which are considered quite rational.

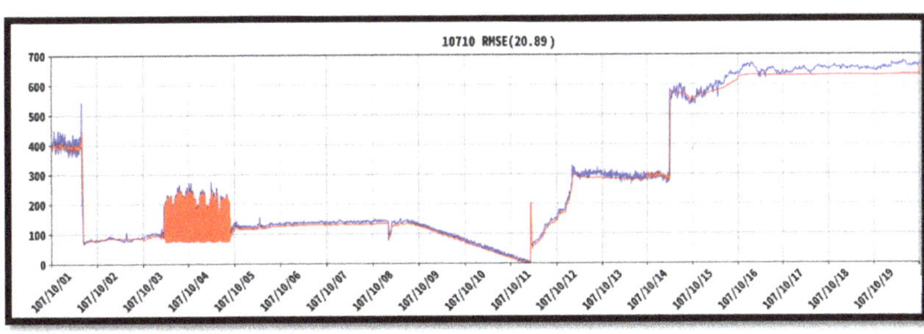

(**a**)

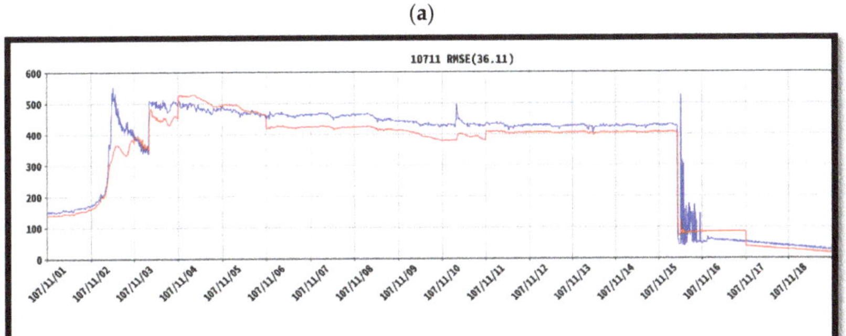

(**b**)

Figure 10. Time–history difference between real (blue) and predicted (red). (**a**) Time–history of October for building of RNN2 model; (**b**) time–history of November for predicting of RNN2 model.

In addition to using time–history analysis, we also employed MAE, *RMSE*, *CV*, and R^2 to observe the performance and accuracy of the model. The performance criteria of RNN2 are presented in Table 1. We illustrated the *MAE*, *RMSE*, and R^2 of October 2018 and November 2018. October represents the training data used to generate the RNN2 model. November represents the verification part, using testing data to examine how the real-time value is displayed.

Table 1. The RNN performance criteria table.

Date	Type	Value
October 2018	*MAE*	13.88
October 2018	*RMSE*	20.89
October 2018	*CV*	0.757
October 2018	R^2	0.993
November 2018	*MAE*	25.20
November 2018	*RMSE*	36.11
November 2018	*CV*	0.768
November 2018	R^2	0.961

5. Summary and Conclusions

The recursive neural network (RNN2) is a type of artificial neural network that uses time-series data to predict a few forward step-wise data items of values. Such deep learning algorithms are usually designed for temporal problems. In addition, RNN2 utilizes training data to learn a learning system, which is similar to its "memory", as it takes information from prior inputs to influence current outputs. Traditional deep neural networks presume that inputs and outputs are independent of each other, as well as the output of recurrent neural networks depending on prior elements within a sequence. When future events help to determine the output of a given sequence, unidirectional recurrent neural networks cannot take account of these events in their predictions. Considering these characteristics of RNN2, the few-step time-series data of water components can be successfully applied by using RNN2 to resolve them.

Turbidity is the most important measure indicator of the relative clarity of water. It is an optical feature of water, and it is a measurement of the amount of light that is scattered by materials in water samples. In general, the higher the intensity of scattered light, the higher the turbidity. However, many materials, such as clay, silt, very tiny inorganic and organic matter, algae, dissolved organic/non-organic compounds, plankton, and other microscopic organisms, cause water to be turbid. We monitored a reservoir's turbidity values using sensors mounted at the waterbed. Simultaneously, a series of images from the UAV were taken to correlate turbidity and the band values of B, G, R, and IR. Remote sensing data were used to present the values of turbidity of water in our selected reservoir. The R-squared value is 0.931, which satisfies the prediction of a regression model. We aimed to construct an ANN system (regular no-rain hours) with an RNN2 system for a subroutine (considering rainy hours) to predict turbidity values using water sensors. Due to the occurrence of rainy hours, the prediction can have huge errors, but our subroutine can adjust the accuracy of results. As part of this study, a deep learning approach using the RNN2 model is presented to forecast time-series data by operating the main program in regular hours. That is, when the data increase dramatically, the subroutine is automatically operated. The main objectives of this work were to design a model that can not only predict the very next step but also generate a sequence of predictions and utilize multiple driving time series together with a set of static (scalar) features as its inputs. The *RMSE* of October is 20.89, which is smaller than that of November (36.11). The predicted R-squared values in RNN2 are 0.993 and 0.941. The differences in *RMSE* for October and November are sufficiently small, which explains why the predictions of outcomes are considered quite rational.

Author Contributions: Conceptualization, S.W.; programing, H.-L.M.; formal analysis, M.-L.Y.; writing—original draft preparation, S.W.; writing—review and editing, T.-Y.C.; project administration, M.-L.Y. All authors have read and agreed to the published version of the manuscript.

Funding: The APC was funded by Ministry of Science and Technology.

Institutional Review Board Statement: Not applicable.

Informed Consent Statement: Not applicable.

Data Availability Statement: The data presented in this study are available on request from the corresponding author.

Acknowledgments: The authors express their gratitude to the National Science Council (MOST 109-2622-M-275-001 -CC3) for sponsoring this work.

Conflicts of Interest: The authors declare no conflict of interest.

References

1. Cero, C.D.L.; Ala, K.J.L.; Arboleda, E.R. Android application for microcontroller-based reservoir water level monitoring. *Indones. J. Electr. Eng. Inform.* **2018**, *6*, 79–85.
2. Karki, S.; Burton, P.; Mackey, B. Climate change adaptation by subsistence and smallholder farmers: Insights from three agro-ecological regions of Nepal. *Cogent Soc. Sci.* **2020**, *6*, 1720555. [CrossRef]
3. Tsiko, R.G.; Haile, T.S. Integrating Geographical Information Systems, Fuzzy Logic and Analytical Hierarchy Process in Modelling Optimum Sites for Locating Water Reservoirs. A Case Study of the Debub District in Eritrea. *Water* **2011**, *3*, 254–290. [CrossRef]
4. Wan, S.; Chang, T.C. The Application of a Hybrid Model of EDBD Algorithms and Grey Relational Analysis: A Solution for Knockout Poor Quality of Water Reservoirs. *J. Grey Syst.* **2005**, *8*, 131–139.
5. Scholten, T.; Hartmann, T.; Spit, T. The spatial component of integrative water resources management: Differentiating integration of land and water governance. *Int. J. Water Resour. Dev.* **2019**, *36*, 800–817. [CrossRef]
6. Chang, K.T.; Wan, S.; Lei, T.C. Development of a spatial decision support system for monitoring earthquake-induced landslides based on aerial photographs and the finite element method. *Int. J. Appl. Earth Obs. Geoinf.* **2010**, *12*, 448–456. [CrossRef]
7. Kosaka, K.; Koike, Y.; Miyabayashi, Y.; Saito, K.; Asami, M.; Sasaki, M.; Sato, S.; Akiba, M. National survey of utilization of continuous water quality monitors in water supply systems in Japan. *Water Supply* **2019**, *19*, 1347–1353. [CrossRef]
8. Chang, B.; He, K.-N.; Li, R.-J.; Sheng, Z.-P.; Wang, H. Linkage of climatic factors and human activities with water level fluctuations in Qinghai Lake in the northeastern Tibetan Plateau, China. *Water* **2017**, *9*, 552. [CrossRef]
9. Chen, Y.-C.; Lei, T.-C.; Yao, S.; Wang, H.-P. PM2.5 Prediction Model Based on Combinational Hammerstein Recurrent Neural Networks. *Mathematics* **2020**, *8*, 2178. [CrossRef]
10. Barzegar, R.; Aalami, M.T.; Adamowski, J. Short-term water quality variable prediction using a hybrid CNN–LSTM deep learning model. *Stoch. Hydrol. Hydraul.* **2020**, *34*, 415–433. [CrossRef]
11. Ferreira, L.B.; da Cunha, F.F. Multi-step ahead forecasting of daily reference evapotranspiration using deep learning. *Comput. Electron. Agric.* **2020**, *178*, 105728. [CrossRef]
12. Granata, F.; Di Nunno, F. Forecasting evapotranspiration in different climates using ensembles of recurrent neural networks. *Agric. Water Manag.* **2021**, *255*, 107040. [CrossRef]
13. Wan, S.; Lei, T.C.; Ma, H.L.; Cheng, R.W. The Analysis on Similarity of Spectrum Analysis of Landslide and Bareland through Hyper-Spectrum Image Bands. *Water* **2019**, *11*, 2414. [CrossRef]
14. Notton, G.; Voyant, C.; Fouilloy, A.; Duchaud, J.L.; Nivet, M.L. Some applications of ANN to solar radiation estimation and forecasting for energy applications. *Appl. Sci.* **2019**, *9*, 209. [CrossRef]
15. Hong, Y.-M.; Wan, S. Information-based system identification for predicting the groundwater-level fluctuations of hillslopes. *Appl. Hydrogeol.* **2011**, *19*, 1135–1149. [CrossRef]
16. Wan, S.; Lei, T.C.; Chou, T.Y. An enhanced supervised spatial decision support system of image classification: Consideration on the ancillary information of paddy rice area. *Int. J. Geogr. Inf. Sci.* **2010**, *24*, 623–642. [CrossRef]
17. Wei, X.; Zhang, L.; Yang, H.-Q.; Zhang, L.; Yao, Y.-P. Machine learning for pore-water pressure time-series prediction: Application of recurrent neural networks. *Geosci. Front.* **2020**, *12*, 453–467. [CrossRef]
18. Li, L.; Jiang, P.; Xu, H.; Lin, G.; Guo, D.; Wu, H. Water quality prediction based on recurrent neural network and improved evidence theory: A case study of Qiantang River, China. *Environ. Sci. Pollut. Res.* **2019**, *26*, 19879–19896. [CrossRef]
19. Bengio, Y.; Simard, P.; Frasconi, P. Learning long-term dependencies with gradient descent is difficult. *IEEE Trans. Neural Networks* **1994**, *5*, 157–166. [CrossRef]
20. Graves, A.; Fernandez, S.; Gomez, F.J.; Schmidhuber, J. Connectionist temporal classification: Labelling unsegmented sequence data with recurrent neural nets. In Proceedings of the 23rd International Conference on Machine Learning, Pittsburgh, PA, USA, 25–29 June 2006.
21. Sak, H.; Senior, A.; Beaufays, F. Long Short-Term Memory based recurrent neural network architectures for large vocabulary speech recognition. *arXiv* **2014**, arXiv:1402.1128.

22. Hochreiter, S.; Bengio, Y.; Frasconi, P.; Schmidhuber, J. Gradient flow in recurrent nets: The difficulty of learning long-term dependencies. In *A Field Guide to Dynamical Recurrent Neural Networks*; Kremer, S.C., Kolen, J.F., Eds.; IEEE Press: Manhattan, NY, USA, 2001.
23. Lin, T.; Horne, B.G.; Tino, P.; Giles, C.L. Learning long-term dependencies in NARX recurrent neural networks. *IEEE Trans. Neural Networks* **1996**, *7*, 1329–1338. [CrossRef] [PubMed]
24. Wang, L.; Zhang, T.; Jin, X.; Xu, J.; Wang, X.; Zhang, H.; Yu, J.; Sun, Q.; Zhao, Z.; Xie, Y. An approach of recursive timing deep belief network for algal bloom forecasting. *Neural Comput. Appl.* **2018**, *32*, 163–171. [CrossRef]
25. Acharya, T.D.; Subedi, A.; Lee, D.H. Evaluation of Machine Learning Algorithms for Surface Water Extraction in a Landsat 8 Scene of Nepal. *Sensors* **2019**, *19*, 2769. [CrossRef]
26. Zhang, Y.; Xiong, R.; He, H.; Pecht, M.G. Long Short-Term Memory Recurrent Neural Network for Remaining Useful Life Prediction of Lithium-Ion Batteries. *IEEE Trans. Veh. Technol.* **2018**, *67*, 5695–5705. [CrossRef]
27. Jain, A.K.; Prasad Rao, P.V.R.D.; Venkatesh Sharma, K. Deep learning with recursive neural network for temporal logic implementation. *Int. J. Adv. Trends Comput. Sci. Eng.* **2020**, *9*, 6829–6833. [CrossRef]
28. Mirikitani, D.T.; Nikolaev, N. Recursive Bayesian Recurrent Neural Networks for Time-Series Modeling. *IEEE Trans. Neural Networks* **2009**, *21*, 262–274. [CrossRef]
29. Møller, M. Supervised learning on large redundant training sets. *Int. J. Neural Syst.* **1992**, *4*, 79–89.
30. Parish, E.J.; Carlberg, K.T. Time-series machine-learning error models for approximate solutions to parameterized dynamical systems. *Comput. Methods Appl. Mech. Eng.* **2020**, *365*, 112990. [CrossRef]
31. Srivastava, N.; Hinton, G.; Krizhevsky, A.; Sutskever, I.; Salakhutdinov, R. Dropout: A simple way to prevent neural networks from overfitting. *J. Mach. Learn. Res.* **2014**, *15*, 1929–1958.

water

MDPI

Article

Performance Assessment of Event-Based Ensemble Landslide Susceptibility Models in Shihmen Watershed, Taiwan

Chun-Yi Wu * and Sheng-Yu Lin

Department of Soil and Water Conservation, National Chung Hsing University, Taichung 402, Taiwan; linshengyu0630@gmail.com
* Correspondence: cywu@nchu.edu.tw; Tel.: +886-4-2284-0381 (ext. 605)

Abstract: While multi-year and event-based landslide inventories are both commonly used in landslide susceptibility analysis, most areas lack multi-year landslide inventories, and the analysis results obtained from the use of event-based landslide inventories are very sensitive to the choice of event. Based on 24 event-based landslide inventories for the Shihmen watershed from 1996 to 2015, this study established five event-based single landslide susceptibility models employing logistic regression, random forest, support vector machine, kernel logistic regression, and gradient boosting decision tree methods. The ensemble methods, involving calculating the mean of the susceptibility indexes (PM), median of the susceptibility indexes (PME), weighted mean of the susceptibility indexes (PMW), and committee average of binary susceptibility values (CA) of the five single models were then used to establish four event-based ensemble landslide susceptibility models. After establishing nine landslide susceptibility models, using each inventory from the 24 event-based landslide inventories or a multi-year landslide inventory, we identified the differences in landslide susceptibility maps attributable to the different landslide inventories and modeling methods, and used the area under the receiver operating characteristic curve to assess the accuracy of the models. The results indicated that an ensemble model based on a multi-year inventory can obtain excellent predictive accuracy. The predictive accuracy of multi-year landslide susceptibility models is found to be superior to that of event-based models. In addition, the higher predictive accuracy of ensemble landslide susceptibility models than that of single models implied that these ensemble methods were robust for enhancing the model's predictive performance in the study area. When employing event-based landslide inventories in modeling, PM ensemble models offer the best predictive ability, according to the Kruskal–Wallis test results. Areas with a high mean susceptibility index and low standard deviation, identified using the 24 PM ensemble models based on different event-based landslide inventories, constitute places where landslide mitigation measures should be prioritized.

Keywords: landslide susceptibility analysis; event-based landslide inventory; ensemble model; Shihmen watershed

Citation: Wu, C.-Y.; Lin, S.-Y. Performance Assessment of Event-Based Ensemble Landslide Susceptibility Models in Shihmen Watershed, Taiwan. *Water* **2022**, *14*, 717. https://doi.org/10.3390/w14050717

Academic Editor: Aizhong Ye

Received: 29 January 2022
Accepted: 21 February 2022
Published: 24 February 2022

1. Introduction

Under the impact of climate change, extreme rainfall events have caused frequent landslides and debris flows in Taiwan's mountainous areas. In order to effectively reduce the losses caused by the landslides and debris flows, it is necessary to employ landslide susceptibility analysis to delineate those areas in watersheds that are susceptible to landslides and use this information as a reference for overall watershed management plans. The chief methods for landslide susceptibility analysis consist of heuristic, statistical, probability, and deterministic methods [1]. Many types of machine learning methods have been broadly applied to landslide susceptibility analysis in recent years and have yielded excellent results; machine learning algorithms can be classified as either parametric or nonparametric [2].

Parametric machine learning algorithms first select a form of the function and then learn the function's coefficients through a training process. The advantage of these algorithms is that the methodology is easy to explain and understand, and the training process is short and does not require the collection of vast amounts of data; their limitation is that the prior selection of a function often constrains the learning process, the method is only suitable for simple problems, and the fit is often relatively poor. Parametric machine learning algorithms may employ logistic regression and linear discriminant analysis, and logistic regression, in particular, is often applied in landslide susceptibility analysis [3–7]. For their part, nonparametric machine learning algorithms do not require prior selection of the functional form and may fit the function of any form through the training process. The advantage of this approach is its versatility and ability to generate good performance for training sample data; its limitation is its need for vast amounts of data, a slow training process, and a higher chance of overfitting occurring [2]. When the training sample is too small, the nonparametric algorithm inevitably leads to inadequate training, which will reduce the accuracy [8]. The nonparametric machine learning algorithms most commonly used in landslide susceptibility analysis include the support vector machine [9–15], random forest [8,9,12,15–18], kernel logistic regression [10,11,19], and boosted regression tree [15–17].

When performing landslide susceptibility analysis, landslide inventories can be classified as either multi-year or event-based, depending on the length of data collection time in the inventory. Apart from establishing a landslide susceptibility model based on a multi-year landslide inventory [20–22], when the research area lacks a multi-year landslide inventory, an event-based landslide inventory and triggering factors can be used to perform susceptibility analysis [3]. When establishing an event-based landslide susceptibility model, a landslide inventory for the event and data concerning the spatial distribution of triggering factors must be available; triggering factors, such as rainfall or earthquake intensity, are taken as independent variables in the model [6,23–25].

When establishing a landslide susceptibility model, the input sample data set is usually divided into a training set and a testing set. After using the training set to establish a model, the testing set is used to assess the performance of the model. The sample data set commonly contains a 50:50 ratio of landslide and no-landslide samples [9,13,26,27], and the ratio of the training set sample to the testing set sample is typically 70:30 [8,12,13,26,28,29]. Furthermore, when establishing a nonparametric machine learning model, the training set is also used to perform hyperparameter optimization. During the optimization process, fivefold cross-validation [9] and tenfold cross-validation [12,30] are often used to tune the hyperparameters.

Because each modeling method has its own advantages and limitations, different models can be used to perform landslide susceptibility analysis for the same research area, but uncertainty associated with the results of these models may exist. The ensemble method can then be used to aggregate the results of different models and can delineate areas with high susceptibility and low uncertainty [16,31]. The five most commonly used ensemble methods [32] involve the calculation of mean of landslide probabilities (PM), confidence interval for the mean of landslide probabilities (CI), median of landslide probabilities (PME), weighted mean of landslide probabilities (PMW), and committee averaging (CA), respectively. The stacking ensemble method, which uses a meta-learning algorithm to combine different single models [33], was also employed to establish ensemble modes [34,35]. In this study, the landslide susceptibility models were constructed by adopting the ensemble methods, such as PM, PME, PMW, and CA.

In order to assess the performance of different event-based ensemble landslide susceptibility models, this study used event-based and multi-year landslide inventories for the Shihmen watershed to establish single and ensemble landslide susceptibility models. To assess the robustness of the ensemble methods, we used numerous landslide inventories, rather than a single one, and compared the predictive accuracy of these single models and ensemble modes, established using the same inventory. Additionally, the rainfall-triggering factors were incorporated as independent variables into the landslide susceptibility models, which contributes to the development and improvement of landslide early-warning systems. Apart from comparing the landslide susceptibilities in the different models, this study

also located those areas with high landslide susceptibility with
can provide a reference for decision-making when planning lan

2. Methods

This study first employed a logistic regression model and learning models to establish single landslide susceptibility mod methods to establish ensemble landslide susceptibility models an event-based landslide susceptibility model, based on an e tory, we also combined 24 event-based landslide inventories inventory, to establish a multi-year landslide susceptibility mo operating characteristics (ROC) curve, Spearman's rank corre Whitney test, and the Kruskal–Wallis test to assess the perforı

2.1. Single Landslide Susceptibility Model

2.1.1. Logistic Regression (LR) Model

Because the goal of landslide susceptibility analysis is to will occur in individual slope units, the dependent variables i binary response variables of "landslide" and "no-landslide" developed by Menard [36] was used to establish a paramet which took the form shown in Equation (1):

$$\ln\left(\frac{p_i}{1-p_i}\right) = \alpha_i + \sum_{j=1}^{k} \beta_{ij} x_{ij}.$$

Here, p_i is the probability of landslide occurrence, α_i, β value of the susceptibility factor, i represents different eve susceptibility factors.

2.1.2. Random Forest (RF) Model

The random forest model proposed by Breiman [37] is ble method and establishes multiple decision trees via the subsets. Because random forest models do not require any the relationship between the independent variables and th model is suitable for the analysis of large datasets with non process of establishing different decision trees, the re-sampli selection of variable subsets increase the diversity of the de Chang et al. [12], there are three reasons for random forest m is a form of nonparametric nature-based analysis; (2) it ca the variables used; and (3) it can provide an algorithm for e method has been extensively used in landslide susceptibili has yielded excellent results [8,9,12,16].

2.1.3. Support Vector Machine (SVM) and Kernel Logistic R

Support vector machines, as proposed by Vapnik [40 sification method. Their special property is their ability to geometric margin and minimize the empirical classificati also referred to as maximum margin classifiers [41]. SVMs the hyperplane with the largest margin between two typ dimensional space. A non-linear kernel function can be u a higher dimensional space, where a hyperplane classify Kernel logistic regression is a kernelized version of line method uses a kernel function to project the input data or space, with the goal of finding a discriminant function of c of landslide and no-landslide.

In the two previous models, the most commonly utilized kernel functions consist of linear kernel functions, polynomial kernel functions, radial basis kernel functions (RBF), and sigmoid kernel functions. Of these types, radial basis kernel functions are the most widely used [11] and offer the best predictive ability in most situations, especially in the case of nonlinear data [14]. Radial basis kernel functions are also a very popular choice for the establishment of landslide susceptibility models [43].

2.1.4. Gradient-Boosting Decision Tree (GBDT) Model

The gradient-boosting decision tree (GBDT) model proposed by Friedman et al. [44] is similar to the gradient-boost regression tree (GBRT) and multiple additive regression tree (MART) algorithms. GBDT models combine boosting and regression trees in a single algorithm [41]. Boosting relies on the minimization of the loss function at each tree spilt to improve the decision trees [45] and represents one of the learning methods offering the greatest improvement of model accuracy [17]. Rather than being fitted without any relationship with adjacent trees, GBDT trees are fitted on top of the previous trees.

2.2. *Ensemble Landslide Susceptibility Model*

Referring to Thuiller et al. [32], this study selected PM, PME, PMW, and CA as the ensemble methods used to aggregate the results of the 5 single models, as shown in Table 1. Among these methods, the PM ensemble model calculates the mean of the susceptibility indexes of the single models; the PME ensemble model calculates the median of the susceptibility indexes of the single models; and the PMW ensemble model calculates the weighted mean of the susceptibility indexes of the single models. To set weights, this study assigned weights to each single model, based on the accuracy calibrated by the training-event data. Additionally, the CA ensemble model first identifies the threshold value of each single model, converts the landslide susceptibility index to a binary value (landslide or no-landslide), and calculates the committee average of binary values of the 5 single models.

Table 1. The ensemble methods to aggregate the results of the selected models.

e Methods	Description
M	Mean of susceptibility indexes. The PM ensemble model calculates the mean of the susceptibility indexes for the selected models.
ME	Median of susceptibility indexes. The PME ensemble model calculates the median of the susceptibility indexes for the selected models.
IW	Weighted mean of susceptibility indexes. The PMW ensemble model calculates the relative importance of the weights based on the accuracies of the selected models, and then calculates the weighted mean of the susceptibility indexes for the models.
A	Committee averaging. After identifying the threshold value of each selected model and converting the susceptibility index to binary value, the CA ensemble model calculates the average of binary values for the selected models.

2.3. *Single Model Establishment Process*

2.3.1. Logistic Regression (LR) Model

All slope units with landslides in each landslide inventory are included in the landslide sample, and the no-landslide samples with the same sample number as the landslide sample are also selected. A 10-fold cross-validation is then used to perform model validation. The cross-validation process is repeated 5 times to reduce the error from the split subsets, which yields the mean test accuracy for the models, established from that sample dataset. The foregoing sampling process is repeated 10 times in order to reduce sampling error, and the model with the best mean test accuracy is selected for use in subsequent analysis.

from 236 m to 3526 m, and a slope gradient ranging from 20° to 50° accounts for 77% of the whole area (Figure 1). Because of the relatively well-defined topographic boundaries and topographic meaning in the Shihmen watershed, slope units were employed as analytical units for landslide susceptibility analysis. According to the subdivision method suggested by Xie et al. [52], this watershed was divided into 9181 slope units (Figure 2).

Twelve topographic factors, such as maximum slope, average slope, slope roughness, highest elevation, total slope height, terrain roughness, average elevation, distance from the road, distance from the fault, distance from the river, average aspect, and lithology, were selected as intrinsic susceptibility variables according to the previous study [6]. The values of highest elevation, total slope height, terrain roughness, average elevation, maximum slope, average slope, slope roughness, and average aspect of each slope unit were calculated by employing ArcGIS programs and by utilizing 5 m digital elevation model produced by the Ministry of the Interior. After obtaining the 1:5000 orthophoto base maps issued by the Aerial Survey Office of the Forestry Bureau, the 1:50,000 geologic maps issued by the Central Geological Survey and a road map overlay from the Soil and Water Conservation Bureau, we calculated the horizontal distances of each slope unit from the river, fault, and road, respectively. The lithologic types of each slope unit, such as argillite, quartzitic sandstone, hard sandstone and shale, sandstone and shale, and terrace deposit and alluvium, were analyzed utilizing the 1:50,000 geological maps. The distribution maps of 12 topographic factors are presented in Appendix A.

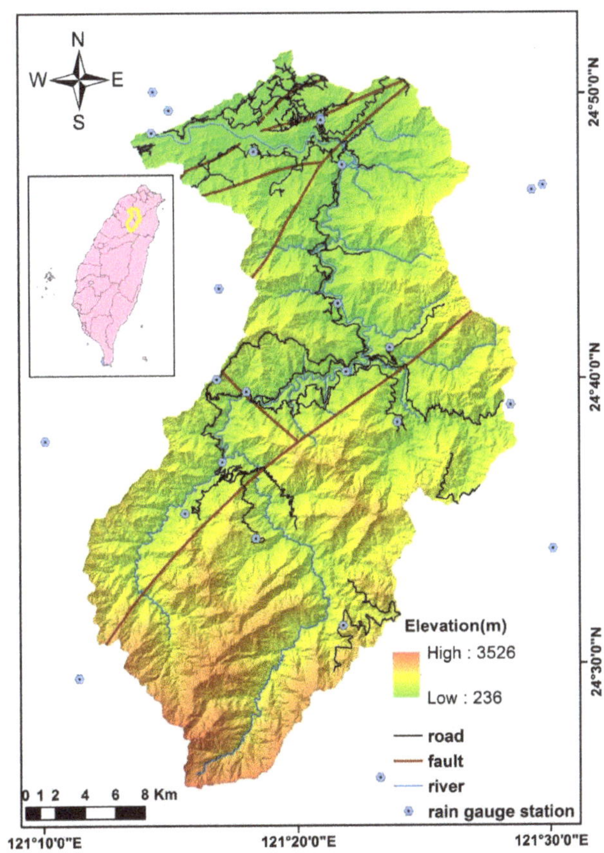

Figure 1. Elevation, road, fault, river system, and rain gauge station in the Shihmen watershed.

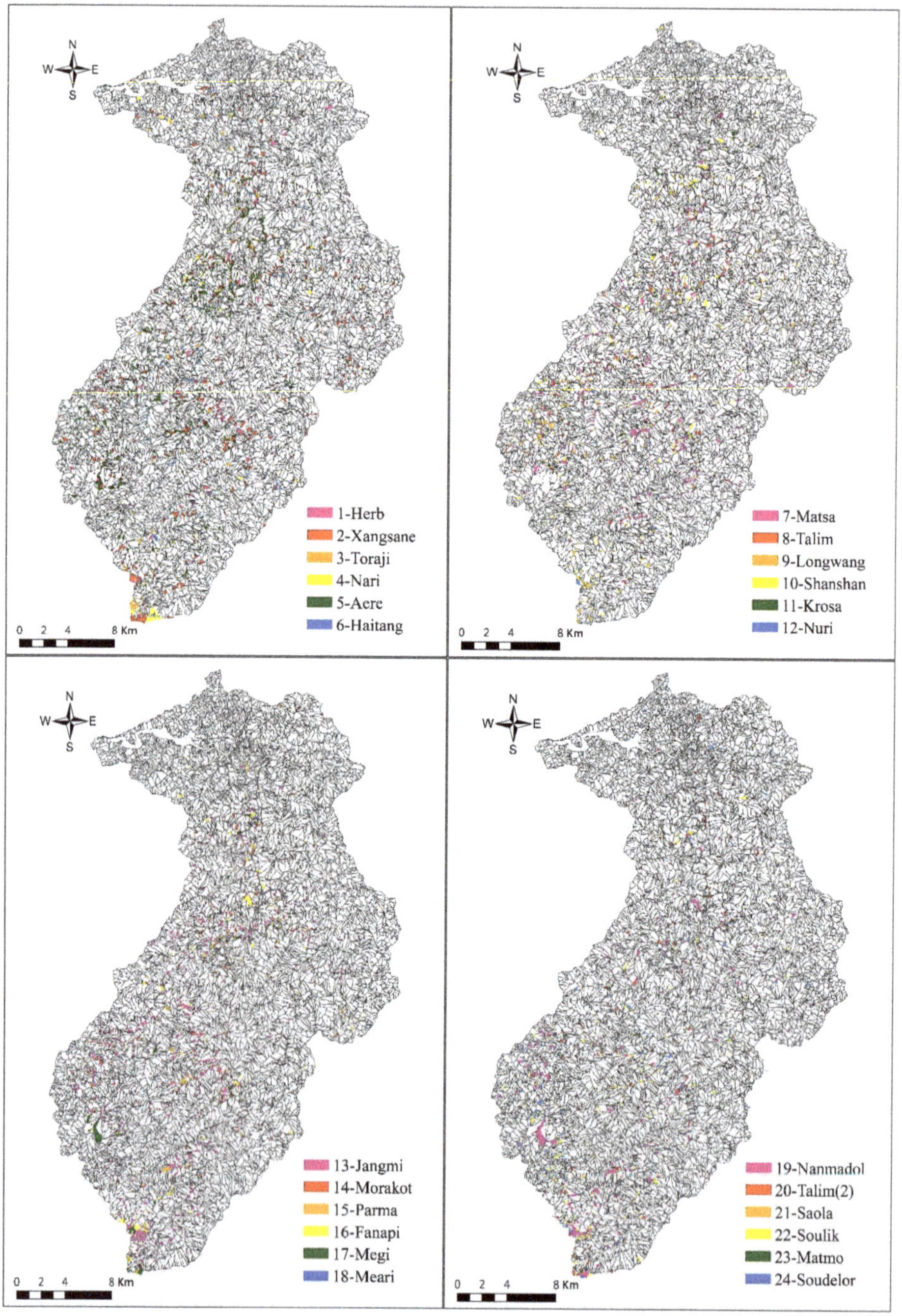

Figure 2. The slope units and landslide inventories, triggered by 24 typhoon events in the Shihmen watershed.

3.2. Landslide Inventory and Rainfall Factor

After collecting 24 sets of satellite images of the Shihmen watershed during the period from 1996 to 2015, landslide inventories triggered by 24 typhoon events were mapped according to the interpretation procedures proposed by Liu et al. [53]. The landslides recorded in the 24 landslide inventories in each slope unit are shown in Figure 2. The number of landslides for each landslide inventory ranged from 59 to 1350 and the total landslide area ranged from 10.19 ha to 577.04 ha (Figure 3).

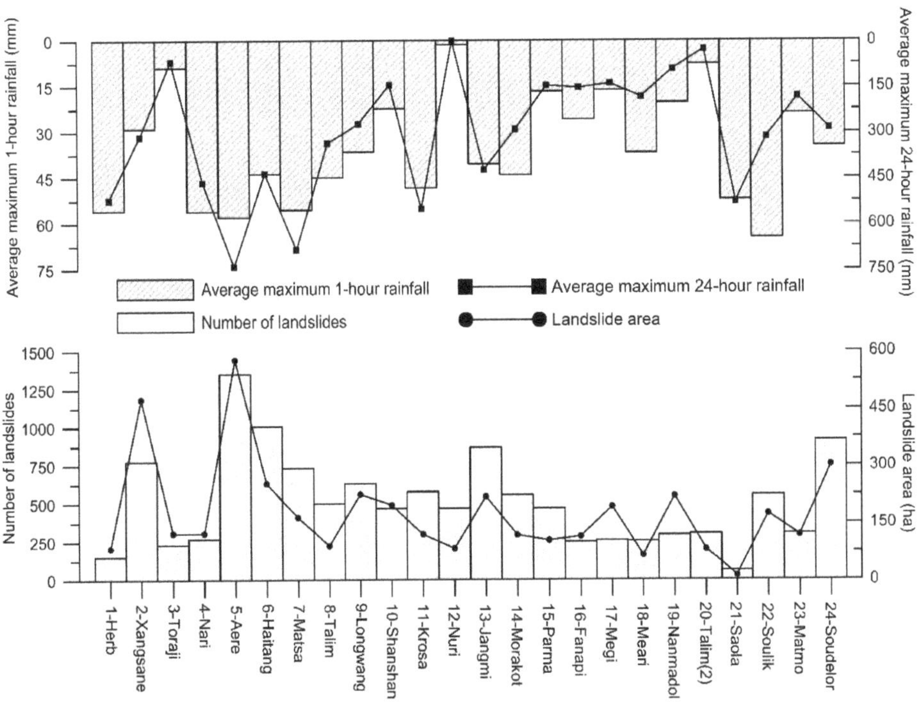

Figure 3. Landslide inventory and rainfall statistics for 24 typhoon events.

Two rainfall factors, namely, maximum 1-h rainfall and maximum 24-h rainfall, were selected as extrinsic triggering variables, according to the previous research [6]. The short-duration rainfall and long-duration rainfall values reflect the rainfall pattern during the typhoon event. After collecting rainfall data from 31 rain-gauge stations (Figure 1), the maximum 1-h rainfall and maximum 24-h rainfall of each station during each typhoon event were analyzed. Then, the rainfall values of each slope unit were calculated, after using the Kriging method to estimate the spatial distribution of rainfall. The average maximum 1-h rainfall and maximum 24-h rainfall for each typhoon event are shown in Figure 3.

4. Results of Analysis

4.1. Results of Single Models

4.1.1. Logistic Regression (LR) Model

This study used LR to establish a parametric landslide susceptibility model. In the modeling process, 10-fold cross-validation was repeatedly used to assess model performance. The repeated application of this process reduced the sampling error and enabled the selection of the model with the best mean test accuracy for subsequent analysis. The test accuracy of 24 event-based logistic regression models (i.e., the AUROC value of the test stage) ranged from 0.740 to 0.862, and the mean accuracy was 0.819 (Table 4). Additionally, the test accuracy of the multi-year logistic regression model was 0.798.

Table 4. Performances of the LR models.

Event	Test Accuracy	Event	Test Accuracy
1-Herb	0.825	14-Morakot	0.832
2-Xangsane	0.776	15-Parma	0.802
3-Toraji	0.740	16-Fanapi	0.815
4-Nari	0.760	17-Megi	0.859
5-Aere	0.824	18-Meari	0.798
6-Haitang	0.789	19-Nanmadol	0.862
7-Matsa	0.832	20-Talim	0.812
8-Talim	0.806	21-Saola	0.815
9-Longwang	0.859	22-Soulik	0.835
10-Shanshan	0.826	23-Matmo	0.835
11-Krosa	0.850	24-Soudelor	0.823
12-Nuri	0.842	Multi-year	0.798
13-Jangmi	0.849		

The 24 event-based logistic regression models established in this study enabled the spatial variation in each event's landslide susceptibility index to be determined. The mean values and standard deviations of the 24 landslide susceptibility indices for each slope unit were then calculated (Figure 4). Similarly, the mean landslide susceptibility indices and standard deviations were calculated for each slope unit in the multi-year logistic regression model.

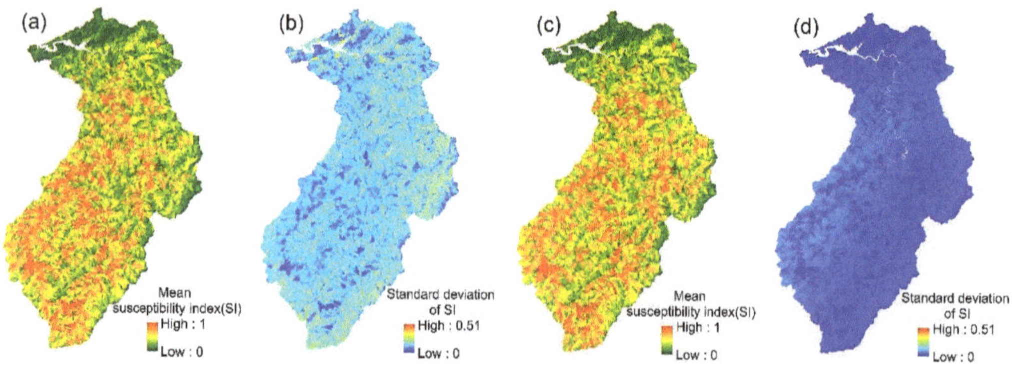

Figure 4. LR models: (**a**,**b**) the mean susceptibility index and standard deviation of 24 event-based models; (**c**,**d**) the mean susceptibility index and standard deviation of the multi-year model.

4.1.2. Random Forest (RF) Model

The hyperparameter tuning results for each event-based model, established using the RF algorithm, are shown in Table 5; it can be seen that the number of trees (numtree) ranged from 100 to 1000 and the number of variables (mtry) ranged from 7 to 14. The test accuracy of the 24 event-based models ranged from 0.772 to 0.944, and the mean was 0.842. Hyperparameter tuning for the multi-year RF model yielded a numtree: 400 and mtry: 14, and the model's test accuracy was 0.789.

Table 5. The tuned hyperparameters and model performances for RF models.

Event	Hyperparameter Tuned		Test Accuracy	Event	Hyperparameter Tuned		Test Accuracy
	Numtree	Mtry			Numtree	Mtry	
1-Herb	100	8	0.797	14-Morakot	300	13	0.815
2-Xangsane	500	14	0.772	15-Parma	1000	14	0.820
3-Toraji	700	11	0.851	16-Fanapi	600	14	0.848
4-Nari	400	13	0.821	17-Megi	200	11	0.872
5-Aere	1000	12	0.820	18-Meari	500	7	0.849
6-Haitang	700	11	0.792	19-Nanmadol	600	14	0.913
7-Matsa	700	12	0.845	20-Talim	600	13	0.824
8-Talim	300	11	0.808	21-Saola	700	13	0.944
9-Longwang	600	13	0.843	22-Soulik	500	11	0.846
10-Shanshan	500	14	0.819	23-Matmo	900	13	0.883
11-Krosa	700	12	0.855	24-Soudelor	400	13	0.827
12-Nuri	1000	14	0.881	Multi-year	400	14	0.789
13-Jangmi	900	13	0.855				

The spatial variation in each event's landslide susceptibility index could be obtained from the 24 event-based RF models. The mean values and standard deviations of the 24 landslide susceptibility indices for each slope unit were then calculated (Figure 5). Similarly, the mean landslide susceptibility indices and standard deviations were calculated for each slope unit in the multi-year RF model.

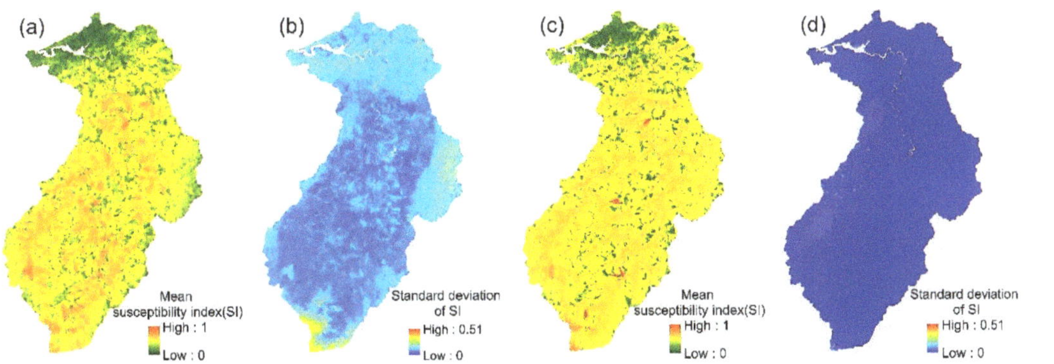

Figure 5. RF models: (**a**,**b**) the mean susceptibility index and standard deviation of 24 event-based models; (**c**,**d**) the mean susceptibility index and standard deviation of the multi-year model.

4.1.3. Support Vector Machine (SVM) Model

The hyperparameter tuning results for each event-based model, established using the SVM algorithm, are shown in Table 6; it can be seen that the penalty parameter (C) ranged from 0.029 to 754.312 and the RBF parameter (γ) ranged from 0.001 to 0.091. The test accuracy of all event-based models ranged from 0.674 to 0.861, and the mean was 0.754. Hyperparameter tuning for the multi-year SVM model yielded a penalty parameter (C) of 0.1 and an RBF parameter (γ) of 0.774, and the model's test accuracy was 0.806.

The spatial variation in each event's landslide susceptibility index could be obtained from the 24 event-based SVM models. The mean values and standard deviations of the 24 landslide susceptibility indices for each slope unit were then calculated (Figure 6). Similarly, the mean landslide susceptibility indices and standard deviations were calculated for each slope unit in the multi-year SVM model.

Table 6. The tuned hyperparameters and model performances for SVM models.

Event	Hyperparameter Tuned		Test Accuracy	Event	Hyperparameter Tuned		Test Accuracy
	C	γ			C	γ	
1-Herb	2.683	0.017	0.721	14-Morakot	0.029	0.029	0.768
2-Xangsane	79.060	0.007	0.717	15-Parma	2.024	0.052	0.766
3-Toraji	0.494	0.091	0.759	16-Fanapi	3.556	0.029	0.772
4-Nari	0.655	0.029	0.755	17-Megi	0.121	0.017	0.786
5-Aere	4.715	0.013	0.732	18-Meari	2.024	0.029	0.787
6-Haitang	33.932	0.001	0.729	19-Nanmadol	1.151	0.017	0.797
7-Matsa	754.312	0.001	0.757	20-Talim	4.715	0.002	0.674
8-Talim	2.024	0.069	0.758	21-Saola	10.985	0.091	0.861
9-Longwang	14.563	0.005	0.754	22-Soulik	19.307	0.002	0.712
10-Shanshan	138.950	0.002	0.741	23-Matmo	1.151	0.017	0.739
11-Krosa	2.024	0.007	0.758	24-Soudelor	0.373	0.007	0.728
12-Nuri	3.556	0.022	0.754	Multi-year	0.1	0.774	0.806
13-Jangmi	184.207	0.001	0.769				

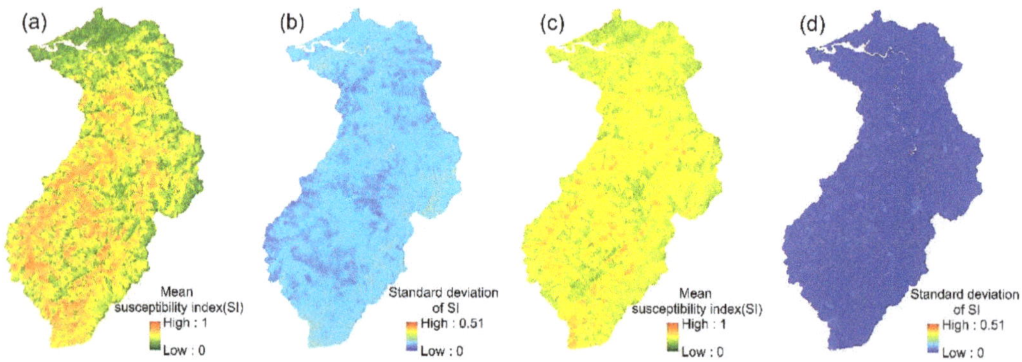

Figure 6. SVM models: (**a**,**b**) the mean susceptibility index and standard deviation of 24 event-based models; (**c**,**d**) the mean susceptibility index and standard deviation of the multi-year model.

4.1.4. Kernel Logistic Regression (KLR) Model

The hyperparameter tuning results for each event-based model established using the KLR algorithm are shown in Table 7; it can be seen that the penalty parameter (C) ranged from 0.017 to 244.205 and the RBF parameter (γ) ranged from 0.002 to 0.281. The test accuracy of every event-based model ranged from 0.712 to 0.833 and the mean was 0.754. Hyperparameter tuning for the multi-year KLR model yielded a penalty parameter (C) of 1.0 and an RBF parameter (γ) of 0.1; the model's test accuracy was 0.812.

The spatial variation in each event's landslide susceptibility index could be obtained from the 24 event-based KLR models. The mean values and standard deviations of the 24 landslide susceptibility indices for each slope unit were then calculated (Figure 7). Similarly, the mean landslide susceptibility indices and standard deviations were calculated for each slope unit in the multi-year KLR model.

Table 7. The tuned hyperparameters and model performances for KLR models.

Event	Hyperparameter Tuned		Test Accuracy	Event	Hyperparameter Tuned		Test Accuracy
	C	γ			C	γ	
1-Herb	244.205	0.005	0.712	14-Morakot	0.017	0.039	0.729
2-Xangsane	59.636	0.005	0.712	15-Parma	2.683	0.017	0.771
3-Toraji	1.151	0.052	0.781	16-Fanapi	33.932	0.013	0.784
4-Nari	10.985	0.017	0.775	17-Megi	2.024	0.069	0.815
5-Aere	1.151	0.022	0.760	18-Meari	33.932	0.007	0.725
6-Haitang	1.526	0.029	0.717	19-Nanmadol	8.286	0.003	0.829
7-Matsa	244.205	0.002	0.747	20-Talim	1.151	0.069	0.712
8-Talim	1.526	0.069	0.744	21-Saola	14.563	0.281	0.833
9-Longwang	3.556	0.039	0.765	22-Soulik	244.205	0.001	0.730
10-Shanshan	0.494	0.029	0.745	23-Matmo	1.526	0.039	0.737
11-Krosa	33.932	0.004	0.746	24-Soudelor	8.286	0.005	0.737
12-Nuri	59.636	0.029	0.723	Multi-year	1.0	0.1	0.812
13-Jangmi	3.556	0.029	0.767				

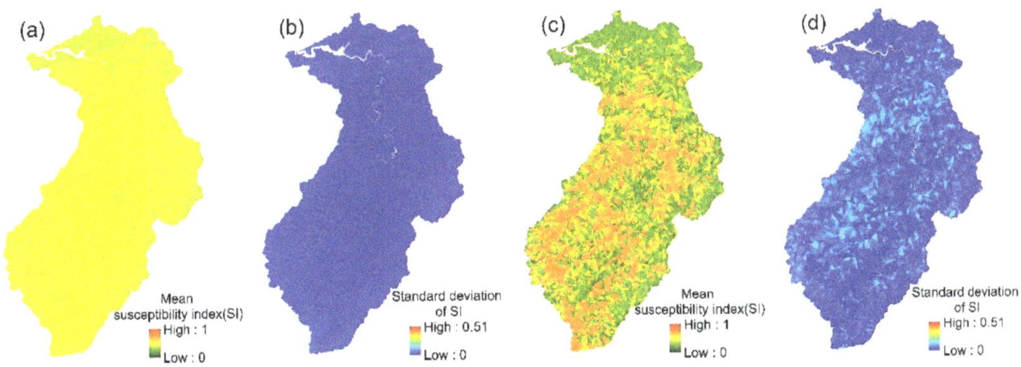

Figure 7. KLR models: (**a**,**b**) the mean susceptibility index and standard deviation of 24 event-based models; (**c**,**d**) the mean susceptibility index and standard deviation of the multi-year model.

4.1.5. Gradient-Boosting Decision Tree (GBDT) Model

The hyperparameter tuning results for each event-based model established using the GBDT algorithm are shown in Table 8, and it can be seen that the number of trees (numtree) ranged from 100 to 1000, the number of variables (mtry) ranged from 6 to 14, and the learning rate ranged from 0.1 to 1.0. The test accuracy of the 24 event-based models ranged from 0.772 to 0.861 and the mean was 0.820. Hyperparameter tuning for the multi-year GBDT model yielded a numtree of 900, an mtry of 7, and a learning rate of 0.1; the model's test accuracy was 0.804.

The spatial variation in each event's landslide susceptibility index could be obtained from the 24 event-based GBDT models. The mean values and standard deviations of the 24 landslide susceptibility indices for each slope unit were then calculated (Figure 8). Similarly, the mean landslide susceptibility indices and standard deviations were calculated for each slope unit in the multi-year GBDT model.

Table 8. The tuned hyperparameters and model performances for GBDT models.

Event	Hyperparameter Tuned			Test Accuracy	Event	Hyperparameter Tuned			Test Accuracy
	Numtree	Mtry	Learning Rate			Numtree	Mtry	Learning Rate	
1-Herb	100	8	0.1	0.800	14-Morakot	200	8	0.1	0.833
2-Xangsane	700	13	0.9	0.772	15-Parma	300	13	0.5	0.823
3-Toraji	100	8	0.1	0.777	16-Fanapi	100	13	0.5	0.815
4-Nari	300	11	0.6	0.807	17-Megi	1000	13	0.8	0.848
5-Aere	600	14	0.3	0.821	18-Meari	1000	14	0.4	0.788
6-Haitang	200	11	0.7	0.772	19-Nanmadol	1000	10	0.7	0.852
7-Matsa	100	12	1.0	0.837	20-Talim	100	13	0.7	0.807
8-Talim	100	9	0.3	0.817	21-Saola	100	7	0.3	0.832
9-Longwang	200	6	0.1	0.845	22-Soulik	200	6	0.2	0.839
10-Shanshan	200	6	0.9	0.815	23-Matmo	100	6	1.0	0.843
11-Krosa	200	11	0.5	0.841	24-Soudelor	100	6	0.1	0.811
12-Nuri	1000	9	0.3	0.827	Multi-year	900	7	0.1	0.804
13-Jangmi	100	12	0.1	0.861					

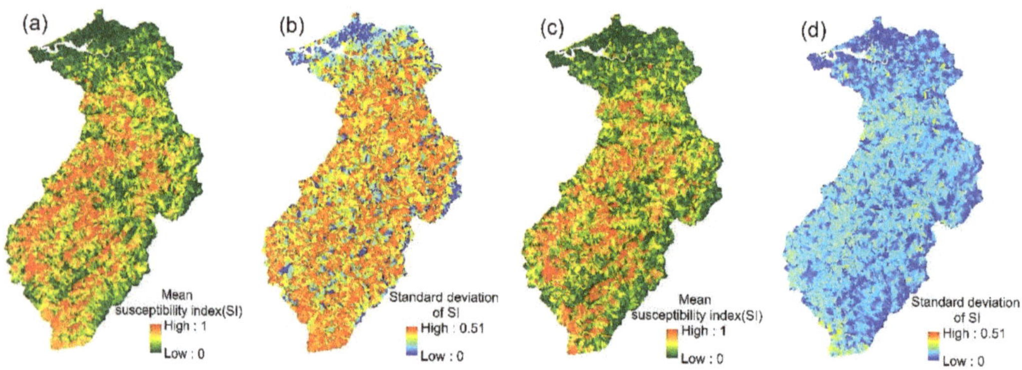

Figure 8. GBDT models: (**a**,**b**) the mean susceptibility index and standard deviation of 24 event-based models; (**c**,**d**) the mean susceptibility index and standard deviation of the multi-year model.

4.2. Results of Ensemble Models

After establishing the single models, this study used the PM, PME, PMW, and CA ensemble methods to aggregate the landslide susceptibility indices of each single model for each event, yielding the landslide susceptibility indices of the 4 ensemble models.

The spatial variation in each event's landslide susceptibility index could be obtained from the 24 event-based PM ensemble models. The mean values and standard deviations of the 24 landslide susceptibility indices for each slope unit were then calculated (Figure 9). The mean landslide susceptibility indices and standard deviations that were calculated for each slope unit in the multi-year PM ensemble model are shown in Figure 9. Similarly, the mean susceptibility index and the standard deviation of the 24 event-based models, obtained using the PME, PMW, and CA ensemble methods, are shown in Figures 10–12.

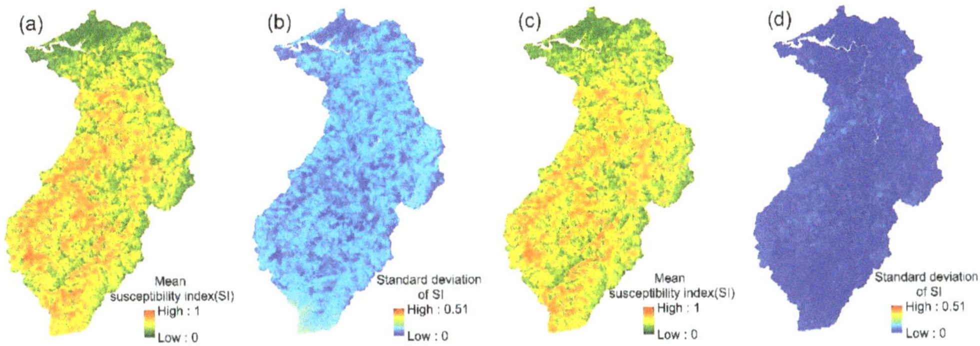

Figure 9. PM ensemble models: (**a**,**b**) the mean susceptibility index and standard deviation of 24 event-based models; (**c**,**d**) the mean susceptibility index and standard deviation of the multi-year model.

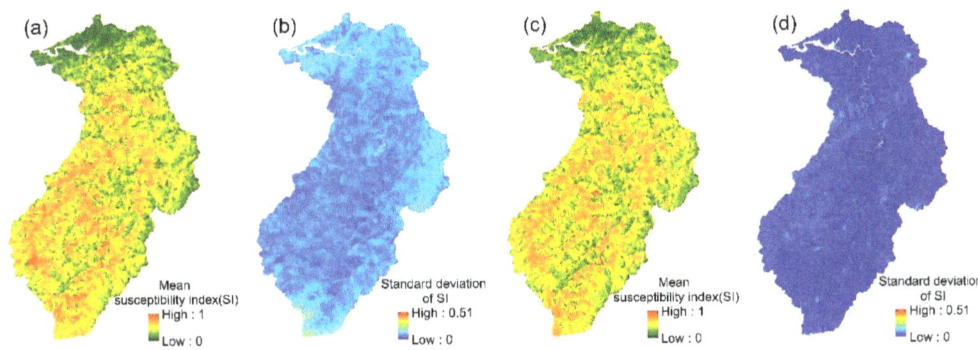

Figure 10. PME ensemble models: (**a**,**b**) the mean susceptibility index and standard deviation of 24 event-based models; (**c**,**d**) the mean susceptibility index and standard deviation of the multi-year model.

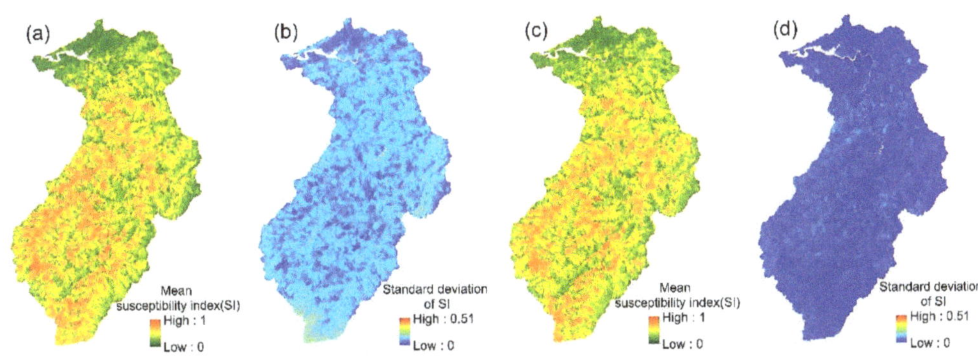

Figure 11. PMW ensemble models: (**a**,**b**) the mean susceptibility index and standard deviation of 24 event-based models; (**c**,**d**) the mean susceptibility index and standard deviation of the multi-year model.

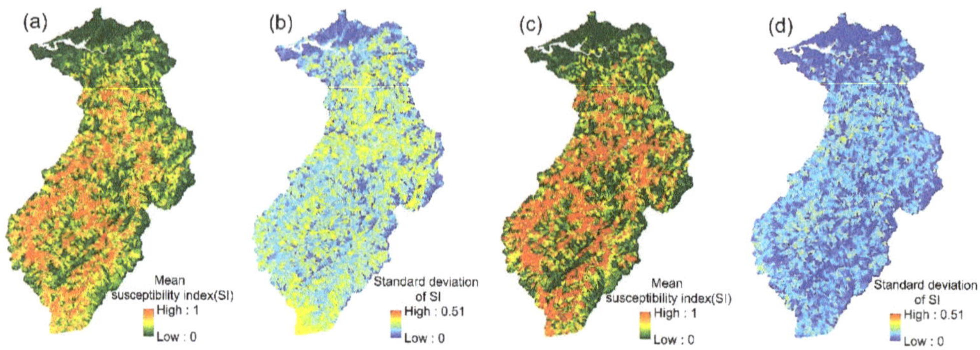

Figure 12. CA ensemble models: (**a**,**b**) the mean susceptibility index and standard deviation of 24 event-based models; (**c**,**d**) the mean susceptibility index and standard deviation of the multi-year model.

4.3. Assessment of Model Accuracy

After establishing all the above-mentioned models, this study assessed the predictive ability of each model established using a specific landslide inventory by examining that model's ability to predict the remaining landslide events. For the sake of clarity, the section below will employ the terminology $a_{i,j,k}$ to indicate the accuracy of the various models established using different landslide inventories and different modeling methods. Here, i = 1–25 indicate the individual landslide inventories used in modeling, where 1–24 are event-based landslide inventories, and 25 is the multi-year landslide inventory; j = 1–24 indicate the predicted events; and k = 1–9 indicate the different modeling methods.

The accuracy of each PM ensemble model is shown in Table 9; AUROC > 75% is indicated in green, AUROC 75–50% is indicated in yellow, and AUROC < 50% is indicated in red. The average predictive accuracy of event-based models (i = 1–24, j = 1–24, $i \neq j$, k = 6) ranged from 70.9% to 77.9% and the mean was 74.8%; the average predictive accuracy of the multi-year model (i = 25, j = 1–24, k = 6) was 91.1%. The average predictive accuracy of the LR models (k = 1), RF models (k = 2), SVM models (k = 3), KLR models (k = 4), GBDT models (k = 5), PME ensemble models (k = 7), PMW ensemble models (k = 8), and CA ensemble models (k = 9) were also obtained.

The average predictive accuracy of the 5 event-based single landslide susceptibility models (k = 1–5) with regard to the other landslide events is shown in Figure 13. In Figure 13, from top down, the various symbols represent the maximum, third quartile, median, first quartile, and minimum of a box plot of average predictive accuracy. This figure also shows the predictive accuracy distribution of the multi-year single landslide susceptibility models. In particular, the average predictive accuracy of the event-based LR models (i = 1–24, j = 1–24, $i \neq j$, k = 1) ranged from 48.8% to 76.1%, and the mean was 71.2%; the multi-year LR model had a mean predictive accuracy of 78.8%. Similarly, the average predictive accuracy of event-based RF models ranged from 57.9% to 74.7%, and the mean was 69.5%; the multi-year RF model had a mean predictive accuracy of 79.5%. The average predictive accuracy of event-based SVM models ranged from 50.0% to 76.1%, and the mean was 68.1%; the multi-year SVM model had a mean predictive accuracy of 88.0%. The average predictive accuracy of event-based KLR models ranged from 50.0% to 76.4% and the mean was 67.4%; the multi-year KLR model had a mean predictive accuracy of 88.6%. The average predictive accuracy of event-based GBDT models ranged from 69.2% to 76.3%, while the mean was 72.8%; the multi-year GBDT model had a mean predictive accuracy of 94.3%.

The average predictive accuracy of the 4 event-based ensemble landslide susceptibility models (k = 6–9) with regard to the other landslide events is shown in Figure 13, which also shows the predictive accuracy distribution of the multi-year ensemble landslide susceptibility models. In Figure 13, the average predictive accuracy of event-based PME models (i = 1–24, j = 1–24, $i \neq j$, k = 7) ranged from 67.0% to 77.5% and the mean was 73.8%; the multi-year PME model had a mean predictive accuracy of 89.2%. The average predictive accuracy of event-based PMW models ranged from 70.9% to 77.8%, and the mean was 74.7%; the multi-year PMW model had a mean predictive accuracy of 91.6%. The average predictive accuracy of event-based CA models ranged from 72.7% to 77.1%, while the mean was 74.8%; the mean predictive accuracy of the multi-year CA model was 89.0%. These results indicate that the event-based ensemble models all had an AUROC > 50% with regard to other landslide events (i = 1–24, j = 1–24, $i \neq j$, k = 6–9).

Table 9. AUROCs (%) of each PM ensemble model for the calibration or prediction of other landslide events. AUROC > 75% is indicated in green, AUROC 75–50% is indicated in yellow.

		Event for Calibration or Prediction																							
		1	2	3	4	5	6	7	8	9	10	11	12	13	14	15	16	17	18	19	20	21	22	23	24
Model trained by event	1	90	71	70	67	79	67	75	74	76	77	75	77	78	75	76	75	78	76	78	73	74	69	76	67
	2	74	90	75	78	71	74	70	76	81	79	78	78	75	73	79	79	83	76	81	75	80	73	78	77
	3	70	75	92	75	71	74	73	74	78	78	76	74	79	75	77	77	81	74	81	72	78	76	78	74
	4	71	75	72	92	65	71	67	70	77	76	78	75	77	69	77	76	79	68	79	70	80	69	73	72
	5	80	77	74	74	91	74	81	78	81	82	78	79	83	79	77	78	83	78	82	74	73	75	80	72
	6	67	73	69	74	70	90	76	82	75	77	78	78	78	76	76	72	79	76	77	74	79	76	75	72
	7	72	73	68	69	79	76	91	78	77	76	79	77	78	77	76	72	76	72	77	70	76	72	72	70
	8	64	70	65	65	73	74	72	91	75	74	76	78	70	73	72	67	68	75	77	72	65	72	72	68
	9	76	75	72	75	78	75	76	78	93	79	82	80	78	74	78	77	78	78	84	75	79	78	80	79
	10	76	75	74	75	80	72	76	75	82	90	80	80	81	77	79	77	84	77	79	73	77	76	79	73
	11	74	74	67	74	74	74	76	77	81	79	92	74	79	77	78	71	79	73	69	68	75	67	69	71
	12	71	70	70	71	72	70	72	72	75	73	73	90	74	71	74	77	79	73	76	70	77	72	76	72
	13	77	77	74	76	78	75	80	79	78	79	82	75	92	79	77	75	75	74	76	71	82	79	77	64
	14	72	70	66	68	78	72	79	78	76	78	78	78	77	90	74	69	76	77	70	71	69	68	73	65
	15	69	76	72	77	69	69	68	68	80	81	78	82	74	69	91	77	84	73	82	75	77	67	77	77
	16	67	71	71	69	74	66	72	63	71	74	74	72	70	69	73	93	81	69	69	67	77	63	76	74
	17	72	73	73	74	76	63	71	69	79	79	72	79	71	75	78	78	93	74	81	73	70	73	76	78
	18	72	71	69	69	68	72	70	74	79	77	71	78	73	71	76	72	74	89	80	75	79	77	79	78
	19	72	72	70	74	74	74	76	74	78	77	72	79	74	72	78	74	81	73	93	74	83	79	80	77
	20	69	70	63	68	71	73	72	73	76	73	71	79	72	70	70	72	75	75	79	89	80	74	76	74
	21	66	74	70	72	69	74	73	75	76	77	74	79	80	73	76	74	80	73	81	73	91	79	77	75
	22	72	72	67	72	75	73	74	76	77	77	71	80	80	76	74	72	78	80	85	75	83	92	81	79
	23	72	73	70	71	76	71	72	73	79	78	73	79	76	75	75	76	80	77	82	74	75	80	92	81
	24	74	74	69	72	75	75	75	77	82	80	77	81	77	77	78	76	83	80	82	76	80	81	82	89
	25	90	88	89	90	92	89	92	90	92	91	91	94	93	90	91	89	94	91	95	91	92	93	92	89

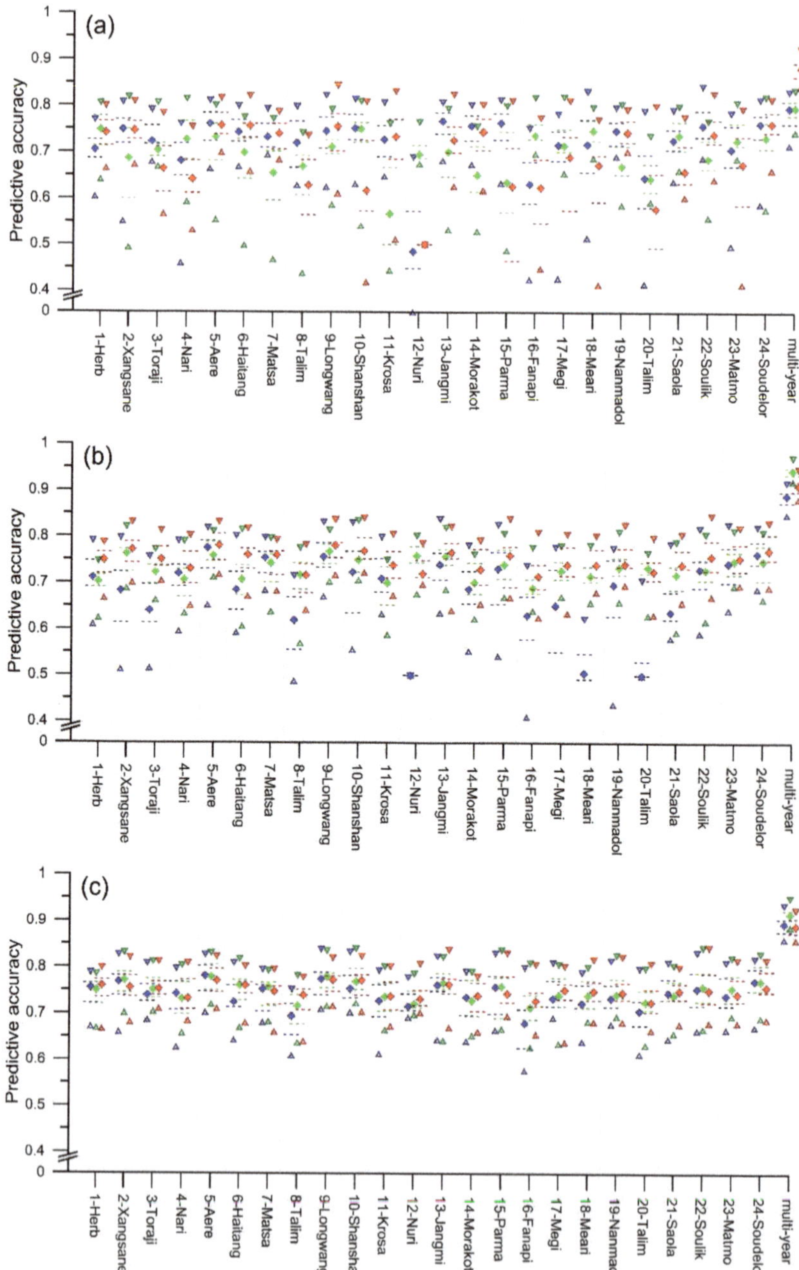

Figure 13. Box plot of the average predictive accuracy of single and ensemble models in the prediction of other landslide events: (**a**) blue, green, and red represent LR, RF, and SVM, respectively; (**b**) blue, green, and red represent KLR, GBDT, and PM, respectively; (**c**) blue, green, and red represent PME, PMW, and CA, respectively.

5. Discussion

5.1. Comparison of the Performance of Single and Ensemble Models

It can be seen from Figure 13 that the predictive accuracy of the ensemble models is superior to that of single models under most circumstances. In particular, the four ensemble models established on the basis of event-based landslide inventories 1, 3, 4, 9, 10, 17, and 21, as well as the multi-year landslide inventory all had greater predictive accuracy than any single models. In addition, the four ensemble models established on the basis of the event-based landslide inventories 2, 5, 7, 11, 12, 13, 15, 16, 18, 20, and 22 had greater predictive accuracy than at least any four single models. In other words, when establishing a model using the same landslide inventory, most ensemble models will offer superior predictive accuracy.

The mean predictive accuracy of different modeling methods (k = 1–9) are compared in Table 10. It can be seen that the mean predictive accuracy of ensemble models (k = 6–9) ranged from 0.738 to 0.748 and was higher than the accuracy range of 0.674–0.728 for single models (k = 1–5). Since the Kolmogorov–Smirnov test indicated that not all datasets were normally distributed, the Kruskal–Wallis test was used to compare the predictive accuracy of different modeling methods (Table 11). The post hoc test indicated that the predictive accuracy of ensemble models is consistently superior to that of single models. Furthermore, the coefficient of variation (CV) of the predictive accuracy of ensemble models ranged from 0.047 to 0.063, which was lower than the CV range for single models. In summary, our results show that ensemble landslide susceptibility models offer superior predictive ability and relatively low uncertainty.

Table 10. Performance assessment of the different modeling methods.

Inventory Type	Metric	LR(1)	RF(2)	SVM(3)	KLR(4)	GBDT(5)	PM(6)	PME(7)	PMW(8)	CA(9)
Event-based	Mean training accuracy	0.813	0.878	0.833	0.856	0.977	0.909	0.883	0.912	0.923
	CV of training accuracy	0.038	0.028	0.054	0.037	0.011	0.013	0.017	0.013	0.019
	Mean predictive accuracy	0.712	0.695	0.681	0.674	0.728	0.748	0.738	0.747	0.748
	CV of predictive accuracy	0.118	0.104	0.146	0.142	0.059	0.055	0.063	0.055	0.047
Multi-year	Mean predictive accuracy	0.788	0.795	0.880	0.886	0.943	0.911	0.892	0.916	0.890
	CV of predictive accuracy	0.040	0.029	0.026	0.021	0.014	0.019	0.022	0.018	0.022

Table 11. Kruskal–Wallis test of the predictive accuracy of different modeling methods.

	N	Mean Rank	d.f.	H	p	Post Hoc Test
LR(1)	576	2481.79	8	514.142	0.000	>2–4
RF(2)	576	2013.05				-
SVM(3)	576	2080.67				-
KLR(4)	576	1924.60				-
GBDT(5)	576	2571.61				>2–4
PM(6)	576	3134.95				>1–5
PME(7)	576	2873.52				>1–5
PMW(8)	576	3120.86				>1–5
CA(9)	576	3131.45				>1–5

Prior studies have demonstrated that the predictive ability of the landslide susceptibility models established by different ensemble methods was superior to that of single landslide susceptibility models [16,31,34,35]. In accordance with the previous study results, we found that most ensemble models were superior in terms of predictive accuracy to the single models developed with the same inventory. Moreover, this study used 24 inventories to establish the corresponding ensemble models. The higher predictive ability of the ensemble models for each inventory implied that the PM, PME, PMW, and CA ensemble methods were robust for enhancing the predictive performance of landslide susceptibility models in the study area.

Among the single models, while LR models had the lowest mean training accuracy (i = 1–24, j = 1–24, i = j, k = 1), their mean predictive accuracy of 0.712 (i = 1–24, j = 1–24, $i \neq j$, k = 1) was higher than that of the RF, SVM, and KLR models. Although the RF, SVM, and KLR models had very good mean training accuracy, their mean predictive accuracy was poor; this may be because these nonparametric models require a greater quantity of data for training and are prone to overfitting [2,8].

5.2. Comparison of the Performance of Event-Based and Multi-Year Models

It can be seen from Figure 13 that among the nine modeling methods, the predictive accuracy of multi-year models is consistently superior to that of the 24 event-based models. Table 10 also reveals that the mean predictive accuracy of multi-year models (i = 25, j = 1–24) ranged from 0.788 to 0.943, which was higher than the values of 0.674–0.748 in the event-based models (i = 1–24, j = 1–24, $i \neq j$). The results of the Mann–Whitney U test (Table 12) indicate that among the nine modeling methods, the predictive performance of models established based on multi-year landslide inventories is uniformly superior to that of event-based models. Furthermore, the CV of the multi-year models' predictive accuracy (0.014–0.040) was lower than that of event-based models. In summary, multi-year landslide susceptibility models offer excellent predictive performance and low uncertainty.

Table 12. Mann–Whitney U test of the predictive accuracy of models, based on different types of landslide inventories.

Modeling Method	Inventory Type	N	Mean Rank	Sum of	U	p
LR (1)	Event-based	552	279.95	154,531.00	1903.000	0.000
	Multi-year	24	485.21	11,645.00		
RF (2)	Event-based	552	277.78	153,332.00	704.000	0.000
	Multi-year	24	535.17	12,844.00		
SVM (3)	Event-based	552	276.50	152,629.00	1.000	0.000
	Multi-year	24	564.46	13,547.00		
KLR (4)	Event-based	552	276.50	152,628.00	0.000	0.000
	Multi-year	24	564.50	13,548.00		
GBDT (5)	Event-based	552	276.50	152,628.00	0.000	0.000
	Multi-year	24	564.50	135,48.00		
PM (6)	Event-based	552	276.50	152,628.00	0.000	0.000
	Multi-year	24	564.50	13,548.00		
PME (7)	Event-based	552	276.50	152,628.00	0.000	0.000
	Multi-year	24	564.50	13,548.00		
PMW (8)	Event-based	552	276.50	152,628.00	0.000	0.000
	Multi-year	24	564.50	13,548.00		
CA (9)	Event-based	552	276.50	152,628.00	0.000	0.000
	Multi-year	24	564.50	13,548.00		

The findings of the current study that the relatively excellent predictive performance and low uncertainty of the landslide susceptibility models established using multi-year landslide inventories verifies the advantage of using a combination of event-based inventories and confirms the previous study results. The relatively high predictive abilities of the landslide susceptibility models, built by the combination of different event-based landslide inventories, have been thought to be related to their bigger landslide sample size and the wider numerical range of rainfall parameters in the training sample [23,54,55], or to their lower concentration of landslides in areas with the same lithology and a lower collinearity between rainfall parameters and lithology [56,57].

It can also be seen from Figure 13 that when applying the same modeling method, the predictive accuracy of event-based models depends on the choice of event. For example, in terms of the average predictive accuracy of the 24 event-based LR models, a maximum of 76.1% appeared when employing the event in Jangmi, and a minimum of 48.8% was obtained when using the event in Nuri, while the range is 27.30%. Similarly, the ranges of the average predictive accuracy of the 24 event-based RF, SVM, KLR, GBDT, PM, PME, PMW, and CA models are 16.80%, 26.10%, 26.40%, 7.10%, 7.00%, 10.50%, 6.90%, and 4.40%, respectively. These results confirmed the findings of previous studies that the choice of event has an influence on the predictive ability of the event-based landslide susceptibility model established [22–24], which may correlate with the event's rainfall intensity range [3] and the event's spatial concentration degree of landslides [56,57].

5.3. Correlations between the Susceptibility Maps of the Optimal Model and Other Models

It can be seen from Figures 4–12 that when a specific modeling method is used, the high susceptibility areas in landslide susceptibility maps, based on multi-year models, are similar to those created using the mean susceptibility indices of 24 event-based models. Nevertheless, there are significant differences in the landslide susceptibility index range and standard deviation among the different modeling methods. Because PM ensemble models have optimal predictive accuracy (Table 11) and multi-year models are superior to event-based models (Table 12), this study considered the multi-year PM ensemble model to be the optimal landslide susceptibility model. This model's landslide susceptibility map is the most representative and can best reflect the probabilities of landslides in different slope units of the research area.

We also compared these landslide susceptibility maps, in order to analyze the correlations in the spatial distribution of susceptibility index between the multi-year PM ensemble model and other models. Rather than performing the mutual subtraction algorithm [6,58–60] or the histogram matching method [55,56], we calculated the Spearman's rank correlation coefficient to assess the degree of difference between the susceptibility maps of the optimal model and other models. As shown in Table 13, the correlation coefficient of the single models ranged from 0.811 to 0.946, with an average of 0.912, which was lower than the 0.940–1.00 correlation coefficient range of the ensemble models. This indicates that there are only relatively small differences between the susceptibility maps of the optimal model and other ensemble models. In addition, when a multi-year landslide inventory is not available, the fact that the average correlation coefficient of the single models was lower than the average correlation coefficient of the ensemble models indicates that ensemble models can effectively reduce the discrepancies between the susceptibility maps of the established models and the optimal model.

Table 13. Degree of difference between the susceptibility maps of the multi-year PM ensemble model and other models.

Susceptibility Map	Spearman's Rank Correlation Coefficient	Susceptibility Map	Spearman's Rank Correlation Coefficient
Event-based LR model	0.924	Multi-year LR model	0.928
Event-based RF model	0.917	Multi-year RF model	0.913
Event-based SVM model	0.938	Multi-year SVM model	0.811
Event-based KLR model	0.946	Multi-year KLR model	0.864
Event-based GBDT model	0.938	Multi-year GBDT model	0.936
Event-based PM ensemble model	0.962	Multi-year PM ensemble model	1.000
Event-based PME ensemble model	0.954	Multi-year PME ensemble model	0.990
Event-based PMW ensemble model	0.963	Multi-year PMW ensemble model	1.000
Event-based CA ensemble model	0.940	Multi-year CA ensemble model	0.950

6. Conclusions

This study collected 24 event-based landslide inventories for the Shihmen watershed and employed logistic regression, random forest, support vector machine, kernel logistic regression, and gradient boosting decision tree methods to establish event-based single landslide susceptibility models. We also used four ensemble methods to aggregate the results of single models, to establish event-based ensemble models. In addition, the 24 event-based landslide inventories were combined to form a multi-year landslide inventory, which was used to establish multi-year single landslide susceptibility models and multi-year ensemble models.

As shown in Tables 10–12, the current study found that an ensemble model based on a multi-year inventory can achieve excellent predictive accuracy. Compared with event-based models, multi-year landslide susceptibility models offer superior predictive ability and lower uncertainty; compared with single models, ensemble landslide susceptibility models have higher predictive ability and lower uncertainty for each inventory, implying that the four ensemble methods are robust for enhancing the model's predictive performance in the study area.

When relying on an event-based landslide inventory instead of a multi-year inventory to establish a model, the predictive accuracy of single models has considerable uncertainty due to differences in the predicted landslide events. The ensemble models can both reduce uncertainty and achieve better predictive accuracy, while the established PM ensemble models are the most effective of all. The susceptibility map created using the 24 PM ensemble models, based on different event-based landslide inventories, revealed areas where landslides are likely to occur. High-priority landslide mitigation measures should be implemented in places with a high mean susceptibility index and a low variation in susceptibility index to effectively reduce the losses caused by the landslides.

We recommend that other modeling methods, such as neural networks and deep learning, be further employed to establish landslide susceptibility models. When there are large numbers of single models, other ensemble methods, such as the confidence interval of the mean susceptibility index, may be used to establish even more effective ensemble landslide susceptibility models. Finally, due to the influence of the choice of event on the predictive ability of an event-based model and the better predictive ability of the models built by the combination of different event-based landslide inventories, future research can investigate possible improvements in predictive ability by combining two different event-based inventories to create an ensemble model when researchers are lacking a multi-year inventory.

Author Contributions: Conceptualization, C.-Y.W.; data curation, S.-Y.L.; formal analysis, S.-Y.L.; methodology, C.-Y.W. and S.-Y.L.; supervision, C.-Y.W.; visualization, S.-Y.L.; writing—original draft, C.-Y.W. and S.-Y.L.; writing—review and editing, C.-Y.W.; funding acquisition, C.-Y.W. All authors have read and agreed to the published version of the manuscript.

Funding: This research was funded by the Ministry of Science and Technology, Taiwan (MOST 108-2625-M-005-003).

Acknowledgments: The authors would like to thank the Soil and Water Conservation Bureau, COA, and Water Resources Agency, MOEA, Taiwan for providing data, and C.L. Hsueh for her work on the event-based landslide inventories.

Conflicts of Interest: The authors declare no conflict of interest.

Appendix A

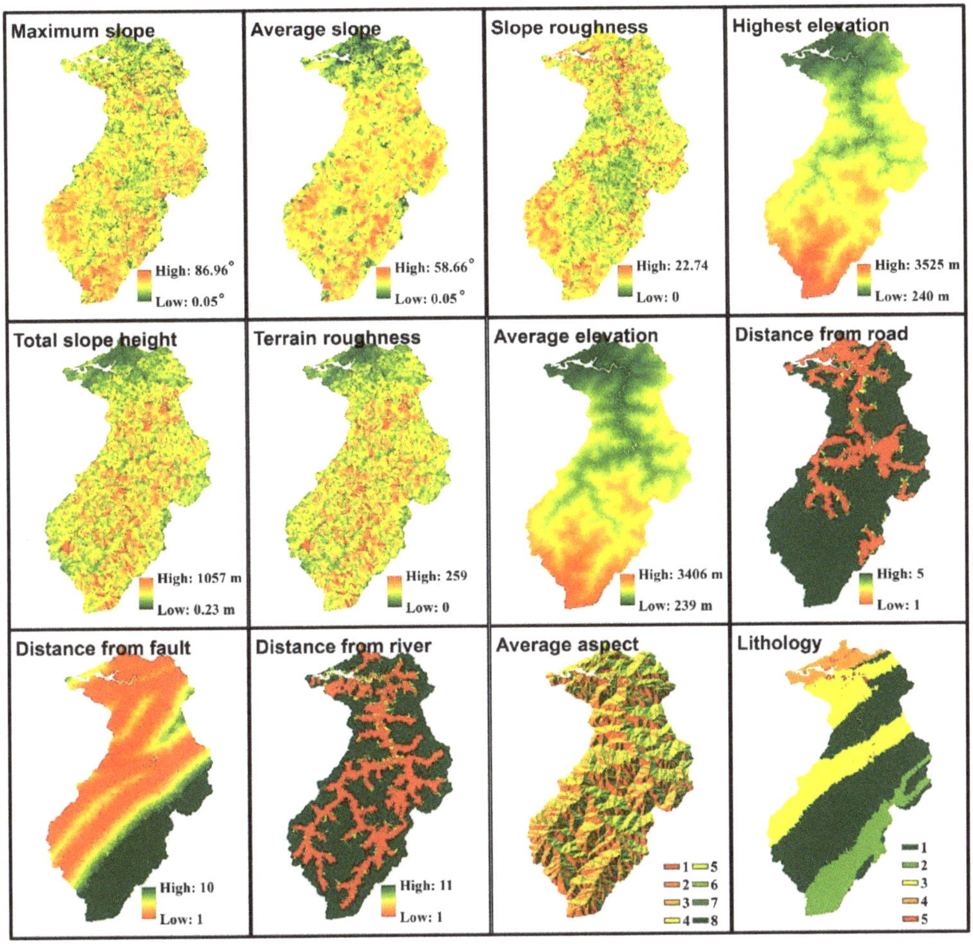

Figure A1. Maps of 12 topographic factors.

References

1. Van Westen, C.; Van Asch, T.W.; Soeters, R. Landslide hazard and risk zonation—Why is it still so difficult? *Bull. Eng. Geol. Environ.* **2006**, *65*, 167–184. [CrossRef]
2. Brownlee, J. Parametric and Nonparametric Machine Learning Algorithms. Machine Learning Mastery. Available online: https://machinelearningmastery.com/parametric-and-nonparametric-machine-learning-algorithms (accessed on 12 January 2021).
3. Lee, C.T.; Huang, C.C.; Lee, J.F.; Pan, K.L.; Lin, M.L.; Dong, J.J. Statistical approach to storm event-induced landslides susceptibility. *Nat. Hazards Earth Syst. Sci.* **2008**, *8*, 941–960. [CrossRef]
4. Lu, A.; Haung, W.K.; Lee, C.F.; Wei, L.W.; Lin, H.H.; Chi, C. Combination of Rainfall Thresholds and Susceptibility Maps for Early Warning Purposes for Shallow Landslides at Regional Scale in Taiwan. In *Workshop on World Landslide Forum*; Casagli, N., Tofani, V., Sassa, K., Bobrowsky, P.T., Takara, K., Eds.; Springer: Cham, Germany, 2020; pp. 217–225.
5. Rossi, M.; Guzzetti, F.; Reichenbach, P.; Mondini, A.C.; Peruccacci, S. Optimal landslide susceptibility zonation based on multiple forecasts. *Geomorphology* **2010**, *114*, 129–142. [CrossRef]
6. Wu, C.Y.; Chen, S.C. Integrating spatial, temporal, and size probabilities for the annual landslide hazard maps in the Shihmen watershed, Taiwan. *Nat. Hazards Earth Syst. Sci.* **2013**, *13*, 2353–2367. [CrossRef]
7. Wu, C. Landslide Susceptibility Based on Extreme Rainfall-Induced Landslide Inventories and the Following Landslide Evolution. *Water* **2019**, *11*, 2609. [CrossRef]

8. Zhao, L.; Wu, X.; Niu, R.; Wang, Y.; Zhang, K. Using the rotation and random forest models of ensemble learning to predict landslide susceptibility. *Geomat. Nat. Hazards Risk* **2020**, *11*, 1542–1564. [CrossRef]
9. Brock, J.; Schratz, P.; Petschko, H.; Muenchow, J.; Micu, M.; Brenning, A. The performance of landslide susceptibility models critically depends on the quality of digital elevation models. *Geomat. Nat. Hazards Risk* **2020**, *11*, 1075–1092. [CrossRef]
10. Bui, D.T.; Le, K.T.T.; Nguyen, V.C.; Le, H.D.; Revhaug, I. Tropical forest fire susceptibility mapping at the Cat Ba National Park Area, Hai Phong City, Vietnam, using GIS-based kernel logistic regression. *Remote Sens.* **2016**, *8*, 347. [CrossRef]
11. Bui, D.T.; Tuan, T.A.; Klempe, H.; Pradhan, B.; Revhaug, I. Spatial prediction models for shallow landslide hazards: A comparative assessment of the efficacy of support vector machines, artificial neural networks, kernel logistic regression, and logistic model tree. *Landslides* **2016**, *13*, 361–378.
12. Chang, K.T.; Merghadi, A.; Yunus, A.P.; Pham, B.T.; Dou, J. Evaluating scale effects of topographic variables in landslide susceptibility models using GIS-based machine learning techniques. *Sci. Rep.* **2019**, *9*, 12296. [CrossRef]
13. Mokhtari, M.; Abedian, S. Spatial prediction of landslide susceptibility in Taleghan basin, Iran. *Stoch. Environ. Res. Risk Assess.* **2019**, *33*, 1297–1325. [CrossRef]
14. Oh, H.J.; Kadavi, P.R.; Lee, C.W.; Lee, S. Evaluation of landslide susceptibility mapping by evidential belief function, logistic regression and support vector machine models. *Geomat. Nat. Hazards Risk* **2018**, *9*, 1053–1070. [CrossRef]
15. Schratz, P.; Muenchow, J.; Iturritxa, E.; Richter, J.; Brenning, A. Hyperparameter tuning and performance assessment of statistical and machine-learning algorithms using spatial data. *Ecol. Model.* **2019**, *406*, 109–120. [CrossRef]
16. Kim, H.G.; Lee, D.K.; Park, C.; Ahn, Y.; Kil, S.H.; Sung, S.; Biging, G.S. Estimating landslide susceptibility areas considering the uncertainty inherent in modeling methods. *Stoch. Environ. Res. Risk Assess.* **2018**, *32*, 2987–3019. [CrossRef]
17. Park, S.; Kim, J. Landslide susceptibility mapping based on random forest and boosted regression tree models, and a comparison of their performance. *Appl. Sci.* **2019**, *9*, 942. [CrossRef]
18. Nsengiyumva, J.B.; Valentino, R. Predicting landslide susceptibility and risks using GIS-based machine learning simulations, case of upper Nyabarongo catchment. *Geomat. Nat. Hazards Risk* **2020**, *11*, 1250–1277. [CrossRef]
19. Chen, X.; Chen, W. GIS-based landslide susceptibility assessment using optimized hybrid machine learning methods. *Catena* **2021**, *196*, 104833. [CrossRef]
20. Gassner, C.; Promper, C.; Beguería, S.; Glade, T. Climate change impact for spatial landslide susceptibility. In *Engineering Geology for Society and Territory*; Christine, G., Catrin, P., Santiago, B., Thomas, G., Eds.; Springer: Cham, Germany, 2015; Volume 1, pp. 429–433.
21. Lucà, F.; D'Ambrosio, D.; Robustelli, G.; Rongo, R.; Spataro, W. Integrating geomorphology, statistic and numerical simulations for landslide invasion hazard scenarios mapping: An example in the Sorrento Peninsula (Italy). *Comput. Geosci.* **2014**, *67*, 163–172. [CrossRef]
22. Ozturk, U.; Pittore, M.; Behling, R.; Roessner, S.; Andreani, L.; Korup, O. How robust are landslide susceptibility estimates? *Landslides* **2021**, *18*, 681–695. [CrossRef]
23. Knevels, R.; Petschko, H.; Proske, H.; Leopold, P.; Maraun, D.; Brenning, A. Event-Based Landslide Modeling in the Styrian Basin, Austria: Accounting for Time-Varying Rainfall and Land Cover. *Geosciences* **2020**, *10*, 217. [CrossRef]
24. Shou, K.J.; Yang, C.M. Predictive analysis of landslide susceptibility under climate change conditions—A study on the Chingshui River Watershed of Taiwan. *Eng. Geol.* **2015**, *192*, 46–62. [CrossRef]
25. Tanyas, H.; Rossi, M.; Alvioli, M.; van Westen, C.J.; Marchesini, I. A global slope unit-based method for the near real-time prediction of earthquake-induced landslides. *Geomorphology* **2019**, *327*, 126–146. [CrossRef]
26. Moayedi, H.; Khari, M.; Bahiraei, M.; Foong, L.K.; Bui, D.T. Spatial assessment of landslide risk using two novel integrations of neuro-fuzzy system and metaheuristic approaches; Ardabil Province, Iran. *Geomat. Nat. Hazards Risk* **2020**, *11*, 230–258. [CrossRef]
27. Zhang, S.; Li, C.; Peng, J.; Peng, D.; Xu, Q.; Zhang, Q.; Bate, B. GIS-based soil planar slide susceptibility mapping using logistic regression and neural networks: A typical red mudstone area in southwest China. *Geomat. Nat. Hazards Risk* **2021**, *12*, 852–879. [CrossRef]
28. Nachappa, T.G.; Kienberger, S.; Meena, S.R.; Hölbling, D.; Blaschke, T. Comparison and validation of per-pixel and object-based approaches for landslide susceptibility mapping. *Geomat. Nat Hazards Risk* **2020**, *11*, 572–600. [CrossRef]
29. Sur, U.; Singh, P.; Meena, S.R. Landslide susceptibility assessment in a lesser Himalayan road corridor (India) applying fuzzy AHP technique and earth-observation data. *Geomat. Nat. Hazards Risk* **2020**, *11*, 2176–2209. [CrossRef]
30. Luo, L.; Lombardo, L.; van Westen, C.; Pei, X.; Huang, R. From scenario-based seismic hazard to scenario-based landslide hazard: Rewinding to the past via statistical simulations. *Stoch. Environ. Res. Risk Assess.* **2021**. [CrossRef]
31. Di Napoli, M.; Carotenuto, F.; Cevasco, A.; Confuorto, P.; Di Martire, D.; Firpo, M.; Pepe, G.; Raso, E.; Calcaterra, D. Machine learning ensemble modelling as a tool to improve landslide susceptibility mapping reliability. *Landslides* **2020**, *17*, 1897–1914. [CrossRef]
32. Thuiller, W.; Georges, D.; Engler, R. Biomod2 Package Manual. 2015. Available online: https://CRAN.R-project.org/package=biomod2 (accessed on 12 July 2021).
33. Wolpert, D.H. Stacked Generalization. *Neural Netw.* **1992**, *5*, 241–259. [CrossRef]
34. Lee, M.J.; Choi, J.W.; Oh, H.J.; Won, J.S.; Park, I.; Lee, S. Ensemble-based landslide susceptibility maps in Jinbu area, Korea. *Env. Earth Sci.* **2012**, *67*, 23–37. [CrossRef]

35. Hu, X.; Zhang, H.; Mei, H.; Xiao, D.; Li, Y.; Li, M. Landslide Susceptibility Mapping Using the Stacking Ensemble Machine Learning Method in Lushui, Southwest China. *Appl. Sci.* **2020**, *10*, 4016. [CrossRef]
36. Menard, S.W. *Applied Logistic Regression Analysis*, 2nd ed.; Sage University Paper Series on Quantitative Application in the Social Sciences, Series no. 106; Sage: Thousand Oaks, CA, USA, 1995.
37. Breiman, L. Random forests. *Mach. Learn.* **2001**, *45*, 5–32. [CrossRef]
38. Olden, J.D.; Kennard, M.J.; Pusey, B.J. Species invasions and the changing biogeography of Australian freshwater fishes. *Glob. Ecol. Biogeogr.* **2008**, *17*, 25–37. [CrossRef]
39. Youssef, A.M.; Pourghasemi, H.R.; Pourtaghi, Z.S.; Al-Katheeri, M.M. Landslide susceptibility mapping using random forest, boosted regression tree, classification and regression tree, and general linear models and comparison of their performance at Wadi Tayyah Basin, Asir Region, Saudi Arabia. *Landslides* **2016**, *13*, 839–856. [CrossRef]
40. Vapnik, V.N. The Nature of Statistical Learning Theory. *IEEE Trans. Neural Netw.* **1997**, *8*, 1564.
41. Foody, G.M.; Mathur, A. A relative evaluation of multiclass image classification by support vector machines. *IEEE Trans. Geosci. Remote Sens.* **2004**, *42*, 1335–1343. [CrossRef]
42. Cawley, G.C.; Talbot, N.L. Efficient approximate leave-one-out cross-validation for kernel logistic regression. *Mach. Learn.* **2008**, *71*, 243–264. [CrossRef]
43. Bui, D.T.; Pradhan, B.; Lofman, O.; Revhaug, I.; Dick, O.B. Landslide susceptibility assessment in the Hoa Binh province of Vietnam: A comparison of the Levenberg–Marquardt and Bayesian regularized neural networks. *Geomorphology* **2012**, *171*, 12–29.
44. Friedman, J.H. Greedy Function Approximation: A Gradient Boosting Machine. *Ann. Stat.* **2001**, *29*, 1189–1232. [CrossRef]
45. Cheong, Y.L.; Leitão, P.J.; Lakes, T. Assessment of land use factors associated with dengue cases in Malaysia using Boosted Regression Trees. *Spat. Spatio-Temporal Epidemiol.* **2014**, *10*, 75–84. [CrossRef]
46. Green, D.M.; Swets, J.A. *Signal Detection Theory and Psychophysics*; Wiley: New York, NY, USA, 1966; Volume 1, pp. 1969–2012.
47. Mann, H.B.; Whitney, D.R. On a test of whether one of two random variables is stochastically larger than the other. *Ann. Math. Statist.* **1947**, *18*, 50–60. [CrossRef]
48. Wilcoxon, F. Some uses of statistics in plant pathology. *Biometrics Bull.* **1945**, *1*, 41–45. [CrossRef]
49. Kruskal, W.H.; Wallis, W.A. Use of ranks in one-criterion variance analysis. *J. Am. Stat. Assoc.* **1952**, *47*, 583–621. [CrossRef]
50. Spearman, C. The Proof and Measurement of Association Between Two Things. *Am. J. Psychol.* **1904**, *15*, 72–101. [CrossRef]
51. Myers, J.L.; Well, A.D. *Research Design and Statistical Analysis*; Routledge: London, UK, 2003.
52. Xie, M.; Esaki, T.; Zhou, G. GIS-based probabilistic mapping of landslide hazard using a three-dimensional deterministic model. *Nat. Hazards* **2004**, *33*, 265–282. [CrossRef]
53. Liu, J.K.; Weng, T.C.; Hung, C.H.; Yang, M.T. Remote Sensing Analysis of Heavy Rainfall Induced Landslide. In Proceedings of the 21st Century Civil Engineering Technology and Management Conference, Hsinchu, Taiwan, 28 December 2001; Minghsin University of Science and Technology: Xinfeng Township, Taiwan, 2001; pp. C21–C31. (In Chinese).
54. Dai, F.; Lee, C. A spatiotemporal probabilistic modelling of storm-induced shallow landsliding using aerial photographs and logistic regression. *Earth Surf. Processes Landf. J. Br. Geomorphol. Res. Group* **2003**, *28*, 527–545. [CrossRef]
55. Lee, C.T.; Chung, C.C. Common patterns among different landslide susceptibility models of the same region. In Proceedings of the Workshop on World Landslide Forum, Ljubljana, Slovenia, 30 May–2 June 2017; Springer: Cham, Germany, 2017; pp. 937–942.
56. Chien, F.C. The Relationship among Probability of Failure, Landslide Susceptibility and Rainfall. Master's Thesis, National Central University, Taoyuan City, Taiwan, 2015.
57. Fu, C.C. Event-Based Landslide Susceptibility and Rainfall-Induced Landslide Probability Prediction Model in the Zengwen Reservoir Catchment. Master's Thesis, National Central University, Taoyuan City, Taiwan, 2017.
58. Lai, J.S.; Chiang, S.H.; Tsai, F. Exploring influence of sampling strategies on event-based landslide susceptibility modeling. *ISPRS Int. J. Geo-Inf.* **2019**, *8*, 397. [CrossRef]
59. Xiao, T.; Segoni, S.; Chen, L.; Yin, K.; Casagli, N. A step beyond landslide susceptibility maps: A simple method to investigate and explain the different outcomes obtained by different approaches. *Landslides* **2020**, *17*, 627–640. [CrossRef]
60. Lei, X.; Chen, W.; Pham, B.T. Performance evaluation of GIS-based artificial intelligence approaches for landslide susceptibility modeling and spatial patterns analysis. *ISPRS Int. J. Geo-Inf.* **2020**, *9*, 443. [CrossRef]

water

MDPI

Article

Using Deep Learning to Formulate the Landslide Rainfall Threshold of the Potential Large-Scale Landslide

Jie-Lun Chiang [1,*], Chia-Ming Kuo [2] and Leila Fazeldehkordi [2]

1 Department of Soil and Water Conservation, National Pingtung University of Science and Technology, Pingtung 912, Taiwan
2 Department of Civil Engineering, National Pingtung University of Science and Technology, Pingtung 912, Taiwan
* Correspondence: jlchiang@mail.npust.edu.tw

Abstract: The complex and extensive mechanism of landslides and their direct connection to climate change have turned these hazards into critical events on a global scale, which can have significant negative influences on the long-term sustainable development of nations. Taiwan experiences numerous landslides on different scales almost every year. However, Typhoon Morakot (2009), with large-scale landslides that trapped people, demonstrated the importance of an early warning system. The absence of an effective warning system for landslides along with the impossibility of its accurate monitoring highlighted the necessity of landslide rainfall threshold prediction. Accordingly, the prediction of the landslide rainfall threshold as an early warning system could be an effective tool with which to develop an emergency evacuation protocol. The purpose of this study is to present the capability of the deep learning algorithm to determine the distribution of landslide rainfall thresholds in a potential large-scale landslide area and to assess the distribution of recurrence intervals using probability density functions, as well as to assist decision makers in early responses to landslides and reduce the risk of large-scale landslides. Therefore, the algorithm was developed for one of the potential large-scale landslide areas (the Alishan D098 sub-basin), Taiwan, which is classified as a Type II Landslide Priority Area. The historical landslide data, maximum daily rainfall, 11 topographic factors from 2004 to 2017, and the Keras application programming interface (API) python library were used to develop two deep learning models for landslide susceptibility classification and landslide rainfall threshold regression. The predicted result shows the lowest landslide rainfall threshold is located primarily in the northeastern downstream of the Alishan catchment, which poses an extreme risk to the residential area located upstream of the landslide area, particularly if large-scale landslides were to be triggered upstream of Alishan. The landslide rainfall threshold under controlled conditions was estimated at 780 mm/day (20-year recurrence interval), or 820 mm/day (25-year recurrence interval). Since the frequency of extreme rainfall events caused by climate change is expected to rise in the future, the overall landslide rainfall threshold was considered 980 mm/day for the entire area.

Keywords: deep learning; rainfall-induced landslide; sediment disaster; Keras; multilayer perceptron; hyperparameter tuning; mountain disaster risk reduction

Citation: Chiang, J.-L.; Kuo, C.-M.; Fazeldehkordi, L. Using Deep Learning to Formulate the Landslide Rainfall Threshold of the Potential Large-Scale Landslide. *Water* **2022**, *14*, 3320. https://doi.org/10.3390/w14203320

Academic Editor: Su-Chin Chen

Received: 26 September 2022
Accepted: 17 October 2022
Published: 20 October 2022

1. Introduction

Landslides are natural hazards that cause loss of life and heavy financial and economic damage to residential areas all over the world. Therefore, investigating the effective factors of landslides for planning and providing management solutions in sensitive areas is essential.

Landslides generally occur under the influence of physical, human, and climatic factors [1]. Among all factors, intense rainfall is highly significant [2], and is thus considered a crucial factor in the implementation of early warning systems for landslides [3]. Formulation approaches used for the landslide rainfall threshold are mainly divided into the empirical and the deterministic approaches. The empirical approach generally is defined based on empirical formulae proposed in previous studies [4,5]. Since empirical formulae are provided for a specific study area, their application is constrained to areas with similar characteristics and properties. In addition, landslide precipitation threshold prediction through empirical formulae with insufficient raw data leads to unreliable results with larger confidence prediction intervals. On the other hand, the deterministic approach has mainly been presented by models, such as slope stability models or artificial intelligence models [3,6–8].

The main point of these models is to require sufficient physical parameters or factors to analyze and obtain optimized results; otherwise, despite multiple analyses at various time points, the result will be suboptimal. The most important challenge in landslide prediction through statistical-based methods is access to the location and date of landslides, factors of landslide occurrence, and controlling factors such as topography, geology, soil, and land cover [9].

The topography of the watershed is diverse, complex, and heterogeneous, and these different factors establish different conditions for landslide occurrence. Geology, topography, and soil type are requisite factors for formulating the landslide rainfall threshold [4]. Research has shown that high-risk landslide areas are mainly located within high altitudes, steep slopes, high curvature, barren lands, loose geology or lithology, and near river systems or roadsides [10–12]. In order to predict high-risk landslide areas, first, the landslide precipitation for the entire area should be determined, which is difficult when there is insufficient information.

Deep learning has evolved gradually in recent years as a result of technological advances in software and hardware computer equipment, such as the advent of tensor and graphics processing units (TPUs and GPUs). Moreover, as landslide control factors limit the capability of traditional machine learning methods [13], deep learning can be a wiser choice due to some of its advantages, including: (i) its capability to be used in multiple fields; (ii) ability to extract inherent and deep features easily; (iii) ability to distinguish effective information and obtain optimal parameters to construct models (iv) without requiring prior knowledge or hypotheses [14]; and (v) capability of accepting a higher sample size without manually constructing and (vi) selecting feature layers [15]. Various studies have applied deep learning techniques for landslide analysis [16–18], landslide susceptibility analysis [19–22], and landslide displacement predictions [23,24], and have proved the accuracy of these methods compared to traditional methods. Similarly, traditional machine learning methods compared to empirical approaches and physical modeling have provided highly efficient and accurate landslide disaster warning models [25]. Multiple studies employed precipitation data and artificial intelligence methods to develop landslide displacement prediction methods [1–3,5–9,13–15,26–33].

Recent studies attempted to enhance the reliability of landslide displacement prediction models through machine learning algorithms [26,27].

Although landslides in Taiwan have been studied for a long time, landslide hazards came into prominence in 2009 coincidentally with Typhoon Morakot, which trapped people. Due to the large population of Taiwan compared to the area of the country, concern was raised about potential property damage and fatalities in residential areas. On the other hand, the lack of landslide prediction and warning systems adds to this concern. Therefore, this paper attempts to assist responsible authorities in taking emergency evacuation decisions by introducing a model of estimation of the landslide rainfall threshold. This study combined the advantages of empirical approaches (with sufficient raw data) and deterministic approaches (with high resolution and high accuracy); in addition, we used a deep learning model to establish the landslide susceptibility and the landslide rainfall threshold map of potential large-scale landslide area for Alishan D098. To indicate the range with a "highly landslide potential and lower rainfall threshold", the critical rainfall threshold should be considered in landslide susceptibility. Therefore, the prediction results of the two models were multiplied to obtain the revised landslide rainfall threshold. The deep learning model parameters were chosen through hyperparameter tuning with a Keras tuner to probe and choose the model with the highest score (low validation loss score). Missing values were interpolated on the basis of historical landslide data. Figure 1 shows the workflow of this study.

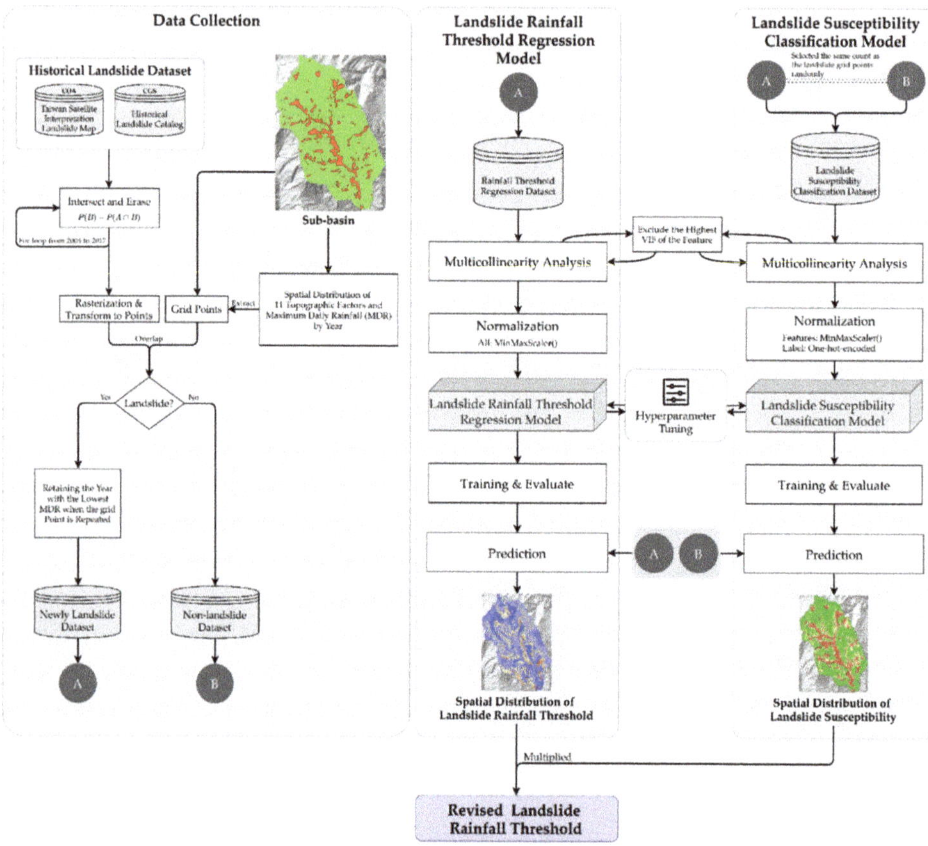

Figure 1. Study workflow for the prediction model.

2. Study Area

The study area was Alishan D098 (Figure 1) in southwestern Taiwan, which is known as the Laitou Landslide Area, located upstream of the Tsengwen Reservoir Watershed. It serves as a reference area for potential large-scale landslide areas affecting Gongxin Village, Fanlu Township, and Shanmei Village in Chiayi County's Alishan Township (Figure 2a). The potential area is approximately 42.9 hectares, with a downstream influence area of 74.2 hectares; landslide occurrence along the river is the debris flow type [34,35]. The geological formation is primarily Changchihkeng, which consists of sandstone and sand-shale interbedding lithology, and the Laitou Fault runs through this area from the northwest to southeast, resulting in relatively fragmented geology [36].

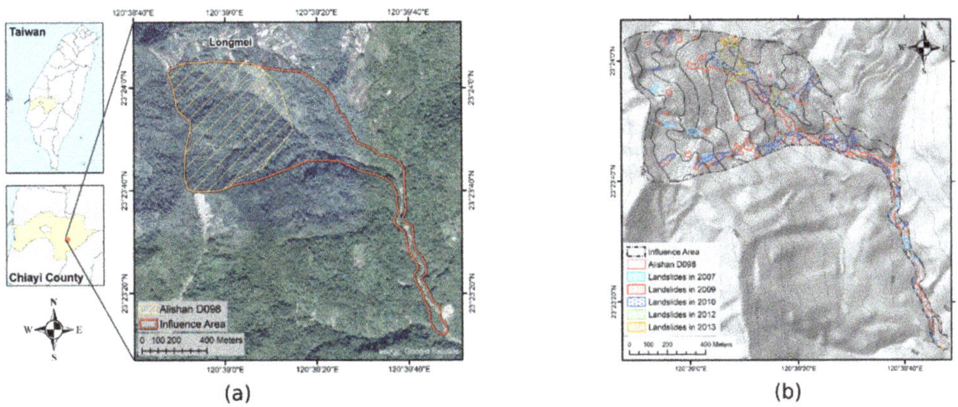

Figure 2. (**a**) Location of the influence area of Alishan D098; (**b**) the historical landslide areas by years.

Landslides have previously occurred in the northeast of Alishan D098, trapping ten residents during Typhoon Morakot in 2009 [37]. The recent multiphase regulation [38] has stabilized this area. The Soil and Water Conservation Bureau classified the area as a Type II Landslide Priority Area, which can be described by potential large-scale landslide objects without overlapping with existing debris-flow-protected targets [35].

3. Materials and Methods

3.1. Data Collection

Historical landslide datasets, including the Taiwan Satellite Interpretation Landslide Map provided by the Council of Agriculture (COA) from 2003 to 2017, and the Historical Landslide Catalog, provided by the Central Geological Survey (CGS) from 2003 to 2010, with all polygon vector data types, were used in this study. First, the landslide area was analyzed in 2003 and 2004. To obtain the overlapping area, the Intersect tool (ArcGIS 10.8) was employed afterward, and the overlapping area in 2004 was eliminated by the Erase tool to establish the new landslide area for the same year. The process was repeated year after year to generate all vector data of landslide areas from 2004 to 2017. After that, we converted it to grid data with the same resolution as the DEM (5 × 5 m) to obtain point data and finally created new landslide grid points.

The analysis result shows the new landslide grid points in Alishan D098 and its influence area were calculated at about 7000 grid points, which is considered insufficient. As a result, new landslide grid points in the sub-basin where the influence area is located were added to augment the dataset (Figure 3a). The bar chart illustrates new landslide grid points in the sub-basin (Figure 3b) from 2004 to 2017, with a total of 36,369 grid points. The highest frequency occurred in 2009, with 23,806 grid points (equivalent to an increase of 59.5 hectares of new landslide area) in the basin as a result of Typhoon Morakot's heavy rainfall. In 2004, 2008, and 2011, there were no landslide increases. Moreover, a significant decline in landslide occurrence has been observed since 2013.

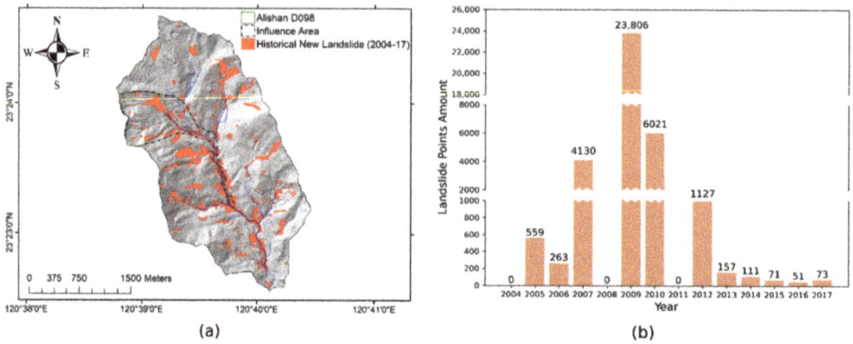

Figure 3. (**a**) The new landslide distribution; (**b**) the number of grid points by year in the sub-basin.

Based on literature research [18,22,39–42], 11 topographic factors were chosen (Table 1). The annual maximum daily rainfall (MDR) was collected from local rain gauge stations around Tsengwen Reservoir Watershed by the Central Weather Bureau (CWB) from 2004 to 2017 and was interpolated using inverse distance weighted (IDW) for conversion into spatial distribution grids (Figure 4a). The values extracted by the Point tool were used to match newly acquired landslide grid points from these grids with the corresponding topographic factors and the MDR. Figures 4b and 5 depict the spatial distribution of 11 topographic factors and the MDR in the sub-basin.

Repeated points in the new landslide grid points dataset represented the landslides that occurred more than once during or after the recovery period due to topography with a high landslide risk. Given that the high-landslide-susceptibility area requires a lower landslide rainfall threshold, the year with the lowest MDR was retained when the grid point was repeated. After filtering, 5243 repeated points were reduced to 31,126 non-repeated points.

The grid points in the sub-basin were used as training and testing datasets for the deep learning models of landslide susceptibility classification and rainfall threshold regression.

The former was the input feature for both the landslide and non-landslide grid points, with the non-landslide grid points having the same count as the landslide grid points chosen at random for dataset balance; the latter was only used for the landslide grid points because the rainfall for the non-trigger landslide was unknown.

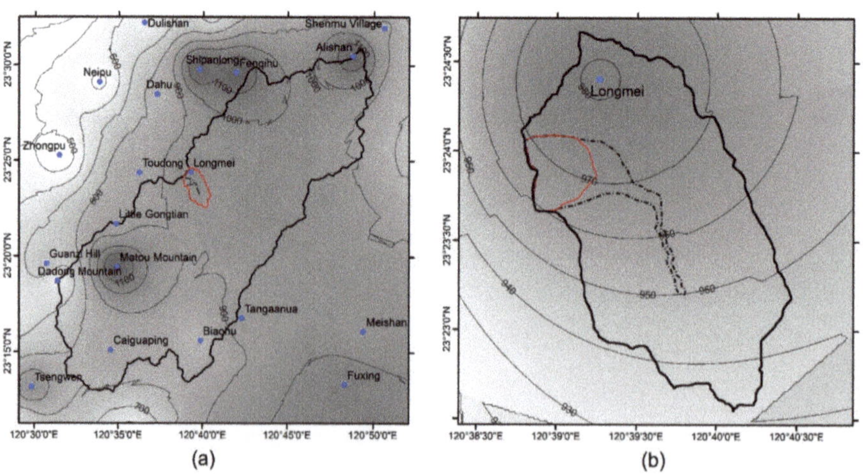

Figure 4. Spatial distribution of maximum daily rainfall in 2009. (**a**) Tsengwen Reservoir Watershed; (**b**) sub-basin where the study area is located.

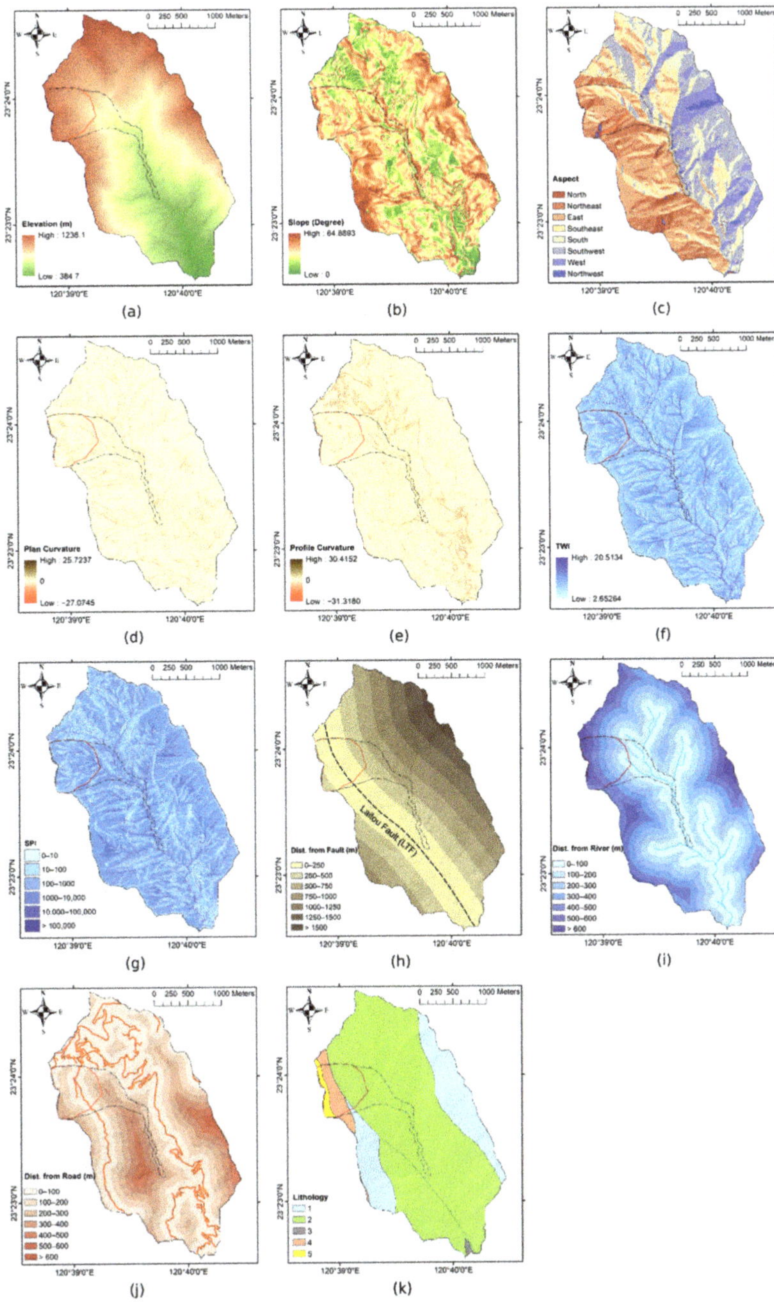

Figure 5. Topographic factors: (**a**) elevation; (**b**) slope; (**c**) aspect; (**d**) plan curvature; (**e**) profile curvature; (**f**) Topographic Wetness Index (TWI); (**g**) Stream Power Index (SPI); (**h**) distance from fault; (**i**) distance from river; (**j**) distance from road; (**k**) lithology. 1. Thick-bedded massive sandstone and argillaceous sandstone; 2. sandstone and sand-shale interbedding; 3. bodies of water; 4. massive shale, occasionally with thin-bedded siltstone; 5. siltstone and shale thin interbedding.

Table 1. The information about factors selected in the study.

Topography Factors	Data Source	Data Type	Value Range
Elevation	DEM	Continuous	384.70–1238.10
Slope	DEM	Continuous	0.00–64.89
Aspect	DEM	Continuous	−1.00–360.00
Plan curvature	DEM	Continuous	−27.07–25.72
Profile curvature	DEM	Continuous	−31.32–30.42
Topographic Wetness Index (TWI)	DEM	Continuous	2.65–20.51
Stream Power Index (SPI)	DEM	Continuous	0–4,388,483
Distance to fault	CGS	Continuous	0.00–1646.15
Distance to river	DEM	Continuous	0.00–680.66
Distance to road	OpenStreetMap	Continuous	0.00–610.98
Lithology	CGS	Categorical	n/a

3.2. Preprocess

Preprocessing of new landslide grid points was required before model training. Only 11 topography factors were left to be assigned as input features, and the new landslide year of occurrence was considered as the label. The scikit-learn *MinMaxScaler* function was applied for feature standardization [43], which scaled all input features into a range of 0 to 1. The transformation serves two purposes: the first is to allow the optimization function to converge earlier during iteration, improving model accuracy, and the second is to combine the landslide susceptibility and rainfall threshold regression models, correcting the rainfall threshold once the model prediction is complete. The label was one-hot-encoded for the landslide susceptibility model. To give the lower *MDR* a large-scale value after transformation in the rainfall threshold regression model, first, the multiplicative inverse of the original *MDR* was used, and then the reciprocal of the *MDR* was scaled to a range from 0 to 1 with *MinMaxScaler*. Therefore:

$$\overline{MDR}_{iscaled} = \frac{\overline{MDR}_i - \overline{MDR}_{min}}{\overline{MDR}_{max} - \overline{MDR}_{min}} \tag{1-1}$$

$$\overline{MDR}_i = \frac{1}{MDR_i} \tag{1-2}$$

3.3. Multicollinearity Analysis

The variance inflation factor (*VIF*) is the ratio of the overall model variance to the variance of a single independent variable model. It is used to determine variable multicollinearity. The *VIF* formula is shown below [44]:

$$VIF_i = \frac{1}{Tolerance} = \frac{1}{1 - R_i{}^2} \tag{2}$$

where $R_i{}^2$ is the coefficient of multiple determination of regression *i*.

Although the *VIF* threshold is still controversial, there is agreement on the maximum threshold. If the *VIF* of a factor exceeds 10.0 (or tolerance is less than 0.1), it is described as higher multicollinearity, which may have a negative impact on predictive power in multiple regression models [44,45]. The statsmodels was used to calculate the *VIF* [46]. In this study, the *VIF* threshold was set to 5 based on previous study recommendations [19–21] and considered the condition that exceeding values would be excluded and then reanalyzed until the *VIF* of all features showed a value of less than 5.

3.4. The Framework of Deep Learning Model

Deep learning is an artificial neural network (ANN) architecture with multiple and continuous layers between the input and output layers, with the number of layers increasing the model's depth. The weight values stored in the layers are interacted with and transformed in this model to determine the rules via the connection of neurons between layers. The predicted values are expected to match the true values.

Keras API is a Python deep learning library for high-level neural network programming, similar to TensorFlow and PyTorch. However, Keras API can build deep learning models faster and easier [47]. Hale [48] evaluated 11 popular deep learning applications based on usage, interest, and popularity. TensorFlow, Keras, PyTorch, Caffe, and Theano topped the preference list, with Keras being preferred most by deep learning novices. According to a Kaggle survey of data scientists and machine learning engineers, Scikit-learn is the most popular machine and deep learning library, followed by Tensorflow and Keras API [49].

The multilayer perceptron (MLP) in Keras API was used to build the deep learning model in this study. MLP architecture is shown in Figure 6. Moreover, the dropout layer was added after each hidden layer to reduce the model's overfitting problem; the input value for the dropout layer indicates that the ratio of neurons will not be used in this layer. The normalized exponential function (*Softmax*), Softplus, Softsign, rectified linear unit (*ReLU*), hyperbolic tangent (tanh), *sigmoid*, and exponential are some activation functions provided by Keras API [47]. As a result, different activation functions are appropriate for various layers and purposes. The *ReLU* function was used for each hidden layer, and the *Softmax* function was employed for the output layer of the landslide susceptibility model. However, the regression model is usually not required to set the activation function of the output layer, with the output value expected to be between 0 and 1; therefore, *Sigmoid* was used for the rainfall threshold regression model's output layer. The *ReLU*, *Softmax*, and *Sigmoid* function formulae are as follows:

$$Softmax(x) = \frac{e^{z_j}}{\sum_{K=1}^{K} e^{z_k}} for\ j = 1, \ldots K \quad (3)$$

$$ReLU(x) = max(0, x) \quad (4)$$

$$Sigmoid(x) = \frac{1}{1 + e^{-x}} \quad (5)$$

where z_j is the raw output value of the layer j, and K is the total number of labels. The total percent of the output value in the classification model must be equal to 1.

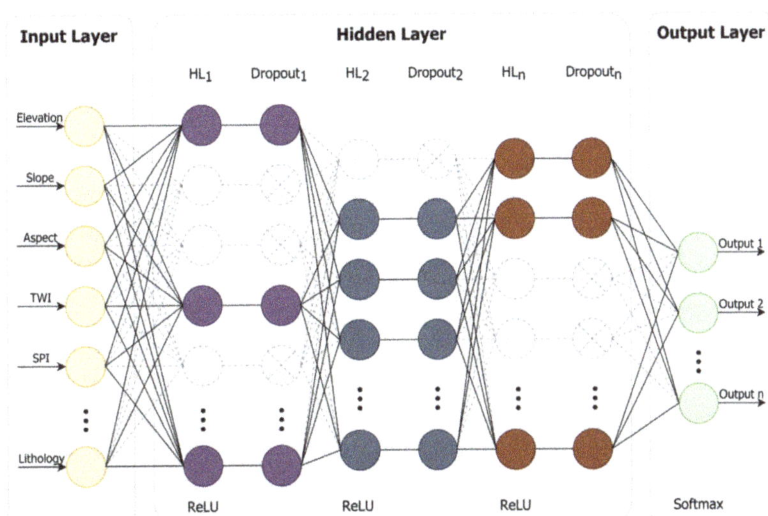

Figure 6. The MLP model architecture.

Stochastic Gradient Descent (SGD), Root Mean Square Propagation (RMSProp), and Adaptive Moment Optimization (Adam) are among the Keras optimizer algorithms. Binary cross-entropy loss (CE, also known as sigmoid cross-entropy loss), mean absolute error (MAE), and mean squared error are among the loss functions (MSE). Setting Adam and MSE as the optimization algorithm and loss function, respectively, yields the highest model predictive accuracy [22]. Therefore, Adam and MSE were assigned to this study.

Adam is an SGD-based algorithm with the added benefits of AdaGrad and RMSProp. It has high computational efficiency and low memory usage. Adam Optimizer's algorithm functions are as follows [50]:

$$\widehat{m}_t = \frac{\beta_1 \cdot m_{t-1} + (1-\beta_1)\cdot g_t}{1-\beta_1^t} \tag{6-1}$$

$$\widehat{v}_t = \frac{\beta_2 \cdot v_{t-1} + (1-\beta_2)\cdot {g_t}^2}{1-\beta_2^t} \tag{6-2}$$

$$w_t = w_{t-1} - \frac{\alpha \cdot \widehat{m}_t}{\sqrt{\widehat{v}_t} + \varepsilon} \tag{6-3}$$

where t is the timestep, m is the first moment vector, v is the second moment vector, g_t is the gradients with respect to the stochastic objective at t, w_t is the resulting parameter, α is the stepsize, β_1, β_2 are the exponential decay rates for the moment estimate, and ε is the value that, to be prevented, must be divided by zero. The final parameter vector is returned when w_t is converged. The suitable parameters for Adam Optimizer are: $\alpha = 0.001$; $\beta_1 = 0.9$; $\beta_2 = 0.999$, and $\varepsilon = 10^{-8}$.

Typically, the epoch is determined by training a model several times and selecting the appropriate epochs from the results. If the epochs are too long or too short, it will result in overfitting for the validation loss or underfitting for the training loss, and the model's performance will suffer as a result. To improve on this, we set the EarlyStopping function to monitor the validation loss and stopped training if the validation loss was no longer reduced after 10 consecutive training sessions.

3.5. Hyperparameter Tuning

Deep learning model parameters are typically chosen via manual selection based on previous experiences and research, trial and error, grid search, and or random search methods. Although the babysitting and random search are simple and quick, these strategies are not always the most appropriate combination. On the other hand, the grid search will always obtain the best combination; however, it will be time-consuming to search in cases with a large number of combinations. The current study used KerasTuner [51], a Keras hyperparameter tuning library, to determine the parameter of each hyperparameter in the model, such as the number of hidden layers, the neurons of each hidden layer, the dropout rate of each dropout layer, and the stepsize (α) of the Adam Optimizer, in order to provide the most suitable combination in a short period of time. In addition, the Hyperband Tuner was selected to obtain the best model parameters. The range of hyperparameters for the model defined in this study is listed in Table 2.

Table 2. The range of hyperparameters for the model defined.

Hyperparameters	Type	Defined Parameters
Counts of Hidden Layer	Integer	2–10
Neurons of Hidden Layer	Integer	4, 8, 12, . . . , 124, 128
Dropout Rate	Real	0.00, 0.05, 0.10, . . . , 0.85, 0.90
Stepsize of Adam Optimizer	Float	0.0001–0.1

3.6. Model Evaluate

The model was tested on validation and test datasets to assess the model's accuracy after training. In this study, the accuracy of the landslide susceptibility model was evaluated using a binary confusion matrix. The model performance evaluation metrics used the OA, which is defined as the ratio of all true predicted data (TP + TN) to all data; precision, which is defined as the ratio of true positive data predicted by the model (TP) to all true data predicted by the model (TP + FP); and recall, which is defined as the ratio of true positive data predicted by the model (TP) to all actual true data (TP + FN):

$$Overall\ Accuracy = \frac{TP + TN}{TP + FP + FN + TN} \tag{7}$$

$$Precision = \frac{TP}{TP + FP} \tag{8}$$

$$Recall = \frac{TP}{TP + FN} \tag{9}$$

In addition, some common indices for evaluation were selected, such as the *F-score*, Cohen's kappa coefficient (κ), and Matthews correlation coefficient (*MCC*). The F-score was proposed by van Rijsbergen in 1979, and introduced by Chinchor [52] in 1992. The formula of the *F-score* is the following:

$$F_{\beta}\ score = \frac{(\beta^2 + 1) \cdot Precision \cdot Recall}{\beta^2 \cdot Precision + Recall} \tag{10}$$

where β is the relative importance of the precision and recall; if $\beta = 1$, the precision is as important as the recall, meaning the F_1 *score*, as well as the harmonic mean of the precision and recall [53].

Matthews introduced the *MCC* classification accuracy index in 1975 (in the range of -1 to 1). If the value is close to 1, it represents the best and most consistent classification accuracy; a score close to 0 indicates poor classification accuracy, and a score close to -1 shows an inconsistent difference between the true and predicted classification accuracy. According to Chicco et al. [54], the *MCC* can provide more realistic and informative results than the commonly used Cohen's kappa coefficient and Brier score. Chicco and Jurman [55] also demonstrated that the *MCC* will receive a higher score when the confusion matrix categories have high precision, whereas the accuracy and *F-score* in imbalanced datasets are prone to over-optimism. The formulae for Cohen's kappa coefficient (κ) and *MCC* are as follows [54]:

$$MCC = \frac{TP \cdot TN - FP \cdot FN}{\sqrt{(TP + FP) \cdot (TP + FN) \cdot (TN + FP) \cdot (TN + FN)}} \tag{11}$$

$$\kappa = \frac{2 \cdot (TP \cdot TN - FN \cdot FP)}{(TP + FP) \cdot (FP + TN) + (TP + FN) \cdot (FN + TN)} \tag{12}$$

The rainfall threshold regression model was evaluated using *MAE*, root mean square error (*RMSE*), and mean absolute percentage error (*MAPE*). *MAE* and *RMSE* are used to compute model error, while *MAPE* is used to calculate the model accuracy. The formulae for the three metrics are as follows:

$$MAE = \frac{1}{n} \sum_{i=1}^{n} |A(i) - P(i)| \tag{13}$$

$$RMSE = \sqrt{\frac{\sum_{i=1}^{n} [A(i) - P(i)]^2}{n}} \tag{14}$$

$$MAPE = \frac{1}{n}\sum_{i=1}^{n}\left|\frac{A(i) - P(i)}{A(i)}\right| \times 100\% \quad (15)$$

where $A(i)$ is the actual value of data i, $P(i)$ is the predicted value of data i, and n is the total of the dataset.

All models in this study were built using Desktop PC with Intel Core i7, Processor 3.20 GHz, 40 GB RAM, and python language 3.9.7 through Spyder IDE 5.1.5 compiler.

4. Results and Discussion

4.1. Correlation and Multicollinearity between Factors

First, the correlation between factors was investigated by using the Pearson correlation coefficients (PCCs). PCCs close to 1 or −1 represent the probability of extreme multicollinearity while PCCs close to 0 indicate almost no multicollinearity between factors. The results (Figure 7) show that the highest PCCs between factors were observed in elevation/distance to river, elevation/distance to fault, SPI/TWI, and aspect/distance to fault, with both PCCs being greater than 0.5 or less than −0.5. Table 3 shows the results of the multicollinearity analysis. The VIFs of 11 terrain factors were all less than 5, with the highest being 2.916 and 3.330 in elevation, and the second being 2.444 and 2.492 in distance from the fault. The distance to the road was the lowest. All factors were used for model training and prediction based on the VIF results.

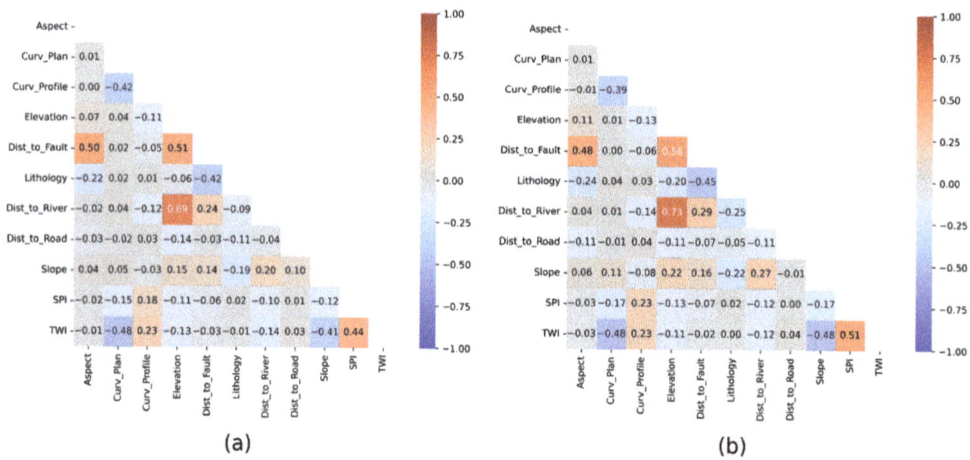

Figure 7. Pearson correlation coefficient between factors for training datasets. (**a**) Landslide susceptibility classification model; (**b**) rainfall threshold regression model.

Table 3. The VIF results of each factor for training datasets of landslide susceptibility and rainfall threshold regression model.

Features	Landslide Susceptibility (n = 62,252)	Rainfall Threshold (n = 31,126)
Elevation	2.916	3.330
Slope	1.359	1.502
Aspect	1.435	1.381
Plan Curvature	1.548	1.548
Profile Curvature	1.246	1.252
TWI	1.963	2.293
SPI	1.279	1.447
Distance to fault	2.444	2.492
Distance to river	2.094	2.491
Distance to road	1.060	1.041
Lithology	1.346	1.423

The variance inflation factor.

4.2. Result of Hyperparameter Tuning

Table 4 shows the top three validation loss score combinations based on model hyperparameter results. The validation loss score of the best combination for the landslide susceptibility classification model was 0.0801; the architecture contained four hidden layers with 108, 84, 100, and 112 neurons, and the dropout rate of the dropout layers was 0.05, 0.25, 0.45, and 0.3, respectively. Additionally, the stepsize of the Adam Optimizer was 0.000874, and the validation accuracy was 90%; this combination was the same as the second. The validation loss score of the best combination for the rainfall threshold regression model was 0.0675, which had four hidden layers with 124, 128, 36, and 72 neurons, respectively, and the dropout rate of the dropout layers was 0.2, 0.2, 0.25, and 0.1, respectively, and the stepsize of the Adam Optimizer was 0.001889. The combination with the best validation loss score was applied as the model parameter in the study based on the results.

Table 4. Top 3 ranking of hyperparameter tuning results for models.

Hyperparameters	Landslide Susceptibility Classification			Rainfall Threshold Regression		
	1st	2nd	3rd	1st	2nd	3rd
Counts of hidden layer	4	4	3	4	4	3
Neurons of hidden layer 1	108	108	108	124	120	120
Rate of dropout layer 1	0.05	0.05	0.30	0.20	0.20	0.45
Neurons of hidden layer 2	84	84	96	128	108	84
Rate of dropout layer 2	0.25	0.25	0.45	0.20	0.70	0.05
Neurons of hidden layer 3	100	100	60	36	28	88
Rate of dropout layer 3	0.45	0.45	0.10	0.25	0.05	0.00
Neurons of hidden layer 4	112	112	-	72	28	-
Rate of dropout layer 4	0.30	0.30	-	0.10	0.40	-
Stepsize of Adam Optimizer	0.000874	0.000874	0.000383	0.001889	0.000669	0.000507
Initial epoch	67	23	67	67	67	67
Epochs	200	67	200	200	200	200
Best step	91	41	127	83	120	99
Validation accuracy	0.8916	0.8683	0.8652	n/a	n/a	n/a
Score (Validation loss)	0.0801	0.0977	0.0998	0.0675	0.0704	0.0710

4.3. Training and Evaluated Results

The total dataset was split into train and test datasets in the ratio of 70% to 30%, respectively. The prediction data were based on all grid points within the study area, with a batch size of 50,200 epochs, and 20% for validation dataset. The model parameters were determined using tuning hyperparameters to produce the best combinations of validation loss scores.

The results for the landslide susceptibility classification model show that the early stopping function ceased during training while the epoch was approaching 155 times (Figure 8a, Table 5). The final accuracy and loss for the training dataset were 0.8816 and 0.0871, respectively, and 0.8973 and 0.0786 for the validation dataset. The overall accuracy, F1 score, Cohen's kappa, and MCC for the training dataset were 0.9094, 0.9092, 0.8187, and 0.8211, respectively, while the results show 0.8959, 0.8957, 0.7918, and 0.7944 for the test dataset. The high model metric scores indicate the model is able to predict reliable results.

During the rainfall threshold regression model training, the early stop function paused when the epoch reached 89 times (Figure 8b, Table 5). The calculated final loss for the training and validation datasets was 0.0734 and 0.0698, respectively. The MAE, RMSE, and MAPE of the training dataset computed 180.09 mm/day, 228.03 mm/day, and 26.31%, respectively; for the validation dataset, these were 185.98 mm/day, 233.97 mm/day, and 27.16%. The reason for the high MSE and RMSE errors is that the actual rainfall at the time of the landslides could not be captured; therefore, this study assumed that the landslides occurred when precipitation reached the maximum daily rainfall for the year; accordingly, the training results for the model still had errors and uncertainties. Although the model's results were not perfect, MAPE was calculated at about 27%, which is within an acceptable range. Figure 9 shows the ROC area under curve of training and test models.

(a)

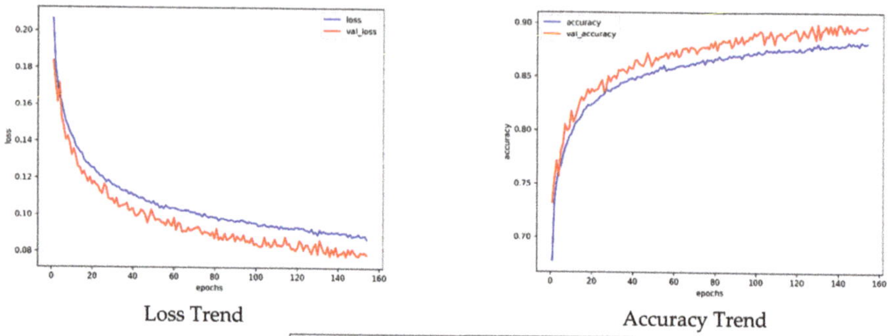

(b)

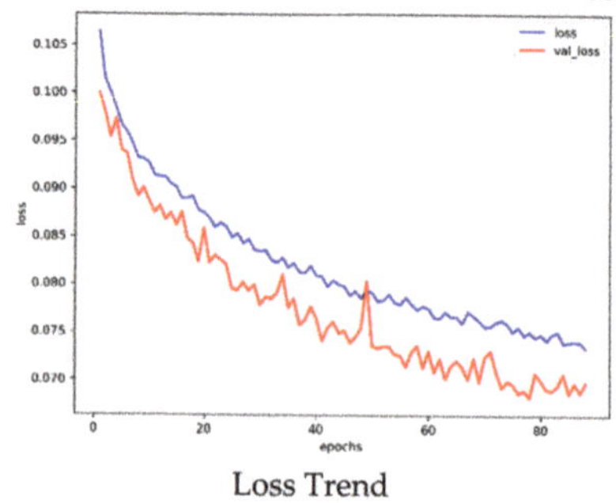

Figure 8. The accuracy and loss trend of training and validation dataset. (**a**) Landslide susceptibility classification model; (**b**) rainfall threshold regression model.

Table 5. The evaluated result for models.

Models	Evaluate	Train Dataset	Test Dataset
Landslide Susceptibility Classification	Overall accuracy	0.9094	0.8959
	Precision	0.9118	0.8985
	Recall	0.9094	0.8959
	ROC AUC	0.9633	0.9550
	F_1 score	0.9092	0.8957
	Cohen's kappa	0.8187	0.7918
	MCC	0.8211	0.7944
Rainfall Threshold Regression	MAE	180.09	185.98
	RMSE	228.03	233.97
	MAPE	0.2631	0.2716

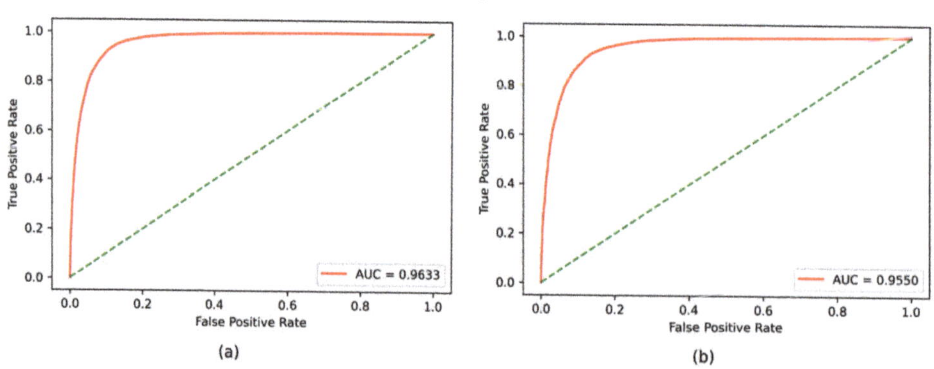

Figure 9. The ROC area under curve. (**a**) Training dataset; (**b**) test dataset.

4.4. Predicting and Revising Landslide Rainfall Threshold

Generally, when the susceptibility factor shows a higher value and the rainfall threshold factor a lower value, the risk of landslide occurrence is elevated. As a result, the emergency evacuation is triggered earlier.

Therefore, the prediction dataset was used as input for the trained model to calculate the landslide susceptibility factor and rainfall threshold factor with values ranging from 0 to 1.

The spatial distribution tagged image file (TIF) for prediction results is from the Geospatial Data Abstraction Library (GDAL) [56] and was mapped in with ArcGIS. Figure 10 shows that the landslide susceptibility factor in the historical landslide area and downstream was higher than in other areas, whereas the high rainfall threshold factor was primarily found in areas with obvious topographic changes.

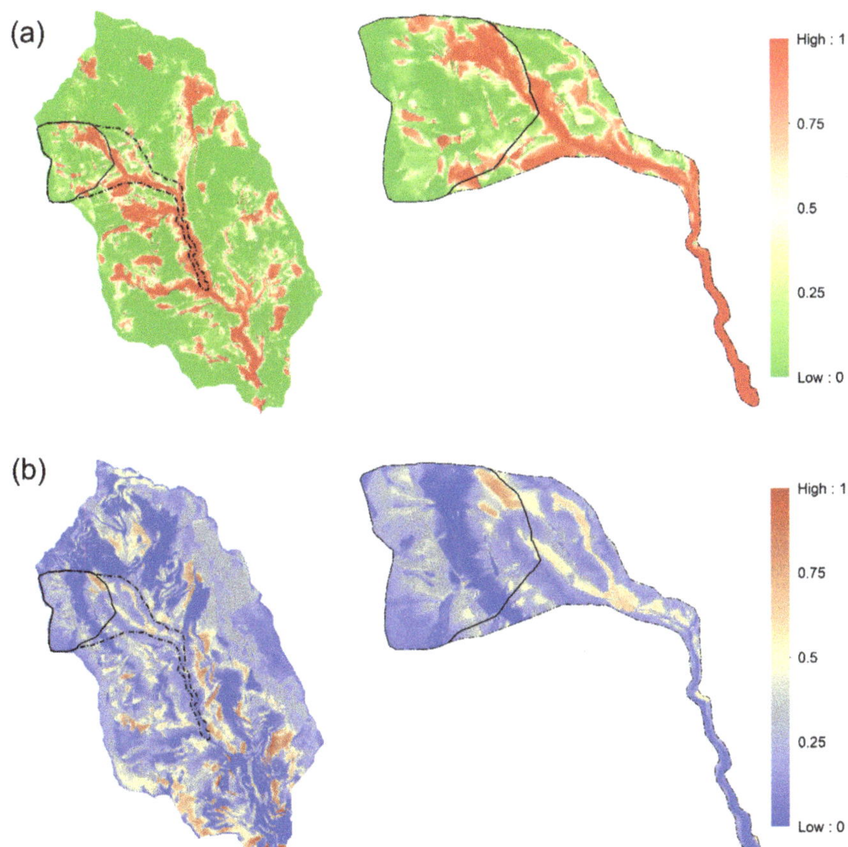

Figure 10. The prediction result of the sub-basin and Alishan D098 influence area. (**a**) Landslide susceptibility classification model; (**b**) rainfall threshold regression model.

Although the rainfall threshold model's input data include topographic factors, these factors were excluded from the training data due to missing landslide rainfall values in the non-landslide area, because the rainfall threshold model was only able to estimate the rainfall threshold data in the landslide areas. Accordingly, the training model eliminated topographic features of the non-landslide zone and the prediction results show a lower rainfall threshold value for non-landslide areas. It was assumed that the landslide occurred when rainfall surpassed the highest daily precipitation of the year because the actual precipitation landslide threshold was not known; therefore there was still uncertainty in the training model's outputs. To obtain the revised landslide rainfall threshold factor, two predictions were multiplied. This method significantly reduces the factor of the area with low landslide susceptibility and low rainfall threshold; as a result, the high coefficient of "high landslide susceptibility and low rainfall threshold" was retained. The result (Figure 11) shows that the lowest landslide rainfall threshold in Alishan D098 was primarily located in the northeast, with a downstream influence area. Due to the area's high landslide susceptibility but lower rainfall threshold in the prediction results, the MDR ranged from 350 to 780 mm/day after revision. Because of the topography, the western area showed a lower rainfall threshold, but the landslide susceptibility was lower. The original landslide rainfall threshold value of about 600 mm/day was increased to about 900 mm/day after the rainfall threshold was revised.

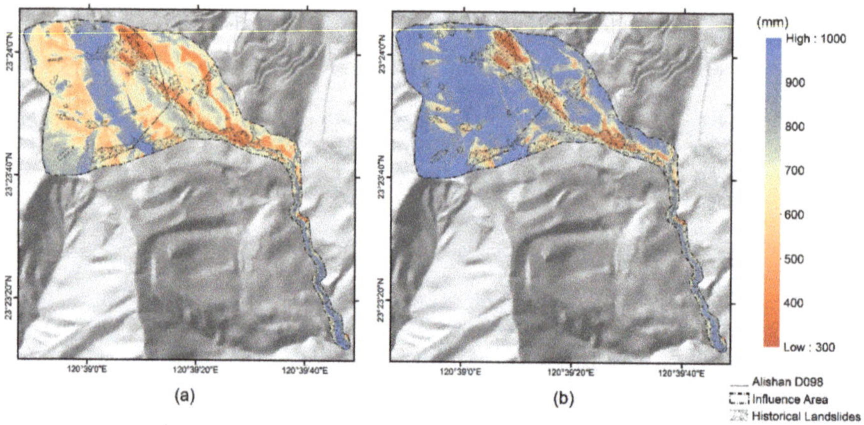

Figure 11. The spatial distribution of landslide rainfall threshold in Alishan D098 influence area. (**a**) Original result, which did not take the landslide susceptibility into account; (**b**) revised result, which did take the landslide susceptibility into account.

4.5. Establishing Recurrence Interval Distribution of Revised Landslide Rainfall Threshold

MDR data (from 1989 to 2021) were collected from the Longmei rain gauge station (C1M390), which is located near Alishan D098. The probability density functions, such as Extreme-value Type I, Normal, Pearson Type III, Log-Normal, and Log-Pearson Type III Distribution frequency analysis, were employed for frequency analysis. Moreover, the Kolmogorov–Smirnov Test (K-S Test) was used to examine the suitability. After that, the lowest RMSE among distribution methods was selected to establish the distribution for each corresponding recurrence interval. Since the extreme values of type I distribution showed the lowest RMSE (Table 6) to determine the corresponding recurrence intervals, the revised landslide rainfall threshold values were inputted into this distribution (Figure 12).

Table 6. Result of KS-Test and RMSE * by distributions.

Recurrence Interval	*p*-Value	RMSE
Extreme-value Type I Distribution	0.9995	0.0424
Normal Distribution	0.8510	0.0555
Pearson Type III Distribution	0.9725	0.0441
Log-Normal Distribution	0.9725	0.0437
Log-Pearson Type III Distribution	0.8510	0.0840

* Root mean square error.

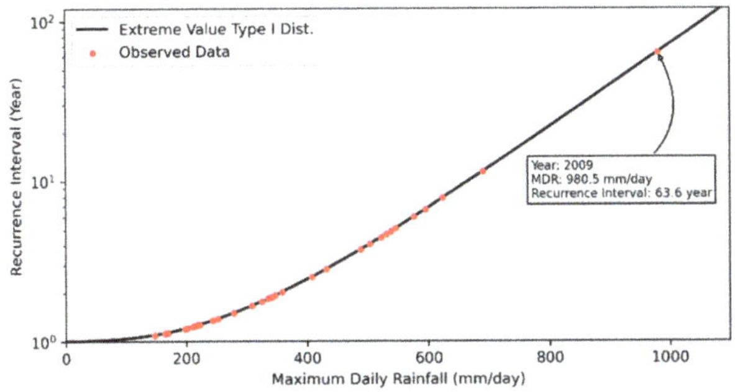

Figure 12. Extreme-value Type I distribution of Longmei rainfall gauge station.

The results (Figure 13) indicate that the lowest landslide rainfall threshold areas are primarily located in the northeastern and downstream in Alishan D098, with a recurrence interval of about 2–20 years, because the area has experienced sporadic landslides since the occurrence of Typhoon Morakot disturbances in 2009. The lowest recurrence interval was observed in the red-colored area (Figure 13), which showed the highest susceptibility to landslides. Thus, landslide management is critical for this area. In addition, the landslide risk in this region will be more crucial when a large-scale landslide occurs in Alishan D098, which could endanger the upstream settlements (Longmei). The MDR of the landslide rainfall threshold was defined as 780 mm/day (about a 20-year recurrence interval) for yellow-colored areas, which eventually increased to 820 mm/day (about a 25-year recurrence interval) for the historical landslide areas that have been managed.

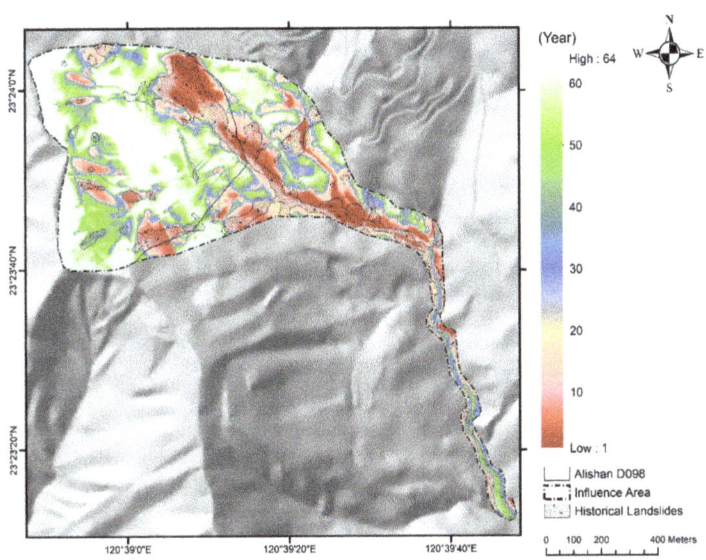

Figure 13. The distribution of recurrence interval of revised landslide rainfall threshold in the influence area.

Except for the historical landslide area in the northeast part, the obvious landslide or small-scale landslide was not observed in areas surrounding Alishan D098. The predicted landslide rainfall threshold was around 980.5 mm/day with a recurrence interval of approximately 63.6 years.

Typhoon Morakot's devastation in 2009 showed climate change impacts in the future will change the MDR. As a result, the landslide rainfall threshold for the entire area was proposed to be 980 mm/day. Table 7 shows the proposed landslide rainfall threshold for the Alishan D098 region.

Table 7. The landslide rainfall threshold of each region as suggested.

Regions	Condition	Rainfall Threshold (mm/day)	Recurrence Interval (Year)
Northeast Alishan D098 Downstream of Alishan D098 Longmei Settlement	Non-managed	780	20
	Has been managed	820	25
Whole Area	n/a	980	64

5. Conclusions

The landslide incidence time series limitation, the limitation of landslide monitoring data, the vagueness of the rainfall threshold for the landslide's occurrence, and the complexity of topographic factors have made it difficult to predict landslide incidence. Despite numerous types of research, most previous methods for determining the landslide threshold have limitations and errors. As a result, with the advancement of artificial intelligence algorithms, new methods can compensate for the lack of data availability and produce satisfactory results. Recent studies which applied deep learning algorithms to predict or map landslide susceptibility [13,14,20–22,39] acknowledged the capability of deep learning and, at the same time, have stated uncertainty in the prediction of landslide hazards due to insufficient data, climate change, and human activities. Moreover, they considered the achievement of satisfactory results to be contingent on future research.

The purpose of the current study was to predict the distribution of landslide susceptibility and rainfall threshold in the influence area of Alishan D098 using a deep learning algorithm. The results indicate the model achieved sufficient efficiency in determining the landslide rainfall threshold. In general, compared to the previous research [19,20], which focused on the effect of precipitation parameters on the occurrence of landslides, this study employed the topographic factors, the historical landslide data, and the maximum daily rainfall. The results show the lowest MDR threshold prediction was about 980.5 mm/day (64-year recurrence interval), decreasing to 820 mm/day in the managed historical landslide area. However, the MDR for the entire Alishan D098 region is anticipated to exceed 980.5 mm/day in the future due to climate change impacts. Therefore, this study proposed the landslide rainfall threshold to be 980 mm/day.

The current research findings are expected to be used to estimate and adjust the landslide rainfall threshold and, as a result, to make an emergency evacuation decision for the region. In addition, the results can be applied for future planning, land management, and policies or strategies for landslide hazard mitigation.

Since this study provides a new model, it faced challenges and limitations. Due to insufficient data in the study area, soil texture was not considered as an input feature in the deep learning models. Furthermore, the amount of rainfall at the time of the landslide could not be accurately calculated. As a result, the maximum daily rainfall for the year was chosen as the rainfall landslide threshold; consequently, the training results of the rainfall regression model were uncertain and showed errors. Therefore, the current study suggests using the model as a reference for other areas with more available data or improving input data through the application of different references, such as satellite data to predict future landslide incidence in this area.

Author Contributions: Conceptualization, J.-L.C.; methodology, J.-L.C. and C.-M.K.; software, J.-L.C. and C.-M.K.; validation, J.-L.C. and C.-M.K.; formal analysis, J.-L.C. and C.-M.K.; investigation, J.-L.C. and C.-M.K.; resources, J.-L.C.; data curation, J.-L.C., C.-M.K. and L.F.; writing—original draft preparation, J.-L.C. and C.-M.K.; writing—review and editing, J.-L.C., C.-M.K. and L.F.; visualization, C.-M.K.; supervision, J.-L.C.; project administration, J.-L.C.; funding acquisition, J.-L.C. All authors have read and agreed to the published version of the manuscript.

Funding: This research was funded by National Science and Technology Council of Taiwan, ROC. grant number MOST 110-2625-M-020-001 and MOST 111-2221-E-002-164.

Data Availability Statement: Taiwan CGS (Central Geological Survey) website provides free access to the geology, lithology, fault and landslide data and CWB (Central Weather Bureau) website provides free access to the rainfall data.

Acknowledgments: The authors gratefully acknowledge the Taiwan SWCB, CGS, and CWB for providing us with free data for use in this study.

Conflicts of Interest: The authors declare no conflict of interest.

References

1. Skilodimou, H.; Bathrellos, G.; Koskeridou, E.; Soukis, K.; Rozos, D. Physical and Anthropogenic Factors Related to Landslide Activity in the Northern Peloponnese, Greece. *Land* **2018**, *7*, 85. [CrossRef]
2. Srivastava, S.; Anand, N.; Sharma, S.; Dhar, S.; Sinha, L.K. Monthly Rainfall Prediction Using Various Machine Learning Algorithms for Early Warning of Landslide Occurrence. In Proceedings of the 2020 International Conference for Emerging Technology (INCET), Belgaum, India, 5–7 June 2020; pp. 1–7.
3. Distefano, P.; Peres, D.J.; Scandura, P.; Cancelliere, A. Brief communication: Introducing rainfall thresholds for landslide triggering based on artificial neural networks. *Nat. Hazards Earth Syst. Sci.* **2022**, *22*, 1151–1157. [CrossRef]
4. Palladino, M.R.; Viero, A.; Turconi, L.; Brunetti, M.T.; Peruccacci, S.; Melillo, M.; Luino, F.; Deganutti, A.M.; Guzzetti, F. Rainfall thresholds for the activation of shallow landslides in the Italian Alps: The role of environmental conditioning factors. *Geomorphology* **2018**, *303*, 53–67. [CrossRef]
5. Weng, M.-C.; Wu, M.-H.; Ning, S.-K.; Jou, Y.-W. Evaluating triggering and causative factors of landslides in Lawnon River Basin, Taiwan. *Eng. Geol.* **2011**, *123*, 72–82. [CrossRef]
6. Teja, T.S.; Dikshit, A.; Satyam, N. Determination of Rainfall Thresholds for Landslide Prediction Using an Algorithm-Based Approach: Case Study in the Darjeeling Himalayas, India. *Geosciences* **2019**, *9*, 302. [CrossRef]
7. Sun, D.; Gu, Q.; Wen, H.; Shi, S.; Mi, C.; Zhang, F. A Hybrid Landslide Warning Model Coupling Susceptibility Zoning and Precipitation. *Forests* **2022**, *13*, 827. [CrossRef]
8. Dou, J.; Yamagishi, H.; Pourghasemi, H.R.; Yunus, A.P.; Song, X.; Xu, Y.; Zhu, Z. An integrated artificial neural network model for the landslide susceptibility assessment of Osado Island, Japan. *Nat. Hazards* **2015**, *78*, 1749–1776. [CrossRef]
9. Tehrani, F.; Santinelli, G.; Herrera, M. A framework for predicting rainfall-induced landslides using machine learning methods Un cadre pour prédire les glissements de terrain induits par les précipitations à l'aide d'un apprentissage automatique. In Proceedings of the XVII ECSMGE-2019 Geotechnical Engineering Foundation of the Future, Reykjavik, Iceland, 1–6 September 2019.
10. Naseer, S.; Haq, T.U.; Khan, A.; Tanoli, J.I.; Khan, N.G.; Qaiser, F.-u.-R.; Shah, S.T.H. GIS-based spatial landslide distribution analysis of district Neelum, AJ&K, Pakistan. *Nat. Hazards* **2021**, *106*, 965–989. [CrossRef]
11. Wu, C.Y.; Yeh, Y.C. A Landslide Probability Model Based on a Long-Term Landslide Inventory and Rainfall Factors. *Water* **2020**, *12*, 937. [CrossRef]
12. Mind'je, R.; Li, L.H.; Nsengiyumva, J.B.; Mupenzi, C.; Nyesheja, E.M.; Kayumba, P.M.; Gasirabo, A.; Hakorimana, E. Landslide susceptibility and influencing factors analysis in Rwanda. *Environ. Dev. Sustain.* **2020**, *22*, 7985–8012. [CrossRef]
13. Nam, K.; Wang, F. An extreme rainfall-induced landslide susceptibility assessment using autoencoder combined with random forest in Shimane Prefecture, Japan. *Geoenvironmental Disasters* **2020**, *7*, 1–16. [CrossRef]
14. Zhu, L.; Huang, L.; Fan, L.; Huang, J.; Huang, F.; Chen, J.; Zhang, Z.; Wang, Y. Landslide Susceptibility Prediction Modeling Based on Remote Sensing and a Novel Deep Learning Algorithm of a Cascade-Parallel Recurrent Neural Network. *Sensors* **2020**, *20*, 1576. [CrossRef] [PubMed]
15. Zhang, T.; Li, Y.; Wang, T.; Wang, H.; Chen, T.; Sun, Z.; Luo, D.; Li, C.; Han, L. Evaluation of different machine learning models and novel deep learning-based algorithm for landslide susceptibility mapping. *Geosci. Lett.* **2022**, *9*, 26. [CrossRef]
16. Yang, R.; Zhang, F.; Xia, J.; Wu, C. Landslide Extraction Using Mask R-CNN with Background-Enhancement Method. *Remote Sens.* **2022**, *14*, 2206. [CrossRef]
17. Lu, H.; Ma, L.; Fu, X.; Liu, C.; Wang, Z.; Tang, M.; Li, N. Landslides Information Extraction Using Object-Oriented Image Analysis Paradigm Based on Deep Learning and Transfer Learning. *Remote Sens.* **2020**, *12*, 752. [CrossRef]
18. Prakash, N.; Manconi, A.; Loew, S. Mapping Landslides on EO Data: Performance of Deep Learning Models vs. Traditional Machine Learning Models. *Remote Sens.* **2020**, *12*, 346. [CrossRef]

19. Park, S.-J.; Lee, D.-k. Predicting susceptibility to landslides under climate change impacts in metropolitan areas of South Korea using machine learning. *Geomat. Nat. Hazards Risk* **2021**, *12*, 2462–2476. [CrossRef]
20. Saha, S.; Sarkar, R.; Roy, J.; Hembram, T.K.; Acharya, S.; Thapa, G.; Drukpa, D. Measuring landslide vulnerability status of Chukha, Bhutan using deep learning algorithms. *Sci. Rep.* **2021**, *11*, 16374. [CrossRef] [PubMed]
21. Azarafza, M.; Azarafza, M.; Akgün, H.; Atkinson, P.M.; Derakhshani, R. Deep learning-based landslide susceptibility mapping. *Sci. Rep.* **2021**, *11*, 24112. [CrossRef] [PubMed]
22. Nhu, V.-H.; Hoang, N.-D.; Nguyen, H.; Ngo, P.T.T.; Thanh Bui, T.; Hoa, P.V.; Samui, P.; Tien Bui, D. Effectiveness assessment of Keras based deep learning with different robust optimization algorithms for shallow landslide susceptibility mapping at tropical area. *Catena* **2020**, *188*, 104458. [CrossRef]
23. Li, X.; Li, S. Large-Scale Landslide Displacement Rate Prediction Based on Multi-Factor Support Vector Regression Machine. *Appl. Sci.* **2021**, *11*, 1381. [CrossRef]
24. Meng, Q.X.; Wang, H.L.; He, M.J.; Gu, J.J.; Qi, J.; Yang, L.L. Displacement prediction of water-induced landslides using a recurrent deep learning model. *Eur. J. Environ. Civ. Eng.* **2020**, 1–15. [CrossRef]
25. Orland, E.; Roering, J.J.; Thomas, M.A.; Mirus, B.B. Deep Learning as a Tool to Forecast Hydrologic Response for Landslide-Prone Hillslopes. *Geophys. Res. Lett.* **2020**, *47*, e2020GL088731. [CrossRef]
26. Ma, J.; Xia, D.; Wang, Y.; Niu, X.; Jiang, S.; Liu, Z.; Guo, H. A comprehensive comparison among metaheuristics (MHs) for geohazard modeling using machine learning: Insights from a case study of landslide displacement prediction. *Eng. Appl. Artif. Intell.* **2022**, *114*, 105150. [CrossRef]
27. Ma, J.; Xia, D.; Guo, H.; Wang, Y.; Niu, X.; Liu, Z.; Jiang, S. Metaheuristic-based support vector regression for landslide displacement prediction: A comparative study. *Landslides* **2022**, *19*, 2489–2511. [CrossRef]
28. Li, S.H.; Wu, L.Z.; Huang, J. A novel mathematical model for predicting landslide displacement. *Soft Comput.* **2020**, *25*, 2453–2466. [CrossRef]
29. Shihabudheen, K.V.; Peethambaran, B. Landslide displacement prediction technique using improved neuro-fuzzy system. *Arab. J. Geosci.* **2017**, *10*, 502. [CrossRef]
30. Ma, J.; Tang, H.; Liu, X.; Wen, T.; Zhang, J.; Tan, Q.; Fan, Z. Probabilistic forecasting of landslide displacement accounting for epistemic uncertainty: A case study in the Three Gorges Reservoir area, China. *Landslides* **2018**, *15*, 1145–1153. [CrossRef]
31. Li, H.; Xu, Q.; He, Y.; Deng, J. Prediction of landslide displacement with an ensemble-based extreme learning machine and copula models. *Landslides* **2018**, *15*, 2047–2059. [CrossRef]
32. Krkač, M.; Špoljarić, D.; Bernat, S.; Arbanas, S.M. Method for prediction of landslide movements based on random forests. *Landslides* **2016**, *14*, 947–960. [CrossRef]
33. Wang, L.; Chen, Y.; Huang, X.; Zhang, L.; Li, X.; Wang, S. Displacement prediction method of rainfall-induced landslide considering multiple influencing factors. *Nat. Hazards* **2022**. [CrossRef]
34. Council of Agriculture. 36 Potential Large-Scale Landslide Areas in 2022. Available online: https://data.coa.gov.tw/open_detail.aspx?id=I20 (accessed on 4 July 2022).
35. Tsai, Y.-J.; Syu, F.-T.; Shieh, C.-L.; Chung, C.-R.; Lin, S.-S.; Yin, H.-Y. Framework of Emergency Response System for Potential Large-Scale Landslide in Taiwan. *Water* **2021**, *13*, 712. [CrossRef]
36. Shao, P.-H.; Kao, M.-C. *Explanatory Text for the Geological Map of TAIWAN Scale 1:50,000—Zhongpu*; Central Geological Survey: Taipei, 2009.
37. Feng, Z.-Y.; Huang, H.-C.; Huang, J.-J.; Lai, H.-Y. Preliminary Discussion of the Factors Causing Laitou Landslide in Jiayi. *J. Soil Water Conserv.* **2012**, *44*, 177–188.
38. Chen, C.-K.; Lin, B.-S.; Chi, S.-Y.; Chien, Y.-D.; Tsai, M.-F. Sediment Stability Assessment for Large-Scale Landslides Induced by Typhoon Morakot After Remediation in an Upstream Reservoir Watershed. *Sinotech Eng.* **2015**, *182*, 25–37.
39. Wang, H.J.; Zhang, L.M.; Yin, K.S.; Luo, H.Y.; Li, J.H. Landslide identification using machine learning. *Geosci. Front.* **2021**, *12*, 351–364. [CrossRef]
40. Vojtek, M.; Vojteková, J.; Pham, Q.B. GIS-Based Spatial and Multi-Criteria Assessment of Riverine Flood Potential: A Case Study of the Nitra River Basin, Slovakia. *ISPRS Int. J. Geo-Inf.* **2021**, *10*, 578. [CrossRef]
41. Nguyen, K.; Chen, W. DEM- and GIS-Based Analysis of Soil Erosion Depth Using Machine Learning. *ISPRS Int. J. Geo-Inf.* **2021**, *10*, 452. [CrossRef]
42. Chen, S.-C.; Wu, C.-Y. Establishment of Landslide Susceptibility Early Warning Model in National Forest Areas Based on Geo-intrinsic and Hydro-extrinsic Factors. *J. Chin. Soil Water Conserv.* **2018**, *49*, 89–97. [CrossRef]
43. Pedregosa, F.; Varoquaux, G.; Gramfort, A.; Michel, V.; Thirion, B.; Grisel, O.; Blondel, M.; Prettenhofer, P.; Weiss, R.; Dubourg, V.; et al. Scikit-learn: Machine Learning in Python. *J. Mach. Learn. Res.* **2011**, *12*, 2825–2830.
44. Kutner, M.H.; Nachtsheim, C.J.; Neter, J.; Li, W. *Applied Linear Statistical Models*, 5th ed.; McGraw-Hill/Irwin: New York, NY, USA, 2004.
45. O'brien, R.M. A Caution Regarding Rules of Thumb for Variance Inflation Factors. *Qual. Quant.* **2007**, *41*, 673–690. [CrossRef]
46. Seabold, S.; Perktold, J. Statsmodels: Econometric and statistical modeling with python. In Proceedings of the 9th Python in Science Conference, Austin, TX, USA, 28 June–3 July 2010.
47. Chollet, F. Keras: The Python deep learning API. Available online: https://keras.io/ (accessed on 20 December 2021).

48. Hale, J. Deep Learning Framework Power Scores 2018. Available online: https://www.kaggle.com/discdiver/deep-learning-framework-power-scores-2018 (accessed on 22 December 2021).
49. Kaggle. State of Machine Learning and Data Science 2021. Available online: https://www.kaggle.com/kaggle-survey-2021 (accessed on 23 August 2022).
50. Kingma, D.P.; Ba, J. Adam: A Method for Stochastic Optimization. In Proceedings of the ICLR 2015, San Diego, CA, USA, 7–9 May 2015; 2015.
51. O'Malley, T.; Bursztein, E.; Long, J.; Chollet, F.; Jin, H.; Invernizzi, L. KerasTuner. Available online: https://github.com/keras-team/keras-tuner (accessed on 4 May 2022).
52. Chinchor, N. MUC-4 Evaluation Metrics. In Proceedings of the MUC4 92: Conference on Message Understanding, Stroudsburg, PA, USA, 16–18 June 1992; pp. 22–29.
53. Taha, A.A.; Hanbury, A. Metrics for evaluating 3D medical image segmentation: Analysis, selection, and tool. *BMC Med. Imaging* **2015**, *15*, 29. [CrossRef] [PubMed]
54. Chicco, D.; Warrens, M.J.; Jurman, G. The Matthews Correlation Coefficient (MCC) is More Informative Than Cohen's Kappa and Brier Score in Binary Classification Assessment. *IEEE Access* **2021**, *9*, 78368–78381. [CrossRef]
55. Chicco, D.; Jurman, G. The advantages of the Matthews correlation coefficient (MCC) over F1 score and accuracy in binary classification evaluation. *BMC Genom.* **2020**, *21*, 6. [CrossRef] [PubMed]
56. GDAL/OGR Contributors. GDAL/OGR Geospatial Data Abstraction Software Library. Available online: https://gdal.org (accessed on 30 August 2022).

water

MDPI

Article

The Effect of Temporal Characteristics on Developing a Practical Rainfall-Induced Landslide Potential Evaluation Model Using Random Forest Method

Yi-Min Huang [1,*] and Shao-Wei Lu [2]

[1] Department of Civil Engineering, Feng Chia University, Taichung 407, Taiwan
[2] True Dreams Construction Company, Taichung 403, Taiwan; leuwilly81304@gmail.com
* Correspondence: ninerh@mail.fcu.edu.tw

Abstract: With the unique rainfall patterns of typhoons, plum rains, and short-term heavy rainfalls, the frequent landslide and debris flow disasters have caused severe loss to people in Taiwan. In the studies of landslide susceptibility, the information of factors used for analysis was usually annual-based content, and it was assumed that the same elements from different years were independent between each year. However, the occurrence of landslides was usually not simply due to the changes within a year. Instead, landslides were triggered because the factors that affected the potential of landslides reached critical conditions after a cumulative change with time. Therefore, this study had well evaluated the influence of temporal characteristics and the ratios of antecedent landslide areas in the past five years in the landslide potential evaluation model. The analysis was conducted through the random forest (RF) algorithm. Additional rainfall events of 2017 were used to test the proposed model's performance to understand its practicality. The analysis results show that in the study area, the RF model had considerably acceptable performance. The results have also demonstrated that the antecedent landslide ratios in the past five years were essential to describe the significance of cumulative change with time when conducting potential landslide evaluation.

Keywords: landslide potential; random forest; antecedent landslides; machine learning

Citation: Huang, Y.-M.; Lu, S.-W. The Effect of Temporal Characteristics on Developing a Practical Rainfall-Induced Landslide Potential Evaluation Model Using Random Forest Method. *Water* **2021**, *13*, 3348. https://doi.org/10.3390/w13233348

Academic Editor: Su-Chin Chen

Received: 22 October 2021
Accepted: 23 November 2021
Published: 25 November 2021

1. Introduction

Taiwan is an island located in the hot zone of the Circurm-Pacific Earthquake Belt, typhoons, and subtropics monsoon climatic region. About 70% or more of the island's area is hillsides and mountains. Because of the steep topography and geological vulnerability, two or more natural disasters are common in Taiwan. Landslide disasters have become a critical issue in recent years due to their increasing frequency of occurrence caused by extreme weather and climate change.

To better understand and deal with the increase in landslide disasters, researchers conducted studies on the potential for landslide and occurrence probability by considering environmental conditions. Most researchers used the landslide susceptibility analysis (LSA) to develop landslide evaluation models. Factors that describe the environmental conditions and triggering behavior of landslides were usually used in the LSA method. The LSA-based models can be divided into qualitative and quantitative methods [1,2]. The approaches of statistical analysis, geotechnical engineering analysis (deterministic and probability), artificial intelligence [3–9], and data mining were commonly used for quantitative LSA modeling. Artificial intelligence has become a popular data analysis method in recent years [8] utilizing machine learning, and deep learning algorithms have become more efficient and reliable in recent years. The machine learning algorithm, Random Forest (RF), was often used for landslide potential evaluation [9–15] and was the study focus of this paper.

Landslide risk analysis (LRA) is another approach to study the landslide disaster. The concept of the risk triangle, proposed by Crichton [16], including hazard, vulnerability, and exposure, is usually used to develop an LRA model, in which probability assessment is performed to describe the risk level. A recently developed landslide evaluation model using the landslide fragility curves (LFC) was a model of LRA [1,2,17]. LFC model used spatial statistics and GIS for data processing on the grid base. Rainfall data over the study areas and landslide cases were obtained and analyzed by constructing a fragility curve to describe the exceeding probability of landslide for given environmental conditions. Fragility curves were built for each combination of factors in the model. With the fragility curves, the critical hazard potential and critical fragility potential were determined to express the probability of exceeding a damage state of landslides under certain conditions of rainfall intensity and accumulated rainfall [2].

Both LSA and LRA models use factors to describe the potential of landslides. These factors, such as rainfall intensity, accumulated rainfall, slope degree, etc., were categorized as environmental factors and triggering factors. The development of the landslide potential evaluation model was usually conducted by considering the effectiveness of each chosen factor for a given period, e.g., the accumulated rainfall of each heavy rainfall event used for developing the model. Among the rainfall-induced researches, the rainfall thresholds are usually the main factor when evaluating the landslide susceptibility [18], and the temporal resolution of rainfall affects the thresholds significantly [19]. In these studies, the rainfall factors were included in the evaluation procedure by yearly-based input, not considering the consecutive influence from previous years.

In addition to rainfall factors, previous studies [1,2,20,21] on landslides usually used other annual factors of each year to establish a potential evaluation model. The variation of the same factors between each year was typically assumed to be independent and uncorrelated. However, landslides were usually caused not simply due to the change that occurred within a year. The cumulative impacts of the factors had affected the landslide potential over time, and the landslide was triggered when the critical condition was reached. Therefore, not only the spatial factors but the temporal consecutive characteristics should be considered in the model. For this purpose, a novelty attempt was made by including the consecutive changes of landslide areas in this study. The antecedent landslide area was adopted as a temporal characteristic factor in the proposed model, and its effect was discussed in this paper.

Finally, the performance of the proposed RF model was evaluated, and two events of heavy rainfalls in 2017 were used for specific event-driven performance tests. The insights from the newly developed RF model were also included in this paper.

2. Study Areas and Environmental Factor Database

2.1. Study Areas

In the middle south of Taiwan, the watershed areas of the ChenYuLan River, the Laonong River, and the Qishan River (Figure 1) were chosen to develop the landslide potential evaluation model. The ChenYuLan River watershed is located in the mountainous area of Nantou County. The mainstream ChenYuLan River is one of the important tributaries of the Zhuoshui River, with a total length of about 42 km. The terrain of the ChenYuLan River is characterized by a significant height difference and steep slope. The watershed area of ChenYuLan River is about 448 km^2, and the average gradient is over 55%. Most of the Laonong River watershed area is located in Kaohsiung City. The mainstream of the Laonong River is the first tributary of the Gaoping River basin, with a total length of about 137 km. The terrain of the Laonong River is characterized as the valley topography, and the downstream is a suddenly widened sediment area. The watershed area of the Laonong River is extensive, with an area of 1408.71 km^2 and an average slope of over 65%. The Qishan River watershed area is located in Kaohsiung City and Chiayi County. The mainstream Qishan River is a tributary of the Gaoping River, with a total length of about 117 km. The terrain of the Qishan River is characterized as the terraces valley.

The river is suddenly widened at the downstream section. The watershed of Qishan River is about 750.79 km^2 and with an average slope of more than 55%.

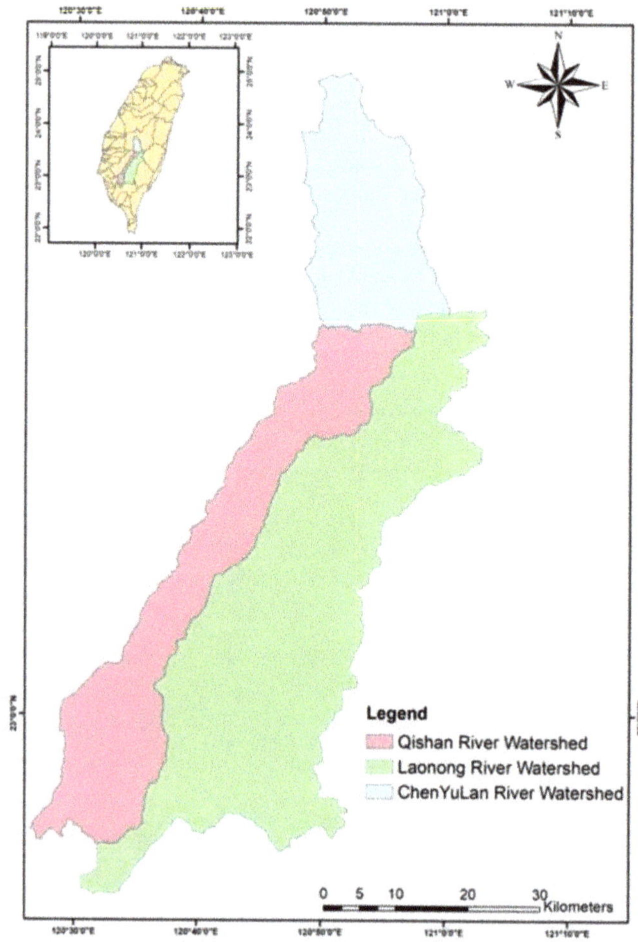

Figure 1. Watersheds of ChenYuLan River, Laonong River, and Qishan River in Taiwan.

To better understand what has occurred in the region, satellite images from 2011 to 2016 were used to build the model factor for this study. The images had undergone a radiometric correction and orthorectification process. The satellite images of 2011 and 2016 are shown as examples in Figure 2.

2.2. Slope Units

To develop the RF models, the area impacted by landslides was first calculated based on slope units. A slope unit is defined as one slope part or the left/right part of a watershed. Slope units can be topologically divided by the watershed divide and drainage line (Figure 3) with the help of the GIS tool [22]. We used the slope unit as the analysis basis to physically and adequately represent the local conditions. Then, the environmental database was applied in accordance with the slope units at the site of interest. The ChenYuLan River, Laonong River, and Qishan River watersheds are 6651, 21,279, and 10,985 units, respectively (Figure 4).

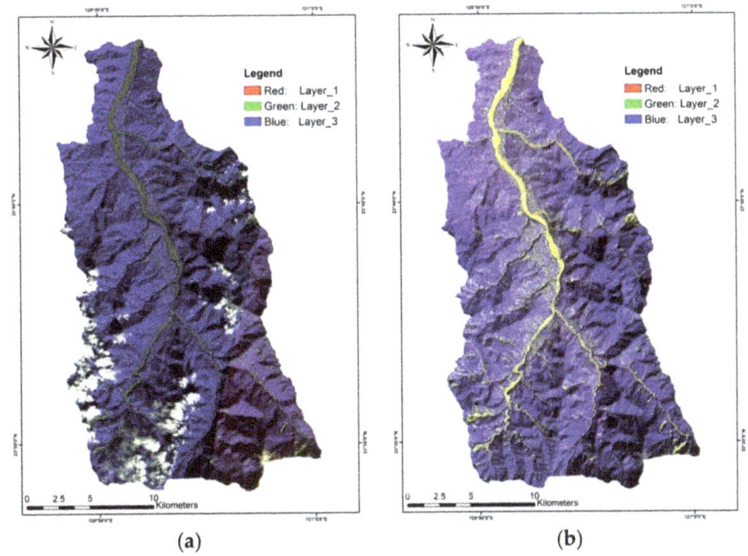

Figure 2. Satellite images of the ChenYuLan River watershed in (**a**) 2011, (**b**) 2016.

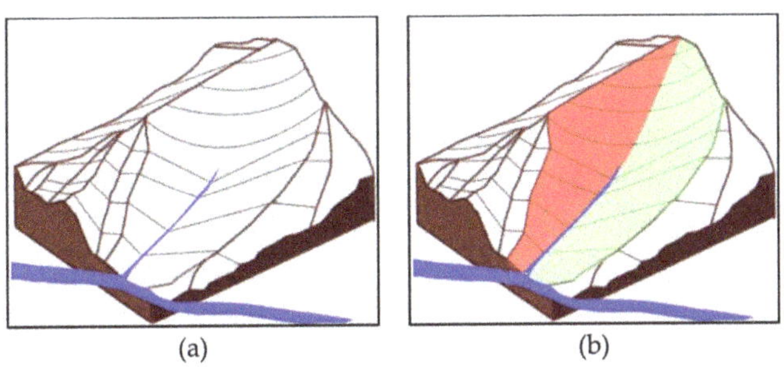

Figure 3. Slope unit delineation, the (**a**) and (**b**) slope units of a watershed.

2.3. *Environmental Database and Model Factors*

From research and studies, the landslide factor used for analysis included slope, lithology, aspect, elevation, land use, river system-related (including distance, density, etc.), vegetation-related (including species, density, age, etc.), geology (including distance, type, geological structure, etc.), soil-related (including type, thickness, content, etc.), slope shape, curvature (including horizontal, vertical, etc.), rainfall (including accumulated rainfall, rainfall intensity), and many more.

Factors directly or indirectly related to landslide occurrence were analyzed by their importance to establish the model for landslide potential evaluation. These factors were generally classified as environmental and triggering factors [23–25]. The rainfall was the primary concern, and the rainfall intensity (maximum hourly precipitation, I_{max}) and effective accumulated rainfall (R_{te}) were used as triggering factors in this study. In addition, elevation, slopes, slope aspects, normalized difference vegetation index (NDVI), distance to the river (stream), geology type, and the ratio of incremental landslide area were used as environmental factors of hillside slope in the study. These factors (Table 1) were collected from satellite images, digital elevation model (DEM), and monitoring records, and a GIS database was created to obtain and process the necessary content of factors. The basic statistical information of the database is shown in Table 2.

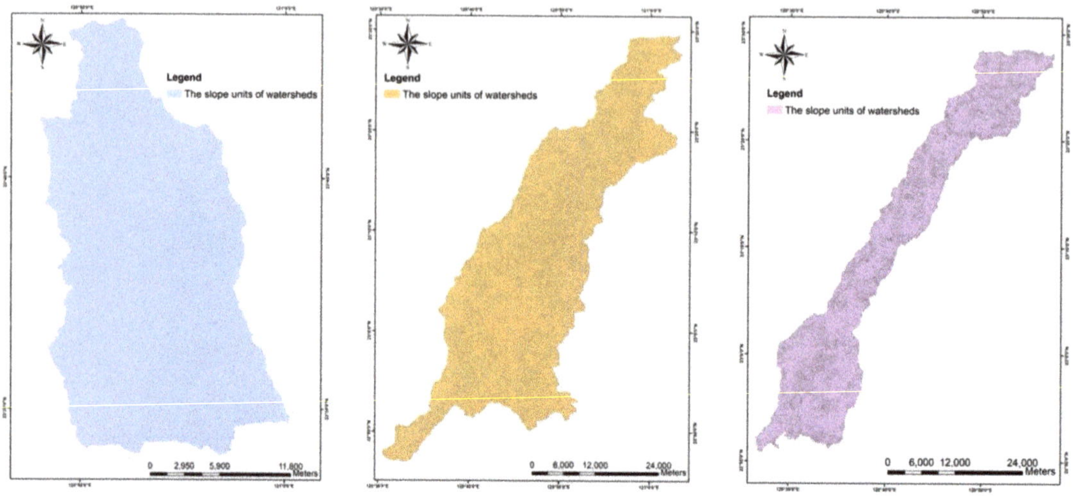

(a) (b) (c)

Figure 4. The slope units of watersheds (**a**) ChenYuLan River, (**b**) Laonong River, (**c**) Qishan River.

Table 1. Environmental database of factors.

Type	Source	Factors	Data Format
Grid	DEM	Elevation Slope (degree) Aspect	Raster, 5 m
	SPOT satellite images	NDVI (from 2011 to 2016)	Raster, 20 m
	Rainfall	Maximum hourly rainfall (I_{max}) of events from 2006 to 2016 Effective accumulated rainfall (R_{te}) of events from 2006 to 2016	Raster, 20 m
Vector	River system	River system Distance to river (stream)	Shapefile, Polygon
	Geology Map	Geology type	
	Landslide [1]	Incremental landslide areas, from 2011 to 2016	

[1] the source of satellite images was from Taiwan's government open data platform (https://data.gov.tw accessed on: 27 August 2021).

Table 2. Statistical information of database.

Watersheds	Year	I_{max} (mm) (Mean, SD)	R_{te} (mm) (Mean, SD)	NDVI (Average)	Slope (Degree) (Average)	Elevation (m) (Average)	Distance to the River (m) (Average)
ChenYuLan River	2011	(46.7, 14.2)	(289, 84)	0.1261	33.7	1591.2	287.3
	2012	(40.1, 9.9)	(223, 46)	−0.1072			
	2013	(53.2, 23.1)	(166, 39)	0.0344			
	2014	(21.4, 4.0)	(202, 75)	0.0497			
	2015	(31.7, 3.5)	(139, 28)	0.0055			
	2016	(55.4, 6.8)	(509, 58)	0.0893			
Laonong River	2011	(50.4, 11.8)	(509, 152)	0.1016	32.2	1508.0	265
	2012	(49.5, 10.7)	(248, 85)	−0.1372			
	2013	(62.9, 18.7)	(289, 73)	0.0225			
	2014	(41.4, 13.4)	(322, 74)	−0.0051			
	2015	(53.7, 13.9)	(347, 179)	0.0235			
	2016	(58.8, 9.8)	(421, 161)	−0.0010			
Qishan River	2011	(49.8, 12.1)	(322, 117)	0.0956	24.7	926.8	190.6
	2012	(49.0, 8.2)	(187, 58)	−0.0376			
	2013	(56.1, 15.1)	(311, 130)	−0.2053			
	2014	(40.9, 6.9)	(278, 74)	0.0799			
	2015	(50.3, 9.9)	(287, 103)	0.0997			
	2016	(64.1, 8.8)	(473, 70)	−0.0062			

In the database, the landslide area, especially the incremental landslide area, was considered significant when describing the landslide situation in each target area. Figure 5 illustrates the difference of landslide areas between two periods of SPOT image. The total area, labeled as No. 3 and No. 4 in the figure, was treated as the new landslide areas during a single period. The ratios related to landslide areas, i.e., the proportions of No. 1, No. 2, No. 3, No.4, and No. 5 in Figure 5 were calculated for each slope unit after satellite image processing and interpretation. The condition of a slope unit would be classified as landslide when the ratio of incremental landslide area (No. 3 + No. 4) was greater or equal to 5%. The landslide ratios on each period from 2006 to 2016 were estimated and used in the proposed model.

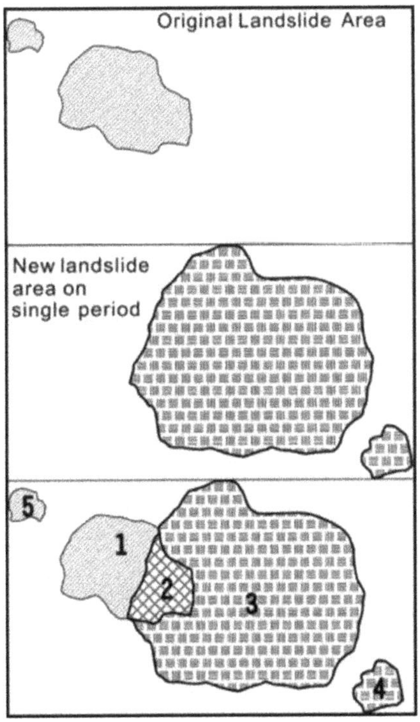

Figure 5. Illustration of landslide area between two periods of a satellite image.

According to the Soil and Water Conservation Bureau (SWCB), Taiwan, a rainfall event starts when the hourly rainfall is greater than 4 mm. The event ends when the accumulated rainfall of continuous 6 h does not exceed 24 mm, and any hourly rain is less than 10 mm during the 6-h period (Figure 6).

The factor, effective accumulated rainfall, R_{te}, was calculated by the expression as follows.

$$R_{te} = R_0 + \sum_{i=1}^{7} (0.7)^i R_i \quad (1)$$

where the R_0 is the current daily rainfall, and R_i represents the daily rainfall i-th day before.

The maximum hourly rainfall, I_{max}, was used to label the timing of landslides by assuming that the landslide occurred when the hourly precipitation reached Imax. First, for a given rainfall event, each grid's I_{max} and R_{te} were calculated by applying Kriging interpolation based on the data of reference weather stations. Then, the I_{max} and R_{te} of a slope unit were determined by averaging the values of grids within the slope unit. Figure 7 shows the processed results of factors in the study areas, and Figure 8 shows an example of the interpolation of rainfall of given rainfall events in 2011.

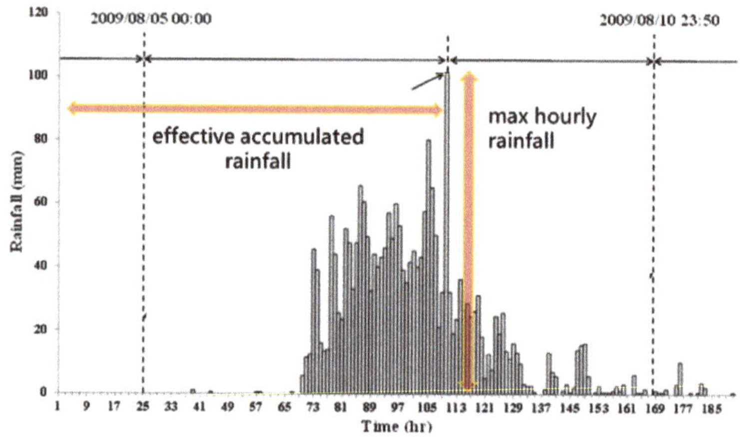

Figure 6. The definition of a rainfall event and rainfall factors.

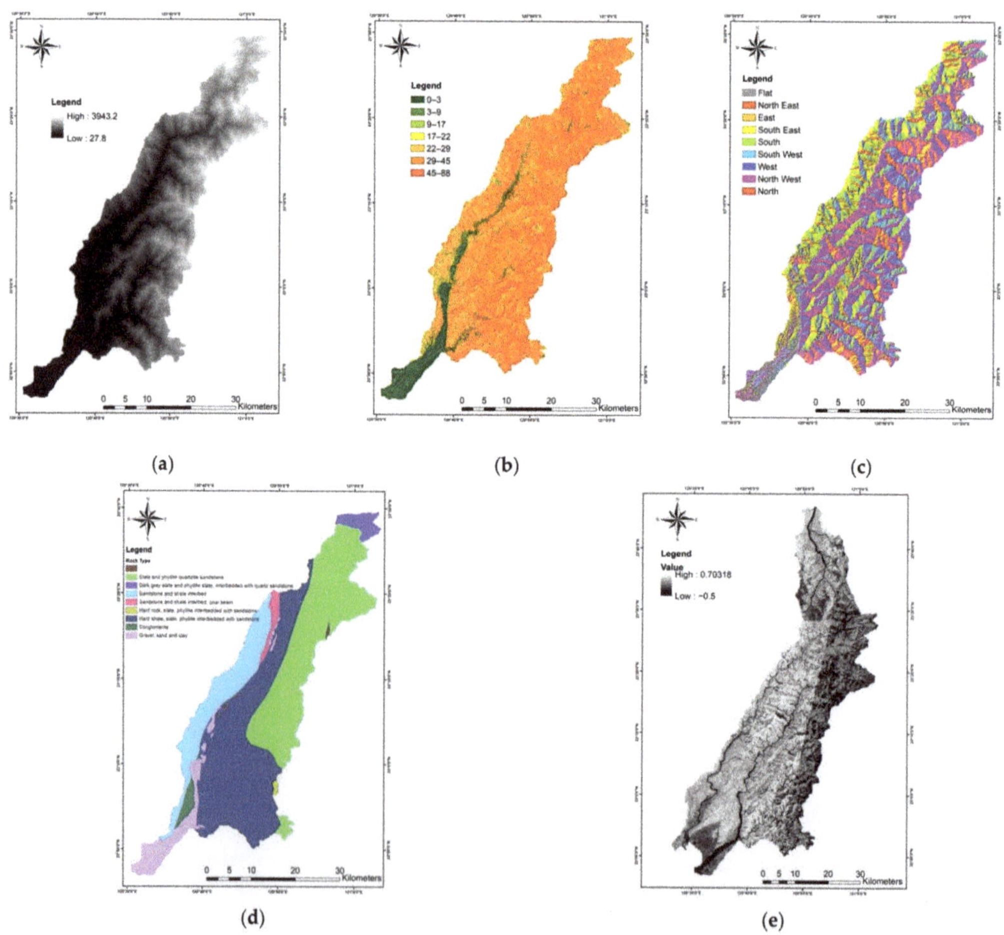

Figure 7. The factors of (**a**) elevation; (**b**) slope; (**c**) slope aspects; (**d**) geology type; (**e**) NDVI (2011) of study areas.

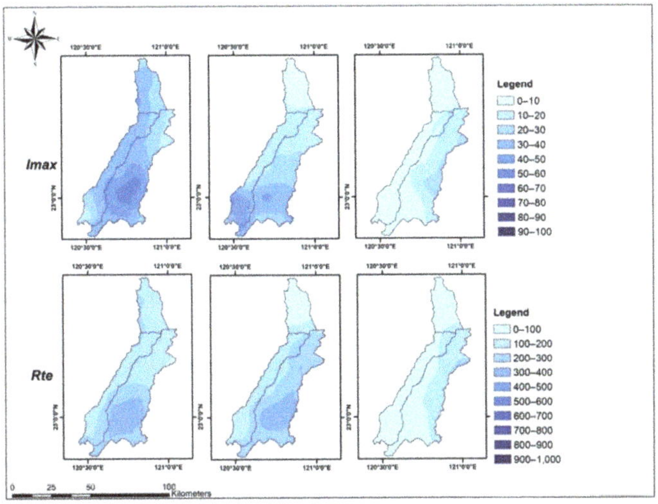

Figure 8. Examples of I_{max} and R_{te} in the study area in 2011. From left to right are events of 0719 Heavy Rainfall, Typhoon Nanmadol, and 1001 Heavy Rainfall.

3. Development of Landslide Potential Evaluation Model

Machine learning, a branch of artificial intelligence (AI), has become popular in various fields. Computer algorithms are necessary and essential in machine learning to robustly analyze and predict information based on learning from training data [22]. Among its applications, the use of machine learning has increased in landslide susceptibility analysis [10]. The methods of artificial neural network (ANN) [26,27], logistic regression [28], support vector machine (SVM) [29], and random forest [11,12] have been popular for landslide-related analysis.

The random forest is a classifier based on the ensemble method and consists of many decision trees. Each tree must meet the basic classification ability and certain accuracy conditions. The built-in classifier performance will eventually be better than the result predicted by a single classifier. The random forest will randomly take the data for training, and the output will be based on the voting in each decision tree. At present, random forests are also widely used due to the advantages of accurate classification, the capability of large amounts of data processing, multi-variable data, and fast calculations.

In the analysis of landslide prediction, rainfalls and landslides in the past years may cause the current site condition to be unstable in the investigated areas. Therefore, the impacts of temporal characteristics should be taken into account in the analysis. Thus, the antecedent rainfalls and landslides were used as the temporal characteristics in the proposed random forest model. The data of the past five years were considered to represent the previous environmental impacts related to current landslide conditions.

In addition to the six environmental factors of elevation, slope, slope aspects, geology type, distance to the river, and NDVI, the temporal factors of annual landslide areas in the past five years were also considered. Therefore, these temporal factors were included in the model in annual area ratios of five-type landslides (Figure 5) in a slope unit. Furthermore, the triggering factors of annual I_{max} and annual R_{te} of a slope unit in the past five years were also treated as temporal characteristics in the model. Therefore, there was a total of 43 factors (6 environmental factors, 25 annual landslide area ratios in the past five years, ten annual factors of I_{max} and R_{te} in the past five years, and current yearly I_{max} and R_{te}) in the model database. The examples of annual landslide areas are shown in Figure 9.

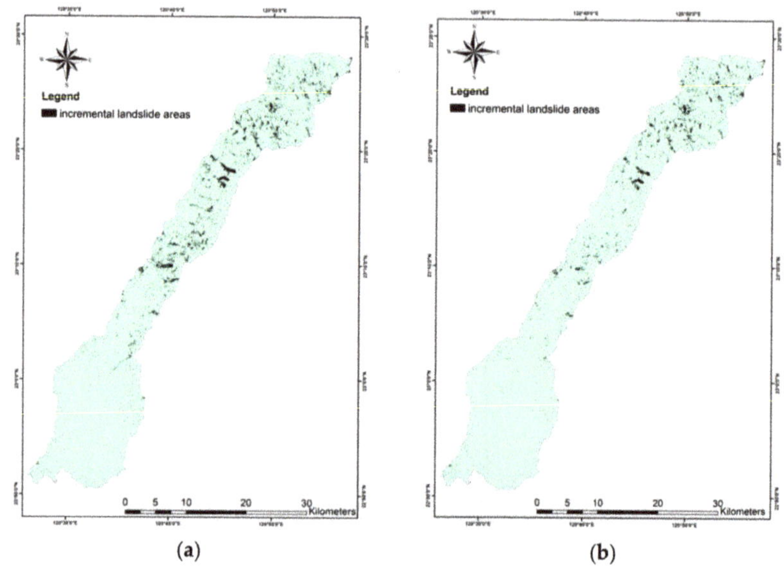

Figure 9. Annual incremental landslide areas of the Qishan River Watershed in (**a**) 2011 and (**b**) 2016.

The program WEKA (Waikato Environment for Knowledge Analysis) was used to develop the random forest model. The data were processed by standardization method, Z-Score, and normalization procedure (values of 0 to 1). From a total of 38,915 slope units, 10,000 units were randomly selected for model development, with a landslide to a non-landslide ratio of 1:1. In total, 50% of data were used for training during the development, and the 50% left were used for internal validation. The supervised classification algorithm was applied in training.

Validation indices are usually used to evaluate the performance of a model. In machine learning, the validation indices can be divided into classification metrics and regression metrics. The proposed model was mainly a binary case. Thus, the confusion matrix of classification metrics (Table 3) was used to validate the model performance.

Table 3. The confusion matrix of landslide evaluation.

Predicted\Actual	Landslide (1)	Non-Landslide (0)
Landslide (1)	True Positive (TP)	False Positive (FP)
Non-Landslide (0)	False Positive (FP)	True Negative (TN)

The validation indices from Table 2 include accuracy (ACC), precision, sensitivity (or recall), Kappa (kappa index of agreement, KIA), and receiver operating characteristic curve (ROC) and area under the curve (AUC). These indices were used to validate the performance of the proposed model.

The random forest model was trained using data from 2006 to 2010 to obtain the 5-year temporal factors (rainfall factors of I_{max} and R_{te}, and ratios of antecedent landslides) and the data from 2011 as the input to represent the current state. After establishing the RF model, the landslide area data from 2012 to 2016 and the annual I_{max} and R_{te} were used as the external validation to analyze the model performance of prediction. In addition to the confirmation of the 43-factor model, the version of the model using only eight factors (elevation, slope, slope aspects, geology type, distance to river, and NDVI, and I_{max} and R_{te}) in 2011, i.e., without factors of antecedent landslide ratios, was also evaluated. The training and validation results were represented by the indices of the confusion matrix, as shown in Table 4. The examples of training and validation results are shown in Figures 10 and 11.

Table 4. RF Model training and validation results.

Training/Validation Year	Accuracy	Precision	Recall	Kappa	AUC
2011 (model training)	0.808	0.791	0.833	0.616	0.889
2012	0.716	0.860	0.515	0.431	0.839
2013	0.706	0.870	0.485	0.412	0.844
2014	0.839	0.773	0.708	0.622	0.901
2015	0.833	0.664	0.828	0.616	0.898
2016	0.834	0.619	0.819	0.592	0.897
2011 (8 factors)	0.702	0.680	0.758	0.406	0.776

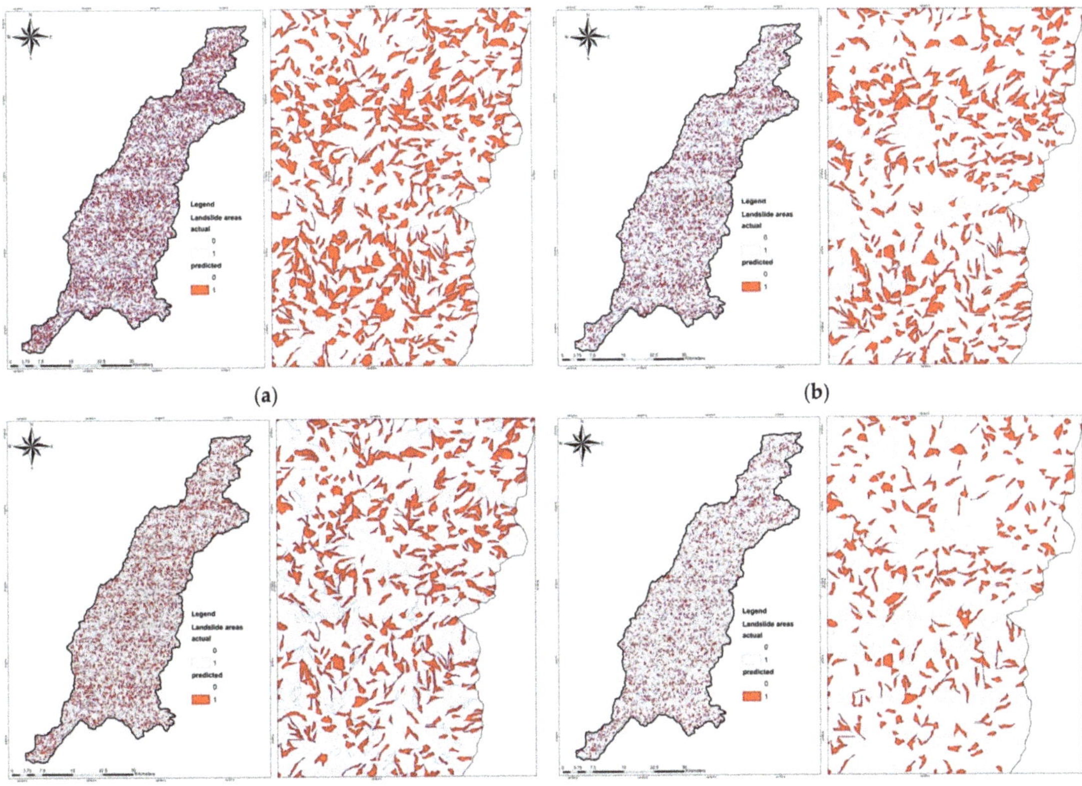

Figure 10. The results of model training and validations of Laonong River watershed (**a**) 2011; (**b**) 2012; (**c**) 2014; (**d**) 2016.

From Table 4, the analysis showed that the accuracy and sensitivity (recall) of the RF model had been greatly improved after adding factors of antecedent landslide ratios, which indicated a noticeable improvement in the model's capability of classification. Although the overall ACC of the model was improved due to a large amount of TN, the lower missing rate, i.e., 1-recall, from the RF model, was critical to denote more landslide samples. The Kappa value showed fair to good accuracy [30], within an acceptable range (0.4~0.74).

It was also noted that the precision values from 2012 to 2014 were at the same level of around 80% as 2011, and the precision from 2015 and 2016 was slightly lower than the level of the trained model. The precision variation from the tests had indicated that the proposed RF model was valuable and referential when applied to predict future landslide potential without re-training the RF model. However, it was also noted that the model might need to update every three years if precision performance was a concern.

In any subsequent year (2012 to 2016), the AUC value of the temporal-factor-added RF model was higher than that of the model that did not include factors of antecedent landslide ratios. The validation results show that the RF model with temporal characteristics had better performance than the model without those temporal factors.

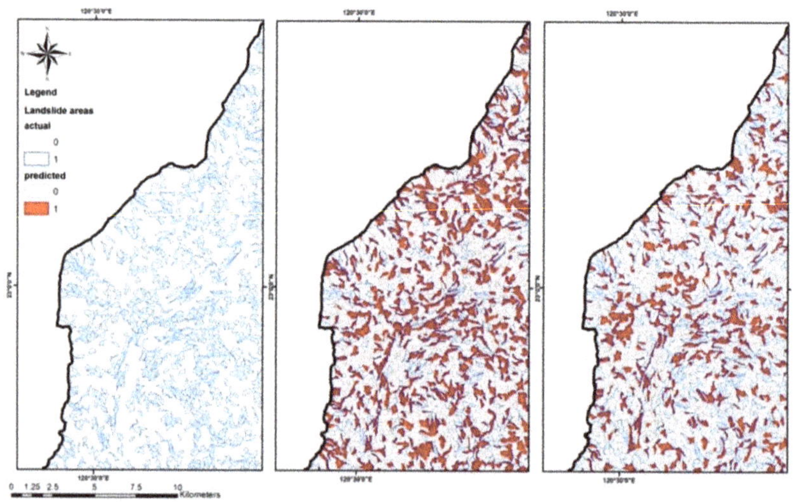

Figure 11. The comparison of model training between 43 and 8 factors using 2011 database of the Qishan River watershed.

4. Model Performance Evaluation for Individual Rainfall Events

Instead of using annual I_{max} and R_{te} for analysis, the selected rainfall events in 2017 (Table 5) were used to test the model's ability of landslide potential prediction if only short-period data were available. The test results would help to understand the capability of the RF model in practical applications during disaster response operations.

Based on the data from the 73 rainfall stations surrounding the study area, the I_{max} and R_{te} of each event (Table 5) were determined and inputted into the model as the factors of current annual rainfall indices. The in-situ rainfall conditions, I and R, disaster overview, and disaster scale were available from the disaster case reports, as shown in Table 5. The proposed RF model obtained the predicted landslide areas using events' rainfall conditions, and the result of the ChenYuLan River watershed was shown in Figure 12. The local government reported the landslide disasters, and three locations were in the study areas. Figure 13 shows the predicted landslide locations and observed ones from the disaster reports of events. It is noted that the reported three locations of landslides are not precisely at the expected locations by the proposed RF model. The model did not capture the two landslide locations at the northern ChenYuLan River watershed. However, the third landslide location at Tongfu Village was close to the predicted area by the model. The results imply that the "hit rate" of the RF model might not perform well when targeting event-based predictions. A rational inference was that the rainfall conditions used in the model input were different from the actual local rainfall conditions since the maximum rainfall intensity was determined from the surrounding rainfall stations, which might not be suitable to represent the local conditions. Another possible explanation of the model's miss-hit was the lack of past five years' data of landslides at the two northern locations in the ChenYuLan River watershed. Nevertheless, the proposed RF model had a promising performance for landslide potential evaluation when applied to rainfall events.

Table 5. Selected landslide disaster cases in 2017 [31,32].

Event	I_{max}/R_{te} (mm) [1]	Disaster Location [2]	Time & Date	Rainfalls at Occurrence [3]	Disaster Overview	Disaster Scale	Hazard Type
0601 Heavy Rainfall (2017/06/02 ~ 2017/06/04)	76.7/749.0	Shuili Township/Xinshan Village (X:236472 Y:2628694)	8:00 on 3 June	I = 43 mm/h R = 270 mm (Rainfall Sta.: XiLuan 01H47)	During the 0601 Heavy Rainfall, the landslide of Renlun Forest Road had caused blockage of about 10 m on Taiwan Provincial Highways 21.	The landslide scale was about 20 m in length, 10 m in width, and 2 m in depth. The landslide area was 200 m^2, and the amount of debris accumulated was about 400 m^3.	landslide
		Xinyi Township/Mingde Village (X:235029 Y: 2622652)	15:00 on 3 June	I = 17 mm/h R = 430 mm (Rainfall Sta.: Xinyi C0I080)	During the heavy rainfall, the slopes on both sides of potential debris flow torrent DF187 collapsed. As a result, existing rivers were piled up with soil and rocks, and a large amount of debris could not be discharged in time, resulting in blockage of 87 k~88 k of Taiwan Provincial Highways 21.	The landslide was about 10 m long, 5 m wide, 2 m deep and covered an area of 100 m^2. The accumulation of debris was about 600 m^3, with the length about 100 m, the width about 3 m, and the depth about 2 m.	landslide
Typhoons Nesat and Haitang	62.2/307.8	Xinyi Township/Tongfu Village (X:239004 Y:2607456)	7:50 on 30 July	I = 0 mm/h R = 150 mm (Rainfall Sta.: Heshe C1I070)	During Typhoons Nesat and Haitang, the slope at 115 k+540 of Taiwan Provincial Highways 21 collapsed and caused road blockage.	The landslide was about 15 m long, 10 m wide, and 2 m deep. The area of soil and debris was 150 m^2, and the accumulated earth volume was about 3500 m^3.	rock-fall, landslide

[1] Reference rainfall station: Nan Tian Chih (0601 event); San Di Men (Typh. Nesat and Haitang). [2] at Nantou County, coordinate using TWD97. [3] The recorded rainfalls when the landslide occurred. R: accumulated rainfall; I: hourly rainfall.

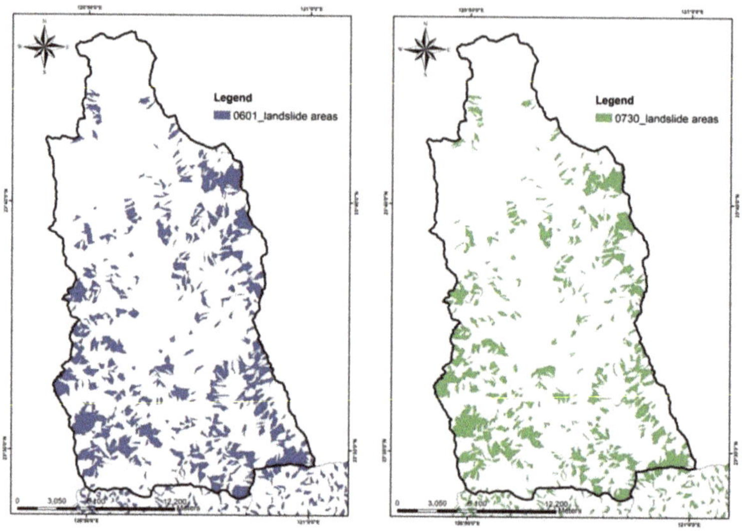

(a) (b)

Figure 12. The predicted landslide areas of ChenYuLan River watershed: (**a**) 0601 Heavy Rainfall; (**b**) Typhoons Nesat and Haitang in 2017.

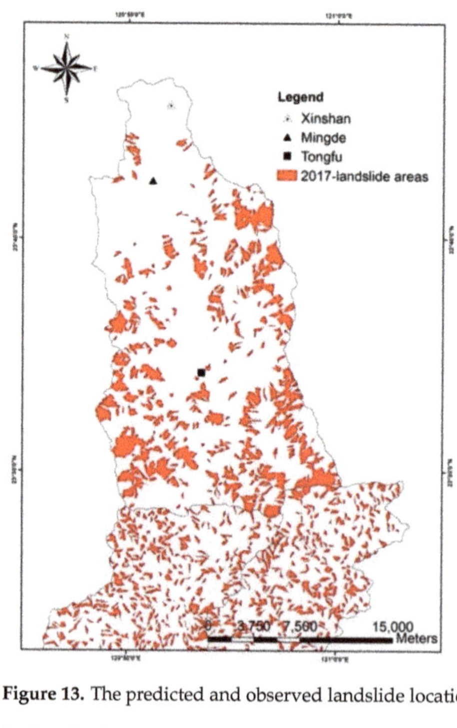

Figure 13. The predicted and observed landslide locations of events in 2017.

5. Conclusions

Based on the analysis results from the study, the importance of antecedent landslide areas in the past five years has been emphasized and discussed. Furthermore, the performance of the proposed RF model was also highlighted by conducting annual validation from 2011 to 2016. As a result, the following conclusions were derived.

1. According to the analysis results, adding temporal characteristics had significantly improved the performance of landslide potential prediction by the proposed random forest model.
2. The contribution of antecedent landslide ratios was significant in improving the model performance. The performance improvement of the model indicated that the time-dependent factors should be taken into consideration, in terms of a series of inputs within a period, such as the five years in this study.
3. The results of better model performance had shown the significance of cumulative change with time. Therefore, a relevant factor, e.g., the antecedent landslide ratios of the past five years, to describe the time effects was significant in landslide potential evaluation.
4. The trained model by considering annual temporal factors provided an angle to estimate the landslide potential in event-type disaster responses.

Author Contributions: Conceptualization, Y.-M.H.; methodology, Y.-M.H.; software, S.-W.L.; validation, Y.-M.H. and S.-W.L.; formal analysis, Y.-M.H. and S.-W.L.; investigation, Y.-M.H.; resources, Y.-M.H.; data curation, Y.-M.H. and S.-W.L.; writing—original draft preparation, Y.-M.H.; writing—review and editing, Y.-M.H.; visualization, Y.-M.H. and S.-W.L.; supervision, Y.-M.H. Huang; project administration, Y.-M.H.; funding acquisition, Y.-M.H. All authors have read and agreed to the published version of the manuscript.

Funding: This research was funded by Soil and Water Conservation Bureau, Council of Agriculture, Executive Yuan, Taiwan, grant number 109AS-10.7.1-SB-S3.

Institutional Review Board Statement: Not applicable.

Informed Consent Statement: Not applicable.

Data Availability Statement: No new data were created or analyzed in this study. Data sharing is not applicable to this article.

Acknowledgments: The authors would like to express their gratitude to research assistant Hsin-Ping Wang, for technical support in generating figures and GIS operations. Authors would also like to thank the editor and anonymous reviewers for the careful review of the original manuscript.

Conflicts of Interest: The authors declare no conflict of interest. The funders had no role in the design of the study; in the collection, analyses, or interpretation of data; in the writing of the manuscript, or in the decision to publish the results.

References

1. Lei, T.C.; Huang, Y.M.; Lee, B.J.; Hsieh, M.H.; Lin, K.T. Development of an empirical model for rainfall-induced hillside vulnerability assessment: A case study on Chen-Yu-Lan watershed, Nantou, Taiwan. *Nat. Hazards* **2014**, *74*, 341–373. [CrossRef]
2. Huang, Y.M.; Lei, T.C.; Lee, B.J.; Hsieh, M.H. Landslide Potential Evaluation Using Fragility Curve Model. In *Landslides—Investigation and Monitoring*, 1st ed.; Ray, R., Lazzari, M., Eds.; IntechOpen: London, UK, 2019. [CrossRef]
3. Chang, F.J.; Lee, S.P. A study of the intelligent control theory for the debris flow warning system. In Proceedings of the First International Conference on Debris Flow Hazards Mitigation, San Francisco, CA, USA, 7–9 August; Philip Chen, C.L., Ed.; American Society of Civil Engineers: Reston, VA, USA, 1997; pp. 109–123.
4. Lee, C.T.; Cheng, C.T.; Liao, C.W.; Tsai, Y.B. Site classification of Taiwan free-field strong-motion stations. *Bull. Seismol. Soc. Am.* **2001**, *91*, 1283–1297.
5. Lee, S.; Choi, J. Landslide susceptibility mapping using GIS and the weight-of-evidence model. *Int. J. Geogr. Inf. Sci.* **2004**, *18*, 789–814.
6. Ermini, L.; Catani, F.; Casagli, N. Artificial neural networks applied to landslide susceptibility assessment. *Geomorphology* **2005**, *66*, 327–343.
7. Wang, H.B.; Sassa, K. Rainfall-induced landslide hazard assessment using artificial neural networks. *Earth Surf. Proc. Land.* **2006**, *31*, 235–247.
8. Tan, L.; Guo, J.; Mohanarajah, S.; Zhou, K. Can we detect trends in natural disaster management with artificial intelligence? A review of modeling practices. *Nat. Hazards* **2021**, *107*, 2389–2417. [CrossRef]

9. Dikshit, A.; Sarkar, R.; Pradhan, B.; Segoni, S.; Alamri, A.M. Rainfall Induced Landslide Studies in Indian Himalayan Region: A Critical Review. *Appl. Sci.* **2020**, *10*, 2466.
10. Park, S.; Kim, J. Landslide Susceptibility Mapping Based on Random Forest and Boosted Regression Tree Models, and a Comparison of Their Performance. *Appl. Sci.* **2019**, *9*, 942. [CrossRef]
11. Pourghasemi, H.R.; Kerle, N. Random forests and evidential belief function-based landslide susceptibility assessment in Western Mazandaran Province, Iran. *Environ. Earth Sci.* **2016**, *75*, 1–17.
12. Youssef, A.M.; Pourghasemi, H.R.; Pourtaghi, Z.S.; Al-Katheeri, M.M. Landslide susceptibility mapping using random forest, boosted regression tree, classification and regression tree, and general linear models and comparison of their performance at Wadi Tayyah Basin, Asir Region, Saudi Arabia. *Landslides* **2015**, *13*, 1–18.
13. Di Napoli, M.; Carotenuto, F.; Cevasco, A.; Confuorto, P.; Di Martire, D.; Firpo, M.; Pepe, G.; Raso, E.; Calcaterra, D. Machine learning ensemble modelling as a tool to improve landslide susceptibility mapping reliability. *Landslides* **2020**, *17*, 1897–1914. [CrossRef]
14. Catani, F.; Lagomarsino, D.; Segoni, S.; Tofani, V. Landslide susceptibility estimation by random forests technique: Sensitivity and scaling issues. *Nat. Hazards Earth Syst. Sci.* **2013**, *13*, 2815–2831. [CrossRef]
15. Chen, W.; Xie, X.; Wang, J.; Pradhan, B.; Hong, H.; Bui, D.T.; Duan, Z.; Ma, J. A comparative study of logistic model tree, random forest, and classification and regression tree models for spatial prediction of landslide susceptibility. *CATENA* **2017**, *151*, 147–160.
16. Crichton, D. The risk triangle. In *Natural Disaster Management*; Ingleton, J., Ed.; Tudor Rose: London, UK, 1999; pp. 102–103.
17. Lee, C.Y.; Lee, B.J.; Huang, Y.M.; Lei, T.C.; Hsieh, M.H.; Wang, H.P.; Fang, Y.M.; Yin, H.Y. Risk Assessment of Landslide by using Fragility Curves—A Case Study in Shenmu, Taiwan. In *Geotechnical Hazard Mitigations: Experiment, Theory and Practice, Proceedings of the 5th International Conference on Geotechnical Engineering for Disaster Mitigation and Rehabilitation*; Taipei, Taiwan, 13–14 September 2017, Lin, M.L., Ed.; Chinese Taipei Geotechnical Society and Ainosco Press Taiwan: Taipei, Taiwan, 2017; pp. 137–148. [CrossRef]
18. Segoni, S.; Piciullo, L.; Gariano, S.L. A review of the recent literature on rainfall thresholds for landslide occurrence. *Landslides* **2018**, *15*, 1483–1501.
19. Gariano, S.L.; Melillo, M.; Peruccacci, S.; Brunetti, M.T. How much does the rainfall temporal resolution affect rainfall thresholds for landslide triggering? *Nat. Hazards* **2020**, *100*, 655–670.
20. Fookes, P.G.; Sweeney, M.; Manby, C.N.D.; Martin, R.P. Geological and Geotechnical Engineering Aspects of Low-cost Roads in Mountainous Terrain. *Eng. Geol.* **1985**, *21*, 1–152.
21. Koukis, G.; Ziourkas, C. Slope Instability Phenomena in Greece: A Statistical Analysis. *Bull. Int. Assoc. Eng. Geol.* **1991**, *43*, 47–60.
22. Xie, M.; Esaki, T.; Zhou, G. GIS-based probabilistic mapping of landslide hazard using a three-dimensional deterministic model. *Nat. Hazards* **2004**, *33*, 265–282. [CrossRef]
23. Pradhan, B.; Lee, S. Delineation of landslide hazard areas on Penang Island, Malaysia, by using frequency ratio, logistic regression, and artificial neural network models. *Environ. Erath Sci.* **2009**, *60*, 1037–1054.
24. Lei, T.C.; Wan, S.; Chou, T.Y.; Pai, H.C. The knowledge expression on debris flow potential analysis through PCA + LDA and rough sets theory: A case study of Chen-Yu-Lan Watershed, Nantou, Taiwan. *Environ. Earth Sci.* **2011**, *63*, 981–997.
25. Wan, S.; Lei, T.C.; Chou, T.Y. A landslide expert system: Image classification through integration of data mining approaches for multi-category analysis. *Int. J. Geogr. Inf. Sci.* **2012**, *26*, 747–770.
26. Gomez, H.; Kavzoglu, T. Assessment of shallow landslide susceptibility using artificial neural networks in Jabonosa River Basin, Venezuela. *Eng. Geol.* **2005**, *78*, 11–27.
27. Nefeslioglu, H.A.; Gokceoglu, C.; Sonmez, H. An assessment on the use of logistic regression and artificial neural networks with different sampling strategies for the preparation of landslide susceptibility maps. *Eng. Geol.* **2008**, *97*, 171–191.
28. Shahabi, H.; Hashim, M.; Ahmad, B.B. Remote sensing and GIS-based landslide susceptibility mapping using frequency ratio, logistic regression, and fuzzy logic methods at the central Zab basin, Iran. *Environ. Earth Sci.* **2015**, *73*, 8647–8668.
29. Kavzoglu, T.; Sahin, E.K.; Colkesen, I. An assessment of multivariate and bivariate approaches in landslide susceptibility mapping: A case study of Duzkoy district. *Nat. Hazards* **2015**, *76*, 471–496.
30. Fleiss, J.L.; Levin, B.; Paik, M.C. *Statistical Methods for Rates and Proportions*, 3rd ed.; Wiley-Interscience: Hoboken, NJ, USA, 2003.
31. Debris Flow Disaster Prevention Information: Major Disaster Events. Available online: https://246.swcb.gov.tw/Achievement/MajorDisasters (accessed on 21 July 2021).
32. Global Disaster Event Book. Available online: https://den.ncdr.nat.gov.tw/ (accessed on 21 July 2021). (In Chinese)

water

MDPI

Article

Using Landslide Statistical Index Technique for Landslide Susceptibility Mapping: Case Study: Ban Khoang Commune, Lao Cai Province, Vietnam

Long Nguyen Thanh [1], Yao-Min Fang [2,*], Tien-Yin Chou [2], Thanh-Van Hoang [2], Quoc Dinh Nguyen [1], Chen-Yang Lee [3], Chin-Lun Wang [3], Hsiao-Yuan Yin [3] and Yi-Chia Lin [3]

1 Vietnam Institute of Geosciences and Mineral Resources, 67 Chien Thang Rd, Van Quan Street, Ha Dong, Hanoi 100000, Vietnam
2 Geographic Information System Research Center, Feng Chia University, 100 Wenhwa Rd, Seatwen, Taichung City 40724, Taiwan
3 Soil and Water Conservation Bureau, Council of Agriculture, No 6, Guanghua Rd, Nantou City 540206, Taiwan
* Correspondence: frankfang@gis.tw; Tel.: +886-4-24516609

Abstract: Ban Khoang is a mountainous commune in Sa Pa district located in the central part of Lao Cai province, Vietnam. Landslides occur frequently in this area and seriously affect the local living conditions. To help the local authority in developing a landslide disaster action plan, the statistical index method for landslide susceptibility mapping is applied. As the result, the landslide susceptibility zonation (LSZ) map was created. The LSZ map indicates that areas of low, moderate, high and very high landslide susceptibility zones are, respectively, 20.3 km^2, 12.4 km^2, 15.4 km^2, and 5.2 km^2; most of the observed landslide areas that are well predicted belong to high or very high landslide susceptibility classes. In detail, 80% observed landslide areas and 78.57% number of observed landslides were well predicted, and the area (AUC) under the receiver operating characteristic (ROC) curve obtained 80.3%. Hence, the high and very high landslide susceptibility classes in the LSZ map can be considered highly believable, and the LSZ map will be reliable to use in the practice.

Keywords: natural hazards; landslide; susceptibility; GIS; Vietnam

Citation: Thanh, L.N.; Fang, Y.-M.; Chou, T.-Y.; Hoang, T.-V.; Nguyen, Q.D.; Lee, C.-Y.; Wang, C.-L.; Yin, H.-Y.; Lin, Y.-C. Using Landslide Statistical Index Technique for Landslide Susceptibility Mapping: Case Study: Ban Khoang Commune, Lao Cai Province, Vietnam. *Water* **2022**, *14*, 2814. https://doi.org/10.3390/w14182814

Academic Editor: Su-Chin Chen

Received: 23 July 2022
Accepted: 6 September 2022
Published: 9 September 2022

Publisher's Note: MDPI stays neutral with regard to jurisdictional claims in published maps and institutional affiliations.

1. Introduction

Ban Khoang is a mountainous commune in Sa Pa district, Lao Cai province of Vietnam, where these landslides occur regularly (Figure 1). In particular, a vast landslide happened in Can Ho A village, Ban Khoang commune in September 2013, causing 14 people loss and severe property damage. Hence, predicting landslide hazards is very important for the inhabitants and local administration of Ban Khoang commune to mitigate landslide damage in this area.

According to the result of a nationwide project "Investigation, assessment and geohazards susceptibility zonation in mountainous areas of Vietnam" [1] recently, Ban Khoang is one of 200 communes with highest level of landslides susceptibility in Vietnam.

Therefore, the LSZ mapping will be very necessary and helpful for local authorities and people in landslide hazard prevention and mitigation, as well as developing a landslide action plan. In addition, the LSZ map will be a technical foundation for practical activities relevant to setting up landslide early warning systems.

The most straightforward initial approach to any study of landslide hazards is the compilation of a landslide inventory and analyzing the relationship with different causative factors to predict landslide-prone areas [2]. In Ref. [3], Carrara (1983) introduced the so-called statistical approach for landslide hazard assessment. This technique has been widely employed and has become one of the most popular approaches for landslide hazard assessment worldwide. Combinations of factors that have led to landslides in the past are identified statistically, and quantitative predictions are made for areas currently free of landslides but with similar conditions. Since then, many other statistical approaches have been proposed and used in landslide susceptibility mapping and analyses.

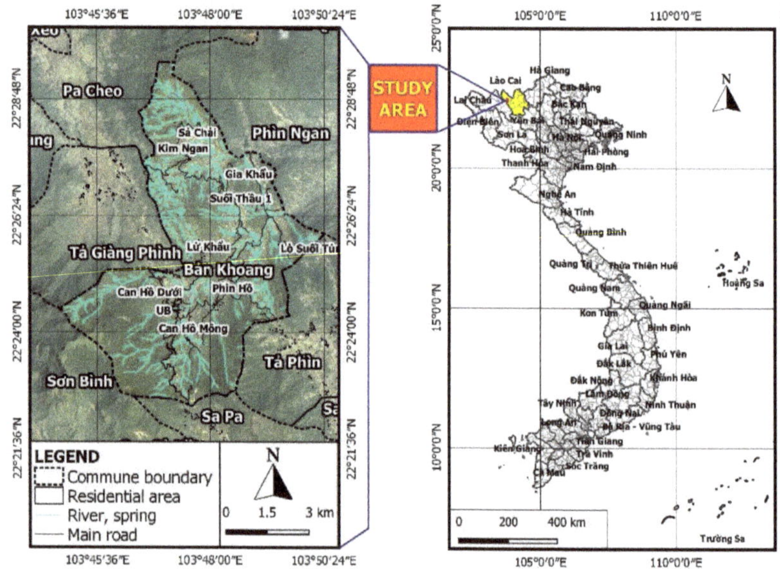

Figure 1. In this study area, 28 landslides, covering 0.262 km^2 (Figure 2), were identified (from 2012 up to May 2022) based on (1) field recognizance to investigate landslide occurrences, (2) collection of historical literature on landslides, and (3) interpretation of available multi-serial google images coupled with field verification.

Basically, statistical landslide susceptibility approaches are based on related spatial information on past landslide activities (i.e., landslide presence/absence) to static geoenvironmental factors (e.g., topography, geology, geomorphology, land use, fault density, soil, and drainage density) using statistical techniques. In Ref. [4], Steger et al. (2016) commented that the generated empirical relation, commonly expressed as a relative susceptibility score, is then applied to each spatial unit of an area (e.g., grid cell, and slope unit) [5–7]. The validation of spatial predictions is commonly evaluated by interpreting inventory-based predictive performance estimates [8–10].

It is obvious that the landslide inventory is a vital component to obtaining high-quality statistical landslide susceptibility models because most analysis steps are dependent on a correct representation of past landslide occurrences [4,9,11–14].

Several studies compared statistical landslide susceptibility models produced from heterogeneous inventories [4,15–19]. However, a differentiated evaluation of the propagation of potential inventory-based errors into landslide susceptibility models was hampered due to the practical inseparability of positional accuracy and inventory completeness as well as the lack of truly accurate reference inventories.

There are many previous works using the statistical approaches for landslide susceptibility assessment (e.g., methods of statistical index, certainty factor, probability, weight of evidence modeling, and logisitic regression). However, the selection of input parameters or causative factors for landslide susceptibility mapping, the method for landslide susceptibility mapping and landslide susceptibility index classification are still confused between many studies.

The statistical index method is considered the simplest and quantitatively suitable method for statistical approaches for landslide susceptibility mapping. However, it has been adopted by various researchers [19–27].

Therefore, in this study, the statistical index method is applied for landslide susceptibility analyses of Ban Khoang commune in Sa Pa district, Lao Cai province of Vietnam. The research result will play an important role for landslide hazard prevention and mitigation in this mountainous commune in Vietnam.

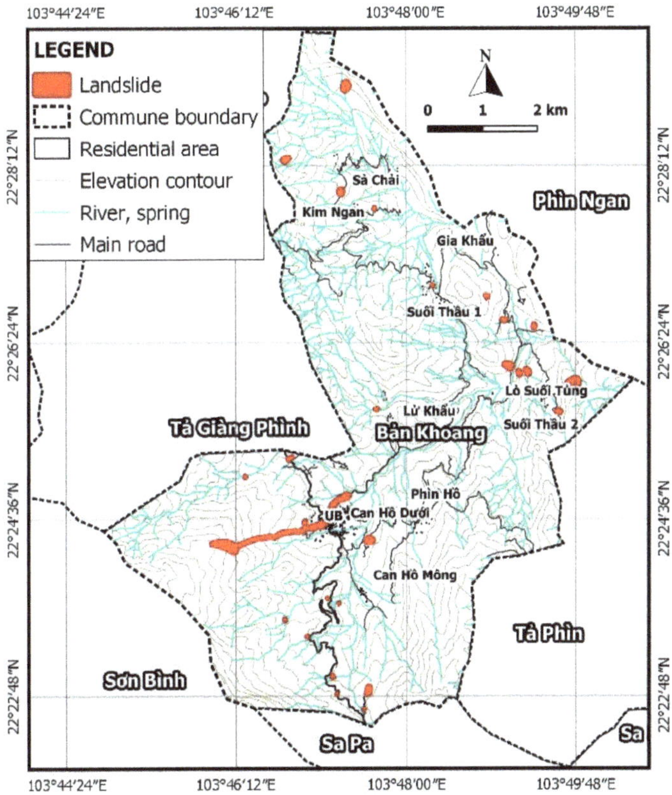

Figure 2. Map of landslide inventory in the study area.

2. Landslide Inventory

The study area, Ban Khoang commune selected for assessment of landslides susceptibility (Figure 1) is about 53.3 km^2.

The average size of the landslides in the study area is approximately 9369 m^2, but the details about width, depth, types, or causes of some landslides were not identified. Some pictures of landslide inventory are displayed in Figure 3.

(A)

(B)

(C)

Figure 3. Some landslide pictures in Ban Khoang commune, Sa Pa district, Lao Cai province of Vietnam. (**A**) Landslide as debris flow occurred near hospital of Ban Khoang commune in 2013. (**B**) Landslides near provincial road 155 in Ban Khoang commune. (**C**) Landslide on the provincial road 155, the section closed to Can Hồ B village in Ban Khoang commune on 22 May 2022.

3. Landslide Causative Factors

The selection of causative factor maps for landslide susceptibility should be considered carefully based on relevance, availability and scale attributes. These are cumbersome in Vietnam, as systematic studies and inventories of spatial characteristics and land cover features have only been initiated recently by different government institutions. Therefore, such data are often lacking, incomplete, or on a scale that is not useful for scientific purposes, especially in remote and rural regions as the present study area. Hampered by such constraints, eight digital causative factors map for landslide analysis could be developed:

- Topography is intrinsically associated with landslides by slope gradient and other factors, such as weathering, precipitation, soil thickness, etc. Hence, topography strongly affects landslides [28,29]. Ban Khoang is a mountainous area where the microclimate is quite predominant. Hence, the aspect is considered an indirect landslide causative factor in this study. A digital elevation map (DEM) of the study area with a pixel size of 10 m by 10 m was obtained by using inverse distance weighted interpolation in QGIS 3.6 from elevation points and contours of a topographic map, scale 1:10,000, published by the Cartographic Publishing House, Vietnamese Ministry of Natural Resources and Environment (2019). Then the aspect map of Ban Khoang commune (Figure 4A) was developed based on the Aspect tool inside QGIS 3.6 software.
- In most landslide studies, slope gradient is considered a principal causative or triggering factor. A slope map was derived from the DEM using the slope function tool of QGIS 3.6. The slope map is in the form of a raster map with the same 10 m pixel size as the DEM, but was converted to vector by separating the slope angles into six classes: (1) flat-gentle slope (<5°), (2) fair slope (5–15°), (3) moderate slope (15–25°), (4) fairly moderate slope (25–35°), (5) steep slope (35–45°), and (6) very steep slope (>45°). The map of slope classes of Ban Khoang commune is displayed in Figure 4B.
- Geology and slope instability are strongly associated [30,31]. Hence, a geological map of Ban Khoang (Figure 4C) was derived from the map of geology and mineral resources of the Lao Cai sheet group, scale 1:50,000 by Lap et al. (2003) [32]. Figure 4C displays the distribution of geological classes in Ban Khoang commune in Sa Pa district, Lao Cai province of Vietnam.
- Geomorphology is considered an essential factor related to landslide occurrence in the study area. Based on the analyses of the topological characteristics, geological structures, neotectonic movements, and morphometries, six geomorphological units can be identified in the study area by [33] (Figure 4D).
- Soil is an essential factor of slope instability in many settings [34,35]. A digital map of soil was derived from previous work in Lao Cai province carried out by the National Institute of Agriculture Planning and production (2019), identifying three types of soil mechanics in the study area, i.e., (1) outcrop, (2) reddish-yellow humus soil on claystone, and (3) reddish-yellow humus soil on magma rocks (Figure 4E). The soil depth map (Figure 4F) was derived based on the soil depth information based on the map of soil mechanics.
- Neotectonics contribute to slope instability by fracturing, faulting, jointing, and deforming foliation structures [36,37]. For this study, faults were extracted from the map of geology and mineral resources scale 1:50,000. Additionally, lineaments were interpreted from free available Landsat 8 captured by NASA in 2020. The fault and lineament density was calculated as the total length of faults and lineament per 1 km^2 (See Figure 4G).
- Studies have shown that the proximity to drainage axes with intensive gully erosion is an important factor controlling the occurrence of landslides [38,39]. A map of river density was derived on the basis of the digitizing river and stream courses on the topographic map and interpolation in QGIS software (version 3.6). A map of the river density class (Figure 4H) was created by subdividing the river density range values into five classes: (1) <1000 m/km^2, (2) 1000–2000 m/km^2, and (3) 2000–3000 m/km^2, (4) 3000–4000 m/km^2, and (5) >4000 m/km^2.
- Vegetation augments slope stability primarily in two ways: (1) by removing soil moisture through evapotranspiration and (2) by providing root cohesion to the soil mantle [40]. A land-use map was obtained from the land-use map of Lao Cai published by the land administration department of the Ministry of Natural Resources and Environment, 2019 [41]. The land use composed of 10 land-use classes is displayed in Figure 4I.

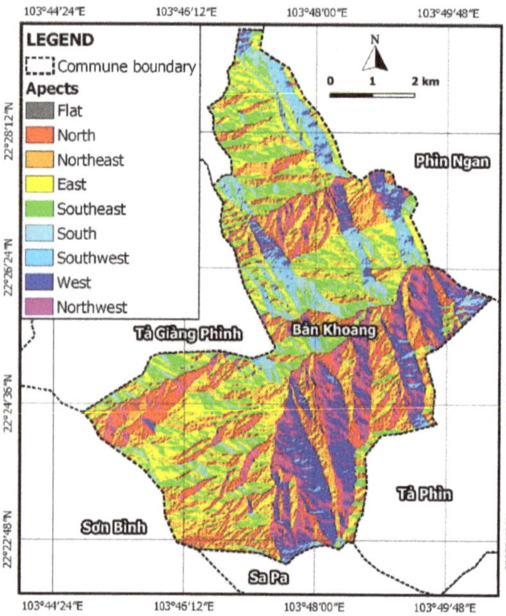

(**A**) Aspect map.

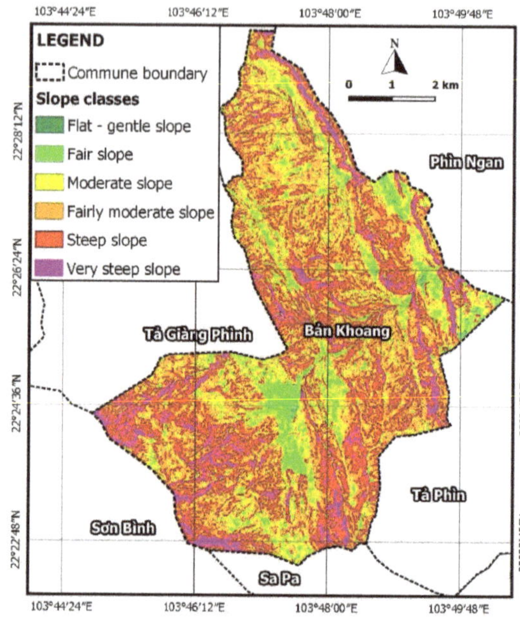

(**B**) Slope map.

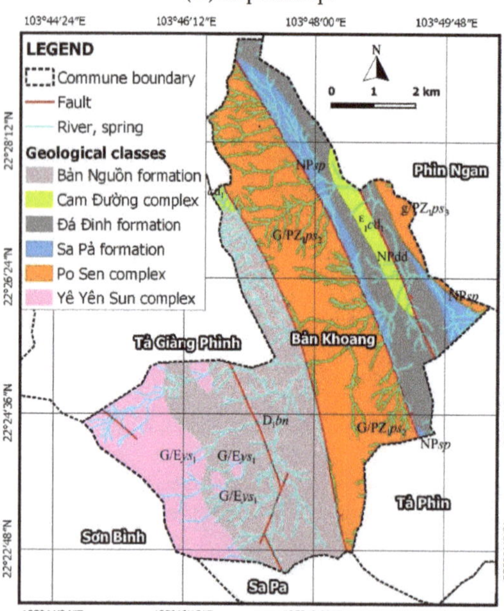

(**C**) Geological map.

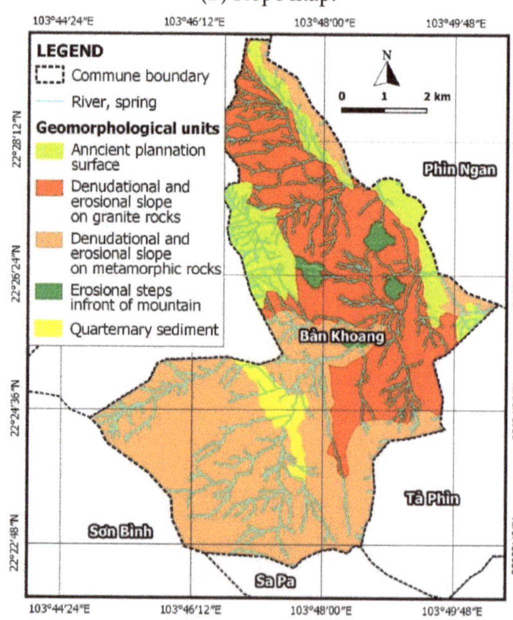

(**D**) Geomorphological map.

Figure 4. *Cont.*

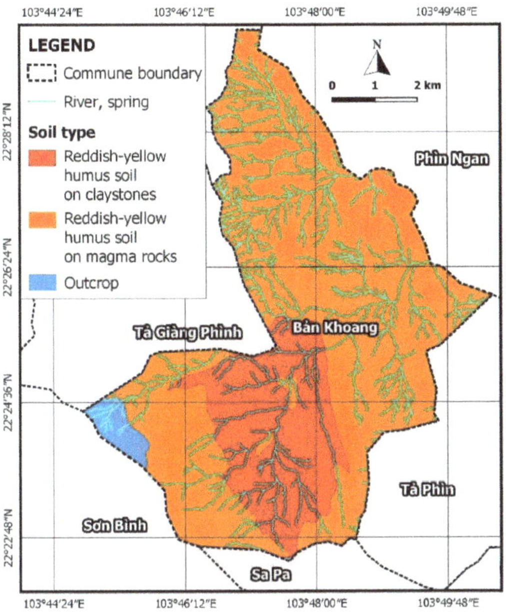

(**E**) Soil-type map.

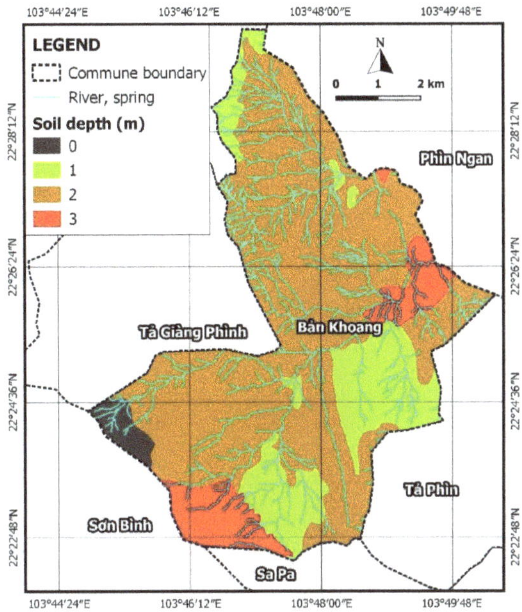

(**F**) Soil-depth map.

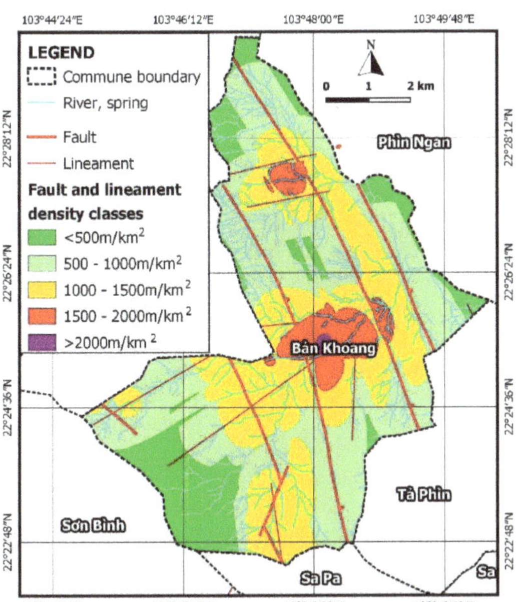

(**G**) Map of fault and lineament density.

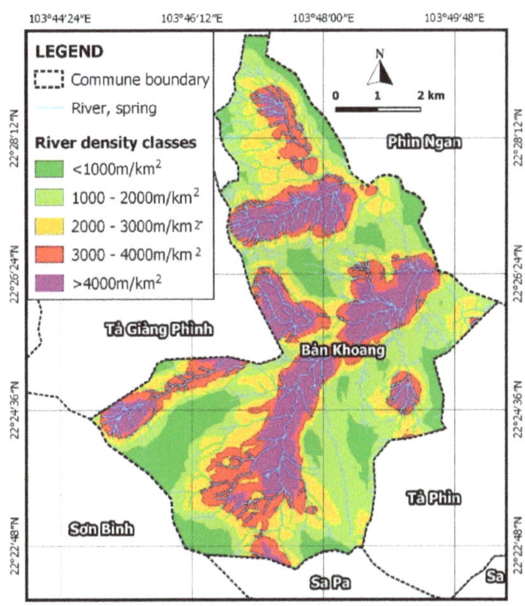

(**H**) Map of river density.

Figure 4. *Cont.*

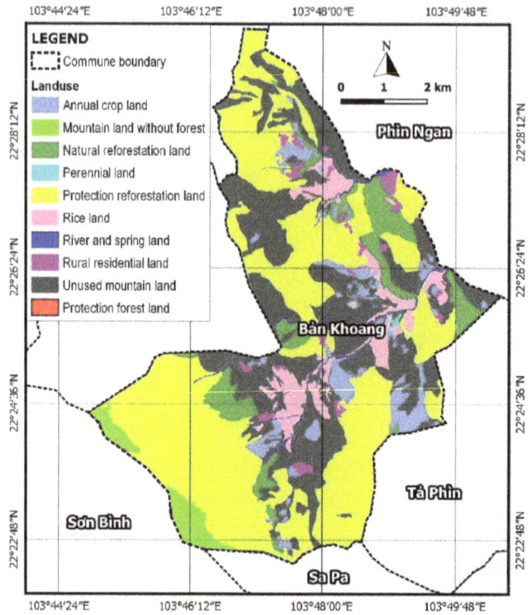

(I) Land-use map.

Figure 4. Landslide causative parameters for landslide susceptibility mapping of Ban Khoang commune.

4. Method for Landslide Susceptibility Analysis

The statistical index method is a bivariate statistical technique introduced by van Westen in 1997 [20] for landslide susceptibility analyses. Other researchers, such as Gebremedhin et al., 2021 [21], Mandal et al., 2018 [22], Wu et al., 2017 [23], Wang et al., 2016 [24], Dieu et al., 2011 [25], Long, 2008 [19], Cevik and Topal, 2003 [27], and Oztekin and Topal, 2005 [26], also applied this technique and termed it the statistical index method. In the statistical index method, a weight value for a parameter class is defined as the natural logarithm of the landslide density in the class divided by the landslide density in the entire map [20].

$$W_{ij} = \ln\left(\frac{f_{ij}}{f}\right) \tag{1}$$

where W_{ij} is the weight of a class i of parameter j, f_{ij} the landslide density within the class i of parameter j, and f the landslide density within the entire map. Hence, the statistical index method is based on statistical correlation of the landslide inventory map with attributes of different parameter maps. The W_{ij} value in Equation (1) is only calculated for classes that have landslide occurrences. If there are no landslide occurrences in a parameter class, the W_{ij} will be assigned to zero [20,30]. This also means that the parameter class having no landslide occurrences will have no correlation with the landslide inventory. Hence, it does not influence the calculation of the landslide susceptibility index.

In this study, nine landslide causative factors, i.e., (1) slope, (2) geology, (3) geomorphology, (4) soil depth, (5) soil type, (6) land use, (7) fault and lineament density, and (8) river density (Figure 4), were used as the layer input for landslide susceptibility index mapping. The workflow for landslide susceptibility mapping in Ban Khoang commune is shown in Figure 5. Every parameter map is crossed with the landslide map, and the density of the landslide in each class is calculated. The distribution of landslides for various data layers and weight wij values are shown in Table 1. The distribution of landslides for various data layers, weight of class (W_{ij}) of all the causative factors in the study area is displayed in Table 2.

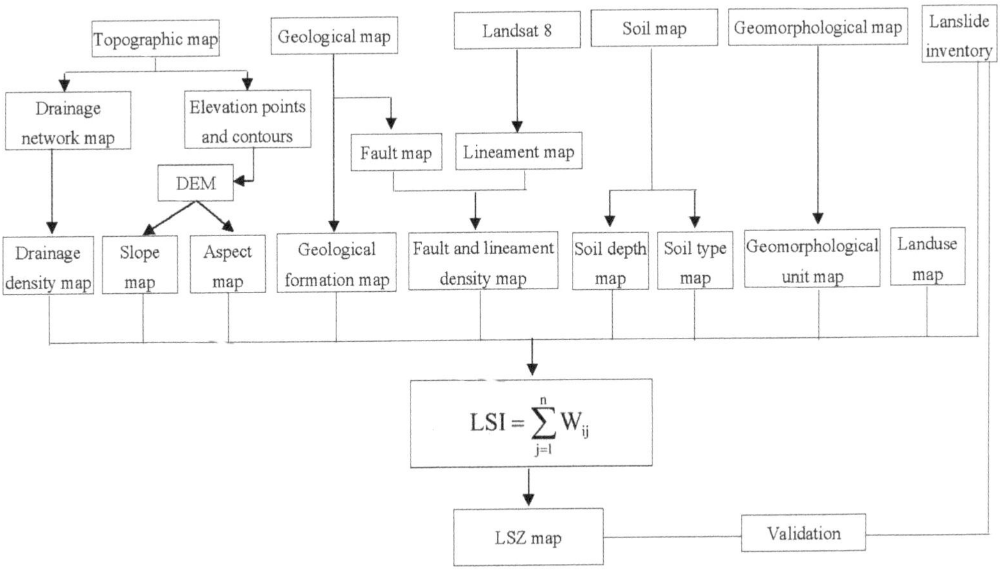

Figure 5. Work flow for landslide susceptibility mapping in the study area.

Table 1. Distribution of landslides for various data layers, weight of class (W_{ij}) of all causative factors in the study area.

Landslide Causative Factors	Landslide Occ. Pixels	% Occ.	No. of Pixels in Domain	% Domain	W_{ij}
Slope					
<5°	11	0.42	2798	0.52	−0.2261
5–15°	537	20.43	45,087	8.46	0.8823
15–25°	580	22.07	114,691	21.51	0.0257
25–35°	508	19.33	182,665	34.26	−0.5723
35–45°	609	23.17	138,605	25.99	−0.1149
>45°	383	14.57	49,356	9.26	0.4539
Fault and lineament density					
<500 m/km^2	348	13.24	81,710	15.32	−0.1461
500–1000 m/km^2	1103	41.97	235,319	44.13	−0.0502
1000–1500 m/km^2	1170	44.52	178,785	33.53	0.2835
1500–2000 m/km^2	7	0.27	35,889	6.73	−3.2296
>2000 m/km^2	0	0.00	1499	0.28	0.0000
River density					
<1000 m/km^2	758	28.84	86,095	16.15	0.5802
1000–2000 m/km^2	603	22.95	135,804	25.47	−0.1044
2000–3000 m/km^2	684	26.03	113,416	21.27	0.2018
3000–4000 m/km^2	250	9.51	96,950	18.18	−0.6478
>4000 m/km^2	333	12.67	100,937	18.93	−0.4014
Soil depth					
0 m	0	0.00	11,686	2.19	0.0000
1 m	175	6.66	94,253	17.68	−0.9763
2 m	2314	88.05	376,835	70.67	0.2198
3 m	139	5.29	50,428	9.46	−0.5812

Table 1. *Cont.*

Landslide Causative Factors	Landslide Occ. Pixels	% Occ.	No. of Pixels in Domain	% Domain	W_{ij}
Soil type					
Reddish-yellow humus soil on magma rocks	1629	61.99	401,940	75.38	−0.1957
Outcrop	0	0.00	11,686	2.19	0.0000
Reddish-yellow humus soil on claystone	999	38.01	119,576	22.43	0.5277
Geomorphology					
Ancient planation surface	93	3.54	59,101	11.08	−1.1417
Denudational and erosional slope on metamorphic rocks	1597	60.77	193,797	36.35	0.5140
Denudational and erosional slope on granite rocks	489	18.61	256,604	48.13	−0.9502
Quaternary sediment	427	16.25	9101	1.71	2.2533
Erosional steps in front of mountain	22	0.84	14,599	2.74	−1.1850
Geology					
Sa Pả formation	262	9.97	32,774	6.15	0.4836
Cam Đường formation	169	6.43	15,573	2.92	0.7893
Yê Yên Sun complex	571	21.73	69,501	13.03	0.5110
Po Sen complex	163	6.20	183,843	34.48	−1.7154
Đá Đinh formation	312	11.87	64,195	12.04	−0.0140
Bản Nguồn formation	1151	43.80	167,316	31.38	0.3334
Landuse					
Protection reforestation land	1089	41.44	252,972	47.44	−0.1353
Rice land	241	9.17	29,879	5.60	0.4926
Annual crop land	2	0.08	33,860	6.35	−4.4242
Natural reforestation land	585	22.26	39,883	7.48	1.0906
Rural residential land	16	0.61	8229	1.54	−0.9302
Unused mountain land	682	25.95	152,292	28.56	−0.0958
Mountain land without forest	0	0.00	12,683	2.38	0.0000
Perennial land	0	0.00	619	0.12	0.0000
River and spring land	13	0.49	2660	0.50	−0.0085
Protection forest land	0	0.00	125	0.02	0.0000
Aspect					
Flat	0	0.00	24	0.00	0.0000
North	422	16.06	80,933	15.18	0.0563
Northeast	904	34.40	103,501	19.41	0.5722
East	586	22.30	102,496	19.22	0.1484
Southeast	285	10.84	64,916	12.17	−0.1157
South	100	3.81	40,339	7.57	−0.6872
Southwest	102	3.88	33,324	6.25	−0.4764
West	58	2.21	54,362	10.20	−1.5303
Northwest	171	6.51	53,307	10.00	−0.4295

Table 2. Distribution of landslides for various data layers, weight of class (W_{ij}) of all causative factors in the study area.

Landslide Causative Factors	LSI_{Min}	LSI_{Max}	LSI_{Range}	LSI_{Dev}
Slope	−0.5723	0.8823	1.4546	−0.5183
Fault and lineament density	−3.2296	0.2835	3.5131	−1.4628
River density	−0.6478	0.5802	1.2280	−0.4851
Soil depth	−0.9763	0.2198	1.1961	−0.5453
Soil type	−0.1957	0.5277	0.7234	−0.3742
Geomorphology	−1.1850	2.2533	2.5047	−0.9107
Geology	−1.7154	0.7893	2.5047	−0.9107
Land use	−4.4242	1.0906	5.5147	−1.5015
Aspect	−1.5303	0.1484	2.1025	−0.6039

From Tables 1 and 2, it can be noted the following:

- For the slope factor, there is an obvious distinction between classes with slope angles 5–15° and >45° compared to other classes. This indicates that landslides in the study area are mainly occurring in areas with slope angles 5–15° and >45°.
- The class of fault density of 1000–1500 m/km^2 has the highest W_{ij} value (0.2835) compared to the remaining classes from all causative factors; hence, it has the highest impact on landslides in the study area.
- Cam Đường formation (W_{ij} = 0.7893) are distinctly more favorable for landslides compared to the other geological formations ($W_{ij} \leq 0.5110$).
- For the geomorphological factor, denudational and erosional slope on metamorphic rocks, Quaternary sediment, also favor landslides.
- For the land-use factor, natural reforestation land is most favorable for landslide occurrence. Other classes seem to have very little or no influence for landslides.

All W_{ij} layers for the different causative factors were constructed with QGIS 3.6 software. Next, these were summed up to obtain a resultant landslide susceptibility index map.

$$LSI = \sum_{j=1}^{n} W_{ij} \quad (2)$$

where LSI is the landslide susceptibility index and n the number of parameters.

As the result, the LSI map of Ban Khoang commune was developed and is displayed in Figure 6.

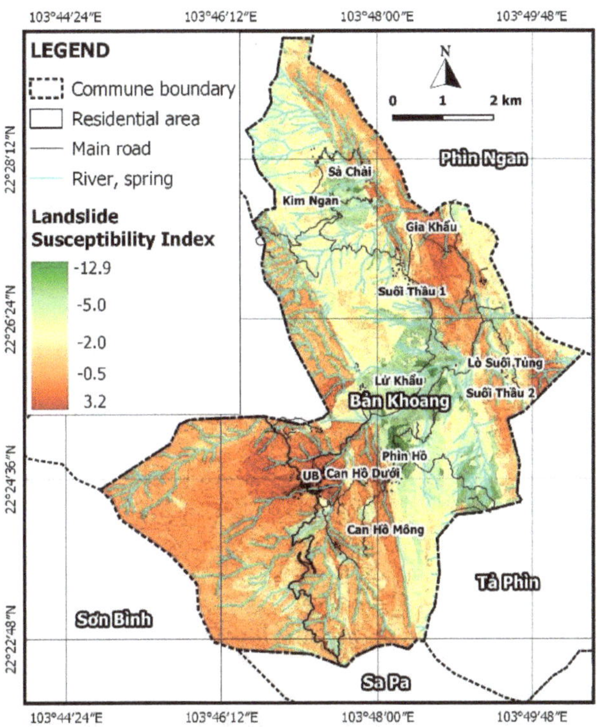

Figure 6. Map of landslide susceptibility index of Ban Khoang commune.

5. Results and Discussion

Classifier methods that have been used in landslide classification are manual classification [26,38,42–48], equal interval classification [49,50], standard deviation classification [51–55]. However, the authors usually do not explain the reasons for using a certain method in previous works.

In this study, the manual classifier method was used to reclassify the LSI values into four different susceptibility zones, according to the classification method that was proposed by Galang (2004) [56]. The susceptibility classes are low, moderate, high, and very high. Ideally, the classification method should satisfy the principle that higher landslide susceptibility classes should capture more or most landslide occurrences. Therefore, it is assumed that the expected number of observed landslide occurrences within a higher landslide susceptibility class equals two times the expected numbers in the next lower landslide susceptibility class. Hence, the expected numbers of observed landslide occurrences in the very high landslide susceptibility class equals two times the expected numbers in the high landslide susceptibility class, and so on. Based on this rule, it can be inferred that the expected percentages of observed landslide occurrences in the low, moderate, high, and very high landslide susceptibility classes are 6.7%, 13.3%, 26.7%, and 53.3% respectively.

Hence, the procedure is as follows. The landslide occurrence map is compared to the LSI map, and the cumulative percentage of observed landslide values versus ranked LSI values is calculated as shown in Figure 6. Three cut-off percentages of observed landslide occurrence in the cumulative curve are used to identify the four landslide susceptibility classes. It is 6.7% for separating the low from the moderate class, 20% for separating the moderate from high class, and 46.7% for separating the high from the very high class, as shown in Figure 7.

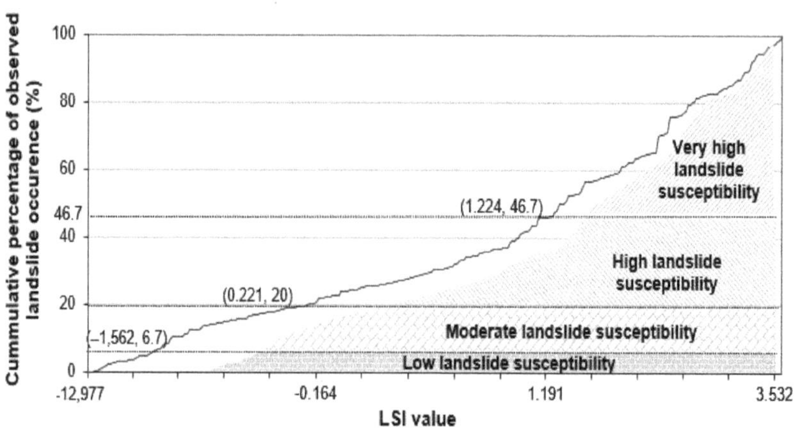

Figure 7. Cumulative percentage of observed landslide occurrence against LSI values.

As the result, the final map of landslide susceptibility zonation is shown in Figure 8. The statistical index shows that areas of low, moderate, high and very high landslide susceptibility zones are, respectively, 20.3 km^2 (38.0%), 12.4 km^2 (23.3%), 15.4 km^2 (28.9%), and 5.2 km^2 (9.8%).

In addition, to minimize the damage caused by natural disasters caused by climate change to people in Lao Cai province, Taiwan's Soil and Water Conservation Bureau (SWCB) and the Vietnam Institute of Science and Mineral Geology (VIGMR) have built a landslide monitoring station in Ban Khoang commune, Lao Cai in November 2019.

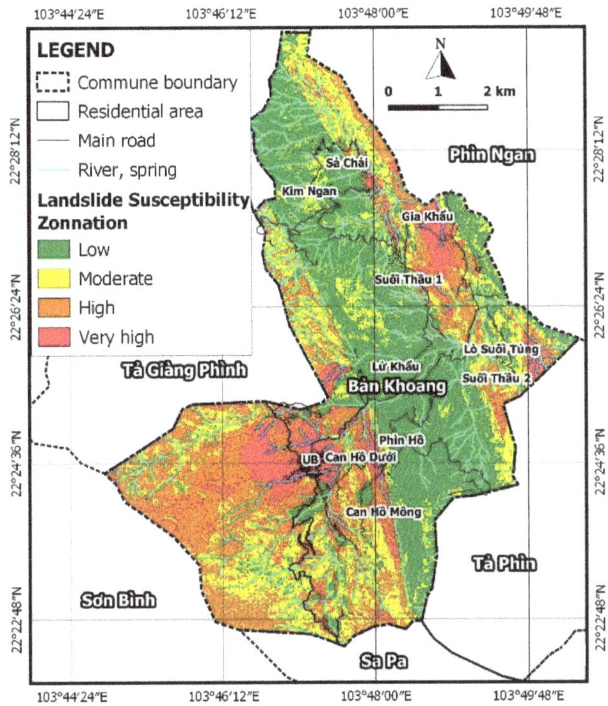

Figure 8. Map of landslide susceptibility zonation of Ban Khoang commune.

According to a survey by the VIGMR, Ban Khoang commune (Lao Cai province) is the area most at risk of landslides in Lao Cai province. At the same time, it is also the place with the highest risk of landslides in Vietnam. Therefore, installing a real-time landslide monitoring station in these two areas is essential.

Within the framework of international cooperation among the SWCB, GIS.FCU (the Geography information System Research Center of Taiwan Feng Chia University), VIGMR, WeatherPlus company (Former is AgriMedia), the projects "Study, develop a pilot debris flow early warning system in real time for mountainous areas of Vietnam" and "Study, develop a pilot debris flow early warning system in real time for mountainous areas of Vietnam" were applied to Ban Khoang area, Sa Pa town, Lao Cai province.

The real-time landslide early warning system deployed and installed in Ban Khoang includes a series of sensors, such as geophone, water level, tensiometer sensor, infrared cameras, and auto-rain gauges installed in three areas (Figure 9) (upstream (Figure 10) and midstream (Figure 11), and downstream (Figure 12)) to observe and record changes in weather conditions, such as precipitation, geology (by geogphone), hydrology (flow, water level) and the surface movement of liquid mud, soil and rock. All data are collected and processed on site (Figure 13) by the Data Processing Center.

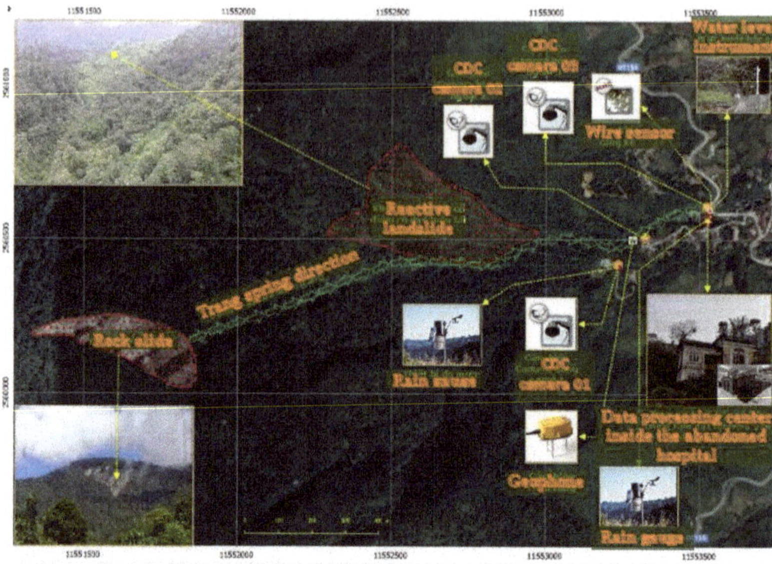

Figure 9. Real-time landslide monitoring station system in Ban Khoang, Lao Cai.

Figure 10. Upstream of the real-time landslide monitoring station.

Figure 11. Middle stream of the real-time landslide monitoring station.

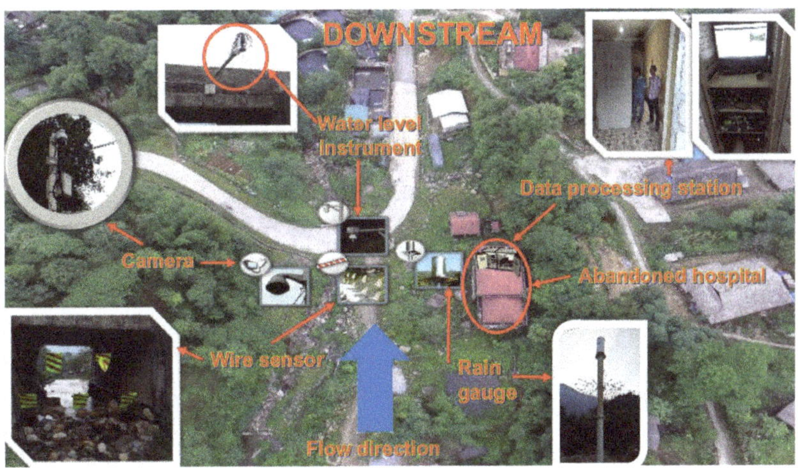

Figure 12. Downstream of the real-time landslide monitoring station.

Figure 13. On-site station at the downstream of the real-time monitoring station.

6. Validation of Landslide Susceptibility Map

The final map of landslide susceptibility zonation for the study area is shown in Figure 8, and the area percentages of landslide susceptibility classes and posterior landslide susceptibility of these classes in the final LSZ map are shown in Figure 14.

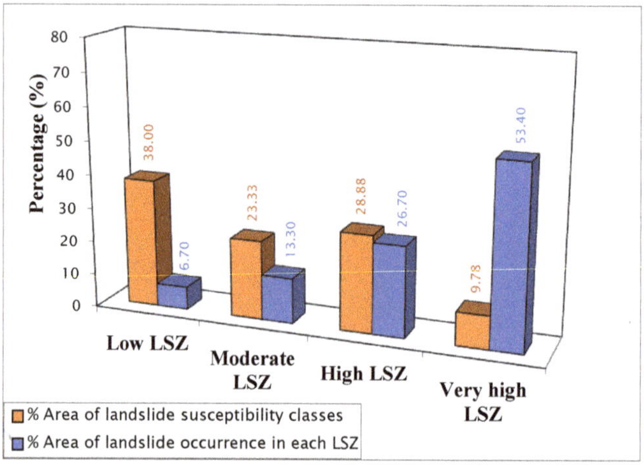

Figure 14. Area percentage of LSZ classes and the observed landslide accumulation in each LSZ class.

The accuracy of the final LSZ map is evaluated based on the observed landslides. First, Figure 13 shows that 80% observed landslide areas belonging to very high and high LSZ classes. Secondly, the final LSZ map is checked by overlaying it with the observed landslide map. In addition, As shown in Figure 14, there are various possibilities of different LSZs coinciding with a landslide polygon. Because in the inventory of the observed landslide, no distinction was made between the initiation part of the landslide and the areas of debris or flows, there can be no complete correspondence between the LSZ classes (Figure 15) that blue line is landslide area, and the complete observed landslide affected area. Hence, we consider a landslide as having "good" prediction when at least part of it is situated in a high or very high susceptibility zone. Otherwise, based on the above criteria, the model predicts 28 landslides in the study area, as shown in Table 3.

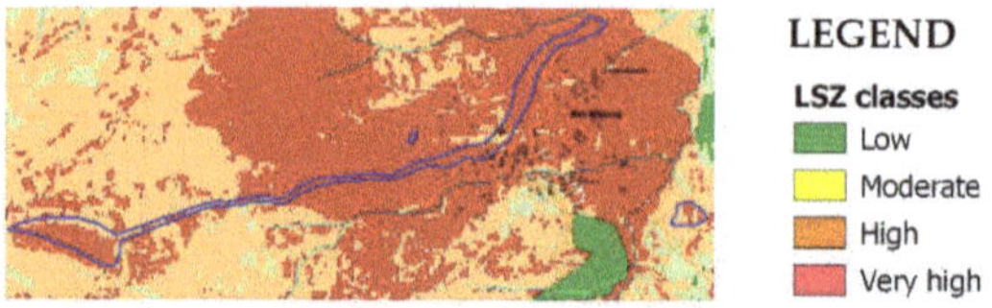

Figure 15. Example of some landslides overlaying the final LSZ map.

Table 3. LSZ validation result with observed landslide.

Accuracy of Prediction	Observed Landslide	
	Number	Percentage (%)
Good	22	78.57
Wrong	6	21.43

Table 3 indicates that 22 of the 28 observed landslides are well predicted (78.57%), and only 6 of the total landslides are wrongly predicted (21.43%). Figure 16 shows the LSZ map with the observed landslides indicating the different levels of prediction.

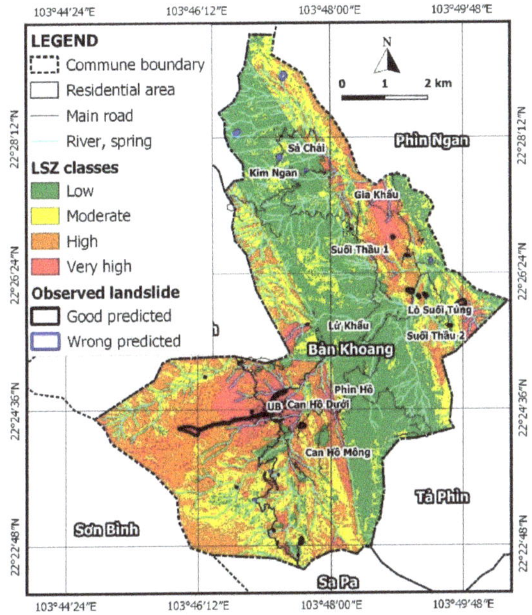

Figure 16. Validation LSZ map with observed landslides of Ban Khoang commune.

The Area Under the Curve (AUC) is used to qualitatively analyze the prediction accuracy of the landslide susceptibility map (Figure 17). The analysis results of the success rate curve indicated that the statistical index model has an approximately high AUC value of 0.803.

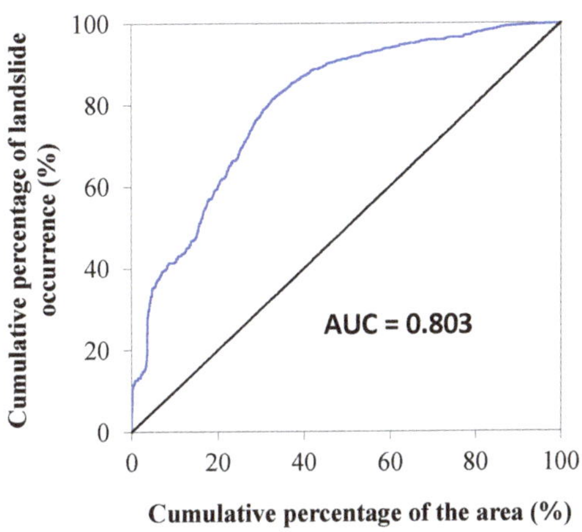

Figure 17. AUC representing quality model a success rate curve.

In terms of model performance, the accuracy of the statistical index method for landslide susceptibility mapping is approximately 80.3%, which is much closer to other studies (e.g., 74%

in the work of Conoscenti et al. (2016) [57], 71% in the work of Camilo et al. (2017) [58], 75% in the work of Youssef et al. (2015) [59], and 73.3% in the work of Shu et al. (2021) [60]). It must be admitted that this accuracy is not superior, which mainly includes the following reasons: one is that the data quality of the inventory is not very high, and the other is associated with the limitation of statistically based methods and assumptions of the landslide classification method.

Because we do not have such another area, a validation of the landslide susceptibility was performed as follows:

- A total of 75% of the observed landslides in the study area is selected at random (see Figure 18). These areas form the training data set. The actual selection was made arbitrarily without considering causative factors. It was only taken into account to spread the training data set as evenly as possible over the study area.
- On the basis of the training data set, a new LSZ map based on the statistical index method for the whole study area was created (see Figure 19).
- The remaining 25% of the observed landslides in the study area is used to evaluate the correctness of the new LSZ map.

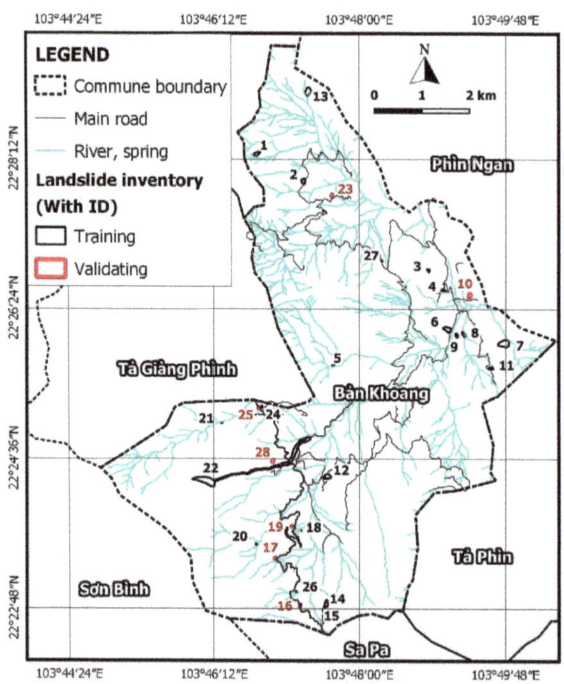

Figure 18. Randomly selected observed landslides in the study area for model validation.

It can be seen that the 80.95% of landslide number has "good" prediction for the new LSZ map based on 75% landslide training data set. Meanwhile, it is a higher value of "good" prediction (83.33%) for the landslide validating data set.

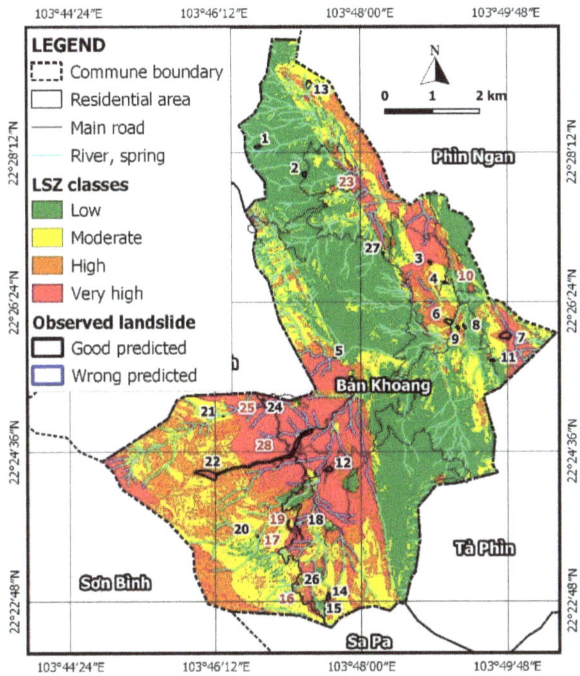

Figure 19. LSZ map based on 75% training landslides of Ban Khoang commune.

The area of landslide belonging to high and very high LSZ classes for the landslide training data set is 80%. Meanwhile the similar value for the landslide validating data set is obtained to be a little bit higher, with 81.2%. The model predictions for training and validating landslides in the study area are shown in Table 4. Generally, the results show that the target data can be predicted well with the modeling approach.

Table 4. LSZ validation result with training and validating landslide.

Accuracy of Prediction	Landslide Training Data Set				Landslide Validating Data Set			
	Number	*Percentage (%)*	**Area (km^2)**	*Percentage %*	*Number*	*Percentage %*	**Area (km^2)**	*Percentage %*
Wrong	4	19.05	0.0512	20	1	16.67	0.0011	18.8
Good	17	80.95	0.2050	80	6	83.33	0.0047	81.2

7. Conclusions

Because most of the observed landslides are well predicted, the high and very high landslide susceptibility classes in the final LSZ map can be considered highly believable.

For all landslides that are wrongly predicted, of course, due to the assumptions of the landslide classification method, 6.7% and 13.3% of the total observed landslide areas fall in the low and medium landslide susceptibility class, respectively. Hence, it is easy to understand that some observed landslides are not correctly predicted.

The causes of these landslides remain unanswered in this study. This probably has to do with some unique local conditions that promote landslides that were not considered in the present analyses or errors or misinterpretations of the data and factor maps.

Finally, the good prediction can be evaluated based on observed landslides belonging to very high and high LSZ classes, it can be seen that 80% observed landslide areas and 78.57% number of observed landslides were well predicted, and AUC obtained 0.803.

Hence, the LSZ map was created, and the real-time landslide monitoring station will be reliable to use in the practice.

Author Contributions: Conceptualization, T.-Y.C. and L.N.T.; data curation, Y.-M.F. and H.-Y.Y.; formal analysis, L.N.T. and T.-V.H.; funding acquisition, C.-Y.L. and C.-L.W.; investigation, T.-Y.C. and C.-Y.L.; methodology, Q.D.N. and C.-L.W.; project administration, L.N.T. and Y.-M.F.; software, T.-V.H. and Y.-C.L.; supervision, T.-Y.C. and C.-Y.L.; validation, T.-V.H. and H.-Y.Y.; visualization, Y.-M.F. and Y.-C.L.; writing—Original draft, L.N.T. and T.-V.H.; writing—Review and editing, T.-V.H. and L.N.T. All authors have read and agreed to the published version of the manuscript.

Funding: This research received the funding from Vietnam Ministry of Natural Resources and Environment, Taiwan Soil and Water Conservation Bureau (SWCB): TNMT.2021.02.11.

Acknowledgments: We acknowledge the support of Vietnam Institute of Geology and Mineral Resources (VIGMR), Taiwan (Soil Water Conservation Bureau (SWCB)), GIS Research Center of Feng Chia University, WeatherPlus company. We would like to express our sincere thanks and appreciations to funded MONRE projects: "Research, design and manufacture a monitoring system for real-time early warning of landslides, debris flow, debris floods, and flash floods in mountainous and midland areas of Vietnam" code TNMT.2019.03.01 and "Study, develop a pilot debris flow early warning system in real time for mountainous areas of Vietnam" code TNMT.2021.02.11. Without the projects' supported data, it would have not been possible to complete our paper.

Conflicts of Interest: The authors declare no conflict of interest.

References

1. Hoa, T.X.; Khanh, N.Q.; Ha, N.D.; Son, P.V. *Project Report of Investigation, Assessment and Geohazards Susceptibility Zonation in Mountainous Areas of VIETNAM*; Vietnam Institute of Geosciences and Mineral Resources, Ministry of Natural Resources and Environment of Vietnam: Hanoi, Vietnam, 2021.
2. Dai, F.C.; Lee, C.F. Landslide characteristics and slope instability modeling using GIS, Lantau Island, Hong Kong. *Geomorphology* **2002**, *42*, 213–228. [CrossRef]
3. Carrara, A. Multivariate models for landslide hazard evaluation. *Math. Geol.* **1983**, *15*, 403–426. [CrossRef]
4. Steger, S.; Brenning, A.; Bell, R.; Glade, T. The influence of systematically incomplete shallow landslide inventories on statistical susceptibility models and suggestions for improvements. *Landslides* **2017**, *14*, 1767–1781. [CrossRef]
5. Rossi, M.; Guzzetti, F.; Reichenbach, P.; Mondini, A.C.; Peruccacci, S. Optimal landslide susceptibility zonation based on multiple forecasts. *Geomorphology* **2010**, *114*, 129–142. [CrossRef]
6. Reichenbach, P.; Busca, C.; Mondini, A.C.; Rossi, M. The Influence of Land Use Change on Landslide Susceptibility Zonation: The Briga Catchment Test Site (Messina, Italy). *Environ. Manag.* **2014**, *54*, 1372–1384. [CrossRef] [PubMed]
7. Goetz, J.N.; Brenning, A.; Petschko, H.; Leopold, P. Evaluating machine learning and statistical prediction techniques for landslide susceptibility modeling. *Comput. Geosci.* **2015**, *81*, 1–11. [CrossRef]
8. Brenning, A. Spatial cross-validation and bootstrap for the assessment of prediction rules in remote sensing: The R package sperrorest. In Proceedings of the 2012 IEEE International Geoscience and Remote Sensing Symposium, Munich, Germany, 22–27 July 2012; pp. 5372–5375. [CrossRef]
9. Guzzetti, F.; Reichenbach, P.; Ardizzone, F.; Cardinali, M.; Galli, M. Estimating the quality of landslide susceptibility models. *Geomorphology* **2006**, *81*, 166–184. [CrossRef]
10. Frattini, P.; Crosta, G.; Carrara, A. Techniques for evaluating the performance of landslide susceptibility models. *Eng. Geol.* **2010**, *111*, 62–72. [CrossRef]
11. Cascini, L. Applicability of landslide susceptibility and hazard zoning at different scales. *Eng. Geol.* **2008**, *102*, 164–177. [CrossRef]
12. Fell, R.; Corominas, J.; Bonnard, C.; Cascini, L.; Leroi, E.; Savage, W.Z. Guidelines for landslide susceptibility, hazard and risk zoning for land use planning. *Eng. Geol.* **2008**, *102*, 85–98. [CrossRef]
13. Harp, E.L.; Keefer, D.K.; Sato, H.P.; Yagi, H. Landslide inventories: The essential part of seismic landslide hazard analyses. *Eng. Geol.* **2011**, *122*, 9–21. [CrossRef]
14. Petschko, H.; Brenning, A.; Bell, R.; Goetz, J.; Glade, T. Assessing the quality of landslide susceptibility maps—Case study Lower Austria. *Nat. Hazards Earth Syst. Sci.* **2014**, *14*, 95–118. [CrossRef]
15. Ardizzone, F.; Cardinali, M.; Carrara, A. Impact of mapping errors on the reliability of landslide hazard maps. *Nat. Hazards Earth Syst. Sci.* **2020**, *2*, 3–14. [CrossRef]
16. Galli, M.; Ardizzone, F.; Cardinali, M.; Guzzetti, F.; Reichenbach, P. Comparing landslide inventory maps. *Geomorphology* **2008**, *94*, 268–289. [CrossRef]
17. Zêzere, J.; Henriques, C.S.; Garcia, R.A.C.; Piedade, A. Effects of Landslide Inventories Uncertainty on Landslide Susceptibility Modelling; RISKam Geographical Research Centre, University of Lisbon, Lisbon, Portugal. 2009. Available online: http://eost.u-strasbg.fr/omiv/Landslide_Processes_Conference/Zezere_et_al.pdf (accessed on 20 July 2022).

18. Fressard, M.; Thiery, Y.; Maquaire, O. Which data for quantitative landslide susceptibility mapping at operational scale? Case study of the Pays d'Auge plateau hillslopes (Normandy, France). *Nat. Hazards Earth Syst. Sci.* **2014**, *14*, 569–588. [CrossRef]
19. Long, N.T. *Landslide Susceptibility Mapping of the Mountainous Area in a Luoi District, Thua Thien Hue Province, Vietnam;* Faculty of Engineering, Department of Hydrology and Hydraulic Engineering, Vrije Universiteit Brussel: Brussel, Belgium, 2008.
20. Van Westen, C. Statistical Landslide Hazard Analysis. In *ILWIS 2.1 for Windows Application Guide;* ITC Publication: Enschede, The Netherlands, 1997; pp. 73–84.
21. Berhane, G.; Tadesse, K. Landslide susceptibility zonation mapping using statistical index and landslide susceptibility analysis methods: A case study from Gindeberet district, Oromia Regional State, Central Ethiopia. *J. Afr. Earth Sci.* **2021**, *180*, 104240. [CrossRef]
22. Mandal, S.; Mandal, K. Bivariate statistical index for landslide susceptibility mapping in the Rorachu river basin of eastern Sikkim Himalaya, India. *Spat. Inf. Res.* **2018**, *26*, 59–75. [CrossRef]
23. Wu, Z.; Wu, Y.; Yang, Y.; Chen, F.; Zhang, N.; Ke, Y.; Li, W. A comparative study on the landslide susceptibility mapping using logistic regression and statistical index models. *Arab. J. Geosci.* **2017**, *10*, 187. [CrossRef]
24. Wang, Q.; Li, W.; Wu, Y.; Pei, Y.; Xie, P. Application of statistical index and index of entropy methods to landslide susceptibility assessment in Gongliu (Xinjiang, China). *Environ. Earth Sci.* **2016**, *75*, 1–13. [CrossRef]
25. Bui, D.T.; Lofman, O.; Revhaug, I.; Dick, O. Landslide susceptibility analysis in the Hoa Binh province of Vietnam using statistical index and logistic regression. *Nat. Hazards* **2011**, *59*, 1413–1444. [CrossRef]
26. Oztekin, B.; Topal, T. GIS-based detachment susceptibility analyses of a cut slope in limestone, Ankara—Turkey. *Environ. Earth Sci.* **2005**, *49*, 124–132. [CrossRef]
27. Cevik, E.; Topal, T. GIS-based landslide susceptibility mapping for a problematic segment of the natural gas pipeline, Hendek (Turkey). *Environ. Geol.* **2003**, *44*, 949–962. [CrossRef]
28. Guzovski, L.A.; Toản, T.X.; Hiến, P.Đ. Some Problems of Study on the Weathering Crust in South Vietnam. *Geol. Miner. Resour.* **1989**, *2*, 29–36.
29. An, P.V.; Bính, H.V.; Hiền, L.V.; Phú, Đ.X.; Dũng, N.T. Geochemical charateristics of the tropical humid weathering crust in Vietnam. *Geol. Miner. Resour. J.* **1990**, *3*, 95–104.
30. Sarkar, S.; Kanungo, D.P.; Mehrotra, G.S. Landslide hazard zoning: A case study in Garhwal Himalaya, India. *Mt. Res. Dev.* **1995**, *15*, 301–309. [CrossRef]
31. Sidle, R.C.; Ochiai, H. *Landslides: Processes, Prediction, and Land Use;* Water Resources Monograph No.18; American Geophysical Union: Washington, DC, USA, 2006; 312p.
32. Lap, D.Q.; Binh, D.T.; Loc, N.V.; Son, P. *Map of Geology and Mineral Resources of Lao Cai Sheet Group, Scale 1:50,000;* Intergeo Geological Section; MONRE: Hanoi, Vietnam, 2003.
33. Thành, N.Q.; Dung, N.P.; Hoàng, N.V.; Hiển, T.T. *Study for Assessments of Landslide, Debris flow and Flash Flood in Focused Area of Lao Cai Province (Districts of Bat Xat & Sa Pa and Lao Cai city) and Propose Remedy Damage Solutions (Branch Project Belonging to National Project KC 01-08);* Hanoi, Vietnam, 2006.
34. Chigira, M.; Nakamoto, M.; Nakata, E. Weathering mechanisms and their effects on the landsliding of ignimbrite subject to vapor-phase crystallization in the Shirakawa pyroclastic flow, northern Japan. *Eng. Geol.* **2002**, *66*, 111–125. [CrossRef]
35. Wakatsuki, T.; Tanaka, Y.; Matsukura, Y. Soil slips on weathering-limited slopes underlain by coarse-grained granite or fine-grained gneiss near Seoul, Republic of Korea. *CATENA* **2005**, *60*, 181–203. [CrossRef]
36. Ibetsberger, H.J. The Tsergo Ri landslide: An uncommon area of high morphological activity in the Langthang valley, Nepal. *Tectonophysics* **1996**, *260*, 85–93. [CrossRef]
37. Pachauri, A.K.; Gupta, P.V.; Chander, R. Landslide zoning in a part of the Garhwal Himalayas. *Environ. Earth Sci.* **1998**, *36*, 325–334. [CrossRef]
38. Gökceoglu, C.; Aksoy, H. Landslide susceptibility mapping of the slopes in the residual soils of the Mengen region (Turkey) by deterministic stability analyses and image processing techniques. *Eng. Geol.* **1996**, *44*, 147–161. [CrossRef]
39. Gokceoglu, C. Discussion on "Landslide hazard zonation of the Khorshrostam area, Iran" by A. Uromeihy and M.R. Mahdavifar. *Bull. Eng. Geol. Environ.* **2001**, *58*, 207–213.
40. Greenway, D.R. Vegetation and slope stability. In *Slope Stability, Geotechnical Engineering and Geomorphology;* Anderson, M.G., Richards, K.S., Eds.; John Wiley & Sons: Chichester, UK, 1987; pp. 187–230.
41. Land Administration Department of the Ministry of Natural Resources and Environment. *Landuse Map of Lao Cai Province, Scale 1:50,000;* Land Administration Department of the Ministry of Natural Resources and Environment: Hanoi, Vietnam, 2019.
42. Yalcin, A. GIS-based landslide susceptibility mapping using analytical hierarchy process and bivariate statistics in Ar-desen (Turkey). *Catena* **2007**, *72*, 1–12. [CrossRef]
43. Van Westen, C.J.; Rengers, N.; Terlien, M.T.J.; Soeters, R. Prediction of the occurrence of slope instability phenomena through GIS-based hazard zonation. *Geol. Rundsch.* **1997**, *86*, 404–414. [CrossRef]
44. Binaghi, E.; Luzi, L.; Madella, P.; Pergalani, F.; Rampini, A. Slope Instability Zonation: A Comparison Between Certainty Factor and Fuzzy Dempster–Shafer Approaches. *Nat. Hazards* **1998**, *17*, 77–97. [CrossRef]
45. Barredo, J.; Benavides, A.; Hervás, J.; van Westen, C.J. Comparing heuristic landslide hazard assessment techniques using GIS in the Tirajana basin, Gran Canaria Island, Spain. *Int. J. Appl. Earth Obs. Geoinform.* **2000**, *2*, 9–23. [CrossRef]
46. David, J.W.; Paul, F.H. Mapping landslide susceptibility in Travis County, Texas, USA. *Geol. J.* **2000**, *51*, 245–253.

47. Saha, A.K.; Gupta, R.P.; Arora, M.K. GIS-based Landslide Hazard Zonation in the Bhagirathi (Ganga) Valley, Himalayas. *Int. J. Remote Sens.* **2002**, *23*, 357–369. [CrossRef]
48. Lan, H.; Zhou, C.; Wang, L.; Zhang, H.; Li, R. Landslide hazard spatial analysis and prediction using GIS in the Xiaojiang watershed, Yunnan, China. *Eng. Geol.* **2004**, *76*, 109–128. [CrossRef]
49. Lee, C.F.; Li, J.; Xu, Z.W.; Dai, F.C. Assessment of landslide susceptibility on the natural terrain of Lantau Island, Hong Kong. *Environ. Earth Sci.* **2001**, *40*, 381–391. [CrossRef]
50. Kanungo, D.; Arora, M.; Sarkar, S.; Gupta, R. A comparative study of conventional, ANN black box, fuzzy and combined neural and fuzzy weighting procedures for landslide susceptibility zonation in Darjeeling Himalayas. *Eng. Geol.* **2006**, *85*, 347–366. [CrossRef]
51. Ayalew, L.; Yamagishi, H.; Ugawa, N. Landslide susceptibility mapping using GIS-based weighted linear combination, the case in Tsugawa area of Agano River, Niigata Prefecture, Japan. *Landslides* **2004**, *1*, 73–81. [CrossRef]
52. Akgün, A.; Bulut, F. GIS-based landslide susceptibility for Arsin-Yomra (Trabzon, North Turkey) region. *Environ. Earth Sci.* **2006**, *51*, 1377–1387. [CrossRef]
53. Mehmet, L.S.; Vedat, D. A comparison of the GIS based landslide susceptibility assessment methods: Multivariate versus bivariate. *Environ. Geol.* **2004**, *45*, 665–679. [CrossRef]
54. Komac, M. A landslide susceptibility model using the Analytical Hierarchy Process method and multivariate statistics in perialpine Slovenia. *Geomorphology* **2006**, *74*, 17–28. [CrossRef]
55. Foumelis, M.; Lekkas, E.; Parcharidis, I. Landslide susceptibility mapping by GIS-based qualitative weighting pro-cedure in Corinth area. Bulletin of 10th International Congress of the Geological Society. *Thessalloniki* **2004**, *34*, 904–912.
56. Galang, J.S. A Comparison of GIS Approaches to Slope Instability Zonation in the Central Blue Ridge Mountains of Virginia. Master's Thesis, Faculty of Virginia Polytechnic Institute and State University, Blacksburg, Virginia, 2004; 99p.
57. Conoscenti, C.; Rotigliano, E.; Cama, M.; Caraballo-Arias, N.A.; Lombardo, L.; Agnesi, V. Exploring the effect of absence selection on landslide susceptibility models: A case study in Sicily, Italy. *Geomorphology* **2016**, *261*, 222–235. [CrossRef]
58. Camilo, D.C.; Lombardo, L.; Mai, P.M.; Dou, J.; Huser, R. Handling high predictor dimensionality in slope-unit-based landslide susceptibility models through LASSO-penalized Generalized Linear Model. *Environ. Model. Softw.* **2017**, *97*, 145–156. [CrossRef]
59. Youssef, A.M.; Al-Kathery, M.; Pradhan, B. Landslide susceptibility mapping at Al-Hasher area, Jizan (Saudi Arabia) using GIS-based frequency ratio and index of entropy models. *Geosci. J.* **2015**, *19*, 113–134. [CrossRef]
60. Shu, H.; Guo, Z.; Qi, S.; Song, D.; Pourghasemi, H.R.; Ma, J. Integrating Landslide Typology with Weighted Frequency Ratio Model for Landslide Susceptibility Mapping: A Case Study from Lanzhou City of Northwestern China. *Remote Sens.* **2021**, *13*, 3623. [CrossRef]

water

Article

Triggering Rainfall of Large-Scale Landslides in Taiwan: Statistical Analysis of Satellite Imagery for Early Warning Systems

Tsai-Tsung Tsai *, Yuan-Jung Tsai, Chjeng-Lun Shieh and John Hsiao-Chung Wang

Department of Hydraulic and Ocean Engineering, National Cheng Kung University, Tainan 701, Taiwan
* Correspondence: victor@dprc.ncku.edu.tw; Tel.: +886-6-2091061 (ext. 10)

Abstract: Typhoon Morakot had a serious impact on Taiwan, especially the uncommon type of landslide called large-scale landslide (LSL), not many in number but serious in effect, the origin of which the study induced. To establish a specific relationship between LSL and triggering rainfall for future applications of LSL early warning predictions, relevant cases from satellite imagery, along with field investigation data, major event reports, and seismic data from 2004 to 2016, were collected. All collected cases are distributed around the mountainous area in Taiwan, and a total of 107 cases which were mainly distributed in the southern part of the mountainous area were finally selected, including 28 occurrence-time-known cases and 79 occurrence-time-unknown cases. In addition, 149 potential areas identified by the Soil and Water Conservation Bureau (SWCB) were used for improving bounding estimates. Based on the concept of safety factor, two dimensionless quantities, rainfall/landslide depth (R/D) and friction angle/slope (ϕ/θ), were analyzed by linear regression. In addition, D was assumed to be nonlinearly dependent on R, θ, and ϕ, and the parameter uncertainties were evaluated by the resampling with bootstrap method. Based on the currently obtained data, there were 8% Type-I errors in the results of the linear regression analysis, and 1% Type-II errors in the results of the nonlinear regression analysis. Through the comparison of statistical indicators, the results of nonlinear regression analysis have a better correlation trend. Based on the needs of early warning operations, more conservative indicators can reduce the risks faced by management operations. Therefore, according to the results of this study, the lower boundary values from nonlinear analysis could be used as the LSL early warning management settings. Incorporated with real-time rainfall forecasts, the variation of statistical indicators will provide the trend information dynamically, and will help to increase the response time for relevant evacuation operations, that will be welcome for the further extended applications to guide the evacuation operations of early warning systems.

Keywords: landslide; large-scale landslide; triggering rainfall; early warning system; linear regression; nonlinear regression; uncertainty

Citation: Tsai, T.-T.; Tsai, Y.-J.; Shieh, C.-L.; Wang, J.H.-C. Triggering Rainfall of Large-Scale Landslides in Taiwan: Statistical Analysis of Satellite Imagery for Early Warning Systems. *Water* **2022**, *14*, 3358. https://doi.org/10.3390/w14213358

Academic Editor: Su-Chin Chen

Received: 14 September 2022
Accepted: 20 October 2022
Published: 23 October 2022

1. Introduction

Typhoon Morakot brought Taiwan a historically record-breaking rainfall in August 2009. Half of Taiwan Island had accumulated rainfalls exceeding 500 mm [1] with some hot areas receiving up to 3000 mm [2]. There were 673 people killed, 26 people missing, and more than TWD 19.5 billion in agricultural losses [3]. Various types of sediment disasters occurred in large numbers during Typhoon Morakot.

Among the sediment disasters, there were 43,519 landslide cases found after this event, but only 259 cases of the uncommon type of landslide called large-scale landslide (LSL), which is not many in number but serious in effect, were found. The most famous case was Xiaolin Village, in Jiaxian District of Kaohsiung City, which was horribly destroyed in this event and 462 people from 180 households were buried in mud and rocks [4]. The tragedy of Xiaolin Village not only shocked all sectors of society, but also made the public realize that we are not familiar with LSL, and knowledge of the key for prevention and mitigation of LSL incidents is urgently needed. This need prompted the present study.

After Typhoon Morakot, the National Science and Technology Center for Disaster Reduction (NCDR) of Taiwan defined an LSL as a landslide area larger than 10 ha, earth volume larger than 100,000 m^3, or a collapse depth deeper than 10 m [5]. Under the situation that epistemic conditions of LSLs are insufficient, in order to find a practical solution for early warning works, this study tries to find the most intuitive indicator, and rainfall and some other factors of LSLs seem to be the best choice. In order to establish the specific relationship between LSLs and triggering rainfall for the future LSL early warning predictions, LSL cases, satellite imagery, rainfall data, seismic data, and other support datasets were collected. In this study, two dimensionless factors, rainfall/landslide depth (R/D) and friction angle/slope (ϕ/θ), were assumed to have a linear relationship, and all factors, R, D, ϕ, θ, were assumed to have a nonlinear relationship, and the both linear and nonlinear regressions were analyzed statistically.

The results of LSL occurrence time and location evaluated from the data of the Broadband Array in Taiwan for Seismology (BATS) were applied in this study for establishing the specific relationship between LSLs and triggering rainfall. Compared with previous research results using only cumulative rainfall of events, the reliability is expected to be considerably improved.

2. Literature Review

Current research topics of LSLs include a wide range of subjects, e.g., occurrence mechanisms, monitoring, and early warning. In order to clearly understand past studies, including the key points and effectiveness, a detailed literature review on the above topics was conducted.

2.1. Occurrence Mechanism-Related Research

The stability of a slope can be considered by the factor of safety (FS), the ratio of shear strength τ_R and shear stress τ_D. When the FS is greater than or equal to 1, it can be regarded as a balanced state [6]. The limit equilibrium method is popular in many studies on slope stability analysis [7]. Slope stability analyses can be roughly divided into three common theories based on the corresponding concept of slope type: infinite slope theory [6,8–11], finite slope theory [6,9,12], and the method of slices [6,9–11].

Numerous numerical models have been developed based on the three simulation functionalities of movement status as the following:

1. Simulate the occurrence of landslide only models, such as SHALSTABLE, SINMAP, TRIGRS, Scoops3D, PLAXIS, and GeoSutio [6,13–17].
2. Simulate the movement of soil on the slope after landslide only models, such as Flo2D, Landslide2D, PFC, and DDA [6,18,19].
3. Simultaneously simulate the occurrence of landslide and the movement of soil on the slope after landslide models, such as FLAC, ABAQUS, and Anura [6,20].

In this study, the most important issue is whether the slope is stable or not, which is the reason why limit equilibrium theory is used, but not numerical simulations of landslides, for the selection of parameters for statistical analysis.

2.2. Monitoring-Related Research

Slope monitoring data is the basis for the LSL research and early warning predictions. Slope monitoring can be carried out through on-site monitoring and remote sensing.

1. On-site monitoring

The purpose of on-site monitoring is mainly to obtain on-site data, including surface changes, underground changes, surface hydrology, groundwater hydrology, and structural deformation. The comparison of various on-site monitoring programs is shown in Table 1 [21]. In recent years, new technologies and equipment have been continuously improved, e.g., GPS [22,23], TDR [23], RGB-D sensors [24,25], and ERT [26]. On-site monitoring has the advantages of high efficiency, high frequency, and accuracy from directly measuring on-site data, and it is quite convenient for subsequent analysis and application. Therefore, the program of on-site monitoring is widely used. However, since the results of on-site monitoring are limited to "points", if wide-range monitoring is to be carried out, a lot of resources have to be invested not only in equipment construction but also in subsequent maintenance because underground monitoring instruments are vulnerable to damage from ground deformations.

Table 1. Comparison table of on-site monitoring programs (modified from [21]).

Investigation Item	Instruments		Investigation Objects	Accuracy
Surface changes	Surface inclinometer		Tilting direction and amount of ground surface	$1''$
	Surface extensometer		Fracture displacement and velocity	0.2 mm
	Surface measurement	Optical measuring instruments	Tilting direction and amount of ground surface	1~10 mm
		GNSS	Displacement of the ground surface	NA
		LiDAR scanner	Terrain 3D variation	NA
Underground changes	In-place inclinometer		Sliding surface position and variation	$5\sim10''$
	Pipe strain gauge		Sliding surface position and variation	1×10^{-6}
	Borehole extensometer		Sliding surface dislocation rate	0.2 mm
	Multipoint borehole extensometer		Sliding surface position and dislocation rate	0.3 mm
Surface hydrology	Rain gauge		Rainfall amount	0.5 mm
Underground hydrology	Water level gauge		Variation of water level in the hole	0.05%FS
	Pore pressure gauge		Variation of water pressure of the sliding surface	0.05%FS
	Soil moisture meter		Variation of soil saturation	NA
	Flowmeter		Variation of discharge	NA
Structures	Earth pressure gauge		Earth pressure acting on retaining walls, deep foundation piles	0.1%FS
	Load cell		Tension acting on the ground anchor	0.1%FS
	Strain gauge		Deformation of the structure	1×10^{-6}
	Rebar gauge		Stress acting on the rebar gauge	0.1%FS
	Inclinometer		Tilt variation of structure	$1\sim10''$
	In-place inclinometer		Bending deformation of steel pipe piles	$5\sim10''$

Note: FS—Factor of Safety; ″ —inch; NA—Not Available.

Some of the occurrence times of LSLs in this study were obtained from analyses using the seismic data from the BATS. Since 1992, 42 stations have been set up in Taiwan and offshore islands. Each station is equipped with a broadband seismograph, which can record a wide frequency range of fluctuations due to the characteristics of its sensitive sensors, which can record rich and high-quality seismic waveforms. It can effectively make up for the insufficiency of seismic wave information recorded by acceleration type or traditional narrow-band velocity type seismographs, thereby improving the evaluation accuracy of earthquake location and scale [27].

2. Remote Sensing

Remote sensing mainly includes optical imagery, airborne LiDAR, and radar data. Optical imagery can be obtained from unmanned aerial vehicle (UAV) imagery [28], aerial photos [29], satellite imagery [29–31], etc., and usually requires special processes [30–32] to derive surface variation or trend information. The airborne LiDAR acquires a large amount of point cloud data through intensive light wave scanning ranging, and can derive high-precision surface deformation after data calculation and spatial matching [33–37]. In recent years, with the gradual miniaturization of LiDAR products, there have been small LiDAR applications that can be matched with UAVs [38]. Radar data analysis evaluates surface variability through radar wave phase changes. In the application of surface variability, the current common applications include differential InSAR (D-InSAR) [39,40], persistent scatterer InSAR (PS-InSAR) [40–48], short baseline subset InSAR (SBAS-InSAR) [49], and temporarily coherent point InSAR (TCP-InSAR) [50,51]. The advantage of landslide monitoring through remote sensing data analysis is to obtain the "plane" data of a landslide. As acquisitions of data have to be coordinated with the schedule of the remote sensing payload, and most of the original data have to be processed by some complex interpretation and analysis, the time frame of data acquisition is slower than for on-site monitoring.

Rainfall data for statistical analysis in this study come from the CWB's on-site monitoring data and radar evaluation data, and all the terrestrial data of LSLs are based on airborne LiDAR and satellite optical imagery.

2.3. Early Warning-Related Research

By comparing the literature of current LSL early warning studies from Taiwan [52], the United States [53], Japan [54], Italy [55], and Canada [56], it is clear that each country has a different approach to establishing warning systems according to the risks to be faced and the technologies to be mastered. The approaches are follows:

1. Early warning indicators

Through the aforementioned on-site monitoring, remote sensing, and other solutions, on-site information is obtained to establish early warning indicators, including:

- Surface or underground deformation, velocity, or amount of deformation [21,30,40,42,44,49–51];
- variation speed of groundwater level or water level changes [21,57–60];
- rainfall amount [59,61];
- comprehensive indicators [59].

2. Early warning management values

The management values of early warning indicators are determined by statistical methods, which are generally divided into warning management values and evacuation management values [62–68].

3. Real-time simulation

In addition, early warning indicators and management values are also introduced into the numerical model of LSLs, and real-time simulation is performed to provide guidance for warning and evacuation.

3. Materials and Methods

The process with major data analysis steps is as follows: case collection and screening (Section 3.1), case confirmation (Section 3.2), occurrence time confirmation (Section 3.3), triggering rainfall analysis (Section 3.4), linear regression analysis (Section 3.5), and nonlinear regression analysis (Section 3.6). Details are delineated in the following subsections.

3.1. Case Collection and Screening

The FORMOSAT-2 satellite images of landslides caused by typhoons and torrential rain events from 2004 to 2016 were collected. The images were compared with each other to confirm each landslide type, location, and size. Based on the NCDR's definition of an LSL, 43,519 collected landslide events were screened out and narrowed down to 259 landslides. In addition, new landslides can be classified into the newborn landslide and the expanded landslide (as shown in Figure 1), and 107 newborn landslides were selected for analysis.

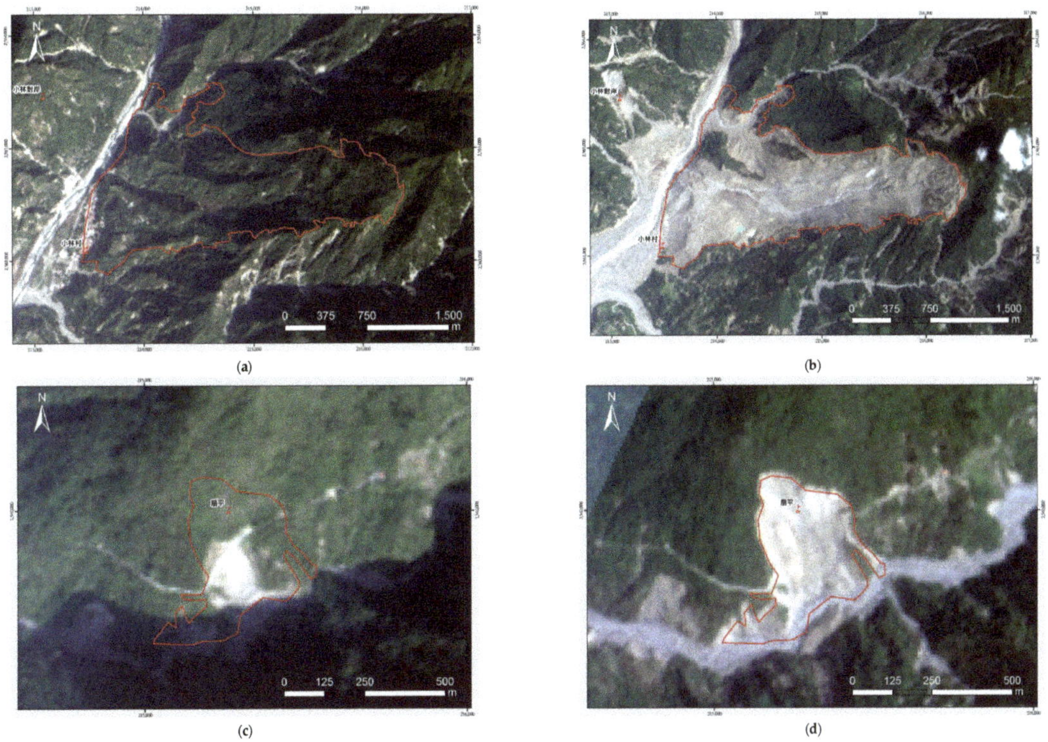

Figure 1. Landslide types of satellite images: newborn landslide (**a**,**b**) and expanded landslide (**c**,**d**). (modified from: [66]). (**a**) Xiaolin Village Landslide (pre-event), (**b**) Xiaolin Village Landslide (post-event), (**c**) Shanping Landslide (pre-event), (**d**) Shanping Landslide (post-event).

3.2. Case Confirmation, Area Size, and Average Slope Identification of LSL

In this study, the practice of the Central Geological Survey, MOEA on the identification of landslides was referred to [69,70], and landslide cases caused by the influence of specific events were obtained through the creation and comparison of landslide catalogues. The LSLs discussed in this study are mainly new landslides. A total of 259 new LSL cases were identified, including 107 newborn landslides and 152 expanded landslides. Only the 107 newborn landslides were used for analysis. In this stage, area size and average slope of LSLs were obtained at the same time. The projected area A (m^2) of an LSL was calculated by the Calculate Geometry function of ESRI ArcMap® software. Based on 5 m resolution DEM data, the slope of each data grid was calculated first with the Slope function of the Surface tool in the Spatial Analyst Tools module of ArcToolbox, and the average slope θ (degree) for the corresponding range of each newborn LSL was calculated by the Zonal Statistics function in the Zonal tool module.

3.3. The Occurrence Time Confirmation of LSL

In order to accurately correlate the triggering rainfall with LSLs, the first task is to confirm the occurrence time of each landslide. The occurrence time data cited in this study mainly come from interviews with local residents [71,72], and the evaluation results of the LSL occurrence time from the BATS [73]. Among the 107 newborn LSL cases, 28 cases were identified with their occurrence time, and for the remaining 79 cases the exact time of occurrence was unknown. The information of the 28 newborn LSLs is shown in Table 2.

Table 2. Information of the 28 newborn LSLs.

No.	ID	Event	Area Size (ha)	Occurrence Time	Cited From
1	SR-3	Typhoon Morakot (200908)	19	09 August 2009 17:00	Interviews with local residents
2	SR-5		238	09 August 2009 17:00	
3	SR-6		142	10 August 2009 12:00	
4	SR-7		130	09 August 2009 02:00	
5	SR-8		88	09 August 2009 02:00	
6	SR-9		74	09 August 2009 04:00	
7	SR-11		40	08 August 2009 16:00	
8	SR-12		32	09 August 2009 07:00	
9	SR-16		26	09 August 2009 07:00	
10	SR-19		23	08 August 2009 15:00	
11	SR-42		15	09 August 2009 07:00	
12	SR-43		15	09 August 2009 07:00	
13	SR-46		15	09 August 2009 05:00	
14	SR-53		14	09 August 2009 00:00	
15	SR-94		351	09 August 2009 10:00	
16	SR-95		249	09 August 2009 06:00	
17	SR-96		81	09 August 2009 10:00	
18	SR-97		61	09 August 2009 09:00	
19	SR-98		52	09 August 2009 06:00	
20	SR-99		15	09 August 2009 04:00	
21	SR-100		11	08 August 2009 10:00	
22	SR-101		10	09 August 2009 09:00	
23	2005_002	Typhoon Haitang (200505)	18	21 July 2005 14:33	The evaluation results of the LSL occurrence time and location from the BATS
24	2006_002	0609 Torrential Rain	12	10 Jun 2006 00:53	
25	2008_002	Typhoon Sinlaku (200813)	89	18 September 2008 02:50	
26	2008_003	Typhoon Kamaegi (200807)	10	19 July 2008 05:30	
27	2012_002	Typhoon Saola (201209)	19	03 August 2012 09:02	
28	2012_004		25	03 August 2012 03:00	

3.4. The Triggering Rainfall Analysis

From previous methods of analyzing the triggering rainfall of landslides and debris flows [62,68,74,75], the triggering rainfall of LSLs can be described as Equation (1). The rain field can be divided as in Figure 2. The analysis of landslide triggering rainfall R should consider R_0 and P. The cumulative rainfall R_0 directly contributes to the landslide event from the beginning of rainfall to the moment the LSL occurred and is called "direct rainfall". The rainfall P is called "antecedent rainfall" that occurred before the start of the current rain event, and is related to the moisture content of soil which also affects the likelihood of having a landslide. The length of time considered for the antecedent rainfall can be adjusted according to the local geological characteristics. In this study, the rainfall of the antecedent seven days is used. The sum of the direct rainfall of the current rainfall event and the antecedent rainfall is the effective cumulative rainfall of the landslide, and is called the triggering rainfall, which can be expressed by Equation (1).

$$R = R_0 + P \approx R_0 + \sum_{i=1}^{N} \alpha^i R_i, \quad (1)$$

where:

R is the triggering rainfall for LSL (mm);

R_0 is the accumulated rainfall from the beginning of the rainfall event that caused the LSL to the moment the landslide occurred (mm);
P is the antecedent rainfall (mm) $\approx \sum_{i=1}^{N} \alpha^i R_i$;
R_i is the rainfall on the i-th day (24 h) before the start of the rain field t_0 (mm);
N is the number of days to consider the antecedent rainfall (), generally N = 7;
α is the daily (24 h) rainfall triggering landslide decay coefficient (), which can be 0.7 or 0.8. At present, α = 0.7 is used in this study to calculate the antecedent rainfall [62].

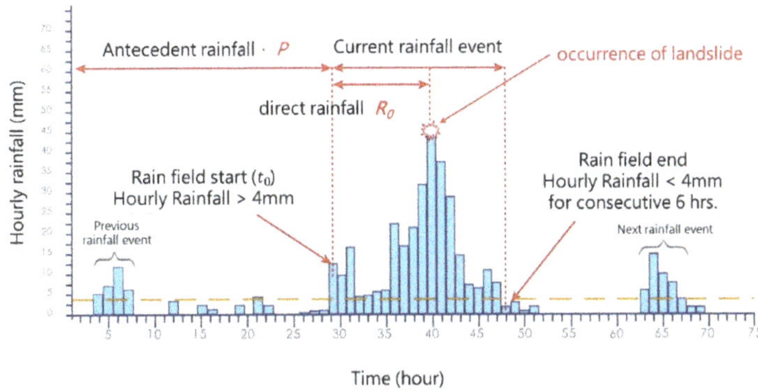

Figure 2. Schematic diagram of rain field cutting (modified from: [76]).

The aforementioned 107 newborn LSL cases and the data of 149 potential LSL areas evaluated by the SWCB were used in this study for the follow-up analysis, with the information of the cases shown in Table 3. Among the 107 newborn LSL cases, the occurrence time of 28 cases was confirmed through local resident interviews and evaluation of the data from the BATS, and then triggering rainfall was also calculated. Since the time of occurrence of the remaining 79 cases could not be confirmed, the total rainfall of the corresponding event was used instead of the triggering rainfall. For the 149 potential LSL areas without landslide, the total rainfall of Typhoon Morakot was used as the triggering rainfall in this study. In summary, three different sets of LSL events were used in this study: 28 occurrence-time-known cases, 79 occurrence-time-unknown cases, and 149 potential LSL areas. As for other relevant parameters, the relevant data were collected and estimated based on the actual collapse area and the potential collapse range delineated by the SWCB.

Table 3. Information of selected cases.

Cases		Landslide or Not	Rainfall Type Used
107 newborn LSLs	28 occurrence-time-known cases	Yes	Triggering rainfall
	79 occurrence-time-unknown cases	Yes	Total event rainfall
149 potential LSL areas		No	Total rainfall of Typhoon Morakot

3.5. The Linear Regression Analysis

When the rain infiltrates the sliding surface and starts to accumulate, the pore water pressure rises. It is assumed that the soil will start to slide when the pore water pressure rises to a critical h_c (m). According to the infinite slope theory, θ (degree) is defined as the average slope of the sliding surface, D (m) is defined as the thickness of the sliding soil layer, and h_c is defined as landslide triggering pore water pressure or critical pore water pressure for soil sliding. According to Towhata et al. [77], the h_c/D is a function as shown in Equation (2).

$$\frac{h_c}{D} = f\left(\frac{internal\ friction\ angle}{slope\ of\ sliding\ surface}\right) = f\left(\frac{internal\ friction\ angle}{\theta}\right) \quad (2)$$

Alternatively, h_c is replaced with the landslide triggering rainfall R, and the internal friction angle is replaced by the equivalent friction angle ϕ (degree). Therefore, Equation (2) can be rearranged in two dimensionless quantities, R/D and ϕ/θ, for a linear statistical regression analysis as the following:

$$\left(\frac{R}{D}\right) = a \times \left(\frac{\phi}{\theta}\right) + b. \tag{3}$$

In Equation (3), R can be derived from Equation (1), θ can be obtained by calculating the average slope by the steps in Section 3.2, and D is the landslide volume V (m^3) divided by the projected landslide area A as Equation (4). With an empirical volume–area relation [78] as Equation (4) from the SWCB, D can be evaluated as Equation (5).

$$V = 0.1025 \times A^{1.401} \tag{4}$$

$$D = \frac{V}{A} = 0.1025 \times A^{0.401} \tag{5}$$

Scheidegger mentioned that the equivalent friction coefficient f [] is a function of the landslide volume V as in Equation (6) [79]. Since the friction coefficient f is equal to $\tan\phi$ based on the force balance (gravity and friction) of an incline plane, the equivalent friction angle ϕ can be calculated by Equation (7).

$$\log_{10} f = -0.1466 \log_{10} V + 0.62419 \tag{6}$$

$$\phi = \frac{\tan^{-1} f \times 180}{\pi} = \frac{\tan^{-1}\left(e^{-0.1466 \log_{10} V + 0.62419}\right) \times 180}{\pi} \tag{7}$$

3.6. The Nonlinear Regression Analysis

To evaluate of the stability of the slope using the concept of FS in Section 2.1, D, R, θ, and ϕ are the four factors to be considered, and a generic nonlinear relationship is assumed as shown in Equation (8).

$$D = \mathrm{f}(R, \theta, \phi; a, b, c, d, e) = \frac{a \times R^b \times \theta^c}{\phi^d} + e, \tag{8}$$

where:
D is a nonlinear function of R, θ, ϕ (m),
a, b, c, d, e are the regression coefficients (). To be dimensionally correct, these coefficients should have certain physical units. However, they are just regressionally fitted values and the actual units are ignored in this paper for simplicity.

To deal with this inverse problem, a nonlinear regression is performed with bootstrap resampling to evaluate parameter uncertainty. The analysis steps are shown in Figure 3. The key steps of the entire nonlinear process are detailed as follows.

Step 1. Initial calculation

In this step, the coefficients of a, b, c, d, and e are initially set to be 1.0, and the initial predicted values can be calculated as Equation (8). The (relative) residual is defined as Equation (9).

$$\Delta_i \equiv 1 - \frac{p_i}{o_i}, i = 1 \cdots n, \tag{9}$$

where:

Δ_i is the residual of the i-th data [],
p_i is the landslide thickness of the i-th prediction (m),
o_i is the landslide thickness of the i-th observation (m),
n is the total number of observations ().

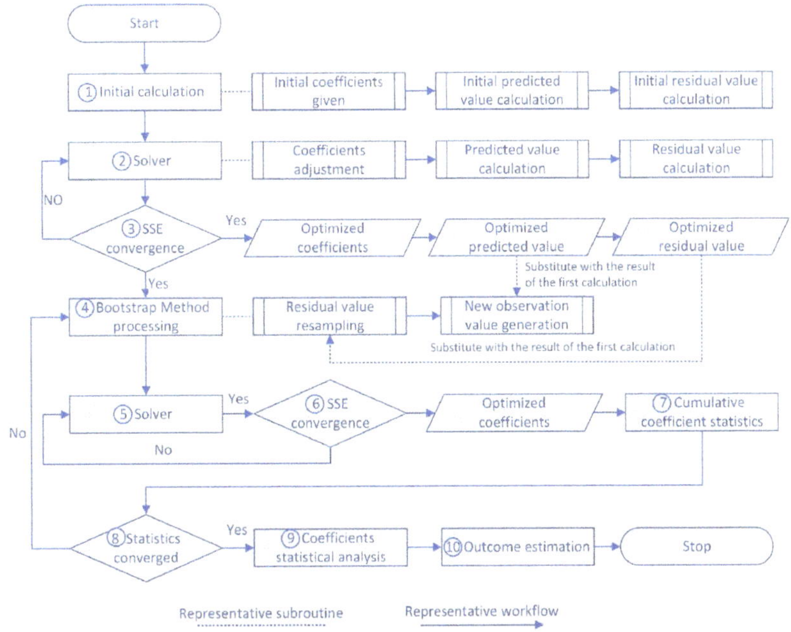

Figure 3. Nonlinear regression workflow. The numbered items are the key steps of the process, detailed in the following text.

Step 2. Solver

For the nonlinear regression analysis, the built-in Solver add-in of Microsoft Excel® is used to find the optimized coefficients a, b, c, d, e. The sum of squared errors (SSE) as in Equation (10) is used as the target function, and the coefficients can be obtained by minimizing the SSE with Solver. Solver uses the GRG nonlinear solving method to solve the problem with an accuracy of the constraint of 0.000001.

$$SSE = \sum_{i=1}^{n} {\Delta_i}^2 = \sum_{i=1}^{n} \left(1 - \frac{p_i}{o_i}\right)^2 \tag{10}$$

Step 3. SSE convergence

To obtain the optimal solution, a convergence criterion of 0.0001 is set in Solver. When the absolute value of the change of the SSE in the last 5 iterations is less than the convergence criteria, the GRG nonlinear solution method will stop. With the final optimized coefficient $\widetilde{a}$, $\widetilde{b}$, $\widetilde{c}$, $\widetilde{d}$, $\widetilde{e}$, the optimized prediction value $\widetilde{p}_i$ and the corresponding optimized residual value $\widetilde{\Delta}_i$ will be used for the following bootstrap procedure (steps 4 to 8).

Step 4. Bootstrap method processing

The bootstrap method was proposed by Efron in 1979 [80] and can be used to estimate the uncertainty of system parameters. The concept of this method is to use the existing data to generate a large number of "phantom samples" through a procedure of resampling with replacement [81]. Since this method is a nonparametric method, there is no need to make assumptions about the data distribution pattern [82].

When carrying out the bootstrap method, the optimal set of residual values $\widetilde{\Delta}_i$ from step 3 is resampled with replacement to obtain a new set of residual value ${\Delta_i}^*$. With the rearrangement of Equation (9), the observation value can be described as Equation (11).

Therefore, with the optimized prediction value $\widetilde{p}_i$ and the bootstrapped residual value $\Delta_i{}^*$, the new observation value $o_i{}^*$ can be obtained as Equation (12).

$$\Delta_i = 1 - \frac{p_i}{o_i} \Rightarrow o_i = \frac{p_i}{1 - \Delta_i}, \quad i = 1 \cdots n \tag{11}$$

$$o_i{}^* = \frac{\widetilde{p}_i}{1 - \Delta_i{}^*}, \quad i = 1 \cdots n \tag{12}$$

Step 5. Solver

This step is essentially the same as step 2, and the main difference is that the new observation value $o_i{}^*$ from step 4 will be used for optimization.

Step 6. SSE convergence

This step is the same as step 3. Different SSEs are solved repeatedly through iterative optimization, and new optimized coefficients $\widetilde{a}^j$, $\widetilde{b}^j$, $\widetilde{c}^j$, $\widetilde{d}^j$, $\widetilde{e}^j$ are obtained for j-th calculation of the bootstrap method.

Step 7. Cumulative coefficient statistics

In this step, the cumulative coefficient of variation (CV) for each optimized coefficient is defined as Equation (13).

$$CV_j = \frac{\sigma_{1\sim j}}{\mu_{1\sim j}}, \quad j - 1 \cdots , \tag{13}$$

where:

CV_j is coefficient of variation of the specific coefficient from the 1st to the j-th bootstrap resampling [],
$\sigma_{1\sim j}$ is the standard deviation of the specific coefficient from the 1st to the j-th bootstrap resampling [],
$\mu_{1\sim j}$ is the average value of the specific coefficient from the 1st to the j-th bootstrap resampling [].

Step 8. Statistics converged

The variation value ΔCV is defined as the absolute value of the difference between two consecutive CV values shown in Equation (14) and it can be used to gauge the trend asymptotically stable solution. The convergence criterion for ΔCV is set to be 0.0001 and steps 4 to 8 are repeated m times until the ΔCV is converged. In the bootstrap procedure, m is the total times of resampling and is not a preset number but can only be determined by the convergence in this step.

$$\Delta CV_j = \left| CV_j - CV_{j-1} \right|, \quad j = 2 \cdots m \tag{14}$$

Step 9. Coefficient statistical analysis

After the iterative outcomes converge to stability, the statistical analysis of $\widetilde{a}^j$, $\widetilde{b}^j$, $\widetilde{c}^j$, $\widetilde{d}^j$, $\widetilde{e}^j$ is performed to obtain the maximum value, minimum value, average value, mode value, and statistical values of 40% and 60% confidence intervals of each coefficient.

Step 10. Outcome estimation

With the nonlinear relationship of D from the previous step, the new predicted D can be calculated with the existing LSL data. By comparing the calculated values with the existing data, the prediction ability of each nonlinear relationship of D can be estimated.

4. Results

Through the data preparation steps in Sections 3.1–3.3, all the data were analyzed according to Sections 3.4 and 3.5, and the results are described in the following sections.

4.1. Linear Regression Analysis Results

With a linear relationship between dimensionless (R/D) and dimensionless (ϕ/θ) according to Equation (3), the regression results are illustrated as Figure 4. In the figure, the black dots are the 28 known occurrence time cases, the red line is the regression trend line of Equation (15), and the green and purple dotted lines are the upper and the lower boundary lines of Equations (16) and (17), respectively.

$$\text{Trend line}: \frac{R}{D} = 0.1347 \times \frac{\phi}{\theta} - 0.0281,\ R^2 = 0.55 \tag{15}$$

$$\text{Upper boundary line}: \frac{R}{D} = 0.1347 \times \frac{\phi}{\theta} + 0.0032 \tag{16}$$

$$\text{Lower boundary line}: \frac{R}{D} = 0.1347 \times \frac{\phi}{\theta} - 0.0543 \tag{17}$$

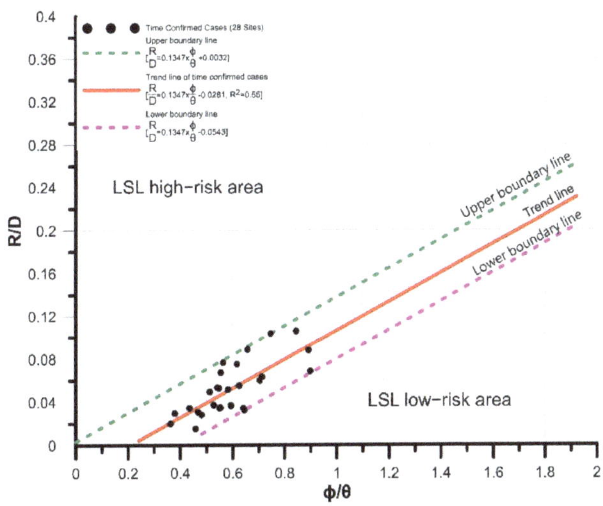

Figure 4. Dimensionless factor relationship of LSL cases with known occurrence time.

In order to examine whether the relationship of Equation (15) is reasonable, this study collected data according to the procedures of Sections 3.1–3.3 for the 79 occurrence-time-unknown cases and 149 potential areas delineated by the Soil and Water Conservation Bureau (SWCB). Since there is no information about the time of landslide for these two types of cases, the effective cumulative rainfall is evaluated alternatively. The total accumulated rainfall values of the corresponding events were used for the 79 occurrence-time-unknown cases, and the maximum total accumulated rainfall recorded over the years was taken for the 149 potential areas. After calculating the two dimensionless quantities, the results were superimposed with the trend line and the boundary lines from Figure 5 to display in Figures 6 and 7.

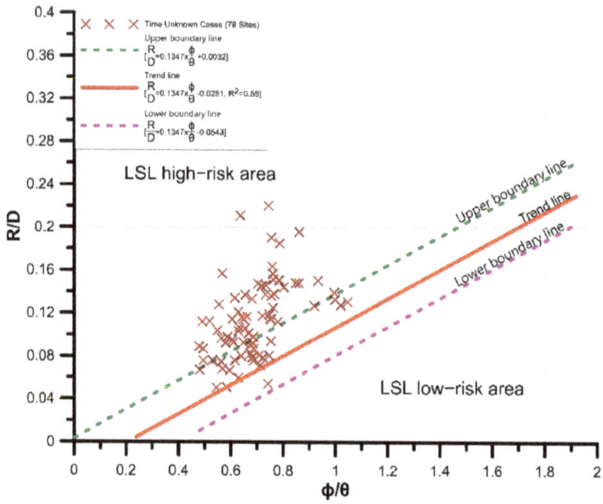

Figure 5. Dimensionless factor relationship of LSL cases with unknown occurrence time.

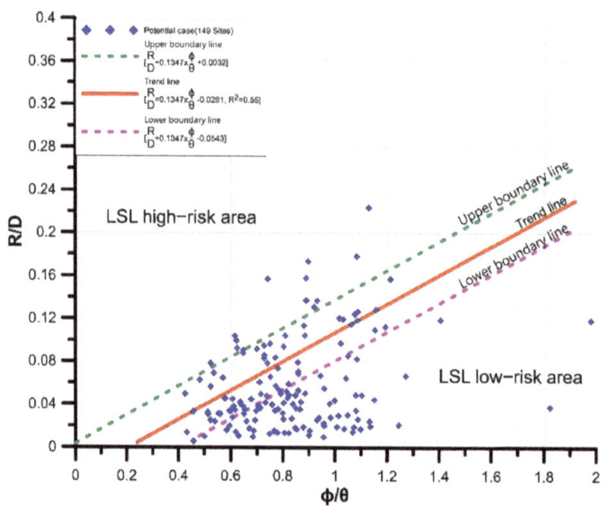

Figure 6. Dimensionless factor relationship of potential LSL area.

In Figure 5, almost all cases are above the trend line, and only one case falls between the trend line and the lower boundary line. Since the total rainfall should cap the actual effective cumulative rainfall, it is reasonably expected that most of the data (if not all) are above the trend line.

Among the 149 records in Figure 8, 67 records are located below the lower boundary line (45%), 112 below the trend line (75%), 135 below the upper boundary line (91%), and 14 above the upper boundary line (9%).

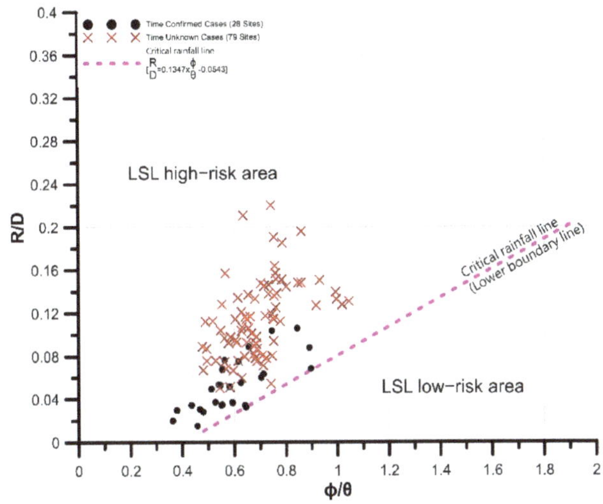

Figure 7. Dimensionless factor relationship of potential LSL area.

To be conservative, the lower boundary line (the purple dotted line in Figure 9) of the occurrence rainfall is suggested to be the management value for LSL evacuation and refuge.

4.2. Nonlinear Regression Analysis Results

In this study, the Analysis ToolPak of Microsoft Excel® was used to perform the nonlinear fit analysis between the predictions and the observations according to Equation (8). The results are shown in Table 4 and Figure 8. From the value of R square and the adjusted R square, it was found that Equation (8) has a good degree of model fit, and F is very different from the significance F (last two columns of ANOVA). This indicates that each parameter has large differences between groups, and small differences within groups. From the results in Figure 8b, it was found that the predicted value and the observed value have similar trends, and the function described by Equation (8) is reasonable.

Table 4. Result of complex data analysis.

Regression Statistics						
Multiple R						0.9569
R Square						0.9157
Adjusted R Square						0.9052
Standard Error						2.8447
Observations						28
ANOVA						
	df	SS	MS	F		Significance F
Regression	3	2109.4427	703.1476	86.8894		4.9903×10^{-13}
Residual	24	194.2187	8.0924			
Total	27	2303.6614				
	Coefficients	Standard Error	*t*-Test	*p*-value	Lower 95%	Upper 95%
Intercept	66.7058	6.1858	10.7838	1.10×10^{-10}	53.9390	79.4725
R(m)	−2.2273	2.8034	−0.7945	0.4347	−8.0133	3.5587
θ(degree)	0.0043	0.1279	0.0334	0.9737	−0.2597	0.2682
ϕ(degree)	−2.2251	0.1456	−15.2793	7.26×10^{-14}	−2.5257	−1.9246

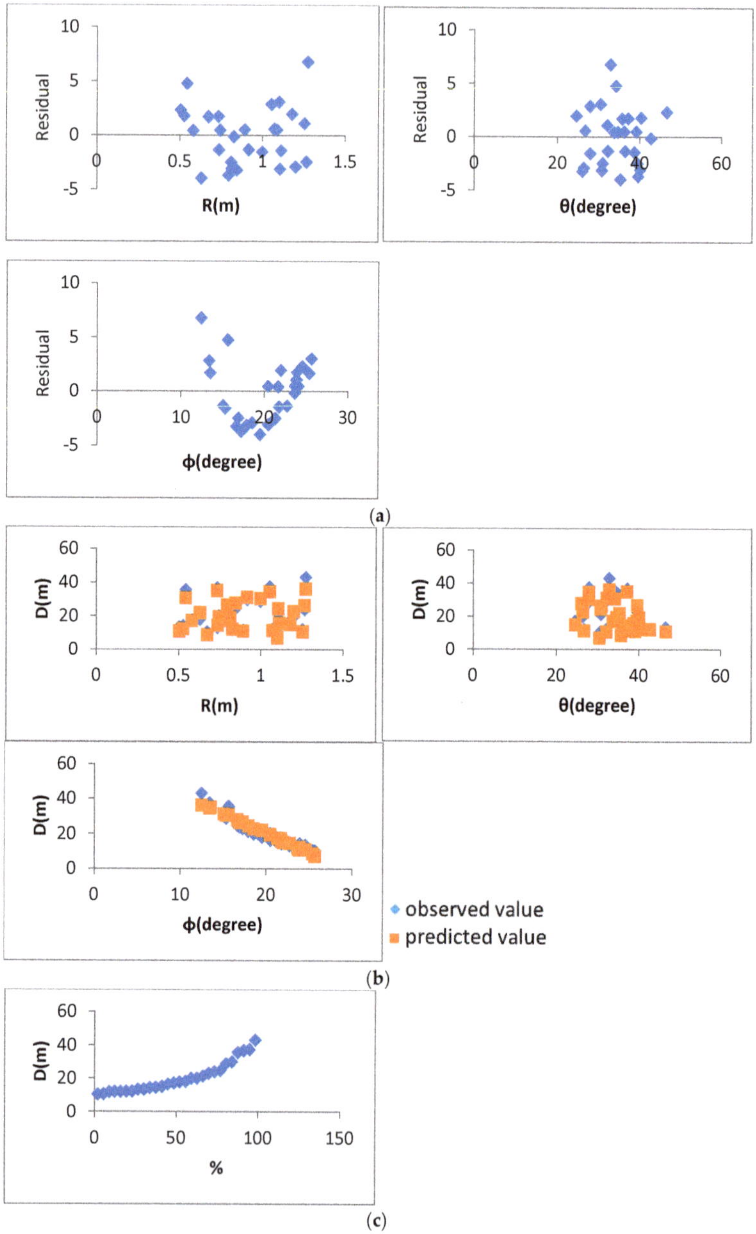

Figure 8. The results of the fit analysis between the nonlinear predicted value and the observed value, where (**a**) is the residual of each parameter, (**b**) is the sample regression line of each parameter, and (**c**) is the normal probability. D is the thickness of the sliding soil layer, R is the landslide triggering rainfall, θ is the slope of sliding surface, and ϕ is the equivalent friction angle.

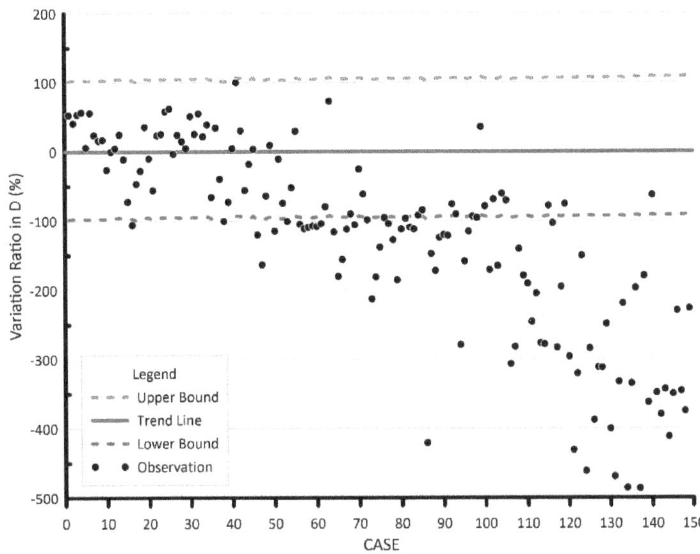

Figure 9. Comparison between predicted and estimated values of collapse thickness in 149 potential LSL areas.

After confirming that Equation (8) was reasonable, the nonlinear regression analysis according to the procedures in Figure 5 was performed. With 5000 iterations of calculation, it was found that the variation of the CV of each coefficient is approaching asymptotic stability as the number of iterations increases. The optimized coefficients of each group were obtained from the statistics in Table 5. Through the results in Table 5, the statistical values of each coefficient were substituted into Equation (8) to calculate the predicted value. The predicted values were compared with the observed values of 28 occurrence-time-known cases and 79 occurrence-time-unknown cases. The results of landslide thickness difference ΔD are shown in Tables 6 and 7.

Table 5. Statistics of nonlinear regression coefficients.

Parameter	Max	Min	Mean	Median	40%	60%
a	24,469.2875	648.5690	6095.1517	4438.0571	3576.8579	5539.8139
b	−0.0229	−0.3621	−0.1647	−0.1604	−0.1744	−0.1460
c	0.1614	−0.4220	−0.1339	−0.1298	−0.1492	−0.1089
d	2.4898	0.8831	1.6260	1.6288	1.5488	1.7067
e	5.0123	−22.1158	−4.4903	−3.5814	−4.7650	−2.5826

Table 6. The landslide thickness of the prediction minus the observation for 28 occurrence-time-known cases. (Unit: m, % relative to the observation).

	ΔD_{max}	ΔD_{min}	ΔD_{mean}	ΔD_{median}	$\Delta D_{40\%}$	$\Delta D_{60\%}$
Standard deviation	9.18 (15%)	6.85 (35%)	3.52 (9%)	1.38 (6%)	1.65 (6%)	1.77 (7%)
Mean of absolute values	15.03 (72%)	31.00 (167%)	8.03 (41%)	1.07 (5%)	2.20 (11%)	2.87 (15%)
Maximum value	42.61 (99%)	−23.20 (−117%)	16.15 (55%)	2.38 (12%)	0.02 (0%)	6.63 (26%)
Mean value	15.03 (72%)	−31.00 (−167%)	8.03 (41%)	0.16 (1%)	−2.20 (−11%)	2.82 (15%)
Median value	11.38 (76%)	−28.18 (−163%)	6.94 (43%)	0.45 (3%)	−1.85 (−10%)	2.84 (17%)
Minimum value	6.38 (37%)	−50.56 (−231%)	3.69 (21%)	−4.00 (−13%)	−7.11 (−23%)	−0.36 (−2%)

Table 7. The landslide thickness of the prediction minus the observation for 79 occurrence-time-unknown cases. (Unit: m, % relative to the observation).

	ΔD_{max}	ΔD_{min}	ΔD_{mean}	ΔD_{median}	$\Delta D_{40\%}$	$\Delta D_{60\%}$
Standard deviation	1.42 (4%)	1.60 (20%)	1.20 (8%)	0.65 (5%)	0.69 (6%)	0.67 (5%)
Mean of absolute values	9.26 (75%)	26.47 (216%)	4.39 (35%)	0.63 (5%)	2.13 (17%)	1.48 (12%)
Maximum value	16.21 (83%)	−22.69 (−150%)	8.71 (61%)	1.62 (15%)	0.13 (1%)	3.36 (30%)
Mean value	9.26 (75%)	−26.47 (−216%)	4.39 (35%)	−0.39 (−3%)	−2.13 (−17%)	1.48 (12%)
Median value	8.81 (75%)	−26.41 (−219%)	4.19 (35%)	−0.43 (−3%)	−2.17 (−18%)	1.42 (12%)
Minimum value	7.46 (62%)	−32.25 (−254%)	1.85 (17%)	−1.72 (−16%)	−3.55 (−32%)	0.02 (0%)

From Tables 6 and 7, Equation (18) was obtained by substituting the median value of each coefficient into Equation (8) which has a relatively good predictive ability in both cases with or without known occurrence time. Assuming a normal distribution of Equation (8), the upper and lower boundaries of the 99.7% range covered by three standard deviations were used. The lower boundary illustrated in Equation (19) was obtained by the 40th percentile values of each coefficient, and the upper boundary illustrated in Equation (20) was obtained by the 60th percentile value of each coefficient.

$$\text{Trend line}: D_{TL} = \frac{4438.0571 \times R^{-0.1604} \times \theta^{-0.1298}}{\phi^{1.6288}} - 3.5814 \tag{18}$$

$$\text{Upper boundary line}: D_{LB} = \frac{3576.8579 \times R^{-0.1744} \times \theta^{-0.1492}}{\phi^{1.5488}} - 4.7650 \tag{19}$$

$$\text{Lower boundary line}: D_{UB} = \frac{5539.8139 \times R^{-0.1460} \times \theta^{-0.1089}}{\phi^{1.7067}} - 2.5826 \tag{20}$$

To better illustrate the data distribution for the 149 potential LSL areas, the trend line and the boundary lines were transformed and normalized. The results from Equation (18) were set as the zero-reference line, and those from Equations (19) and (20) were scaled to be +100% (upper boundary) and −100% (lower boundary), respectively. The three lines of the 149 observation data are plotted in Figure 9. Of the 149 records, 83 records (56%) are below the lower boundary line, 115 records (77%) below the trend line, and all 149 records (100%) are below the upper boundary line.

5. Discussion

According to the aforementioned information and results, some issues are discussed in the following sections.

5.1. Prediction Ability

In order to directly compare the predictive capabilities (rainfall is the key indicator for early warning systems) of the results from linear regression analysis and nonlinear regression analysis, Equations (15) to (20) can be rewritten with R as the dependent variable as Equations (21) to (26).

$$R_{LT} = D \times (0.1347 \times \frac{\phi}{\theta} - 0.0281), \tag{21}$$

$$R_{LU} = D \times (0.1347 \times \frac{\phi}{\theta} + 0.0032), \tag{22}$$

$$R_{LL} = D \times (0.1347 \times \frac{\phi}{\theta} - 0.0543), \tag{23}$$

$$R_{NT} = \left[\frac{(D + 3.5814) \times \phi^{1.6288}}{4438.0571 \times \theta^{-0.1298}}\right]^{\frac{1}{-0.1604}}, \tag{24}$$

$$R_{NU} = \left[\frac{(D + 2.5826) \times \phi^{1.7067}}{5539.8139 \times \theta^{-0.1089}}\right]^{\frac{1}{-0.1460}}, \tag{25}$$

$$R_{NL} = \left[\frac{(D + 4.7650) \times \phi^{1.5488}}{3576.8579 \times \theta^{-0.1492}}\right]^{\frac{1}{-0.1744}}, \tag{26}$$

where:

R_{LT} is the predicted rainfall by the linear regression trend line (m),
R_{LU} is the predicted rainfall by the linear regression upper boundary line (m),
R_{LL} is the predicted rainfall by the linear regression lower boundary line (m),
R_{NT} is the predicted rainfall by the nonlinear regression trend line (m),
R_{NU} is the predicted rainfall by the nonlinear regression upper boundary line (m),
R_{NL} is the predicted rainfall by the nonlinear regression lower boundary line (m).

With the 79 occurrence-time-unknown cases and the 149 potential areas, the predicted rainfalls were calculated by Equations (21) to (26), the trend lines and boundary lines were transformed and scaled to be 0% and ±100%, similar to Figure 9, and the results are shown in Figures 10–13.

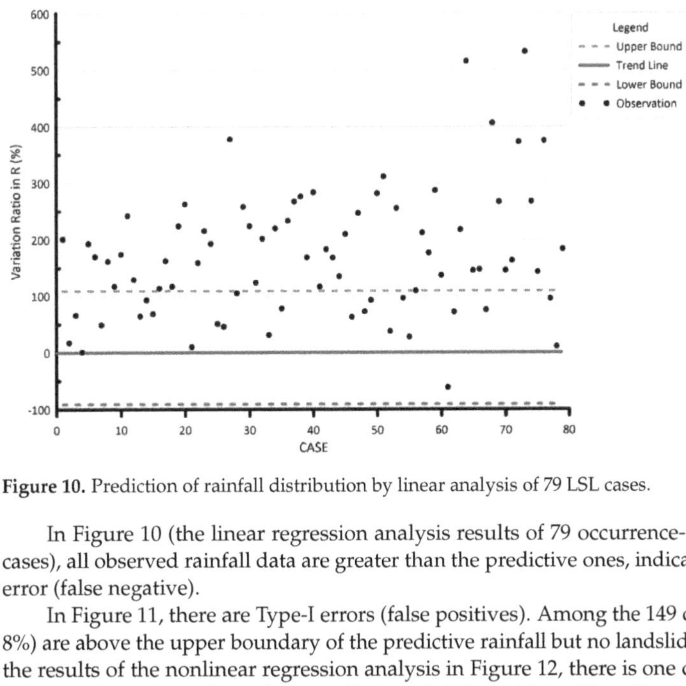

Figure 10. Prediction of rainfall distribution by linear analysis of 79 LSL cases.

In Figure 10 (the linear regression analysis results of 79 occurrence-time-unknown cases), all observed rainfall data are greater than the predictive ones, indicating no Type-II error (false negative).

In Figure 11, there are Type-I errors (false positives). Among the 149 cases, 12 (about 8%) are above the upper boundary of the predictive rainfall but no landslide occurred. For the results of the nonlinear regression analysis in Figure 12, there is one case (about 1%) below the lower boundary of the predictive rainfall, which is a Type-II error. According to the results in Figure 13, all of the 149 cases are below the upper boundary of the predictive rainfall, which means no Type-I error.

In summary, based on the limited data used in this study, there is no Type-II error from linear regression and no Type-I error from nonlinear regression. If different situations occur in the future, the new data should be updated and included for improving risk probability assessment.

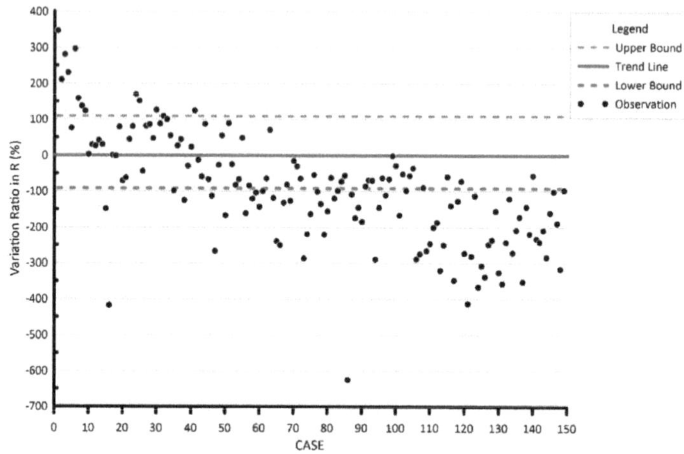

Figure 11. Prediction of rainfall distribution by linear analysis of 149 potential LSL areas.

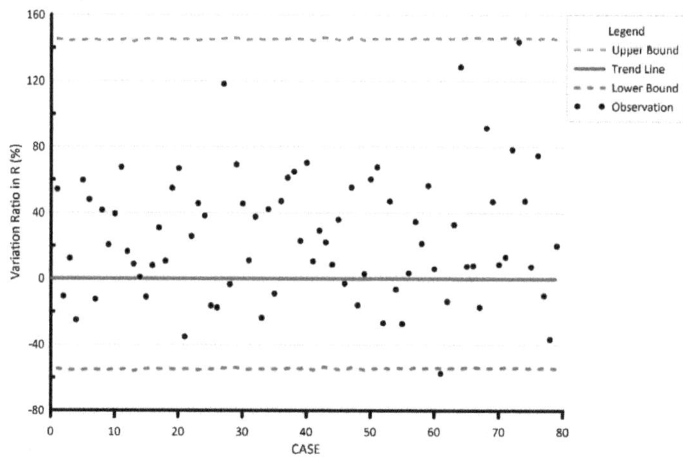

Figure 12. Prediction of rainfall distribution by nonlinear analysis of 79 LSL cases.

With Equations (21) to (26) for the 28 occurrence-time-known cases, the predicted rainfalls were compared with the observed triggering rainfalls, and error sums of squares and the root mean square errors are shown in Table 8. By examining the error sum results, the nonlinear trend (R_{NT}) has the best predictive capacity. Following the same procedures, 79 occurrence-time-unknown cases and 149 potential areas were evaluated and the results are shown in Tables 9 and 10. It is also found that the nonlinear trend line has the best predictive capacity. Nevertheless, the conservative lower boundary value should be used regarding the evacuation management value of the early warning operation. From the comparison of the error sums in Tables 8 and 9, the estimates by the lower boundary line of nonlinear analysis do provide better predictions.

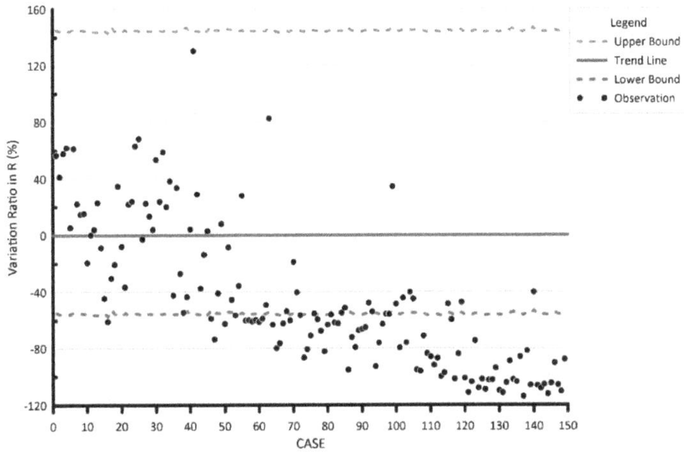

Figure 13. Prediction of rainfall distribution by nonlinear analysis of 149 LSL potential areas.

Table 8. Statistics of rainfall predictions for 28 occurrence-time-known LSL cases.

	R	R_{LT}	R_{LL}	R_{LU}	R_{NT}	R_{NL}	R_{NU}
Maximum value (mm)	1271.3	1599.1	1147.8	2554.0	1425.8	804.3	3069.4
Minimum value (mm)	501.6	571.4	−198.5	955.9	277.0	178.7	507.4
Error sum of squares (m^2)		2.3922	9.0971	19.3008	2.1096	4.3647	55.0174
Normalized error sum of squares		4.1646	9.9850	34.2229	3.2921	4.8297	70.8433
Standard deviation (mm)		297.7	580.5	845.5	279.5	402.1	1427.5
Normalized standard deviation		0.3927	0.6081	1.1258	0.3492	0.4229	1.6198

Table 9. Statistics of rainfall predictions for 79 occurrence-time-unknown LSL cases.

	R	R_{LT}	R_{LL}	R_{LU}	R_{NT}	R_{NL}	R_{NU}
Maximum value (mm)	1994.4	1544.9	1174.1	1987.9	1688.7	950.8	3589.9
Minimum value (mm)	779.2	464.8	132.2	862.1	872.7	470.0	1966.4
Error sum of squares (m^2)		11.2251	26.5031	3.7198	4.1568	18.3132	40.8837
Normalized error sum of squares		4.5775	12.7829	1.9193	1.5839	7.6505	31.4102
Standard deviation (mm)		644.8	990.8	371.2	392.4	823.6	1230.5
Normalized standard deviation		0.4117	0.6881	0.2666	0.2422	0.5323	1.0786

Table 10. Statistics of rainfall predictions for 149 potential LSL areas.

	R	R_{LT}	R_{LL}	R_{LU}	R_{NT}	R_{NL}	R_{NU}
Maximum value (mm)	2267.9	3016.4	2685.2	3412.0	2739.7	1575.6	5647.8
Minimum value (mm)	822.3	804.5	511.1	1037.4	1144.4	591.5	2583.9
Error sum of squares (m^2)		10.0854	20.0943	8.5405	6.1552	23.6371	94.8223
Normalized error sum of squares		3.1882	5.8767	3.4506	2.3811	6.6163	44.9751
Standard deviation (mm)		611.2	862.7	562.4	477.5	935.7	1874.0
Normalized standard deviation		0.3436	0.4665	0.3575	0.2970	0.4950	1.2906

5.2. Limitations

In order to establish the specific relationship between LSLs and triggering rainfall for the future LSL early warning predictions, this study collected LSL case data for modeling. As aforementioned, under the situation that epistemic conditions of LSLs are insufficient, coupled with the fact that LSL is not common, the data that can be collected are very limited. Although the model verified the rationality, it does not mean this model could be used directly in another region outside Taiwan, and those who want to apply this model should follow the procedures mentioned in this study to retrieve the suitable parameters before practice.

5.3. Early Warning Application

When the evacuation of early warning systems is announced, it is important to take into account not only the number of evacuees, the length of the evacuation route, and the opening of evacuation shelters, but also the time required for the activation of the evacuation mechanism, agency communication, evacuation notification and enforcement, evacuation status confirmation, etc. It is recommended that the operation should be carried out 3 to 6 h before the cumulative rainfall value reaches the evacuation management value. For this reason, it is suggested to use the quantitative rainfall forecast of 3 to 6 h, such as: ensemble model-based typhoon quantitative precipitation forecast (ETQPF) or blended quantitative precipitation forecast (BQPF). When the forecast of accumulated rainfall reaches the critical value, the evacuation mechanism should be activated immediately for an adequate response time.

6. Conclusions

When a large-scale landslide (LSL) disaster occurs, it could have a great impact on people's lives and properties. In response to the potential threat caused by LSLs, this study collected most relevant data of LSL cases, analyzed them to quantify the relationship between the LSL and the triggering rainfall, and proposed to apply this relationship to LSL warning predictions.

The satellite imagery and additional information of landslides from 2004 to 2016 were collected in this study. After screening of 43,519 landslide records, 107 newborn landslides were selected for analysis, including 28 occurrence-time-known cases and 79 occurrence-time-unknown cases. In addition, 149 potential LSL areas evaluated by the Soil and Water Conservation Bureau (SWCB) were also used for improving the lower boundary line.

This study employs two methods of linear and nonlinear regression analysis to assess the relationship between LSL and rainfall. The results show that there are 8% Type-I errors (false positives) in the linear regression analysis, and 1% Type-II errors (false negatives) in the nonlinear regression analysis. With the comparison of statistical indicators, the trend line of nonlinear regression analysis shows better predictive power. Considering the response time required for the early warning operation, it is suggested that the nonlinear lower boundary line can be used as the evacuation and refuge management value. Combined with the numerical rainfall forecast of 3 to 6 h, it can be applied to the evacuation and refuge operations of LSLs.

LSL is not common, but always causes serious impacts. In this study, although a reasonable model for Taiwan was successfully established through limited data, it does not mean that it can be directly applied to other regions. In order to make this model more widely used, testing through foreign cases and expanding the factors considered, such as: geological type, slope aspect, water content, land use, NDVI, time error effect, etc., would be the future topics for further understanding and applications of LSL-related issues.

Author Contributions: Conceptualization, C.-L.S., T.-T.T., Y.-J.T. and J.H.-C.W.; methodology, T.-T.T., Y.-J.T., C.-L.S. and J.H.-C.W.; data curation, Y.-J.T.; writing original draft, T.-T.T.; visualization, T.-T.T. and Y.-J.T.; supervision, J.H.-C.W.; project administration, C.-L.S. All authors have read and agreed to the published version of the manuscript.

Funding: This research was funded by the Soil and Water Conservation Bureau, Council of Agriculture, Executive Yuan, Taiwan, grant number SWCB-109-153, and the APC was funded by Disaster Prevention Education Center, National Cheng-Kung University.

Data Availability Statement: The data that support the findings of this study are available from the author Y.-J.T., upon reasonable request.

Acknowledgments: We thank the Soil and Water Conservation Bureau of Council of Agriculture under Executive Yuan for providing information regarding the project, large-scale landslide disaster prevention, and mitigation under climate change.

Conflicts of Interest: The authors declare no conflict of interest.

References

1. Shieh, C.-L.; Wang, C.-M.; Chen, Y.-S.; Tsai, Y.-J.; Tseng, W.-H. An overview of disasters resulted from Typhoon Morakot in Taiwan. *J. Disaster Res.* **2010**, *5*, 236–244. [CrossRef]
2. Chen, Y.-S.; Kuo, Y.-S.; Lai, W.-C.; Tsai, Y.-J.; Lee, S.-P.; Chen, K.-T.; Shieh, C.-L. Reflection of typhoon morakot—The challenge of compound disaster simulation. *J. Mt. Sci.* **2011**, *8*, 571–581. [CrossRef]
3. Central Weather Bureau. Typhoon Database. Available online: https://rdc28.cwb.gov.tw/TDB/public/typhoon_detail?typhoon_id=200908 (accessed on 7 March 2022).
4. Xiaolin Pingpu Cultural Museums. Typhoon Morakot Caused Flood. Available online: https://xiaolin.khcc.gov.tw/index.php?option=module&lang=cht&task=pageinfo&id=4&index=1 (accessed on 8 March 2022).
5. National Science and Technology Center for Disaster Reduction. *Action Plan for the Prevention and Control of Large-Scale Collapse Disasters*; National Science and Technology Center for Disaster Reduction: Taipei, China, 2015. (In Chinese)
6. Lee, W.-L. Research on Pore Water Pressure Changes during Infiltration and Its Impact on Slope Stability. Ph.D. Thesis, National Cheng Kung University, Tainan, Taiwan, 2020. (In Chinese)
7. Kalatehjari, R.; Ali, N. A review of three-dimensional slope stability analyses based on limit equilibrium method. *Electron. J. Geotech. Eng.* **2013**, *18*, 119–134.
8. Duncan, J.M.; Wright, S.G.; Brandon, T.L. *Soil Strength and Slope Stability*; John Wiley & Sons: Hoboken, NJ, USA, 2014.
9. Lu, N.G.; Jonathan, W. *Hillslope Hydrology and Stability*; Cambridge University Press: Cambridge, UK, 2013.
10. Newmark, N.M. Effects of earthquakes on dams and embankments. *Geotechnique* **1975**, *15*, 139–160. [CrossRef]
11. Wilson, R.C.; Keefer, D.K. Dynamic analysis of a slope failure from the 6 August 1979 Coyote Lake, California, earthquake. *Bull. Seismol. Soc. Am.* **1983**, *73*, 863–877. [CrossRef]
12. Culmann, K. *Die Graphische Statik*; Meyer & Zeller (A. Reimann): Zürich, Switzerland, 1875; Volume 1. (In German)
13. William, E.; Dietrich, D.R.M. Shalstab. Available online: http://calm.geo.berkeley.edu/geomorph/shalstab/index.htm (accessed on 15 March 2022).
14. Pack, R.T.; Tarboton, D.G.; Goodwin, C.; Prasad, A. SINMAP 2.0—A Stability Index Approach to Terrain Stability Hazard Mapping, User's Manual. Available online: https://digitalcommons.usu.edu/cee_facpub/16/ (accessed on 15 September 2022).
15. Baum, R.L.; Savage, W.Z.; Godt, J.W. TRIGRS—A Fortran program for transient rainfall infiltration and grid-based regional slope-stability analysis. In *US Geological Survey Open-File Report*; US Geological Survey: Reston, VA, USA, 2002.
16. Survey, U.S.G. Scoops3D. Available online: https://www.usgs.gov/software/scoops3d (accessed on 27 May 2022).
17. Brinkgreve, R.B.J. *Plaxis Finite Element Code for Soil and Rock Analyses 2D—Version 8*; Balkema Rotterdam: Rotterdam, The Netherlands, 2008.
18. Miyamoto, K. Two dimensional numerical simulation of landslide mass movement. *J. Eros. Control Eng.* **2002**, *55*, 5–13.
19. Itasca Consulting Group, Inc. PFC. Available online: https://www.itascacg.com/software/PFC (accessed on 27 May 2022).
20. Itasca Consulting Group, Inc. FLAC. Available online: https://www.itascacg.com/software/FLAC (accessed on 27 May 2022).
21. Wang, W.-N. *Geological Characteristics of Landslide and Disaster Prevention*; Taiwan Disaster Prevention Society: Tainan, China, 2016. (In Chinese)
22. Coveney, S.; Fotheringham, A.S.; Charlton, M.; McCarthy, T. Dual-scale validation of a medium-resolution coastal DEM with terrestrial LiDAR DSM and GPS. *Comput. Geosci.* **2010**, *36*, 489–499. [CrossRef]
23. Su, M.-B.; Chen, I.-H.; Liao, C.-H. Using TDR cables and GPS for landslide monitoring in high mountain area. *J. Geotech. Geoenviron. Eng.* **2009**, *135*, 1113–1121. [CrossRef]
24. Caviedes-Voullième, D.; Juez, C.; Murillo, J.; García-Navarro, P. 2D dry granular free-surface flow over complex topography with obstacles. Part I: Experimental study using a consumer-grade RGB-D sensor. *Comput. Geosci.* **2014**, *73*, 177–197. [CrossRef]
25. Juez, C.; Caviedes-Voullième, D.; Murillo, J.; García-Navarro, P. 2D dry granular free-surface transient flow over complex topography with obstacles. Part II: Numerical predictions of fluid structures and benchmarking. *Comput. Geosci.* **2014**, *73*, 142–163. [CrossRef]
26. Tsai, W.-N.; Chen, C.-C.; Chiang, C.-W.; Chen, P.-Y.; Kuo, C.-Y.; Wang, K.-L.; Lin, M.-L.; Chen, R.-F. Electrical Resistivity Tomography (ERT) monitoring for landslides: Case study in the Lantai Area, Yilan Taiping Mountain, Northeast Taiwan. *Front. Earth Sci.* **2021**, *9*, 737271. [CrossRef]
27. Broadband Array in Taiwan for Seismology, BATS. Available online: https://gdms.cwb.gov.tw/gdms-broad.php (accessed on 18 May 2022).
28. Wang, K.-L.; Lin, J.-T.; Lee, Y.-H.; Chen, T.-W. Study on research on simulation of shallowcollapse behavior with particle flow—Case study of Hongye landslide, Taiwan. In Proceedings of the XIII International Symposium on Landslides, Cartagena, Colombia, 15–19 June 2020.
29. Lo, C.-M. Evolution of deep-seated landslide at Putanpunas stream, Taiwan. *Geomat. Nat. Hazards Risk* **2017**, *8*, 1204–1224. [CrossRef]
30. Hölbling, D.; Betts, H.; Spiekermann, R.; Phillips, C. Identifying spatio-temporal landslide hotspots on North Island, New Zealand, by analyzing historical and recent aerial photography. *Geosciences* **2016**, *6*, 48. [CrossRef]
31. Lahousse, T.; Chang, K.; Lin, Y.; Günther, A. Landslide mapping with multi-scale object-based image analysis–a case study in the Baichi watershed, Taiwan. *Nat. Hazards Earth Syst. Sci.* **2011**, *11*, 2715–2726. [CrossRef]

32. Wang, K.-L.; Lin, M.-L. Initiation and displacement of landslide induced by earthquake—A study of shaking table model slope test. *Eng. Geol.* **2011**, *122*, 106–114. [CrossRef]
33. Booth, A.M.; Roering, J.J.; Perron, J.T. Automated landslide mapping using spectral analysis and high-resolution topographic data: Puget Sound lowlands, Washington, and Portland Hills, Oregon. *Geomorphology* **2009**, *109*, 132–147. [CrossRef]
34. Dewitte, O.; Jasselette, J.-C.; Cornet, Y.; Van Den Eeckhaut, M.; Collignon, A.; Poesen, J.; Demoulin, A. Tracking landslide displacements by multi-temporal DTMs: A combined aerial stereophotogrammetric and LiDAR approach in western Belgium. *Eng. Geol.* **2008**, *99*, 11–22. [CrossRef]
35. Lee, C.-F.; Lo, C.-M.; Chou, H.-T.; Chi, S.-Y. Landscape evolution analysis of large scale landslides at Don-Ao Peak, Taiwan. *Environ. Earth Sci.* **2016**, *75*, 29. [CrossRef]
36. Lin, M.-L.; Chen, T.-W.; Lin, C.-W.; Ho, D.-J.; Cheng, K.-P.; Yin, H.-Y.; Chen, M.-C. Detecting large-scale landslides using LiDAR data and aerial photos in the Namasha-Liuoguey area, Taiwan. *Remote Sens.* **2014**, *6*, 42–63. [CrossRef]
37. Ventura, G.; Vilardo, G.; Terranova, C.; Sessa, E.B. Tracking and evolution of complex active landslides by multi-temporal airborne LiDAR data: The Montaguto landslide (Southern Italy). *Remote Sens. Environ.* **2011**, *115*, 3237–3248. [CrossRef]
38. Lin, M.-L.; Chen, Y.-C.; Tseng, Y.-H.; Chang, K.-J.; Wang, K.-L. Investigation of geological structures using UAV LiDAR and its effects on the failure mechanism of deep-seated landslide in Lantai Area, Taiwan. *Appl. Sci.* **2021**, *11*, 52. [CrossRef]
39. Wang, K.-L.; Lin, J.-T.; Zhang, J.-W.; Lin, Y.-H.; Hung, J.-S.; Hung, J.-Y.; Ciou, C.-F.; Fu, K.-L. A debris estimation method combining high accuracy UAV, DInSAR and NDSI analysis. *J. Taiwan Agric. Eng.* **2019**, *65*, 26–39.
40. Hung, W.-C.; Hwang, C.; Chen, Y.-A.; Chang, C.-P.; Yen, J.-Y.; Hooper, A.; Yang, C.-Y. Surface deformation from persistent scatterers SAR interferometry and fusion with leveling data: A case study over the Choushui River alluvial fan, Taiwan. *Remote Sens. Environ.* **2011**, *115*, 957–967. [CrossRef]
41. Su, P.-T. A PSInSAR Method with Redundant Observations. Master's Thesis, National Cheng Kung University, Tainan, China, 2012.
42. Champenois, J.; Fruneau, B.; Pathier, E.; Deffontaines, B.; Lin, K.-C.; Hu, J.-C. Monitoring of active tectonic deformations in the Longitudinal Valley (Eastern Taiwan) using Persistent Scatterer InSAR method with ALOS PALSAR data. *Earth Planet. Sci. Lett.* **2012**, *337*, 144–155. [CrossRef]
43. Chou, F.-M. Measurement of Surface Deformation in Northern Taiwan by Using PSInSAR Techniques. Master's Thesis, National Central University, Taoyuan, Taiwan, 2009. (In Chinese).
44. Greif, V.; Vlcko, J. Monitoring of post-failure landslide deformation by the PS-InSAR technique at Lubietova in Central Slovakia. *Environ. Earth Sci.* **2011**, *66*, 1585–1595. [CrossRef]
45. Lu, P.; Casagli, N.; Catani, F.; Tofani, V. Persistent Scatterers Interferometry Hotspot and Cluster Analysis (PSI-HCA) for detection of extremely slow-moving landslides. *Int. J. Remote Sens.* **2011**, *33*, 466–489. [CrossRef]
46. Lu, C.-H. *Faults Activities and Crustal Deformation near Hualien City, Eastern Taiwan Analysed by Persistent Scatterer InSAR*; University of Taipei: Taipei, China, 2009. (In Chinese)
47. Siao, Y.-F. Permanent Scatter Radar Interferometry Applied to Surface Change Detection. Master's Thesis, National Central University, Taoyuan, China, 2010. (In Chinese).
48. Tung, H.; Hu, J.-C. Assessments of serious anthropogenic land subsidence in Yunlin County of central Taiwan from 1996 to 1999 by Persistent Scatterers InSAR. *Tectonophysics* **2012**, *578*, 126–135. [CrossRef]
49. Zhao, C.; Lu, Z.; Zhang, Q.; de la Fuente, J. Large-area landslide detection and monitoring with ALOS/PALSAR imagery data over Northern California and Southern Oregon, USA. *Remote Sens. Environ.* **2012**, *124*, 348–359. [CrossRef]
50. Wang, K.-L.; Lin, J.-T.; Wu, C.-C.; Chen, C.-F.; Chiu, T.-W.; Hsieh, M.-H. Study on sliding displacement classification of slope land—An example of NeiHu District, Taipei City. *J. Chin. Inst. Civ. Hydraul. Eng.* **2021**, *33*, 139–149. (In Chinese) [CrossRef]
51. Chen, R.-F.; Lee, C.-Y.; Yin, H.-Y.; Huang, H.-Y.; Cheng, K.-P.; Lin, C.-W. *Monitoring the Deep-Seated Landslides by Using ALOS/PALSAR Satellite Imagery in the Disaster Area of 2009 Typhoon Morakot, Taiwan*; Springer International Publishing: Cham, Switzerland, 2017.
52. Soil and Water Conservation Bureau. *Large-Scale Landslide Disaster Prevention and Mitigation under Climate Change*; Soil and Water Conservation Bureau: Nantou, China, 2020. (In Chinese)
53. Del Bene, S.K.; Kilmer, D.; Heck, D.; Smith, A.; Cartwright, M.; Bonamici, S.; Cohen, S. *National Landslide Preparedness Act*; the U.S. Government Publishing Office: Washington, DC, USA,, 2021.
54. Kambara, J.; Uchida, T. *Technical Guideline for Countermeasures against Deep-Seated Catastrophic (Rapid) Landslide*; National Institute for Land and Infrastructure Management: Tsukuba City, Ibaraki Prefecture, Japan, 2014.
55. National System for Environmental Protection. *Linee Guida per il Monitoraggio Delle Frane*; National System for Environmental Protection: Roma, Italy, 2021. (In Italian)
56. Jackson, L.; Bobrowsky, P.; Bichler, A. *Identification, Maps and Mapping Canadian Technical Guidelines and Best Practices Related to Landslides: A National Initiative for Loss Reduction*; Geological Survery of Canada: Ottawa, ON, Canada, 2012. [CrossRef]
57. Chae, B.-G.; Choi, J.; Seo, Y.-S. Suggestion of a landslide early warning method using a gradient of volumetric water content. In *Landslide Science for a Safer Geoenvironment*; Springer: Berlin/Heidelberg, Germany, 2014; pp. 545–550.
58. Kimura, T.; Yokoyama, N. Monitoring of landslide movement and forecasting of slope failure occurrence. *J. Jpn. Landslide Soc.* **2006**, *43*, 172–179.

59. Wang, J.-S.; Zhong, M.-J.; Ji, S. Discussion on monitoring feedback analysis and early warning application of rainfall-induced landslide. *Sinotech Eng.* **2011**, *110*, 27–40. (In Chinese) [CrossRef]
60. Lin, D.-G.; Chang, K.-C.; Su, M.-B. Numerical assessment of slope stability at Li-San landslide during typhoon rainfall. *J. Chin. Soil Water Conserv.* **2008**, *39*, 57–81. (In Chinese)
61. Gian, Q.A.; Tran, D.-T.; Nguyen, D.C.; Nhu, V.H.; Tien Bui, D. Design and implementation of site-specific rainfall-induced landslide early warning and monitoring system: A case study at Nam Dan landslide (Vietnam). *Geomat. Nat. Hazards Risk* **2017**, *8*, 1978–1996. [CrossRef]
62. Jan, C.-D. *Rainfall Warning for Debris Flow Disaster Prevention*; Sinotech Foundation for Research & Development of Engineering Sciences & Technologies: Taipei, China, 2016. (In Chinese)
63. Guzzetti, F.; Peruccacci, S.; Rossi, M.; Stark, C.P. Rainfall thresholds for the initiation of landslides in central and southern Europe. *Meteorol. Atmos. Phys.* **2007**, *98*, 239–267. [CrossRef]
64. Liao, J.-T.; Chen, C.-W.; Chi, C.-C.; Lin, H.-H. Study of slope displacement for management from landslide monitoring cases in Taiwan. *Sino-Geotech.* **2013**, *136*, 59–70. (In Chinese)
65. Martelloni, G.; Segoni, S.; Fanti, R.; Catani, F. Rainfall thresholds for the forecasting of landslide occurrence at regional scale. *Landslides* **2012**, *9*, 485–495. [CrossRef]
66. Tsai, T.-T.; Tsai, Y.-J.; Shieh, C.-L. Critical rainfall analysis of large-scale landslide occurrence. In Proceedings of the Interpraenent 2018, Toyama, Japan, 1–4 October 2018; pp. 91–96.
67. Zêzere, J.; Vaz, T.; Pereira, S.; Oliveira, S.; Marques, R.; Garcia, R.A. Rainfall thresholds for landslide activity in Portugal: A state of the art. *Environ. Earth Sci.* **2015**, *73*, 2917–2936. [CrossRef]
68. Shieh, C.-L.; Luh, Y.-C.; Yu, P.-S.; Chen, L.-J. The critical line of debris flow occurrence. *J. Chin. Soil Water Conserv.* **1995**, *26*, 167–172.
69. Tsai, T.-T.; Li, C.-C.; Tsai, H.-C.; Lin, Y.-F.; Chiu, T.-W.; Lin, C.-W. *An Analysis of Slopeland Disasters with Formosat-2 Imagery*; Central Geological Survey: Taipei, China, 2010; pp. 18–27. (In Chinese)
70. Tsai, T.-T. The review of FOMOSAT-II applications in disaster prevention and mitigation. *NARL Quart.* **2009**, *23*, 53–60. (In Chinese) [CrossRef]
71. Sinotech Engineering Consultants Inc. *Dynamic Susceptibility Assessment and Warning System of Rainfall Thresholds for Debris Slide*; Central Geological Survey: Taipei, China, 2013. (In Chinese)
72. Soil and Water Conservation Bureau. Majon Disaster Events. Available online: https://246.swcb.gov.tw/Achievement/MajorDisasters (accessed on 22 March 2022).
73. Lin, G.-W. *Detecting Large Scale Landslides by Taiwan Broadband Seismic Network and the Study of Triggering Rainfall for Landsliding*; Soil and Water Conservation Bureau: Nantou, China, 2015. (In Chinese)
74. Huang, T.-H. Characteristics of Rains Triggering Debris Flows in the Watershed of Chenyoulen Stream. Master's Thesis, National Cheng Kung University, Tainan, Taiwan, 2002. (In Chinese).
75. Jan, C.-D.; Lee, M.-H. A debris-flow rainfall-based warning model. *J. Chin. Soil Water Conserv.* **2004**, *35*, 275–285.
76. Soil and Water Conservation Bureau. *Soil and Water Conservation Handbook: Debris-Flow Chapter*; Soil and Water Conservation Bureau: Taiwan, 2021.
77. Towhata, I.; Goto, S.; Goto, S.; Akima, T.; Tanaka, J.; Uchimura, T.; Wang, G.; Yamaguchi, H.; Aoyama, S. Mechanism and future risk of slope instability induced by extreme rainfall event in Izu Oshima Island, Japan. *Nat. Hazards* **2021**, *105*, 501–530. [CrossRef]
78. Soil and Water Conservation Bureau. *Development and Application of Disaster Prevention and Mitigation Technologies for Large-Scale Landslides*; Soil and Water Conservation Bureau: Nantou, China, 2015. (In Chinese)
79. Scheidegger, A.E. On the prediction of the reach and velocity of catastrophic landslides. *Rock Mech.* **1973**, *5*, 231–236. [CrossRef]
80. Efron, B. Bootstrap methods: Another look at the jackknife. *Ann. Stat.* **1979**, *7*, 1–26. [CrossRef]
81. Singh, K.; Xie, M. *Bootstrap: A Statistical Method*; Department of Statistics, Rutgers University: New Brunswick-Piscataway, NJ, USA, 2008; (unpublished work).
82. Hu, W.; Xie, J.; Chau, H.W.; Si, B.C. Evaluation of parameter uncertainties in nonlinear regression using Microsoft Excel Spreadsheet. *Environ. Syst. Res.* **2015**, *4*, 4. [CrossRef]

water

MDPI

Article

Self-Sustaining Landslide Mitigation Strategy through Long-Term Monitoring

Filippo Tommaso Catelan, Giulia Bossi * and Gianluca Marcato

Italian National Research Council (CNR), Research Institute for Geo-Hydrological Protection (IRPI), Corso Stati Uniti 4, 35127 Padova, Italy
* Correspondence: giulia.bossi@irpi.cnr.it

Abstract: In mountain areas, anthropic pressure is growing while, concurrently, landslide frequency in most of the mountain regions of the world is increasing due to a more extreme precipitation pattern and permafrost deglaciation. Because of budget constraints, the need to investigate innovative and low-cost countermeasures for landslide risk is becoming more and more pressing. In this context, the Passo della Morte area (North-East Italy) is a perfect benchmark case. It comprises an extensive, long-term database of monitoring data that allows for testing hypotheses and validating them. Based on this data, a strong correlation between the velocity of a displacement of a landslide and the discharge of the Rio Verde stream was found. According to this evidence, local authorities have started the construction of a completely innovative mitigation strategy. It is focused on the triggering factor by identifying a significant component of the flow rate of the stream that cuts through the landslide. In addition, aiming to reduce the cost of construction and maintenance, this mitigation strategy is coupled with a micro-hydropower plant that can provide economic revenue by exploiting the discharge itself to produce electricity. Considering the active monitoring system that will be used to verify the effectiveness of the countermeasure, the Passo della Morte case study could become a starting point for implementing this pioneering and low-cost mitigation solution in similar morphologies.

Keywords: landslide mitigation; monitoring; micro-hydro plant; costs reduction; gray-box model

Citation: Catelan, F.T.; Bossi, G.; Marcato, G. Self-Sustaining Landslide Mitigation Strategy through Long-Term Monitoring. *Water* **2022**, *14*, 3824. https://doi.org/10.3390/w14233824

Academic Editor: Su-Chin Chen

Received: 17 October 2022
Accepted: 22 November 2022
Published: 24 November 2022

Publisher's Note: MDPI stays neutral with regard to jurisdictional claims in published maps and institutional affiliations.

1. Introduction

Structural mitigation of landslides was an expenditure item on which there was substantial investment in the 20th century, especially in associated high-risk areas such as in Europe, North America, and Far East Asia [1]. However, many of these structures are now increasingly old and damaged, significantly decreasing their capability to reduce risk. In Japan, for example, within 20 years, 60% of the Sabo facilities will exceed 50 years of age [2]. The major concern about the deterioration of these structures is that this will result in a potential risk even higher than the one prior to the implementation of the countermeasure, such as the Barlin dam (Taiwan) failure in August 2007 that caused the sudden release of over 7.5 Mm^3 of stored sediment [3,4].

Three elements must be considered when facing the optimization problem regarding maintenance and/or replacement of old countermeasure works. First, risk aversion and risk tolerance are changing in several countries globally [5]. Secondly, global population growth is pushing people to expand the road network. As a result, the urbanized areas in new locations may not be as safe as the historically built ones, especially near active debris flow and debris flood fans in mountain areas [6]. At the same time, the population is becoming older, which means that more and more people require assistance during rapid evacuations. Consequently, the probability of landslides causing disasters and fatalities is increasing since it is also increasing (and changing) the number and types of exposed elements [7,8]. Thirdly, the modification of the precipitation patterns that most scientists attribute to climate change has a significant impact on the type and frequency of floods and landslide events [9], and that should be taken into account in the assessment of the efficacy of structural mitigation works that have been designed 50 years ago considering a different climate.

An excellent example of this is well illustrated in the research of Junichi and Naoki 2020 [10] about the Japan situation. According to the study, over the last 20 years, landslide processes caused 678 human fatalities in Japan, encompassing 21% of the total deaths due to natural disasters (excluding the Great East Japan Earthquake of 2011), and the most impacted age class is the elderly. Moreover, between 2000 and 2019, the number of landslides increased due to heavy rains and extreme events.

In this framework, it is evident that new countermeasure structures, reinforcements, and continuous maintenance of the old ones are needed. However, before that, each country would have to convert landslide damage into economic terms to evaluate the potential budget that would allow paying for all these works [11,12]. Further, some nations assign jurisdiction and responsibility for risk reduction control measures to local authorities, such as regions or even municipalities; in this case, the overall costs of all the required mitigation actions percolate through different institutions and are challenging to assess. One crucial example is Italy. In fact, despite the presence in its territory of two of the most important mountain ranges in Europe, rarely are the costs associated with countermeasure works for landslide stability taken into account by national or regional authorities.

Consequently, mountain municipalities and local communities must assume the critical economic burden of financing infrastructures designed to reduce landslide risk [13,14]. However, the annual cost that those small municipalities can bear is often minimal. Therefore, when faced with the issues of funding kindergartens or landslides' countermeasures, the choice is evident and understandably the first. Moreover, in addition to degradation and sediment deposition in mitigation structures, which requires continuous use of economic resources for maintenance, local communities are often against extensive stabilization intervention that would lead to the appropriation of some of their properties or significant disturbance to the landscape [15].

A possible strategy to limit the impact on mitigation structures is acting on the triggering factors of the landslide process. Unfortunately, these are not always so easy to identify. The other option is to produce very sophisticated geotechnical models of the landslide risk to design countermeasure works [16]. However, these require a deep knowledge of the geotechnical and geological characteristics, such as subsoil reconstruction, groundwater regime, and rock/soil parameters burdened by high uncertainties. For this reason, numerical simulations, despite their solid theoretical basis, are not always the best option for producing a landslide model that is functional for mitigation.

On the other hand, careful analysis of an already present long-term series of recorded monitoring datasets such as inclinometric measures, Global Navigation Satellite System (GNSS) surveys, piezometers' data, acoustic emissions, discharge rate, rainfall, and temperature reports can lead to significant statistical testing for correlation between slope movement and other variables [17–19]. If the link is statistically significant, then it can be modeled using a gray-box approach [20]. The latter requires mainly identifying a correlation between input and output with limited knowledge of the intrinsic process that produces the relationship. Another critical aspect is that the monitoring period should be long enough to catch a few high-intensity events to record the duration and magnitude of a landslide occurrence under these circumstances. This not only reduces the time and cost of modeling but also increases its reliability and predictive value as it is based on actual signals provided by the investigated landslides without introducing uncertainties regarding the rheological property of the landslide mass [21,22]. In addition, acting on the triggering factors rather than barring a landslide through steel and concrete deployment is generally cheaper [23,24].

Since water is one of the main triggering factors for landslides in the various forms of rainfall, snowmelt, groundwater, or surface runoff, water sequestration is usually one of the most successful forms of "soft countermeasure" [25–27]. Given that landslides occur on steep mountain slopes, it is not so futuristic to think about using the potential energy of water runoff and velocity to produce power employing micro-hydropower plants. This has the potential to encourage municipalities to implement this mitigation strategy as it may become self-paying or, in some contexts, even profitable. This approach will be more and more feasible in the forthcoming future as research into and investments for designing efficient micro-hydro power plants are booming [28].

The case study presented here is an example of this approach. The area is located in the upper Tagliamento Valley, where a main national road crosses through it. A medium-size landslide monitored for more than a decade intercepts the latter. However, despite the considerable risk of a paroxysmal event that would hinder the road's usability, thus causing major disruption in the life of the valley inhabitants, only monitoring actions have been implemented since the Spring of 2022. This was due to the lack of resources of the municipality and jurisdictional disputes between the national and local authorities. In this context, we proposed a new mitigation strategy in 2019. It consists of pulling out the water that flows above and within the slope and funneling it away from the landslide to reduce the landslide's movement.

Additionally, the water would not be directly returned to the Rio Verde downstream of the landslide. Thus, the ultimate plan calls for exploiting the available 130 m vertical drop to produce power by installing a micro hydropower plant. This solution would provide a long-term financial return that would pay for the stabilization costs, offer funds for maintenance, and possibly economic revenue.

2. Study Area

A large sector of the Eastern Alps falls within the borders of the Friuli Venezia Giulia region in North-East Italy. The morphology generated by these high reliefs and tectonics favors the existence of numerous areas with an associated high risk of landslides [29,30]. Additionally, the abundance of evaporites causes the formation of sinkholes [31].

The most dominant hydrologic element in the region is the Tagliamento river. Its course is 178 km long with a basin of 2916 km^2 [32]. Part of its mountain path in the Eastern Alps is closely linked to the Tagliamento fault's presence [33]. This relation originated in a 45 km E–W oriented valley traversed by a major national road called the SS52 "Carnica" that connects the upper plains of the Friuli Venezia Giulia Region with the industrial areas of the Cadore (Veneto Region).

The road is located in an area with confined morphology and high relief potential energy. Widespread landslides, such as rockfalls, slides, and snow avalanches, are frequent [34,35].

One of the most hazardous segments of the road is located on the left flank of the upper Tagliamento valley between Ampezzo and Forni di Sotto municipalities. It is called Passo della Morte (46.396 N, 12.715 E) (Figure 1). In this tortuous tract, the avalanche and rockfall relative risk was so high that the construction of a bypass tunnel was needed. The works started in 1994, and finally opened in 2008. In fact, in 1996, during the construction, an extreme precipitation event disrupted part of the infrastructure. The extreme rainfall event allowed the discovery of two major slow slope movements, causing incapacitating road infrastructure problems. The first is a 24 $\times$ 10^6 m^3 slow-moving landslide called "Frana 3" [36]. Frana 3 intercepts the road's tunnel 300 m from the eastern entrance and in 1996, caused the collapse of the tunnel in this segment [37]. The same portal and 200 m of the SS52 were affected by another 2.1 $\times$ 10^6 m^3 slope movement called "Frana 1-2" [38]. The latter comprises two different rotational landslide bodies in the upper part that converge in the toe with a translational behavior.

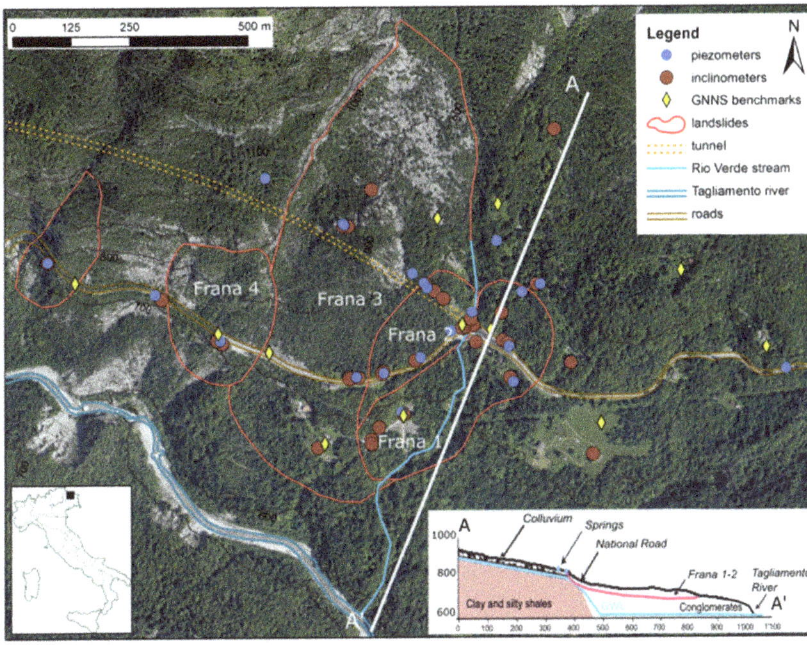

Figure 1. The Passo della Morte area with the monitoring devices installed during the last 20 years. In the right-hand corner, a hydrological cross-section of the slope.

Frana 1-2 is cut across by the Rio Verde torrent. This same stream crosses the new and old national roads below a concrete box culvert. The retaining walls along the SS52 and the road pavement suffer continuous damage due to unremitting movements of heart-shaped landslide. To mitigate the slope movements of both the Frana 3 and the Frana 1-2, a drainage tunnel was drilled underneath the road tunnel in 2007. Despite that, this remediation strategy proved insufficient for a complete solution to the problem.

2.1. Hydrological Context

The Rio Verde basin covers an area of approximately 2 km^2 from 2122 m a.s.l. of Mt. Tinisa to the Tagliamento River at 578 m a.s.l. The upper part of the basin lies over the tectonized dolomitic limestone of the Monte Tiarfin Formation, which is affected by karst phenomena. In a different way, the lower domain is characterized by colluvium and back scarp deposit of Frana 3 consisting of dolomitic boulders. The runoff coming from the upper part of the basin, except for extremely intense rainfall events, is completely infiltrated into the groundwater circulation of the right bank of the Rio Verde and almost does not participate in the water discharge of the lower part of the torrent [39]. In fact, the flow rate of the stream is mainly sustained by the water flowing from the springs located north of Frana 1-2. The stream guarantees an average "dry base flow" of 40 L/s related to the deep groundwater circulations, with long residence times set in the dolomitic limestone on the hydrographic western area of the Rio Verde Torrent [40]. This dry baseflow represents an emergency source of water for the aqueduct of the nearby village of Ampezzo, when other sources of freshwater fail to provide enough discharge during arid summers. During extreme rainy events, the discharge from the springs can increase substantially, thanks to the rapid infiltration into the high-permeability debris at the base of the Tinisa massif [20]. The geological context thus induces the water table to stay shallow near the contact between the clay and silty shales and the colluvium until the scarp of Frana 1-2 (Figure 1). Then, all the groundwater that did not manage to surface through the springs and reach the

Rio Verde, is again rapidly infiltrated into the conglomerates forming the landslide and then dispersed.

2.2. The Vaia Storm (2018) and Other Intense Meteorological Events

The Vaia storm was an extreme meteorological event that affected the Italian Northeast from 26 to 30 October 2018. The event originated following a perturbation of Atlantic origin. In the context of a strong wave of bad weather over Italy (also affecting the neighboring regions of Switzerland, Austria, and Slovenia), it brought strong winds and persistent, intense rainfall in the region, inducing floods and landslides. At the same time, the wind, blowing between 100 and 200 km/h for several hours, caused the felling of millions of trees, with the consequent destruction of tens of thousands of hectares of alpine coniferous forests. Several peripheral valleys were isolated due to road blocks caused by fallen trees or landslides, and the electric grid was out of operation for days. Vaia was a large-scale multi-hazard natural disaster in the Veneto and Friuli Venezia Giulia mountain areas.

In the Passo della Morte area, the Vaia storm brought a cumulated rainfall of 800 mm, almost 700 mm in the initial three days. Vaia was a hot windstorm, so all the precipitation in the first days was in the form of rain, even at 2000 m above sea level (Figure 2). To provide a reference measure, this area's average yearly accumulated rainfall amount is around 1700 mm.

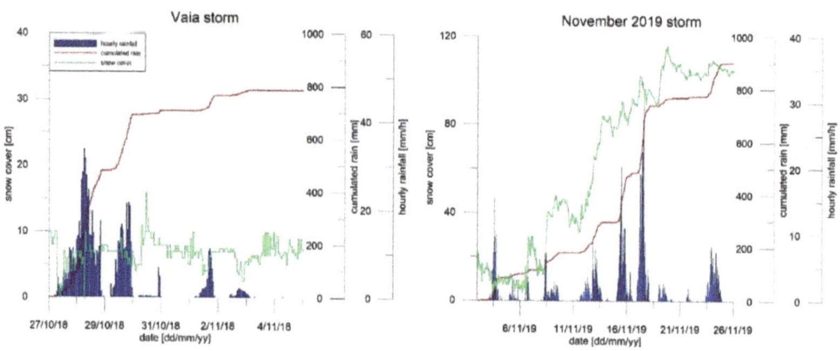

Figure 2. Daily rainfall, cumulated rainfall, snow.

In November 2019, another extreme event hit the Carnian Alps with an accumulated rainfall amount of 908 mm within a week in Passo della Morte. The event was intense; the peak daily accumulated rainfall was 250 mm on 17 November but lasted longer with constant and unremitting precipitation amount for days that were, however, in the form of snow above 1500 m a.s.l.

These two events induced significant disruption in the study area and are presented here since they are essential to interpret the 2018–2019 data and in understanding the following analysis and results.

2.3. Societal Background

As mentioned, the landslides in the Passo della Morte are constantly moving and damaging the road infrastructure [37]. The Italian National Road Authority (ANAS) manages the new tunnel's damage and associated risk. On the other hand, the jurisdiction is disputed over the 200 m of the road before the eastern entrance of the tunnel. In fact, several institutions could take charge of the slope's stabilization. Among them, the Civil Protection of Friuli Venezia Giulia decided to act based on the monitoring data, given the high societal risk linked to the possibility of the failure of the only road serving the valley. In case of road disruptions, the logistical impact would be severe, with traffic detours through another valley and, consequently, damaging the local economy (the extra travel time for the local commuters might reach two hours).

The municipality of Forni di Sotto is also the most invested in resolving the problem. However, the municipality of Forni di Sotto is small (less than 600 inhabitants); consequently, the funds to finance a big slope stabilization plan are difficult to gather. On the other hand, the Civil Protection of the Friuli Venezia Giulia region can act and finance simple, environmentally friendly, and "light" solutions, but in any case, the municipality would be in charge of the maintenance of any countermeasure.

In this framework, the National Research Council of Italy (CNR-IRPI) has been consulted to design a low-cost, low-impact remediation strategy that would at least mitigate, if not stop, the movements of Frana 1-2. This was completed based on a long-lasting monitoring campaign and several modeling tests.

3. Materials and Methods

Monitoring Network

The landslides in Passo della Morte have been monitored since 2002. One of the primary purposes was to control the slope's movement and characterize the kinematics of each landslide in the area. In addition,, through the continuous maintenance of the monitoring network, it is possible to assess, now and in the future, the effectiveness of each countermeasure and provide data on which to design additional mitigation strategies.

A complex system was deployed over 20 years of monitoring (Figure 3). It consists of inclinometers (periodic and in-place, IPI), Periodical Global Navigation Satellite System (GNSS) surveys, piezometers, a rain gauge, and temperature sensors [16,30]. The IPI series namely I21 (I21, I21bis, I21ter, I21quater), PC1 (PC1, PC1bis, PC1ter) and I15 (I15, I15bis) have been deployed along the slip surfaces and each time backed up with another one drilled nearby after a couple of years in order to follow the evolution of the landslides for several years and obtain a large, continuous dataset. The location of the IPI series along the landslide cross sections can be seen in Figure 4. In addition, a fundamental element joined the monitoring network in 2012: a sharp-crested thin-plate weir that can measure the water flow rate of the Rio Verde torrent. The weir was deployed in a regular 6 × 5 m concrete culvert that acts as a bridge for the National Road over the stream (Figure 5). The rating curve for the measuring station was assessed through periodic measurements with different flow rates and then interpolated using simple hydraulic equations considering the free fall setting of the streamflow.

For Frana 1-2, 18 inclinometers have been installed since 2002. In addition, for three locations considered of higher interest, a backup borehole was drilled nearby every 3–4 years to be able to follow continuously the movements of the landslides through IPI (In-Place Inclinometers), minimizing the gaps of measurements in the database and consequently gaining a complete and large dataset that covers several years. This series of inclinometers, namely I21, PC1, and I15, are still active today and have been recording since August 2010 [20].

The information collected through GNSS provided quantitative data about the surface movement of Frana 1-2. These measurements revealed an average annual displacement more significant on the Frana 1 side compared to the Frana 2 side. Benchmark PM18 and PM22 have recorded values of 4.2 cm/year and 3.7 cm/year, respectively. On the other hand, PM21 shows a 1.2 cm/year rate of movement.

Lastly, but most importantly, in 15 years of continuous monitoring in all piezometers installed in Frana 1-2 located along the main road or below the road, the water table location was way below the slip surface.

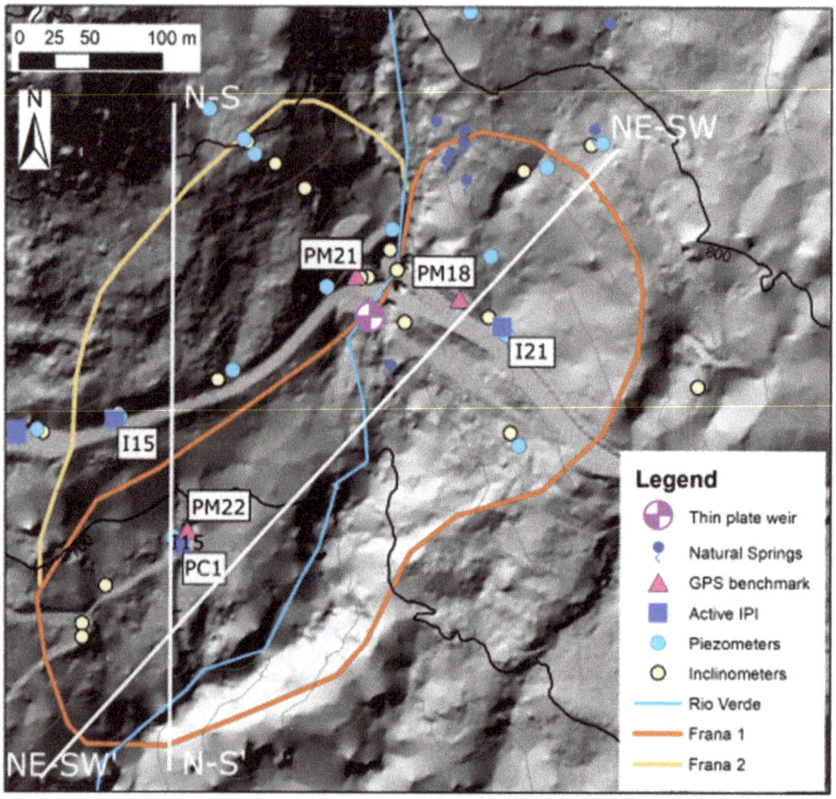

Figure 3. The monitoring network deployed in Frana 1-2.

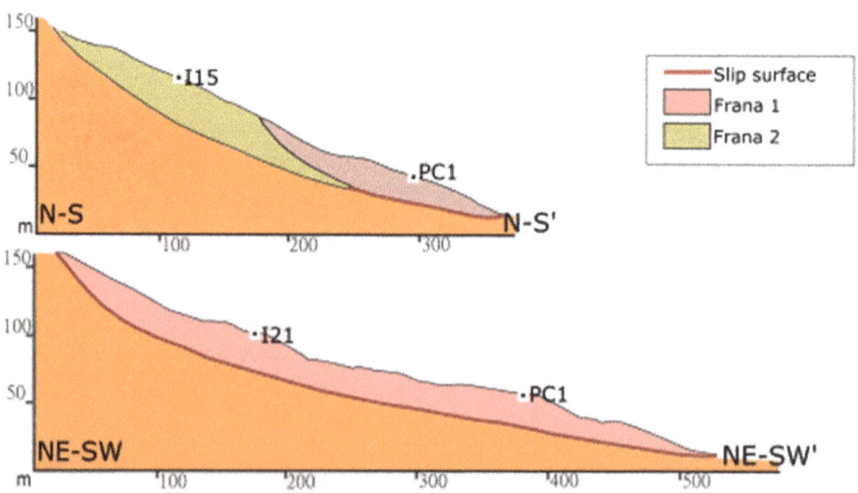

Figure 4. Cross sections of Frana 1-2 as seen in the map in Figure 3.

Figure 5. Thin plate weir on the Rio Verde torrent seen from upstream (**left**) and downstream (**right**).

4. Gray Box Modelling

In 2019, Bossi and Marcato [20] presented a gray-box model that could reproduce the velocity of displacement recorded in IPI I21Ter in function of the discharge of the Rio Verde torrent. The data presented in the paper stopped on 15 July, 2018. Of course, the system continued recording, and until the Vaia storm (end of October 2018), the correlation between the modeled velocities and the actual recorded ones produced a high fit to the model (R^2 = 0.82).

During the Vaia Storm, the peak discharge rate of the Rio Verde reached about 1000 L/s; we cannot be sure of the actual peak discharge of the flow since it was not confined in the small *adit* of the measuring section, and in this condition our rating curve becomes imprecise. For the 2019 event, the estimated peak discharge was about 700 L/s. The lower discharge of the 2019 event is linked with the temperature regime that caused snow precipitation above 1300 m a.s.l. (Figure 6).

Figure 6. Rio Verde Discharge—(**A**): ordinary days; (**B**): during the Vaia storm; (**C**): during the 2019 event.

During the Vaia Storm, Frana 1-2 and Frana 3 landslides showed a marked peak acceleration never observed before in more than 12 years of continuous monitoring. It also caused damage to the emergency aqueduct due to the sudden displacement of Frana 1-2.

After Vaia, we also observed a marked change in the baseline velocity of inclinometer I21Ter (Figure 7) linked to the storm's effect that increased the landslide's overall mobility.

Therefore, a new equation to model the velocity change in the Rio Verde discharge was developed:

$$0.000095\overline{Q_{5g}} + 0.001 = \overline{V_{10g}}$$

where ($\overline{Q_{5g}}$) is the average flow calculated over the previous 5 days; $\overline{V_{10g}}$ is the average velocity calculated over 10 days. The fit is linear as before [20] and produces a coefficient of determination (R^2 = 0.86) that indicates a high fit of the data to the model, however, the coefficients differ from the pre-event gray-box model. The base-velocity was estimated based on the data from a dry period (from 1 March 2020 to 1 June 2020), when the discharge was lower and optimized to reduce the systematic error that is the average of the residuals. The new equation that fits the post-Vaia data can also reproduce the landslide dynamic during the 2019 event (Figure 8). However, some data about the discharge needed to be discarded (from 19 March 2019 to 1 August 2019) as there were reliability issues for the measurements since the sensor was experiencing problems with limescale and sediment accumulation in the stilling basin. In July 2019, part of the accumulated sediment was removed, and the sensor was cleaned and repaired. On 1 August 2019, the rating curve for the discharge was re-tested and validated.

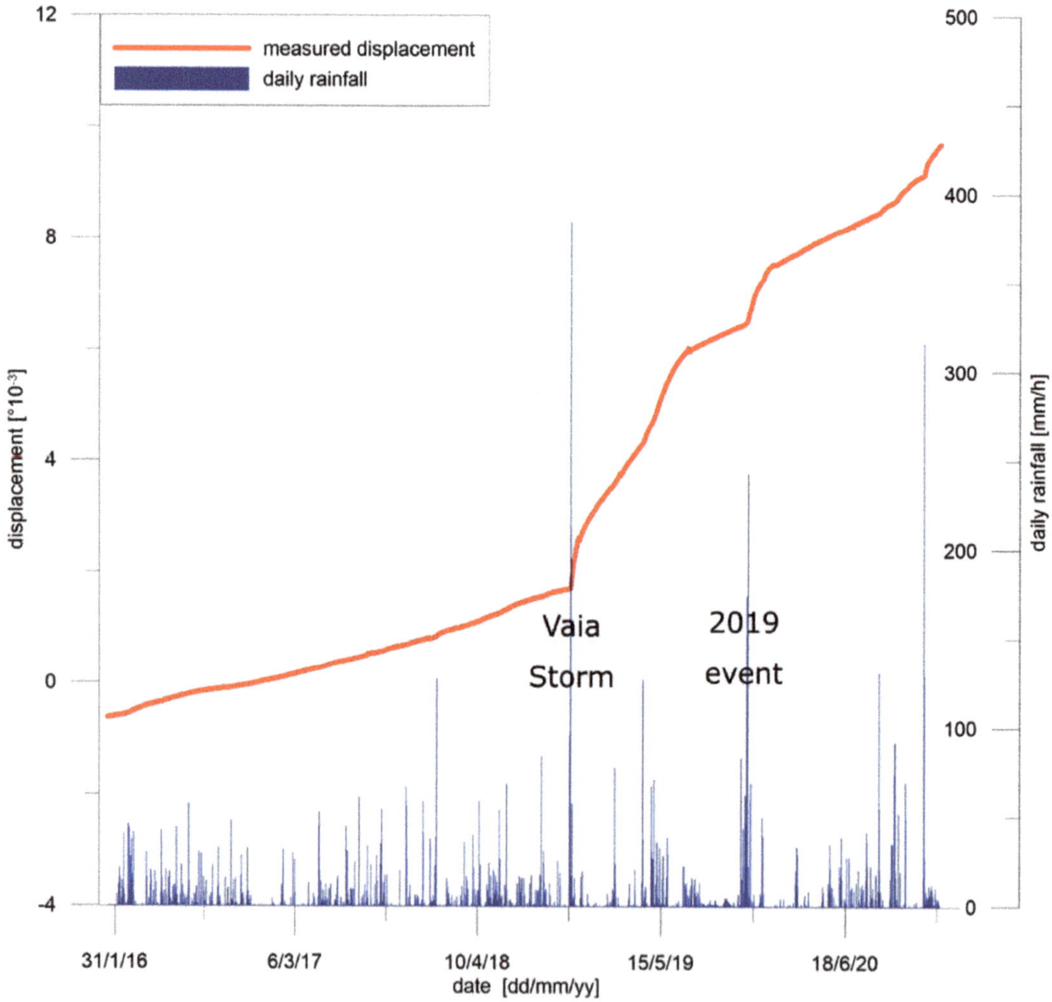

Figure 7. Data from the IPI I21Ter compared with the daily precipitation.

These strong results, showing a long-term direct correlation between the Rio Verde discharge and the displacements, convinced the local authorities (Protezione Civile Friuli Venezia Giulia) to start the realization of remediation works based on the reduction in the flow discharge in the Rio Verde stream.

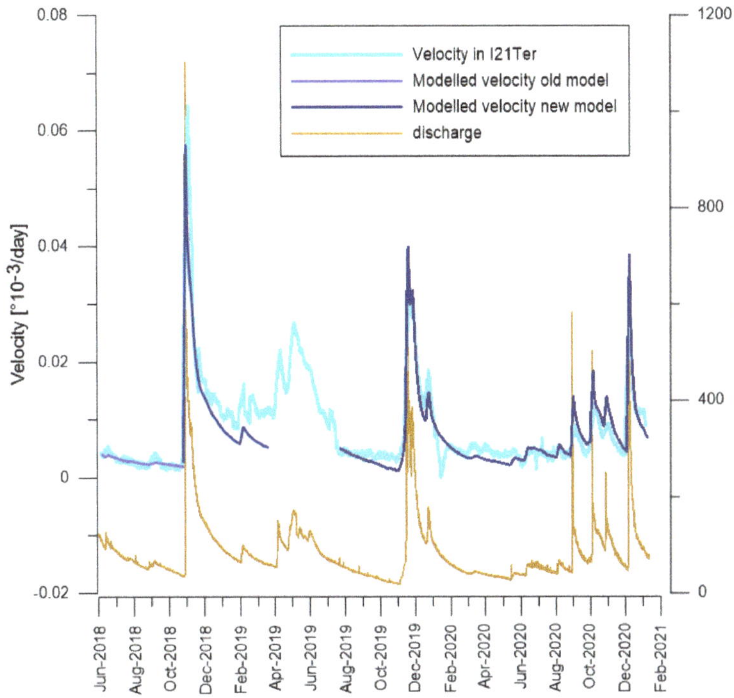

Figure 8. Modeled velocities based on the gray-box models (pre- and post- Vaia storm) and the Rio Verde discharge.

5. Implementation of the Risk Mitigation Strategy

A cost–benefit analysis was performed to decide the optimum value of discharge to be extracted. The elements that were considered were: (1) the expected reduced displacements; (2) the expected cost of the construction works; (3) the expected economic revenue from the hydro plant; and (4) the environmental impact both in terms of bioenvironmental effects on the stream and from the visual impact on the landscape.

If we calculate, for the period June 2018–January 2021, the expected displacement considering sequestration of 100 L/s with the linear model, and we confront them with the measured displacements, the decrease in the movements reaches about 70% (Figure 9). The value of 100 L/s represents an optimum considering both the Rio Verde flow regime and the possible countermeasures' size.

The project calls for constructing two small (10 × 2 m) transversal sedimentation/intake basins in the upper part of the slope (Figure 10). The need for sediment removal is not so crucial during dry periods when the spring water component is prevalent, but it is fundamental during extreme rainy events.

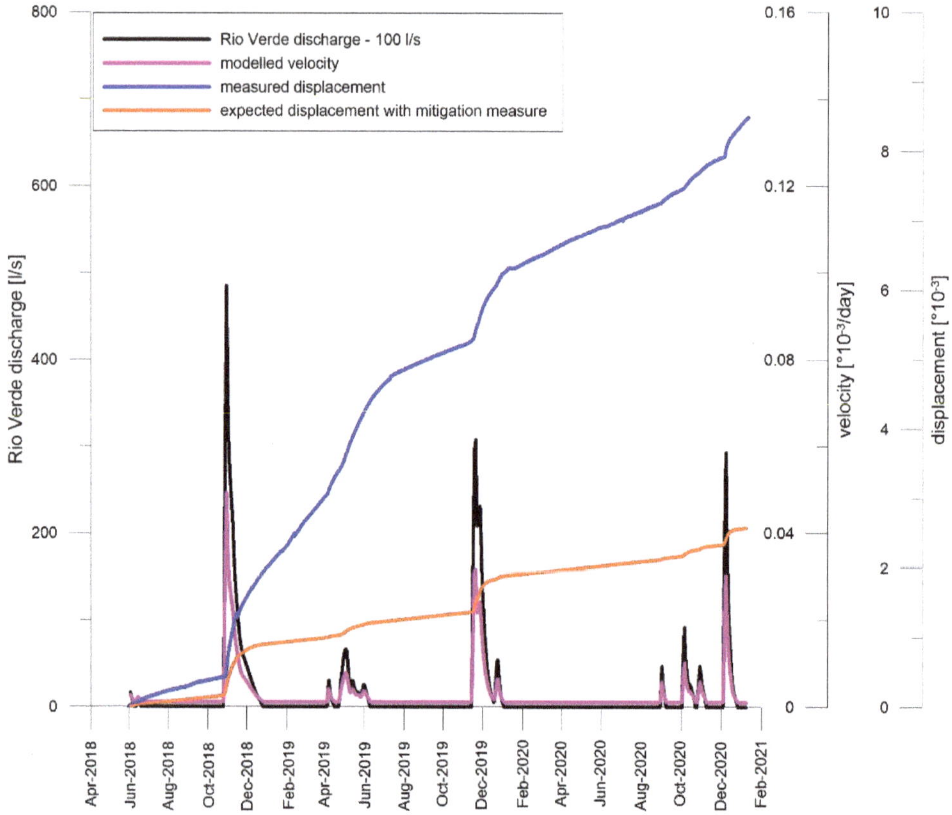

Figure 9. Recorded displacement after the Vaia storm vs. expected displacement with the proposed mitigation measure.

The first tank will be placed on the Rio Verde upstream of the road bridge. Here, the water from the natural springs above the road (except those serving the emergency aqueduct) will be conveyed through sub-horizontal drains and little spillways. Then, with a pipe, the water will reach the second intake basin, placed downstream of the bridge, where all the discharge from the drainage network of the road tunnel will be collected along with the water from lower springs. Funneling the water below the bridge through a pipe is necessary due to large cracks in the culvert below the bridge that could cause water loss through infiltration.

A 900 m long HDPE (High-Density Polyethylene) penstock is currently under construction. It will gather the discharge to the Tagliamento river at an altitude of 590 m a.s.l. The total cost of these works for the Friuli Venezia Giulia Civil Protection will be around 800,000 EUR.

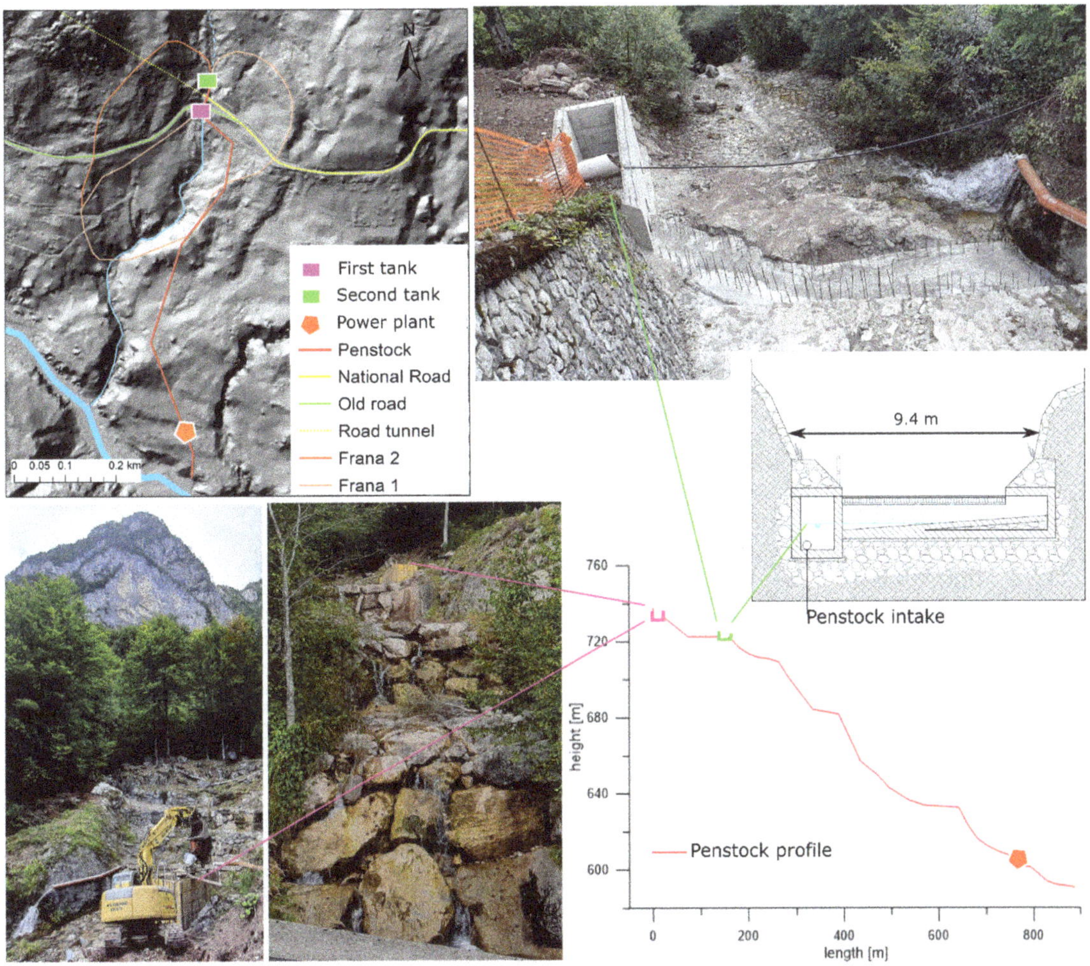

Figure 10. Design and construction of the tanks and penstock.

6. Discussion Regarding Costs and Self-Sustainability

As emphasized in the introduction, it is essential to consider economic revenue for the municipalities to incentivize them to implement mitigation strategies and maintain the countermeasure works. That is why, in this study, the integration of the remediation system with a micro-hydro plant has been proposed. In Italy, the latter (5 kW to 100 kW) is subsidized on a national basis, giving 0.22 EUR/kWh. Consequently, for the Forni di Sotto municipality, the micro-hydro plant option is an excellent solution for exploiting the discharge of the Rio Verde stream by producing electric power and selling it to the national network.

For the hydroelectric exploitation of the system, the horizontal axis Pelton turbine is optimal, considering the available 130 m drop and low average flow rates of the torrent. The turbine should be integrated with a three-phase asynchronous generator with a rated power of 100 kW. This generator is cheaper and requires less maintenance than a synchronous generator of the same deliverable power. Additionally, considering the quality of the water entering the hydro plant, which is very poor in terms of the level of sediments, it is believed that the conservation of the turbine will be excellent. This power plant will be located over an old fluvial terrace of the Tagliamento river (603 m a.s.l.). The position is isolated and not subject to landslides or erosion. This will reduce the landscape impact and the risk of damage. In addition, the building will be more than 10 m above the active course of the Tagliamento river to avoid any chance of damage by the river's floods. To further mitigate the impact on the landscape, the penstock will be buried in the soft colluvium present in the area which is easy to excavate.

The minimum predicted discharge of water funneled into the plant would be around 15 L/s, while the maximum will be 100 L/s with an estimated average of 45 L/s. Considering the 130 m drop, an average nominal power at 58.27 kW has been calculated: the yearly mean production will be 378,000 kWh. This will allow the municipality of Forni di Sotto a profit of 65,745 EUR every year, considering a selling price of 219 EUR/MWh, while the total cost of the hydro-plant will be around 300,000 EUR. As shown in Table 1, this relevant financial intake will pay for the power plant in less than five years. After that, the additional profit will finance the maintenance of the mitigation strategy of Frana 1-2 and probably provide economic income for the municipality.

Table 1. Cost and benefit of micro-hydro power plant implementation.

	Cost	Expected Revenue
1 year	300,000 EUR	65,745 EUR
5 years		328,725 EUR
20 years	maintenance	1,314,900 EUR

Given the relatively low cost of the operation and the possibility of revenue, the local authorities deemed the project worthy of a try. During several meetings the expected output, and the unlikely possibility that the project as it is would not work, were presented. What was crucial from the general risk assessment point of view was the fact that it would be simple to reinstate the fluvial system and the landslide area to its original condition if the expected reduction in displacement targets were not met. Moreover, it would be, in any event, a much easier option than with other structural mitigation works that were considered over the years [41]. Lastly, the project would not interfere with secondary structural mitigation solutions implemented in a stepwise manner if in the future the target would be stopping the landslide altogether rather than reducing its displacements. Actually, it would have decreased for some time the stresses within the slope, thus allowing smaller "bolts and nails" countermeasures.

The Civil Protection of Friuli Venezia Giulia was so involved in this project that it also financed an additional monitoring campaign consisting of two permanent GNSS stations deployed on the Frana 1-2 landslide. In this way, the reduction in the superficial movements will be assessed with a state of the art technology.

7. Conclusions

A constant flow of money is needed to maintain old landslide mitigation structures and build new ones, given the increasingly risk-averse attitudes from the exposed population. This leads to developing countermeasures that are easier to restore and less expensive.

Most of the time, when dealing with small-to-medium landslides, the usual pipeline followed by engineering geologists is to reduce the monitoring phase to a minimum (due to time and budget constraints), test some soil samples, develop a geotechnical model of the landslide based on laboratory results and on this model design the structural countermeasure works. Unfortunately, due to the high uncertainties linked with this process, the countermeasures are usually over-dimensioned and thus overpriced [42].

Acting on the triggering factor, when possible, is way more focused and surgical. However, the monitoring activity needs to be pursued in more extended periods than usual and with several different instruments for robustness. While the time factor is in itself a cost, we cannot forget that these slow-moving landslides are usually chronic phenomena that are, in any case, bureaucratically taking a long time to pass through all the phases from the identification to the actual mitigation. Moreover, from the mere actual cost point of view, the cost of a well-designed monitoring network is a tiny fraction of the cost of a classic "walls and nails" mitigation structure.

Thanks to long-term monitoring data in Passo della Morte, a strong correlation and model fit between the velocity of the displacement of Frana 1-2 and the discharge of the Rio Verde stream was found. The gray-box model presented by Bossi 2019 [20] was validated by the data gathered in the months after publication until the Vaia storm. Then, the landslide dynamic changed, showing an increase in the baseline displacements. In any case, even this new behavior is still strongly linearly correlated with the water discharge in the Rio Verde torrent, as previously stated.

Considering this result, local authorities have started constructing a series of tanks and penstocks to redirect the majority of the stream's discharge. The cost of this type of mitigation action for the Civil Protection is around 800,000 EUR. With an additional 300,000 EUR from the municipality of Forni di Sotto, this solution could be associated with a micro-hydropower plant that has the potential to payback its cost in five years. The future profits will guarantee the maintenance of the mitigation system itself and probably some economic income for the municipality. Moreover, this kind of countermeasure would drastically reduce the visual impact on the landscape compared to extensive engineering slope stabilization or large Sabo works.

Another advantage of this type of design is that it is scalable. The monitoring system will be maintained and enhanced with a permanent GNSS receiver along with the tanks and penstocks. This will allow for the precise quantification of the displacements, confronting them with the estimate from the gray-box model, and checking to see if the reduction in movements is in line with what was expected. In this case, the Passo del Morte case study serves as a benchmark for implementing this type of low-cost remediation solution.

Author Contributions: G.M. is the PI of the monitoring project; G.B. and F.T.C. analyzed data; G.B. produced the statistics and the gray-box model; G.M. conceived the mitigation strategy; F.T.C. wrote the paper; F.T.C. and G.B. designed the figures; G.M. and G.B. revised the paper. All authors have read and agreed to the published version of the manuscript.

Funding: This research received no external funding.

Acknowledgments: We thank Protezione Civile della Regione Autonoma Friuli Venezia Giulia and in particular Gianni Burba and Gabriele Peressi. Protezione Civile della Regione Autonoma Friuli Venezia Giulia funded the deployment of the monitoring system. Other resources already available at CNR-IRPI provided the funds for the research activity. Protezione Civile della Regione Autonoma Friuli Venezia Giulia was not involved in the manuscript design and production.

Conflicts of Interest: The authors declare no conflict of interest.

References

1. Chanson, H. Sabo check dams—Mountain protection systems in Japan. *Int. J. River Basin Manag.* **2004**, *2*, 301–307. [CrossRef]
2. Watanabe, H.; Toshio, M. Actions for the Maintenance and Lifespan Prolongation of SABO Facilities. In Proceedings of the INTERPRAEVENT 2016, Lucerne, Switzerland, 30 May–2 June 2016; pp. 862–870.
3. Tullos, D.; Wang, H.-W. Morphological responses and sediment processes following a typhoon-induced dam failure, Dahan River, Taiwan. *Earth Surf. Process. Landforms* **2013**, *39*, 245–258. [CrossRef]
4. Wang, H.; Kondolf, G.M. Upstream Sediment-Control Dams: Five Decades of Experience in the Rapidly Eroding Dahan River Basin, Taiwan. *JAWRA J. Am. Water Resour. Assoc.* **2013**, *50*, 735–747. [CrossRef]
5. Ben Sim, K.; Lee, M.L.; Wong, S.Y. A review of landslide acceptable risk and tolerable risk. *Geoenviron. Disasters* **2022**, *9*, 3. [CrossRef]
6. Marcato, G.; Bossi, G.; Rivelli, F.; Borgatti, L. Debris flood hazard documentation and mitigation on the Tilcara alluvial fan (Quebrada de Humahuaca, Jujuy province, North-West Argentina). *Nat. Hazards Earth Syst. Sci.* **2012**, *12*, 1873–1882. [CrossRef]
7. Eisbacher, G.H. Slope Stability and Land Use in Mountain Valleys. *Geosci. Can.* **1982**, *9*.
8. Blaikie, P.; Wisner, B.; Cannon, T.; Davis, I. *At Risk—Natural Hazards, People's Vulnerability and Disasters*, 2nd ed.; Routledge: London, UK, 2014.
9. Gariano, S.L.; Guzzetti, F. Landslides in a changing climate. *Earth Sci. Rev.* **2016**, *162*, 227–252. [CrossRef]

10. Junichi, K.; Naoki, I. Outline of measures for sediment disaster by the Sabo department of MLIT, Japan. *Landslides* **2020**, *17*, 2503–2513. [CrossRef]
11. Hilker, N.; Badoux, A.; Hegg, C. The Swiss flood and landslide damage database 1972–2007. *Nat. Hazards Earth Syst. Sci.* **2009**, *9*, 913–925. [CrossRef]
12. Klose, M.; Maurischat, P.; Damm, B. Landslide impacts in Germany: A historical and socioeconomic perspective. *Landslides* **2015**, *13*, 183–199. [CrossRef]
13. Prenger-Berninghoff, K.; Cortes, V.J.; Sprague, T.; Aye, Z.C.; Greiving, S.; Głowacki, W.; Sterlacchini, S. The connection between long-term and short-term risk management strategies for flood and landslide hazards: Examples from land-use planning and emergency management in four European case studies. *Nat. Hazards Earth Syst. Sci.* **2014**, *14*, 3261–3278. [CrossRef]
14. Scolobig, A.; Linnerooth-Bayer, J.; Pelling, M. Drivers of transformative change in the Italian landslide risk policy. *Int. J. Disaster Risk Reduct.* **2014**, *9*, 124–136. [CrossRef]
15. Scolobig, A.; Thompson, M.; Linnerooth-Bayer, J. Compromise not consensus: Designing a participatory process for landslide risk mitigation. *Nat. Hazards* **2015**, *81*, 45–68. [CrossRef]
16. Titti, G.; Bossi, G.; Zhou, G.G.; Marcato, G.; Pasuto, A. Backward automatic calibration for three-dimensional landslide models. *Geosci. Front.* **2021**, *12*, 231–241. [CrossRef]
17. Angeli, M.-G.; Pasuto, A.; Silvano, S. A critical review of landslide monitoring experiences. *Eng. Geol.* **2000**, *55*, 133–147. [CrossRef]
18. Corominas, J.; Moya, J.; Ledesma, A.; Lloret, A.; Gili, J. Prediction of ground displacements and velocities from groundwater level changes at the Vallcebre landslide (Eastern Pyrenees, Spain). *Landslides* **2005**, *2*, 83–96. [CrossRef]
19. Pfeiffer, J.; Zieher, T.; Schmieder, J.; Rutzinger, M.; Strasser, U. Spatio-temporal assessment of the hydrological drivers of an active deep-seated gravitational slope deformation: The Vögelsberg landslide in Tyrol (Austria). *Earth Surf. Process. Landforms* **2021**, *46*, 1865–1881. [CrossRef]
20. Bossi, G.; Marcato, G. Planning Landslide Countermeasure Works through Long Term Monitoring and Grey Box Modelling. *Geosciences* **2019**, *9*, 185. [CrossRef]
21. Corsini, A.; Borgatti, L.; Caputo, G.; De Simone, N.; Sartini, G.; Truffelli, G. Investigation and monitoring in support of the structural mitigation of large slow moving landslides: An example from Ca' Lita (Northern Apennines, Reggio Emilia, Italy). *Nat. Hazards Earth Syst. Sci.* **2006**, *6*, 55–61. [CrossRef]
22. Mantovani, M.; Bossi, G.; Dykes, A.P.; Pasuto, A.; Soldati, M.; Devoto, S. Coupling long-term GNSS monitoring and numerical modelling of lateral spreading for hazard assessment purposes. *Eng. Geol.* **2021**, *296*, 106466. [CrossRef]
23. Zapico, I.; Molina, A.; Laronne, J.B.; Castillo, L.S.; Duque, J.F.M. Stabilization by geomorphic reclamation of a rotational landslide in an abandoned mine next to the Alto Tajo Natural Park. *Eng. Geol.* **2019**, *264*, 105321. [CrossRef]
24. Islam, A.; Islam, M.S.; Islam, T. Landslides in Chittagong Hill Tracts and Possible Measures. In Proceedings of the International Conference on Disaster Risk Mitigation, Dhaka, Bangladesh, 23–24 September 2017.
25. Pepe, G.; Cevasco, A.; Piazza, M.; Macciò, R.; Arrighetti, F.; Casagli, N. On the efficiency and effectiveness of automatic deep drainage systems during an extreme rainfall event: The Mendatica landslide case study (western Liguria, Italy). *Landslides* **2021**, *18*, 3799–3820. [CrossRef]
26. Mikoš, M.; Fazarinc, R.; Pulko, B.; Petkovšek, A.; Majes, B. Stepwise mitigation of the Macesnik landslide, N Slovenia. *Nat. Hazards Earth Syst. Sci.* **2005**, *5*, 947–958. [CrossRef]
27. Troncone, A.; Pugliese, L.; Conte, E. Rainfall Threshold for Shallow Landslide Triggering Due to Rising Water Table. *Water* **2022**, *14*, 2966. [CrossRef]
28. Kishore, T.; Patro, E.; Harish, V.; Haghighi, A. A Comprehensive Study on the Recent Progress and Trends in Development of Small Hydropower Projects. *Energies* **2021**, *14*, 2882. [CrossRef]
29. Pasuto, A. The Vajont Valley (Eastern Alps): A Complex Landscape Deeply Marked by Landsliding. In *World Geomorphological Landscapes*; Springer: Cham, Switzerland, 2017.
30. Schopper, N.; Mergili, M.; Frigerio, S.; Cavalli, M.; Poeppl, R. Analysis of lateral sediment connectivity and its connection to debris flow intensity patterns at different return periods in the Fella River system in northeastern Italy. *Sci. Total Environ.* **2018**, *658*, 1586–1600. [CrossRef]
31. Calligaris, C.; Devoto, S.; Zini, L. Evaporite sinkholes of the Friuli Venezia Giulia region (NE Italy). *J. Maps* **2017**, *13*, 406–414. [CrossRef]
32. Tockner, K.; Ward, J.V.; Arscott, D.B.; Edwards, P.J.; Kollmann, J.; Gurnell, A.M.; Petts, G.E.; Maiolini, B. The Tagliamento River: A model ecosystem of European importance. *Aquat. Sci.* **2003**, *65*, 239–253. [CrossRef]
33. Carulli, G.B. Geological Map of the Friuli Venezia Giulia. Regione Autonoma Friuli Venezia Giulia, Direzione Centrale Ambente e Lavori Pubblici, S.E.L.C.A. 2006. Available online: https://www.regione.fvg.it/rafvg/export/sites/default/RAFVG/ambiente-territorio/geologia/FOGLIA01/allegati/Geofvg.pdf (accessed on 18 November 2022).
34. Bossi, G.; Zabuski, L.; Pasuto, A.; Marcato, G. Capabilities of Continuous and Discontinuous Modelling of a Complex, Structurally Controlled Landslide. *Geotech. Geol. Eng.* **2016**, *34*, 1677–1686. [CrossRef]
35. Codeglia, D.; Dixon, N.; Bossi, G.; Marcato, G. Alpine landslide risk scenario: Run-out modelling using a 3D approach. *Rend. Online Soc. Geol. Ital.* **2017**, *42*, 14–17. [CrossRef]

36. Marcato, G. Valutazione Della Pericolosità Da Frana in Località Passo Della Morte. Ph.D. Thesis, University of Modena and Reggio Emilia, Modena, Italy, 2007.
37. Bossi, G.; Schenato, L.; Marcato, G. Structural Health Monitoring of a Road Tunnel Intersecting a Large and Active Landslide. *Appl. Sci.* **2017**, *7*, 1271. [CrossRef]
38. Sinigardi, G.; Bossi, G.; Scuri, A.; Marcato, G.; Borgatti, L. Geological and numerical models as a tool to manage landslide risk: The Passo della Morte case study (UD, Italy). *Rend. Online Soc. Geol. Ital.* **2015**, *34*, 46–53. [CrossRef]
39. Petronici, F.; Borgatti, L.; Cervi, F.; Piccinini, L.; Bonaga, G.; Marcato, G. Hydrogeological monitoring and modelling in the S. Lorenzo road tunnel area (Passo della Morte, Udine) for the design of countermeasure works. *Rend. Online Soc. Geol. Ital.* **2016**, *39*, 93–96. [CrossRef]
40. Cervi, F.; Borgatti, L.; Dreossi, G.; Marcato, G.; Michelini, M.; Stenni, B. Isotopic features of precipitation and groundwater from the Eastern Alps of Italy: Results from the Mt. Tinisa hydrogeological system. *Environ. Earth Sci.* **2017**, *76*, 410. [CrossRef]
41. Zabuski, L.; Bossi, G.; Marcato, G. Influence of the Geometry Alteration of the Landslide Slope on its Stability: A Case Study in the Carnian Alps (Italy). *Arch. Hydro-Eng. Environ. Mech.* **2017**, *64*, 101–114. [CrossRef]
42. Bossi, G.; Borgatti, L.; Gottardi, G.; Marcato, G. Quantification of the uncertainty in the modelling of unstable slopes displaying marked soil heterogeneity. *Landslides* **2019**, *16*, 2409–2420. [CrossRef]

water

MDPI

Article

Analysis of Socio-Economic Factors That Influence Loss of Life in Sediment-Related Disasters

Miki Chiba [1,*], Hiromichi Furuido [2], Shigemitsu Shibasaki [2] and Kazuki Haga [2]

1 SABO Frontier Foundation, Tokyo 1020093, Japan
2 Department of Forest Science, Graduate School of Agriculture and Life Sciences, The University of Tokyo, Tokyo 1138657, Japan; afuru@g.ecc.u-tokyo.ac.jp (H.F.); shiba8888@g.ecc.u-tokyo.ac.jp (S.S.); khaga@g.ecc.u-tokyo.ac.jp (K.H.)
* Correspondence: sff-chiba@sff.or.jp

Abstract: Non-structural measures, including relocation from a hazard zone, land development regulations, and evacuation, are important sediment-related countermeasures. Such measures depend on the behaviour of residents and are affected by socio-economic conditions. In Japan, the declining birth rate and ageing population are expected to result in rapid changes in socio-economic conditions; accordingly, there is a need to evaluate the impact of such changes. However, there is no established methodology for the establishment of non-structural measures that considers the socio-economic conditions of all areas. Therefore, this study analysed the regional characteristics of disaster-affected areas with high numbers of casualties to elucidate the impact of socio-economic conditions. The results imply that severe losses occurred under various topographical and socio-economic conditions, such as in urban areas with high population densities and in mountainous areas with increasing depopulation. More data are needed, especially regarding socio-economic conditions. Importantly, the community-based analytical method used in this study enables a comparative analysis of disasters in different regions at different times.

Keywords: sediment-related disaster; non-structural measurement; socio-economic conditions; human loss; community-based analysis

Citation: Chiba, M.; Furuido, H.; Shibasaki, S.; Haga, K. Analysis of Socio-Economic Factors That Influence Loss of Life in Sediment-Related Disasters. *Water* **2022**, *14*, 2408. https://doi.org/10.3390/w14152408

Academic Editors: Yu Huang and Ian Prosser

Received: 27 June 2022
Accepted: 28 July 2022
Published: 3 August 2022

1. Introduction

More than 1000 sediment-related disasters occur annually in Japan; they occur in both urban and mountainous areas under different socio-economic conditions, and their causes differ among areas. Sediment-related disasters in this context are disasters that cause direct damage to human lives and houses as a result of sediment movement, and are broadly classified as debris flows, slope failures and landslides [1]. Because Japan is a mountainous and densely populated country, many areas are vulnerable to sediment-related hazards. Despite ongoing preventive structural measures being undertaken annually, many sediment-related casualties occur. Therefore, the importance of non-structural measures, such as land development regulations and evacuation warnings, is regaining attention. However, these measures require the cooperation of residents, and it is important to consider socio-economic factors and to select appropriate measures for both urban and mountainous areas.

According to the FY2020 White Paper on Forest and Forestry [2], mountain villages, most of which are located in mid-mountainous areas (mountainous areas on the outer edges of plains), experienced a significant outflow of population, particularly of young people, during the recent period of rapid economic growth in Japan (ca. the 1960s). Additionally, depopulation and ageing are rapidly progressing. The report also indicated that many villages face problems, such as fewer households, a large proportion of older residents, difficulties in maintaining village functions, the potential for village disappearance, and an absence of newcomers. These problems are worse in mountain villages than in flatlands or mid-mountainous areas. Furthermore, Karasaki [3] described a situation in which social changes (e.g., agricultural decline, depopulation, and widespread ageing) are causing the ties between local community members to fade; thus, the nature of local communities is changing. Such changes threaten livelihoods, particularly in mid-mountainous areas.

Kanai [4] focused on sediment-related disasters in urban areas, in a study that traced the relationship between the conditions of residential land and major sediment-related disasters from the Meiji period (1868–1912) to the collapse of the bubble economy. He concluded that post-war housing policies that prioritised single-family homes, combined with urban planning that prioritised the economy, led to residential land development that drastically changed the topography. After the collapse of the bubble economy, the number of housing starts declined, and the housing industry was restructured. These changes resulted in the emergence of a business model in which large numbers of houses with land were built for sale at low prices. A related issue is the inappropriate disposal of construction soil. Although this was a problem during the period of rapid economic growth and in construction related to the 1964 Tokyo Olympics, there was no shortage of locations to dispose of the soil because of strong demand for soil and sand as aggregate for coastal reclamation, road construction, and building construction. However, during the 1990s, a period of low economic growth, the need for dumping sites increased, resulting in mountainous areas being chosen for the disposal of construction soil. This situation resulted in disaster-prevention structures that were established under conventional socio-economic conditions no longer functioning because of the changes in the socio-economic conditions, which led to sediment-related disasters.

Differences in socio-economic conditions among regions are also affected by the intensification of disasters due to climate change and population decline; these factors are relevant to the nature of future sediment-related disaster countermeasures. However, to date, research has not sufficiently considered socio-economic conditions and the causes of sediment-related disasters. Therefore, this study analysed the socio-economic conditions of major sediment-related disaster sites where casualties occurred, considering the relationship between the occurrence of casualties and socio-economic conditions. It provides key information for the assessment of sediment-related disaster countermeasures in the context of socio-economic conditions.

2. Literature Review of Human Suffering Caused by Sediment-Related Disasters

In one case, the mortality rate from sediment-related disasters was higher among people in their 40s and 50s than among older adults. This can be explained by social norms that influenced the living spaces of different age groups using the data from 38 cases occurring between 1881 and 2019 [5]. Additionally, Jason [6] reported that disaster preparedness was negatively associated with renter status, although the reasons for this were unclear. Socio-economic conditions can affect loss of life as a result of sediment-related disasters. Václav et al. [7] demonstrated that the memory and experience of past disasters in a community affect where people live, indicating that by reducing the number of residents in sediment-disaster-related areas, past disasters have a significant socio-economic impact on subsequent losses from sediment-related disasters.

In terms of non-structural measures based on the Sediment-related Disaster Prevention Act (enacted in 2001), 41% of respondents to a survey that covered 47 prefectures (up to three items selected; percentage of total points when first, second, and third choice scored 3, 2, and 1 points, respectively; same below) indicated 'concern about a decrease in land prices' as the reason why residents opposed the designation of areas based on the Sediment-related Disaster Prevention Act. The top responses were 'dissatisfaction with structural restrictions on building' (25%) and 'dissatisfaction with the lack of implementation of structural measures, if designated' (17%) [8]. These results imply that the designation of areas and the development of controls and structural regulations under the Sediment-related Disaster Prevention Law are affected by the possibility of land sales (e.g., the degree of population increase or decrease), the type of homeownership (e.g., owner-occupied or rental housing), and conditions that affect the implementation of structural measures (e.g., financial status of government agencies and population size).

Of these, population size and the rate of increase or decrease are among the socio-economic conditions that require the most attention. In the previous section, we discussed the population declines in agricultural and mountain villages as a socio-economic condition that influences sediment-related-disaster countermeasures. In some villages, ageing residents cannot drive and have problems moving to evacuation centres [9]. In addition, population size can affect the ability to provide assistance: if the number of residents in a village with a declining population decreases further because of a sediment-related disaster, assistance becomes more difficult, thus increasing the risk of a disaster and potentially producing a vicious cycle.

The initial budget for erosion-control projects declined continuously from 1972 to 2006, with a peak decline in 1998 [10]. Since then, the rate of increase or decrease has been approximately 1.0 [11]. Data from 27 prefectures show that, from FY2009 to FY2013, the cost of non-structural measures was less than 10% of that of structural measures annually [12]. Thus, each prefecture primarily promoted structural measures, while providing less financial support to non-structural measures. In addition, considering the large number of designated zones, it would be difficult to allocate sufficient financial resources toward explaining the conditions in each zone. Therefore, the financial situations of municipalities located closest to residents at risk and responsible for building-evacuation precautions under the Sediment-related Disaster Prevention Law may directly affect the human loss resulting from a lack of sediment-related disaster countermeasures. Mizuyama [13] described the need for a cost–benefit analysis of evacuation alerts and provided a discussion of the costs of gathering information and recommending evacuation, the cost of providing food at evacuation sites, and other expenses to support evacuation. It has been estimated that the financial and human resources of municipalities that conduct such activities affect human loss.

In this study, we presumed that socio-economic conditions (e.g., the population size and its increase or decrease, the financial status of administrative agencies, and the type of home ownership) were related to the occurrence of casualties resulting from sediment-related disasters. We analysed these and other conditions at disaster sites.

3. Method

In our study, we selected study areas based on the magnitude of human loss resulting from sediment-related disasters. A characteristic of sediment-related disasters, as distinct from other natural disasters, is that the forces generated by water and rocks are extremely large and have considerable potential to impact people's lives. In addition to its significant effect of further decreasing resident numbers in mountainous areas with continuing population decline, loss of life is an important indicator that can both maintain the memory of disasters and aid in the implementation of disaster-prevention activities.

We collected study cases from the annual reports of major sediment-related disasters listed by the Erosion Control Department of the Ministry of Land, Infrastructure, Transport and Tourism, Japan, from 2004 to August 2021 [14]. All study cases were defined as community units, smaller than municipalities. The cases selected included fatalities or missing persons. We supplemented these cases with reports from national or prefectural verification committees and academic papers related to disasters. Based on these results, 28 cases with $\geq$5 people missing or dead per location were included (Table 1, Figure 1). In Table 1, the term 'meteorological phenomenon' is based on the definition provided by the Japan Meteorological Agency [15]. The term 'agricultural community' is based on units from the agricultural census [16].

Table 1. Study areas where losses occurred between 2004 and 2011.

Years	Natural Phenomena	Prefecture	Agricultural Community (Study Area)
2004	TY Meari (0421)	Mie	Takiya
	TY Tokage (0423)	Okayama	Uno
2006	Heavy rain of July 2006	Nagano	Hanaoka
2008	The Iwate-Miyagi Nairiku Earthquake in 2008	Miyagi	Koei
2009	Heavy rain in Chugoku-Kyushu of July 2009	Yamaguchi	Nango
2011	The 2011 off the Pacific coast of Tohoku Earthquake	Fukushima	Iizawa
	STS Talas (1112)	Wakayama	Ichinono Iseki-1 Nagano-6
		Nara	Ui Nojiri
2012	Heavy rain in Kyushu of July 2012	Kumamoto	Fukuoka
2013	Heavy rain of August 2013	Akita	Sendatsu
	TY WIPHA (1326)	Tokyo	Motomachi
2014	Heavy rain of August 2014	Hiroshima	Yashiki Kobara Kamirakuchi Muroya
2016	The 2016 Kumamoto Earthquake	Kumamoto	Kurokawa
2017	Heavy rain in Kyushu of July 2017	Fukuoka	Ishizume Tachi
2018	–	Ooita	Kajigaharu-yukihiro
	Heavy rain of July, 2018	Hiroshima	Kawasumi Tenjin Oonishi Koyaura
	Hokkaido Eastern Iburi Earthquake, 2018	Hokkaido	Yoshino
2021	Heavy rain of July 2021	Shizuoka	Kidani

The disaster phenomena included in the 28 cases were 13 cases of debris flow (reported in terms of the agricultural community names: Sendatsu, Motomachi, Hanaoka, Ui, Ichinono, Iseki-1, Uno, Muroya, Kamirakuchi, Ohara, Yashiki, Nango, and Fukuoka), 3 cases of landslides (Iizawa, Nagano-6, and Kurokawa), 2 cases of slope failure (Yoshino, and Kajigaharu-yukihiro), and 10 unknown cases (3 cases described as 'debris flow and other': Kidani, Kawasumi, and Koyaura; 5 cases without a description of the disaster in the annual reports: Kouei, Tenjin, Onishi, Tachi, and Ishizume, and 2 cases without a description of the disaster phenomenon: Takiya and Nojiri). The most common disaster among these cases was a debris flow that damaged a wide area.

Published data were collected to elucidate the characteristics of the 'place' and 'people' in the study area affected by the targeted sediment-related disaster, including conditions that affected losses during the sediment-related disaster. We used the agricultural community unit as the unit of the study area. According to the Census of Agriculture and

Forestry [17], an agricultural community is 'a community based on agriculture in part of a municipal area,' and 'specifically, a production community closely connected to the maintenance and management of agricultural roads and water facilities, common forests, marriage, funerals, and other aspects of life. In addition, an agricultural community functions as a unit of self-governance and administration'. As such, an agricultural community is considered similar to the unit used in normal times and in times of disaster. However, the sub-regional units from the census [18] were used in cases where data were not organised into units of agricultural community.

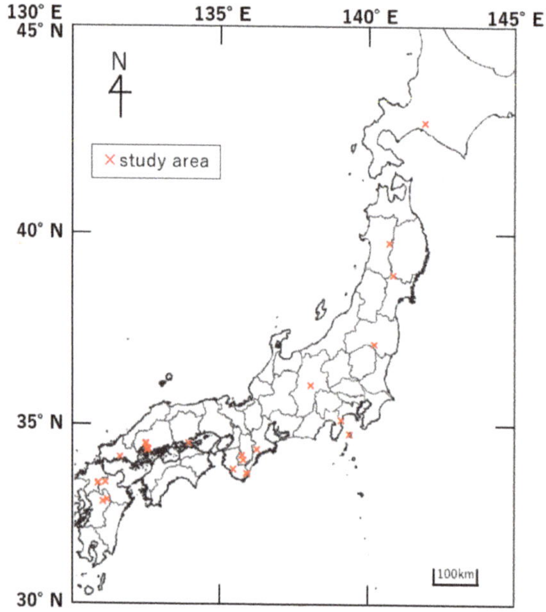

Figure 1. Location of study areas. (Partially edited from [19]).

4. Results

4.1. Topography, Geology, and Rainfall Conditions

In this study, the topographical conditions of the study areas can be summarised as follows: average elevation of the centre and average slope of the centre [20], geology [21], and annual precipitation [22] (Table 2).

None of the study areas had an average elevation <10 m above sea level; however, 17.9% and 24.9% of the areas had an average elevation of 30–50 m or ≥200 m, respectively, while the corresponding national averages are 10.2% and 14.9%. Thus, although the affected areas were generally higher in elevation compared to the nation as a whole, many areas were located at relatively low elevations of 30–50 m. No bias suggesting an impact on the amount of human harm could be identified.

The average slopes of the central part of the areas were <5° for 11 (42.9%), 5–15° for 16 (53.6%), and >15° for 1 (27.1°). The areas with an average central land slope of ≥5°, which constituted half of the settlements, are presumably located in areas with poor agricultural production; accordingly, many of them tend to have low agricultural incomes and an ageing population. The average slope does not directly affect the number of human casualties, although it is necessary to consider the impact of social conditions as well.

Table 2. Topography, geology, and rainfall conditions of the study areas.

Study Area	Elevation (m)	Slope (°)	Geology	Annual Precipitation (mm)
Yoshino	33.4	3.6	Sedimentary rock	1028.4
Koei	578.5	7.4	Igneous rock	2124.9
Sendatsu	266.8	2.5		2354.1
Iizawa	366.1	4.5		1456.7
Motomachi	54.9	4.0		2858.9
Hanaoka	798.7	12.5		1301.5
Kidani	193.8	13.6		2012.7
Takiya	183.8	10.5	Accretionary wedge	3369.0
Ui	418.4	5.8	Sedimentary rock, Accretionary wedge	2538.2
Nojiri	329.0	27.1	Accretionary wedge	2538.2
Nagano-6	219.6	7.8	Aedimentary rock, Accretionary wedge	2581.3
Ichinono	72.6	8.8	Sedimentary rock	3784.7
Iseki-1	30.8	5.2		3332.9
Uno	11.1	4.4	Igneous rock	1038.5
Muroya	48.3	10.6		1678.3
Kamirakuchi	17.2	2.2		1678.3
Kobara	32.4	8.2		1678.3
Yashiki	14.0	3.1		1572.2
Tenjin	80.7	8.7		1572.2
Oonishi	27.9	6.6		1417.2
Kawasumi	227.4	1.4		1417.2
Koyaura	19.3	3.6		1417.2
Nango	44.9	3.5		1653.7
Tachi	103.6	8.0		1876.3
Ishidume	195.2	8.6		1876.3
Fukuoka	605.8	13.5		3009.6
Kurokawa	449.3	3.8		3009.6
Kajigaharu-yukihiro	217.0	11.3		1944.9
Average	201.4	7.5	—	2075.8

Note: Date sources: Elevation and Slope [20], Geology [21], Annual Precipitation [22].

The geological classification of the sediment movement that caused damage to each area was established. The number of movements classified was larger than the number of target areas because two areas included two geological classifications. Based on the results, we classified the geological units into sedimentary rock, igneous rock, and accretionary complexes; igneous rock constituted the largest proportion (70%), followed by sedimentary rock (17%) and accretionary complexes (13%). Among igneous rock, granite was the most common rock type, constituting 30% (nine sites) of the total, because of the large number of disasters in Hiroshima Prefecture. The frequency of slope failures and debris flows caused by heavy rainfall is once every 10–1000 years based on 34 reports of such events in Japan, but the frequency is as high as once every 10–100 years for siliceous and granitic rocks and as low as once every 1000 years for accretionary sedimentary rocks of the Mesozoic to Paleozoic boundaries [23]. Therefore, the geological features of an area may lead to different disaster experiences. It is possible that differences in the frequency of incidents due to geology may also affect the number of human casualties, but this cannot be confirmed due to the bias in the cases collected in this study.

We compiled annual rainfall data from the nearest rainfall stations for each area. Japan, which is located on the eastern edge of monsoon Asia (one of the world's rainiest regions), receives an average annual precipitation of 1718 mm, which is approximately twice the global average (880 mm) [24]. Thus, half of the study areas generally receive rainfall above the average for Japan, while the remaining areas receive below-average rainfall. Some of the affected areas had experienced very high rainfall of as much as 3784.7 mm/year, while others had low rainfall of 1301.5 mm/year (the lowest value excluding earthquake-damaged areas). However, no direct relationship between the amount of rainfall and the number of human casualties was observed.

4.2. Land Use

The land-use status of each area was summarised based on the agricultural area classification [25] and changes in land-use status determined from aerial photographs [26].

- Type of agricultural area

The study areas were organised based on agricultural area type, which is one of the main land-use categories. In accordance with the Ministry of Agriculture, Forestry and Fisheries of Japan [27], agricultural areas were classified on the basis of the land-use characteristics in each municipality. This classification is based on the conditions that define the regional agricultural structure and is used as a fundamental consideration in analyses of agricultural and forestry statistics, as well as in the promotion of agricultural policy. The classification comprises the following categories: urban area, flat agricultural area, intermediate agricultural area between flat and mountain areas, and mountain agricultural area. Comparison of agricultural area categories in the study areas and the nation as a whole [28] revealed that none of the study areas was 'flatland agricultural area'; the proportions of the other categories (urban area, intermediate agricultural area, and mountain agricultural area) were all higher in the study areas than nationwide (Figure 2). The ratio of the number of fatalities in 'urban areas' in the study areas was lower than the ratio of population in Japan, whereas the ratio in 'intermediate agricultural areas' and 'mountainous agricultural areas' was fourfold and sixfold higher, respectively (Figure 3). The 'mountainous agricultural areas' have had the highest rate of population decline, with a consistent decline since 1970, and had the highest ageing rate (38.5%) [28]. Sediment-related disasters occurred in various areas, but the effects of loss of life were greater in mid-mountain areas and mountainous areas than in urban areas.

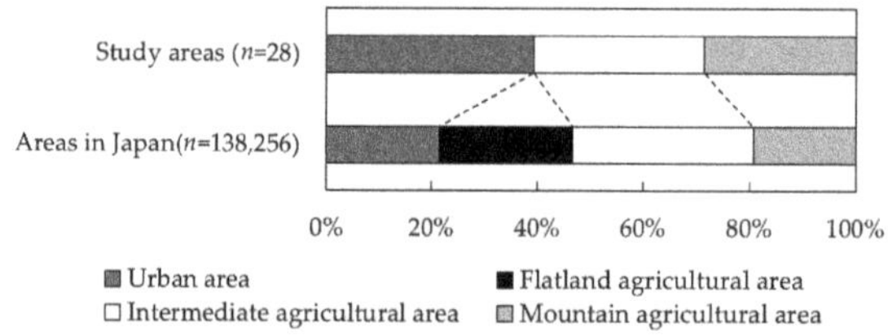

Figure 2. Proportions of the numbers of study areas and nationwide areas stratified according to agricultural area type. (Date sources: [25,28]).

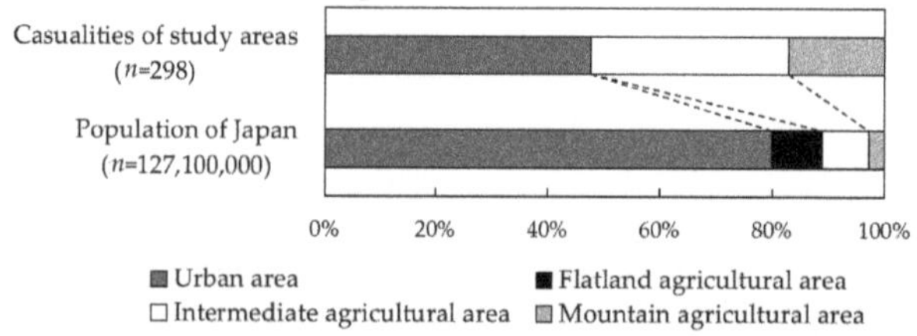

Figure 3. Proportion of casualties among the study areas and proportion of nationwide population, stratified according to agricultural area type. (Date sources: [14,25,28]).

As the distribution of the number of human casualties differs significantly from that of the population in each agricultural area category, it is expected that land use and living patterns in each agricultural area category have an impact on the occurrence of human casualties.

- Changes in land use

Among the tertiary-mesh land-use data of the National Land-use Data, the oldest data (1976) were compared with the newest (2016). The average values of the mesh units comprising each set of data were as follows, in descending order: 'building sites', 'forests', 'saltwater areas', 'other sites (including golf courses)', and 'land for transportation'. The types of land that decreased were as follows, in descending order: 'rice paddies', 'wastelands', 'other agricultural land', 'beaches', and 'river and lakes' (Table 3). Based on the changes in the land-use classification of Japan [29] (Figure 4), 'residential land' and 'roads' have increased continuously during the same period, while 'agricultural land' has decreased continuously; 'forests' have increased and decreased at regular intervals. The study areas were unique in that 'forests' continuously increased in these areas.

Table 3. Changes in land use (average of the study areas).

Land Use	Change of Study Area (m^2)
Building sites	44,555.6
Forests	19,666.5
Saltwater areas	16,740.4
Other sites (including golf courses)	2299.1
Land for transportation	1457.8
River and lakes	1082.0
Beaches	−300.8
Other agricultural land	−7669.5
Wastelands	−27,168.8
Rice paddies	−30,706.3

Note: Date source: [26].

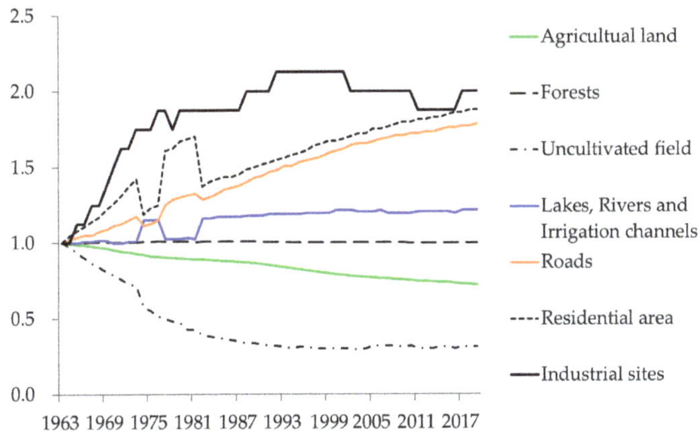

Figure 4. Changes in Land Use Classification in Japan; the baseline value of 1963 was set as 1 (Date source: [29]).

The increase in 'forest' was likely the result of a decrease in 'wasteland' in mountainous areas because of afforestation and other factors. Furthermore, the increase in 'building sites' was generally consistent with the decrease in 'rice paddies' and some 'wasteland', implying that the overall trend in the target villages was for rice paddies and wasteland to be used for residential areas.

Analysis of the changes in land use (only building sites, forests, rice paddies, and wasteland with large increases or decreases were extracted) in each study area revealed that the changes varied among areas, but the main patterns were that 'rice paddies' decreased and 'building sites' increased, while 'wasteland' decreased, and 'forest' increased. Where 'rice paddies' decreased and 'building sites' increased, the expansion of residential areas was one cause of human suffering. Where 'wasteland' decreased and 'forest' increased, it is possible that the increase in forested areas, which at first glance would not be expected to cause landslides, led to a decline in awareness of landslide hazards.

The results indicate that land-use change and the associated expansion of settlements tend to increase human suffering.

4.3. Population and Its Change

We checked the population size (total population, sex ratio, and population aged $\geq$ 65 years) and population trends [30], together with the long-term status of population trends aggregated according to municipality [31].

- Population size and population trends in recent years

Table 4 shows the population size and recent population trends in the study areas. The total populations ranged from 40 to 5472 in 2005 and from 18 to 4771 in 2015, with considerable variation. The populations of all study areas except one (Muroya) declined; the average rate of decline was 23.8%. Sex ratios (here, the ratio of the female population to the male population) remained virtually unchanged at 1:1 in 2005 and 2015, but Yoshino and Nango experienced declines of >0.5 (the female population declined more than the male population). The proportion of the population aged $\geq$ 65 increased from 27.8% in 2005 to 36.7% in 2015. Based on the median values in 2015, the average area had a total population of 559, a male-to-female ratio of 1.1, a population in which 38.2% were aged $\geq$ 65 years, and a population decline rate of 19.6% over 10 years. Population size and composition, as well as population trends in recent years, did not directly affect the number of disasters with high numbers of human casualties.

- Long-term population trends (1995–2015)

The population growth rates of the 20 municipalities included in the study areas tended to decline from 1985 to 2015, similar to the national trend (Figure 5). However, the degree of change during this period was −4.8%, which was slightly larger than the national rate of −4.2%. It was also 3.3% lower on average for each year, compared to the nation as a whole. Over approximately the past 30 years, disasters with large human losses have occurred in municipalities with large population losses.

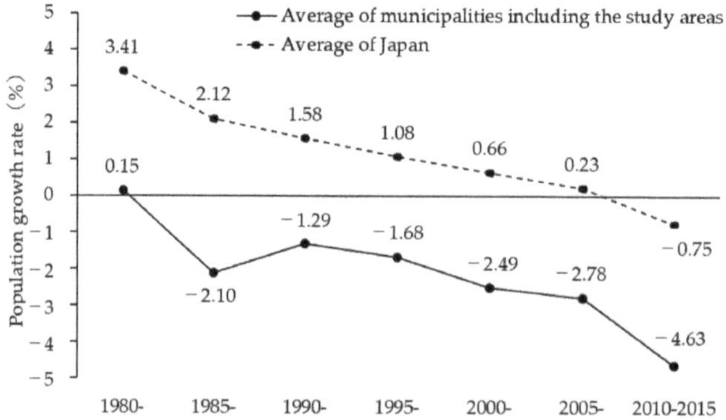

Figure 5. Population growth rate in the municipalities including the study areas. (Date sources: [31]).

Table 4. Comparison of 2005 and 2015 populations in the study areas.

	2005			2015		
	Population	Ratio of Female to Male	Ratio of Aged 65 and Over	Population	Ratio of Female to Male	Ratio of Aged 65 and Over
Yoshino	54	1.35	0.37	32	0.78	0.47
Koei	126	0.77	0.47	64	0.64	0.45
Sendatsu	145	1.13	0.35	117	1.17	0.48
Iizawa	1098	1.03	0.18	979	1.05	0.31
Motomachi	2789	1.04	0.26	2461	0.97	0.32
Hanaoka	1551	1.09	0.28	1304	1.05	0.36
Kidani	1024	1.15	0.27	898	1.12	0.40
Takiya	64	1.29	0.42	30	1.14	0.47
Ui	108	1.16	0.30	26	1.00	0.58
Nojiri	40	0.90	0.20	18	1.25	0.44
Nagano-6	177	1.16	0.34	115	1.21	0.45
Ichinono	775	1.25	0.28	632	1.32	0.40
Iseki-1	664	1.14	0.31	467	1.22	0.41
Uno	5472	1.12	0.29	4771	1.04	0.38
Muroya	1093	0.81	0.17	1231	0.88	0.19
Kamirakuchi	1540	1.06	0.17	1168	1.02	0.25
Kobara	1137	1.08	0.17	832	1.04	0.25
Yashiki	2355	1.08	0.16	2330	1.13	0.20
Tenjin	1598	1.07	0.20	1396	1.09	0.38
Oonishi	535	1.25	0.24	486	1.20	0.38
Kawasumi	4556	1.12	0.15	4465	1.09	0.24
Koyaura	2333	1.12	0.26	1871	1.09	0.39
Nango	173	1.54	0.61	83	1.02	0.42
Tachi	102	1.17	0.19	80	1.29	0.26
Ishidume	77	1.41	0.29	57	1.85	0.33
Fukuoka	40	1.11	0.28	35	1.19	0.37
Kurokawa	1222	0.89	0.23	1149	0.89	0.28
Kajigaharu-Yukihiro	78	1.05	0.33	59	1.03	0.42
Average	1104.5	1.12	0.28	969.9	1.10	0.37

Note: Date sources: [30].

4.4. Residency Status

The residential status in each area was summarised based on the length of residence [32] and housing type (proportion of households living in owner-occupied houses) [33].

- Length of residence

We compared the average values of the study areas to those for Japan (Figure 6). The term "since birth" included people who had lived in an area for various lengths of time.

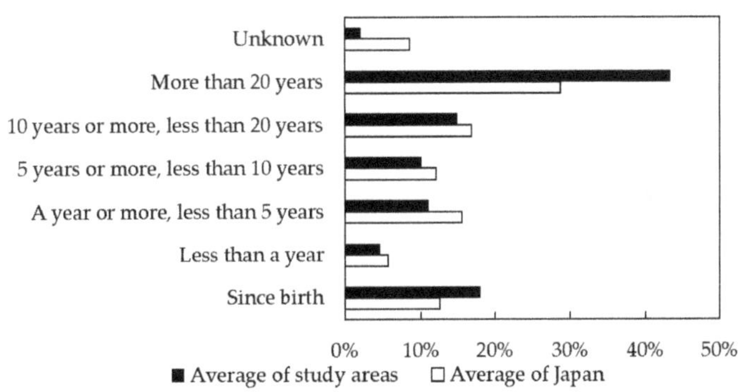

Figure 6. Length of residence. (Date sources: [32]).

The percentages of the 'since birth' and 'more than 20 years' categories were higher in the study areas than in the whole of Japan. This suggests that disasters with large human losses have occurred in areas of low turnover of people. In addition, the proportion of women in the study areas was higher than the proportion of men in the 'more than 20 years' category and lower than the proportion of men in the 'since birth' category, compared to the national proportions. In many cases, there was minimal migration of people in the study areas. Many women presumably entered the study areas at marriage.

- Housing type

The proportion of households living in 'owner-occupied' housing as a percentage of the general number of households living in housing was analysed for each study area (we used sub-region units because of data limitations). Here, housing other than 'owner-occupied' homes included 'public, the Urban Renaissance Agency, public corporation rented housing', 'private rented housing', 'salaried housing', and 'rented rooms'. The data were not disclosed because of confidentiality for one study area (Motomachi), and one sub-region included two study areas; thus, 26 sub-regions were covered. The results showed that the average proportion of households living in 'owner-occupied housing' was 83.2% of the total number of general households, which was higher than the national average of 62.3%. Disasters with high human casualty numbers occur in areas with high numbers of people in owner-occupied housing, who generally do not move as frequently.

4.5. Characteristics of Municipal Administration

The numbers of staff members and the financial statuses of 22 municipalities including the study areas were summarised.

- Number of staff members in municipalities

The numbers of staff members were based on data (as of 15 July 2021; 2005–2020) published in the 'Capacity Management of Local Public Organisations' [34]. The data for 2006 and 2020 were used. We used the 2006 data to take into account the impact of the merger of many municipalities that took place in 2005 throughout the country. We excepted Hiroshima City from some analyses because it is a 'designated city' (only 20 such cities exist in Japan).

In the results, the trend in the average number of staff members was higher in the study areas than the average in Japan in each case, whether 'designated cities' or not. The gap between the study areas and Japan narrowed when 'designated cities' were not included, and the average number of staff members in the study areas between 2006 and 2020 decreased by >100 (Figure 7). However, the sediment-related hazard areas were designated inhabitable areas and a warning evacuation system would have been established by municipalities. Therefore, the amount of inhabitable area could have affected the workload of municipal staff. For this reason, we determined the number of staff members per inhabitable area (using 2020 data; however, the inhabitable area used here is for 2019 [35]); this value was 10.1 persons/km^2 in Japan overall and 12.5 persons/km^2 in the study areas. Furthermore, when 'designated cities' were excluded, this value was 7.7 persons/km^2 in Japan (excluding the 'designated cities') and 6.6 persons/km^2 in the study areas (excluding Hiroshima City).

These results indicate that the target municipalities had slightly more staff members than the nationwide average in Japan. However, the number of staff members was 1.1 persons/km^2 lower in the study areas than in Japan overall, considering the greater proportion of 'designated cities' in the study areas and the numbers of staff members per inhabitable area including sediment-related-disaster hazard warning areas. Disasters with high human casualties occur in municipalities where the number of staff is small in relation to the size of the area.

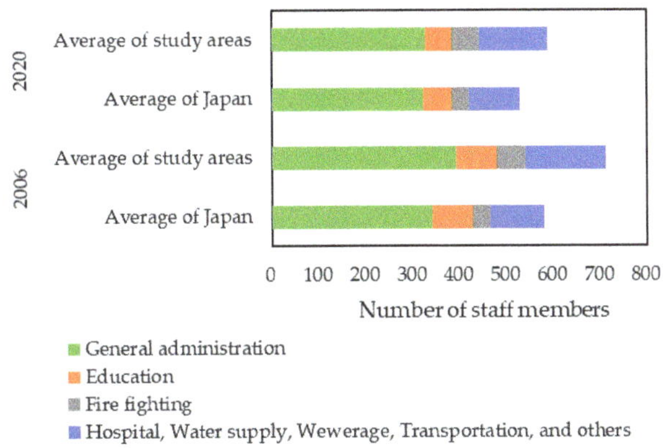

Figure 7. Average number of staff members per municipality (excluding designated cities) (Date sources: [34]).

- Financial status of municipalities

The standard fiscal scale was collected from the 'Municipal Accounts Card' [36] and the 'List of Major Financial Indicators of Local Public Organisations' [37], and the financial indicators of the study areas were compared with the financial indicators of Japan. With the exception of one municipality (Miyagawa Village; merged with Odai Town in 2006), which includes data from Takiya from FY2005, the earliest year for which data were available for all target municipalities, and FY2019, the most recent year, were used. Since Miyagawa Village was not listed in FY2005 or FY2019, we used the 2004 values for Miyagawa Village and the 2019 values for Odai Town.

The results are shown in Table 5. A comparative analysis between the study areas and Japan showed that the 'fiscal capacity index', which indicates the affordability of financial resources, remained almost identical. Although the differences were limited, the following indicator values were larger in the study areas than in the nation as a whole: 'current account balance ratio', which indicates the ratio of fixed expenditures; the 'real bond cost ratio', which indicates the degree of cash flow; and the 'future burden ratio', which indicates the possibility of future financial pressures. In addition, the 'Laspeyres Index', which indicates salary, was smaller. These results imply that disasters with large human casualties occur in municipalities that have financial difficulties.

Table 5. Fiscal indicators according to municipality.

		Fiscal Capacity Index	Current Account Balance Ratio	Real Bond Cost Ratio	Future Burden Ratio	Laspeyres Index
FY2005	Average of Japan	0.52	90.2	14.8	11.4 *	98.0
	Average of Study areas	0.52	93.0	15.2	12.5 *	95.2
FY2019	Average of Japan	0.51	93.6	5.8	27.4	99.1
	Average of Study areas	0.50	94.1	8.4	62.7	97.4

Note(s): Date sources: [36,37]. * Instead of the 'Future Burden Ratio', 'Debt Limit Ratio' is published.

5. Discussion

There is no obvious bias in the natural conditions (altitude, slope, geology, precipitation) of the areas where human-casualty disasters have occurred, but social conditions, as opposed to natural conditions, are likely to have an influence. Disasters with large human losses have occurred in areas where there have been land-use changes, such as the expansion of settlements, areas where people have lived for a relatively long period of time, areas where many people live in owner-occupied housing, and municipalities with a continuously high rate of population decline and difficulties in terms of staff number and finances. However, no effects of total population, sex ratio, and population aged 65 and over were identified.

These results imply the following relationships between socio-economic factors and large human losses in sediment-related disasters.

- The study showed that development activities in hazard areas can have an impact even after a considerable number of years of development. On the other hand, sediment-related disasters with significant human losses have also occurred in areas that have been afforested. The main reason for not evacuating is the lack of anticipation of landslides (see, for example, [38]). Forests may lead people to believe that the threat of landslides is low and that there is no need to evacuate.
- The total population, sex ratio, and population over 65 years of age in the units examined in this study were not relevant. This is inconsistent with previous reports, e.g., [39], which have concluded that older people are more likely to be victims. The combined effect of total population and sex ratio, such as the presence or absence of people to support the elderly, may have influenced the results.
- The fact that more landslides occur in areas where more people live in owner-occupied houses is consistent with reports [6] that people living in rented houses, who probably do not plan to live there for a long time, are less likely to make individual household emergency plans. It may be that people who do not plan to remain in their residence for a long time perceive the need to be prepared less strongly than those who intend to stay in their residence for an extended period of time. However, disasters with significant human suffering have occurred in areas where people have lived for relatively long periods of time. A report concluded that memory of floods depends on living witnesses, and that it diminishes within two generations [7]. Given the frequency of sediment-related disasters, which may occur only once every few hundred years, the relatively long residence period of more than 20 years may still have been too short for people to have experienced a disaster, which would explain the result that a long period of residence did not result in a reduction in human losses.
- The positive relationship between the number of disasters in a municipality and high rates of population decline and small staff and financial resources relative to the area is consistent with examples of existing disasters in which inadequate disaster management systems led to damage [40]. Differences in municipal disaster management systems may affect the number of human casualties.

These discussions indicate the importance of sediment-related-disaster countermeasures. The study areas were located in municipalities where the populations have long been declining and are expected to continue to decline in the future. The most serious casualties occurred both in areas with generally flat land, where the population density was the main cause, and in areas with mountainous terrain and small populations. In areas with generally flat land, although development regulations under the Sediment-related Disaster Prevention Act would have reduced the population density in hazard areas, considerable damage still occurred, implying the existence of pre-regulation development. In areas with mountainous terrain and small populations, it was difficult to implement building restrictions because such restrictions could have led to a further population decline. In addition, although many of the study areas had few people moving in or out, many did not acquire the knowledge and experience related to disasters that would have been gained by a person living in the area for a long time. Furthermore, for households residing in owner-occupied housing, it would have been effective to encourage them to relocate to a safer location, such as at the time of rebuilding in response to the ageing of residents; unlike households residing in housing complexes, these households did not have to go through the trouble of confirming the intentions of many residents. However, the study areas had more staff-related and financial difficulties than other parts of Japan, and therefore efforts and subsidies may not have been sufficiently implemented.

Therefore, although development regulations need to be continued, emphasis should be placed on hard and soft measures in areas that had already been developed before the regulations and are inhabited by a large number of people. On the other hand, in mountainous areas, where the population density and the number of human casualties in this study was small, disasters occur in large numbers and account for a very high proportion of the number of inhabitants. Because of the low number of residents in these areas, it is not realistic to spend a lot of money on hard measures, so it is suggested that detailed soft measures be implemented in each district. Non-structural measures include the accumulation and utilisation of empirical knowledge, as well as teaching and discussion opportunities to encourage relocation. However, support from the national and prefectural governments is necessary in municipalities that do not have sufficient staff or financial resources. To realise these measures, a classification method focusing on the social conditions of each district in addition to the threat of landslides should be considered and measures should be implemented accordingly.

6. Conclusions

In this study in Japan, natural and socio-economic conditions at the community level were determined at the sites of recent sediment-related disasters that resulted in significant human suffering. The results indicate that sediment-related disasters associated with extensive damage occurred under various natural conditions; the extensive human suffering caused by sediment-related disasters was not limited to specific natural conditions. In addition, although sediment-related disasters occur more frequently in populated areas, they also occur in less populated areas. In areas where it is difficult to justify large expenditures on sediment-related disaster countermeasures, non-structural measures that require lower costs than structural measures may be more important for reducing human suffering.

We attempted to understand the characteristics of each area using published data; however, the results should be confirmed by more detailed surveys in each area to determine whether the characteristics identified from the data are indicative of the characteristics that are influenced by a sediment-related disaster in each area. Furthermore, the limited number of cases in this study does not allow verification of its generalisability. The relationship between disaster cases and social factors should also be investigated under different conditions, such as different periods and countries with very different social conditions. Furthermore, alternative data items and complementary methods should be considered for items and areas for which no data have been obtained. Additionally, deepening our knowledge of the relationships between a wider range of socio-economic factors and the number of human casualties is also an important issue.

Author Contributions: Conceptualisation, M.C.; methodology, M.C., K.H. and S.S.; Investigation, M.C.; data curation, M.C.; writing—original draft preparation, M.C.; writing—review and editing, K.H., S.S. and H.F.; visualisation, M.C.; supervision, H.F. All authors have read and agreed to the published version of the manuscript.

Funding: This research received no external funding.

Institutional Review Board Statement: Not applicable.

Informed Consent Statement: Not applicable.

Data Availability Statement: The data presented in this study are available on request from the corresponding author.

Acknowledgments: I am grateful to the member who attend the seminar of Forest Research laboratory Department of Forest Science, Graduate School of Agriculture and Life Sciences, The University of Tokyo, for their insightful comments.

Conflicts of Interest: The authors declare no conflict of interest.

References

1. Ministry of Land, Infrastructure, Transport and Tourism, Japan. Development of Warning and Evacuation System against Sediment Disasters in Developing Countries. Available online: https://www.mlit.go.jp/sogoseisaku/inter/keizai/gijyutu/pdf/sediment_e.pdf (accessed on 17 July 2022).
2. Annual Report on Forest and Forestry in Japan (Fiscal Year 2020). Available online: https://www.rinya.maff.go.jp/j/kikaku/hakusyo/R2hakusyo/ (accessed on 22 April 2022). (In Japanese)
3. Karasaki, T. Research on Communication in Community making Activities. *Bull. Natl. Inst. Rural. Eng. Japan.* **2012**, *51*, 195–257. (In Japanese) [CrossRef]
4. Kamai, T. *Disaster Mitigation Studies of Residential Lots-Urban and Slope Modern and Contemporary*, 1st ed.; Kyoto University Press: Kyoto, Japan, 2020; 326p. (In Japanese)
5. William, P.; Joseph, W. Human Vulnerability to Landslides. *GeoHealth* **2020**, *4*, e2020GH000287. [CrossRef]
6. Jason, D.R. The likelihood of having a household emergency plan: Understanding factors in the US context. *Nat. Hazards* **2020**, *104*, 1331–1343. [CrossRef]
7. Václav, F.; Miroslav, Š.; Petr, S. How long do floods throughout the millennium remain in the collective memory? *Nat. Commun.* **2019**, *10*, 1105. [CrossRef]
8. Ministry of Land, Infrastructure, Transport and Tourism, Japan. Draft of Policy Review on Landslide Prevention Law (2012). Available online: https://www.mlit.go.jp/river/sabo/dosyahou_review/03/120130_shiryo3.pdf (accessed on 4 April 2022). (In Japanese)
9. Aging Population and Increasing Number of "Marginalized Communities" Preventing "Mutual Aid", Nails of the Torrential Rains in Western Japan—Kyoto Newspaper. Available online: https://www.kyoto-np.co.jp/politics/article/20190711000077 (accessed on 12 July 2019). (In Japanese).
10. Ministry of Land, Infrastructure, Transport and Tourism, Japan. Sediment-related Disaster Warning Evacuation Guidelines Study Committee (First Meeting) Explanatory Material (Ref-1, January 15, 2007). Available online: https://www.mlit.go.jp/common/001024528.pdf (accessed on 3 November 2021). (In Japanese)
11. Information on Erosion Control-Related Budgets, etc., National Flood Control and Erosion Control Association. Available online: http://www.sabo.or.jp/seimei-zaisan.htm (accessed on 15 June 2022). (In Japanese)
12. Board of Audit of Japan. Report to the Diet and the Cabinet on the Project Cost for Sediment-Related Disaster Countermeasures and the Implementation Status of Projects for the Countermeasures (Pursuant to Article 30-2 of the Act on the Board of Audit, September 2015). Available online: https://report.jbaudit.go.jp/org/h26/ZUIJI2/2014-h26-Z2000-0.htm (accessed on 29 October 2021). (In Japanese)
13. Mizuyama, T. Evacuation from sediment-related disasters. *J. Jpn. Soc. Eros. Control. Eng.* **2015**, *68*, 61. (In Japanese) [CrossRef]
14. Ministry of Land, Infrastructure, Transport and Tourism, Japan. Past Disaster Response (2004–2021). Available online: https://www.mlit.go.jp/river/sabo/saigai_sokuhou.html (accessed on 22 April 2021). (In Japanese)
15. Japan Meteorological Agency. List of Weather, Earthquake, and Volcanic Phenomena Named by the Japan Meteorological Agency. Available online: https://www.jma.go.jp/jma/kishou/know/meishou/meishou_ichiran.html (accessed on 3 November 2021). (In Japanese)
16. Ministry of Agriculture, Forestry and Fisheries, Japan. 2015 Agricultural Community Boundary Data. Available online: https://www.maff.go.jp/j/tokei/census/shuraku_data/2015/ma/index.html (accessed on 3 November 2021). (In Japanese)
17. Ministry of Agriculture, Forestry and Fisheries, Japan. The Concept of an Agricultural Community, Report on the Survey of Agricultural and Rural Areas, 2015 Census of Agriculture and Forestry. Available online: https://www.e-stat.go.jp/stat-search/files?page=1&layout=datalist&toukei=00500209&tstat=000001032920&cycle=7&year=20150&month=0&tclass1=000001077437&tclass2=000001077396&tclass3=000001085316 (accessed on 3 February 2021). (In Japanese)
18. Ministry of Internal Affairs and Communications, Japan. Boundary Data by Town, Street, etc. from the 2015 National Census, General Contact for Government Statistics (e-Stat). Available online: https://www.e-stat.go.jp/gis/statmap-search?page=1&type=2&aggregateUnitForBoundary=A&toukeiCode=00200521&toukeiYear=2015&serveyId=A002005212015&datum=2000 (accessed on 3 November 2021). (In Japanese)
19. Geospatial Information Authority of Japan. White Map of Japan. Available online: https://maps.gsi.go.jp/#6/38.004820/140.119629/&base=blank&ls=blank&disp=1&vs=c0g1j0h0k0l0u0t0z0r0s0m0f0&d=m (accessed on 17 June 2022).
20. Ministry of Agriculture, Forestry and Fisheries, Japan. Average Elevation of Central Land and Average Slope of Central Land, National Land Numerical Information (Legal Area Designation, etc.), DB to See, Know, and Make Use of Local Agriculture. Available online: https://www.maff.go.jp/j/tokei/census/shuraku_data/2015/ka/index.html (accessed on 3 November 2021). (In Japanese)
21. National Institute of Advanced Industrial Science and Technology. 1:200,000 Seamless Geological Map of Japan. Available online: https://maps.gsi.go.jp/ (accessed on 3 November 2021). (In Japanese)
22. Japan Meteorological Agency. Details (Precipitation), Annual and Monthly Normal Value, Historical Weather Data Search. Available online: https://www.data.jma.go.jp/obd/stats/etrn/index.php?prec_no=21&block_no=0124&year=&month=&day=&view=a1 (accessed on 3 November 2021). (In Japanese)
23. Nishiyama, K.; Tsuyoshi, W. Frequency of Slope Failure and Debris Flow Caused by Heavy Rainfall on Mountain Slopes in Japan. *J. Jpn. Soc. Eng. Geol.* **2015**, *55*, 325–333. (In Japanese) [CrossRef]

24. Ministry of Land, Infrastructure, Transport and Tourism, Japan. Flood Control Measures (Chapter 3-3-1 Flood-prone Land). Available online: https://www.mlit.go.jp/river/pamphlet_jirei/bousai/saigai/kiroku/suigai/suigai_3-1-1.html (accessed on 3 November 2021). (In Japanese)
25. Ministry of Agriculture, Forestry and Fisheries, Japan. Agricultural Region Typology, Regional Indicators, DB to See, Know, and Make Use of Local Agriculture. Available online: https://www.maff.go.jp/j/tokei/census/shuraku_data/2020/ia/index.html (accessed on 3 November 2021). (In Japanese)
26. Ministry of Land, Infrastructure, Transport and Tourism, Japan. Land Use Tertiary Mesh Data, National Land Information—Agriculture and Rural Areas in Japan. Available online: https://nlftp.mlit.go.jp/ksj/gml/datalist/KsjTmplt-L03-a.html (accessed on 14 July 2021). (In Japanese)
27. Ministry of Agriculture, Forestry and Fisheries, Japan. About Agricultural Area Types. Available online: https://www.maff.go.jp/j/tokei/chiiki_ruikei/setsumei.html (accessed on 6 May 2021). (In Japanese)
28. Ministry of Agriculture, Forestry and Fisheries, Japan. Current Situation and Future Projections for Rural Areas and Villages, Section 1: Current Situation of Rural Areas and Movement of Rural Development, FY2019 Summary of the Annual Report on Food-Agriculture and Rural Areas in Japan. Available online: https://www.maff.go.jp/j/wpaper/w_maff/r1/r1_h/trend/part1/pdf/c3_1_00.pdf (accessed on 29 April 2022). (In Japanese)
29. Ministry of Land, Infrastructure, Transport and Tourism, Japan. National Land Plan: Area of National Land by Use Category—Agriculture and Rural Areas in Japan. Available online: https://www.mlit.go.jp/kokudoseisaku/kokudoseisaku_fr3_000033.html (accessed on 3 November 2021). (In Japanese)
30. Ministry of Agriculture, Forestry and Fisheries, Japan. Population by Sex and Age (Census of 2015, 2010, 2005), Census, DB to See, Know, and Use Local Agriculture. Available online: https://www.maff.go.jp/j/tokei/census/shuraku_data/2020/sb/index.html (accessed on 3 December 2021). (In Japanese)
31. Ministry of Internal Affairs and Communications, Japan. Population Change Rate (FY1985-FY2015, Every 5 Years), Population, Households, Indicator Data, Prefectural and Municipal Statistics (Social and Demographic Statistics System), General Contact for Government Statistics (e-Stat). Available online: https://www.e-stat.go.jp/regional-statistics/ssdsview (accessed on 3 November 2021). (In Japanese)
32. Ministry of Internal Affairs and Communications, Japan. Population by Period of Residence (6 Categories), Sex, Town, Street, Character, etc., Tabulation Related to Sex, Age, etc. Tabulation of Migratory Population, 2015 Census. Available online: https://www.e-stat.go.jp/stat-search/files?page=1&query=%E5%B9%B3%E6%88%9027%E5%B9%B4%E5%9B%BD%E5%8B%A2%E8%AA%BF%E6%9F%BB%E3%80%80%E5%B0%8F%E5%9C%B0%E5%9F%9F%E9%9B%86%E8%A8%88%E3%80%80%E5%B1%85%E4%BD%8F%E6%9C%9F%E9%96%93&layout=dataset&metadata=1&data=1 (accessed on 3 November 2021). (In Japanese)
33. Ministry of Internal Affairs and Communications, Japan. Number of General Households and Number of Persons in General Households and Persons per Household by Type of Dwelling and Relationship of Ownership of Dwelling (6 Categories), Town, Street, etc., Tabulations Related to Population and Other Basic Tabulations, Census 2015. Available online: https://www.e-stat.go.jp/stat-search/files?page=1&query=%E5%B9%B3%E6%88%9027%E5%B9%B4%E5%9B%BD%E5%8B%A2%E8%AA%BF%E6%9F%BB%E3%80%80%E5%B0%8F%E5%9C%B0%E5%9F%9F%E3%80%80%E4%BD%8F%E5%B1%85%E3%81%AE%E7%A8%AE%E9%A1%9E&layout=dataset&metadata=1&data=1 (accessed on 3 November 2021). (In Japanese)
34. Ministry of Internal Affairs and Communications, Japan. List of Staff by Department, Designated City Data and Municipalities (Excluding Designated Cities) Data, Local Government Capacity Management Related. Available online: https://www.soumu.go.jp/main_sosiki/jichi_gyousei/c-gyousei/teiin/ (accessed on 3 November 2021). (In Japanese)
35. Ministry of Internal Affairs and Communications, Japan. Area of Inhabitable Land, Natural Environment, Basic Data, Statistics on Municipalities, 2021 Social and Demographic Statistics System, e-Stat (e-Stat). Available online: https://www.e-stat.go.jp/stat-search/files?page=1&query=%E5%8F%AF%E4%BD%8F%E5%9C%B0%E9%9D%A2%E7%A9%8D&layout=dataset&stat_infid=000032093332 (accessed on 11 November 2021). (In Japanese)
36. Ministry of Internal Affairs and Communications, Japan. Municipal Balance Sheet, Materials Related to Local Financial Condition Surveys. Available online: https://www.soumu.go.jp/iken/zaisei/card.html (accessed on 3 December 2021). (In Japanese)
37. Ministry of Internal Affairs and Communications, Japan. List of Major Financial Indicators for Local Governments. Available online: https://www.soumu.go.jp/iken/shihyo_ichiran.html (accessed on 4 December 2021). (In Japanese)
38. Chiba, M.; Kono, Y.; Fujimoto, O. Effects of non-structural measures on regional warning and evacuation during sediment-related disasters. *J. Jpn. Soc. Eros. Control Eng.* **2022**, *74*, 14–25.
39. Ushiyama, M.; Honma, M.; Yokomaku, S.; Sugimura, K. Characteristics of human suffering due to the heavy rainfall disaster of 30 July 2008. *J. Jpn. Soc. Nat. Disaster Sci.* **2019**, *38*, 29–54. [CrossRef]
40. Chiba, M. Warning and evacuation in response to sediment-related disasters. *Nat. Hazards* **2011**, *56*, 499–507. [CrossRef]

water

MDPI

Article

Disaster-Resilient Communities on Flood Plains and Their Agricultural Regeneration: A Case Study in Meinong Plain, Taiwan

Shu-Chen Tsai [1] and Su-Hsin Lee [2,*]

[1] College of Arts and Design, Jimei University, Xiamen 361021, China; tsaishuchen@jmu.edu.cn
[2] Department of Geography, National Taiwan Normal University, Taipei 10644, Taiwan
* Correspondence: shlee@ntnu.edu.tw

Abstract: Taiwan's near-mountain alluvial plain is a high-risk area with frequent disasters, and residents have become more tolerant of the compound disasters that occur with overall environmental changes associated with the development of urbanization in recent years. This paper presents a case study of a near-mountainous alluvial plain in Southern Taiwan. The Hakka ethnic group is the main community in the study area and also the main research object. This case study illustrates the disaster resilience of the community to natural and artificial disasters. This study adopted two research approaches, namely historical geography and political economic geography, as well as community resilience theory. Research methods including case study, secondary literature analysis, fieldwork, and interviews were used. Through text analysis, it was found that (1) the community's awareness of disaster avoidance was rooted in the experience of reclamation in the early 17th century; (2) communities have experienced artificial disasters caused by political and economic intervention, which have been transformed into disaster awareness and community resilience; (3) cumulative artificial disasters have a greater impact on communities than unpredictable natural disasters; and (4) the energy of community resilience and agricultural regeneration is based on the duality of disasters.

Keywords: actions; reorganization; artificial disaster; young farmer; resilience community

Citation: Tsai, S.-C.; Lee, S.-H. Disaster-Resilient Communities on Flood Plains and Their Agricultural Regeneration: A Case Study in Meinong Plain, Taiwan. *Water* **2022**, *14*, 1736. https://doi.org/10.3390/w14111736

Academic Editor: Su-Chin Chen

Received: 4 April 2022
Accepted: 24 May 2022
Published: 28 May 2022

1. Introduction

There is an ecological network between humans and the environment, and in the face of environmental threats, humans use culture to adapt in order to protect themselves in this network [1]. This concept of cultural adaptation results in reflexive thinking and has opened up research into environmental cognition. For example, human civilization must be close to water and be able to prevent water disasters, which is part of the relationship between reaction and behavior when flooding occurs [2]. International academic circles have begun to focus on the importance of traditional disaster-prevention knowledge—especially in the area of small regional, local and community-based knowledge research—and believe that it should be regarded as a form of disaster-prevention technology [3–6]. Different societies also have different levels of disaster awareness. For example, coastal residents know how to interpret changes in walrus behavior and the environment to predict tsunamis and earthquakes [7,8], while mountain residents in areas with frequent orogeny carefully select their sites and test their bases during the monsoon season before settling down in order to prevent landslide disasters [9]. One study showed that disasters are the result of the interaction between the natural environment, socio-demographic characteristics, and man-made facilities [10]. However, with the rapid development of human society, there will be sudden changes when the natural load capacity exceeds the threshold [11]. Therefore, research on community resilience with regard to living environments in the face of natural disasters, environmental degradation and toxic waste is being actively carried out [12,13].

According to IPCC reports, human societies do not have sufficient resilience under extreme climatic phenomena [14,15]. The latest research work is committed to contributing effective methods to deal with the impact of floods, e.g., applying big data to assist disaster risk management and emergency response sharing [16], and a framework entailing AI/ML methods for identifying the safest route to destinations using UAV and path planning has been proposed for the timely disaster response and evacuation of the residents of aged care facilities [17], as well as for the accurate identification of flood-affected areas after floods for effective flood management [18]. The accurate assessment of flood sensitivity with DEM in arid regions [19,20] can provide prediction models for flash flood disasters in small catchment areas [21]. A study on community resilience to floods proposed local residents' resilience practices that can be applied to flood management measures and focused on rehabilitation, community participation, and local indigenous practices [22], as well as developing more accurate forecasting methods for the flood sensitivity of the catchment area to enhance the resilience of catchment residents to natural disasters [23]. However, changes in hydrological conditions within a catchment not only affects communities but also causes significant social change. Therefore, in addition to leveraging technology for flood protection, the main question of this study was how can communities develop resilience under the impact of long-term environmental disasters? This paper provides a case study of an area located downstream of a catchment, where environmental conditions are wet, rainy, feature floodplains that are often disturbed by flooding, and allow water to infiltrate the area within a short period of time. This case provides reflections on the impact and transformation of repeated flooding on community resilience.

Taiwan is a long mountainous island with a long north–south and narrow east–west mountain; the plain area accounts for only 30% of the island [24]. Taiwan's topographic characteristics have created short river basins and steep slopes. The geology of the upstream mountainous areas is young, and the clay slate is fragile. In addition, the daily rainfall in the rainy season is 1030 mm, and sudden heavy rainfall often causes large-scale landslides in the upstream areas [25]. Therefore, in order to stabilize the mountainous area and ensure the safety of the downstream regions, most earth and rock disaster studies have focused on the catchment region in the mountainous area, as secondary cascade disasters are caused when the catchment region cannot be loaded [26].

However, small catchments are highly sensitive to land use and short rainfall, which affect the safety of the downstream alluvial plain [27]; every heavy rain event may cause serious disasters in the plain along the river [28–30] due to the limitations of the terrain. More than 90% of Taiwan's population lives in the western plain, and only 5% reside in the mountains [31]. The near-mountain alluvial plain in the valley mouth area is marginal land with a high flood sensitivity, and it has no urban disaster-prevention system or mountain-monitoring system. Recently, the Taiwan government announced potential streams of debris flow so that residents could take the necessary refuge and evacuation measures before a disaster in order to maintain the safety of settlements on slopes and valleys.

However, research has shown that the lack of spatial coverage still puts communities near mountains at high risk [32]. Another study confirmed that the gap between urban and rural areas makes the disaster risk in rural areas much greater than that in urban areas [33]. With the development of urbanization, mountain areas must bear other risks associated with overall environmental changes in addition to natural disasters, e.g., the output of water resources and environmental pollution caused by the input of urban waste. It was pointed out that the rapid development of Taiwan's metropolitan areas in recent years has affected the discharge of runoff and changed the hydrological cycle, resulting in increases in flood frequency and water volume, as well as losses of life and property.

Therefore, the increase in downstream population and water demand should be considered to be water resource factors [27,34]; according to early research data, the groundwater resources in near-mountain areas also bear a high risk of pollution from the development of metropolitan areas [35]. From this point of view, disasters should not only be regarded as natural events caused by natural driving forces, as they may also result from social problems. Therefore, researchers should also seek solutions in community development context [34].

In this study, five cases were used to explore the relationship between disasters and community resilience. The first case comprised irrigation waterways built in the Ching Dynasty, mainly located in the north of the Meinong Plain. At that time, the irrigation waterway was completely excavated by local residents in accordance with natural conditions. The second to the fifth cases comprised modern environmental events, which are cases of cross-regional water allocation. The geographic area spanned two counties and cities along the Gaoping Stream and the time range was from 1975 to 2019, which makes them indicator cases of artificial disasters.

This study argue that natural disasters and social forces are intertwined, and the phenomenon that communities near mountains bear the dual risks of natural and artificial environmental disasters should have more attention paid to it. Therefore, based on the dual points of view of nature and society, this study used the flood plain in Southern Taiwan as the research object. A case study on a regional scale was used to explore the process of disaster resistance and the agricultural regeneration action of environmental awareness transformation. This research mainly contributes to bridge the research gap between physical and human geography. In contrast to disaster research that only considers natural factors, social factors can also be considered when looking at the man-made causes behind natural phenomena. In terms of research innovation, we propose an approach to escape a single disaster event or single point in time and integrated two research methods and theoretical perspectives in an attempt to understand the interrelationship between disasters and resilience.

2. Materials and Methods

2.1. Study Framework and Design

This study adopted two research approaches, namely historical and political economic geography, as well as community resilience theory. Research methods including case studies, secondary literature analysis, fieldwork, and interviews were used. This study was focused on five cases in the north floor area of the Pingtung Plain, covering the period of 1680–2020.

(1) Historical geography of natural disasters: from the historical perspective, we analyzed how a balance between man and nature was achieved with agricultural technology in the face of natural disasters during the process of reclamation by the well-cultivated Hakka ethnic group in the Meinong Plain.

(2) Political and economic geography of modern disasters: four cases were used to explain the change in the nature of disaster and community anti-disaster action in the study area.

(3) Community resilience theory: knowledge reorganization and the social self-organization process of residents facing disasters in the study area were examined.

The simultaneous application of the two approaches allowed us to effectively compare the community's response to the same issues. The selection of the research method was based on the results of exploratory research [36,37] and the data of the study cases, which have the main characteristics of fragmentation and heterogeneity. A limitation of this study is that the political economy and the stakes involved in civic movements will keep some of the data hidden. In the late 1970s and early 1980s, actor network theory was able to effectively explain the relationship between actors and space [38]. However, the spatial and temporal relationships of the five cases were relatively scattered, and although there was a common issue, there was no intersection between the actors; therefore, we advocate for the concept of underground rhizomes as a research strategy [39]. The phenomenon of data dispersion was found in the exploratory research [36,37]. This research suggested that the effective integration of data is the most important link in the research process.

Therefore, we used case study as the main method, treated the plains community as a unique phenomenon, and collected data from different sources as the basis for the analysis.

An important step in analyzing secondary literature is to interpret the data collected by original studies from another perspective [40].

The inclusion of multiple viewpoints in this study could have helped to prevent the researchers' personal values from affecting the research results because a wide variety of data, different methods, and perspectives of researchers in different fields can be neutralized [41,42]. In response to two research approaches, this study firstly collected official and quasi-official public information and secondly analyzed media materials and interviewed reporters and interviewees on the subject in the process. Third, Triangulation for the study to cross-check three different types of data (Figure 1). The reliability of the research originated from the use of various methods and different data sources. The data were verified at all levels of time (long-term), space (specific), and people (key roles). The scientific philosophical framework of this research comprised a multi-method approach of considering "retrospections data components back and forth", "triangulation to seek the most applicable interpretation", "descriptive laws from all levels", and "breakthrough awareness of research limitations" [43], with the aim of generating more trustworthy research interpretations [42].

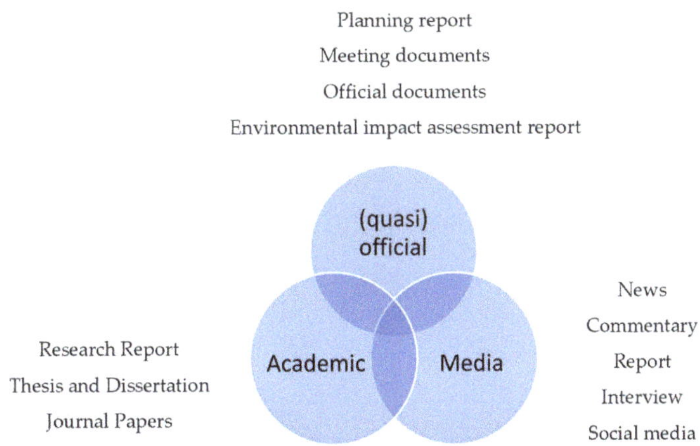

Figure 1. Data triangulation.

2.2. Study Area

The study area is located at the northernmost end of the Pingtung Plain in Southern Taiwan, in the center of Kaohsiung City and the northernmost plain of Pingtung County. As the plain is located between a fault zone and a trench, it forms a complete geographical area [44–46]. The geographical scope is to the east of Chishang River, to the north of the Ligung Embankment, to the west of Dawu Mountain cliff, and to the south of Moonlight Mountain. It is an independent plain isolated by the natural environment. The administrative area includes Ligung Township in Pingtung County, Meinong District, and the Chishang District in Kaohsiung City (Figure 2). The total area of the Meinong Plain is approximately 106 km^2, as shown in Figure 2. The formation of its range comes from the coupling of special geographical conditions, which are the Laonong Stream, Chishang River, and Meinong Stream alluvial fan and alluvial plain.

The area of this study has received attention from the humanities and nature perspectives, but no previous studies have explored the relationship between the two simultaneously. In terms of humanities, the Hakka people's reclamation and development have formed a special social group. As many as 93.5% of the Hakka people in Meinong District (Table 1) are characterized by a high level of education, a strong sense of national identity, and strong political and cultural influence [47,48]. The geographical space where the settlement is located is also a high disaster-risk area. This area has a maximum water output of surface and groundwater (including underlying water), as well as natural energy due to the intersection of a fault zone, uneven rainfall, and typhoons [49].

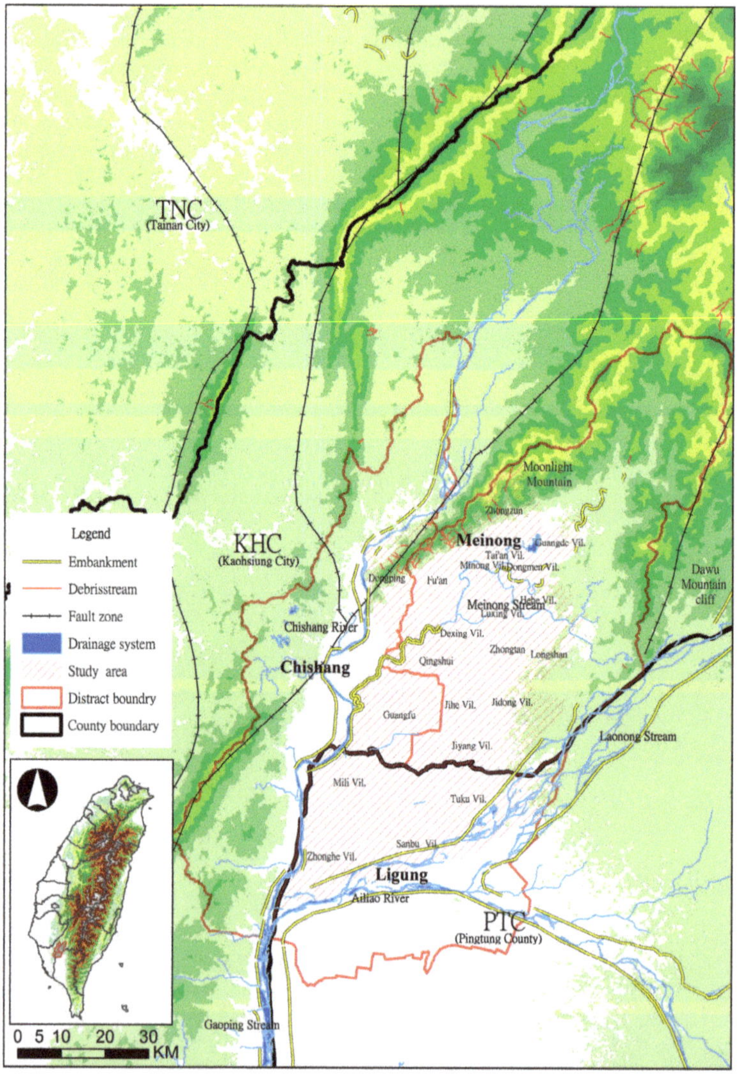

Figure 2. Study area: Meinong Plain, Taiwan.

Table 1. Population and ethnic groups in the study area.

District	Name of Village	Population	Main Ethnic
Meinong District	Hehe vil.; Luxing vil.; Zhongtan vil.; Dexing vil.; Longshan vil.; Guangde vil.; Dongmen vil.; Tai'an vil.; Minong vil.; Jidong vil.; Fu'an vil.; Zhongzun vil.; Qingshui vil.; Jiyang vil.; Jihe vil. (Total 15 vil.)	28,359	Hakka
Chishang District	Dongping vil.; Guangfu vil. (Total 2 vil.)	2786	Hakka
Ligung Township	Sanbu vil.; Zhonghe vil.; Mili vil.; Tuku vil. (Total 4 vil.)	4549	Hokkien and unnan–Burmese Veterans

2.3. Data Collection and Analysis of Secondary Literature

According to the research design, this study began with an investigation of: (1) natural disaster data in historical documents (mainly water and soil disasters); (2) how regional residents altered nature to meet their survival needs (mainly irrigation waterways); (3) artificial disasters (mainly public pollution and environmental movements, such as the anti-Meinong reservoir and the anti-Jiyang artificial lake (after the artificial lake project was rejected in 2013, the government reintroduced the new project and named it Gaoping Great lake; in this study, it is called Jiyang artificial lake) movements in 1992); (4) water conflicts (such as the well sinking in Ligung in 1975, the well closure and power shut-off in Ligung in 1990, and the anti-deep-water well in Meinong District in 2015–2017); and (5) environmentally friendly actions (such as young men returning home for farming, organic farming and environmental protection). All the collected data were classified and coded with Excel software. There were two main sources of data. The first was secondary literature on natural disasters, including data on (1) and (2), which were mainly from research reports and research papers; the second source comprised data on (3), (4), and (5), which came from official and institutional information, official statements, internal meeting minutes, research reports, planning reports, survey reports, etc. Historical water conflict events and management organization data were also collected. Sources included newspapers, magazines, and social media, as well as various planning reports and statistical data from the Ministry of Economic Affairs (MEA) and the Taiwan Water Company (TWC) and its related sub-organizations. In addition, announcement information, planning reports, and meeting records were collected from the Pingtung County government. The interview records from various agencies represented unofficial opinions and descriptions. Data analysis was conducted with Excel statistics software on 54 natural disaster incidents, 35 adjusted natural incidents, 11 water conflict incidents, 26 artificial disaster incidents, and 7 environmental action incidents (Figure 3).

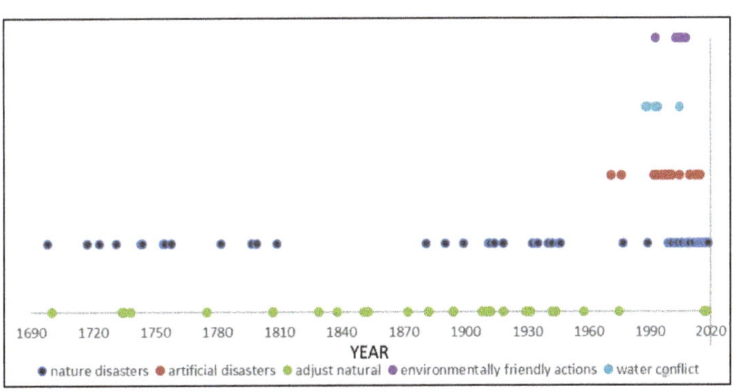

Figure 3. Number and time of water events (1680–2020).

2.4. Case Studies

Five cases were used to explore the relationship between disasters and community resilience (Table 2). Objects included residents, farmers, religious groups, organizations, and county governments. A network of nodes and connections, including attitudes, experiences, knowledge, and awareness, was constructed. This network illustrates the experience of disaster in life and in the environment, as well as culture, ordinary people's knowledge of the local hydrological environment, the formation of environmental risk awareness and the resilience of life, and the process of organization.

Table 2. Basic descriptions of five study cases.

Case	Location	Related Parties	Claim	Situation
Waterway in Ching Dynasty	Meinong District, KHC	1. Hakka ethnic group 2. Hakka lineage 3. Irrigation waterway owner	Survival and settlement	Settlement
Ligung Well Closure and Power Shut-Off	Ligung Township, PTC	1. Ligung villagers' congress 2. Pingtung County Government	Well closure and compensation	Complete well closure and monetary compensation
Anti-Meinong reservoir	Meinong District, KHC	1. Group 7 2. Meinong People's Association	Stop	Whole project is cancelled
Anti-Jiyang artificial lake	Chishang and Meinong District, KHC	1. Anti-Jiyang artificial lake self-rescue association 2. Meinong People's Association 3. Kaohsiung teachers' Association Ecological Education Centre 4. Pingtung teachers' Association Ecological Education Centre 5. Pingtung Environmental Protection Alliance	Stop	Whole project is cancelled
Meinong anti-deep-water well	Meinong District, KHC	1. Anti-deep-water well self-rescue association 2. Chishang, Meinong and Ligung villagers	Stop	Project is on hold

The first case was an irrigation waterway built in the Ching Dynasty, mainly located in the north of the Meinong Plain. At that time, the irrigation waterway was completely excavated by local residents in accordance with natural conditions. There was no official waterway until the mid-19th century. The second to the fifth cases covered modern environmental events, which are cases of cross-regional water allocation. These involved compensation for power shuts-off for closed wells in the Ligung Township, the anti-Meinong reservoir movement, the anti-Jiyang artificial lake movement, and the Meinong anti-deep-water well movement. The geographic area spanned two counties and cities along the Gaoping Stream, and the time range was from 1975 to 2019, which makes them indicator cases of artificial disasters.

2.4.1. Irrigation Waterway in Ching Dynasty

In 1736, the Hakka ethnic group crossed the Laonong Stream from Pingtung to the Meinong alluvial plain for reclamation. At the beginning, a few people travelled back and forth for reclamation every day. There were two types of irrigation waterways for agriculture. One was used to divert water from the depression on the plain (called Zhongzhenpi) to the rice field, and the other was used to divert the water left in the valley to irrigate dry fields when floods subsided [50]. After the cultivated land was slightly enlarged, Meinong Stream water was introduced into the irrigation waterway, and the rice field environment was completed towards the end of the Ching Dynasty, with some areas already having the ability to harvest two crops a year [51]. Due to the fear of invasion by the indigenous people near the mountains and the difficulty of fetching water from the plain, settlement did not take shape in the Meinong Plain until the middle of the 18th century, i.e., after the irrigation waterways were roughly completed [52]. In the Ching Dynasty, all irrigation waterways on the Meinong Plain were located in areas that were not easily flooded. A total of six irrigation waterways formed the prototype of the settlements. At the end of the 19th century, in order to expand the demand for land development, the Japanese government built the Lion Head waterway in flooded areas. However, the waterway has been damaged by floods year after year, so the government has further built the Ailiao River embankment on a larger scale to protect the developed land and waterways, which has prompted changes in land use and industry in Meinong Plain [25,28,29].

2.4.2. Ligung Well Closure and Power Shut-Off

Since 1974, Kaohsiung City and Gangshan Township have relied on 13 deep-water wells in Ligung Township. They extract 127,700 cubic meters per day, supplying 420,000 people with domestic and industrial water [53,54]. In 1987, the water company dug five deep wells in the Shoujinliao area of Kaohsiung County, which caused panic among residents in Ligung, who jointly opposed it [53] (in 1989, the Ligung Township Congress predicted that the groundwater level would drop, with accompanying reductions in crop yield and demand and the cessation of pumping). The Pingtung government (PtCG) shut down the power of its own accord when the water rights of two wells expired, and the water company appealed to the Provincial Construction Department [54]. The PtCG implemented a water cut-off, and the Kaohsiung County government responded by cutting off road access, preventing people from entering Kaohsiung [55–58]. In 1988, the Ligung Township Congress petitioned the PtCG and requested that the TWC carry out evaluation work. In 1989, the TWC agreed to conduct an evaluation and appraisal with Fengjia University; in 1990, the TWC signed a consultation resolution. The conclusions included a deadline to stop pumping, funding subsidies and pipeline compensation, and water conservancy facility subsidies. The well closure and power shut-off incident in Ligung ended in 1994 after the well shut-off was completed.

2.4.3. Anti-Meinong Reservoir

The goal of the Meinong reservoir is to meet the long-term water demand in Kaohsiung, Tainan and Pingtung [59]. Due to safety concerns caused by the fact that the dam site of Meinong reservoir is located above a fault [60] and will seriously damage the ecological environment of the tropical forest in Yellow Butterfly Valley [61] (as well as the impact on the traditional Hakka culture that has long been present in Meinong village), at the end of 1992, after the villagers learned that the reservoir was about to be built, local residents formed a Group 7 and then established the Meinong People's Association to take action against the Meinong reservoir [62,63]. In 1993, they led Meinong's residents to take a night bus to Taipei to protest the construction of the reservoir. Through effective mobilization and lobbying by Congress, the Legislative Yuan removed the project budget for the Meinong reservoir in 1993 and 1994. In 2000, the government announced that it would not build the Meinong reservoir, although it still proposed to build it in 2015, causing residents to protest again [64]. It can be seen that the water resources on the Meinong Plain are still a hot spot of industrial demand.

2.4.4. Anti-Jiyang Artificial Lake

The Jiyang artificial lake was proposed by the WRA in 2000 and was intended to address the daily 500,000 tons shortages of long-distance and industrial water in Kaohsiung [59,65]. In other words, it is part of a new scheme to replace the function of the Meinong reservoir [66].

This project planned to use the surplus water from Laonong Stream during the wet season for introduction into the artificial lake for regulation and storage in order to increase the water supply in the Kaohsiung and Pingtung areas. The excavated earthwork could also provide a source of sand and stone materials, which is one of the reasons for the prohibition of sand and gravel mining in the Gaoping River. In this project, the existing farm would have been excavated into a lake, surrounded by earth embankments, and arranged into five lake areas (A, B, C, D and E), with a total area of 770 hectares. The project is located at the Shoujinliao farm and the Tuku farm at the confluence of the Laonong Stream and the Chishang River [67–72]. Residents from the Jiyang, Mili, Guangfu and Sanbu villages in the Chishang and Meinong Districts, as well as Ligung Township, comprised the main members of the self-rescue association, while the Meinong People's Association, the Kaohsiung Teachers' Association Ecological Education Centre, the Pingtung Teachers' Association Ecological Education Centre, and the Pingtung Environmental Protection Alliance jointly launched an appeal. The main reason for the appeal was that the development of the artificial lake may cause groundwater resource issues, flood safety concerns, and disputes over the government's expropriation of a large amount of soya bean agricultural land. After 13 years of continuous negotiation and resistance, the plan was terminated at the Environmental Impact Assessment stage in 2013 [73–75].

2.4.5. Meinong Anti-Deep-Water Well

In the case of the Meinong anti-deep-water well movement, the conflict directly involves the water corporation on the front line. Based on the support of residents and non-profit organizations, local and central representatives hold the same view. They also formed a self-rescue association that can be expanded. Since the Ligung incident in 1979, Meinong's anti-deep-water well movement has engendered a sense of opposition. In the local area, water conflicts have forced the suspension of a strong water distribution policy.

2.5. *Field Investigations and Interviews*

Field investigations and interviews were conducted from 2014 to 2019. In 2014, the TWC planned to dig deep-water wells in Sojingliao in Chishan District, which triggered mass resistance from residents who set up a self-rescue association. Between 2015 and 2017, the association had several violent conflicts with TWC; as of 2019, the association was still in operation because it thinks the TWC is just suspending the plan and that if the well-sinking plan was enforced, it would have a negative outcome for the 2020 elections.

The main interviewees were local people familiar with the situation (school principals, members of NGOs, public opinion representatives, farmers, residents, and members of self-rescue associations), local officials (the Taiwan Agricultural Research Institute, the Seventh River Management Office, the Ministry of Economic Affairs, the water resource authority (WRA), the Water Resource Planning Institute of the WRA, the Water Resources Bureau of Kaohsiung City, the Department of Urban and Rural Development of Pingtung County, and the Department of Agriculture of Pingtung County), and professionals (a professor of hydrology). A total of 20 interviews were conducted, and the responses were snowballed to connect related objects.

2.6. Community Resilience

Resilience is a social–ecological system concept [76] that is complex, nonlinear, and unpredictable and that feeds back to other systems [12,77,78]. It differs from natural disasters in that the system emphasizes slow-onset hazards associated with slow drivers of change [79,80]. However, the social–ecological system view emphasizes that resilient communities can be identified and have development advantages, establish resilience through self-organization, and pay attention to the connection between humans and land, values and beliefs, knowledge and learning, social networks, collaborative governance, economic diversification, infrastructure, leadership, and outlook [12]. By applying the concept of resilience to space science, a living environment that will not collapse in the face of the impacts of disaster can be created [81,82]. Scholars in Taiwan have long paid attention to the relationship between human settlements and disasters [83], emphasizing disaster prediction, especially the use of disaster sensing before early warning monitoring systems, such as the more sensitive detection of stratum subsidence from all aspects [84], slope disasters [85,86], flooding [87], earthquakes [88,89], and learning recovery and adjustment after disasters [88–94]. In recent years, people's participation in disaster prevention and traditional indigenous disaster-prevention knowledge have been further studied [34,95,96].

However, the disaster data in this study area showed that, because the three watersheds affect the alluvial plain, they are no longer a single catchment, which may explain the causes of the natural disasters. Moreover, the types of disasters in the study area have changed, and the factors of the disasters are complex and feed back into each other. The current research results also lack insight into how social–ecological systems learn to resist, absorb or adapt to disasters from the diachronic time axis. Therefore, it is necessary to study the actions and feedback of communities experiencing disasters from the perspective of social–ecological systems. Scholars have proposed four aligned suggestions: learning to live with change and uncertainty, nurturing diversity for reorganization and renewal, combining different types of knowledge for learning, and release and reorganization [77].

3. Results

3.1. Natural Disasters and Changes on the Plain

The central mountain range on the east side of the Pingtung Plain rises from 100 to 3000 m above sea level within a horizontal distance of 12 km. When the highlands on the east side of the plain are violently eroded by rivers due to the high source and long flow, a strong scouring force and a large amount of accumulation form a wide plain [97]. The Meinong Plain, which is alluvial and formed by three drainage systems, has received the largest concentration of heavy rainfall in recent years [98]. The geological conditions of the plain are fragile, and the erosion under the flow path is rapid, able to carry rock debris and sediment and to promote the resurgence of tributary river erosion. Moreover, this area also has a rainfall form with summer rain and winter drought. Showers often cause large-scale landslides in the upstream, and a large amount of debris forms debris flows, resulting in vertical and lateral erosion. Due to accumulation, it is often diverted or overflowed, which is also the main cause of disasters [99].

According to historical data, Taiwan experienced 223 typhoons in the Ching Dynasty from 1684 to 1887, 31 of which were disasters that stuck Southern Taiwan [100,101]. From 1897 to 1945, the country experienced a total of 178 typhoons, 13 of which caused serious disasters to Southern Taiwan [102–104]. The disaster events in the study area in recent years have mainly been caused by typhoons. Heavy rainfall caused the collapse of the upper reaches of the Chishang River and the Laonong Stream, resulting in the flooding of the plain [30,98,105–107]. Disasters in the study area have been the focus of research in recent years [107–109].

In 1736, the Hakka ethnic group reclaimed land from the high ground at the foot of the mountain and irrigated it with Zhongzhenpi on the plain. After it was stable, it was gradually expanded to the Moonlight Mountain at the upstream of the Meinong Stream to form Meinong and other villages. The Meinong Plain was maintained in a stable state by the Hakka ethnic group. In the early 20th century, a large number of immigrants entered the Meinong Plain after the Japanese government built the Ailiao River embankment [50]. Before the embankments of Ailiao and the Laonong Stream were built, the streams flowed to the southwest. One can imagine that if heavy rain caused flooding, it would be difficult for the whole north Pingtung area to survive. After the embankment was built, the waterway of the Ailiao River was restrained to the north while the braided river of the Laonong Stream was restrained to the south (Figure 4), converging to Gaoping River [110]. After the construction of the embankment, the riverbed of Laonong Stream became reclaimed land, which is now Sojingliao Farm and Tuku Farm (also the main area in of cases 4 and 5), Kaohsiung Farm, and the Shingo area.

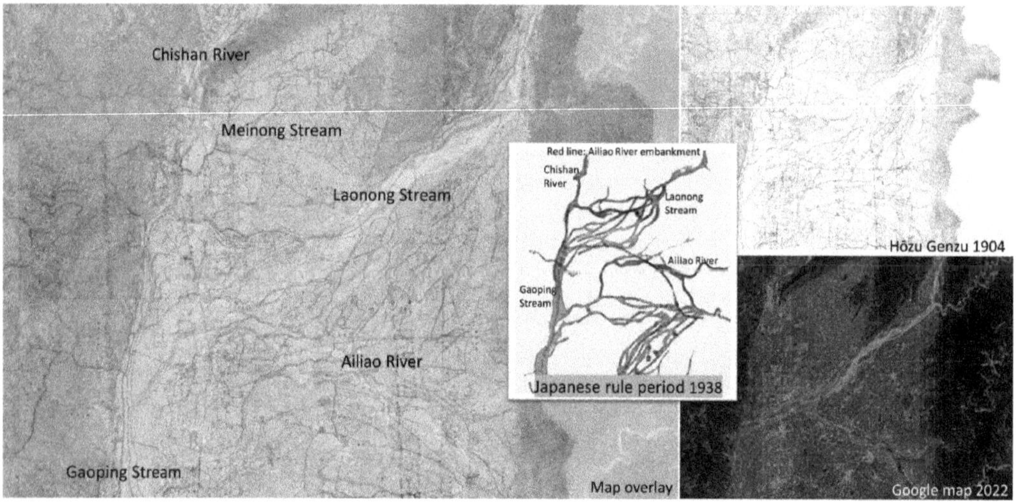

Figure 4. Overlay map of drainage system changes in Meinong Plain (1904–1938).

Since the Japanese rule of law, the construction of embankments has become the most important policy to protect both banks of the rivers [25,29]. However, recent studies have shown that typhoons can induce gully-type earth rock flow. Taking the 2009 Morakot typhoon as an example, the farmland that was expanded after the construction of embankments after 1948 was submerged during the Morakot typhoon, while the farmland reclaimed before 1948 was outside the area of flood overflow [30]; This study has confirmed that the farmland reclaimed on the flood plain before 1948 is an area with disaster-prevention awareness.

3.2. *Artificial Disasters in the Ecological-Social System: Reservoirs, Artificial Lakes, Deep-Water Wells and Public Pollution*

In 1921, the Japanese government built the Ailiao River embankment to control the flood disaster on the Pingtung Plain, resulting in the gradual drying up of the original river due to the closure of the upstream embankment. Later, due to man-made reclamation and cultivation, the appearance of the original river gradually disappeared on the surface, though there was abundant groundwater [111]. Industrial development in Southern Taiwan, including traditional and science and technology industries, totally depends on the groundwater resources in the plain. Industrial water grabbing is another disaster for communities that have been farming for generations. As early as 1990, the Ligung Township had an environmental recession due to groundwater over-pumping, and the following was announced by the PtCG in the media:

> *"Over the past decade, a large amount of water from Pingtung has been pumped to Kaohsiung, with 500,000 tons a day. For many years, thousands of wells have been*

scrapped because they can't get water; The loss of irrigation water in the countryside is serious and farming is affected." [112]

Residents recounted the disappearance of water and species:

"In the past, Tuku area were swamps on the other side of the soil bank. Now there is no water and it has become a dry field. I remember that in summer, the water was knee deep and there were loaches and eels in the water. I had several classmates there, this is our common experience. Now it is dry."

In the 1960s, Taiwan's economy was booming, and most of the infrastructure was reinforced concrete structures, so a lot of sand and gravel were needed as building materials. The sand and gravel in Ligung, Tuku and other areas at the northern end of the Gaoping River alluvial fan were of good quality and had good traffic conditions, making it the most important material supplier at that time. In particular, many low-use lands, such as Taiwan sugar company land and retired farm land, are highly suitable for sand and gravel development. The integrity, large area, and low utilization of this kind of land make it suitable for use as a sand and gravel resource area [113]. The construction of the Ailiao River embankment stabilized the land and brought economic benefits, but it has also brought stagnant river sand. A large amount of sand and gravel accumulated in the center of the river, forming a high ridge. Heavy rainfall in summer resulted in the river sand washing into the two embankments, causing the collapse of the embankment and the erosion and loss of fields near the river. The annual siltation volume of the Laonong Stream (1975–1990) is 4.07 million cubic meters, and that of the adjacent the Ailiao River is approximately 60,800 cubic meters [53]. Therefore, with the early development of artificial mining and the establishment of special gravel areas by the government, the north and south banks of the Laonong Stream have become a special area with the highest density of gravel in Taiwan [49,99,114].

While meeting the needs of infrastructure construction, the government continues to plan the remote development of water resources in the Gaoping Stream basin to meet future needs. Under the premise of giving priority to economic development, all resources must fully contribute to economic policies [66,67,115,116]. Therefore, implementation plans have been formulated, as with cases two to five in our study; however, a few elite-led programs that lack a local environmental perspective have become a form of artificial disaster. One study showed that the north of Ailiao River has the largest water yield, the highest soil yield/value, and the highest number of gravel fields in Taiwan [49]. This phenomenon is known to be caused by the gravel pits (locals call it the "Grand Canyon") and waste pollution caused by illegal gravel mining on river banks and on agricultural land (Figure 5). For example, agricultural land in Chishang was backfilled with waste slag that polluted the Chishang River in 2013, and other rivers were also polluted by industrial waste in 1999, 2000, 2013, 2016 and 2019 [117–119]. The gravel pits formed by illegal mining led the government to establish the artificial lake project and directly use a large number of gravel pits in this area as recharge pools for groundwater recharge [120].

Early research results have revealed that the groundwater resources in this study area bear high pollution risk from the development of Kaohsiung and Pingtung [35,121]. Recent studies have also shown that politics and economic integrated regional scale strategies to carry out cross-regional water distribution in the study area have resulted in fierce conflicts [122]. The export of water resources in the study area contributes to economic development, but industrial waste returns to the area, with the whole social–ecological system linked through the negative feedback of artificial disasters. The main reason for this is that economic factors distort the allocation of resources, and scholars have pointed out that this can be regarded as interference. The stronger the interference, the weaker the resilience and the lower the stability of the system [123].

3.3. Social System and Action of Residents in Disaster Events

In the case of Ligung, it was found that there was no solution to residents' complaints about the declining groundwater level, the reduced agricultural production, and the subsidence of the land. In Meinong, the self-rescue association used eddy currents, starting from Jiyang at the center, and sent a message to the Ligung and Chisang areas to use street sweeping, virtual and physical gatherings, and collective appeals to the power center; that is, the association directed the areas to fight against lobbying with knowledge and materials, reject compensation and exchange, and use every means available to organize social cohesion.

In 2017, the deep-water well project was restarted in Meinong. Mainly, the Kaohsiung City Government actively encouraged the TSMC (Taiwan Semiconductor Manufacturing Company) to set up a factory in Luzhu Park, Southern Science Park. The aim was to provide stable water resources as promised. In addition, in order to maintain the advantages of the semiconductor industry supply chain, the central government had to restore regional integration and dynamic balance [122].

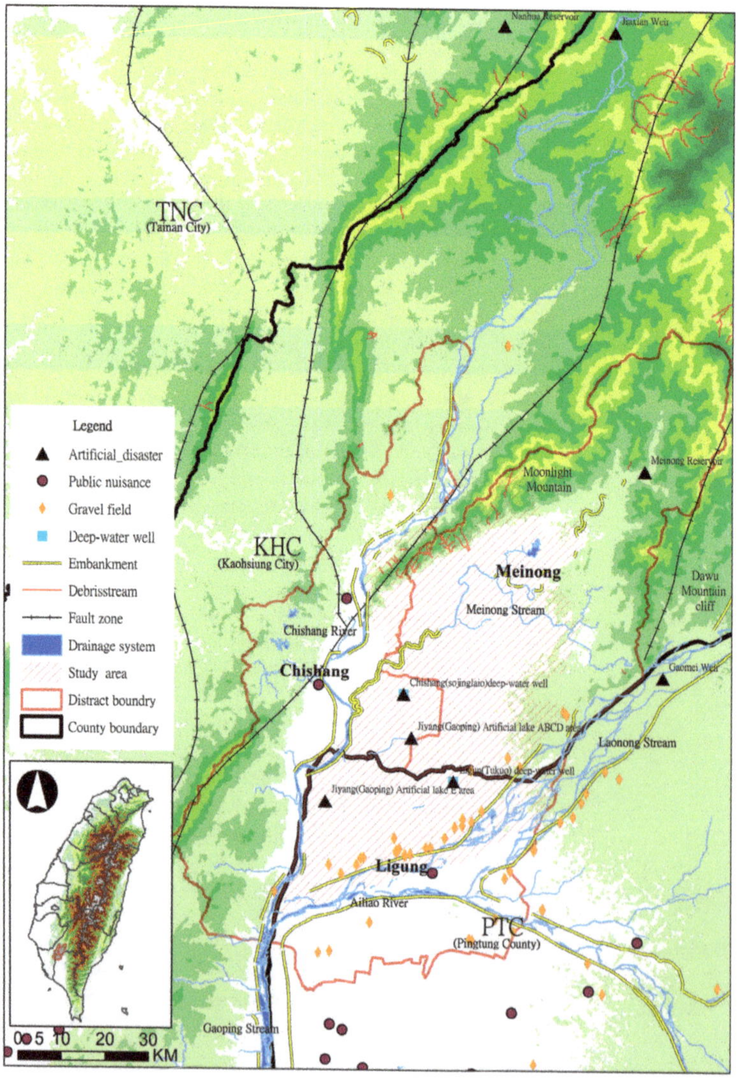

Figure 5. Spatial distribution of artificial disasters on the Meinong Plain.

At the upper reaches of the Gaoping Stream is the intersection of Tainan, Kaohsiung, and Pingtung. In 1992, the location of the Meinong reservoir was quite controversial. At that time, it resulted in the most influential water environment movement in Taiwanese history and led to the formation of an important environmental sustainability group [61,124,125]. After these projects failed, the central government began to develop adaptation plans. The plans were to shift the reservoir model to a distribution model by using a combination of surface water and groundwater abstraction. For example, the Gaoping River has multiple water abstraction options [125–127], and deep-water wells in the underground aquifer are designed to increase pure raw water to supply the south-moving high-tech industrial park [122]. However, after the plan to drill new deep wells was resisted, there was proposal to rent the existing water wells from the Taiwan Sugar Corporation. Using the old wells to improve and clean the wells to make up for the original plan because the two could together increase the daily water supply by 100,000 cubic meters [128]. Based on five deep wells dug in Sojiliao, Kaohsiung County in 1987 [53], the TWC initiated negotiation and compensation [128]. Finally, in 2011, 14 more deep-water wells were dug in the Sojiliao area. The daily water intake was raised 100,000 tons to fill the gaps in Kaohsiung's livelihood and industrial needs for water, as well as to maintain the dynamic balance of the region's current situation by means of negotiated diffusion.

3.4. Risk Awareness and Community Resilience

The main impact of adaptation plans is that the Meinong Plains are full of water-intake facilities and the expected reservoir space has turned into water-intake weirs, radiation wells, deep-water wells and artificial lakes. More facilities have brought more disasters and risks to local residents, such as lower agricultural production (R15), water pump motors being damaged due to groundwater level drops (R13), the general feeling of residents that the fountain disappeared (R6), groundwater level drops (R7), and the illegal mining of sand and gravel (R12) (Figure 6). The residents' experience of water shortage in their lives directly formed their awareness of disasters, which not only prompted the establishment of self-rescue associations but also led to anti-deep-water well operations. Because these previous disaster experiences were often flooding or prolonged drought rather than the decline of the groundwater level, they were different from the experience of extreme natural disasters that promotes the formation of risk awareness.

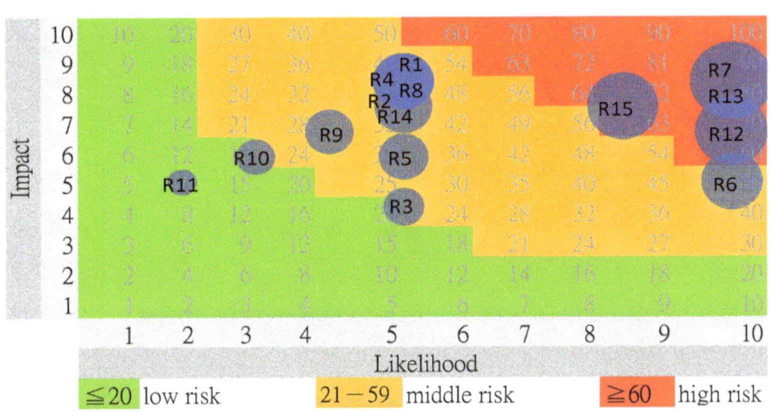

Figure 6. Risk matrix.

The personal experience of residents slowly suffering from disasters in their lives is the main reason for the formation of environmental awareness. Residents in the study area were found to have different reactions to sudden disasters and disasters accumulated year after year. When economic losses are caused by natural disasters (such as floods that wash away farmland), residents can apply for tax reduction. However, when residents find that the main cause of the disaster is man-made, their main response has been to regroup as a crowd to demand that all human intervention be stopped. The specific methods have included self-organization, active information collection, external connection, knowledge learning, showing that the community can learn from and adapt to disaster risks to strengthen community resilience (Table 3).

Table 3. Five cases and community resilience.

Resilience Transformation \ Five Cases	Water Tunnel in Ching Dynasty	Ligung Well Closure and Power Shut-Off	Meinong Anti-Reservoir	Jiyang Anti-Artificial Lake	Meinong Anti-Deep-Water Well
Learning change and uncertainty in disasters	• Typhoon (R1) • Floods (R2)	• Groundwater level drops (R7) • Declining agricultural production (R15)	• Reservoir collapse (R4) • Earthquake (R3) • Breaking the embankment (R5)	• Edamame fields were expropriated (R14) • Illegal mining of sand and gravel (R12)	• The fountain disappeared (R6) • Groundwater level dropped (R7) • The water pump motor was damaged due to groundwater level drops (R13)
Water sensitivity nurturing diversity for reorganization and renewal	• Hakka ethnic group • Hakka lineage • Irrigation waterway owner	• Farmer • Ligung villagers' congress • Pingtung County Government	• Group 7 • Self-rescue association • Meinong People's Association	• Self-rescue association	• Self-rescue association
Combining different types of knowledge for learning	• Feng Shui • Belief • Tradition • Culture	• Agriculture • Water resources • Disappearance of species	• Earthquake • Fault zone • Hakka culture, • Yellow Butterfly ecology • Food self-sufficiency food safety	• Water resources • Education • Ecology • Agriculture	• Agriculture • Groundwater pollution (R8) • River water pollution (R9) • Hazardous waste (R10) • Factory waste (R11)
Release and reorganization	• Meinong consciousness • Hakka consciousness	• Ligung, Tuku, and Sanbu	• Chimei Community College • Agricultural regeneration	• Kaohsiung teachers' Association Ecological Education Centre • Pingtung teachers' Association Ecological Education Centre • Pingtung Environmental Protection Alliance	• Chishang, Meinong, and Ligung

4. Discussion

4.1. Learning Change and Uncertainty in Disasters

The spatial sequence of reclamation in the Meinong Plain was related to natural disasters. During the Ching Dynasty, avoiding disasters was the first choice. The Hakka ethnic group, who have skilled farmers, knew how to conduct the reclamation so that disasters could be avoided in the environment (Lingshan). The top of the fan area has almost no settlements and or farming due to the flooding of the Laonong Stream in the rainy season (from the south of Jingualiao to Hengshan). Moreover, residents know how to adjust crops for the flood season and temporarily leave the cultivated land (Beishangtang), as well as how to adjust crops according to the dry land (Jingualiao), as follows: select the fertile field after floods to plant rice (Longdu); plant seedlings according to the season; use rainwater, stream water, and waterway water in a planned way (Longdu, Dafen, Xiaofen, Hobienliao, and Hengshanwei); set up cultural symbols in key places of water (Shuikou-lisher-jenkuan-bogong); and name the space with regard to water (Sanjia means three water clips, Xiaoshuigou means gate for distributing water, and Longque means hole for distributing water).

After the irrigation waterway was built, the community residents used the processes of digging culverts and wooden boards to adjust the water volume in order to prevent disasters [50,129]. Even after torrential water from the Laonong Stream washed away land and caused collapse, the affected households successfully applied to the government for rent reduction (Nanlong) [130]. At this time, spatial development was restricted by natural conditions. Even if the Japanese government had intended to set up an immigrant village here, it would still have failed. The main causes of failure were flooding, lack of water supply and drainage facilities, and ant damage [50].

Building embankments was the starting point of environmental awareness. During the initial reclamation period, the rules for allocating crops were based on the amount of water. Even if it was necessary to divert water, the residents still chose to avoid the flooded areas. However, after the Japanese government built the official irrigation waterway, large and flat agricultural areas were developed in the south of the Meinong Plain, and they planted a larger number of high-economic-value crops, such as sugarcane and tobacco, than that on the Meinong Plain [50].

4.2. Water Sensitivity Nurturing Diversity for Reorganization and Renewal

The changes in the irrigation system on the Meinong Plain in the Ching Dynasty were mostly funded by the lineage or Kenshou (capitalists). Farmers did not need to pay grain rent, but they had to pay the costs of management and maintenance. The mayor in the irrigation area was responsible for the maintenance of the irrigation waterway and water rent. If the mayor was too busy, he was able to employ more staff [50].

The organization of waterways in the Ching Dynasty was large-scale and stable; however, the Japanese government began to designate private irrigation waterways as public waterways, mainly because the irrigation waterways in the Ching Dynasty had poor equipment and were particularly vulnerable to damage in the case of rain. Joining the public irrigation waterways meant owners could obtain bank loans to build waterways, obtain legal protection, and manage themselves under the supervision of the government [131,132]. In 1912, the Japanese government announced the Taiwan water conservancy combination rule, under which all irrigation waterways now belonged to the government, thus allowing the waterways to enter a complete water conservancy organization system [133].

The scale of disasters and development has prompted the continuous renewal and reorganization of waterway organizations, and the residents on the plain have adjusted their corresponding abilities for hundreds of years. This means that when facing issues such as reservoirs, artificial lakes, and deep wells, local residents have been able to organize the Group 7 and the Meinong People's Association and to invite various groups from Kaohsiung and Pingtung to challenge the government's economic and environmental policies in terms of safety, ecology, culture and industry [63,134].

In the cases of Meinong and Ligung, there were attempts to negotiate a water allocation plan after 2006 [135,136]. Even though the central budget was prepared, it was still unsuccessful [137,138]. Residents in this area formed a social network across counties and cities, and they refused monetary compensation. This community acted like a network of underground rhizomes strung together [122]. Ultimately, the TWC removed its budget and changed its allocation strategy [139]. This also proves that even small-scale political bodies, such as local associations and communities [140], have the ability to enact water negotiations with the central government.

4.3. Combining Different Types of Knowledge for Learning

The development of disaster-prone areas needs to incorporate more environmental awareness. Disaster awareness on the Meinong Plain combines different types of knowledge including geology, hydrology, ecology, culture, anthropology and agriculture. For example, regarding the issue of the Meinong reservoir being located on the fault belt, the residents continue to uncover academic data that have become important evidence for the anti-reservoir movement. Academic attention has also deepened the geological study of the

Meinong Plain [60]. The development of artificial lakes and deep-water wells has helped us gain knowledge of groundwater hydrology in flood plains [141].

The anti-Meinong reservoir issue not only strengthens the environmental awareness of residents but also creates a special memory landscape [62,142,143]. In the process of local consolidation, combined with different areas of knowledge such as hometown, land and environment, the agricultural characteristics of "Meinong consciousness" and "hometown consciousness" have been created [144]. There are also Meinong's local writers, who continue to create and pass literature on from generation to generation, spreading the meaning of local environmental characteristics to the community [145]. On the other hand, the unique Hakka ethnic group and the preservation and development of Hakka local characteristics have highlighted the implications of Hakka culture [146]. Following the environmental movement, Chimei Community University has become a local knowledge institution, holding agricultural experience education activities such as the Rice Transplanting Festival, the Harvest Festival, and a Farmers' Market; cultivating rural issues; promoting agricultural knowledge; and offering courses such as environmentally friendly farming and agricultural machinery repair in order to shorten the distance between people the land. They have also published a large number of stories about the lives of organic farmers [117,147].

4.4. Release and Reorganization

The anti-Meinong reservoir movement promoted the formation of a Meinong consciousness and a Hakka consciousness. The artificial lake and deep-water well strengthened the characteristics of local environmental activism, while the multi-faceted view of released water was encouraged by the agency of the society. Taiwan's joining to the World Trade Organization strengthened the protective emotions around local agriculture, encouraged small farmers to enter the countryside, and created a rural regenerative social phenomenon of young people returning home that is referred to as Meinong's new rural movement [147]. It has also been found that the environmental awareness formed by the environmental movement resulted in young people choosing to connect the land with agriculture and becoming more willing to take action to respond to climate change and environmental change issues [117].

The environmental movement on the Meinong Plain became the foundation for unleashing awareness of environmental disasters. Returning young farmers share their farming skills and experience at Chimei Community University, further cooperating with local farmers and participating in the social movement of contemporary agricultural regeneration [147]. With the aid of media communication, the organic farming of young farmers returning home has gradually influenced the local farmers to adopt environmentally friendly agricultural methods [148–150] and has promoted the establishment of special agricultural production areas [151,152]. The young farmers have also encouraged the reorganization of local agricultural groups, improvements in farming methods, and the renewal of agricultural products, thus forming a social–ecological system with a positive feedback loop.

4.5. Summary: Community Resilience Developed from the Duality of Disasters

The resilience of the community in this study area can be further explained on the basis of the achievements of research [78]. The resilience in this study area is interdependent with agriculture and economy; coupled to geographical conditions, ethnic groups, agriculture, economy, and so on; and coevolving in the duality of disasters. In addition, the agency and reorganization of this area involve some concepts of vortex concentration of community resilience [12], but their particularity lies in the fact that their community resilience is generated in a dual loop of positive and negative directions (Figure 7).

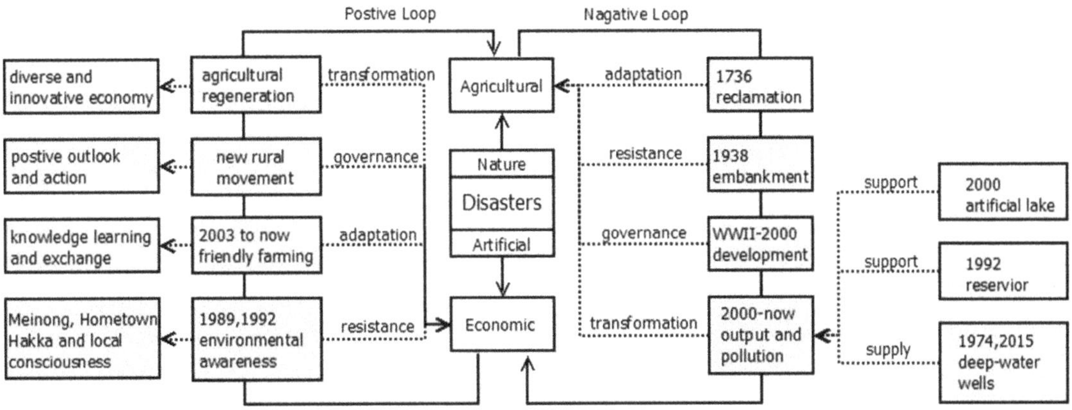

Figure 7. Two loops of disaster duality.

Under the interaction of natural and artificial disasters, different stages produce different responses of adaptation, resistance, governance, and transformation. The study found that under the threat of natural disasters, agriculture communities should first develop the ability to adapt until the big government brings in technology to generate the ability to resist natural disasters, and then these communities should develop governance and accordingly transform their agriculture industries. However, the resistance of big government to natural disasters is also the key point in the negative loop, because the economic benefit of land development is the main purpose of this resistance. These results confirm the idea that the nature of disturbing the stability of resilience is an economic factor [11]. However, in artificial disasters, resistance to the economy is what first arises, followed by the development of adaptation, governance, and transformation processes. Developing risk-awareness resistance actions is also key to turning community resilience into a positive loop, because the purpose of changing farming practices is to become environmentally friendly.

5. Conclusions

This study focuses on the interrelationship between disaster disturbance and reorganization within the study area. The main findings of this study are as follows. (1) The disaster resilience in the study area is rooted in the experience of land reclamation in the early 17th century, and its main factors are the Hakka ethnic group, culture, feng shui, and belief. (2) The community has experienced artificial disasters caused by political and economic interventions that have formed residents' awareness of risk and community resilience. The main events have been anti-reservoir and anti-deep-water well movements. (3) Cumulative artificial disasters have a greater impact on communities than unpredictable natural disasters, and their main factors are groundwater level drops, the disappearances of species, declines in agricultural production, farmland expropriation, and illegal mining. (4) The duality of disasters forms positive and negative loops that generate the energy of community resilience and agricultural regeneration in resistance, adaptation, governance, and transformation. One limitation of this research is that the natural disaster data in the Qing Dynasty currently only have descriptive text. With accurate spatial positioning, more complete research could be conducted on the sources and subsequent impacts of disasters. Additionally, studies of social resilience in this case are suggested in the future to obtain comparable findings.

Author Contributions: Conceptualization, S.-C.T. and S.-H.L.; methodology, S.-C.T. and S.-H.L.; software, S.-C.T.; validation, S.-H.L.; investigation, S.-C.T. and S.-H.L.; resources, S.-C.T.; data curation, S.-C.T.; writing—original draft preparation, S.-C.T.; writing—review and editing, S.-H.L.; visualization, S.-C.T.; supervision, S.-H.L. All authors have read and agreed to the published version of the manuscript.

Funding: This research was partially funded by the Ministry of Science and Technology of Taiwan, grant number MOST 110-2321-B-003-001, and an award for doctoral candidates in the humanities and social sciences to write doctoral dissertations, grant number MOST 108-2420-H-003-006-DR.

Institutional Review Board Statement: Not applicable.

Informed Consent Statement: Not applicable.

Data Availability Statement: Not applicable.

Conflicts of Interest: The authors declare no conflict of interest.

References

1. Barrows, H.H. Geography as human ecology. *Ann. Assoc. Am. Geogr.* **1923**, *13*, 1–14. [CrossRef]
2. White, G. *Human Adjustment to Floods*; University of Chicago Press: Chicago, IL, USA, 1942.
3. Agrawal, A. Dismantling the divide between indigenous and scientific knwoledge. *Dev. Chang.* **1995**, *20*, 413–439. [CrossRef]
4. Kalland, A. Indigenous knowledge: Prospects and limitations. In *Indigenous Environmental Knowledgeand Its Transformations: Critical Anthropological Perspectives*; Ellan, R., Parkes, P., Bicker, A., Eds.; Routledge: London, UK, 2000; pp. 319–331.
5. Dudgeon, R.C.; Berkes, F. Local understanding of the land: Traditional ecological knowledge and indigenous knowledge. *Nat. Across Cult.* **2003**, *4*, 75–96. [CrossRef]
6. Kameda, H. (Ed.) Disaster reduction hyperbase: Asian application. In Proceedings of the DRH Idea Workshop, Delhi, India, 26–28 July 2007; pp. 19–20.
7. Guan, K.; Ohara, M.; Mahoro, S.; Seino, S.; Naoya, F. *Disaster and Biodiversity*; Nishida, H., Yokoyama, J., Wagstaff, S.J., Callomon, P., Eds.; International Union of Biological Sciences: Paris, France, 2017; Available online: https://www.iubs.org/ (accessed on 19 March 2022).
8. Seino, S.; Kojima, A.; Hinata, H.; Magome, S.N.; Isob, A. Multi-sectoral research on East China sea beach litter based on oceanographic methodology and local knowledge. *J. Coast. Res.* **2009**, *4*, 1289–1292.
9. Rautela, P. Indigenous technical knowledge inputs for effective disaster management in the fragile Himalayan ecosystem. *Disaster Prev. Manag. Int. J.* **2005**, *4*, 233–241. [CrossRef]
10. Mileti, D.S. *Disasters by Design: A Reassessment of Natural Hazards in the United States*; Joseph Henry Press: Washington, DC, USA, 1999.
11. Scheffer, M. *Critical Transitions in Nature and Society*; Princeton University Press: Princeton, NJ, USA, 2009.
12. Berkes, F.; Ross, H. Community resilience: Toward an integrated approach. *Soc. Nat. Resour.* **2013**, *26*, 5–20. [CrossRef]
13. Paton, D.; Johnston, D. Disasters and communities: Vulnerability, resilience and preparedness. *Disaster Prev. Manag. Int. J.* **2001**, *10*, 09653562. [CrossRef]
14. IPCC. *Climate Change 2017*; IPCC: Geneva, Switzerland, 2017.
15. IPCC. *Climate Change 2013*; IPCC: Geneva, Switzerland, 2013.
16. Munawar, H.S.; Qayyum, S.; Ullah, F.; Sepasgozar, S. Big data and its applications in smart real estate and the disaster management life cycle: A systematic analysis. *Big Data Cogn. Comput.* **2020**, *4*, 4. [CrossRef]
17. Munawar, H.S.; Mojtahedi, M.; Hammad, A.W.; Ostwald, M.J.; Waller, S.T. An ai/ml-based strategy for disaster response and evacuation of victims in aged care facilities in the Hawkesbury-Nepean Valley: A perspective. *Buildings* **2022**, *12*, 80. [CrossRef]
18. Munawar, H.S.; Hammad, A.; Ullah, F.; Ali, T.H. (Eds.) After the flood: A novel application of image processing and machine learning for post-flood disaster management. In Proceedings of the 2nd International Conference on Sustainable Development in Civil Engineering (ICSDC 2019), Jamshoro, Pakistan, 5–7 December 2019; pp. 5–7.
19. Avand, M.; Kuriqi, A.; Khazaei, M.; Ghorbanzadeh, O. DEM resolution effects on machine learning performance for flood probability mapping. *J. Hydro-Environ. Res.* **2022**, *40*, 1–16. [CrossRef]
20. Avand, M.; Moradi, H. Spatial modeling of flood probability using geo-environmental variables and machine learning models, case study: Tajan watershed, Iran. *Adv. Space Res.* **2021**, *67*, 3169–3186. [CrossRef]
21. Costache, R.; Tin, T.T.; Arabameri, A.; Crăciun, A.; Ajin, R.; Costache, I.; Islam, A.R.M.T.; Abba, S.; Sahana, M.; Avand, M. Flash-flood hazard using deep learning based on H_2O R package and fuzzy-multicriteria decision-making analysis. *J. Hydrol.* **2022**, *609*, 127747. [CrossRef]
22. Munawar, H.S.; Khan, S.I.; Anum, N.; Qadir, Z.; Kouzani, A.Z.; Mahmud, P. Post-flood risk management and resilience building practices: A case study. *Appl. Sci.* **2021**, *11*, 4823. [CrossRef]
23. Avand, M.; Moradi, H.R.; Ramazanzadeh Lasboyee, M. Spatial prediction of future flood risk: An approach to the effects of climate change. *Geosciences* **2021**, *11*, 25. [CrossRef]

24. Ou, S. Ayasaka Ichiro and the Development of Geology Research in Modern Taiwan. *J. Taiwan Hist.* **2020**, *71*, 53–102.
25. Ma, C. *The Study of Taiwan's River Policy in the Period of Japanese Rule—The Focus of Water Control*; National Cheng Kung University: Tainan, Taiwan, 2015.
26. Chen, J.-C.; Huang, W.-S. Evaluation of Rainfall-Triggered Debris Flows under the Impact of Extreme Events: A Chenyulan Watershed Case Study, Taiwan. *Water* **2021**, *13*, 2201. [CrossRef]
27. Chang, M. Flood Control in Small Forested Watershed. *J. Chin. Soil Water Conserv.* **1979**, *10*, 1–12.
28. Jian, Y. Taiwan River Flood Control in the Early Japanese Occupation and the Contribution of Civil Engineer Kataro Tokawa. *J. Chin. Inst. Civ. Hydraul. Eng.* **2019**, *46*, 38–44.
29. Ma, C. *Research on Taiwan's Water Control during the Japanese Occupation Period*; National Central University: Taoyuan, Taiwan, 2005.
30. Weng, Y.; Shen, S.; Chuang, Y. Flooding and Debris Flow Triggered by Typhoon Morakot and their Geomorphic Implications, Nanzisian Stream. *Bull. Geogr. Soc. China* **2010**, *45*, 59–74.
31. Wen, Z.; Huang, Q.; Ye, G. *Atlas of Taiwan's Postwar Census*; National University of Population and Taiwan Gender Research Center, Department of Geography, Environment and Resources, National Taiwan University: Taipei, Taiwan, 2012.
32. Lin, S. Underestimation of the Number of Streams of Potential Debbris Flow and Households for Preservation According to the General Survey of Slope Settlements in Taiwan. *J. Eng. Environ.* **2014**, *33*, 17–25.
33. Chang, Y.; Lin, T. Unequal Disaster: Hazard and Social Stratification in an Earthquake. *J. Soc. Sci. Philos.* **2012**, *24*, 193–231.
34. Kuan, D. Indigenous Ecological Knowledge and Contemporary Disaster Management a Case Study on the Tayal Communities' Experience in the Watershed of Shih-Men Reservoir. *J. Geogr. Sci.* **2015**, *76*, 97–132. [CrossRef]
35. Ting, C.; Kan, T. Preliminary Study on Combination of DRASTIC Model and the Sampling Frequency for Regional Groundwater Quality Monitoring. *J. Chin. Agric. Eng.* **2005**, *51*, 57–68.
36. Tsai, S. Water and Space: The Carrier Changes and Political Economic Governance of Donggang River. In Proceedings of the 20th International Geographical Conference of Taiwan, Taipei, Taiwan, 20–22 May 2016.
37. Tsai, S. Are Pingtung water shortage? An interrogate of water transfer in the space. In Proceedings of the 19th International Geographical Conference of Taiwan, Taipei, Taiwan, 23–24 May 2015.
38. Murdoch, J. *Post-Structuralist Geography-Aguide to Lelational Space*; SAGE Publications Ltd.: London, UK, 2006.
39. Deleuze, G.; Guattari, F. *Capitalisme et Schizophrenie 2. Mille Plateaux*; Shanghai Bookstore Press: Shanghai, China, 2010.
40. Stewart, D.W.; Kamins, M.A. *Secondary Research*; Hurng-Chih Book Co., Ltd.: Taipei, Taiwan, 2000.
41. Jick, T.D. Mixing Qualitative and Ouantitative Methods: Triangulation in Action. *Adm. Sci. Q.* **1979**, *24*, 602–611. [CrossRef]
42. Hu, Y. *Qualitative Research: Theory, Method and Examples of Indigenous Female Research*, 2nd ed.; Chuliu Book Co., Ltd.: Taipei, Taiwan, 2008.
43. Brody, H. Philosophic Approaches. In *Doing Qualitative Reasearch*; Miller, B.F.C.W.L., Ed.; Sage Publication: London, UK, 1992.
44. Shyu, J.H.; Sieh, K.; Chen, Y.; Lee, C.; Wang, Y.; Ota, Y.; Lee, J.; Sung, Q.; Rubin, C. A Preliminary Neotectonic Map of Taiwan, and its Implications for Future Destructive Earthquakes. In Proceedings of the AGU Fall Meeting Abstracts, San Francisco, CA, USA, 9–13 December 2019.
45. Shyu, J.B.H.; Chuang, Y.R.; Chen, Y.L.; Lee, Y.R.; Cheng, C.T. A new on-land seismogenic structure source database from the Taiwan earthquake model (TEM) project for seismic hazard analysis of Taiwan. *Terr. Atmos. Ocean. Sci.* **2016**, *27*, 311–323. [CrossRef]
46. He, C. *Introduction to Geology of Taiwan—Description of Geological Map of Taiwan*; Central Geological Survey; Ministry of Economic Affairs: Taipei, Taiwan, 1986.
47. Kaohaiung City Government. *Demographics*; Bureau, C.A., Ed.; Kaohaiung City Government: Kaohaiung City, Taiwan, 2022.
48. Chung, Y. *The Study of the Local Writings of Meinung Writers*; National Central University: Taoyuan, Taiwan, 2014.
49. Tsai, S. *Theory on Adaptive Mode and Creative Destruction under the Shift-Reset: A Case Study of Water Distribution in Pingtung Plain*; National Taiwan Normal University: Taipei, Taiwan, 2020.
50. Liao, S. Formation and Transition of Irrigated System. In *Mei Nong (1736–1976)*; National Cheng Kung University: Tainan, Taiwan, 2008.
51. Li, Y. *The Study of the Built Environment of Mei Nung Settlement, South Taiwan, during the Period of Late Ching Dynasty to Japanese Occupancy*; Chung Yuan Christian University: Taoyuan, Taiwan, 1989.
52. Liang, Z. *"Collection of Historical Materials of Taiwan Aborigines, Vol. 7" Mid-Range Memorial of the Ching Palace*; The Historical Research Commission of Taiwan Province: Nantou, Taiwan, 1998.
53. Chen, Q.; Wu, G. *The History of Ligang Township*; Ligang Township: Pingtung, Taiwan, 2005.
54. Taiwan Provincial Government. *Cause of Action: The Construction Department Signed a Case for the Application of the Pingtung County Government to Stop the Water Company's Handling of the Fourth and Fifth Deep Well Water Rights Cases in Ligang Township in the County*; Taiwan Provincial Government: Nantou, Taiwan, 1990.
55. China Times. The event of water cut off, the people of Ligung threatened to control the export of sand and gravel to Kaohsiung County. *China Times*, 20 February 1990.
56. China Times. Dissatisfied with the villagers. *China Times*, 10 February 1990.
57. Huang, J. *The Politics of "Bad Water"—The Crisis of Urban Water Governance and the Controversy over Public Drinking Water Quality in the Kaohsiung City*; National Taiwan University: Taipei, Taiwan, 2014.

58. Xiao, C. Discussion on the dispute over well closure and power shutdown in Ligang. *Taiwan Water Mon.* **1990**, *7*. Available online: https://blog.xuite.net/hsiaoch3/twblog/137538536 (accessed on 19 June 2020).
59. China Engineering Consultants. *Preliminary Assessment Report on Environmental Impact of Meinong Reservoir Project in Gaoping river Basin*; Water Resources Unified Planning Committee of the Ministry of Economic Affair: Taipei, Taiwan, 1984.
60. Chen, Y. *Study of Continue Monitoring to Fault Activity in the Mei-Nong Area*; National Cheng Kung University: Tainan, Taiwan, 2004.
61. Li, Y. The battle between water and butterfly- Discussion on Taiwan's environmental resources policy from Meinong reservoir. *Landsc. Archit.* **1993**, *13*, 84–85.
62. Hung, H. A community movement that started against the reservoir but never ends: Meinong's local training team—Meinong people's Association. *Cent. Hakka Stud. Newsl.* **2001**, *4*, 186–193.
63. Meinong people's Assoication. *Return to Meinong: Taiwan's First Anti-Reservoir Movement Documentary*; Morning Star Publishing Inc.: Taiching, Taiwan, 1994.
64. Pan, Z. After 20 Years of Competition for Water Resources, the Development of the Meinong Reservoir and the Gaoping Great Lake Has Resumed, and Residents Have Taken the Night Train to the North to Protest. Taichung: News&Market 2015. Available online: https://www.newsmarket.com.tw/blog/69782/ (accessed on 18 October 2018).
65. Su, J. *Management Strategy Analysis of Sustainable Utilization of Groundwater Resources—A Case Study of Pingdong Plain*; Sun Yat-Sen University: Kaohsiung, Taiwan, 2001.
66. Li, J.; Li, Z. Policy analysis on construction of Jiyang artificial lake. *Natl. Policy Forum* **2003**, *92*, 10.
67. Tsai, M. *Risk Discourses and Democratizing Science: The Study of the Controversies on the KaoPing Great Lakes Project*; National Yang-Ming University: Shinju, Taiwan, 2014.
68. SEC Inc. *Jiyang Artificial Lake Feasibility Planning III, Comprehensive Topic 2. Economic and Financial Analysis*; Water Conservancy Experiment Research Institute, Ministry of Economic Affairs: Taipei, Taiwan, 2013.
69. Water Resources Planning and Testing Institute of Water Resources Department of the Ministry of Economy. *Feasibility Planning of Jiyang Artificial Lake I, Special Project*; Water Resources Planning and Testing Institute of Water Resources: Taichung, Taiwan, 2000.
70. SEC Inc. *Feasibility Planning of Jiyang Artificial Lake III. Comprehensive Topic 3 Feasibility Supplementary Investigation Plan (1) Final Version of Environmental Impact Statement*; Water Conservancy Experiment Research Institute, Ministry of Economic Affairs: Taipei, Taiwan, 2000.
71. SEC Inc. *Feasibility Planning of Jiyang Artificial Lake III. Comprehensive Topic 3 Feasibility Supplementary Investigation Plan (1) Feasibility Review Plan*; Water resources planning and Testing Institute of Water Resources Ministry of Economic Affairs: Taipei, Taiwan, 2002.
72. Water Resources Bureau of Southern District, Ministry of Economic Affairs. *Gaoping Great Lake Project Environmental Current Situation Difference Analysis and Countermeasure Review Report and Change Content Comparison Table*; Water Resources Bureau of Southern District, Ministry of Economic Affairs: Tainan, Taiwan, 2010.
73. Yang, H. The "Jiyang Lake" that has been rejected is renamed Gaoping great lake. It is proposed again, and it be rejected again at the stage of environmental impact assessment. *Merit Times*, 14 February 2012. Available online: https://www.merit-times.com/NewsPage.aspx?unid=252580(accessed on 5 March 2019).
74. Lu, Y. Escorted by the Water Resources Department, Jiyang Artificial Lake was renamed Gaoping Great Lake. *Lihpao* **2012**. Available online: http://www.lihpao.com/?action-viewnews-itemid-115336 (accessed on 1 April 2022).
75. Water Resources Administration. *Minutes of the Local Briefing Meeting of "Gaoping Great Lakes Project" in Pingtung County*; Water Resources Bureau of Southern District: Tainan, Taiwan, 2012.
76. Norberg, J.; Cumming, G. *Complexity Theory for a Sustainable Future*; Columbia University Press: New York, NY, USA, 2008.
77. Berkes, F.; Colding, J.; Folke, C. *Navigating Social–Ecological Systems: Building Resilience for Complexity and Change*; Cambridge University Press: Cambridge, UK, 2003.
78. Berkes, F.; Folke, C. *Linking Social and Ecological Systems. Management Practices and Social Mechanisms for Building Resilience*; Cambridge University Press: Cambridge, UK, 1998.
79. Wilson, G. *Community Resilience and Environmental Transitions*; Routledge: Oxfordshire, UK, 2012.
80. Haque, C.E.; Etkin, D. *Disaster Risk and Vulnerability*; McGill-Queens University Press: Montreal, QC, Canada; Kingston, ON, Canada, 2012.
81. Chiang, Y.; Tsai, F.; Chang, H.; Chen, C.; Huang, Y. Adaptive society in a changing environment: Insight into the social resilience of a rural region of Taiwan. *Land Use Policy* **2014**, *36*, 510–521. [CrossRef]
82. Latai-Niusulu, A.; Binns, T.; Nel, E. Climate change and community resilience in Samoa. *Singap. J. Trop. Geogr.* **2020**, *41*, 40–60. [CrossRef]
83. Chang, C. Research on Environmental Awareness and Natural Disasters. *Bull. Geogr. Soc. China* **1977**, *5*, 57–60.
84. Chen, M. *A Study of Residents' Hazard Perception for Land Subsidence in the Kouhu Township, Yunlin*; National Chiayi University: Chiayi, Taiwan, 2013.
85. Lin, H. *Disaster and Environmental Risk Perception Research to He-Zuo Villager, Ren-Ai Township, Nantou County*; University of Taipei: Taipei, Taiwan, 2009.
86. Hung, C. *Using Regional Disaster System to Access Multiple Spatial Scale Disaster Risk: An Example of Hillside Environment, Taitung County, Taiwan*; National Taiwan Normal University: Taipei, Taiwan, 2013.
87. Chang, T. *Environmental Perception and Adjustment Behaviors for Flood Disasters: An Example from Shalu District in Taichung City*; Feng Chia University: Taichung, Taiwan, 2018.

88. Tsao, C.; Chang, C. The Residents' Experience and Behavior of the Earthquake Hazard: A Comparative Study in Paiho and Tungshih. *Hwa Kang Geogr. J.* **2008**, *21*, 52–75. [CrossRef]
89. Wu, M.; Su, S.B. Residents' Hazard Perception and Life Adjustment in Pu-Li after Chi-Chi Earthquake. *Environ. Worlds* **2003**, *8*, 55–84. [CrossRef]
90. Yeh, K. *Learning from Disaster for Resilience: A Study on Songhe Village*; National Taiwan Normal University: Taipei, Taiwan, 2017.
91. Lin, Y. *The Public's Adaptation Behavior Model on the Risk and Disaster of Typhoon and Flood*; National Taichung University of Education: Taichung, Taiwan, 2021.
92. Lin, W.; Liaw, S. Retailers' Hazard Perception and Adjustment Behaviors for Flooding Disasters Occurring in the Sijhih City. *J. Chin. Soil Water Conserv.* **2005**, *36*, 413–427. [CrossRef]
93. Wu, Z.P.; Wu, T.M.; Chuang, P.C. A Study on the Refuge Behavior in Remote Mountain Areas. *J. Crisis Manag.* **2012**, *9*, 1–14.
94. Liou, Y.; Wu, G.; Lin, P.S. Landslide Leads to Different Futures: A Comparison of Disaster Perception and Adaptation Behavior of Two Indigenous Atayal Tribal Communities. *J. Geogr. Res.* **2020**, 35–54.
95. Chang, S. *Evaluating Public Participation of Flood Management in Flood-Prone Areas:The Case of Meinong, Kaohsiung, Taiwan*; National Sun Yat-Sen University: Kaohsiung, Taiwan, 2013.
96. Song, Y. *Indigenious Ecological Knowledge Related to Typhoon Resilience: A Case Study of The Settlements along The Laonong River in Taoyuan, Kaohsiung*; National Taipei University: New Taipei City, Taiwan, 2020.
97. Hou, D. *Study on Regional Geography of Pingtung Plain*; Chinese Cultural Institute: Taipei, Taiwan, 1970.
98. Chen, S.; Chu, Y.; Wu, C. The Long Term Rainfall Trend and Landslide Distribution in Chishan River Watershed. *J. Chin. Soil Water Conserv.* **2012**, *43*, 293–301.
99. Pan, M. *The Study of Historical Transitions in Settlement for the North of Laonong River in Ligang Township in Ping-Tung County*; National Tainan University: Tainan, Taiwan, 2011.
100. Chen, G. The path and intensity of typhoon invading Taiwan. *Natl. Taiwan Norm. Univ. Geogr. Res. Rep.* **1981**, *7*, 61–74.
101. Shu, H. *Source Materials On The History Of Natural Disasters in Ching*; National Science Council: Taipei, Taiwan, 1983; pp. 35–105.
102. Tsai, S. The measures adopted to Relieve and prevent typhoon disasters in Taiwan during Japan-ruling period. *J. Taiwan Hist.* **2009**, *60*, 253–298.
103. Tan, P.; Tu, H.; Wu, J.; Meng, X.; Huang, W.; Chen, J. Construction of database of meteorological disaster events in 1895–1945 and application on disaster prevention. In Proceedings of the 2020 Symposium on Weather Analysis and Forecasting, Taipei, Taiwan, 13–15 October 2020.
104. Dai, Y. *A Study of Classical Poetry on Natural Disasters in Taiwan—Ching Dynasty to Japanese Occupation Period*; National Chengchi University: Taipei, Taiwan, 2002.
105. Chen, C.; Lin, C.; Chen, M.; Chang, W.; Liu, S. A Landslide Study by Using Multi-temporal Satellite Images: An Example from Lao-Lung River Watershed. *J. Taiwan Disaster Prev. Soc.* **2011**, *3*, 25–38.
106. Weng, M.; Huang, W.; Chen, R. Exploring Failure Mechanisms of Levee System During the Extreme Rainfall Event-Case Study in Laonong River Basin. *J. Eng. Environ.* **2013**, *31*, 17–36. [CrossRef]
107. Chen, T.; Hung, C. Simple One-Dimensional Dam Breach and Flood Route Models for Tributary-Dammed Lake: A Case Study of the Laonong River in Southern Taiwan. *J. Chin. Soil Water Conserv.* **2019**, *50*, 131–142. [CrossRef]
108. Lin, C.; Chen, C.; Fu, G. Deep-seated landslide area extraction in Laonong River Watershed using environmental index. *Water Conserv. Technol.* **2016**, *9*, 1–9.
109. Huang, P. *Analysis of Spatiotemporal Characteristics of Typhoon Rainfall for Laonong River Basin*; Tamkang University: New Taipei, Taiwan, 2014.
110. Chen, Z. *The Topography of Taiwan*; SMC Publishing Inc.: Taipei, Taiwan, 1997.
111. Tsai, C. *Effect of Ai-Liao Paleoflume on Groundwater Flow Regime within Ping-Tung Alluvial Plain*; National Cheng Kung University: Tainan, Taiwan, 2006.
112. PingtungCounty. Dear Kaohsiung Folks. *China Times*, 3 September 1990.
113. Institute ITR. *Report on the Construction Plan of the Distribution Map of Onshore Sand and Gravel Resources in Southern Taiwan (Project Scope: Chiayi, Tainan, Kaohsiung and Pingtung Areas)*; Ministry of Economy Mining Division: Taipei, Taiwan, 2002; *Unpublished*.
114. Hou, Q. <South> 1.4 billion port gravel truck lane, approved by the central government. In *Liberty Times*; Liberty Times: Pingtung, Taiwan, 2010.
115. Hwu, T. *Water Resources Policy in Taiwan—Case Study of Meei Nong Reservoir*; National Cheng Kung University: Tainan, Taiwan, 2007.
116. Chung, Y. *The Interaction between Meinung Anti-dam Movement and the Public Policy*; National Sun Yat-Sen University: Kaohsiung, Taiwan, 2003.
117. Chen, I. *Escaping Strategy or Renaissance—Observation on The Return to Farming Movement of Youth in Taiwan*; National Kaohsiung Normal University: Kaohsiung, Taiwan, 2017.
118. Liao, J. Slag Invades Agricultural Land and Poisons into the Water and Soil Conservation Area, But It Can't Be Managed. Environmental Information Cente. 2014. Available online: https://e-info.org.tw/node/99633 (accessed on 8 March 2022).

119. Alliance, T.W.R.C. The Quality of Drinking Water in Kaohsiung City is Poor, and the Waste Slag of Sinosteel in Chishang Dalin Has Polluted Groundwater (Tap Water) Sources Such As Chishang Stream, Sojingliao, and Meinong Pingtung. Mayor Han Guoyu Asked to Remove It. Kaohsiung: Coolloud. 2019. Available online: https://www.coolloud.org.tw/node/93553 (accessed on 28 November 2019).
120. Wang, B.; Lin, J.; Hou, Z. "Dian" Peiliu "Li"—People and Lands and Boundary Transfer in the Burmese Community of Meinong, Ligang. *Bull. Geogr. Soc. China* **2017**, *58*, 83–138. [CrossRef]
121. Huang, H. *Groundwater Vulnerability Study within the Plain Region of Kao-Hsiung District*; National Cheng Kung University: Tainan, Taiwan, 2007.
122. Tsai, S.; Lee, S.; Zou, Z.; Chu, T. The Competitions, Negotiations, and Collaborations of Regional Integration: A Perspective on Sustainable Management of Water Resources in Pingtung Plain, Taiwan. *Sustainability* **2022**, *14*, 40. [CrossRef]
123. Common, M.; Perrings, C. Toward an Ecological Economics of Sustainability. *Ecol. Econ.* **1992**, *6*, 7–34. [CrossRef]
124. Cheng, Y. *The Collective Action and the Reconstruction of the Locality—The Case of the Anti-Dam Movement in Meinong*; National Taiwan Normal University: Taipei, Taiwan, 2000.
125. Yau, S. *The Contruction Policy of Reservior and the Interation of Environmental Groups in Taiwan-Case Study on Meilong Reservior*; Chinese Cultural University: Taipei, Taiwan, 2000.
126. Ruiwen, C.; Qizheng, C. Brief introduction to the project of drawing water from the river in the upper reaches of Gaoping weir. *J. Prof. Geotech. Eng.* **2019**, *19*, 29–33.
127. Tian, Q.; Zhou, Q.; Cheng, G. Combined Utilization of Surface and Groundwater in Pingtung Plain. *Water Resour. Manag. Q.* **2001**, *3*, 18–32.
128. Nengbang Technology Consulting Co., Ltd. *Report on the Commissioned Technical Service Achievement Report on the Review of the Kaohsiung Area's Multiple Water Source Plans and Promotion of Strategic Analysis*; Water Conservancy Bureau of Kaohsiung Municipal Government: Kaohsiung, Taiwan, 2015.
129. Water Resources Administration; Ministry of Economic Affairs; Taiwan Water Corporation. *Zengwen Nanhua Wushantou Reservoir Regulation and Stable Water Supply Plan in Southern Region (The 3rd Revision of the Implementation Plan)*; Water Resources Administration, Ministry of Economic Affairs, Taiwan Water Corporation: Taipei, Taiwan, 2013.
130. Zhong, F.; Zhong, S. *Longzhuang Ancient Period Memory*; 1965; *Unpublished.*
131. Xiao, S. *The Formation and Development of a Hakka Settlement: A Case Study of Meinong Town, Kaohsiung County*; National Taiwan Normal University: Taipei, Taiwan, 2004.
132. Chen, H. *History of Water Conservancy in Taiwan*; Wu-Nan Book Inc.: Taipei, Taiwan, 2009.
133. Chen, H. *Water Conservancy Development and the Development of the Jianan Plain in Ching Dynasty*; Academia Historical: Taipei, Taiwan, 1996.
134. Chung, M. *The Research of Territoriality of Space of Irrigated System—A Study at the Chia-Nan Irrigated System Area*; Chung Yuan Christian University: Taoyuan, Taiwan, 2004.
135. News Market. Citizen Group Opinion of Kaohsiung Water Resources Forward Looking. Forum Unpublished. 2017. [updated 20190827]. Available online: https://www.newsmarket.com.tw/blog/99658/ (accessed on 19 June 2020).
136. Corporation, T.W. *The Chairman Visited the Mayor Lu of Ligang Township, Pingtung County to Communicate and Explain the Matter of the Re-Extraction Case of Ligangshenjing*; District Seven Management Office: Pingtung, Taiwan, 2006.
137. Corporation, T.W. *Minutes of the 37th General Meeting of Shareholders of Taiwan Water Co., Ltd.*; Taiwan Water Co., Ltd.: Taichung, Taiwan, 2011.
138. Chu, S. Kaohsiung deserves to be thirsty when the Great Lakes in Gaopinga are rejected? *Bus. Weeks* **2013**, *1326*. Available online: http://www.businessweekly.com.tw/KWebArticle.aspx?id=50038 (accessed on 1 April 2022).
139. Ministry of Economic Affairs. *Construction Plan for Disaster Prevention and Backup Wells (Amendment 1) (Approved Version)*; Ministry of Economic Affairs: Taipei, Taiwan, 2019.
140. Gyamfi, C.; Diabene, P.Y.; Odai, S.N.; Anornu, G.K.; Annor, F.O. Conflict prevention and resolution mechanisms in water resources management: A perspective from the Black Volta Basin—Ghana. *Int. J. Dev. Sustain.* **2013**, *2*, 1346–1356.
141. Huang, C. *Groundwater Recharge and Exploitative Potential Zone Mapping and Exploitable Yield in the Upstream of Kaoping River Basin*; National Cheng Kung University: Tainan, Taiwan, 2018.
142. Hung, H. "Return to Meinong" and its strategy—Community University as another social movement in Meinong. In Proceedings of the Hongkong Symposium, Annual Conference on International Social Development, Hong Kong Polytechnic University, Hongkong, China, 4–6 January 2007.
143. Hung, H. The dynamic generation of memory landscape as the preservation and community building of Hakzhuang: A discussion on the "Meinong Yellow Butterfly Festival" as an example. *Kaohsiung Hist.* **2012**, *2*, 6–31.
144. Lin, X. A Study on Agricultural Transformation in Taiwan's Rural Areas—Case Study of Meinong, Kaohsiung. *Cross-Strait Grassroots Elect. Local Gov. Res. Newsl.* **2016**, *3*, 32–58.
145. Kuo, T. From "Rain" to "After Rain"-Transformation of Agricultural Production and Rural Life in Meinong Literature. *J. Taiwan Hist.* **2021**, *11*, 83–118.
146. Chung, H.; CWang, h.; Chao, C. The Reconstruction of the Place Characteristics of Meinong in Responding to Glocalization. *Hwa Kang Geogr. J.* **2006**, *19*, 23–37. [CrossRef]

147. Lo, K. *Building an Organic Rice Co-Op under Contemporary Political-Economic Structure of Meinung Township*; Shih Hsin University: Taipei, Taiwan, 2008.
148. Chiu, T. *The Research of Management of Organic Farm in Traditional Hakka Settlement—The Study of Meinong Ren—Hue Organic Farm*; National Kaohsiung Normal University: Kaohsiung, Taiwan, 2011.
149. Chung, C. *Analysis of Business Model on Environmental-Friendly Agriculture in Meinong Dist., Kaohsiung City, Taiwan*; National Dong Hwa University: Hualian, Taiwan, 2018.
150. Peng, T. *An Analysis on the Appropriate Technological View of the Development of Organic Agriculture in Meinung District*; National Kaohsiung Normal University: Kaohsiung, Taiwan, 2010.
151. Lin, Y. Research and Analysis on the promotion mechanism and benefits of agricultural operation zone~Taking the business model and operation mechanism of Meinong agricultural operation zone as an example. *Kaohsiung Dist. Agric. Res. Ext. Stn.* **2019**, *107*, 59–60.
152. Lin, Y. Research on the Operation Mechanism and Operational Benefit of Large-scale Special Zones—Taking Meinong Agricultural Management Zone as an Example. *Kaohsiung Dist. Agric. Res. Ext. Stn.* **2018**, *106*, 79–82.

water

MDPI

Article

Evaluating the Landslide Stability and Vegetation Recovery: Case Studies in the Tsengwen Reservoir Watershed in Taiwan

Chun-Hung Wu

Department of Water Resources Engineering and Conservation, Feng Chia University, Taichung 40724, Taiwan; chhuwu@fcu.edu.tw; Tel.: +886-424-517-250-3223

Abstract: The sediment yield from numerous landslides triggered in Taiwan's mountainous regions by 2009 Typhoon Morakot have had substantial long-term impacts on the evolution of rivers. This study evaluated the long-term evolution of landslides induced by 2001 Typhoon Nari and 2009 Typhoon Morakot in the Tsengwen Reservoir Watershed by using multiannual landslide inventories and rainfall records for the 2001–2017 period. The landslide activity, vegetation recovery time, and the landslide spatiotemporal hotspot analyses were used in the study. Severe landslides most commonly occurred on 35–45° slopes at elevations of 1400–2000 m located within 500 m of the rivers. The average vegetation recovery time was 2.29 years, and landslides with vegetation recovery times exceeding 10 years were most frequently retrogressive landslide, riverbank landslides in sinuous reaches, and the core area of large landslides. The annual landslide area decline ratios after 2009 Typhoon Morakot in Southern Taiwan was 4.75% to 7.45%, and the time of landslide recovery in the Tsengwen reservoir watershed was predicted to be 28.48 years. Oscillating hotspots and coldspots occupied 95.8% of spatiotemporal patterns in the watershed area. The results indicate that landslides moved from hillslopes to rivers in the 2001–2017 period because the enormous amount of sediment deposited in rivers resulted in the change of river geomorphology and the riverbank landslides.

Keywords: landslide evolution; landslide activity; vegetation recovery time; spatiotemporal hotspot

Citation: Wu, C.-H. Evaluating the Landslide Stability and Vegetation Recovery: Case Studies in the Tsengwen Reservoir Watershed in Taiwan. *Water* **2021**, *13*, 3479. https://doi.org/10.3390/w13243479

Academic Editor: Monica Papini

Received: 22 October 2021
Accepted: 2 December 2021
Published: 7 December 2021

1. Introduction

Several serious landslide disaster events were caused by heavy rainfall events between 2001 and 2010, including 2001 Typhoon Toraji [1], 2004 Typhoon Mindulle [2], 2004 Typhoon Aere, and 2009 Typhoon Morakot [3], in the mountainous areas of central and southern Taiwan. The landslide events are believed to be related to the 1999 Chichi earthquake [4]. The return period of heavy rainfall brought by Typhoon Morakot (5–10 August 2009), which caused approximately 45,000 landslides concentrated in central and southern Taiwan, was estimated to be over 200 years [3]. Moreover, loose deposits from the numerous landslides in mountainous areas continues to affect watershed evolution and landslide recovery. As of 2021, mountainous areas in southern Taiwan remain under high risk of landslides and debris flows.

The analysis of landslide recovery or landslide evolution had been widely used in the long-term observation of landslide distribution after large earthquake events, including the 1999 Chichi earthquake [5], 2005 Kashimir earthquake [6], 2008 Wenchuan earthquake [7,8], and 2015 Gorkha earthquake [9]. Landslide evolution in the years following a large earthquake or extreme rainfall event in a watershed with dense landslide cases is the key determinant of watershed management and the mitigation of secondary geohazards [10]. Some artificial factors, including land use [11,12] and road development [13], were the significant factors for the evolution and reoccurrence of landslide. Some studies have been conducted on the characteristics of landslide evolution, including landslide activity [6,7], the spatiotemporal distributions of landslides [8,14], landslide recovery characteristics [15], and landslide evolution trends [10,16], with regard to severe landslides induced by large earthquakes or extreme rainfall. However, few studies have explored landslide evolution by examining spatiotemporal hotspots. Lin et al. (2017) [17] explained the distribution of landslide hotspots and coldspots on a catchment scale by using multiannual records of landslides, rainfall, and earthquakes for the 2003–2012 period in Taiwan. Conventional hotspot analysis can only describe the clustering pattern and intensity of a particular spatial location for a specific time interval; it cannot explain underlying trends over time [18]. The spatiotemporal evolution of landslide distribution patterns is challenging to investigate but merits scholarly attention.

New analysis methods by using the multi landslide inventories, including spatiotemporal hotspot and landslide activity, had been used to explore the characteristic of landslide evolution after large earthquake or extreme rainfall events [6,7,19]. The emerging hot spot analysis in ArcGIS Pro software has been employed in tracing the spread of COVID-19 [20], locating the source area of pollutant emissions [16], and assessing spatiotemporal changes in fisheries [21]. Emerging hotspot analysis was used in an assessment of long-term landslide evolution in Taiwan after 2009 Typhoon Morakot [19]. The results explained the spatiotemporal pattern and distribution of landslide hotspots and coldspots. Landslide-concentrated areas in the upper reaches of watersheds were determined to have poor landslide recovery and be highly susceptible to further landslides [19]. Emerging hotspot analysis can both explain changes in the spatiotemporal pattern and location of landslide hot spots and evaluate the intensity and location of landslide clustering. Landslide activity data have been employed in assessing the rates of and differences in landslide recovery after large earthquake-induced landslides [6,7] but not after extreme rainfall-induced landslide events.

The present study analyzed characteristics of landslide evolution by assessing landslide activity, estimating vegetation recovery time, and detecting the pattern and spatiotemporal distribution of landslides by using multiannual landslide inventories and long-term rainfall records from 2001 to 2017. Moreover, the characteristics of landslide evolution after large earthquakes were compared with the corresponding characteristics after extreme rainfall events. The Tsengwen Reservoir Watershed (*TRW*) suffered the most serious landslide disasters of all reservoir watersheds in Taiwan following 2009 Typhoon Morakot. The landslide evolution characteristics in the watershed with dense rainfall-induced landslides are poorly understood, but these characteristics represent information essential for watershed management and disaster prevention.

2. Research Area

The *TRW* (Figure 1) is a watershed located in the upper reaches of the Tsengwen River in southwestern Taiwan. The average elevation in the *TRW* is 959 m, and the average slope is 29.2°; 49.5% of the total area has a slope of greater than 30°. According to 1/5000 basin geological maps of Taiwan [22], the main stratigraphical formations in the *TRW* (Figure 2 and Table 1) include the Miocene-era Changchihkeng formation, Pliocene-era Ailiaochiao formation, and Miocene- to Pliocene-era Tangenshan sandstone (occupying 34.2%, 16.6%, and 16.4% of the watershed, respectively) [17]. The *TRW* is also surrounded by nine faults (Figure 2). Regarding land use types, forest, agriculture, development, rivers, and bare land account for 80.6%, 10.4%, 1.4%, 4.3%, and 3.3% of the total area based on the land use maps produced in 2008 by National Land Surveying and Mapping Center in Taiwan.

The average (standard deviation) annual precipitation from 2001 to 2017, based on the records of Matoushan Rainfall Station (Figure 1), was 2998.8 (934) mm. Rainfall characteristics in the *TRW* are non-uniform in space and time. The 3 years with the most precipitation were 2005 (4892 mm), 2008 (4118 mm), and 2001 (3764 mm). Those with the least precipitation were 2003 (1533 mm), 2013 (1735 mm), and 2002 (1878 mm; Figure 3). The difference between the precipitation in 2003 and that in 2005 is 1.12 times the average annual precipitation. The total precipitation in the rainy season (May–October; 2749 mm) was 91.7% of the average annual precipitation (2998.2 mm). The precipitation in August 2009, the largest monthly total in the considered period, was 2425 mm—70.2% of the annual precipitation.

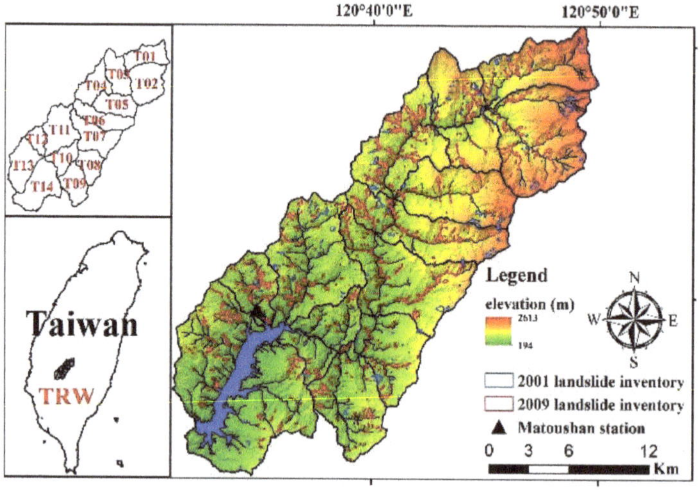

Figure 1. The distribution of elevation, rivers, and landslide after 2001 Typhoon Nari and 2009 Typhoon Morakot in the Tsengwen Reservoir watershed (Abbreviated as *TRW*). The down-left figure shows the location of *TRW* in Taiwan, and the up-left figure shows the names of sub-watersheds in the *TRW*.

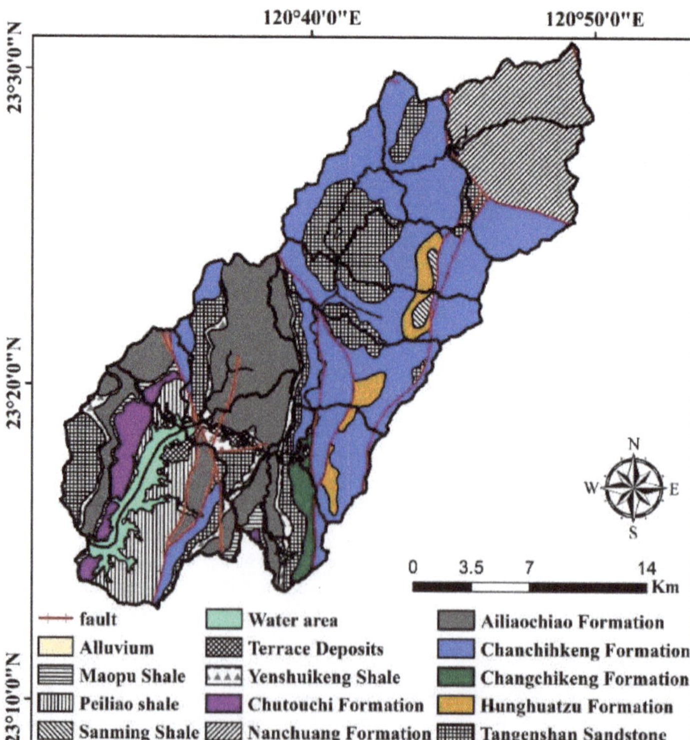

Figure 2. The distribution of geological settings and faults in the *TRW*.

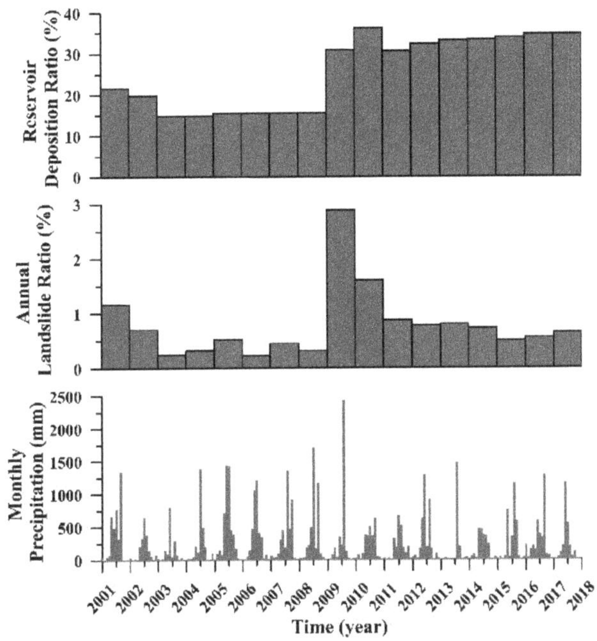

Figure 3. The distribution of monthly precipitation, landslide ratio, and reservoir deposition ratio from 2001 to 2017 in the *TRW*. The up, middle, and down figures are the distribution of reservoir deposition ratio, annual landslide ratio, and monthly precipitation from 2001 to 2017 in the *TRW*.

Landslide and debris flow were the two main geohazards in the *TRW*, and the 1999 Chichi earthquake and heavy rainfall events during the 2000s were the main landslide triggers. Typhoon Nari (9–10 September 2001) and Typhoon Morakot (5–10 August 2009) caused the two main rainfall-induced landslide events during the 2000s. In the *TRW*, the accumulated precipitation brought by 2001 Typhoon Nari and 2009 Typhoon Morakot was 990.5 and 2158.5 mm (62.5% of the annual precipitation in 2009), respectively. The landslide ratios (i.e., the ratio of landslide area to watershed area) after 2001 Typhoon Nari and 2009 Typhoon Morakot were estimated to be 1.17% and 2.88%, respectively. Events with a landslide ratio exceeding 1.0% are considered serious disaster events in Taiwan [3]; thus, these two typhoons are regarded as serious disaster events within the *TRW*. The sediment deposition volume in the Tsengwen Reservoir increased by approximately 1.28×10^8 m^3 from 2001 to 2017 (Figure 3). According to the field survey results, the majority of landslides can be classified as shallow landslide [23].

Table 1. Statistical data of landslide induced by 2001 Typhoon Nari and 2009 Typhoon Morakot in each stratigraphical division in the *TRW*.

Stratigraphical Division	Lithology	Occ. Per. (%)	Landslide Ratio (%)	
			In 2001	In 2009
Water Area	-	3.07	0.00	0.27
Alluvium	Gravel, sand, silt, and clay	0.70	0.80	1.96
Ailiaochiao Formation	Thin alternation of siltstone and shale	16.65	0.80	2.52
Changchihkeng Formation	Sandstone, sandstone interbedded with shale	34.16	1.37	3.48
Chutouchi Formation	Muddy sandstone, interbedded muddy sandstone and shale	2.76	0.36	5.43
Hunghuatzu Formation	Thick-bedded fine sandstone or siltstone, thick alternation of fine sandstone and siltstone	2.48	0.74	4.13

Table 1. *Cont.*

Stratigraphical Division	Lithology	Occ. Per. (%)	Landslide Ratio (%)	
			In 2001	In 2009
Maopu Shale	Shale with thin-bedded sandstone	0.99	0.17	5.69
Nanchuang Formation	Alternation of sandstone and shale	12.44	1.93	2.91
Peiliao Shale	Shale and sandy shale	5.58	0.34	1.00
Sanming Shale	Shale, shale intercalated with thin-bedded siltstone	0.49	4.72	5.30
Terrace Deposits	Sand, silt, mud and gravel	1.19	0.02	0.00
Tangenshan Sandstone	Thick-bedded massive sandstone and muddy sandstone	16.39	1.45	2.65
Yenshuikeng Shale	Massive shale, occasionally intercalated with thin-bedded siltstone	3.11	0.44	2.13

Note: The Occ. Per. means the occupied percentage of each stratigraphical division area to the watershed area.

3. Materials and Method

3.1. Materials

A digital elevation model (DEM) and multiannual landslide inventories were the main study materials. The DEM had a spatial resolution of 5 m, and the 5 m × 5 m grid was the basic analysis unit in the study. The landslide inventories, corresponding to the 2001–2017 period, were produced by Taiwan's Forestry Bureau, and landslides were identified from images captured by the Formosat-2 satellite with spatial resolutions of 2–5 m. Multiannual inventories of landslides in the *TRW* (for the years 2001–2017) were examined, and the minimum landslide area in the inventories was 100 m^2. Landslides with an area of greater than 10 hectares (ha) were considered large landslides. The main types of landslide [24] included in the annual landslide inventories from 2001 to 2017 in Taiwan were fall, slide, and flow [19]. The deep-seated landslide cases, similar to the deep-seated gravitational slope deformation in central Taiwan [25], were not listed in the annual landslide inventories. Data on daily rainfall from 2001 to 2017 were collected from Matoushan Rainfall Station (Figure 1). The *TRW* contains 14 subwatersheds (Figure 1). The subwatersheds in the upper, middle, and lower reaches of the river are labeled as T01–T03, T04–T08, and T09–T14, respectively.

A recovery characteristics comparison of landslides induced by large earthquake and extreme rainfall events was conducted to explore the characteristic of landslide evolution. The large earthquake-induced landslides considered were the landslide cases after the 2005 Kashmir Earthquake (Mw = 7.6) in Pakistan [9] and the 2008 Wenchuan Earthquake (Mw = 7.9) in China [7,8,16]. The extreme rainfall-induced landslides used in the study were landslides occurring in the Chishan River Watershed [10], Ailiao River Watershed [19], and Taimali River Watershed [19] after 2009 Typhoon Morakot, and the *TRW* before and after 2009 Typhoon Morakot.

3.2. Landslide Activity

Landslide activity, which is useful in analyzing the spatiotemporal evolution of landslides, has been assessed by using the presence or absence of landslides in specific years [7,8] or by examining the ratio of the active landslide area to the total landslide area [26]. Following the landslide activity assessment methods used in the evolution discussion of earthquake-induced landslide cases [6,14,27], the criteria for the five types of landslide activity considered are presented in Table 2. The meaning of landslide activity used in a single hillslope scale was the current moving condition of the hillslope [25], but that used in a watershed scale was the vegetation recovery and the stability of loose materials deposited on the hillslopes [6,14,27]. The meanings of extremely active, very active, and active landslide were that the loose material deposited on the hillslope was instability or the vegetation recovery was too poor to stabilize the hillslope. The inventories of landslides in the *TRW* (2001–2017) were appropriate materials for assessing and comparing the activity of landslides induced by 2001 Typhoon Nari and 2009 Typhoon Morakot. The landslide activity induced by the typhoons was evaluated using multiannual landslide inventories for the 2001–2008 and 2009–2017 periods.

Table 2. The criteria of landslide activity in the study.

Activity Type	Criteria
Extremely Active	Landslides present in the annual inventories from 2001 to 2008 or from 2009 to 2017.
Very Active	Landslides present in the annual inventories from 2005 to 2008 or from 2013 to 2017.
Active	Landslides present in the annual inventories of 2007 to 2008, or only in the annual inventory of 2008 for 2001 Typhoon Nari. Landslides present in the annual inventories of 2016 to 2017, or only in the annual inventory of 2017 for 2009 Typhoon Morakot.
Dormant	Landslides present in one or more annual inventories of 2001 to 2007 or 2009 to 2016 but absent in the annual inventory of 2008 or 2017.
Inactive	Landslides present in the annual inventory of 2001 or 2009 but absent in the annual inventories from 2002 to 2008 or 2010 to 2017.

3.3. Landslide Frequency and Vegetation Recovery Time

Landslide frequency in the study was defined as the total occurrence of landslide identified in each grid from 2001 to 2017. The area that had been identified as landslide at least once from 2001 to 2017 in the study had been named as the landslide-identified area. The total occurrence of landslide-identified from 2001 to 2017 can be considered as the maximum vegetation recovery time (in years). The landslide recovery time had been estimated by using different factors, including landslide density [28], landslide rate [29], number of landslides [30], and landslide area [6]. Vegetation recovery is a significant factor to assess the landslide stability in long-term landslide evolution analysis [31] because the reinforcement of vegetation root contributes to the hillslope stability, especially in the shallow landslide cases [32,33]. The total occurrence of identified landslide of a grid was three, which meant that the grid was identified as landslides three times over the 17-year period. If the three landslide-identified years were discontinuous (e.g., 2005, 2008, and 2012), the vegetation recovery time was considered as 1 year. However, if the three landslide-identified years were continuous (e.g., 2002–2004), the vegetation recovery time was considered as 3 years. Herein, the maximum recovery time was taken as the vegetation recovery time in the study. Vegetation recovery time can be employed as an index to assess the difficulty of vegetation recovery in a watershed and in individual landslide cases.

3.4. Emerging Hotspot Analysis

The analysis of landslide spatiotemporal hotspots was based on the landslide spatiotemporal cube (*STC*) model, which represented landslide clustering patterns in each location at various time intervals. The STC model was composed of numerous 5 m × 5 m landslides-identified grids in the TRW from 2001 to 2017 in chronological order. The time series of landslide clustering intensity, represented by the combination of basic units at the same location, described the temporal evolution of landslides.

The emerging spatiotemporal mining method in ArcGIS Pro software was used in the *STC* model to explore the mechanism of temporal landslide evolution in the *TRW*. The Getis-Ord Gi statistic [34] was used to estimate the clustering intensity and to classify hot spot patterns (Table 3). The hotspot classification was based on the landslide clustering intensity in the neighborhood of specific 5 m × 5 m grids in both time and space, and specific basic units were designated as hotspots or coldspots if they featured high and low values of landslide clustering, respectively. Under the emerging spatiotemporal mining method, the consistency and intensity of landslide clustering in each time step was calculated, as was the significance of the autocorrelation and dependence in the vicinity of specific 5 m × 5 m grids. The method also used a space-time implementation of the Mann-Kendall statistic [35] to measure the intensity of landslide clustering in the neighborhood of specific basic units. The time step was set as 1 year, and the neighborhood distance was set as 5, 25, 50, 100, and 200 m for further comparison in the study.

Table 3. The classifications and definition of emerging landslide hotspots and coldspots in the study.

Pattern	Definition
Consecutive (CHS or CCS) *	A landslide grid with a single uninterrupted run of statistically significant hotspot or coldspot grids in the final year during the research time period. The landslide grid has never been a statistically significant hotspot or coldspot before the final hotspot or coldspot run.
Diminishing (DHS or DCS)	A landslide grid that has been a statistically significant hotspot or coldspot for 90% of the research time period, including the final year. In addition, the clustering intensity of landslides in each year is decreasing (increasing) overall, and that decrease (increase) is statistically significant.
Historical (HHS or HCS)	The most recent year is not hotspot or coldspot, but at least 90% of the research time period has been a statistically significant hotspot or coldspot.
Intensifying (IHS or ICS)	A landslide grid that has been a statistically significant hotspot or coldspot for 90% of the research time period. In addition, the clustering intensity of landslide for each year increased (decreased) overall, and that increase (decrease) was statistically significant.
New (NHS or NCS)	A landslide grid identified as a statistically significant hotspot or coldspot since the first year of the research time period but was not previously identified as a statistically significant hotspot or coldspot.
Oscillating (OHS or OCS)	A statistically significant hotspot or coldspot for the final year that has a history of also being a statistically significant coldspot or hotspot during a prior year. Less than 90% of the research time period has been a statistically significant hotspot or coldspot.
Persistent (PHS or PCS)	A landslide grid that has been a statistically significant hotspot or coldspot for 90% of the research time period with no discernible trend indicating an increase or decrease in the clustering intensity of landslide over time.
Sporadic (SHS or SCS)	A landslide grid that is an on-again then off-again hotspot or coldspot. Less than 90% of the research time period has been a statistically significant hotspot or coldspot, and none of the time-step intervals have been a statistically significant coldspot or hotspot.
No pattern detected (No)	The analysis grid does not fit any definition of hotspot or coldspot classifications

*: CHS and CCS are the abbreviations of consecutive hotspot and consecutive coldspot. The regulation of abbreviation was applied to every hotspot and coldspot pattern in the study.

4. Results

4.1. Multiannual Rainfall and Landslide Data

The Matoushan Rainfall Station (Figure 1) was used as the representative rainfall station in the *TRW*. The average annual rainfall recorded by the station from 2001 to 2017 (2998 mm) is 1.06 times that of the corresponding level from 1969 to 2020 (2835 mm) [36]. Table 4 presents data on daily and annual rainfall and on annual landslides in the *TRW* from 2001 to 2017. The annual rainfall distribution in the *TRW* was clearly non-uniform in time. The difference between the lowest annual rainfall (1533 mm in 2003) and the highest annual rainfall (4892 mm in 2005) was 3359 mm, which is 1.12 times the average annual rainfall. The daily rainfall corresponding to return periods of 1.11, 2, 5, 10, 20, 50, 100, and 200 years, based on station records from 1969 to 2020, were estimated to be 144.5, 300.6, 447.9, 545.4, 639.0, 760.1, 850.9, and 941.3 mm, respectively [33]. The maximum daily rainfall from 2001 to 2008 of 545.5 mm was observed twice, once during 2001 Typhoon Nari and once during 2005 Typhoon Haitang, and the return period was estimated to be 10–20 years. The maximum daily rainfall during Typhoon Morakot of 1206.0 mm was the highest amount of rainfall recorded at Matoushan Rainfall Station. The return period of daily rainfall from 2010 to 2017 was <10 years, and the maximum daily rainfall of 502.5 mm was recorded during 2015 Typhoon Soudelor.

The landslide area in 2001 (565.2 ha) was the highest for 2001–2008, and that in 2009 (1392.9 ha) marked the historical high recorded in the *TRW*. The average annual landslide area corresponding to the 2009–2017 period was 2.25 times that of the 2001–2008 period. The annual average number of landslides from 2009 to 2017 was 2.29 times larger than that from 2001 to 2008.

Southern Taiwan suffered serious landslide disaster during 2009 Typhoon Morakot, and the *TRW* was located in this region. The number of landslides in 2009 was 5.3 times larger than that in 2008. The number of the landslide cases with area greater than 100,000 m^2, 1000–100,000 m^2, and <1000 m^2 in 2009 were 29, 1114, and 7, respectively. The average landslide length/width ratio in all landslide cases induced by 2009 Typhoon Morakot was 7.2, and this information showed that the majority type of landslide was rainfall-triggered slide.

Table 4. The statistical data of annual rainfall and landslide from 2001 to 2017 in the *TRW*.

Year	Annual Rainfall (mm)	Daily Rainfall (mm)	Return Period (Years)	Landslide Area (ha)	Landslide Number
2001	3764	545.5	10 to 20	565.2	324
2002	1878	96.5	<1.11	375.3	241
2003	1533	233.0	1.11 to 2	117.4	194
2004	2593	502.5	5 to 10	154.4	294
2005	4892	545.5	10 to 20	252.1	375
2006	3774	436.5	5 to 10	108.0	169
2007	3917	267.0	1.11 to 2	216.0	255
2008	4112	501.5	5 to 10	153.5	217
2009	3454	1206.0	>200	1392.9	1150
2010	2492	379.5	2 to 5	775.6	1091
2011	2225	241.5	1.11 to 2	422.4	434
2012	3569	433.0	2 to 5	375.5	513
2013	1736	381.0	2 to 5	388.3	565
2014	2077	172.5	1.11 to 2	352.0	461
2015	3050	502.5	5 to 10	240.8	345
2016	3478	491.5	5 to 10	270.4	407
2017	2428	334.0	2 to 5	312.6	425

Note: The daily rainfall meant the maximum daily rainfall in the specific year, and the return period was based on the daily rainfall.

4.2. Landslide Activity

Landslide activity data provide insights into the stability of rainfall-induced landslides for the analysis of the spatial distribution of landslide evolution. Figure 4 presents the spatial distribution of landslide activity after 2001 Typhoon Nari and 2009 Typhoon Morakot in the *TRW*. Areas of extremely active, very active, active, dormant, and inactive landslides after 2001 Typhoon Nari constituted 0.04, 40.8, 277.9, 939.5, and 561.4 ha, respectively. The corresponding areas after 2009 Typhoon Morakot were 54.0, 82.0, 424.7, 1577.0, and 762.2 ha, respectively. The accumulated area of extremely active, very active, and active landslides after 2009 Typhoon Morakot was 1.76 times that after 2001 Typhoon Nari, indicating that the landslides induced by Typhoon Morakot were more difficult to recover. The extremely active landslide area in subwatersheds in the upper reaches of the *TRW* increased substantially from 0 ha after 2001 Typhoon Nari to 42 ha after 2009 Typhoon Morakot, reflecting the difficulty of landslide recovery in the upstream subwatersheds.

A comparison of landslide activity after the 2005 Kashmir Earthquake [6] and 2009 Typhoon Morakot revealed that the proportions of unstable (extremely active, very active, or active) landslides and stable landslides (dormant or inactive landslides) were 28% and 72% in 2018 (13 years after the 2005 Kashmir earthquake). However, the corresponding proportions 8 to 9 years after typhoons in the TRW were 17.5–19.3% and 80.7–82.5%, respectively. These results indicate that vegetation recovery of landslides induced by 2009 Typhoon Morakot in the TRW was easier than vegetation recovery of landslides induced by 2005 Kashimir earthquake in Pakistan [6].

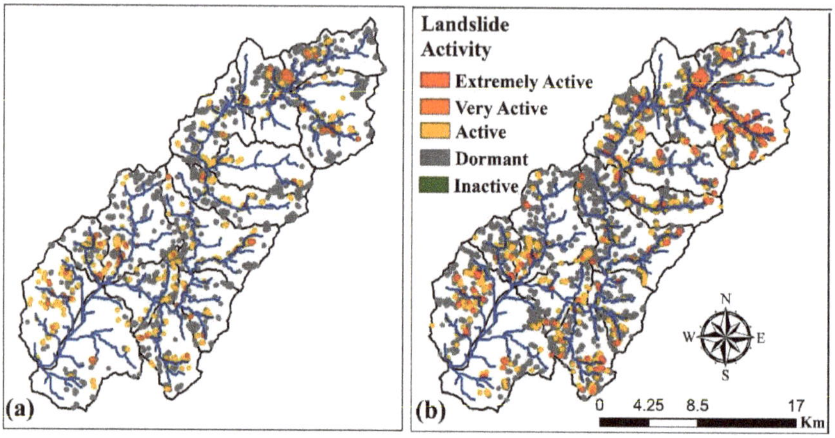

Figure 4. The distribution of landslide activity in the *TRW*. (**a**) is the distribution of landslide activity after 2001 Typhoon Nari and (**b**) is after 2009 Typhoon Morakot.

4.3. Relationship between Landslide Ratio and Geomorphological Factors

The relationship between earthquake-induced landslide area and geomorphological control factors, including elevation, slope, and distance to the river, had been discussed in analyses of the evolution of earthquake-induced landslides [6–8,18]. The temporal distribution of landslide ratios with respect to elevation, slope, and distance to the river from 2001 to 2017 in the *TRW* is shown in Figure 5. The landslide ratio distribution for the 2001–2007 period was used to observe the evolution of landslides induced by 2001 Typhoon Nari, and that for the 2009–2017 period was used to examine the evolution of landslides induced by 2009 Typhoon Morakot.

The landslide ratios were concentrated at 1800 to 2200 m elevation intervals after 2001 Typhoon Nari. However, from 2003 to 2007, they were concentrated at 800–1400 m elevation intervals. The landslide ratios at elevations below 1000 m and above 1600 m decreased notably during this period. After 2009 Typhoon Morakot, the landslide ratios were concentrated at 600 m to 1400 m elevation intervals. From 2010 to 2017, they were concentrated at 1200 m to 1800 m elevation intervals. The landslide ratios at elevations below 1200 m decreased notably, and those at elevations above 1800 m exhibited small changes from 2010 to 2017.

Only the distribution of extremely active landslides after 2009 Typhoon Morakot is plotted herein (Figure 5) because the area of extremely active landslides after 2001 Typhoon Nari was quite small (0.04 ha). Extremely active landslides were clustered at 1400 to 2000 m elevation intervals after Typhoon Morakot. The extremely active or very active landslides were clustered at moderately high elevations after large earthquakes. For example, they were clustered from 1000 to 1600 m after the 2005 Kashmir Earthquake [6] and below 2000 m after the 2008 Wenchuan Earthquake [7]. Overall, the clustered area and locations of extremely active or very active landslides after large earthquakes or extreme rainfall events was very similar.

The landslide ratios were concentrated at two slope intervals, 35° to 45° and 55° to 60°, after 2001 Typhoon Nari and 2009 Typhoon Morakot. The landslide ratios were clustered at the same intervals from 2003 to 2007 and from 2010 to 2017. The extremely active landslides were concentrated at 35–45° slope intervals after 2009 Typhoon Morakot. Extremely active or very active landslides were also clustered at 30–40° after the 2005 Kashmir Earthquake [6] and at 30° to 50° after the 2008 Wenchuan Earthquake [7].

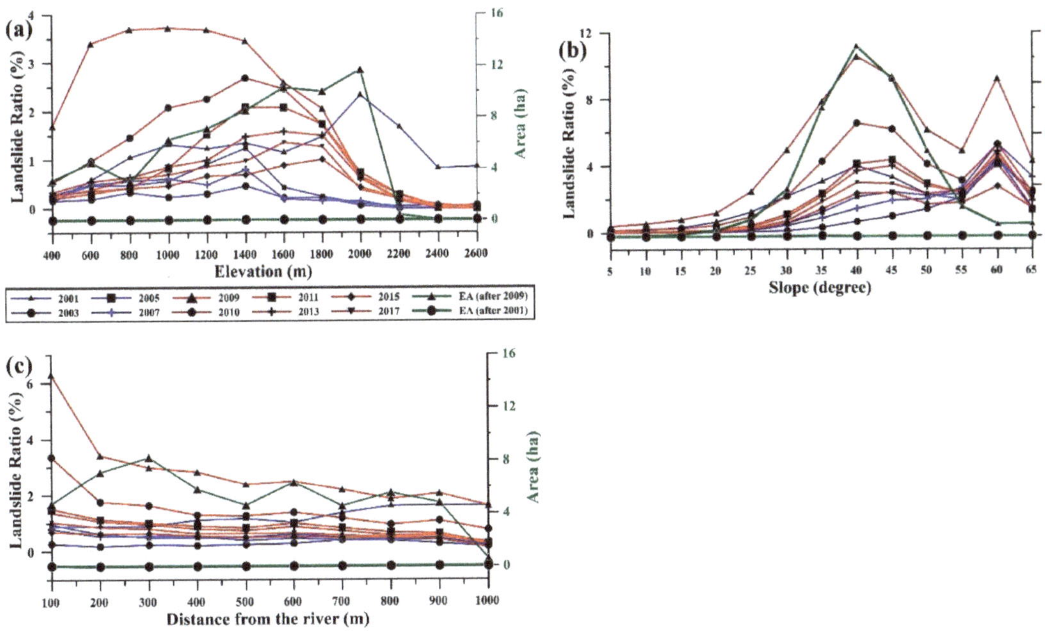

Figure 5. The relationship between landslide ratios and geomorphological factors, including elevation in (**a**), slope in (**b**), and distance to the rivers in (**c**), from 2001 to 2017 in the *TRW*. The EA line (green) means the extremely active landslide after 2009 Typhoon Morakot.

In Taiwan, loose debris from landslides after heavy rainfall events was generally deposited in rivers and gullies [10,15]. The *TRW* was divided into 10 buffer areas with 100 m distance intervals from the rivers. The landslide ratios in all river buffer intervals after 2001 Typhoon Nari ranged from 0.88% to 1.63%, but those after 2009 Typhoon Morakot were all greater than 1.86%. The highest landslide ratio in all river buffer intervals after 2001 Typhoon Nari was 1.63% within 1000 m of the river, and that after 2009 Typhoon Morakot was 6.3% within 100 m of the river.

Extremely active landslides were clustered along the fault line after the 2005 Kashmir Earthquake [6] but were concentrated along rivers in the *TRW* after 2009 Typhoon Morakot. This indicates that large pieces of loose material from landslides were transported by excess surface runoff and flood discharge and deposited in the rivers, thereby inducing riverbank landslides and change of river geomorphology.

4.4. Landslide Frequency and Vegetation Recovery Time

The distribution of landslide frequency from 2001 to 2017 in the *TRW* is shown in Figure 6. The total landslide-identified area was 26.4 km^2 (5.5% of the total area), and the average landslide frequency was 2.29. The area with landslide frequency = 1 constituted 16.28 km^2, which was 61.8% of the landslide-identified area. The area with landslide frequency greater than 10 constituted 0.394 km^2. Those landslides were most commonly retrogressive landslide or riverbank landslides in the sinuous reaches and the core area of the large landslide cases.

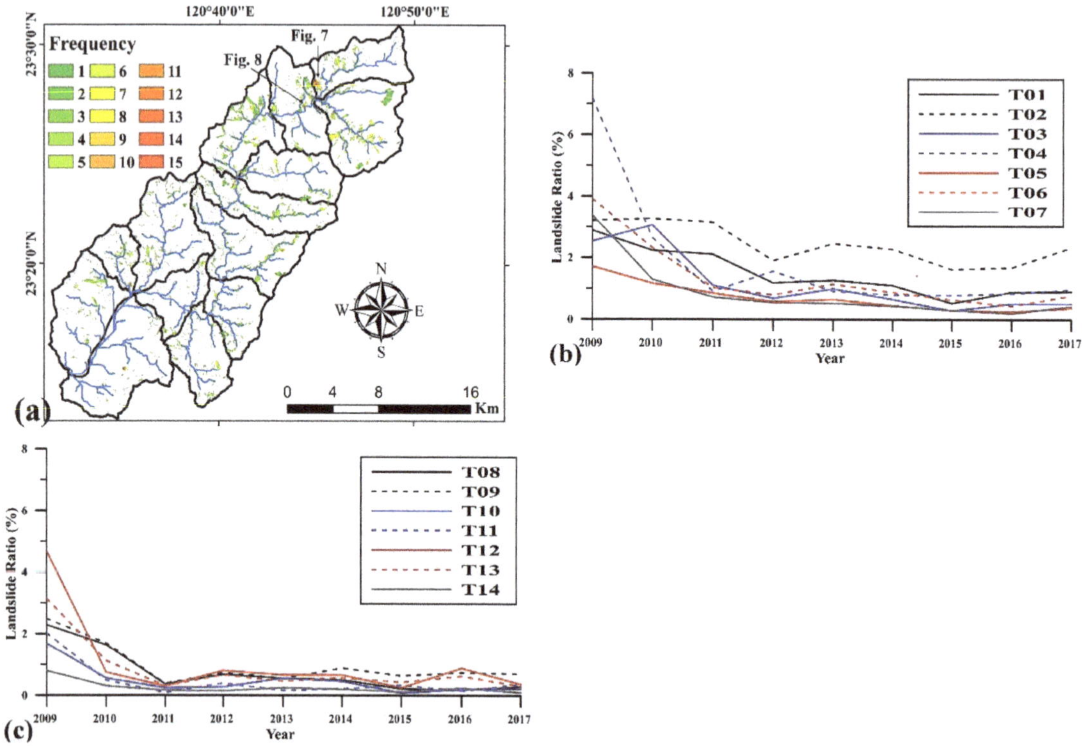

Figure 6. The landslide frequency from 2001 to 2017 (**a**) and the landslide ratio in each subwatershed from 2009 to 2017 (**b**,**c**) in the *TRW*.

The landslide frequency in the subwatersheds in the upper reaches of the *TRW* (2.28–3.73) differed substantially from the distributions in the subwatersheds in the middle and lower reaches (1.82–1.97 and 1.72–2.35, respectively). The average vegetation recovery time in the subwatersheds in the upper reaches was 1.41 to 1.59 times longer than that in the subwatersheds in the middle and lower reaches. The subwatersheds in the upper reaches had the highest landslide ratio (Figure 6b,c) and smallest catchment area, indicating that the large amount of sediment yield from numerous landslides was deposited in the narrow upper reaches, and that resulted in the change of river flowing path and the riverbank landslides.

Large landslides and riverbank landslides in the sinuous reaches had the longest vegetation recovery time of all landslides in the *TRW*. The 2009 landslide inventory contained 22 large landslide cases, the mean vegetation recovery time of which was 3.29 years. Two large landslide cases with mean vegetation recovery times of greater than 6 years occurred in the sinuous upper reaches of the subwatersheds. The large landslide occurring in a gully source area in the T01 subwatershed (Figure 7) can be a typical example to explain the vegetation recovery condition. The area and relief of the large landslide were 35.7 ha and 503.2 m, and the average slope was 42.1°. The stratigraphical formation and lithology of the large landslide were the Nanchuang formation and the sandstone and shale. The activity type of the large landslide was very active. The mean vegetation recovery time of the large landslide was 10.25 years. The vegetation recovery time in the boundary area of the large landslide was less than 3 years, indicating that landslides were easily re-induced in this boundary area. However, the vegetation recovery time in the core area of the large landslide case was over 7 years, and vegetation recovery in the core area of the large landslide was difficult.

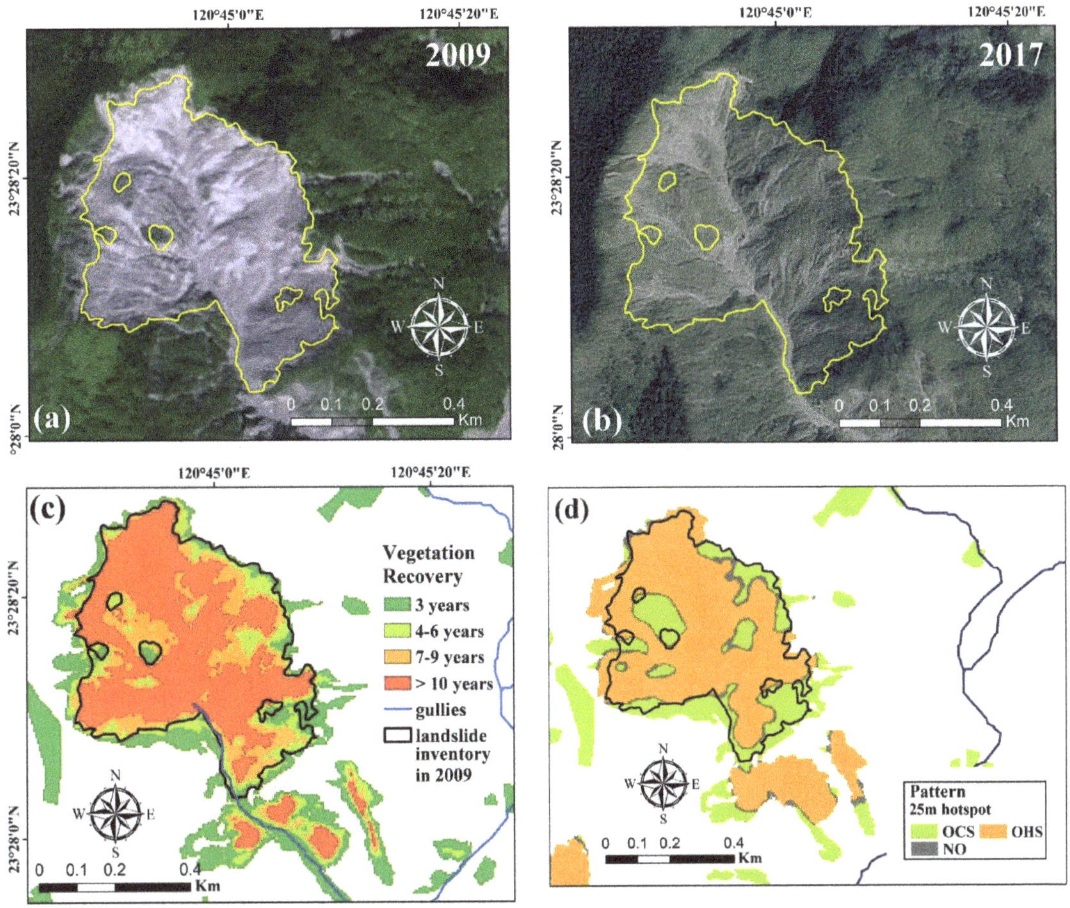

Figure 7. The aerial images in 2009 (**a**) and 2017 (**b**), the distribution of vegetation recovery time (**c**), and the spatiotemporal hotspots (**d**) in the large landslide in the T01 subwatersheds (Figure 6).

The river buffer area was defined as the area within 300 m of the river. The landslide-identified area occupied 7.5% of the total area in the river buffer area in the *TRW*, and the mean vegetation recovery time in the river buffer area was 2.65 years. The proportion of the landslide-identified area and the landslide frequency in the river buffer area were slighter larger than those in the *TRW*. The enormous amount of loose material from landslides was deposited in the confluence of river waters and sinuous reaches, resulting in riverbank landslides, especially in the upper reaches of landslide-concentrated subwatersheds. The riverbank landslide occurring in the confluence downstream of the T01 and T02 subwatersheds (Figure 8) is a typical example demonstrating the impacts of large sediment deposits in sinuous reaches on landslide recovery. The area and relief of the large landslide were 0.92 ha and 269.7 m, and the average slope was 32.5°. The stratigraphical formation and lithology of the large landslide were the Changchihkeng formation and the sandstone and shale. The activity type of the large landslide was active and very active. The riverbank landslide was located in a sinuous reach with the sinuosity index = 1.62, and the sediment deposition depth from 2009 to 2011 in this sinuous reach was estimated as 4.32 m.

The mean vegetation recovery time of the riverbank landslide (Figure 8) was 6.79 years (2.56 times longer than that in the *TRW*). The considerable amount of sediment deposition in the river in the vicinity of the riverbank landslide resulted in the formation of river meanders and several new riverbank landslides.

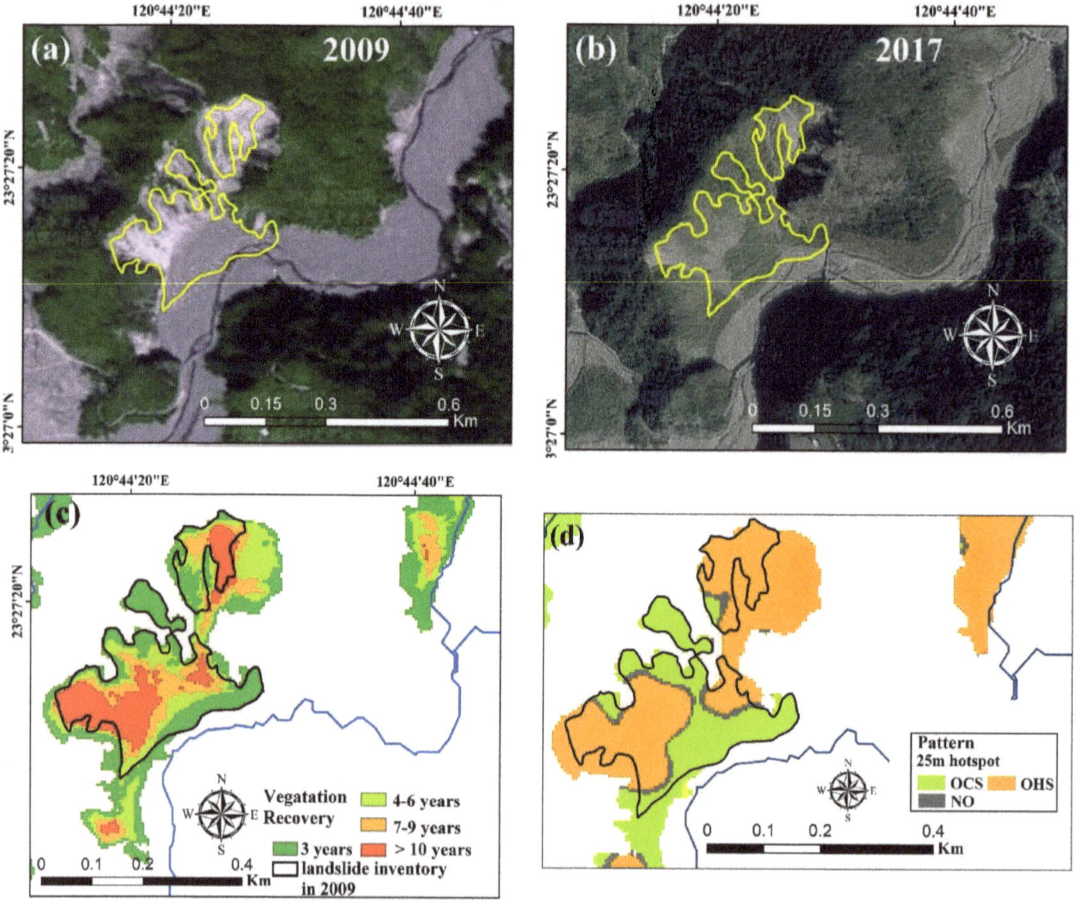

Figure 8. The aerial images in 2009 (**a**) and 2017 (**b**), the distribution of vegetation recovery time (**c**), and the spatiotemporal hotspots (**d**) in the riverbank landslide in the T03 subwatershed (Figure 6).

4.5. Landslide Spatiotemporal Hotspot Distribution and Trend

The total number of 5 m × 5 m grids in the landslide STC model, collected from the aggregated spatiotemporal data corresponding to all landslide grids from 2001 to 2017 in the *TRW*, was 17,861,645. The spatiotemporal hot spot distribution in the *TRW* is displayed in Figure 9. Neighborhood distance settings in analyses of spatiotemporal landslide hotspots have not yet been suggested. Therefore, these settings were based on a performance comparison of results obtained through analyses performed using various neighborhood distances. Specifically, the distances were set to 5, 25, 50, 100, and 200 m.

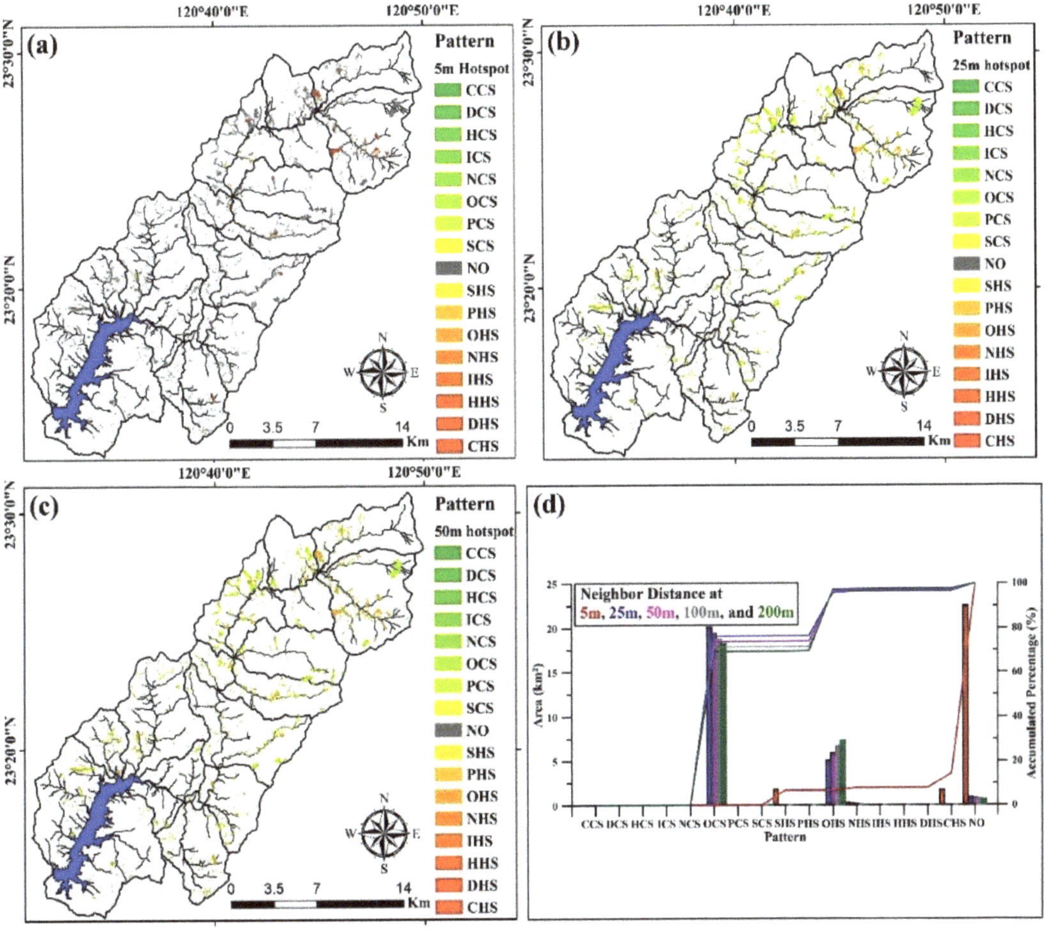

Figure 9. The landslide spatiotemporal hotspot distributions (**a**–**c**) and the occupied percentage (**d**) in the *TRW*.

The no pattern area in the analysis with a 5 m neighborhood distance occupied 85.6% of the *TRW*, but that with a neighborhood distance of ≥25 m dropped to 2.4–3.4% (Figure 9d). This result indicates that the 5 m neighborhood distance setting in the analysis was excessively short for detecting landslide hotspots. Furthermore, the 5 m setting yielded only five spatiotemporal patterns for the entire *TRW*. By contrast, when neighborhood distances of ≥25 m were used, 10–12 spatiotemporal patterns were generated (Figure 9a–c). The main hotspot patterns in the results obtained using a 5 m neighborhood distance were sporadic hotspot (SHS) and consecutive hotspot (CHS). The corresponding patterns in the results obtained using a neighborhood distance of ≥25 m were oscillating hotspot (OHS) and oscillating coldspot (OCS). The spatiotemporal distribution of hotspots in the results obtained using 25, 50, 100, and 200 m neighborhood distances were similar yet distinct from that in the results obtained using the 5 m neighborhood distance. Therefore, the neighborhood distance setting in spatiotemporal analyses of landslide hotspots was suggested to be ≥25 m from the result in the study.

The numbers of 5 m × 5 m grids corresponding to coldspot trends in the analyses performed using neighborhood distances of 200 and 25 m were 2.46 and 3.80 times the number of bins corresponding to hotspot trends. These results indicate that the landslide clustering intensity in the *TRW* was low. The summed proportion of OCS and OHS in analyses with neighborhood distances of ≥25 m in the *TRW* was greater than 95.8%. In other words, the dominant spatiotemporal landslide pattern in the *TRW* was oscillating. The spatiotemporal hot spot pattern of the large landslide in the T01 subwatershed and the riverbank landslide in the T03 subwatershed are presented in Figures 7d and 8d, respectively. Oscillating hotspots and coldspots characterized these two landslides.

5. Discussion

The landslide recovery rate and landslide activity are the keys to understanding the evolution of watersheds with frequent and dense landslides. Studies have addressed the rate of landslide recovery induced by large earthquake events [6–8], but few have examined the rate of recovery from landslides induced by extreme rainfall events [19]. The present study discussed and compared differences in the distribution rate of recovery from active landslides induced by large earthquakes and extreme rainfall events.

The majority of landslide types after the 2005 Kashimir earthquake were rock falls and rock slides [6], and those after the 2008 Wenchuan earthquake were debris slides [7]. The majority of landslide types after 2009 Typhoon Morakot in Taiwan were debris slides. Herein, the landslide area decline ratio was defined as the landslide area in the specific year to the landslide area in the year a large earthquake or extreme rainfall event occurred. For example, the landslide areas in the *TRW* in 2009 and 2010 were 1392.9 and 775.6 ha, and the landslide area decline ratio in 2010 was 55.7%. The landslide area decline ratios for the large earthquake- and extreme rainfall-induced landslides are shown in Figure 10. For each landslide event, the linear fitting equation was estimated. The slope coefficient of this equation can be regarded as the average annual landslide area decline ratio, which was used in the comparison of landslide recovery.

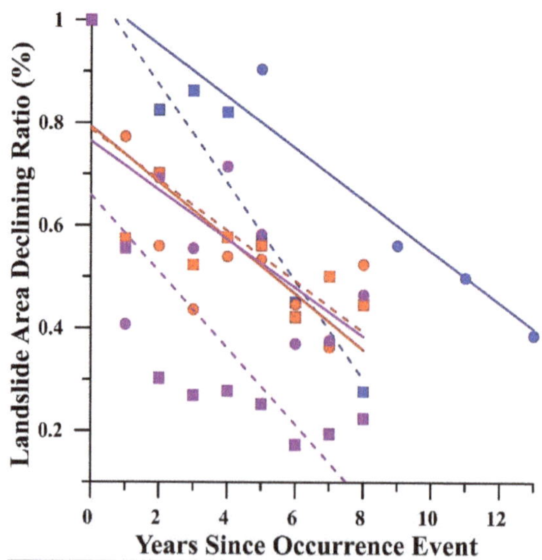

Events	Data	Fitting Equ.	R^2	Fitting lines
2005 Kashimir	●	Y = -0.0502 × X + 1.0520	0.947	——
2008 Wenchuan	■	Y = -0.0956 × X + 1.0650	0.926	– – –
2009 Morakot-CRW	●	Y = -0.0543 × X + 0.7927	0.580	——
2009 Morakot-ARW	■	Y = -0.0493 × X + 0.7871	0.600	– – –
2009 Morakot-TaRW	●	Y = -0.0475 × X + 0.7641	0.403	——
2009 Morakot-TRW	■	Y = -0.0745 × X + 0.6593	0.595	– – –

Figure 10. The area declining ratio fitting equation of large earthquake-induced and extreme rainfall-induced landslide events.

The annual landslide area decline ratios corresponding to the large earthquake events (5.02–9.56%) were slightly larger than those corresponding to the extreme rainfall events (4.75–7.45%). The R^2 values of the linear fitting equations for large earthquake-induced landslides exceeded those of the corresponding equations for extreme rainfall-induced landslides. This indicates that recovery from large earthquake-induced landslides was more stable than recovery was from extreme rainfall-induced landslides. If the declining landslide area in the *TRW* after Typhoon Morakot follows the average annual ratio of landslide area decline, 28.48 years would be required for a return to the pretyphoon landslide area.

The characteristic and evolution comparison of landslide induced by large earthquake events in Northern Pakistan and Sichuan, China and 2009 Typhoon Morakot in *TRW* are shown in Table 5. The main reasons for the difference of landslide evolution between the large earthquake-induced or extreme rainfall-induced landslide cases were the landslide-inducing factors and the deposition location of loose materials yield from landslide. The inducing factor for the earthquake-induced slope instability was the ground shaking force to reduce the shear strength of slope [37], while that for the rainfall-induced slope instability was the infiltration water [38]. The extreme active, very active, and active landslide after large earthquake-induced landslide events were located in the neighboring area of faults, including the Muzaffarbad fault in North Pakistan [6] and the Longmenshan fault in China [7,16], but those after extreme rainfall-induced landslide events in the *TRW* were located in the neighboring area of rivers and gullies (Figure 4). The percentage of the extreme active, very active, and active landslide areas located in the area within 500 m of rivers or the source area of large landslide cases after 2001 Typhoon Nari in the *TRW* was 74.9%, and those after 2009 Typhoon Morakot was 85.3%. The pattern and distribution of large earthquake-induced [6,7] or extreme rainfall-induced landslide spatiotemporal hotspot demonstrated that the landslide hotspot had been moving from mountain regions to the neighborhood area of rivers, but the time needed was the main difference between the large earthquake-induced or extreme rainfall-induced landslide events. The landslide ratio in the area within 500 m to rivers in the first year after 2009 Typhoon Morakot in the *TRW* ranged from 2.4% to 6.3% (Figure 5c), which was much larger than that in the first year after the 2008 Wenchuan earthquake-induced landslide events in China [16]. The evolution of disaster types after extreme rainfall-induced landslide events in the *TRW* was from numerous landslide cases in the hillslope to the river geomorphological changes and riverbank landslide due to much sediment deposition in the rivers, especially in the upstream watersheds. This was also the main reasons for the difficulty of landslide recovery in the T01 and T02 subwatersheds. The key to reduce the disaster in the following years after extreme rainfall-induced landslide events based on the results in the study was how to control the huge amount of loose material yield from numerous landslides and that deposited in the rivers.

Table 5. The characteristic and evolution comparison of large earthquake-induced and extreme rainfall-induced landslide events.

Events	2005 Kashimir Earthquake	2008 Wenchuan Earthquake	2009 Typhoon Morakot
Research area	Northern Pakistan	Sichuan, China	*TRW* in the study
Scale	$M_L = 7.6$	$M_L = 7.9$	Heavy rainfall with return period greater than 200 years
Landslide triggers	ground shaking force to reduce the shear strength of slope		infiltration water to reduce the shear strength of slope
Landslide types	debris fall, debris flow, debris slide, rock falls (majority), rock topple and rock slides (majority) [6]	Rock falls, rock slides, rock flows, debris falls, debris slides (majority), and debris flows [7,16,26]	debris falls, debris slide (majority), and debris flows
Sediment deposition location	concentrated in the down-hillslope to rivers near the fault lines		concentrated in the down-hillslope to rivers
Landslide volume (m^3)	2.16×10^9 [6]	$5–15 \times 10^9$ [39]	6.65×10^6 [40] *

Table 5. *Cont.*

Events	2005 Kashimir Earthquake	2008 Wenchuan Earthquake	2009 Typhoon Morakot
	Distribution of extremely active and very active landslide		
Elevation	1000m to 1600 m	lower than 2000 m	1200 m to 1800 m
Slope	30° to 40°	30° to 50°	30° to 40°
Fault or rivers	close to the Muzaffarbad faults	close to the Longmenshan faults	-
Rivers	-	100 to 400 m to the rivers	<300 m to the rivers
	Long-term landslide evolution		
Annual landslide area decline ratios	5.02%	9.56%	7.45%

*: The landslide volume induced by 2009 Typhoon Morakot in the *TRW* was estimated by using the empirical equations [40].

6. Conclusions

This study evaluated landslide activity and vegetation recovery time and detected the spatiotemporal hotspots of extreme rainfall-induced landslides in the *TRW* by using annual landslide inventories and long-term rainfall records from 2001 to 2017. Areas of extreme rainfall-induced landslides and landslides in general decreased consistently following 2001 Typhoon Nari and 2009 Typhoon Morakot. The return period of rainfall events caused a slight increase in the area of number of landslides; during 2002–2008, it was greater than 10 years. In the 2010–2017 period, it was greater than 2 years. The area of extremely active landslides in the *TRW* after 2009 Typhoon Morakot was notably larger than that after 2001 Typhoon Nari, and extremely active landslides were clustered in the subwatersheds in the upper reaches of the *TRW*. The study also discussed the relationship between temporal landslide distribution and geomorphological factors, including elevation, slope, and distance to the river. Landslides in the years following typhoon events were concentrated at 1400 to 2000 m elevations on 35° to 45° slopes within 500 m of the river. The average vegetation recovery time in the *TRW* was 2.29 years, and landslides with vegetation recovery times of greater than 10 years were commonly retrogressive landslide, riverbank landslides in the sinuous reaches, and landslides in the core area of large landslides. The time required to recover from landslides in the subwatersheds in the upper reaches was 1.41 to 1.59 times longer than the time required to recover from landslides in subwatersheds in the middle and lower reaches. The main spatiotemporal pattern of landslides in the *TRW*, including in the subwatersheds in the upper reaches, was characterized by oscillating hotspots and coldspots. The annual landslide area decline ratio in the *TRW* after Typhoon Morakot was estimated to be 94.08%, and approximately 28.5 years is required for a return to the pretyphoon landslide area. Findings on the characteristics of landslide recovery, the distribution of landslide activity, and the spatiotemporal patterns of landslides are useful for watershed management and disaster prevention in areas where rainfall-induced landslides commonly occur.

Funding: This research was funded by the Ministry of Science and Technology of Taiwan (R.O.C.), grant number MOST 109-2625-M-035-005 and the Project Research in Feng Chia University (grant number: 20H00710 and 21H00710).

Institutional Review Board Statement: Not applicable.

Informed Consent Statement: Not applicable.

Data Availability Statement: Not applicable.

Conflicts of Interest: The authors declare no conflict of interest.

References

1. Cheng, J.D.; Haung, Y.C.; Wu, H.L.; Yen, J.L.; Chang, C.H. Hydrometeorological and landuse attributes of debris flows and debris floods during typhoon Toraji, July 29–30, 2001 in central Taiwan. *J. Hydrol.* **2005**, *306*, 161–173. [CrossRef]
2. Milliman, J.D.; Lee, T.Y.; Huang, J.C.; Kao, S.J. Impact of catastrophic events on small mountainous rivers: Temporal and spatial variations in suspended- and dissolved-solid fluxes along the Choshui River, central western Taiwan, during typhoon Mindulle, July 2–6, 2004. *Geochim. Cosmochim. Acta* **2017**, *205*, 272–294. [CrossRef]
3. Wu, C.H.; Chen, S.C.; Chou, H.T. Geomorphologic characteristics of catastrophic landslides during typhoon Morakot in the Kaoping Watershed, Taiwan. *Eng. Geol.* **2011**, *123*, 13–21. [CrossRef]
4. Lin, C.W.; Liu, S.H.; Lee, S.Y.; Liu, C.C. Impacts of the Chi-Chi earthquake on subsequent rainfall-induced landslides in central Taiwan. *Eng. Geol.* **2006**, *86*, 87–101. [CrossRef]
5. Lin, C.Y.; Chuang, C.W.; Lin, W.T.; Chou, W.C. Vegetation recovery and landscape change assessment at Chiufenershan landslide area caused by Chichi earthquake in central Taiwan. *Nat. Hazards* **2010**, *53*, 175–194. [CrossRef]
6. Shafique, M. Spatial and temporal evolution of coseismic landslides after the 2005 Kashmir earthquake. *Geomorphology* **2020**, *362*, 107228. [CrossRef]
7. Yang, W.; Qi, W.; Wang, M.; Zhang, J.; Zhang, Y. Spatial and temporal analyses of post-seismic landslide changes near the epicentre of the Wenchuan earthquake. *Geomorphology* **2017**, *276*, 8–15. [CrossRef]
8. Li, C.R.; Wang, M.; Liu, K. A decadal evolution of landslides and debris flows after the Wenchuan earthquake. *Geomorphology* **2018**, *323*, 1–12. [CrossRef]
9. Kincey, M.E.; Rosser, N.J.; Robinson, T.R.; Densmore, A.L.; Shrestha, R.; Pujara, D.S. Evolution of coseismic and post-seismic landsliding after the 2015 Mw 7.8 Gorkha earthquake, Nepal. *J. Geophys. Res. Earth Surf.* **2021**, *126*, e2020JF005803. [CrossRef]
10. Wu, C.H. Landslide susceptibility based on extreme rainfall-induced landslide inventories and the following landslide evolution. *Water* **2019**, *11*, 2609. [CrossRef]
11. Glade, T. Landslide occurrence as a response to land use change: A review of evidence from New Zealand. *Catena* **2003**, *51*, 297–314. [CrossRef]
12. Van Beek, L.P.H.; Van Asch, T.W. Regional assessment of the effects of land-use change on landslide hazard by means of physically based modelling. *Nat. Hazards* **2004**, *31*, 289–304. [CrossRef]
13. Phillips, C.; Hales, T.; Smith, H.; Basher, L. Shallow landslides and vegetation at the catchment scale: A perspective. *Ecol. Eng.* **2021**, *173*, 106436. [CrossRef]
14. Valenzuela, P.; Domínguez-Cuesta, M.J.; García, M.A.M.; Jiménez-Sánchez, M. A spatio-temporal landslide inventory for the NW of Spain: BAPA database. *Geomorphology* **2017**, *293*, 11–23. [CrossRef]
15. Wu, C.H. Comparison and evolution of extreme rainfall-induced landslides in Taiwan. *ISPRS Int. J. Geo-Inf.* **2017**, *6*, 367. [CrossRef]
16. Chen, M.; Tang, C.; Xiong, J.; Shi, Q.Y.; Gong, L.F.; Wang, X.D.; Tie, Y. The long-term evolution of landslide activity near the epicentral area of the 2008 Wenchuan earthquake in China. *Geomorphology* **2020**, *367*, 107317. [CrossRef]
17. Lin, S.C.; Ke, M.C.; Lo, C.M. Evolution of landslide hotspots in Taiwan. *Landslides* **2017**, *14*, 1491–1501. [CrossRef]
18. Cheng, S.; Zhang, B.; Peng, P.; Yang, Z.; Lu, F. Spatiotemporal evolution pattern detection for heavy-duty diesel truck emissions using trajectory mining: A case study of Tianjin, China. *J. Clean. Prod.* **2020**, *244*, 118654. [CrossRef]
19. Wu, C.H.; Lin, C.Y. Spatiotemporal Hotspots and Decadal Evolution of Extreme Rainfall-Induced Landslides: Case Studies in Southern Taiwan. *Water* **2021**, *13*, 2090. [CrossRef]
20. Purwanto, P.; Utaya, S.; Handoyo, B.; Bachri, S.; Sari, I.; Sastro, K.; Utomo, B.; Aldianto, Y.E. Spatiotemporal analysis of COVID-19 spread with emerging hotspot analysis and space–time cube models in East Java, Indonesia. *ISPRS Int. J. Geo-Inf.* **2021**, *10*, 133. [CrossRef]
21. Everett, B.I.; Fennessy, S.T.; van den Heever, N. Using hotspot analysis to track changes in the crustacean fishery off KwaZulu-Natal, South Africa. *Reg. Stud.* **2021**, *41*, 101553. [CrossRef]
22. Sinotech Consultants. *Geological Investigation and Database Construction for the UpstreamWatershed of Flood-Prone*; Central Geological Survey, Ministry of Economic Affairs, R.O.C: Taipei, Taiwan, 2007.
23. Tsai, K.J.; Lee, M.H.; Chen, Y.R.; Chen, X.Y. A Study on the Sediment Quanity Variation of Large Scale Landslides Induced by Extremely Heavy Rainfall Occurred at Zengwen Watershed in Southern Taiwan. *Min. Metall.* **2017**, *61*, 8–24. (In Chinese)
24. Varnes, D.J. Slope Movement Types and Processes. In *Special Report 176: Landslides: Analysis and Control*; Schuster, R.L., Krizek, R.J., Eds.; Transportation and Road Research Board, National Academy of Science: Washington, DC, USA, 1978; pp. 11–33.
25. Lin, H.H.; Lin, M.L.; Lu, J.H.; Chi, C.C.; Fei, L.Y. Deep-seated gravitational slope deformation in Lushan, Taiwan: Transformation from cleavage-controlled to weakened rockmass-controlled deformation. *Eng. Geol.* **2020**, *264*, 105387. [CrossRef]
26. Tang, C.X.; VanWesten, C.J.; Tanyas, H.; Jetten, V.G. Analysing post-earthquake landslide activity using multi-temporal landslide inventories near the epicentral area of the 2008 Wenchuan earthquake. *Nat. Hazards Earth Syst. Sci.* **2016**, *16*, 2641–2655. [CrossRef]
27. Hervás, J.; Barredo, J.I.; Rosin, P.L.; Pasuto, A.; Mantovani, F.; Silvano, S. Monitoring landslides from optical remotely sensed imagery: The case history of Tessina landslide, Italy. *Geomorphology* **2003**, *54*, 63–75. [CrossRef]
28. Marc, O.; Behling, R.; Andermann, C.; Turowski, J.M.; Illien, L.; Roessner, S. Long-term erosion of the Nepal Himalayas by bedrock landsliding: The role of monsoons, earthquakes and giant landslides. *Earth Surf. Dyn.* **2019**, *7*, 107–128. [CrossRef]

29. Barth, S.; Geertsema, M.; Bevington, A.R.; Bird, A.L.; Clague, J.J.; Millard, T. Landslide response to the 27 October 2012 earthquake (MW 7.8), southern Haida Gwaii, British Columbia, Canada. *Landslides* **2020**, *17*, 517–526. [CrossRef]
30. Saba, S.B.; van der Meijde, M.; van derWerff, H. Spatiotemporal landslide detection for the 2005 Kashmir earthquake region. *Geomorphology* **2010**, *124*, 17–25. [CrossRef]
31. Yang, W.; Qi, W.; Zhou, J. Decreased post-seismic landslides linked to vegetation recovery after the 2008 Wenchuan earthquake. *Ecol. Indic.* **2018**, *89*, 438–444. [CrossRef]
32. Masi, E.B.; Segoni, S.; Tofani, V. Root Reinforcement in Slope Stability Models: A Review. *Geosciences* **2021**, *11*, 212. [CrossRef]
33. Saito, H.; Uchiyama, S.; Teshirogic, K. Rapid vegetation recovery at landslide scars detected by multitemporal high-resolution satellite imagery at Aso volcano, Japan. *Geomorphology* **2021**, *398*, 107989. [CrossRef]
34. Getis, A.; Ord, J.K. The analysis of spatial association by use of distance statistics. *Geogr. Anal.* **1992**, *24*, 189–206. [CrossRef]
35. Mann, H.B. Nonparametric tests against trend. *Econometrica* **1945**, *13*, 245. [CrossRef]
36. Sheen, S.W. Frequency Analysis of Annual 1-day Maximum Rainfall in the Upstream Drainage Area of Zeng-Wen River. Crop. *Environ. Bioinform.* **2016**, *13*, 1–12. (In Chinese)
37. Kamp, U.; Growley, B.J.; Khattak, G.A.; Owen, L.A. GIS-based landslide susceptibility mapping for the 2005 Kashmir earthquake region. *Geomorphology* **2008**, *101*, 631–642. [CrossRef]
38. Chiu, Y.J.; Lee, H.Y.; Wang, T.L.; Yu, J.; Lin, Y.-T.; Yuan, Y. Modeling Sediment Yields and Stream Stability Due to Sediment-Related Disaster in Shihmen Reservoir Watershed in Taiwan. *Water* **2019**, *11*, 332. [CrossRef]
39. Korup, O.; Wang, G.H. Chapter 8-Multiple Landslide-Damming Episodes. In *Landslide Hazards, Risks, and Disasters*; Elsevier: Amsterdam, The Netherlands, 2015; pp. 241–261.
40. Chan, H.C.; Chang, C.C.; Chen, S.C.; Wei, Y.S.; Wang, Z.B.; Lee, T.S. Investigation and analysis of the characteristics of shallow landslides in mountainous areas of Taiwan. *J. Chin. Soil Water Conserv.* **2015**, *46*, 19–28. (In Chinese)

water

MDPI

Article

Nature-Based Solutions for Disaster Reduction and Improving Ecosystem Services in the Hutoubi Watershed, Taiwan

Yen-Yu Chiu [1], Yi-Hung Wu [2], Kuei-Lin Fu [3], Tsung-Cheng Lai [4], Hung-En Chen [1] and Su-Chin Chen [1,5,*]

1 Department of Soil and Water Conservation, National Chung Hsing University, Taichung 40227, Taiwan; yenyucc@email.nchu.edu.tw (Y.-Y.C.); hechen@email.nchu.edu.tw (H.-E.C.)
2 Sin-Hua Forest Area Experimental Forest Management Office, National Chung Hsing University, Tainan 71249, Taiwan
3 Tainan Branch, Soil and Water Conservation Bureau, Tainan 701017, Taiwan
4 Lee Cheng Engineering Consultant Co., Ltd., Tainan 71842, Taiwan
5 Innovation and Development Center of Sustainable Agriculture, National Chung Hsing University, Taichung 40227, Taiwan
* Correspondence: scchen@nchu.edu.tw; Tel.: +886-4-2285-1558

Abstract: The Hutoubi Reservoir and its mainstream, Huyuan Stream, in the southern mountainous region of Taiwan, have experienced riverbed sedimentation and flood disasters for the past 150 years. In addition to climate change, it is necessary to scientifically consider its regulation for the next hundred years. This study adopted a collaborative approach, involving industry, government, and academia, using Nature-based Solutions (NbS) to enhance ecosystem services. The solution layout is constructed by widening the channel and constructing additional farm ponds and wetlands. An hydraulic simulation indicated that flood control was addressed. The restoration project would create diverse aquatic habitats by simulating and evaluating the distribution of ecological biotopes, using porous materials as revetments. It provided urban residents with forest leisure and recreational sites and supported the local agricultural and forestry products. The restoration has propagated local culture and created environmental and professional education. Therefore, ecological services are enhanced regarding regulation, support, provision, and culture. This pilot study, led by researchers, aimed to promote comprehensive management concepts considering all stakeholders and their active participation. We integrated NbS into the watershed and its river system as a pathway for facing the challenges of rapid urbanization and climate change and improving ecosystem services.

Keywords: mountain stream facilities; Nature-based Solutions (NbS); ecosystem services; climate change; disaster reduction; industry-government-academia collaboration

Citation: Chiu, Y.-Y.; Wu, Y.-H.; Fu, K.-L.; Lai, T.-C.; Chen, H.-E.; Chen, S.-C. Nature-Based Solutions for Disaster Reduction and Improving Ecosystem Services in the Hutoubi Watershed, Taiwan. *Water* **2023**, *15*, 2527. https://doi.org/10.3390/w15142527

Academic Editors: Brian Kronvang and Richard Smardon

Received: 31 May 2023
Revised: 24 June 2023
Accepted: 9 July 2023
Published: 10 July 2023

1. Introduction

The United Nations adopted the "2030 Agenda for Sustainable Development" in 2015, which includes the 15th Sustainable Development Goal (SDG) that urgently calls on all countries in global partnerships to take action on address climate change and strive to protect our terrestrial ecosystems and forest resources while stimulating economic growth. In this context, Nature-based Solutions (NbS) and related concepts are viewed as pathways to achieving the SDGs in various fields. NbS are defined as *"actions to protect, sustainably manage, and restore natural or modified ecosystems, that address societal challenges effectively and adaptively, simultaneously providing human well-being and biodiversity benefits"* [1]. Recently, there has been a noticeable surge in attention directed towards NbS, which are progressively being employed as a viable alternative to conventional engineering methods.

Several cases have been recognized as practical demonstrations of NbS. For instance, the Elwha River Restoration Project in the United States implemented measures like revegetation and removal of dams to restore the river's natural processes and ecosystems [2,3]. The Netherlands's Room for the River program combines traditional measures with NbS, such as creating floodplains and wetlands to manage flood risk [4,5]. These examples illustrate the importance of integrating NbS into river management strategies, promoting sustainable and resilient systems that benefit humans and the environment [6]. NbS offers several advantages, such as enhancing biodiversity, improving water quality, and mitigating the impacts of floods and droughts [1,7]. Additionally, it can provide recreational opportunities and contribute to the local economy through eco-tourism [8,9]. Moreover, NbS typically have lower maintenance costs than traditional engineering methods and can enhance the long-term sustainability of river management [10].

The international trend towards the adoption of NbS is increasingly evident. However, one of the current challenges is translating global policies into national and local policies and legal frameworks to further mainstream and upscale NbS [11]. This challenge is particularly pronounced in Asia where many developing regions are facing environmental degradation due to economic, population, and cultural factors [12]. According to data from the Asian Development Bank, to sustain population and infrastructure development and address climate change, approximately USD 53 billion per year in climate-adjusted water and sanitation infrastructure investments are required in the Asia region from 2016 to 2030 [13]. However, the competition for limited government funding and resources for NbS varies in different countries. This administrative obstacle poses a difficulty in promoting the adoption of NbS [7,9,14,15].

On the other hand, Taiwan's suburban mountain environment has both ecological diversity and environmental sensitivity, as well as diverse characteristics of industry and landscape. However, under the **pressures** of rapid urbanization and climate change, it is facing degradation. In order to maintain the forest's ecological environment and avoid using gray measures, a systematic layout (**action**) of gray-green hybridization and adaptive NbS is necessary to promote an increase in ecosystem services (**response**) but this also poses a technical difficulty for every planning project [10,16].

We aim to share the restoration and regulation project in the Hutoubi watershed and address the existing efforts and future potential of this project from the perspective of ecosystem services. In this context, the actual case experiences serve as valuable learning objectives and are highly important for the mainstreaming and customized design of NbS. The Hutoubi Watershed is a watershed with historical significance, encompassing various historical periods, such as the Dutch colonial period, the Qing Dynasty, the Japanese colonial period, and the post-World War II era. The cultural developments during these periods are closely linked to the history of the Hutoubi Reservoir. However, the watershed and reservoir environment has been gradually deteriorating. Therefore, the watershed in southern Taiwan's mountainous area has undergone watershed management based on the NbS principles. This work has overcome the difficulties as mentioned above and combined the cooperation of the government, industry, and academia, and has also been awarded as a gold engineering example of stream regulation in Taiwan. Therefore, in this study, we exhibit how this project established a consensus, conducted a value analysis, was evaluated using modeling, and designed an NbS systematic layout to implement ecosystem services in the pilot.

2. Materials and Methods

2.1. Case Description

Due to the steep rivers and fragile geology in Taiwan, both streams and reservoirs have problems with sedimentation. The sediment yield from upstream watersheds affects the lifespan of reservoirs, affecting their usage efficiency. The main geological formations in Taiwan's mountainous regions are mudstone and sandstone. Due to their loose geological structure, the soil becomes sticky and slippery after being washed by rainwater for a long time, which can easily flow down slopes and form deep gullies on the mountain surface. Therefore, the deposition ratio of reservoirs in Taiwan is generally high.

The Hutoubi Reservoir is located in southern Taiwan (Figure 1) and was built in 1868, over 150 years ago, primarily to provide irrigation water. The watershed has an average slope of 19.37% and covers an area of approximately 681 hectares. The site is located in a transitional zone of a slope gradient where the phenomenon of sedimentation is explicit [17], with a gentler slope in the west and a steeper slope in the east and elevations ranging from 13 to 154 m above sea level. Natural boundaries form the watershed, and the main source of water in the watershed comes from the Huyuan Stream, which has a total length of approximately 19.1 km.

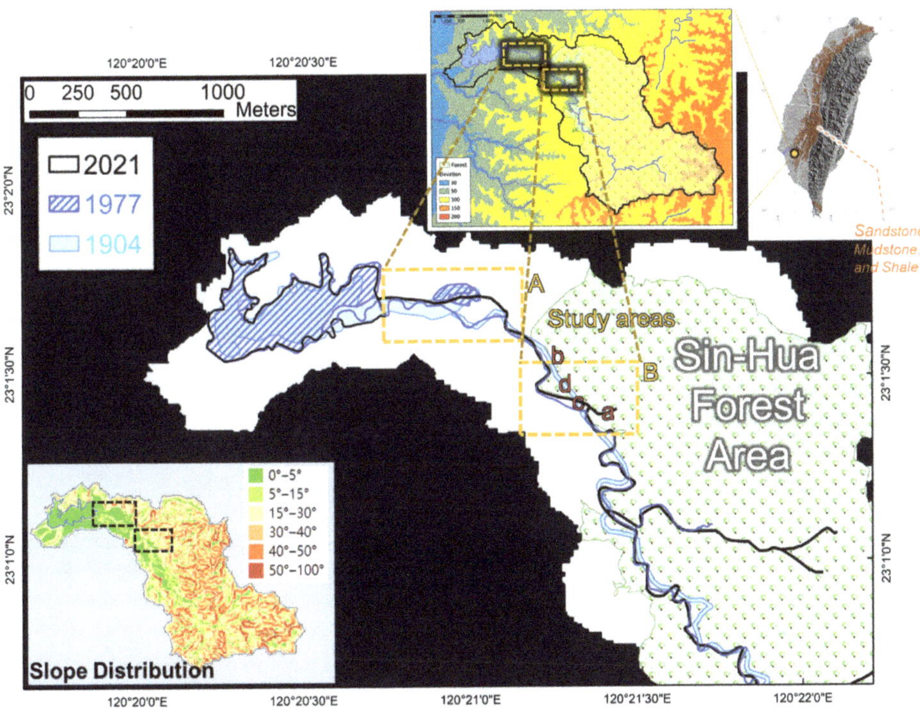

Figure 1. Geographical Location of the Hutoubi Watershed, Sin-Hua Forest, and the study areas. A and B are the study areas. The flood disaster photos of the study, shown in Figure 2, correspond to the locations marked as "a", "b", "c", and "d" in this figure.

The earliest topographic survey of the Hutoubi Watershed can be traced back to 1904, and the range and storage capacity of the Hutoubi Reservoir before that is unclear. Due to the sandstone and mudstone geology of the watershed and no forest cover in the early years, it is highly susceptible to erosion; thereby, it can be inferred that the storage capacity and area at the time of its construction in 1868 was much greater than in 1904. In 1920, National Chung Hsing University established the Sin-Hua Experimental Forest in the Hutoubi Watershed (see Figure 1). In addition to being used for education and training, afforestation was progressively carried out throughout the forest from 1940 onwards to prevent soil erosion. The forest area has reached 503 hectares, accounting for approximately 74% of the watershed (as shown in Table 1). The increase in forest area gradually shows good soil and water conservation function. The Huyuan Stream has become one of the few permanent mountain streams in southern Taiwan. Additionally, the Hutoubi Reservoir is the only reservoir in south Taiwan that can easily survive a century drought [18]. However, although the sedimentation problem has been moderated, the watershed's fine-grained geology still causes the reservoir's tail end to silt up gradually. The volume of the Hutoubi Reservoir has gradually shrunk over the past hundred years, from 1904 to 2021 (Figure 1, study area A), and the tail end of the reservoir has completely silted up.

Figure 2. The post-disaster photographs before restoration: (**a**) serious siltation of tributaries with insufficient flood discharge capacity, (**b**) high water level during rainfall, (**c**) the silting channel and concrete embankment, and (**d**) frequent flooding inundating the building area. The locations of a, b, c, and d are marked in Figure 1.

Table 1. The distribution of land cover types in the Hutoubi watershed.

Land Cover	Area Percentage
Forest	73.7%
Cropland	11.3%
Barren land	5.7%
Water bodies	5.3%
Built-up areas	3.7%
Grassland	0.1%

The rainy season in Taiwan usually falls within the monsoon season, which extends from May to October. The major natural disasters in this watershed are caused by typhoons and heavy rains, especially in the downstream section of the Huyuan Stream connecting the reservoir. In a 2 year return period, the 24 h rainfall amount is approximately 260 mm. Intense rainfall resulted in high discharge that exceeded the capacity of the river channel for the given recurrence interval (Table 2). The occurrence of landslides and soil erosion accompanying typhoons leads to a high sediment yield in the watershed and results in channel siltation. Due to the siltation of the channel, the flood discharge capacity is insufficient, and the river bed gradually aggregates, thereby meandering. Among them, study area B in Figure 1 is a severely silted section, which has caused frequent flooding disasters in the past decade (Figure 2), and the aquatic ecosystem has degraded. For example, during the 2016 Typhoon Meranti, the 2017 Typhoon Haitang, and the heavy rain on 23 August 2018 (332 mm/day), the overflowing channel caused severe flooding in the building area. Soil erosion caused stream banks and embankments to collapse, affected the timbers, and resulted in the loss of farmland and road interruption. The riverbank collapse spans a length of 600 m, impacting an area of 11 hectares of forest. Under the influence of adverse environmental factors, the diversity of biological composition in the stream is low. Non-native species, such as *Tilapia* in large numbers, have caused ecological pollution and impacted the habitat environment, resulting in human avoidance of the aquatic environment.

Table 2. The flood discharge of different return periods.

(m^3/s)	$Q_{1.1}$	Q_2	Q_5	Q_{10}	Q_{25}	Q_{50}
Hutoubi main stream	51.6	76.8	91.3	102.4	117.0	128.0
tributary	2.5	4.9	5.9	6.5	7.5	8.2

Hutoubi is Taiwan's first reservoir, and the Sin-Hua Forest, with its broad-leaved forest cover, is the closest ecological green space to the Tainan metropolitan area. However, river functions have degraded due to sedimentation and frequent flood disasters, leading to ecological decay. With the gradual exacerbation of climate change in the future, addressing the next stage of watershed management and ecological restoration has become an urgent problem to solve. Therefore, the research team has focused on regulating the A and B study areas to maintain the Hutoubi reservoir and watershed functions.

2.2. *Study Process*

This study process can be broken down into three stages, as shown in Figure 3. Firstly, the issues related to the watershed should be clarified and then focused on crucial river reaches (as described in Section 2.1). During this stage, the consensus among all stakeholders is fostered. Next, various strategies and plans are formulated, and the most suitable option is selected. Numerical models serve as tools in this step to assess flood mitigation capabilities. Then, strategic layout, in conjunction with the geographical environment, is employed to achieve multiple objectives. Finally, an initial evaluation of ecosystem services is conducted. Ecosystem services encompass a wide range of aspects, including provisioning, supporting, regulating, and cultural phases. Flood regulation and habitat diversity within the phases of regulating and supporting are initially quantified using numerical models. On the other hand, the assessments of supplying and cultural phases are more indirect. However, the overall improvement in ecosystem services can be indirectly validated through metrics, such as visitor numbers and forest revenue. Research methods available for estimating the economic value of ecosystem services is not considered in this case as the aim is to present compelling specific facts or visions instead of a virtual price.

2.3. *Collaboration Pattern among the Academia, Government, and Forest Management Office*

The Sin-Hua Forest management office collaborated with National Chung Hsing University to enhance the ecosystem services. It proposed to the Soil and Water Conservation Bureau (SWCB) to use this case as a pilot for NbS in the Taiwan mountainous watershed and to jointly develop improvement measures. The research team, government, and forest management office represented the roles of guidance, execution, and post-management, respectively (Figure 4). In this project, the research team proposed strategies and professional recommendations for enhancing ecosystem services, and the SWCB funded the contract with the engineering agency. The completed project was then handed over to the forest management office for operation. This collaboration model allows all three parties to benefit. The forest management office directly benefits from improved river health for recreational activities and reducing flood risks; the research team gains a teaching and research field; and the SWCB gains a valuable pilot project for advanced measure and treatment performance. The collaboration of the three parties has overcome the uncertainties of ownership and maintenance responsibilities, as well as institutional, regulatory, and governance barriers that NbS commonly faced [14]. Moreover, the crucial aspect is to promote comprehensive management concepts that consider all stakeholders and encourage their active participation. Given that government water policies do not typically prioritize ecological improvement as a standalone goal, it is essential to leverage the management initiatives of diverse stakeholders [19].

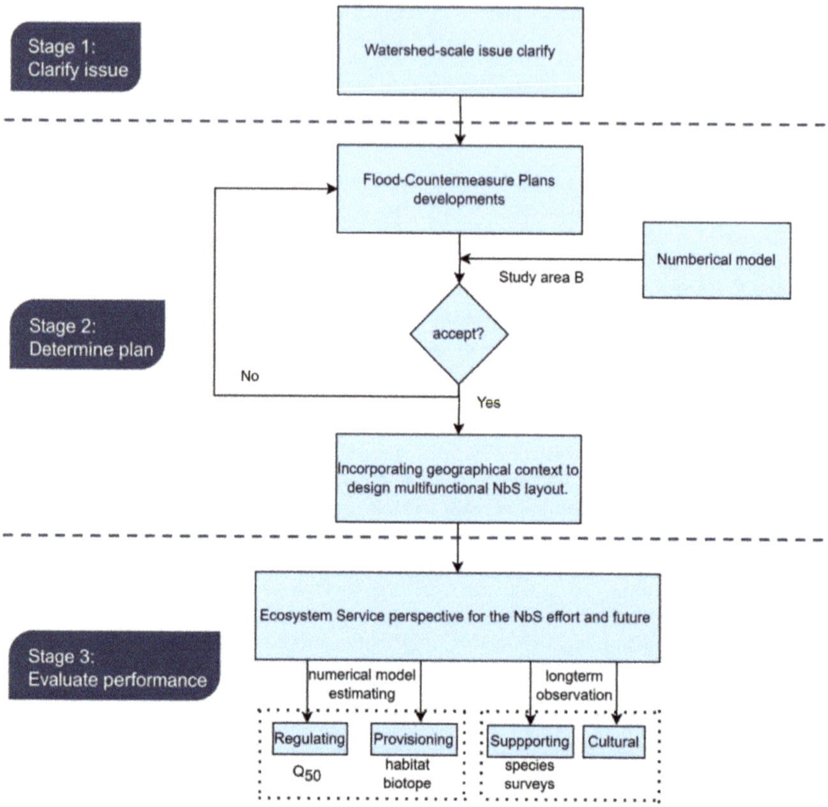

Figure 3. The flowchart of this study.

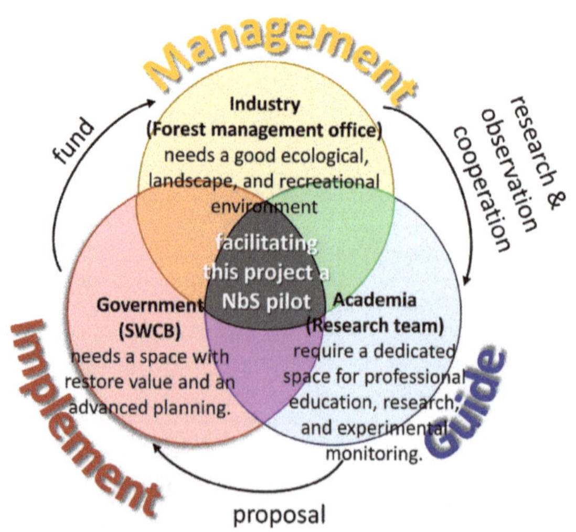

Figure 4. The motivation and interest of stakeholders (management, academia, and government).

2.4. Restoration Plans Developments

During the process, the research team had several different tasks at different stages. Firstly, the three parties jointly explored the main and potential benefits and ecosystem services that could be obtained through restoration by conducting multiple field surveys. Subsequently, based on this, the research team proposed professional advice to the engineering agency for design. In this process, other stakeholders (such as environmental NGO/Non-profit) were invited to enhance the communication and friendly measures for ecological maintenance. During the implementation, recording and monitoring were necessary to collect valuable data for analyzing the expected benefits.

The designed plans were led and proposed by the research team. Initially, in study area B where the channel lacked flood capability, the design of an embankment elevation or channel width was determined through numerical modeling, and the required flood capacity (Q_{50}). Once the plans were proposed, they were discussed with stakeholders, and this stage involves diverse perspectives, including considerations of landscape aesthetics, ecology, and more. These opinions were addressed by incorporating multifunctional NbS layouts and evaluating the diversity of hydraulic biotopes. Ultimately, the adoption of a particular plan was decided jointly by the three groups.

2.5. Numerical Model

Riverflow2D hydraulic numerical is one of the most advanced two-dimensional combined hydraulic and hydrologic flexible-mesh models. Its simulation was used to quantitatively analyze the flood control capacity and understand the types of ecological habitats during low water periods. The model calculates shallow water flows using depth-averaged mass and momentum conservation equations with all the associated assumptions [20]. That system of partial differential equations will be formulated here in a conservative form as follows:

$$\frac{\partial U}{\partial t} + \frac{\partial F(U)}{\partial x} + \frac{\partial G(U)}{\partial y} = S(U, x, y) \tag{1a}$$

where $U = (h, q_x, q_y)^T$ is the vector of the conserved variables with h representing the water depth; $q_x = uh$ and $q_y = vh$ the unit discharges with (u, v) the depth-averaged components of the velocity vector u along the (x, y) coordinates, respectively. The flux vectors are given by:

$$F = \left(q_x, \frac{q_y^2}{h} + \frac{1}{2}gh^2, \frac{q_x q_y}{h} \right)^T \tag{1b}$$

$$G = \left(q_y, \frac{q_x q_y}{h}, \frac{q_y^2}{h} + \frac{1}{2}gh^2 \right)^T \tag{1c}$$

where g is the acceleration of gravity. The terms $\frac{1}{2}gh^2$ in the fluxes have been obtained after assuming a hydrostatic pressure distribution in every water column, usually accepted in shallow water models. The source term vector incorporates the effect of pressure force over the bed, and the tangential forces generated by the bed stress:

$$S = \left(0, gh\left(S_{0x} - S_{fx}\right), gh\left(S_{0y} - S_{fy}\right)\right)^T \tag{1d}$$

where the bed slopes of the bottom level z_b. The bed stress contribution is modeled using the Manning friction law.

The advanced finite-volume engine of this model provides exceptional performance, ensuring precise and volume-conserving calculations. It excels in handling even the most challenging flood modeling scenarios, such as simulating dam-break and levee-break situations on initially dry landscapes. Therefore, it can be used to calculate the low-water condition for hydraulic biotope evolution.

The upstream boundary is the steady flow discharge, and the downstream boundary takes the normal depth boundary condition. Manning *n* is 0.035. In the study, Q_{50} (127 m^3/s, see Table 2) condition is the target flood capacity, we thereby use it as a standard to simulate the condition before regulation and different regulation plans. Additionally, the discharge used as the baseflow to simulate the diversity of hydraulic biotope is 1 m^3/s.

2.6. Hydraulic Biotope

The Froude number (**Fr**) is a reliable hydraulic parameter differentiating the hydraulic ecological units [21–23]. Therefore, this study used hydraulic models to evaluate the distribution and proportion of hydraulic ecological units to ensure the creation of diverse ecological habitats. Allocating different hydraulic parameter ranges to define habitats is a widely accepted assessment method (e.g., [21,22]). At the same time, it can quantify habitat area and diversity, and serve as a basis for quantifying ecological quality, biodiversity, and ecological resilience. This study used a 1 m^3/s baseflow and divided hydraulic biotopes into six categories [21] for simulation and analysis:

$$\mathbf{Fr} = \frac{v}{\sqrt{gh}} \tag{2}$$

These six categories are pool, glide, run, unbroken standing wave, broken standing wave, and chute. They are characterized by different ranges of **Fr**: 0.0–0.04, 0.04–0.15, 0.15–0.24, 0.24–0.49, 0.49–0.70, and 0.70–1.5, respectively [21].

3. Numerical Simulation Results (in Study Area B)

3.1. Condition before Regulation

First, Riverflow 2D was used to realize the flood capacity before regulation. The result is illustrated, which shows that the flood does not remain confined to the channel (red lines in Figure 5). Under Q_{50} conditions, it was observed that the shallow channel cross-section and gentle slope at the confluence caused floods to overflow the embankments and inundate the neighboring terraces and conservation objects in the area. The water depth distribution showed that the flooding inundated the surrounding areas beyond the stream, reaching a height of about 1 to 2 m. The embankments were old and prone to damage during floods, as evidenced by previous events, while the flat slope of the river channel (approximately 0.39% to 1.1%) could lead to silting after floods.

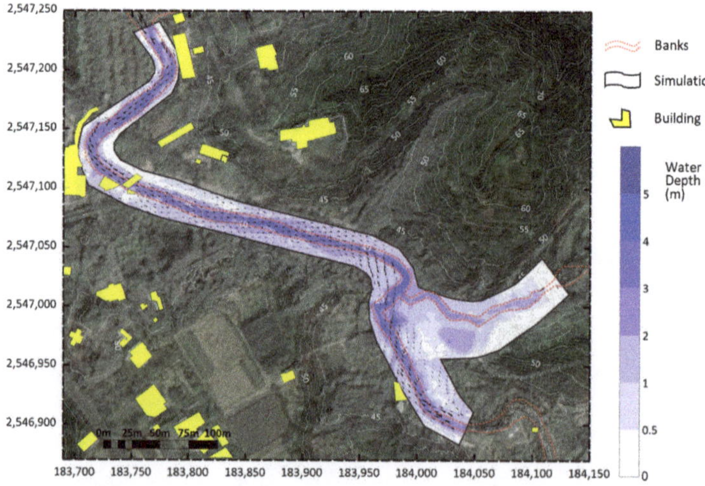

Figure 5. Simulation results of Q_{50} flood before engineering.

3.2. Simulating the Plans for Flood Control and Diversity Habitats

Two flood control plans were developed in response to the problem of insufficient flood control capacity: Plan I, the Raising Dikes Plan and Plan II, the Widening Plan. The engineering details for both plans were evaluated using Riverflow 2D. The results of Plan I are presented in Figure 6a. To meet the 50 year return period flood control standard, the dikes and embankments in the area must be raised by approximately 4–5 m. However, since the river width remains unchanged, the water depth before and after the confluence could reach up to 7 m during the flow peak. Furthermore, damage to the dikes by the high velocity could significantly harm the surrounding areas, indicating that Plan I may have low resilience. The raised dikes may also hinder wildlife passage, interrupt ecological continuity, and harm the landscape.

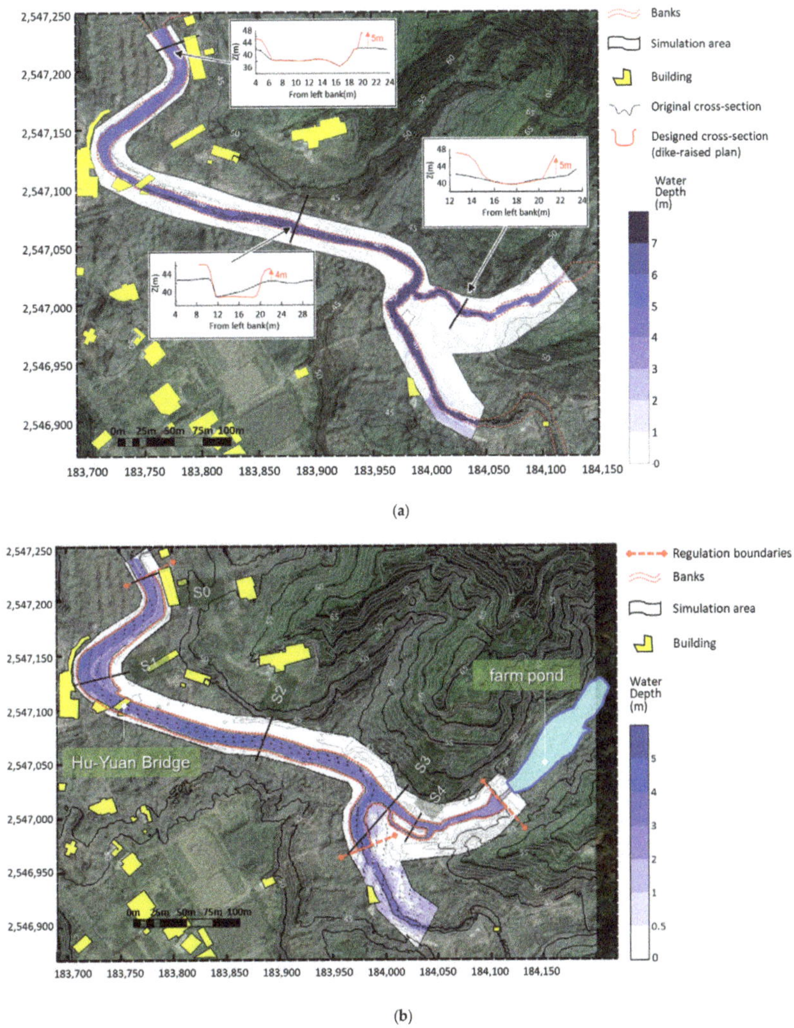

Figure 6. Simulation results of Q_{50} flood for different design plans: (**a**) Plan I: Raising Dikes Plan; (**b**) Plan II: Widening Plan.

The results of Plan II, the Widening Plan, are shown in Figure 6b. Rather than raising the dikes, this plan widens the river to increase the flood control capacity and retention space. The water depth in the widened channel generally ranges from 3 to 4 m, which is lower than the simulation before the engineering scenario (Figure 5). The river widening and channel improvement can significantly improve the flooding situation, and the water flow can be appropriately directed.

Considering the multifunctional aspects of planning and layout, the space above the confluence can be used as a wetland and floodplain to regulate extreme floods. Furthermore, multi-level revetments are planned to be established at the convex bank downstream of the Huyuan Bridge, serving various functions in response to different discharge levels. These revetments address the issue of high velocity and provide an ecological habitat during low-flow periods. After comparing the two plans, the stakeholders agreed that Plan II was more favorable. Furthermore, a farm pond was planned in the northern tributary to address the multiple functions of droughts, sediment retention, and flood control; thereby enhancing the adaptability of the plan to climate change.

The hydraulic biotope diversity simulation (Figure 7) showed that for Plan II, the unbroken standing wave accounted for 36% of the river range, glide accounted for 24%, run accounted for 15%, pool accounted for 10%, broken standing wave accounted for 10%, and chute accounted for 5%. Overall, there is a diverse range of biotope types, and after the disturbance caused by the engineering work stabilizes, there is potential for developing a diverse ecological composition.

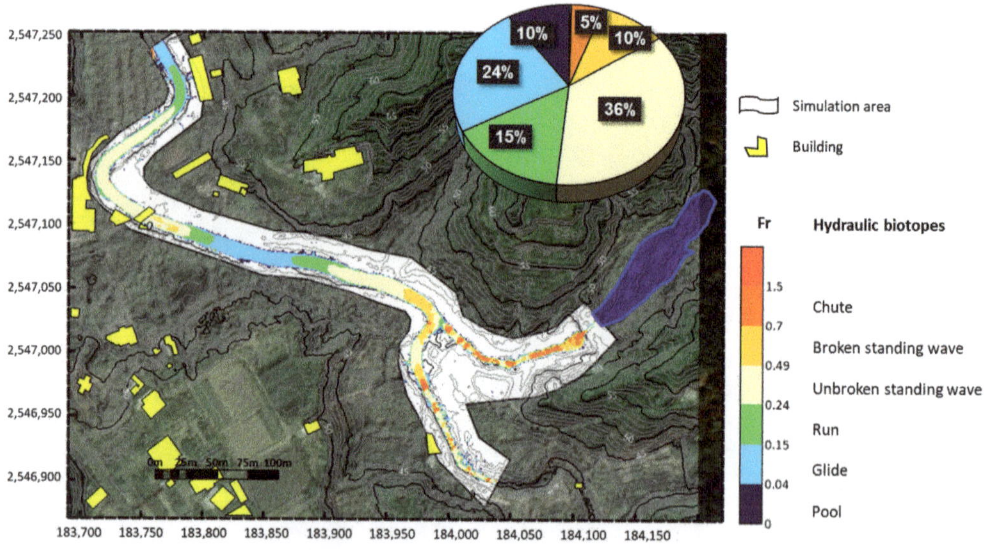

Figure 7. Hydraulic biotopes simulation and the area proportions.

4. NbS Strategies

This study considers improving the river environment during both flood and low-flow periods. The former corresponds to flood control and retention, mainly achieved by providing space for the watercourse. To create diverse hydraulic biotopes, the latter corresponds to the water source, habitat conservation, and landscape considerations. This section summarizes the overall NbS strategies adopted in the watershed.

A farm pond is planned to be set up before the confluence of the northern tributary (blue area in Figure 8a), which has excellent potential for providing ecosystem services. During flood periods, it can reduce peak flow and retain sediments, while during low-flow periods, it can provide water storage, supplement irrigation water, replenish groundwater, regulate microclimates, provide recreational opportunities, and activate the landscape.

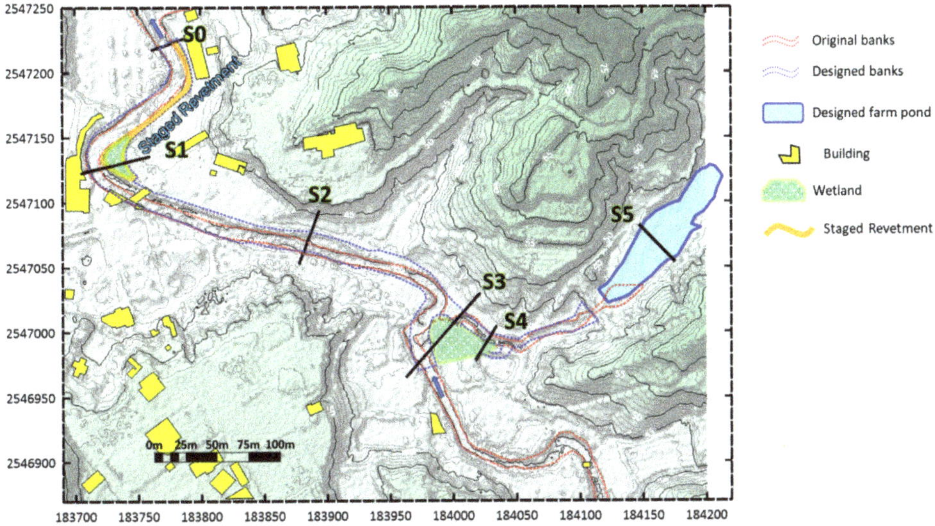

(a)

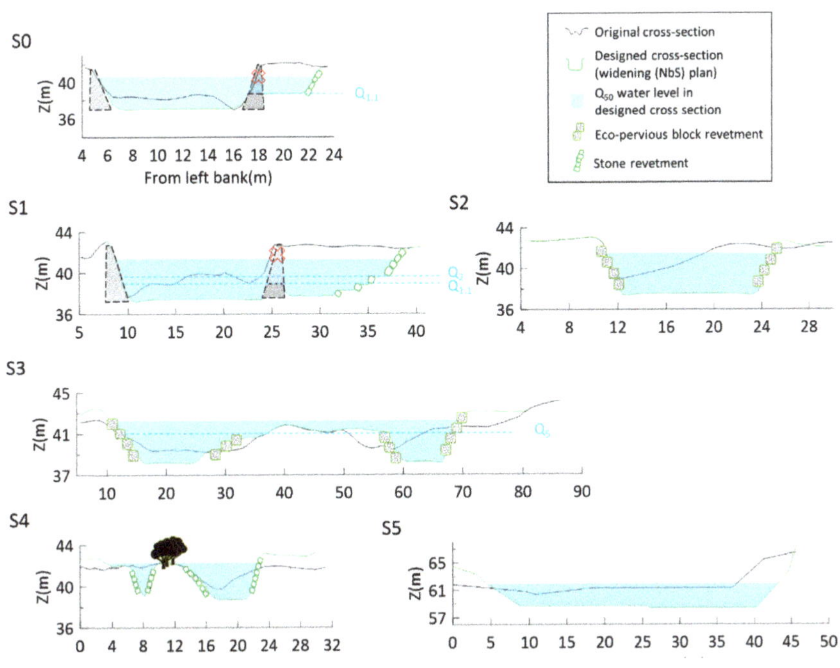

(b)

Figure 8. NbS layout of Huyuan Stream in study area B: (**a**) the plan layout, (**b**) the cross-section designs.

A river islet was set up at the confluence of the main and tributary streams downstream of the farm pond (S4 in Figure 8a) to divert floodwaters. This allows the trees in the center of the river islet to be preserved, and the wetland can serve as a floodplain (S3 in Figure 8a,b) at the confluence where the discharge reaching the maximum. Additionally, this creates an undisturbed ecological habitat where organisms can move between the river and floodplain, helping to spread plant seeds and fragments, nutrients, and carbon and providing opportunities for feeding and reproduction for waterfowl, fish, and amphibians.

Moreover, the flooded space of the river should be significantly increased. Hydraulic modeling planning (Section 3) widened the original mainstream from 6 to 12 m (as in S2 in Figure 8a,b). This design can reduce the flow velocity and lower the risk of scouring the bank and revetment damage. Two types of revetments that belong to ecological methods were used. Stone revetments were used instead of high and vertical concrete dikes, and the pores among the stones facilitate creature habitat and plant growth. In the downstream reach of the confluence, the front yard section of the Sin-Hua Forest, eco-pervious blocks were constructed as another type of ecological revetments. Precast concrete could reduce on-site construction and dust; soil and stones were filled in the hollow parts for planting as integration of gray-green engineering. The river revetments were designed with a mild side slope to enhance the landscape quality.

On the convex side of the downstream bank, the original vertical dike was downgraded and replaced with a multi-level stone revetment (as in S0 and S1 in Figure 8b). The multi-level revetment can serve as a buffer zone during flood periods and as a waterfront space during low-flow periods to enhance the landscape and recreation quality. The multi-level revetment design can also provide shelter for biodiversity during floods and create another wetland area (as in S1 in Figure 8a) in this research area. Aerial photos and pictures after the completion are shown in Figure 9a. Since the vegetation in this pilot area is dense, the principle of protecting the original forest as much as possible was applied during planning. After completion, the trees and vegetation on the slopes can help stabilize the soil and prevent erosion, thus contributing to the watershed management.

(a)

Figure 9. *Cont.*

(b)

Figure 9. The results of the NbS engineering: (**a**) study area B, (**b**) study area A.

Additionally, another pond was set in study area A (Figure 9b) at the tail end of the Hutoubi Reservoir to serve as a detention pond for floodwater and sediment. The pond has a capacity for approximately 22,000 m^3. This pond has no grey (concrete) engineering banks, reshapes the natural landscape, and is beneficial for maintaining the reservoir and facilitating regular dredging work. The sedimentation rate in the reservoir can be slowed, and the dredging goals become more clearly defined.

The selected study areas, study area A and B, were originally located in the Hutoubi Reservoir and Sin-Hua Forest Area, respectively. The restoration in these two areas connected the two tourist areas, creating a more cohesive recreational region. Additionally, these efforts integrated the water environment downstream of the watershed, starting from the restorations of local river reaches.

5. The Review of Ecosystem Services in Hutoubi Watershed

The enhancement of the ecosystem services requires long-term maintenance and development, and this restoration project's primary and potential objectives can be seen as parts of the ecosystem services. The pilot showcased the strategy of using stream regulation to enhance ecosystem services. Figure 10 suggests that improving ecosystem services should follow a logical progression from short-term to long-term objectives. The adaptable regulation of flood and drought can be directly addressed by constructing several restoring measures in a systemic layout (regulating aspect). Instead of using traditional gray measures, the planning adopts a gray-green mixed method that meets the principles of NbS. This approach can achieve not only the goals of flood detention and water resource regulation but also maintain habitats during low-flow conditions and help habitats recover faster from disturbances [24–26] (support aspect). As a result, it gradually restores the ecosystem of native species and creates more opportunities to develop characteristic products (providing aspect). The improved aesthetic value of the riparian zones with a friendly environment is expected to enhance the park's tourism quality. The Sin-Hua Forest could create fresh elements, thereby enriching both professional and public education and maintaining local cultures (cultural aspect).

In contrast, if the river function deteriorates, it will directly affect water resources and water safety, as well as the ecological environment and habitats of aquatic organisms, which causes regional ecological degradation. This can affect organisms higher up in the food chain and soil and water conservation, leading to the deterioration of overall ecosystem services. Therefore, stream regulation can be crucial in maintaining ecosystem services by creating a healthy aquatic environment.

5.1. Regulating

Based on the numerical modeling results, the adopted restoration plan will increase the flood control capacity from the current level (approximately Q_2) to Q_{50}. Additionally, the farm pond with a maximum storage capacity of 6000 m^3 can solve the water shortage problem in the nursery during the dry season and have functions, such as irrigation, disaster prevention, and landscape. Additionally, the sediment detention pond (22,000 m^3) can significantly reduce the deposition rate in the reservoir and help maintain its functionality. Furthermore, the enhanced flood and drought adaptation capacity could regulate the microclimate and hydrology. By widening the river and settling the agricultural pond, the humidity of the forest is regulated. In addition to reducing urban heat, the forest remains green even in the dry season.

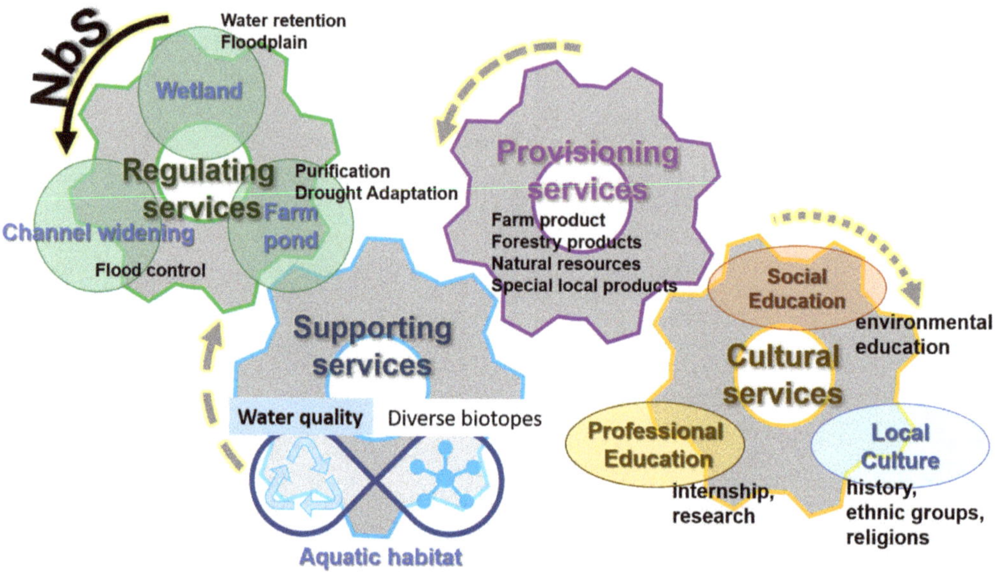

Figure 10. The progression order of improving ecosystem services.

5.2. *Supporting*

The measures to enhance the supporting services include providing eco-friendly environments and diverse river habitats. Although the diversity of biological species is a providing aspect, it can indirectly indicate the improvement of the habitat status. The results of the species composition of the ecological survey in the adjacent river section show that *Formosan stripe dace*, *Opsariichthys pachycephalus*, *Redspotted goby*, and *Caridina pseudodenticulata* are the dominant species in the other adjacent watersheds. In contrast, in the past ecological surveys conducted in the study area in 2020, these native species were not observed, and instead mainly consisted of invasive species.

An ecological species survey was conducted in 2022 after the completion of the engineering works (as shown in Table 3 and Table S1). As the project had just been completed, the ecological recovery of the study area still required time. However, compared to the records in 2020, it was found that there were more native species. For example, in terms of fish, the survey found the presence of the indigenous species *Opsariichthys pachycephalus* (endemic species), which prefers the environment of pools or shallow zones. The survey also found the presence of *Rhinogobius rubromaculatus* (*Redspotted goby*, endemic species), which prefers the environment of glides, and *Pseudorasbora parva*, which prefers muddy and sandy still waters. The survey indicated that hydraulic biotopes (Figure 7) provide diverse habitat environments for aquatic fish species. As for crustaceans and mollusks, two endemic species were found, including *Neocaridina saccam* and the *Geothelphusa olea*. Furthermore, the presence of the *Tarebia granifera* indicates that the river's organic matter has been chiefly decomposed, maintaining high dissolved oxygen levels and rich aquatic insects. *Neocaridina denticulate* and *Neocaridina saccam* prefer clean environments with flowing and oxygenated water, indicating that the aquatic habitats have gradually improved and recovered. In addition, there are also firefly populations in the restored wetlands. *Fireflies* require a clean and unpolluted environment and are considered indicator species of environmental health.

Table 3. The comparison of the species number statistics.

	Total Number of Species		Number of Endemic Species	
	2020	2022	2020	2022
Crustaceans	2	9	0	2
Fishes	5	12	1	2
Amphibians	8	8	1	1
Reptiles	6	8	1	2

However, as the state of the stream is still unstable and the riparian vegetation has not fully recovered, it has also affected the composition of the aquatic organisms. Long-term monitoring is recommended to observe changes in the biological community and to continue investigating and tracking during the different seasons of high and low flows, and to understand the influence of the restoration on the local ecology.

5.3. *Provisioning*

An increase in native and conservation species has been observed through ecological surveys, as described in Section 5.2 and shown in Table 3 and detailed in Table S1. The ecological species surveys focused on aquatic species (including fish and crustaceans) and amphibians and reptiles, considering their sensitivity to improvements in the aquatic environment. The statistics results are in Table 3. While the changes observed in amphibians and reptiles were relatively small, there was a notable increase in the number of fishes and crustaceans species recorded in 2022 compared to 2020. This indicates an enhancement in species diversity and richness in these groups. The diverse biological environment and the activities, such as the restoration of fireflies, the revenue from local timber, agricultural products, and surrounding goods in the Hutoubi Watershed has also increased. The *Honduras Mahogany* from the forest is also used for building wooden houses. The water resources of the Hutoubi Reservoir and ponds also supply the needs of people's livelihoods and agriculture.

5.4. *Cultural*

The Hutoubi Reservoir is the first reservoir in Taiwan and has cultural remains from different regimes periods, including those of indigenous peoples, the Qing Dynasty, the Japanese colonial era, and the post-World War II period. Maintaining the aesthetic value is not only to preserve the aesthetic value of the riparian zone, but also the historical and cultural significance of the waterway. The cohesive restoration of study areas A and B has also enhanced the connectivity and richness of the tourist area. Furthermore, given the success of the cooperative mode used in the Hutoubi Watershed restoration project, similar experimental NbS studies are ongoing. The Sin-Hua Forest has become more diverse in its functions, including serving as a soil and water conservation classroom and an outdoor laboratory for NbS experiments. For example, due to the geological characteristics of the mudstone and sandstone in the Huyuan Stream, there are often gullies caused by heavy rainfall on the slopes. The research team is currently studying innovative methods to use the tree roots left in the gullies and the sediment from the farm ponds to conduct NbS erosion control studies. In the past, traditional gray infrastructure was difficult to replace because it was easy for engineers to control but NbS implementation urgently needs outdoor laboratories as mechanisms for observation and case data verification to overcome obstacles. This restoration project's solutions, strategies, and experiences have also attracted attention from the domestic ecology and soil and water conservation fields, impacting these subcultures. Figure 11 shows that during the phased construction period from 2020 to 2022, the number of tourists increased significantly (despite the impact of the COVID-19 epidemic). The number of tourists is used to evaluate its function as a recreational environment.

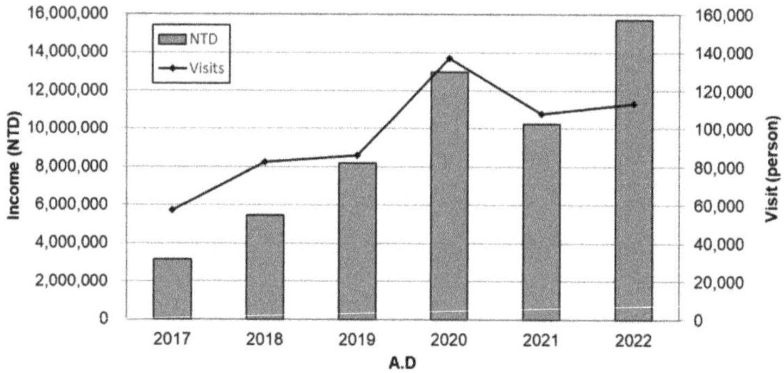

Figure 11. Visitor numbers and revenue statistics for the forest recreation area.

6. Conclusions

This study focused on the Hutoubi Watershed to preserve the reservoir's functions and improve flood and drought adaptations through two study areas, A and B, while improving the overall ecological services by rehabilitating the hostile water environment. Firstly, historic documents were used to trace the changes in the reservoir and river system over the past century to understand the watershed's river path evolution process. Different functions are expected for the basin during dry and flood seasons, so corresponding measures should be taken.

Study area A, originally located at the tail of the reservoir based on historical maps from over a century ago, had been lost due to prolonged sedimentation. A sedimentation pond was established in this area to restore the original state and maintain the reservoir's functionality. In study area B, flood control standards were considered. Regarding flood prevention, two contrasting plans were proposed and compared through numerical simulations to highlight the differences in flood control, landscape, and ecological continuity. The results indicated that to meet the Q_{50} flood control standard, Plan I required raising the embankment by 4–5 m, which would significantly impact the landscape and ecology. Plan II, which involved widening the mainstream channel from 6 to 12 m to provide additional flood space, was ultimately adopted. In addition, the distribution of hydraulic biotopes evaluated using the Froude number was simulated under low water conditions to create diverse ecological habitats. Based on the numerical simulation results, NbS layouts that are sustainable and resilient to extreme weather conditions were designed, with farm ponds and wetlands, and ecological-friendly porous materials were used for bank protection. The data from ecological species surveys and tourist statistics indicated an increase in species richness and higher returns in the park area after the completion of the project.

This case study demonstrates a strategy of using river regulation to continuously improve ecosystem services by chain reaction, adopting gray-green hybrid planning that meets the NbS criteria to address flood and weather issues (regulating) directly. The management of diverse biotopes is operated to maintain the ecological habitats (supporting), make the forest an environmentally friendly environment, and enhance the quality of landscape and agricultural products (provisioning). The Sin-Hua Forest, originally a teaching and experimental area, could maintain the function and local cultures and also create new possibilities (cultural). The Hutoubi Watershed pilot project has established a model with active involvement from all stakeholders, including the forest management office, researchers, and the government, promoting cooperation among them. A customized solution for the watershed can be developed, and long-term research and management projects can be conducted. The restoration of the Hutoubi Watershed is a response to the 15th SDG and represents a resilient approach to addressing climate change impacts. It achieves flood control objectives and enhances biodiversity habitats, species, and benefits to the tourist area. The strategies implemented in this pilot case can serve as guidance for the restoration of other watersheds facing functional degradation due to climate change. However, the importance of subsequent management should not be overlooked.

Supplementary Materials: The following supporting information can be downloaded at: https://www.mdpi.com/article/10.3390/w15142527/s1, Table S1: The species survey in Hutoubi Watershed.

Author Contributions: Conceptualization, Y.-Y.C. and S.-C.C.; Formal analysis, Y.-Y.C. and H.-E.C.; Funding acquisition, S.-C.C.; Investigation, Y.-H.W.; Methodology, T.-C.L. and S.-C.C.; Project administration, S.-C.C.; Resources, Y.-H.W., K.-L.F. and T.-C.L.; Supervision, K.-L.F. and T.-C.L.; Visualization, H.-E.C.; Writing—original draft, Y.-Y.C. and S.-C.C.; Writing—review & editing, S.-C.C. All authors have read and agreed to the published version of the manuscript.

Funding: This research was financially supported by the Nation Science and Technology Council NSTC-111-2321-B-005-007, the Higher Education Sprout Project by the Ministry of Education and Soil and Water Conservation Bureau of Taiwan, grant number SWCB-111-023.

Data Availability Statement: The data presented in this study are available on request from the corresponding author.

Acknowledgments: The authors would like to thank the Public Construction Commission, Executive Yuan for awarding the Public Construction Golden Quality Awards in the Civil Engineering and Hydraulic Engineering categories in recognition of our work. This is the highest honor in Taiwan's engineering industry.

Conflicts of Interest: The authors declare no conflict of interest.

References

1. Cohen-Shacham, E.; Walters, G.; Janzen, C.; Maginnis, S. *Nature-Based Solutions to Address Global Societal Challenges*; IUCN: Gland, Switzerland, 2016; p. 97.
2. Magirl, C.S.; Hilldale, R.C.; Curran, C.A.; Duda, J.J.; Straub, T.D.; Domanski, M.; Foreman, J.R. Large-scale dam removal on the Elwha River, Washington, USA: Fluvial sediment load. *Geomorphology* **2015**, *246*, 669–686. [CrossRef]
3. Randle, T.J.; Bountry, J.A.; Ritchie, A.; Wille, K. Large-scale dam removal on the Elwha River, Washington, USA: Erosion of reservoir sediment. *Geomorphology* **2015**, *246*, 709–728. [CrossRef]
4. Rijke, J.; van Herk, S.; Zevenbergen, C.; Ashley, R. Room for the River: Delivering integrated river basin management in the Netherlands. *Int. J. River Basin Manag.* **2012**, *10*, 369–382. [CrossRef]
5. Wolsink, M. River basin approach and integrated water management: Governance pitfalls for the Dutch Space-Water-Adjustment Management Principle. *Geoforum* **2006**, *37*, 473–487. [CrossRef]
6. Chiu, Y.-Y.; Raina, N.; Chen, H.-E. Evolution of flood defense strategies: Toward nature-based solutions. *Environments* **2021**, *9*, 2. [CrossRef]
7. Barbier, E.B.; Hacker, S.D.; Kennedy, C.; Koch, E.W.; Stier, A.C.; Silliman, B.R. The value of estuarine and coastal ecosystem services. *Ecol. Monogr.* **2011**, *81*, 169–193. [CrossRef]
8. Davis, M.; Krüger, I.; Hinzmann, M. *Coastal Protection and Suds-Nature-Based Solutions*; Policy Brief; European Union: Luxembourg, 2015; Volume 4, pp. 1–14.
9. Temmerman, S.; Meire, P.; Bouma, T.J.; Herman, P.M.J.; Ysebaert, T.; De Vriend, H.J. Ecosystem-based coastal defence in the face of global change. *Nature* **2013**, *504*, 79–83. [CrossRef] [PubMed]
10. De Vriend, H.J.; van Koningsveld, M.; Aarninkhof, S.G.; de Vries, M.B.; Baptist, M.J. Sustainable hydraulic engineering through building with nature. *J. Hydro-Environ. Res.* **2015**, *9*, 159–171. [CrossRef]
11. Estrella, M.; Renaud, F.G.; Sudmeier-Rieux, K.; Nehren, U. Defining new pathways for ecosystem-based disaster risk reduction and adaptation in the post-2015 sustainable development agenda. In *Ecosystem-Based Disaster Risk Reduction and Adaptation in Practice*; Springer: Cham, Switzerland, 2016; pp. 553–591.
12. Wesenbeeck, B.V.; IJzendoorn, C.V.; Sanchez, A.N. *Regional: Protecting and Investing in Natural Capital in Asia and the Pacific*; Asian Development Bank: Mandaluyong, Philippines, 2019.
13. ADB (Asian Development Bank). 2018. *Annual Report 2017*. Manila: ADB. Available online: https://dx.doi.org/10.22617/FLS189307 (accessed on 9 March 2023).
14. Kabisch, N.; Korn, H.; Stadler, J.; Bonn, A. *Nature-Based Solutions to Climate Change Adaptation in Urban Areas: Linkages between science, Policy and Practice*; Springer Nature: Berlin/Heidelberg, Germany, 2017.

15. Narayan, S.; Beck, M.W.; Reguero, B.G.; Losada, I.J.; Van Wesenbeeck, B.; Pontee, N.; Sanchirico, J.N.; Ingram, J.C.; Lange, G.-M.; Burks-Copes, K.A. The effectiveness, costs and coastal protection benefits of natural and nature-based defences. *PloS ONE* **2016**, *11*, e0154735. [CrossRef] [PubMed]
16. Kabisch, N.; Frantzeskaki, N.; Pauleit, S.; Naumann, S.; Davis, M.; Artmann, M.; Haase, D.; Knapp, S.; Korn, H.; Stadler, J. Nature-based solutions to climate change mitigation and adaptation in urban areas: Perspectives on indicators, knowledge gaps, barriers, and opportunities for action. *Ecol. Soc.* **2016**, *21*, 39. [CrossRef]
17. Kuo, C.W.; Tfwala, S.; Chen, S.C.; An, H.P.; Chu, F.Y. Determining transition reaches between torrents and downstream rivers using a valley morphology index in a mountainous landscape. *Hydrol. Process.* **2021**, *35*, e14393. [CrossRef]
18. Wen, Z.-H. Water Shortage in Southern Taiwan, Only the Hutobi Reservoir Still Has 1.01 Million Tons of Water Available for Irrigation of Over 50 Hectares of Rice Paddies. Available online: https://news.pts.org.tw/article/516691 (accessed on 9 March 2023). (In Chinese).
19. Schiemer, F. Building an eco-hydrological framework for the management of large river systems. *Ecohydrol. Hydrobiol.* **2016**, *16*, 19–25. [CrossRef]
20. Vreugdenhil, C.B. *Numerical Methods for Shallow-Water Flow*; Kluwer Academic Publishers: Dordrecht, The Netherlands, 1994.
21. Heritage, M.; Kim, J.; Vendlinski, T.; Herman, J. From evidence to action: A seamless process in formative assessment? *Educ. Meas. Issues Pract.* **2009**, *28*, 24–31. [CrossRef]
22. Kemp, J.L.; Harper, D.M.; Crosa, G.A. The habitat-scale ecohydraulics of rivers. *Ecol. Eng.* **2000**, *16*, 17–29. [CrossRef]
23. Wadeson, R. A geomorphological approach to the identification and classification of instream flow environments. *S. Afr. J. Aquat. Sci.* **1994**, *20*, 38–61. [CrossRef]
24. Brown, B.L. Spatial heterogeneity reduces temporal variability in stream insect communities. *Ecol. Lett.* **2003**, *6*, 316–325. [CrossRef]
25. Dunbar, M.J.; Pedersen, M.L.; Cadman, D.; Extence, C.; Waddingham, J.; Chadd, R.; Larsen, S.E. River discharge and local-scale physical habitat influence macroinvertebrate LIFE scores. *Freshw. Biol.* **2010**, *55*, 226–242. [CrossRef]
26. Townsend, C.R.; Hildrew, A.G. Species traits in relation to a habitat templet for river systems. *Freshw. Biol.* **1994**, *31*, 265–275. [CrossRef]

water

MDPI

Article

Evaluation of Rainfall-Triggered Debris Flows under the Impact of Extreme Events: A Chenyulan Watershed Case Study, Taiwan

Jinn-Chyi Chen [1,*] and Wen-Shun Huang [2]

1 Department of Hydraulic Engineering, Fujian College of Water Conservancy and Electric Power, Yongan 366000, China
2 Ecological Soil and Water Conservation Research Center, National Cheng Kung University, Tainan 70101, Taiwan; n8894103@gmail.com
* Correspondence: chenjinnchyi@gmail.com

Abstract: This study examined the conditions that lead to debris flows, and their association with the rainfall return period (T) and the probability of debris flow occurrence (P) in the Chenyulan watershed, central Taiwan. Several extreme events have occurred in the Chenyulan watershed in the past, including the Chi-Chi earthquake and extreme rainfall events. The T for three rainfall indexes (i.e., the maximum hourly rainfall depth (I_m), the maximum 24-h rainfall amount (R_d), and RI ($RI = I_m \times R_d$)) were analyzed, and the T associated with the triggering of debris flows is presented. The P–T relationship can be determined using three indexes, I_m, R_d, and RI; how it is affected and unaffected by extreme events was developed. Models for evaluating P using the three rainfall indexes were proposed and used to evaluate P between 2009 and 2020 (i.e., after the extreme rainfall event of Typhoon Morakot in 2009). The results of this study showed that the P-T relationship, using the RI or R_d index, was reasonable for predicting the probability of debris flow occurrence.

Keywords: debris flow occurrence; rainfall index; rainfall return period; probability

Citation: Chen, J.-C.; Huang, W.-S. Evaluation of Rainfall-Triggered Debris Flows under the Impact of Extreme Events: A Chenyulan Watershed Case Study, Taiwan. *Water* **2021**, *13*, 2201. https://doi.org/10.3390/w13162201

Academic Editor: Renato Morbidelli

Received: 30 May 2021
Accepted: 5 August 2021
Published: 12 August 2021

1. Introduction

In recent years, the frequency and magnitude of disasters associated with climate extremes have increased [1–6]. Extreme rainfall has given rise to massive sediment transport in mountainous areas; in particular, debris flows have caused significant damage in many countries [7–14].

Studies of rainfall that triggers debris flows, for the purpose of issuing hazard warnings, have focused on empirical thresholds based on rain gauge observations [15]. The properties of cumulative rainfall depth, rainfall intensity, and rainfall duration have been widely used to develop an empirical threshold for triggering of debris flows [16–21]. However, debris flow initiation depends not only on local rainfall properties, but also on sediment availability in the catchment area and local terrain conditions, such as topography, lithology, and soil cover [15,18,22]. The sediment supply can change significantly after extreme events, such as a major earthquake and extreme rainfall, and the triggering threshold or critical rainfall for debris flow may change accordingly [23–28]. This leads to several uncertainties when determining rainfall thresholds, and makes debris flow monitoring and prediction a challenge [29,30]. However, most of the existing empirical models were developed for real-time warning or monitoring of debris flows. In the face of climate change and increasing extreme events, there are few studies on how to reflect the long-term variation in rainfall characteristics of a region (composed of multiple debris flow gullies, townships, and villages) and the impact of extreme rainfall events on debris flow monitoring. After extreme rainfall events, the rainfall conditions that can trigger a debris flow will change, and this unstable period lasts for several years [26,28]. Thus, it is necessary to establish an empirical model for critical rainfall which can be combined with the effects of long-term rainfall changes and extreme events (on an annual scale) for debris flow monitoring and prediction. There are many rainfall indexes, such as cumulative rainfall depth and rainfall intensity, which researchers have used to develop empirical models of critical rainfall. This study aims to integrate these rainfall indicators and establish a model for evaluation of the probability of debris flow occurrence.

The Chenyulan watershed, located in central Taiwan, has experienced extreme events including the Chi-Chi earthquake (CCE) and extreme rainfall, and was therefore selected as the study area. The conditions that lead to debris flows, and their association with the rainfall return period (T) and the probability of debris flow occurrence (P), were examined. This study had two main purposes: (1) to develop an empirical model for how the P-T relationship was affected by extreme rainfall and the CCE, using different rainfall indexes, and (2) to apply the P–T relationship to evaluate P after the recent extreme event of Typhoon Morakot.

2. Study Area and Extreme Events

The study area was the Chenyulan watershed (from 23°24′30″–23°53′30″ N latitude and 120°42′10″–121°02′30″ E longitude) in central Taiwan, as shown in Figure 1. This area is conducive to the study of debris flows caused by extreme events owing to its weak geological characteristics (many faults accompanied by fracture zones; enormous landslides and an abundant source of rock debris [31]), abundant rainfall (the average annual rainfall in the watershed is approximately 3500 mm), as well as the occurrence of the Chi-Chi earthquake (a moment magnitude of M_W 7.6, in 1999) [25,32] and many other extreme events [26,28]. Thus, many studies associated with landslides and debris flows have been conducted in this watershed [25,26,28,32].

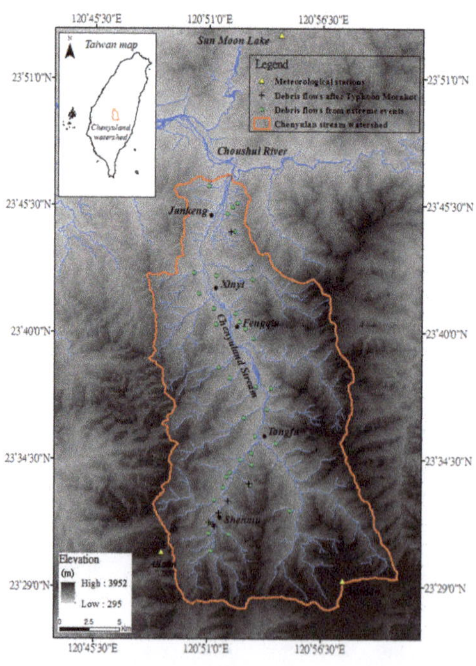

Figure 1. Map showing the study area in the Chenyulan watershed in central Taiwan, and locations of debris flows triggered by five extreme rainfall events (green circles), as well as recent debris flows (crosses) that occurred between 2009–2020 after the extreme event of Typhoon Morakot (TT) in 2009. The Chenyulan stream is a tributary of the Choushui River. The river is approximately 42 km long, the average stream bed gradient is 4°, and the watershed area is 449 km^2 with an elevation of 310–3952 m [6].

Extreme rainfall events in Taiwan have not yet been clearly defined; however, they generally can be considered as rainfall events with accumulated rainfall (R) > 130 mm within 3 h, R > 200 mm within 6 h, or R > 350 mm within 24 h [33]. The rainfall index (RI) is defined as the product of the maximum hourly rainfall depth I_m and the maximum 24-h rainfall amount R_d (i.e., RI = $R_d \times I_m$). Extreme rainfall events with an RI of >365 cm^2/h have triggered multiple debris flows in the Chenyulan watershed [26]. The five most extreme rainfall events in the Chenyulan watershed were Typhoon Herb (TH) in 1996, Typhoon Toraji (TT) in 2001, Typhoon Mindulle (TMi) in 2004, a heavy rainstorm (HR) in 2006, and Typhoon Morakot (TM) in 2009.

3. Debris Flow Occurrence Associated with Rainfall Return Period

As shown in Figure 1, there are three meteorological stations in the study area, namely Sun Moon Lake, Yushan, and Alisan, which have recorded long-term hourly rainfall since 1963. The hourly rainfall recorded from the three meteorological stations was used to determine the representative rainfall for the Chenyulan watershed, which is the regional or average rainfall for the entire watershed. The regional hourly rainfall in the Chenyulan watershed can be determined [6] using the following equation:

$$I = 0.099 I_1 + 0.387 I_2 + 0.514 I_3 \quad (1)$$

where I_1, I_2, and I_3 represent the hourly rainfall records from the Sun Moon Lake, Yushan, and Alisan meteorological stations, respectively. Equation (1) calculates the areal average rainfall at the centroid of the Chenyulan watershed area, using the reciprocal-distance-squared method [34]; the weighting factors in Equation (1) represent a distance function, being inversely proportional to the square of the distance measured between the meteorological station and the centroid point. The reciprocal-distance-squared method may not actually reflect the rainfall characteristics at specific locations where local rainfall varied significantly due to elevation changes [6]. Although the different locations of meteorological stations may influence the average rainfall result, this is a simple method which can be used to directly compute the regional average rainfall characteristics of a watershed [6]. Moreover, Equation (1) can easily represent the long term variation trend in the Chenyulan watershed [6] and it was therefore used to determine the regional hourly rainfall in this study. However, it should be noted that the estimated average rainfall contains high uncertainty. Thus, three rainfall indexes, described in Section 3.1, and the probability concept were introduced in this study in the analysis of debris flow occurrence. Rainfall events in this study were identified by the following criteria: a rainfall event occurred when the hourly rainfall depth was >4 mm and ended when this value remained at <4 mm continuously for 6 h. These criteria have generally been used in official Taiwanese warning models to identify rainfall events that trigger debris flows [35].

3.1. Rainfall Return Period

Three rainfall indexes associated with debris flow occurrence, I_m, R_d, and RI, have been proposed and discussed by [26,28], and were used to analyze T in the Chenyulan watershed. The rainfall return period can be evaluated using many methods, such as the Weibull, Jenkinson, and Gringorten formulae, various computational methods, and the modified Gumbel method [34]. The Weibull formula was used in this study because it is the most efficient and commonly used for most sample data [36]; moreover, it can predict rainfall events with much shorter return periods than the other methods [37]. T can be estimated using the Weibull formula:

$$T = (n+1) / m \quad (2)$$

where n refers to the number of years in the record and m is the rank of the observed rainfall value in a list arranged in descending order. The rainfall index of the annual maximum series collected in the Chenyulan watershed between 1963–2017 (with the exception of

1964; no rainfall record) was used to determine T using Equation (2). A total of 53 years of rainfall index data (I_m, R_d, or RI) and T are shown in Figure 2. The relationship between the rainfall index and T can be expressed in the following form:

$$x_i = a\,(T - b)^{c} \tag{3}$$

$$\text{or } T = (x_i\,/\,a)^{1/c} + b \tag{4}$$

where x_i is the variable of the rainfall index (i.e., I_m, R_d, or RI); and a, b, and c are empirical coefficients, which can be determined by regression analysis based on x_i and T data. Table 1 lists the empirical coefficients a, b, and c, and the coefficient of determination R^2 for rainfall indices I_m, R_d, and RI. All rainfall indexes have high R^2 to T.

Table 1. Empirical coefficients a, b, and c for the rainfall indices: maximum hourly rainfall depth (I_m), maximum 24-h rainfall amount (R_d), and RI ($I_m \times R_d$).

x_i	a	b	c	R^2
I_m	40	0.94	0.23	0.91
R_d	410	0.96	0.26	0.95
RI	180	0.97	0.42	0.93

Equations (3) and (4) represent the long-term rainfall characteristics of the Chenyulan watershed. The return period for the critical rainfall needed to trigger a debris flow can be determined using Equation (4). The x_i - T relationship (Equations (3) or (4)) was developed based on 54 years of data, and the annual maximum x_i was selected. Thus, the optimal range for T was 1–54 years.

3.2. Debris Flow Events Related to the Rainfall Return Period

There are no field investigation data or scientific reports on debris flows in the Chenyulan watershed prior to 1985; therefore, debris flow events prior to 1985 were not considered in the analysis of P. Figure 3 shows rainfall events, using the rainfall indices of I_m, R_d, or RI, and rainfall-triggered debris flow events between 1985–2017. The blue lines in Figure 3 indicate various return periods (T), corresponding to three rainfall indies (I_m, R_d, or RI), which were determined by Equation (3). Data on debris flow events triggered by the five extreme rainfall events (TMi, TT, HR, TH, and TM) and the CCE are also shown in Figure 3.

The extreme rainfall events that triggered numerous debris flows had T values > 5 years. Within approximately 5 years of the CCE, debris flows could be triggered at lower I_m, R_d, or RI values, corresponding to rainfall events with lower T values. After the CCE, the critical rainfall for debris flows decreased significantly, and the critical rainfall returned to the pre-earthquake level within approximately 5 years of the earthquake. This result is consistent with those of previous studies [25,26,28]. In this paper, the period encompassing approximately 5 years after CCE is referred to as the CCE-affected period (CEAP) and is shown in Figure 3. After the CCE, debris flows could be triggered at low RI values with a T of <1 year, which was much smaller than after the other extreme rainfall events. Excluding debris flow events in the CEAP, the majority of debris flow events occurred with T > 1 year. Many rainfall-triggered debris flow events coincided with maximum annual I_m, R_d, or RI values.

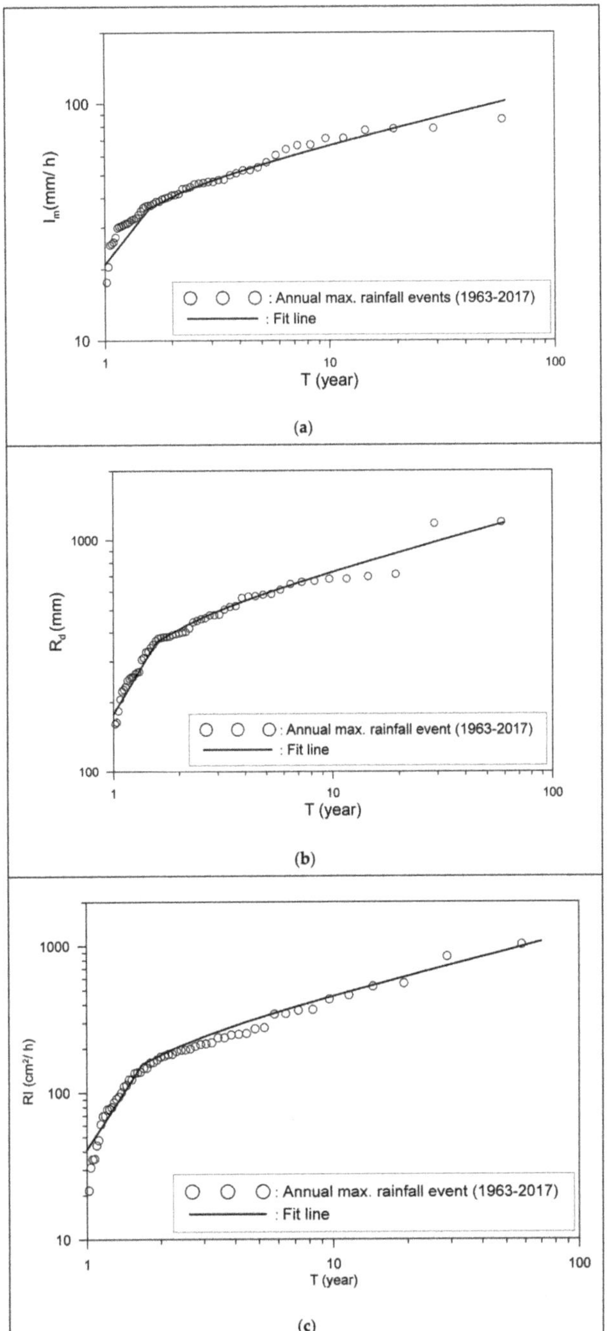

Figure 2. Return period (T) of rainfall index in the Chenyulan watershed. Three rainfall indexes, I_m, R_d, and RI, were used. (**a**) Annual maximum hourly rainfall depth (I_m) between 1963–2017; (**b**) annual maximum 24-h rainfall amount (R_d) between 1963–2017; and (**c**) annual maximum RI ($I_m \times R_d$) between 1963–2017.

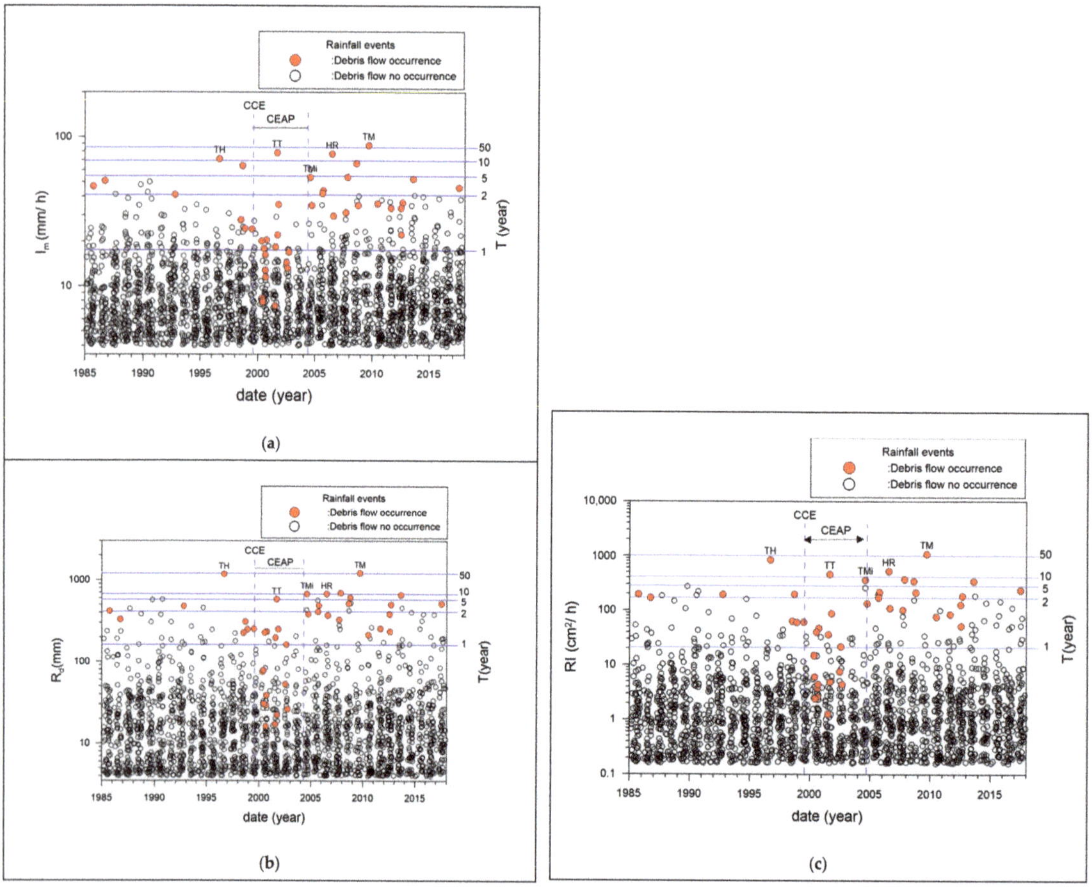

Figure 3. Rainfall events, using rainfall index of: (**a**) I_m, (**b**) R_d, and (**c**) RI, and rainfall-triggered debris flow events between 1985–2017. Data points show debris flow events triggering after the five extreme rainfall events TMi, TT, HR, TH, and TM, Chi-Chi earthquake (CCE), and during the CCE-affected period (CEAP).

4. Models for Evaluating Probability of Debris Flow Occurrence

The rainfall return period T is the average interval in years between events equaling or exceeding a certain magnitude and represents the long-term rainfall characteristics of an area. The probability of debris flow occurrence P considers rainfall events that do not trigger debris flow and represents the uncertainty of debris flow occurrence. The development of an empirical model for the relationship between P and T is presented in this section.

4.1. Probability of Debris Flow Occurrence (P)

Debris flow initiation depends on local rainfall properties, sediment availability, lithology, topography, and soil cover. These properties have a nonuniform spatial distribution and may lead to uncertainty in the evaluation of debris flow occurrence. The average hourly rainfall in the study area was obtained from the long-term rain records of the three meteorological stations. The rainfall was assumed to have a uniform distribution. The benefits of ignoring the spatial variability of hydrogeological properties are as follows: (1) more debris flow events can be identified and analyzed over the whole watershed, and (2) it assists in determining the long-term rainfall characteristics (or T) associated with the triggering of debris flows. However, ignoring spatial distribution characteristics can result in uncertainties. Thus, to reflect the resulting uncertainty, the probability concept was introduced in this study to analyze the probability of a rainfall event that triggered one or more debris flows in the entire watershed.

The P at a rainfall threshold of x_i is referred to as the probability that one or more debris flows are triggered when the rainfall amount is > x_i. P can be calculated using the following equation:

$$P = N_D/N_R \tag{5}$$

where N_R is the number of rainfall events for rainfall amounts > x_i. N_D is the number of those rainfall events that triggered debris flows. P is then determined at various rainfall thresholds x_i, from the maximum annual I_m, R_d, or RI or their corresponding T values. For example, if there were eight rainfall events and all of them triggered debris flows when RI > 285 cm^2/h or T > 5 years (see Figure 3c), thus, $N_D = N_R = 8$ and P = 1.

4.2. Relationship between Probability of Debris Flow Occurrence (P) and Rainfall Return Period (T)

After obtaining datasets for the historical maximum annual I_m, R_d, or RI and their corresponding T, the empirical relationships between P and T can be determined for I_m, R_d, or RI at different periods.

Several extreme events occurred in the Chenyulan watershed during the study period, such as the CCE and five extreme rainfall events. The critical rainfall for debris flows drops significantly after extreme events [25,26,28]; however, it does recover within a certain period thereafter. Thus, periods affected by the CCE and extreme rainfall events were separated in this analysis. The CEAP in the Chenyulan watershed was defined as the 5 years after the CCE [23,32]. The maximum recovery period of critical rainfall is approximately 3 years after extreme rainfall [28]. Rainfall events occurring within the 3 years after extreme rainfall events, were selected to determine the relationship between P and T during the extreme rainfall-affected period (ERAP). Such events included the rainfall events for 3 years after Typhoon Herb and Typhoon Morakot, 2 years after the heavy rainstorm in 2006, and 1 year after Typhoon Mindulle.

Furthermore, the relationship between P and T for the whole period (WP) between 1985–2017 and the unaffected by extreme events period (UEEP) (i.e., the WP excluding CEAP and ERAP) can also be developed. The four periods (i.e., WP, CEAP, ERAP, and UEAP) and their analysis range of rainfall events are summarized and listed in Table 2. Figure 4 shows the relationships between P and T during WP, CEAP, ERAP, and UEEP. Whether using I_m, R_d, or RI indices, the data describing the relationship between P and T showed a similar distribution in all periods. The P-T relationship can be expressed in the form of a Weibull distribution, i.e.;

$$P = 1 - e^{-\alpha\left(\frac{T}{\beta}\right)^{\gamma}} \tag{6}$$

The empirical coefficients α, β, and γ for the four periods were determined by fitting the given data and are listed in Table 2. The P–T relationship models of the four periods are compared in Figure 5. The WP model used long-term data between 1985–2017, and also included data that were affected by extreme events, that is, five extreme rainfall events and the CCE. The UEEP model excluded the influence of extreme events, and the P value predicted by the UEEP model was slightly smaller than that predicted by the WP model at the same T. Compared to the UEEP model, P increased significantly after an extreme rainfall event or the CCE, under the same rainfall conditions or at the same T. In particular, the P value affected by the CCE was markedly higher than that affected by extreme rainfall.

Table 2. Empirical coefficients α, β, and, γ of Equation (6) during the four periods.

Period	Range of Analyzed Rainfall Events	α	β	γ	R^2
I. Whole period (WP)	Between 1985–2017	0.47	0.78	0.85	0.89
II. Chi-Chi earthquake-affected period (CEAP)	Between CCE and TMi (the period of approximately 5 years after the CCE)	0.23	0.87	8.65	0.56
III. Extreme rainfall-affected period (ERAP)	3 years after TH and TM, 2 years after HR, and 1 year after TMi	0.32	0.79	2.89	0.84
IV. Unaffected by extreme events period (UEEP)	WP excluding CEAP and ERAP	0.20	0.46	0.99	0.82

Notes: R^2 = coefficient of determination.

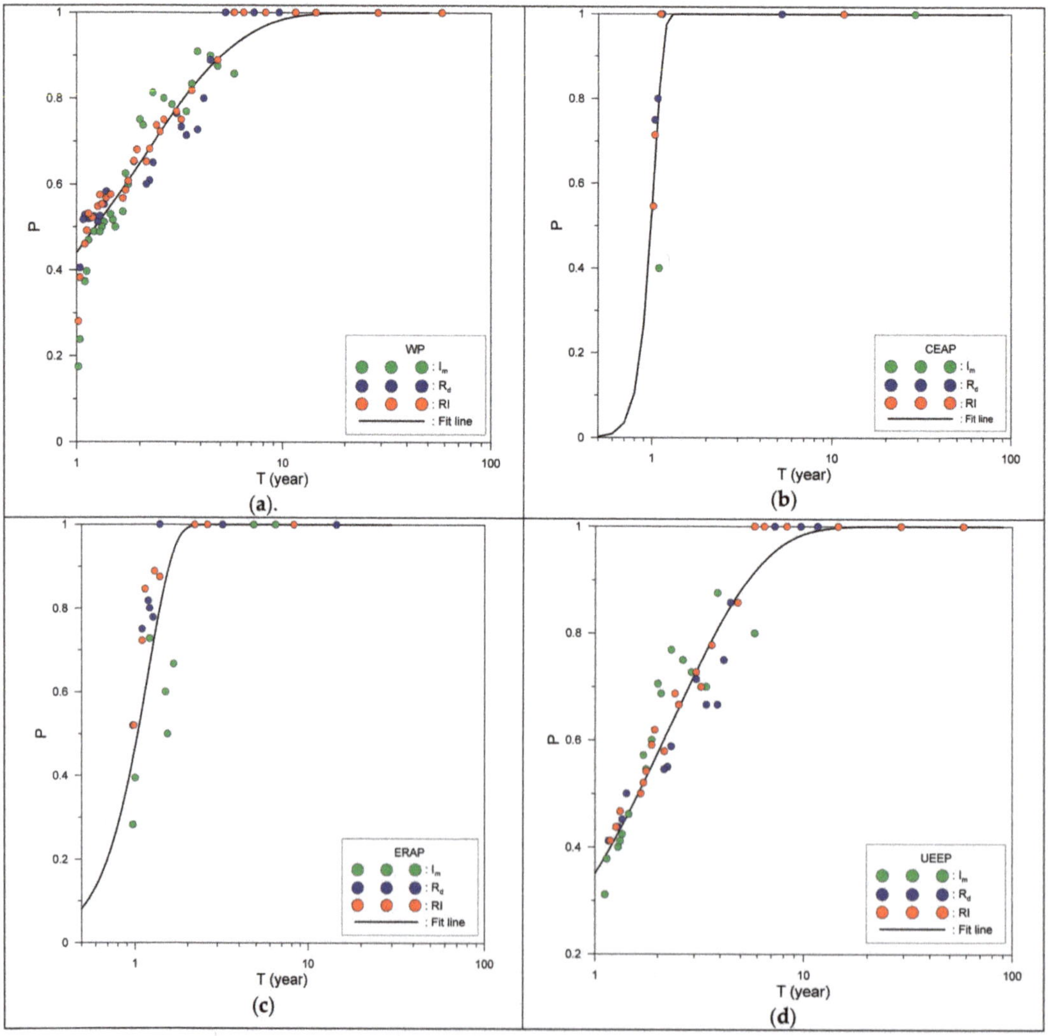

Figure 4. Four models describing the relationship between the probability of debris flow occurrence (P) and the rainfall return period (T). Data for T were determined from historical maximum annual hourly rainfall (I_m), maximum 24-h rainfall (R_d), or $R_d \times I_m$ (RI). (**a**) Whole period (WP); (**b**) Chi-Chi earthquake-affected period (CEAP); (**c**) extreme rainfall-affected period (ERAP); and (**d**) unaffected by extreme events period (UEEP).

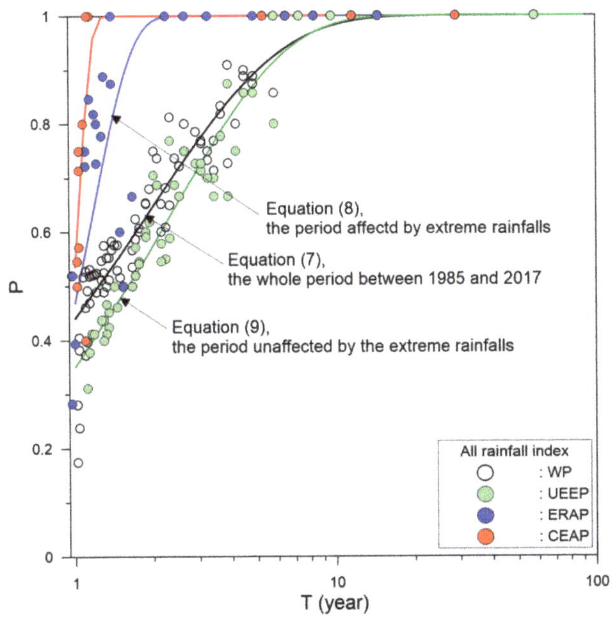

Figure 5. Comparison of the four models describing the relationship between the rainfall return period (T) and the probability of debris flow occurrence (P), i.e., the P–T relationship. The four models are for the Chi-Chi earthquake-affected period (CEAP), the extreme rainfall-affected period (ERAP), the whole period (WP), and the unaffected by extreme events period (UEEP).

The benefits of determining the P–T relationship (Figure 5) are as follows: (1) all rainfall indexes can be integrated by the P–T relationship, and (2) the P values during periods affected by the CCE or extreme rainfall can be evaluated at various T values (or different rainfall conditions). For example, P is 48% at T = 1.5, when the area of the Chenyulan watershed is unaffected by extreme events (see light green curve in Figure 5), whereas P increases to 87% after an extreme rainfall event (see blue curve in Figure 5), and P = 100% after the CCE (see red curve in Figure 5). When affected by the CCE, the P value can reach up to approximately twice that of a period unaffected by extreme events, and P during a period affected by the extreme rainfall events can be approximately 80% higher than during a period unaffected by extreme events at the same T.

5. Discussion

The heavy rainfall caused by Typhoon Morakot in August 2009 was one of the extreme rainfall events in the Chenyulan watershed and had an I_m of 85.5 mm, R_d of 1192.6 mm, and RI = 1019.7 cm^2/h. It caused numerous debris flows that buried more than 20 houses in Shenmu, Tongfu, and Xinyi villages. After Typhoon Morakot, there were seven debris flow events triggered by rainstorms or typhoons between 2009–2020, as shown in Table 3. The relationship between P and T was used to evaluate P after the rainfall events of Typhoon Morakot.

Table 3. Debris flow event and related rainfall characteristics in the Chenyulan watershed between 2009–2020, after the extreme rainfall events of Typhoon Morakot (TM).

No.	Year	Date	Name of Event	I_m (mm/h)	R_d (mm)	RI (cm^2/h)	Remark
0	2009	6–11 August	Typhoon Morakot (TM)	85.5	1192.6	1019.7	Extreme rainfall event
1	2010	23–24 May	Rainstorm	35.8	227.2	81.3	Events within the period affected by extreme rainfall TM
2	2011	17–20 July	Rainstorm	33.6	256.2	86.1	
3	2012	9–12 June	Rainstorm	33.6	384.6	129.2	
4	2012	18–21 June	Typhoon Talim	22.2	243.4	54.0	
5	2012	1–3 August	Typhoon Saola	36.4	502.2	182.8	
6	2013	12–13 July	Typhoon Soulik	52.4	661.7	346.7	Events unaffected by TM
7	2017	1–4 June	Rainstorm	46.1	515.7	237.7	

Notes: I_m (mm/h), maximum hourly rainfall in each rainfall event; R_d (mm), maximum 24-h rainfall in each rainfall event; RI, rainfall index for each rainfall event, which can be determined using the equation RI, $R_d \times I_m$.

5.1. Procedures

The empirical model for evaluating P was applied through the following steps:

1. The hourly rainfall data from three metrological stations was input and the regional hourly rainfall was evaluated using Equation (1).
2. I_m, R_d, and RI were determined from regional rainfall data.
3. T was determined using Equation (4) with I_m, R_d, or RI.
4. P was determined using the P–T relationship. There are three P–T relationships with different empirical coefficients based on various periods (Equation (6) and Table 2). That is, the P–T relationship of the WP is:

$$P = 1 - e^{-0.47\left(\frac{T}{0.78}\right)^{0.85}}; \tag{7}$$

the P–T relationship of the ERAP is:

$$P = 1 - e^{-0.32\left(\frac{T}{0.79}\right)^{2.89}}; \tag{8}$$

and the P–T relationship of the UEEP is:

$$P = 1 - e^{-0.20\left(\frac{T}{0.46}\right)^{0.99}}, \tag{9}$$

where T can be determined using Equation (4). The coefficients in Equation (4) depend on the parameters I_m, R_d, or RI, i.e., $T = (I_m/40)^{4.35} + 0.94$, $T = (R_d/410)^{3.85} + 0.96$, and $T = (RI/180)^{2.38} + 0.97$. P can be determined when I_m, R_d, or RI is given according to Equations (7), (8) or (9).

However, the three equations were developed based on different periods and datasets, and the valid conditions for the three equations may not be identical.

Equation (7) predominantly reflects the long-term characteristics of debris flow occurrence and may not reflect the short-term characteristics caused by extreme events. In contrast, Equation (8) focuses on the influence of extreme rainfall events. Equation (9) reflects P during periods that are unaffected by extreme events, such as significant earthquakes and extreme rainfall. Hence, field data of debris flow occurrence and rainfall events between 2009-2017 were used to assess the suitability of the proposed equations and their adopted rainfall parameters (I_m, R_d, or RI).

5.2. Results

The model for the WP based on the return period of different rainfall indices was used to evaluate P. Figure 6 shows the variation in the predicted P derived from Equation (7) for rainfall events with return periods greater than 1 year (T > 1 year) from 2012–2020, and the associated debris flow events are labelled (note: red dots and those labeled No. 1-7 correspond to the events shown in Table 3).

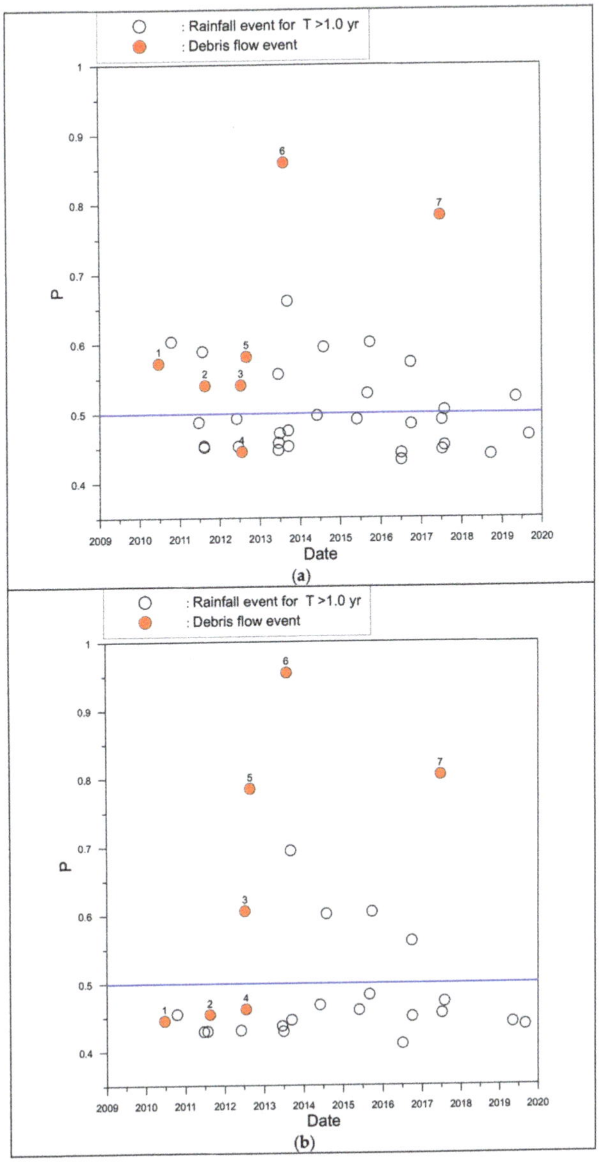

Figure 6. *Cont.*

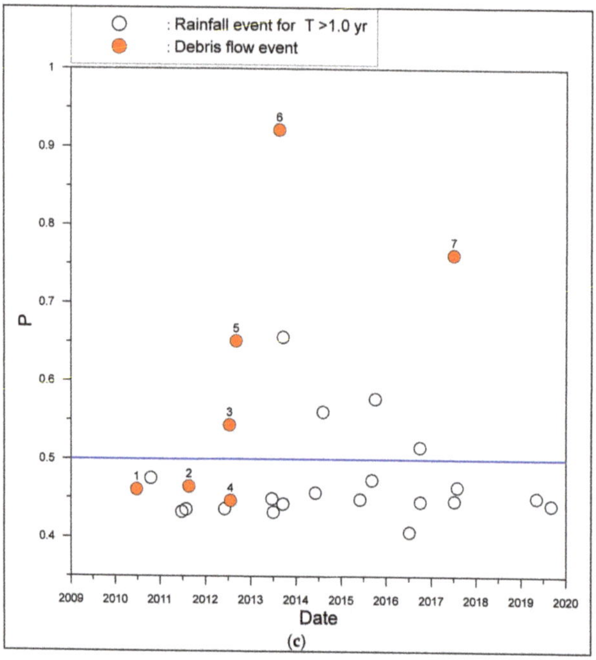

Figure 6. Probability of debris flow occurrence (P) evaluated using the model representing the whole period (WP) and Equation (7) after Typhon Morakot. The rainfall return period (T) in Equation (7) was determined using three rainfall indexes: I_m, R_d, or RI. (**a**) T determined by I_m: six of seven debris flow events were predicted with $P \geq 0.5$. Ten rainfall events that did not trigger debris flows were predicted with $P \geq 0.5$; (**b**) T determined by R_d: four of seven debris flow events were predicted with $P \geq 0.5$. Four rainfall events that did not trigger debris flows were predicted with $P \geq 0.5$; and (**c**) T determined by RI: four of seven debris flow events were predicted with $P \geq 0.5$. Four rainfall events with no triggering of debris flows were predicted with $P \geq 0.5$.

In Figure 6a, using the T of I_m to evaluate P, 10 rainfall events that did not trigger debris flows were predicted with $P \geq 0.5$, although most of the debris flow events (i.e., six of seven) were reasonably predicted by P values > 0.5. In Figure 6b,c, using the return periods of R_d and RI to evaluate P, respectively, only four of the seven debris flow events were predicted with occurrence probabilities of $P \geq 0.5$. The occurrence probability P for the four rainfall events that did not trigger debris flows was overestimated with values of $P \geq 0.5$.

The model for the WP mainly presents the long-term characteristics for P and is unable to respond to the occurrence of a debris flow that is affected or unaffected by extreme events. Thus, the predicted P for the period of 3 years after Typhoon Morakot was determined using the ERAP model (Equation (8)), and the predicted P after the ERAP was determined using the UEEP model (Equation (9)). As shown in Figure 7, three rainfall indexes (I_m, R_d, and RI) were used to evaluate the T values that were associated with the ERAP and UEEP models.

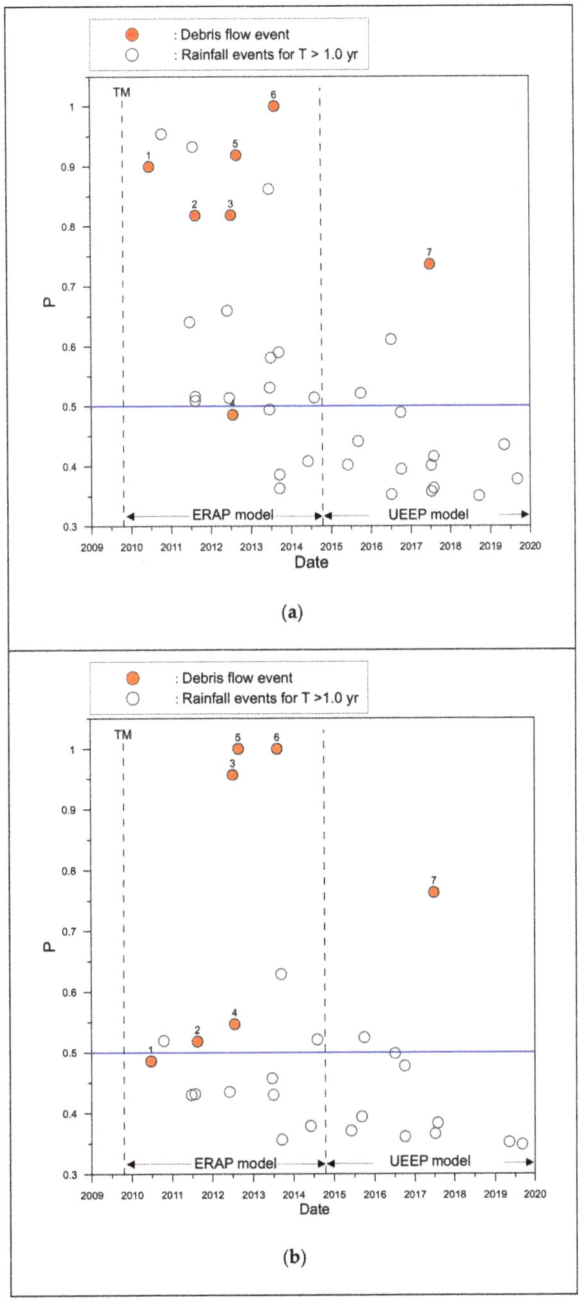

Figure 7. *Cont.*

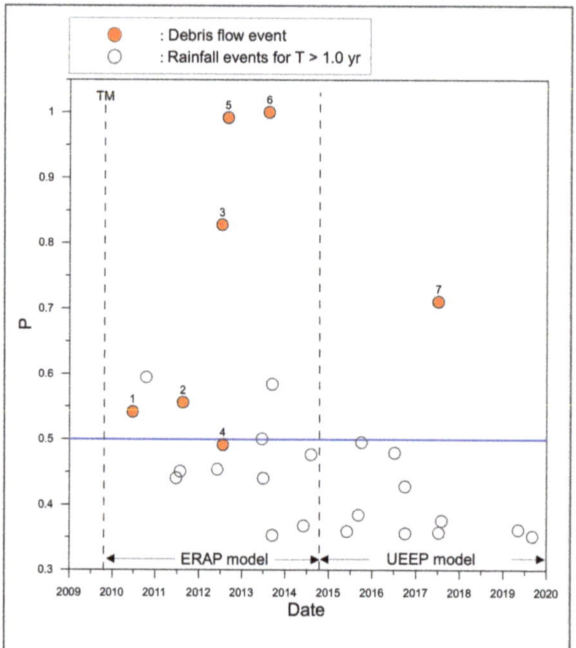

(c)

Figure 7. Probability of debris flow occurrence (P) evaluated using the model for the extreme rainfall-affected period (ERAP) within the 3 years after Typhoon Morakot (Equation (8)), and the model for the unaffected by extreme events period (UEEP) after the ERAP (Equation (9)). The rainfall return period (T in Equation (8) or ((9)) was determined using three rainfall indexes: I_m, R_d, or RI. (**a**) T determined by I_m: six debris flows were predicted within twenty rainfall events when $P \geq 0.5$ and one debris flow events with $P \geq 0.5$; (**b**) T determined by R_d: six debris flows were predicted within ten rainfall events when $P \geq 0.5$ and one debris flow events within $P \geq 0.5$; and (**c**) T determined by RI: Six debris flows were predicted within a total of eight rainfall events when $P \geq 0.5$, and one debris flow event when $P \geq 0.5$.

Using the return period T of I_m, R_d, and RI (Figure 7a–c), six of the seven debris flow events were predicted with $P \geq 0.5$. One debris flow event occurred at $P \geq 0.5$. Most rainfall events that did not trigger debris flows were predicted with $P \geq 0.5$. There were six debris flows within a total of ten rainfall events for T of I_m, six debris flows within eight rainfall events for T of R_d, and six debris flows within eight rainfall events for T of RI when $P \geq 0.5$. Figure 7 shows that 30% (Figure 7a), 60% (Figure 7b), and 75% (Figure 7c) of rainfall events triggered debris flows for T of I_m, R_d, and RI, respectively, when predicted with $P \geq 0.5$. That is, the predicted P using the T of RI and R_d would be more accurate than that using T of I_m. Furthermore, [25] examined the period affected by extreme rainfall that triggers debris flow and the modification of critical rainfall for debris flows after extreme events in the Chenyulan watershed. Three rainfall indices, I_m, R_d, and RI, were used. The modifications of critical rainfall and recovery period have higher correlations with RI driven by extreme rainfall than with I_m and R_d. Thus, the RI index was suggested for convenience. When applying the model, one must initially check whether extreme rainfall (i.e., with RI > 365 cm^2/h) has occurred in the Chenyulan watershed within 3 years. If so, the model that responded to the critical reduction RI (Equation (8)) is suggested; if not, Equation (9) should be used.

6. Conclusions

This study investigated the occurrence of debris flows after extreme events such as the CCE and extreme rainfall in a 449 km^2 region of the Chenyulan watershed. The probability of debris flow occurrence was analyzed to determine the uncertainty of the hydrogeological conditions in the study area. The innovation of this study was the combining of the concepts of T and P to establish a probabilistic model of debris flow occurrence that can be adjusted following extreme events. The findings of this study are as follows:

1. The rainfall indices of I_m, R_d, and RI associated with T were analyzed. The advantages of using T are as follows: (1) it can reflect the long-term rainfall variation characteristics of a region and adjust with rainfall variability, and (2) the concept of T can be easily combined with hydrological analysis to facilitate the subsequent engineering design. Most debris flow events occurred with T > 1 year, excluding debris flow events affected by the CCE, while the T of extreme events triggering numerous debris flows could exceed approximately 5 years. T affected by the CCE for debris-flow-triggering rainfall was < 1 year, (i.e., much smaller than that affected by other extreme rainfall events).
2. The rainfall indices of I_m, R_d, and RI can be used to determine the relationship between P and T. Four empirical models of the P-T relationship were developed (i.e., CEAP, ERAP, WP, and UEEP models). The P values, affected or unaffected by extreme events (such as the CCE or extreme rainfall), can be evaluated at various T values (or different rainfall conditions) using the P-T relationship.
3. P increased significantly after extreme rainfall events or the CCE, with the same T. In particular, the P value of periods influenced by CCE was higher than that during periods influenced by extreme rainfall events. The P during a period affected by the CCE can reach up to approximately twice that of a period that is unaffected by extreme events, while P of periods affected by extreme rainfall events can be approximately 80% higher than the P of a period that is unaffected by extreme events at the same T.
4. A model relating P and T was applied to estimate P during recent rainfall events (2009–2020) after the extreme rainfall of Typhoon Morakot, which showed that a model using the R_d or RI index was reasonably accurate at predicting debris flow occurrence.
5. An empirical model for evaluating P was developed based on the regional characteristics of the Chenyulan watershed, and may not be applicable to areas with different hydrogeological properties. The suitability of this model must be assessed, and empirical coefficients will most likely be required for calibration if the model is applied to other areas. In addition, the RI index was derived empirically, and further studies on the association of the RI index with physical mechanisms are needed.

Author Contributions: Conceptualization, J.-C.C. and W.-S.H.; methodology, J.-C.C.; software, J.-C.C. and W.-S.H.; validation, J.-C.C. and W.-S.H.; formal analysis, J.-C.C. and W.-S.H.; investigation, J.-C.C. and W.-S.H.; resources, J.-C.C. and W.-S.H.; data curation, J.-C.C. and W.-S.H.; writing—original draft preparation, J.-C.C.; writing—review and editing, J.-C.C.; supervision, J.-C.C.; project administration, J.-C.C.; funding acquisition, J.-C.C. All authors have read and agreed to the published version of the manuscript.

Funding: This study received financial support from the Ministry of Science and Technology, Taiwan (MOST 104-2625-M-211-001 and MOST 107-2625-M-211-001).

Acknowledgments: We appreciate the data analysis assistance of graduate student, C. K. Jiang, at Huafan University, Taiwan.

Conflicts of Interest: The authors declare no conflict of interest. The funders had no role in the design of the study; in the collection, analyses, or interpretation of data; in the writing of the manuscript, or in the decision to publish the results.

Abbreviations

CCE	Chi-Chi earthquake
CEAP	Chi-Chi earthquake-affected period
ERAP	Extreme rainfall-affected period
HR	Heavy rainstorm in 2006
TH	Typhoon Herb
TM	Typhoon Morakot
TMi	Typhoon Mindulle
TT	Typhoon Toraji
UEEP	Unaffected by extreme events period
WP	Whole period, 1985–2017
List of symbols	
a, b, and c	Empirical coefficients in Equations (3) and (4).
I	Region hourly rainfall.
I_1, I_2, and I_3	Hourly rainfall record from the Sun Moon Lake, Yushan, and Alisan meteorological stations, respectively.
I_m	Maximum hourly rainfall during each rainfall event.
M_W	Moment magnitude.
m	Rank of a value in a list ordered by descending magnitude.
N_D	Number of rainfall events that have triggered debris flows.
N_R	Number of rainfall events.
n	Number of years in the record.
P	Probability of debris flow occurrence.
R_d	Maximum 24-h rainfall amount during each rainfall event.
R^2	Determination coefficient.
RI	Rainfall index, $RI = R_d\ I_m$.
T	Return period of rainfall.
α, β, and γ	Empirical coefficients in Equation (6).

References

1. Evans, S.G.; Clague, J.J. Recent climatic change and catastrophic geomorphic processes in mountain environments. *Geomorphology* **1994**, *10*, 107–128. [CrossRef]
2. Clarke, M.; Rendell, H. Climate, Extreme Events and Land Degradation. In *Climate and Land Degradation, Environmental Science and Engineering*; Springer: Berlin/Heidelberg, Germany, 2007; pp. 137–152.
3. Petley, D.N.; Hearn, G.J.; Hart, A.; Rosser, N.J.; Dunning, S.A.; Oven, K.; Mitchell, W.A. Trends in landslide occurrence in Nepal. *Nat. Hazards* **2007**, *43*, 23–44. [CrossRef]
4. Hilker, N.; Badoux, A.; Hegg, C. The Swiss flood and landslide damage database 1972–2007. *Nat. Hazards Earth Syst. Sci.* **2009**, *9*, 913–925. [CrossRef]
5. Djalante, R.; Thomalla, F. Disaster risk reduction and climate change adaptation in Indonesia: Institutional challenges and opportunities for integration. *Int. J. Disaster Resil. Built Environ.* **2012**, *3*, 166–180. [CrossRef]
6. Chen, J.C.; Huang, W.S.; Jan, C.D.; Yang, Y.H. Recent changes in the number of rainfall events related to debris flow occurrence in the Chenyulan Stream Watershed, Taiwan. *Nat. Hazards Earth Syst. Sci.* **2012**, *12*, 1539–1549. [CrossRef]
7. Mechler, R.; Bouwer, L.M.; Linnerooth-Bayer, J.; Hochrainer-Stigler, S.; Aerts, J.C.J.H.; Surminski, S.; Williges, K. Managing unnatural disaster risk from climate extremes. *Nat. Clim. Chang.* **2014**, *4*, 235–237. [CrossRef]
8. Joshi, V.; Kumar, K. Extreme rainfall events and associated natural hazards in Alaknanda Valley, Indian Himalayan Region. *J. Mt. Sci.* **2006**, *3*, 228–236. [CrossRef]
9. Borga, M.; Stoffel, M.; Marchi, L.; Marra, F.; Jakob, M. Hydrogeomorphic response to extreme rainfall in headwater systems: Flash floods and debris flows. *J. Hydrol.* **2014**, *518*, 194–205. [CrossRef]
10. Wang, F.; Wu, Y.H.; Yang, H.; Tanida, Y.; Kamei, A. Preliminary investigation of the 20 August 2014 debris flows triggered by a severe rainstorm in Hiroshima City, Japan. *Geoenviron. Disasters* **2015**, *2*, 17. [CrossRef]
11. Joe, E.J.; Tongkul, F.; Roslee, R. Relationship Between Rainfall and Debris Flow Occurrence in the Crocker Range of Sabah, Malaysia. *Malays. J. Geosci.* **2018**, *2*, 18–29. [CrossRef]
12. Nam, D.H.; Lee, S.H.; Kim, B.S. Development of Nomogram for Debris Flow Forecasting Based on Critical Accumulated Rainfall in South Korea. *Water* **2019**, *11*, 2181. [CrossRef]
13. Gong, X.L.; Chen, K.T.; Chen, X.Q.; You, Y.; Chen, J.G.; Zhao, W.Y.; Lang, J. Characteristics of a Debris Flow Disaster and Its Mitigation Countermeasures in Zechawa Gully, Jiuzhaigou Valley, China. *Water* **2020**, *12*, 1256. [CrossRef]

14. Chen, J.C.; Chuang, M.R. Discharge of landslide-induced debris flows: Case studies of Typhoon Morakot in southern Taiwan. *Nat. Hazards Earth Syst. Sci.* **2014**, *14*, 1719–1730. [CrossRef]
15. Destro, E.; Marra, F.; Nikolopoulos, E.I.; Zoccatelli, D.; Creutin, J.D.; Borga, M. Spatial estimation of debris flows-triggering rainfall and its dependence on rainfall return period. *Geomorphology* **2017**, *278*, 269–279. [CrossRef]
16. Caine, N. The rainfall intensity-duration control of shallow landslides and debris flow. *Geogr. Ann. Ser. A Phys. Geogr.* **1980**, *62A*, 23–27.
17. Wieczorek, G.F. Effect of rainfall intensity and during in debris flows in central Santa Cruz Mountain, California. *Eng. Geol.* **1987**, *7*, 93–104.
18. Jan, C.D.; Chen, C.L. Debris flow caused by Typhoon Herb in Taiwan. In *Chapter 21, Debris-Flow Hazards and Related Phenomena*; Jakob, M., Hungr, O., Eds.; Praxis: Chichester, UK, 2005; pp. 539–563.
19. Shieh, C.L.; Chen, Y.S.; Tsai, Y.J.; Wu, J.H. Variability in rainfall threshold for debris flow after the Chi-Chi earthquake in central Taiwan, China. *Int. J. Sediment Res.* **2009**, *24*, 177–188. [CrossRef]
20. Chen, J.C.; Huang, W.S.; Jan, C.D.; Tsai, Y.F. Rainfall conditions for the initiation of debris flows during Typhoon Morakot in the Chen-Yu-Lan watershed in central Taiwan. In Proceedings of the 5th International Conference on Debris-Flow Hazards Mitigation: Mechanics, Prediction, and Assessment, Padua, Italy, 14–17 June 2011; pp. 31–36.
21. Guzzetti, F.; Peruccacci, S.; Rossi, M.; Stark, C.P. Rainfall thresholds for the initiation of landslides in central and southern Europe. *Meteorog. Atmos. Phys.* **2007**, *98*, 239–267. [CrossRef]
22. Chen, J.C.; Jan, C.D. Probabilistic Analysis of Landslide Potential of an Inclined Uniform Soil Layer of Infinite Length—Application. *Environ. Geol.* **2008**, *54*, 1175–1183. [CrossRef]
23. Chen, N.S.; Yang, C.L.; Zhou, W.; Hu, G.S.; Li, H.; Hand, D. The Critical Rainfall Characteristics for Torrents and Debris Flows in the Wenchuan Earthquake Stricken Area. *J. Mt. Sci.* **2009**, *6*, 362–372. [CrossRef]
24. Tang, C.; Zhu, J.; Li, W.L. Rainfall-triggered debris flows following the Wenchuan earthquake. *Bull. Eng. Geol. Environ.* **2009**, *68*, 187–194. [CrossRef]
25. Chen, J.C. Variability of impact of earthquake on debris flow triggering conditions: Case study of Chen-Yu-Lan Watershed, Taiwan. *Environ. Earth Sci.* **2011**, *64*, 1787–1794. [CrossRef]
26. Chen, J.C.; Jan, C.D.; Huang, W.S. Characteristics of rainfall triggering of debris flows in the Chenyulan watershed, Taiwan. *Nat. Hazards Earth Syst. Sci.* **2013**, *13*, 1015–1023. [CrossRef]
27. Pan, H.L.; Jiang, Y.J.; Wang, J.; Ou, G.Q. Rainfall threshold calculation for debris flow early warning in areas with scarcity of data. *Nat. Hazards Earth Syst. Sci.* **2018**, *18*, 1395–1409. [CrossRef]
28. Chen, J.C.; Huang, W.S.; Tsai, Y.F. Variability in the characteristics of extreme rainfall events triggering debris flows: A case study in the Chenyulan watershed, Taiwan. *Nat. Hazards* **2020**, *102*, 887–908. [CrossRef]
29. Guzzetti, F.; Peruccacci, S.; Rossi, M.; Stark, C.P. The rainfall intensity–duration control of shallow landslides and debris flows: An update. *Landslides* **2008**, *5*, 3–17. [CrossRef]
30. Nikolopoulos, E.I.; Crema, S.; Marchi, L.; Marra, F.; Guzzetti, F.; Borga, M. Impact of uncertainty in rainfall estimation on the identification of rainfall thresholds for debris flow occurrence. *Geomorphology* **2014**, *221*, 286–297. [CrossRef]
31. Lin, M.L.; Jeng, F.S. Characteristics of hazards induced by extremely heavy rainfall in central Taiwan—Typhoon Herb. *Eng. Geol.* **2000**, *58*, 191–207. [CrossRef]
32. Lin, C.W.; Shieh, C.L.; Yuan, B.D.; Shieh, Y.C.; Liu, S.H.; Lee, S.Y. Impact of Chi-Chi earthquake on the occurrence of landslides and debris flows: Example from the Chenyulan River watershed, Nantou, Taiwan. *Eng. Geol.* **2013**, *71*, 49–61. [CrossRef]
33. NCDR (National Science and Technology Center for Disaster Reduction). *The Rank of Extreme Rainfall Events over Taiwan*; NCDR: Taipei, Taiwan, 2015. (In Chinese)
34. Chow, V.T.; Maidment, D.R.; Mays, L.W. *Applied Hydrology*; McGraw-Hill Inc.: Singapore, 1988.
35. Jan, C.D.; Lee, M.H. A debris-flow rainfall-based warning model. *J. Chin. Soil Water Conserv.* **2004**, *35*, 273–283. (In Chinese)
36. Viessman, W.; Lewis, G.L.; Knapp, J.W. *Introduction to Hydrology*, 3rd ed.; Harper & Row: New York, NY, USA, 1989.
37. Makkonen, L. Plotting positions in extreme value analysis. *J. Appl. Meteorol. Climatol.* **2006**, *45*, 334–340. [CrossRef]

water

Article

Visual Language Translation Analysis and Scenic Beauty Estimation of Mountain Stream Facilities

Szu-Hsien Peng [1], Chun-Yu Huang [2] and Su-Chin Chen [2,3,*]

1 Department of Spatial Design, Chienkuo Technology University, Changhua City 50020, Taiwan
2 Department of Soil and Water Conservation, National Chung Hsing University, Taichung 40227, Taiwan
3 Innovation and Development Center of Sustainable Agriculture, National Chung Hsing University, Taichung 40227, Taiwan
* Correspondence: scchen@nchu.edu.tw

Abstract: The aesthetic design of mountain stream facilities has not received much attention in Taiwan. In recent years, in addition to safety concerns, the focus has increasingly been on landscape assessment or the sense of integration with the environment. This research is the first attempt to use visual language translation in qualitative research analysis for the landscape assessment of mountain stream facilities. This method is different from a traditional qualitative narrative analysis. It also addresses the shortcomings of previous quantitative analysis methods, in which the topic discussions are too limited. First, mountain stream engineering projects are selected as the research objects. This study uses questionnaire analysis and on-site surveys to summarize the elements, representations, and perceptions of the mountain stream facilities of the subjects concerned to examine their preferences for the visual system. Furthermore, we also employ the scenic beauty estimation (SBE) method for a comparison between the qualitative and quantitative analyses. This study proposes a new method using visual language translation and SBE that combines the features of qualitative research and quantitative analysis. However, the potential limitations include an inability to have a large sample number and the biases caused by the cultural, regional, or personal characteristics of the subjects.

Keywords: visual language translation; mountain stream facilities; perception; qualitative analysis; scenic beauty estimation (SBE); caption evaluation method (CEM)

Citation: Peng, S.-H.; Huang, C.-Y.; Chen, S.-C. Visual Language Translation Analysis and Scenic Beauty Estimation of Mountain Stream Facilities. *Water* **2022**, *14*, 3605. https://doi.org/10.3390/w14223605

Academic Editor: Richard C. Smardon

Received: 7 October 2022
Accepted: 7 November 2022
Published: 9 November 2022

1. Introduction

Taiwan is rich in topography and landscape features. Therefore, in recent years, engineering projects, such as check dams and groundsills, have viewed landscape assessments and integration with the environment as being important, in addition to safety concerns. In mountainous areas, stream erosion control facilities should be considered to reduce the environmental impact, as well as related eco-friendly designs, to increase the harmony of the landscape and environmental sustainability.

Aimed at the design concept of check dam aesthetics, a previous study [1] proposed that the three independent factors of form, color, and texture, as well as sense factors such as harmony, rhythm, and simplicity, could be used for the aesthetic assessment. Subsequently, a fuzzy logic system for landscape assessment was established using the analytic hierarchy process (AHP), and these principles were used to analyze the relationship of the various factors [2].

Furthermore, one study [3] used visual preference and four cognitive factors—naturalness, harmony, vividness, and closeness—to discuss the visual indicators, visual aesthetic experiences, and applications of the empirical relationship. Another study [4] adopted a preference-based psychophysical landscape assessment method (scenic beauty estimation (SBE) method) to explore the aesthetic preferences and differences in various artificial structures in a catchment. The SBE method is a psychophysiological experimental method for assessing the beauty of forest landscapes. The concept derives from the stimulus and response models of behavioral psychology, and is based on the signal-detection method and the Thurstone measurement model, improved for assessing the beauty of landscapes [5–9].

Participant-generated image methods, often used in social science research, present a long application history in social science [10]. Research on participant-generated images (PGIs) in social science can be traced back to 1970. Volunteer participants were offered cameras to take pictures representing specific topics, which might be related to their specific life experiences or places visited. Several qualitative research analysis methods were then used to analyze the pictures taken by the subjects, who became the "participants" in the research.

As type of visual language translation method, the caption evaluation method (CEM), proposed by Koga et al. [11], is a typical PGI method. Using this method, citizens in Minato, Tokyo, Japan, were requested to freely stroll in Minato and take various environmental pictures with a positive or negative impression. Meanwhile, they were invited to write down their feelings about each facility. The participants were further asked to record their reasons for being satisfied with or adverse to each facility.

Many Japanese scholars then began applying the CEM to evaluate various historical monuments, tourist attractions, and urban open spaces [12–17]. For instance, Naoi et al. [18] applied the CEM to evaluate visitors who selected historical areas as tourism destinations. A total of 30 Japanese college students and 27 Japanese adults interested in architecture and urban planning visited facilities in Japan's historical and cultural areas and stated their perceptions of the selected facilities. The results revealed the relationship between the elements in the areas and the participants' perceptions and desires. Furthermore, it also implied the participants' positive assessment of old buildings and adverse perception of modern architecture.

Chen et al. [19] collected comments regarding two hydropower dams in Canada from social media platforms such as Instagram. Each picture, and the title for each topic, was coded as a topic category for statistics. Each comment (e.g., dam construction would affect the perceived aesthetics and sense of home, or dam removal would cause lifestyle changes) might affect the relevant value and activity to predict the group items. For instance, the construction of a dam might affect the local citizens' perceived aesthetics and sense of home, or the deconstruction of a dam might result in lifestyle changes. This type of research could be an extended research method of the CEM.

There are limited studies using visual language translation or the CEM in Taiwan; however, hydraulic engineers need a reference for aesthetic design to meet the landscape requirements. This study uses the CEM as a qualitative research analysis tool for the landscape evaluation of mountain stream facilities. Selecting award-winning projects as the research objects, case studies are conducted to evaluate their design styles and compare them with other types of mountain stream facilities. Meanwhile, the quantitative analysis, using methods such as SBE, is compared with the qualitative results. Finally, qualitative analysis is used to explore the correlations among the characterization factors, perception factors, and preferences. The research goal is to establish a foundation for subsequent aesthetic engineering design in the future. The entire research process is shown in Figure 1.

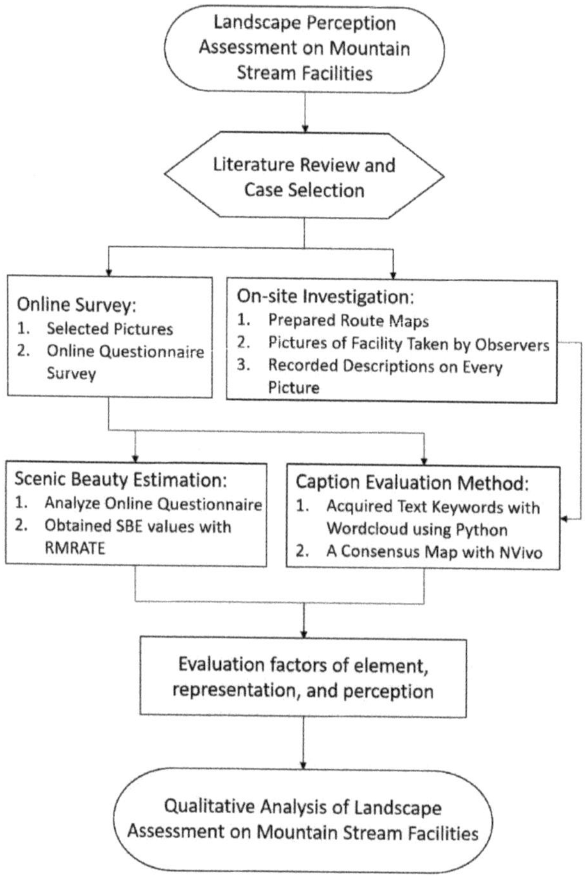

Figure 1. Research flow chart.

2. Materials and Methods

2.1. Caption Evaluation Method (CEM)

This study applies the caption evaluation method (CEM) to evaluate the landscape assessment of mountain stream engineering facilities in Taiwan. The CEM, first developed from architectural psychology and proposed and practiced by Koga et al. [11], is a participatory qualitative research method. The method attempts to investigate the reasons why the subjects take pictures at target locations, and the subjects are asked to state their opinions about the images. With such a method, the pictures of the targets taken by the participants are regarded as the trigger factors in their evaluation of the locations, and the acquired pictures are considered clues to their reactions.

In the CEM, three points to caption the pictures are proposed:

- To which elements do the subjects pay attention?
- What are the features of such elements noticed by the subjects?
- How do the subjects treat the noticed features?

Among the three questions, the first and the second questions are designed to gain an understanding of the features of the elements in the environment. The third question is aimed at inducing the subjects' thoughts regarding the features of such elements. For instance, one subject provided the caption, "It was cool because there were few cars. It would be better when there was no height difference between sidewalks and traffic lanes" [18]. In the example, "elements", such as "cars", "features" of the elements, such as "few", and "perceptions" of the subject, such as "cool", are mentioned. Such data could be used for various positive or negative qualitative analyses.

The CEM presents some advantages: (1) such a method could help the subjects focus on various facilities and elements, and state their opinions; (2) the pictures taken by the subjects could reveal the relationship between such elements and the observers' perceptions, and the elements might be used to manage the destinations to satisfy the interviewers' demands and provide ample opinions; and (3) the subjects can evaluate the facilities they experience.

On the other hand, the CEM also contains some disadvantages: (1) previous research has revealed that the sample size for such a method is relatively small, possibly because it takes a long time to complete taking pictures and recording, and it would be challenging to gather photographers at the same location for a long period; (2) in addition to the small sample size, the subjects' attributes may need to be excluded from the investigation, such as demographic and psychological features, as they would affect the coverage of the results; and (3) in comparison with an in-depth individual interview, the CEM results may lack depth with regard to the topic [18]. However, compared with the traditional questionnaire survey, more perspectives from different users' experiences could be investigated through the CEM, allowing us to think about problems from various perspectives.

2.2. Analysis of Qualitative Data

Qualitative analysis, a modern method used in academic research, is more complicated, with multiple layers and distinct data analyses, than traditional questionnaires or data quantitative research. This method does not simply respond to a questionnaire or discuss the difference, correlations, or predictions among variables. Since the mid-1980s, the development of computer-assisted qualitative data analysis software (CAQDAS) has allowed researchers to properly record intuition, ideas, searches, and analyses to simplify qualitative analyses [20]. The captions acquired through the CEM further preceded qualitative data analysis. Faced with former CEM survey results, many unstructured texts were generated that could not be quickly processed or perceived by humans or computers. For the next step, effective technology and algorithms are needed to mine and extract the most meaningful information [21,22].

The captions obtained by the CEM need to be processed through natural language processing (NLP) before advanced analyses are performed. Natural language processing (NLP) is one of the essential applications of machine learning, for example, text-to-speech and sentence-to-sentence clauses, etc. With advancements in computing speed, the accuracy of natural language processing is gradually being accepted. Segmenting the sentences into a file is the first task in research on text analysis using a computer. "Words" are then used for analyzing and organizing the results; therefore, "word segmentation" can be regarded as the most basic word analysis task.

The present study uses the Jieba module as a Chinese word-segmentation tool. The Chinese meaning of "Jieba" is "stutter," and the default word break principle of the Jieba module is simplified Chinese. Therefore, when using it in Taiwan, the traditional Chinese thesaurus was required to be downloaded in this study [23]. After the word-segmentation process, there was still too much text information; furthermore, more important phrases needed to be chosen.

Among the distinct computer-assisted qualitative data analysis software, NVivo is currently the most popular qualitative research software with complete functions. Data sources of text records, relevant literature, records, videos, and social networking sites after interviews are often used in qualitative research. NVivo helps users systematically organize such information with context and in a mutual relationship. Such collected "Sources" are integrated contextually, and the contents are coded and defined for transformation into "Nodes." When the "attribute value" is added into the coding process, it becomes "Cases." In other words, "Cases" can be regarded as "Nodes with attribute data." Set "Cases" with the same attribute value can be gathered together with the coding function in NVivo queries [24]. The most meaningful and representative labels are obtained through cluster analysis. These labels are then used for encoding CAQDAS to establish the so-called "Case" and "Node" to clarify the research topic's organizational and secondary levels. Through topic coding and classification, as well as comparison and analyses, an organizational hierarchy and sub-hierarchy were established to clarify the research structure and provide a reference for successive research.

2.3. Scenic Beauty Estimation (SBE)

Scenic beauty, an abstract concept, was previously studied using qualitative methods. Prior to the psychophysical paradigm phase, Daniel et al. [25–27] developed scenic beauty estimation (SBE) to quantitatively analyze scenic beauty. The SBE method, a psychophysiological experiment to evaluate the scenic beauty of forests, originated from the stimulus and respondence model in psychology and was improved according to the signal detections method and the Thurston scaling model. In addition, the viewers' preference for landscape or beauty was represented by the perceived scenic beauty distribution of the evaluators.

For the experimental procedure, the subjects viewed representative color slides (stimulus) and gave responses according to the preference scale of 1 to 5, revealing low-quality to high-quality scenic beauty. Statistics was used to standardize the evaluators' values using distinct evaluation criteria to solve the possible differences caused by different evaluation scale baselines. This was expected to accurately measure the public perceived preference for various landscapes. When the sample size was large enough, the randomly sampled perception value would become a normal distribution.

In summary, the SBE method presents the following advantages [26]: (1) It can include the intangible value of resources into the quantitative evaluation, and combine psychology and statistics to exclude individual subjective judgments of managers and planning designers, but adopt the public "perception preference" for different landscapes to respond to the viewers' perceived preference for landscapes. (2) In terms of the validity test, the SBE method provides similar results for the randomly shot slides or photos of the subjects being evaluated by the on-site evaluation. It could therefore save human resources and time. (3) Regarding the reliability test, the SBE method proves the consistency of photos taken at the same site but at different times, excluding special event factors. Therefore, the method shows high reliability. (4) In addition to forest landscape assessments, the SBE method can be applied to various landscapes. The results prove that it is a beauty-estimation method with high reliability and validity. (5) The SBE method eliminates evaluation errors caused by differences in individual aesthetic concepts. (6) The SBE index can be applied to managing large-area landscapes. (7) With regard to evaluation, the listed landscape factors appear to have positive and negative effects on recreation users that could be a reference for future management decision-making.

Nonetheless, the SBE method also presents the following disadvantages [26]: (1) It is arguable whether the selection of landscape samples could represent general landscape groups. (2) Regarding framing, a photographer's techniques and angles affect the evaluation results. (3) It is not easy to view the exact evaluated areas and special regions from photos. (4) The SBE method cannot distinguish whether a viewer's evaluation preference is landscape perception or cognition. (5) The calculated coefficients in the analysis lack definite explanations, would change with different people, and are comparatively subjective.

3. Results

3.1. Study Areas

This study collected relevant cases that have obtained gold awards for water conservancy over the years, selecting two cases according to the completeness of the data (planar layout, disasters experienced, photos before and after remediation, etc.) as the site surveys for analysis.

(1) Upstream Renovation Project of Toubiankeng Stream in Zhongpu

Located in Dongbian Village, Taiping District, Taichung City, Toubiankeng Stream is a tributary of the upstream Dali River. A major flood on 2 July 2004 resulted in debris flow blocking the stream and washing out the surrounding farmland. A large amount of sediment deposited on both sides of the stream at bridge No. 10 of Zhongpu affected the safety of city highway No. 136 and the surrounding houses. "Semi-circle groundsill work" and "arched hollow groundsill work" are the characteristics of this engineering project, with boulders reserved on site to avoid canalization and to direct low-water revetment flow and for the groundsill design of the high riverbank training (Figure 2).

Figure 2. On-site survey picture of the "Upstream Renovation Project of Toubiankeng Stream in Zhongpu" for the 17th Honorable Mention of the Public Construction Golden Quality Award for water conservancy (25 March 2020).

(2) Renovation Project of Sijiaolin Stream in Dongshi Forest Garden

Located in Dongshi District, Taichung City, the Dongshi Forest Garden presents potentially unstable earth next to Dongxin Village and the Da-an River. The existing upstream/downstream check dams appear seriously hollowed out and are almost disabled. A large check dam water body reduction resulted in an ecological barrier, and the serious bank scour caused sediment loss. The original upstream check dam was converted into "water storage type energy dissipation ladders," which has the added function of being a fishway. In the downstream check dam, a "circuitous fishway" was added. Furthermore, a "water and land ecological corridor" using five different types of embankment connecting the slope and stairway channels is a characteristic of this engineering project (Figure 3).

Figure 3. An on-site survey picture of the "Renovation Project of Sijiaolin Stream" for the 19th Excellent Award of the Public Construction Golden Quality Award for water conservancy (20 September 2019).

The selection process considered the influence of weather factors in the photos taken on site. It was a requirement that they were taken and recorded on the same day. The driving distances and the on-site walking paths of the two sample areas are appropriate. Moreover, the second case is located in the Dongshi Forest Garden with good walking paths. The anonymous subjects can learn about different mountain stream facilities with diversified construction methods.

3.2. Analysis Results of the CEM

Regarding the use of pictures, another way could be to have the researcher prepare the images as research materials. However, the images selected by the researcher might not completely capture the subjects' opinions about the target destination. In this case, as in Garrod [28], although the subjects complete the picture evaluation, the picture selection is not controlled. Such a method, therefore, may not accurately interpret the subject's evaluation of the elements. Both methods, including an online survey and an on-site questionnaire survey, are used in this study. As a result, the subjects still take pictures on site for the analyses.

3.2.1. Online Survey Results

The online questionnaire survey period was from 1 May to 30 May 2020. The researchers presented the selected digital photos to the test subjects through a slide presentation so that the test subjects could understand the function of photo structures and evaluate them by filling in an online questionnaire. The subjects were mainly undergraduate students and a small number of graduate students, totaling 212 students.

Aimed at two sample areas, to analyze and explain the effect on the overall perception, this study lists the top three factors from the analysis of the questionnaire results. From the factors acquired in the CEM results of the "Upstream Renovation Project of Toubiankeng Stream in Zhongpu" in Table 1, it is understood, due to year differences, that vegetation densely covers the site. This dense distribution of vegetation affects the overall preference score, with significant differences. For instance, in pictures 2 and 3 of Table 1, the embankments show the score dropping from 4.01 to 3.64; the biological channels in pictures 6 and 7 also drop the score from 3.9 to 3.04; and the average preference for the sample areas is 3.64. This reveals that the images of the sample areas show differences due to year differences, but the form and overall visual perception are compatible with the skyline, presenting high public acceptance. The "Renovation Project of Sijiaolin Stream" in Table 2, shot in the closing year, increases the preference to 3.77 after using the CEM, with multiple factors matching the form and arrangement of multiple embankments and multiple biological channels on site.

Table 1. CEM results of the "Upstream Renovation Project of Toubiankeng Stream in Zhongpu".

No.	Picture	Description of Top 3 Appearance			Preference (Full Score 5)
		Element	Representation	Perception	
1		check dam groundsill embankment	form arrangement stone	relaxing oppressive uneasy	3.15
2		embankment ecology groundsill	form arrangement green	relaxing pressure-relieving comfortable	3.64
3		embankment ecology stone	stone form arrangement	relaxing pressure-relieving comfortable	4.01
4		check dam groundsill stone	form arrangement stone	relaxing pressure-relieving oppressive	3.58
5		ecology embankment channel	form arrangement color	relaxing pressure-relieving comfortable	4.14
6		ecology channel embankment	green color form	relaxing Oppressive Uneasy	3.04

Table 1. *Cont.*

No.	Picture	Description of Top 3 Appearance			Preference (Full Score 5)
		Element	Representation	Perception	
7		ecology channel embankment	arrangement form color	relaxing pressure-relieving comfortable	3.90
8		embankment ecology channel	form arrangement color	uneasy oppressive nervous	3.07
9		ecology channel embankment	arrangement form color	relaxing pressure-relieving comfortable	3.76
10		embankment stone ecology	form arrangement color	relaxing pressure-relieving comfortable	3.71
11		ecology groundsill channel	form arrangement stone	relaxing pressure-relieving comfortable	4.10

Table 2. CEM results of the "Renovation Project of Sijiaolin Stream.".

No.	Picture	Description of Top 3 Appearance			Preference (Full Score 5)
		Element	Representation	Perception	
1		ecological channel check dam embankment	form arrangement stairs	relaxing pressure-relieving comfortable	3.59
2		check dam ecological channel embankment	form arrangement green	pressure-relieving oppressive relaxing	3.08
3		check dam embankment groundsill	form color arrangement	relaxing uneasy nervous	3.77

Table 2. *Cont.*

No.	Picture	Description of Top 3 Appearance			Preference (Full Score 5)
		Element	Representation	Perception	
4		groundsill ecological channel stone	form arrangement stone	relaxing pressure-relieving Pleasant	3.93
5		embankment ecological channel groundsill	form arrangement tidy	relaxing pressure-relieving pleasant	3.93
6		embankment groundsill ecological channel	form arrangement wood	relaxing pressure-relieving comfortable	3.87
7		embankment ecological channel stone	arrangement form stone	relaxing pressure-relieving pleasant	3.86
8		ecological channel check dam embankment	form arrangement tidy	pressure-relieving relaxing pleasant	4.10
9		embankment ecological channel stone	form arrangement stone	relaxing pressure-relieving pleasant	3.92
10		ecological channel embankment channel	form green color	pleasant relaxing uneasy	3.38
11		ecological channel groundsill embankment	form arrangement stone	relaxing pressure-relieving pleasant	4.03

The overall analysis results reveal that the subjects primarily pay attention to the factors of water and plants. Natural texture factors confer the feeling of being close to nature, so the subjects find it relaxing and stress relieving. When a body of water is reduced, the subjects' feeling of being close to nature is diminished, and they feel nervous and uneasy. Regarding structural factors, the subjects first pay attention to form and color. Arc forms and meandering stairs enhance the subjects' preference for artificial structures. Using wood and stone for the structure, and reducing concrete, could minimize the visual impact.

3.2.2. On-Site Investigation Results

Before the on-site investigation, the subjects were offered a map with designated routes or coverage. The subjects were requested to walk for approximately an hour taking 10–20 pictures of the facility, with either negative or positive meanings. Meanwhile, the location of each facility was marked on the map (Figure 4), and software such as JotForm was utilized for taking and recording their opinions about each facility. These descriptions included three captions: attention element, characteristics of the element, and perception of the noticed feature. From these captions, the related analysis and discussion were carried out.

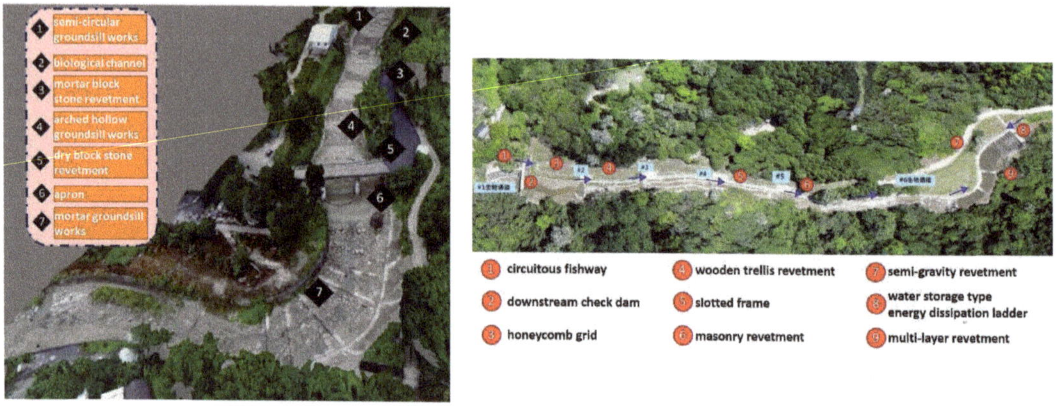

(a) (b)

Figure 4. On-site investigation map: (**a**) Upstream Renovation Project of Toubiankeng Stream in Zhongpu; and (**b**) Renovation Project of Sijiaolin Stream (source: Soil and Water Conservation Bureau, Taiwan).

This study conducted the first on-site investigation on 19 July, the second on 23 September, and the third on 17 October 2020 to establish a more credibly comprehensive evaluation result. A total of 24 volunteer participants visited these two study areas for investigation, taking 231 photographs and recording descriptions of these photos.

Tables 3 and 4 demonstrate part of the text descriptions from the on-site investigation. After acquiring each interviewee's information, Python was used for word-segmentation processing to present the acquired text keywords as a word cloud, including the online questionnaires and on-site investigations (Tables 5 and 6). After coding with NVivo, it was discovered that the online questionnaire representation, as the major direction, merely aims at descriptions of form, color, and arrangement, as the researcher selects the pictures for which the respondents' opinions are limited. The factors acquired in the on-site investigation questionnaire might be more accurate, as the subjects determine the images to better induce the opinions expressed, e.g., descriptions aimed at the stone size and arc form. In addition to static descriptions, dynamic descriptions are covered in the on-site investigation, e.g., water flow, to present the immersive perception, which cannot be realized from images.

Table 3. On-site investigation of the "Upstream Renovation Project of Toubiankeng Stream in Zhongpu".

Picture	On-Site Investigation Text
	This seems to be a semi-circle channel. The planning of guard rails is excellent and beautiful, and it would not be dangerous for fish and animals to drink water here. It feels good to extend the ecology. Look! Egrets are flying around.
	The river bed engineering also pays attention to safety. It seems to slow down the water velocity without special beautification. It would be prettier to tidy the stones on both sides.
	There is a boulder to prevent low water from washing aimlessly and to control the area. Their existence also alerts people that more such boulders could fly down without erosion control. Furthermore, there is no extra money or construction required for the design. So I think that it is a good design.
	The arrangement of stones seems to be a little messy, making it visually uncomfortable. In addition, the color is rather dull, without anything special; so I don't like it very much.

Table 4. On-site investigation of the "Renovation Project of Sijiaolin Stream".

Picture	On-Site Investigation Text
	The feeling of flow is like getting into an original forest, and the stone stacking is original. There does not seem to be embankment burst or fish swimming.
	Here are water storage-type energy dissipation ladders. The design is good with the sense of design but not too much. The dark color selected is a bit like a leopard print. It is pretty good.
	Here is a multi-layer embankment. It does not look pretty, but it shows the convenience. It is designed for Reeves's muntjac and for small animals to drink water. Although it is full of weeds, it is integrated with the background so that small animals, when drinking water, cannot be discovered by predators or humans.
	There is the chirping of cicadas among the trees in Dongshi. The green feeling is like a forest. There are not many visitors in the area to threaten the ecology or the environment. It is pretty comfortable to take a walk here listening to the cicadas chirping. The landscape is also suitable for holidays.

Table 5. Word cloud analysis of the "Upstream Renovation Project of Toubiankeng Stream in Zhongpu".

Word cloud of the Online Questionnaire—Representation

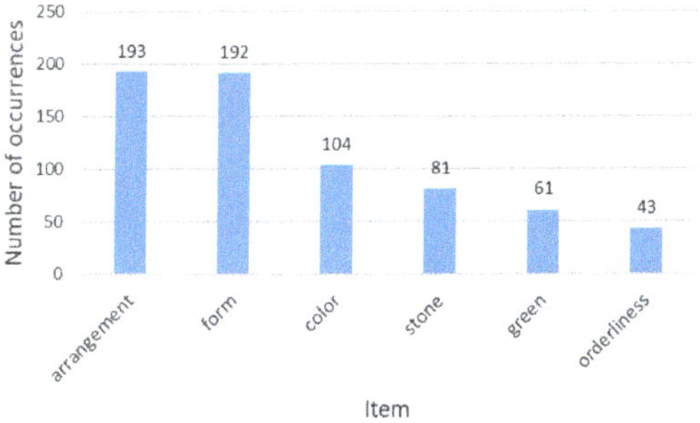

Word cloud of the Online Questionnaire—Perception

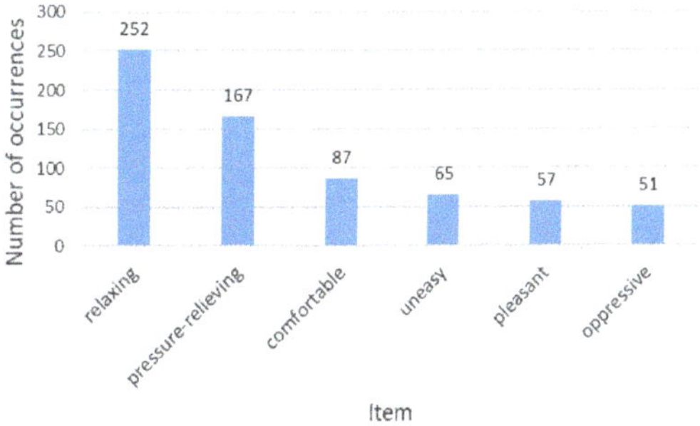

Word cloud of the On-Site Investigation—Representation

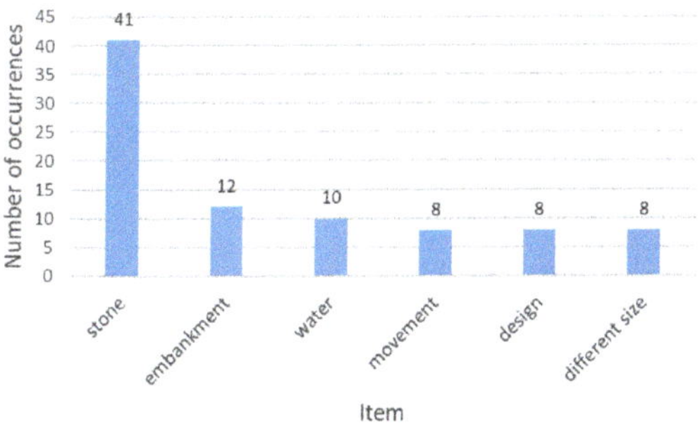

Word cloud of the On-Site Investigation—Perception

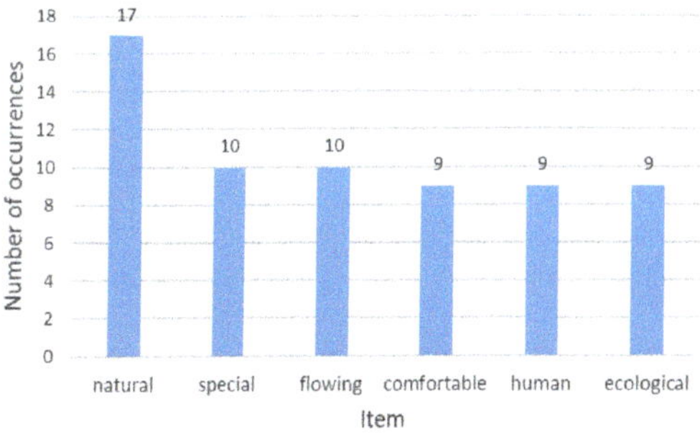

Table 6. Word cloud analysis of the "Renovation Project of Sijiaolin Stream".

Word cloud of the Online Questionnaire—Representation

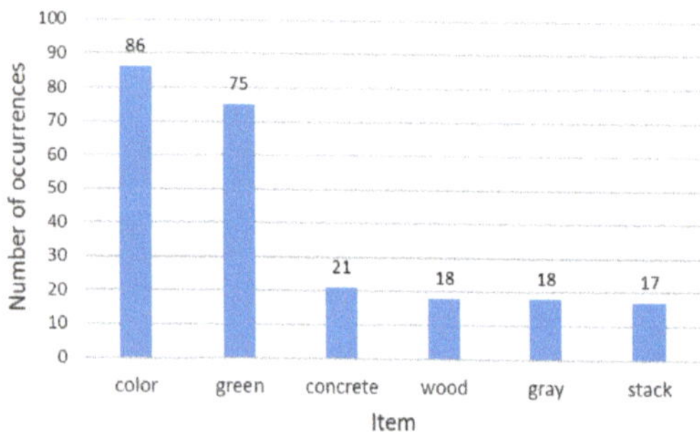

Word cloud of Online Questionnaire—Perception

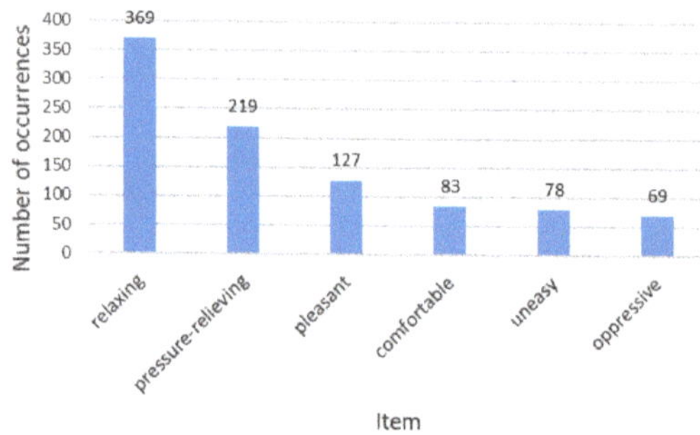

Word cloud of the On-Site Investigation—Representation

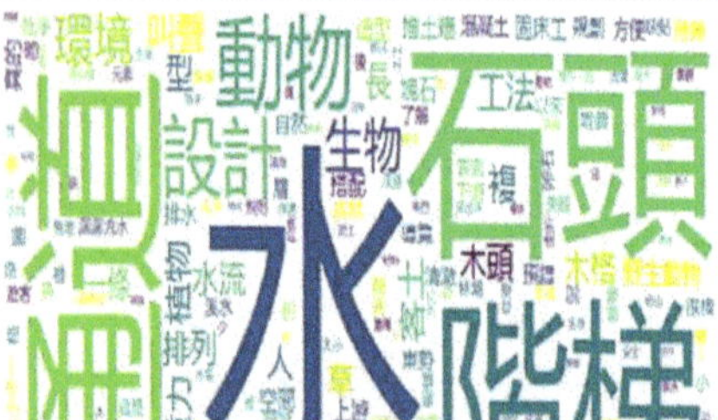

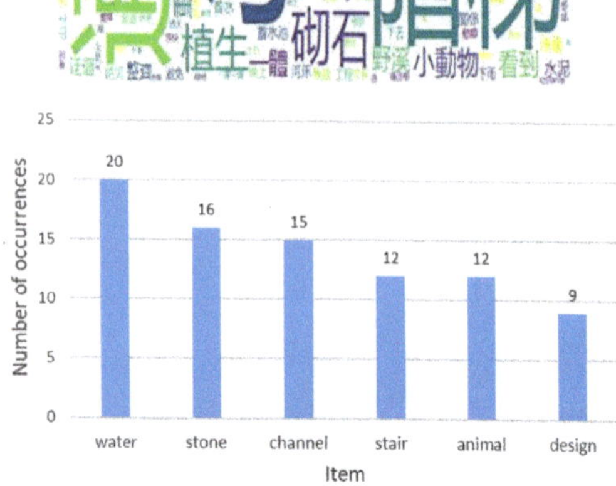

Word cloud of the On-Site Investigation—Perception

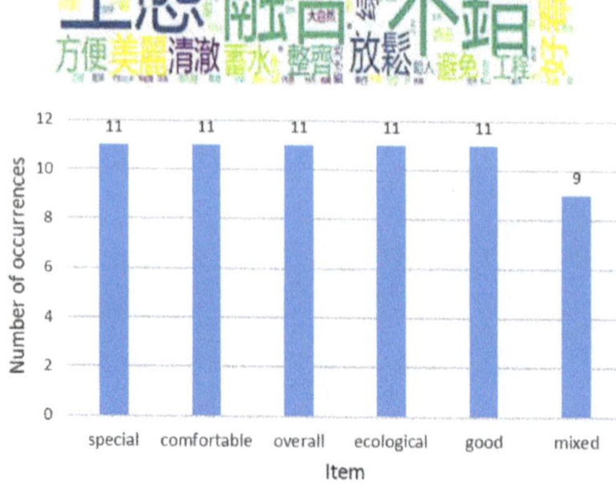

Taking the "Renovation Project of Sijiaolin Stream" as an example to explain the application of the CEM, multiple engineering forms and textures were used in the sample area. The word cloud reveals different online questionnaire representations from it in the other sample area; the factors are multiple and dispersed, but factors of form, arrangement, and color could still be acquired. After the on-site investigation, the word cloud clearly shows the subjects' descriptions of on-site water, stone, biological channels, and stairs. Using NVivo for coding, the multiple textures of the embankment and the layered stairs form were acquired, and actions and sounds were added. For instance, the subjects see flying dragonflies and butterflies and hear bugs chirping and birds singing during the on-site investigation. The on-site flow conditions would not appear visually different due to the picture-taking skills. The overall perception is positive. Moreover, the completion of the sample area is close, environmental education is taken into account in the planning stage, and the Dongshi Forest Garden is well maintained so the perception is positive.

The subjects' feedback information, especially the self-selected shooting pictures, provides rich information and multiple materials. For example, subjects with different backgrounds observe from different perspectives. The ones with an engineering background are accustomed to observing engineering structures, so they describe the form and functions. The students without an engineering background more easily observe from the perspective of life, noticing things such as insects and vegetation in the surrounding environment along with the sound of flowing water, bugs chirping, and birds singing in the background, in their perception of the environment. Referring to the concept of the Zaltman metaphor elicitation technique (ZMET) [29–31], using the text keywords extracted by this study and NVivo, a consensus map of the respondents is produced, as shown in Figure 5. Furthermore, the factors are organized from the above analyses, as shown in Table 7.

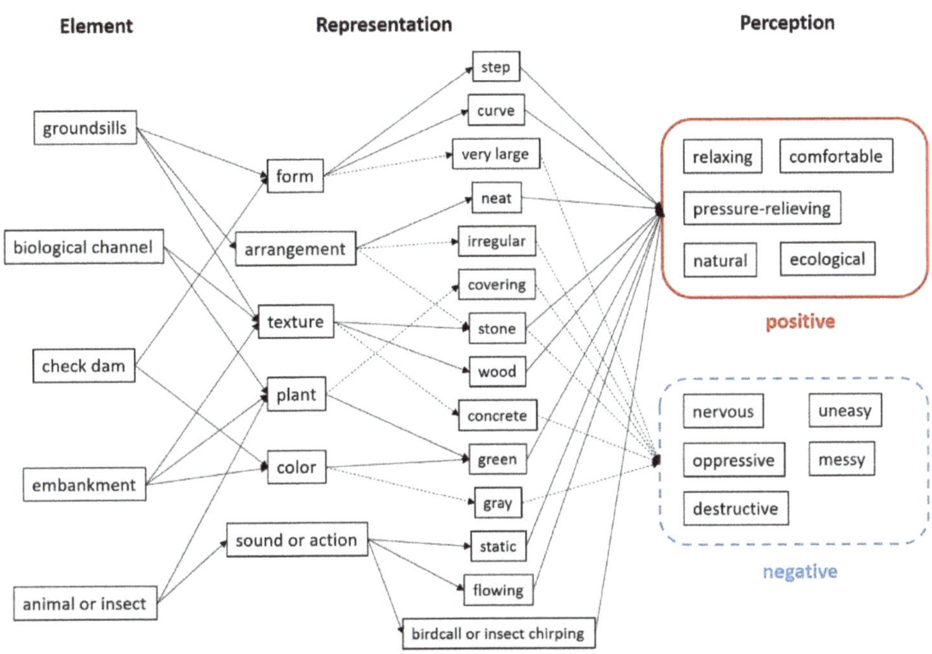

Figure 5. Interviewee consensus map (solid lines represent positive; dashed lines represent negative).

Table 7. Evaluation factors of element, representation, and perception.

Evaluation Factor	Classification	Specific Description
element	natural facility	stream, river, animal, insect, plant
	artificial facility	check dam, groundsill, embankment, biological channel
representation	form	arc, concave, curve, meandering, size, rectangle
	texture	stone, wood, plant, concrete
	arrangement	layer-by-layer as stairs, irregular, tidy, cross
	sound and action	flowing, still, flying around, birdcall, insect chirping
perception	positive	cozy, comfortable, pretty, natural, friendly, practical, graceful, integrative, attractive, safe
	negative	messy, destructive

3.3. Online Questionnaire SBE Analysis Result

The scenic beauty estimation (SBE) method, proposed by Daniel and Boster [26], was applied in this study to analyze the online questionnaire. The subjects evaluated the pictures with their preferences. The online questionnaire was collected via Google for one month (1–20 May 2020); a total of 212 valid copies were acquired. The preference evaluation data were processed using RMRATE [7,9], provided by the United States Department of Agriculture (USDA). The mean of 11 pictures was regarded as the baseline for both sample areas, and the baseline-adjusted Z score, which is similar to the normal distribution, of each image was converted into scenic beauty estimates (SBEs). The results are shown in Tables 8 and 9.

Table 8. SBE results of the "Upstream Renovation Project of Toubiankeng Stream in Zhongpu.".

No.	Picture	SBE Value	Rank
1		−131.21	9
2		−3.79	7
3		99.19	3
4		−20.59	8

Table 8. *Cont.*

No.	Picture	SBE Value	Rank
5		118.73	1
6		−149.75	11
7		61.73	4
8		−135.34	10
9		34.41	5
10		15.06	6
11		111.57	2

Table 9. SBE results of the "Renovation Project of Sijiaolin Stream.".

No.	Picture	SBE Value	Rank
1		45.39	5
2		−51.15	9

Table 9. *Cont.*

No.	Picture	SBE Value	Rank
3		−219.1	11
4		−5.63	8
5		51.29	4
6		23	7
7		29.61	6
8		113.93	1
9		66.85	3
10		−137.39	10
11		83.2	2

3.3.1. CEM and SBE Comparison of the "Upstream Renovation Project of Toubiankeng Stream in Zhongpu"

In the conversion of preferences in the online questionnaire into SBE (Table 8), three images, each with the highest and the lowest scores, were extracted. The top three images are as follows: No. 5, the semi-circle groundsill work; No. 11, the arched hollow-out groundsill work; and No. 3, the embankment with ecological engineering. The former two pictures represent the common point of a rich waterscape, with the on-site stones used in the structure, showing a sense of harmony with the environment. They conform to the aesthetic principle of integrating structures into the environment [1]. Regarding the water conservation facilities in No. 5 and No. 11, the dam body presents a similar shape, with higher immersion into the embankment on both sides, and the natural river shoreline or skyline enhances the harmony and visual preference [3]. The CEM reveals that the subjects with a high-level evaluation show a higher perception of the arrangement of stone as the structure (Table 10). The visual harmony of the layered stone stacks contributes to increasing the subjects' preference [32].

Table 10. CEM comparison of the highest and lowest SBE value images of the "Upstream Renovation Project of Toubiankeng Stream".

No.	Picture	Subject Response Text
5		The overall staircase shape and the integration with nature are relaxing. The idea of the water drop structure is excellent, but more stacks would be fantastic.
6		Messy weeds in a biological channel is not beautiful. Spreading vegetation is messy.

No. 6, No. 8, and No. 1 appear to have the lowest evaluation on the SBE. The former two structures are covered by vegetation, and No. 1 shows incoordination due to a large amount of irregular rocks. The green visual ratio in the image is enhanced to increase the sense of nature. However, the increasing sense of nature diminishes when the green visual ratio is higher than 40%, but the closure property appears to be an increasing trend. The increase in closure property decreases the visual preference [3]. The CEM reveals that the top three subjects show a comparatively oppressive and uneasy perception of the vegetation in the structures with the lowest scores.

3.3.2. CEM and SBE Comparison of the "Renovation Project of Sijiaolin Stream"

The SBE evaluation in Table 9 shows the three images with the highest and lowest scores. The top three images are as follows: No. 8, a roundabout biological channel; No. 11, stone masonry groundsill work; and No. 9, a compound biological channel. The former two images represent the common point of a rich waterscape. The water body in the image with water conservation facilities enhances the vitality, softens the water conservation facilities, reduces the oppressive sense of sand control facilities, and reduces the closure property. When the water visual ratio is lower than 30%, the visual preference of the image is proportional to harmony; the higher harmony, the greater the visual preference [3]. The

CEM reveals that the subjects with a higher evaluation appear to prefer curve or stair forms over traditional straight-line designs (Table 11).

Table 11. CEM comparison of the highest and lowest SBE value images of the "Renovation Project of Sijiaolin Stream".

No.	Picture	Subject Response Text
8		This is a great design to slow down water power and velocity, and the aesthetics are excellent. The light gray color looks comfortable, and the simple form is also comfortable.
3		The river is almost dry, which is worrying. It looks like it lacks vitality without water. Lots of sediment deposits seem to be desolate, and there is no water or fish.

Furthermore, No. 3, No. 10, and No. 2 appear to have the lowest scores on the SBE, and the common point is the absence of a water body. The sample area is a torrent with significant water difference in the dry season—the increasing closure property decreases the visual preference. The qualitative analysis reveals that the top three subjects have nervous and uneasy perceptions regarding the lack of water.

The above findings show that the top three structures in the SBE evaluation are mainly ecological engineering and groundsill structures with a rich waterscape; vegetation is the typical landscape in mountainous areas, and distinct vegetation in forests appears with different landscape aesthetics. Nevertheless, vegetation differences are reduced when there is both a structure and waterscape. The CEM results reveal the significant effects of the structure's form and texture, in addition to the waterscape and vegetation, on the entire landscape, followed by the vegetation and the structure's volume and form.

This study has potential limitations. There are few studies on the landscape assessment of mountain stream facilities using qualitative analysis. Some documents [33,34] have described a detailed narrative analysis of the cultural heritage of facilities in river basins or their relationship with human social development. The use of visual language translation is a new concept, but this method may affect the sample number size and choice because of the complex process. A qualitative study with a large sample size would not be possible in this study. At the same time, it was also limited by time constraints and the impact of the COVID-19 pandemic. Moreover, this study can only be evaluated for case-by-case projects, and there may be biases caused by the cultural or personal characteristics of the subjects. However, the aim of this study is to provide a new evaluation method, and subjects' deeper meanings may be revealed through this method. Future endeavors will be expanded to more projects and subjects, including experts and the general public.

4. Conclusions

Traditional landscape qualitative research is usually based on narrative analysis methods that are too subjective, although discussed in depth. However, the commonly used quantitative analysis method often fails to deeply explore the reasons behind the data. Therefore, this study proposes a new method, namely, visual language translation, or the CEM, which combines the features of qualitative research and quantitative analysis. This has also rarely been mentioned in previous studies, and this is considered a new attempt.

Different from the traditional research method, visual language translation, or the CEM used in this study, clearly understands the subjects' perception of the images and sample areas. Factors are accurately acquired through the subjects' descriptions, and such factors, through qualitative analysis, are coded with NVivo to find the best ones. However, too much open information can easily result in information dispersion and difficulties with induction and analysis. For this reason, both the CEM and SBE are selected to acquire complete information for analysis and comparison. In this study, these methods were established and performed well with good results.

Using SBE, we can acquire the subjects' preferences for images and sample areas, which, corresponding with the CEM, accurately comprehends the subjects' factor weights. In this way, in addition to obtaining a quantitative landscape preference, it is also possible to further understand the possible reasons for the preference. For instance, wooden lattice-framed embankment-matching stairs reduce the visual impact on the entire environment and enhance the subjects' preference. A stone embankment could reduce the sense of artificiality to make the subjects feel safe and stable. Different from individual analysis in the past, two methods are utilized in this study, and an on-site investigation is integrated to understand the sample areas more deeply and to accurately acquire factors for subsequent research and design.

However, this study has limitations. It is mainly based on case studies, of which we still need more varied research objects and a wider survey sample. It is suggested that the CEM could be continuously used for discussing the correlations among representation, perception, and preference. Furthermore, experts' interviews and questionnaires could be combined to build a list of the possible factors in landscape evaluation and to facilitate discussion of the basic concept of aesthetic design in mountain stream engineering.

Author Contributions: Conceptualization, S.-C.C. and S.-H.P.; methodology, C.-Y.H.; software, S.-H.P.; validation, C.-Y.H. and S.-H.P.; formal analysis, S.-C.C.; investigation, C.-Y.H.; writing—original draft preparation, S.-H.P.; writing—review and editing, S.-C.C. All authors have read and agreed to the published version of the manuscript.

Funding: This research was funded by the Soil and Water Conservation Bureau, Council of Agriculture, Taiwan, grant number 109-11.1-01-06-001(8).

Data Availability Statement: The datasets generated during and/or analysed during the current study are available from the corresponding author on reasonable request.

Acknowledgments: The authors thank the anonymous volunteers who participated in this study.

Conflicts of Interest: The authors declare no conflict of interest.

References

1. Chen, S.-C.; Lin, H.-C. Visual aesthetic design principles of check dams. *J. Chin. Soil Water Conserv.* **2010**, *41*, 34–49. (In Chinese)
2. Peng, S.-H.; Tang, C. Blending the analytic hierarchy process and fuzzy logical systems in scenic beauty assessment of check dams in streams. *Water* **2015**, *7*, 6983–6998. [CrossRef]
3. Chen, J.-C.; Cheng, C.-Y.; Huang, C.-L.; Chen, S.-C. Assessment of the visual quality of sediment control structures in mountain streams. *Water* **2020**, *12*, 3116. [CrossRef]
4. Peng, S.-H.; Han, K.-T. Assessment of aesthetic quality on soil and water conservation engineering using the scenic beauty estimation method. *Water* **2018**, *10*, 407. [CrossRef]
5. Brown, T.C.; Daniel, T.C. *Modeling Forest Scenic Beauty: Concepts and Application to Ponderosa Pine*; US Department of Agriculture, Forest Service, Rocky Mountain Forest and Range Experiment Station: Washington, DC, USA, 1984.
6. Brown, T.C.; Daniel, T.C. Predicting scenic beauty of timber stands. *For. Sci.* **1986**, *32*, 471–487.

7. Brown, T.C.; Daniel, T.C. *Scaling of Ratings: Concepts and Methods*; USDA Forest Service, Rocky Mountain Forest and Range Experiment Station: Fort Collins, CO, USA, 1990; Volume 293.
8. Brown, T.C.; Daniel, T.C. Landscape aesthetics of riparian environments: Relationship of flow quantity to scenic quality along a wild and scenic river. *Water Resour. Res.* **1991**, *27*, 1787–1795. [CrossRef]
9. Brown, T.C.; Daniel, T.C.; Schroeder, H.W.; Brink, G.E. *Analysis of Ratings: A Guide to RMRATE*; US Department of Agriculture, Forest Service, Rocky Mountain Forest and Range Experiment Station: Washington, DC, USA, 1990.
10. Balomenou, N.; Garrod, B. A review of participant-generated image methods in the social sciences. *J. Mix. Methods Res.* **2016**, *10*, 335–351. [CrossRef]
11. Koga, T.; Taka, A.; Munakata, J.; Kojima, T.; Hirate, K.; Yasuoka, M. Participatory research of townscape, using "Caption Evaluation Method"-Studies of the cognition and the evaluation of townscape, part 1. *J. Archit. Plan. Environ. Eng.* **1999**, *517*, 79–84.
12. Koga, T.; Sumeragi, T.; Kojima, T.; Munakata, J.; Hirate, K. User's evaluation of living environment in welfare facility for the elderly with caption evaluation method: User's evaluation of living environment in welfare facility for the elderly part 1. *J. Archit. Plan. (Trans. AIJ)* **2006**, *71*, 33–39. [CrossRef]
13. Kojima, T.; Koga, T.; Munakata, J.; Hirate, K. Multivariate analysis on verbal data of 'Caption Evaluation Method'. Studies of the cognition and the evaluation of townscape. Part 2. *J. Archit. Plan. Environ. Eng.* **2002**, *67*, 51–58. [CrossRef]
14. Naoi, T.; Airey, D.; Iijima, S.; Niininen, O. Visitors' evaluation of an historical district: Repertory Grid Analysis and Laddering Analysis with photographs. *Tour. Manag.* **2006**, *27*, 420–436. [CrossRef]
15. Naoi, T.; Iijima, S. Applying the caption evaluation method to research historical districts as tourism destinations. In Proceedings of the CAUTHE 2008: Tourism and Hospitality Research, Training and Practice, Surfers Paradise, Australia, 11–14 February 2008; Volume 122.
16. Oi, N.; Okita, T.; Hiraoka, Y.; Takahashi, H. Attributes of participants and their evaluation tendencies in a research of townscape, using "Caption Evaluation Method". *J. Environ. Eng. (Trans. AIJ)* **2008**, *73*, 393–399. [CrossRef]
17. Ueda, M.; Kawamura, Y.; Oi, N.; Takahashi, H. A Study on cognition of landscape elements in station plazas using Caption Evaluation Method: Landscape survey in station plazas in and around Fukuoka city. *AIJ J. Technol. Des.* **2016**, *22*, 147–151. [CrossRef]
18. Naoi, T.; Yamada, T.; Iijima, S.; Kumazawa, T. Applying the caption evaluation method to studies of visitors' evaluation of historical districts. *Tour. Manag.* **2011**, *32*, 1061–1074. [CrossRef]
19. Chen, Y.; Parkins, J.R.; Sherren, K. Leveraging social media to understand younger people's perceptions and use of hydroelectric energy landscapes. *Soc. Nat. Resour.* **2019**, *32*, 1114–1122. [CrossRef]
20. Gibbs, G.R. The nature of qualitative analysis. In *Analyzing Qualitative Data*; SAGE: New York, NY, USA, 2007; pp. 1–9.
21. Allahyari, M.; Pouriyeh, S.; Assefi, M.; Safaei, S.; Trippe, E.D.; Gutierrez, J.B.; Kochut, K. A brief survey of text mining: Classification, clustering and extraction techniques. *arXiv* **2017**, arXiv:1707.02919. [CrossRef]
22. Hashimi, H.; Hafez, A.; Mathkour, H. Selection criteria for text mining approaches. *Comput. Hum. Behav.* **2015**, *51*, 729–733. [CrossRef]
23. Deng, W.-Y. *Python Machine Learning and Deep Learning Special Training Course: Introduction and Application*; GOTOP Information Inc.: Taipei, Taiwan, 2019. (In Chinese)
24. Bazeley, P.; Jackson, K. *Qualitative Data Analysis with NVivo*; Sage Publications Limited: New York, NY, USA, 2013.
25. Anderson, L.M.; Schroeder, H.W. Application of wildland scenic assessment methods to the urban landscape. *Landsc. Plan.* **1983**, *10*, 219–237. [CrossRef]
26. Daniel, T.C.; Boster, R.S. *Measuring Landscape Esthetics: The Scenic Beauty Estimation Method*; Rocky Mountain Forest and Range Experiment Station: Fort Collins, CO, USA, 1976.
27. Schroeder, H.W.; Daniel, T.C. Predicting the scenic quality of forest road corridors. *Environ. Behav.* **1980**, *12*, 349–366. [CrossRef]
28. Garrod, B. A snapshot into the past: The utility of volunteer-employed photography in planning and managing heritage tourism. *J. Herit. Tour.* **2007**, *2*, 14–35. [CrossRef]
29. Zaltman, G.; Coulter, R.H. Seeing the voice of the customer: Metaphor-based advertising research. *J. Advert. Res.* **1995**, *35*, 35–51.
30. Ramjaun, T.A. Experimenting with ZMET: Issues and adaptions. *Qual. Rep.* **2021**, *26*, 1633–1641. [CrossRef]
31. da Rosa, C.; Baggio, D.K.; Araujo, C.F. Eating: It's the best way to grow: A study on the food consumption of children using an adapted ZMET. *REMark* **2022**, *21*, 577. [CrossRef]
32. Chen, J.-C.; Huang, C.-L.; Chen, S.-C.; Tfwala, S.S. Visual harmony of engineering structures in a mountain stream. *Water* **2021**, *13*, 3324. [CrossRef]
33. Fajer, M. Watermills—A Forgotten River Valley Heritage—Selected examples from the Silesian voivodeship, Poland. *Environ. Socio-Econ. Stud.* **2015**, *2*, 1–9. [CrossRef]
34. Szkurłat, E. Human-river relations at various stages of development of water front factory settlements in Łódź. *Environ. Socio-Econ. Stud.* **2016**, *3*, 47–56. [CrossRef]

MDPI
St. Alban-Anlage 66
4052 Basel
Switzerland
www.mdpi.com

Water Editorial Office
E-mail: water@mdpi.com
www.mdpi.com/journal/water